ICL Contribution to Landslide Disaster Risk Reduction

Series Editor

Kyoji Sassa, The International Consortium on Landslides, ICL, Kyoto, Japan

The ICL Contribution to Landslide Disaster Risk Reduction book-series publishes integrated research on all aspects of landslides. The volumes present summaries on the progress of landslide sciences, disaster mitigation and risk preparation. The contributions include landslide dynamics, mechanisms and processes; volcanic, urban, marine and reservoir landslides; related tsunamis and seiches; hazard assessment and mapping; modeling, monitoring, GIS techniques; remedial or preventive measures; early warning and evacuation and a global landslide database.

More information about this series at http://www.springer.com/series/16332

Željko Arbanas • Peter T. Bobrowsky •
Kazuo Konagai • Kyoji Sassa • Kaoru Takara
Editors

Understanding and Reducing Landslide Disaster Risk

Volume 6 Specific Topics in Landslide Science and Applications

Springer

Editors
Željko Arbanas
Faculty of Civil Engineering
University of Rijeka
Rijeka, Croatia

Kazuo Konagai
International Consortium on Landslides
Kyoto, Japan

Kaoru Takara
Graduate School of Advanced Integrated
Studies in Human Survivability (Shishu-kan)
Kyoto University
Kyoto, Japan

Peter T. Bobrowsky
Geological Survey of Canada
Sidney, BC, Canada

Kyoji Sassa
International Consortium on Landslides
Kyoto University
Uji Campus
Kyoto, Japan

Associate Editors
Amin Askarinejad
Technische Universiteit Delft
Delft, The Netherlands

Giovanna Capparelli
Universita degli Studi della Calabria
Rende, Italy

Yifei Cui
Tsinghua University
Beijing, China

Sabatino Cuomo
University of Salerno
Fisciano, Italy

Stefano Luigi Gariano
CNR IRPI
Perugia, Italy

Hans-Balder Havenith
Universite de Liege
Liege, Belgium

Katsuo Sasahara
Kochi University
Nangoku, Japan

Hendy Setiawan
Universitas Gadjah Mada
Yogyakarta, Indonesia

Faraz Tehrani
Deltares
Delft, The Netherlands

Ryosuke Uzuoka
Kyoto University
Kyoto, Japan

Gonghui Wang
Kyoto University
Kyoto, Japan

Fawu Wang
Tongji University
Shanghai, China

ISSN 2662-1894 ISSN 2662-1908 (electronic)
ICL Contribution to Landslide Disaster Risk Reduction
ISBN 978-3-030-60712-8 ISBN 978-3-030-60713-5 (eBook)
https://doi.org/10.1007/978-3-030-60713-5

Cover illustration: Historical Grohovo Rock Avalanche near Rijeka, Croatia, that buried the Grohovo Village in 1885 (International Consortium on Landslides. All Rights Reserved).

This Springer imprint is published by the registered company Springer Nature Switzerland AG
The registered company address is: Gewerbestrasse 11, 6330 Cham, Switzerland

ICL and Springer created a new book series “ICL Contribution to Landslide Disaster Risk Reduction” in 2019 which is registered as ISSN 2662-1894 (print version) and ISSN 2662-1908 (electronic version). The first books in this series are six volume of books “Understanding and Reducing Landslide Disaster Risk” containing the recent progress of landslide science and technologies from 2017 to 2020.

Editor-in-Chief: Kyoji Sassa
Assistant Editor-in-Chief: Željko Arbanas

Organizational Structure of the Fifth World Landslide Forum

Organizers

International Consortium on Landslides (ICL)
Global Promotion Committee of International Programme on Landslides (IPL-GPC), including: United Nations Educational, Scientific and Cultural Organization (UNESCO), World Meteorological Organization (WMO), Food and Agriculture Organization (FAO), United Nations Office for Disaster Risk Reduction (UNDRR), United Nations University (UNU), International Science Council (ISC), World Federation of Engineering Organizations (WFEO), International Union of Geological Sciences (IUGS), International Union of Geodesy and Geophysics (IUGG) Kyoto University (KU), Japan Landslide Society (JLS), Japanese Geotechnical Society (JGS), Japan Society for Natural Disaster Science (JSNDS) and Japan Association for Slope Disaster Management (JASDiM)

Co-sponsors

Cabinet Office (Disaster Management Bureau) of Japan; Ministry of Foreign Affairs of Japan (MOFA); Ministry of Education, Culture, Sports, Science and Technology-Japan (MEXT); Ministry of Land Infrastructure, Transport and Tourism (MLIT); Ministry of Agriculture, Forestry and Fisheries (MAFF); Science Council of Japan (SCJ); Japan International Cooperation Agency (JICA); Japan Society of Civil Engineers (JSCE); Japanese Society of Irrigation, Drainage and Rural Engineering (JSIDRE); Japan Society of Erosion Control Engineering; Japan Society of Engineering Geology.

Supporting Organizations with Finance

Tokyo Geographical Society
International Union of Geological Sciences (IUGS)
Association for Disaster Prevention Research, Kyoto, Japan

Organizing Committee

Honorary Chairpersons
Audrey Azoulay, Director-General of UNESCO*
Mami Mizutori, Special Representative of the United Nations Secretary-General for Disaster Risk Reduction*

Petteri Taalas, Secretary-General of WMO*
Qu Dongyu, Director-General of FAO*
David Malone, Under-Sectretary General of the Unitred Nations and Rector of UNU
Daya Reddy, President of ISC
Gong Ke, President of WFEO
Qiuming Cheng, President of IUGS
Kathryn Whaler, President of IUGG
Qunli Han, Executive Director of Integrated Research on Disaster Risk (IRDR)
Walter Ammann, President and CEO of Global Risk Forum GRF Davos, Switzerland
Juichi Yamagiwa, President of Kyoto University, Japan
Angelo Borrelli, Head of the National Civil Protection Department, Italian Presidency of the Council of Ministers, Italy
Darko But, Director General of the Administration for Civil Protection and Disaster Relief of the Republic of Slovenia, Slovenia
Akifumi Nakao, Director, International Cooperation Division, Disaster Management Bureau, Cabinet Office, Japan
Kazuyuki Imai, Director General of Sabo Department, Ministry of Land Infrastructure, Transport and Tourism, Japan*
Chungsik Yoo, President of the International Geosynthetics Society
Rafig Azzam, President of the International Association for Engineering Geology and the Environment
(*to be confirmed)

Chairpersons
Kyoji Sassa, Professor Emeritus, Kyoto University; Secretary General of ICL
Peter T. Bobrowsky, Geological Survey of Canada; President of ICL
Kaoru Takara, Kyoto University, Japan; Executive Director of ICL

Members
Željko Arbanas (University of Rijeka, Croatia)
Snježana Mihalić Arbanas (University of Zagreb, Croatia)
Nicola Casagli (University of Firenze, Italy)
Fausto Guzzetti (Department of Civil Protection, Italy)
Matjaž Mikoš (University of Ljubljana, Slovenia)
Paola Reichenbach (Research Institute for Geo-Hydrological Protection, National Research Council, Italy)
Shinji Sassa (Port and Airport Research Institute, Japan)
Alexander Strom (Geodynamics Research Center LLC, Russia)
Binod Tiwari (California State University, Fullerton, USA)
Veronica Tofani (University of Firenze, Italy)
Vít Vilímek (Charles University in Prague, Czech Republic)
Fawu Wang (Tongji University, China)

Chairpersons of Local Organizing Committee
Kaoru Takara (Kyoto University)
Daisuke Higaki (Japan Landslide Society)
Ikuo Towhata (Japanese Geotechnical Society)

Secretary Generals
Ryosuke Uzuoka (Disaster Prevention Research Institute, Kyoto University)
Kazuo Konagai (International Consortium on Landslides)
Khang Dang (International Consortium on Landslides)

International Scientific Committee
Beena Ajmera, North Dakota State University, USA
Snježana Mihalić Arbanas, University of Zagreb, Croatia
Željko Arbanas, Faculty of Civil Engineering, University of Rijeka, Croatia
Amin Askarinejad, Technische Universiteit Delft, Delft, The Netherlands
Peter T. Bobrowsky, Geological Survey of Canada, Sidney, Canada
Michele Calvello, University of Salerno, Italy
Giovanna Capparelli, Universita degli Studi della Calabria, Rende, Italy
Nicola Casagli, University of Florence, Italy
Yifei Cui, Tsinghua University, Beijing, China
Sabatino Cuomo, University of Salerno, Fisciano, Italy
Khang Dang, International Consortium on Landslides, Kyoto, Japan
Elias Garcia-Urquia, National Autonomous University of Honduras, Tegucigalpa, Honduras
Stefano Luigi Gariano, Research Institute for Geo-Hydrological Protection, CNR, Perugia, Italy
Daniele Giordan, Research Institute for Geo-Hydrological Protection, CNR, Italy
Fausto Guzzetti, Department of Civil Protection, Italy
Baator Has, Asia Air Survey, Tokyo, Japan
Hans-Balder Havenith, Universite de Liege, Liege, Belgium
D. P. Kanungo, Central Building Research Institute (CBRI), Roorkee, Uttarakhand, India
Oded Katz, Geological Survey of Israel, Jerusalem, Israel
Kazuo Konagai, International Consortium on Landslides, Kyoto, Japan
Doan Huy Loi, International Consortium on Landslides, Kyoto, Japan
Ping Lu, Tongji University, Shanghai, China
Olga Mavrouli, University of Twente, Enschede, The Netherlands
Matjaž Mikoš, Faculty of Civil and Geodetic Engineering, University of Ljubljana, Slovenia
Alessandro C. Mondini, Research Institute for Geo-Hydrological Protection, CNR, Italy
Veronica Pazzi, Department of Earth Science, University of Florence, Florence, Italy
Dario Peduto, Department of Civil Engineering, University of Salerno, Fisciano, Italy
Paola Reichenbach, Research Institute for Geo-Hydrological Protection, CNR, Italy
Paola Salvati, Research Institute for Geo-Hydrological Protection, CNR, Italy
Katsuo Sasahara, Kochi University, Japan
Kyoji Sassa, International Consortium on Landslides, Kyoto, Japan
Shinji Sassa, Port and Airport Research Institute, Japan
Andrea Segalini, University of Parma, Italy
Hendy Setiawan, Universitas Gadjah Mada, Yogyakarta, Indonesia
Alexander Strom, Geodynamics Research Center LLC, Moscow, Russia
Kaoru Takara, Kyoto University, Japan
Faraz Tehrani, Deltares, Delft, The Netherlands
Binod Tiwari, California State University, Fullerton, California, USA
Veronica Tofani, University of Florence, Italy
Ryosuke Uzuoka, Kyoto University, Kyoto, Japan
Vít Vilímek, Faculty of Science, Charles University, Prague, Czech Republic
Fawu Wang, College of Civil Engineering, Tongji University, Shanghai, China
Gonghui Wang, Kyoto University, Kyoto, Japan
Mike Winter, Winter Associates Limited, Kirknewton, UK
Hiromitsu Yamagishi, Hokkaido Research Center of Geology (HRCG), Sapporo, Japan

Local Organizing Committee
Shinro Abe, Okuyama Boring Co., Ltd.
Kiminori Araiba, Fire and Disaster Management College
Shiho Asano, Forestry and Forest Products Research Institute
Has Baator, Asia Air Survey Co., Ltd.

Hiromu Daimaru, Forestry and Forest Products Research Institute
Khang Dang, International Consortium on Landslides
Mitusya Enokida, Japan Conservation Engineers & Co., Ltd.
Kumiko Fujita, International Consortium on Landslides
Kazunori Hayashi, Okuyama Boring Co., Ltd.
Daisuke Higaki, The Japan Landslide Society
Kiyoharu Hirota, Kokusai Kogyo Co., Ltd.
Kazuo Konagai, International Consortium on Landslides
Taketoshi Marui, MARUI & Co., Ltd.
Satoshi Nishikawa, Nagoya University
Keisuke Oozone, OYO Corporation
Katsuo Sasahara, Kochi University
Kyoji Sassa, International Consortium on Landslides
Shinji Sassa, Port and Airport Research Institute
Go Sato, Teikyo Heisei University
Nobuyuki Shibasaki, Nippon Koei Co., Ltd.
Nobuo Sugiura, Japan Association for Slope Disaster Management
Kaoru Takara, Kyoto University
Keisuke Takimoto, GODAI KAIHATSU Corporation
Yoko Tomita, Public Works Research Institute
Ikuo Towhata, The Japanese Geotechnical Society
Kenichi Tsukahara, Kyushu University
Ryosuke Tsunaki, Sabo & Landslide Technical Center
Taro Uchida, Saitama University
Mie Ueda, International Consortium on Landslides
Ryosuke Uzuoka, Kyoto University
Fawu Wang, Tongji University
Hiroshi Yagi, Yamagata University
Hiromitsu Yamagishi, Shin Engineering Consultants Co., Ltd.
Maki Yano, OSASI Technos Inc.

Foreword by Mami Mizutori

More landslides can be expected as climate change exacerbates rainfall intensity. The long-term trend of the last 40 years has seen the number of major recorded extreme weather events almost double, notably floods, storms, landslides, and wildfires.

Landslides are a serious geological hazard. Among the host of natural triggers are intense rainfall, flooding, earthquakes or volcanic eruption, and coastal erosion caused by storms that are all too often tied to the El Niño phenomenon. Human triggers including deforestation, irrigation or pipe leakage, and mine tailings, or stream and ocean current alteration can also spark landslides. Landslides can also generate tsunamis, as Indonesia experienced in 2018.

Globally, landslides cause significant economic loss and many deaths and injuries each year. Therefore, it is important to understand the science of landslides: why they occur, what factors trigger them, the geology associated with them, and where they are likely to happen.

Landslides with high death tolls are often a result of failures in risk governance, poverty reduction, environmental protection, land use and the implementation of building codes. Understanding the interrelationships between earth surface processes, ecological systems, and human activity is the key to reducing landslide risk.

The Sendai Framework for Disaster Risk Reduction, the global plan to reduce disaster losses adopted in 2015, emphasizes the importance of tackling these risk drivers through improved governance and a better understanding of disaster risk.

One important vehicle for doing that is the Sendai Landslide Partnerships 2015–2025 for global promotion of understanding and reduction of landslide risk facilitated by the International Consortium on Landslides (ICL) and signed by the leaders of 22 global stakeholders, including the UN Office for Disaster Risk Reduction (UNDRR), during the Third UN World Conference on Disaster Risk Reduction in Sendai, Japan.

The Sendai Landslide Partnerships—featured on the Sendai Framework Voluntary Commitments online platform—helps to provide practical solutions and tools, education, and capacity building, to reduce landslide risks.

The work done by the Sendai Partnerships can be of value to many stakeholders including civil protection, planning, development and transportation authorities, utility managers, agricultural and forest agencies, and the scientific community.

UNDRR fully supports the work of the Sendai Landslide Partnerships and ICL and looks forward to an action-oriented outcome from the 5th World Landslide Forum to be held in November 2020 in Kyoto, Japan. Successful efforts to reduce disaster losses are a major contribution to achieving the overall 2030 Agenda for Sustainable Development.

Mami Mizutori
United Nations Special Representative of the
Secretary-General for Disaster Risk Reduction

Foreword by the Assistant Director-General for the Natural Sciences Sector of UNESCO for the Book of the 5th *World Landslide Forum*

As the world slowly recovers from the COVID-19 global pandemic, and looking back at the way this crisis developed, it becomes evident that as a global community we were not prepared for an event of this scale. Although not commonly perceived as such, biological hazards such as epidemics are included in the Sendai Framework for Disaster Risk Reduction 2015–2030. In that sense, the preparedness approach for a pandemic is very similar to that of a geophysical natural hazard such as landslides.

Although natural hazards are naturally occurring phenomena, the likelihood of their occurrence and of associated disasters is rising. Climate change, urban pressure, under-development and poverty and lack of preparedness are increasingly transforming these natural hazards into life-threatening disasters with severe economic impacts. Therefore, Disaster Risk Reduction (DRR) is gaining momentum on the agenda of the UN system of Organizations including UNESCO. While the Sendai Framework for Disaster Risk Reduction 2015–2030 is the roadmap for DRR, other global agendas including the Sustainable Development Goals, the Paris Climate Agreement and the New Urban Agenda have targets which cannot be attained without DRR.

In shaping its contribution to those global agendas, UNESCO is fully committed in supporting its Member States in risk management, between its different mandates and disciplines and with relevant partners. The International Consortium on Landslides (ICL) is UNESCO's key partner in the field of landslide science. The Organization's support to the Consortium is unwavering. Since ICL was established in 2002, the two organizations have a long history of cooperation and partnership and UNESCO has been associated with almost all of ICL activities. I am very glad that ICL and UNESCO are mutually benefitting from their collaboration.

The 5th World Landslide Forum (WLF5) is expected to represent a milestone in the history of landslide science particularly for scientists and practitioners. One of the major outcomes of WLF5 will be the Kyoto 2020 Commitment for global promotion of understanding and reducing landslide disaster risk (KLC2020). This commitment is expected to strengthen and expand the activities of the Sendai Landslide Partnership 2015–2025. With UNESCO already engaged as a partner, the adoption of this international commitment will raise global awareness on landslide risk and mobilize wider partnerships that draw together stakeholders from all levels of society, across different regions, sectors and disciplines.

It is my great pleasure to congratulate the organizers for holding this event and assure you that UNESCO is fully committed in contributing to its success. As part of that contribution, our Organization is proud to host a session on landslides and hazard assessment at UNESCO-designated sites such as natural World Heritage sites, biosphere reserves and UNESCO Global Geoparks. This session aims to assess landslide impacts on our shared cultural and natural heritage, providing the best opportunity to generate public awareness and capacity development for landslide disaster reduction.

I am confident that WLF5 will contribute to further advance the knowledge of both scientists and practitioners regarding landslide disaster risk reduction. This book paves the way for the science, knowledge and know-how which will feature in the deliberations of the Forum. UNESCO commends all of the contributors to this publication. I look forward to an enhanced collaboration between UNESCO and ICL in future activities and undertakings.

Shamila Nair-Bedouelle
Assistant Director-General for Natural Sciences
UNESCO

Preface I

Understanding and Reducing Landslide Disaster Risk

Book Series: ICL Contribution to Landslide Disaster Risk

The International Consortium on Landslides (ICL) was established in pursuance of the 2002 Kyoto Declaration "Establishment of an International Consortium on Landslides," with its Statutes adopted in January 2002. The Statutes define the General Assembly of ICL as follows: in order to report and disseminate the activities and achievements of the Consortium, a General Assembly shall be convened every 3 years by inviting Members of the International Consortium on Landslides, individual members within those organizations, and all levels of cooperating organizations and individual researchers, engineers, and administrators. The General Assembly developed gradually prior to, during and after its first meeting in 2005. In the light of the 2006 Tokyo Action Plan, the Assembly was further facilitated at, and following the First World Landslide Forum held in November 2008. On the occasion of each of its triennial forums, ICL publishes the latest progress of landslide science and technology for the benefit of the whole landslide community including scientists, engineers, and practitioners in an understandable form. Full color photos of landslides and full color maps are readily appreciated by those from different disciplines. We have published full color books on landslides at each forum. In 2019, ICL created a new book series "ICL Contribution to Landslide Disaster Risk Reduction" ISSN 2662-1894 (print version) and ISSN 2662-1908 (electronic version). Six volumes of full color books *Understanding and Reducing Landslide Disaster Risk* will be published in 2020 as the first group of books of this series.

The Letter of Intent 2005 and the First General Assembly 2005

The United Nations World Conference on Disaster Reduction (WCDR) was held in Kobe, Japan, 18–22 January 2005. At this Conference, ICL organized session 3.8 "New international Initiatives for Research and Risk Mitigation of Floods (IFI) and Landslides (IPL)" on 19 January 2005 and adopted a "Letter of Intent" aimed at providing a platform for a holistic approach in research and learning on 'Integrated Earth System Risk Analysis and Sustainable Disaster Management'. This Letter was agreed upon and signed, during the first semester of 2005, by heads of seven global stakeholders including the United Nations Educational, Scientific and Cultural Organization (UNESCO), the World Meteorological Organization (WMO), the Food and Agriculture Organization of the United Nations (FAO), the United Nations International Strategy for Disaster Risk Reduction (UNISDR-currently UNDRR), the United Nations University (UNU), the International Council for Science (ICSU-Currently ISC), and the World Federation of Engineering Organizations (WFEO).

The first General Assembly of ICL was held at the Keck Center of the National Academy of Sciences in Washington D.C., USA, on 12–14 October 2005. It was organized after the aforementioned 2005 World Conference on Disaster Reduction (WCDR). ICL published the first full color book reporting on Consortium activities for the initial 3 years, 2002–2005 titled "Landslides-Risk analysis and sustainable disaster management." In the preface of this book, the Letter of Intent for Integrated Earth System Risk Analysis and Sustainable Disaster Management was introduced. Results of the initial projects of the International Programme on Landslides (IPL) including IPL C101-1 Landslide investigation in Machu Picchu World Heritage, Cusco, Peru and previous agreements and MoU between UNESCO, ICL and the Disaster Prevention Research Institute of Kyoto University including UNESCO/KU/ICL UNITWIN Cooperation programme were published as well in this book.

The 2006 Tokyo Action Plan and the First World Landslide Forum 2008

Based on the Letter of Intent, the 2006 Tokyo Round-Table Discussion—"Strengthening Research and Learning on Earth System Risk Analysis and Sustainable Disaster Management within UN-ISDR as Regards Landslides"—toward a dynamic global network of the International Programme on Landslides (IPL) was held at the United Nations University, Tokyo, on 18–20 January 2006. The 2006 Tokyo Action Plan—Strengthening research and learning on landslides and related earth system disasters for global risk preparedness—was adopted. The ICL exchanged Memoranda of Understanding (MoUs) concerning strengthening cooperation in research and learning on earth system risk analysis and sustainable disaster management within the framework of the United Nations International Strategy for Disaster Reduction regarding the implementation of the 2006 Tokyo action plan on landslides with UNESCO, WMO, FAO, UNISDR (UNDRR), UNU, ICSU (ISC), and WFEO, respectively, in 2006. A set of these MoUs established the International Programme on Landslides (IPL) as a programme of the ICL, the Global Promotion Committee of IPL to manage the IPL, and the triennial World Landslide Forum (WLF), as well as the concept of the World Centres of Excellence on Landslide Risk Reduction (WCoE).

The First World Landslide Forum (WLF1) was held at the Headquarters of the United Nations University, Tokyo, Japan, on 18–21 November 2008. 430 persons from 49 countries/regions/UN entities were in attendance. Both Hans van Ginkel, Under Secretary-General of the United Nations/Rector of UNU who served as chairperson of the Independent Panel of Experts to endorse WCoEs, and Salvano Briceno, Director of UNISDR who served as chairperson of the Global Promotion Committee of IPL, participated in this Forum. The success of WLF1 paved the way to the successful second and third World Landslide Forum held in Italy and China, respectively.

The Second World Landslide Forum 2011 and the Third World Landslide Forum 2014

The Second World Landslide Forum (WLF2)—Putting Science into Practice—was held at the Headquarters of the Food and Agriculture Organization of the United Nations (FAO) on 3–9 October 2011. It was jointly organized by the IPL Global Promotion Committee (ICL, UNESCO, WMO, FAO, UNDRR, UNU, ISC, WFEO) and two ICL members from Italy: the Italian Institute for Environmental Protection and Research (ISPRA) and the Earth Science Department of the University of Florence with support from the Government of Italy and many Italian landslide-related organizations. It attracted 864 participants from 63 countries.

The Third World Landslide Forum (WLF3) was held at the China National Convention Center, Beijing, China, on 2–6 June 2014. A high-level panel discussion on an initiative to create a safer geoenvironment toward the UN Third World Conference on Disaster Risk Reduction (WCDRR) in 2015 and forward was moderated by Hans van Ginkel, Chair of Independent Panel of Experts for World Centers of Excellence (WCoE). In a special address to this high-level panel discussion, Irina Bokova, Director-General of UNESCO, underlined that countries should be united to work against natural disasters and expressed commitment that UNESCO would like to further deepen cooperation with ICL. Ms. Bokova awarded certificates to 15 World Centres of Excellence.

The Sendai Landslide Partnerships 2015 and the Fourth World Landslide Forum 2017

The UN Third World Conference on Disaster Risk Reduction (WCDRR) was held in Sendai, Japan, on 14–18 March 2015. ICL organized the Working Session "Underlying Risk Factors" together with UNESCO, the Japanese Ministry of Land, Infrastructure, Transport and Tourism (MLIT) and other competent organizations. The session adopted ISDR-ICL Sendai Partnerships 2015–2025 (later changed to Sendai Landslide Partnerships) for global promotion of understanding and reducing landslide disaster risk as a Voluntary Commitment to the World Conference on Disaster Risk Reduction, Sendai, Japan, 2015 (later changed to Sendai Framework for Disaster Risk Reduction). After the session on 16 March 2015, the Partnerships was signed by Margareta Wahlström, Special Representative of the UN Secretary-General for Disaster Risk Reduction, Chief of UNISDR (UNDDR), and other representatives from 15 intergovernmental, international, and national organizations. Following the Sendai Landslide Partnerships, the Fourth World Landslide Forum was held in Ljubljana, Slovenia from 29 May to 02 June in 2017. On that occasion, five volumes of full color books were published to disseminate the advances of landslide science and technology. The high-level panel discussion on 30 May and the follow-up round table discussion on 31 May adopted the 2017 Ljubljana Declaration on Landslide Risk Reduction. The Declaration approved the outline of the concept of "Kyoto 2020 Commitment for global promotion of understanding and reducing landslide disaster risk" to be adopted at the Fifth World Landslide Forum in Japan, 2020.

The Fifth World Landslide Forum 2020 and the Kyoto Landslide Commitment 2020

The Fifth World Landslide Forum was planned to be organized on 2–6 November 2020 at the National Kyoto International Conference Center (KICC) and the preparations for this event were successfully ongoing until the COVID-19 pandemic occurred over the world in early 2020. The ICL decided to postpone the actual Forum to 2–6 November 2021 at KICC in Kyoto, Japan. Nevertheless, the publication of six volumes of full color books *Understanding and Reducing Landslide Disaster Risk* including reports on the advances in landslide science and technology from 2017 to 2020 is on schedule. We expect that this book will be useful to the global landslide community.

The Kyoto Landslide Commitment 2020 will be established during the 2020 ICL-IPL Online Conference on 2–6 November 2020 on schedule. Joint signatories of Kyoto Landslide Commitment 2020 are expected to attend a dedicated session of the aforementioned Online Conference, scheduled on 5 November 2020 which will also include and feature the Declaration of the launching of KLC2020. Landslides: Journal of the International Consortium on Landslides is the common platform for KLC2020. All partners may contribute and publish news and reports of their activities such as research, investigation, disaster reduction administration in the category of News/Kyoto Commitment. Online access or/and hard copy of the Journal will be sent to KLC2020 partners to apprise them of the updated information from other partners. As of 21 May 2020, 63 United Nations, International and national organizations have already signed the KLC2020.

Call for Partners of KLC2020

Those who are willing to join KLC2020 and share their achievements related to understanding and reducing landslide disaster risk in their intrinsic missions with other partners are invited to inform the ICL Secretariat, the host of KLC2020 secretariat (secretariat@iclhq.org). The ICL secretariat will send the invitation to the aforementioned meeting of the joint signatories and the declaration of the launching of the KLC2020 on 5 November 2020.

Eligible Organizations to be Partners of the KLC2020

1. ICL member organizations (full members, associate members and supporters)
2. ICL supporting organization from UN, international or national organizations and programmes
3. Government ministries and offices in countries having more than 2 ICL on-going members
4. International associations/societies that contribute to the organization of WLF5 in 2021 and WLF6 in 2023
5. Other organizations having some aspects of activities related to understanding and reducing landslide disaster risk as their intrinsic missions.

Kyoji Sassa
Chair of WLF5/
Secretary-General of ICL
Kyoto, Japan

Peter T. Bobrowsky
President of ICL
Sidney, Canada

Kaoru Takara
Executive Director of ICL
Kyoto, Japan

Appendix: World Landslide Forum Books

WLF	Place/participants	Title	Editors	Publisher/pages
WLF0 (1st General Assembly) 2005	Washington D.C., USA 59 from 17 countries/UNs	Landslides-Risk Analysis and Sustainable Disaster Management	Kyoji Sassa, Hiroshi Fukuoka, Fawu Wang, Goghui Wang	Springer/377 pages ISBN: 978-3-540-2864-6
WLF1 2008	Tokyo, Japan 430 from 49 countries/regions/UNs	Landslides-Disaster Risk Reduction	Kyoji Sassa, Paolo Canuti	Springer/649 pages ISBN: 978-3-540-69966-8
WLF2 2011	Rome, Italy 864 from 63 countries	Landslide Science and Practice Vol. 1 Landslide inventory and Sustainability and Hazard Zoning	Claudia Margottini, Paolo Canuti, Kyoji Sassa	Springer/607 pages ISBN: 978-3-642-31324-0
		Vol. 2 Early Warning, Instrumentation and Monitoring		Springer/685 pages ISBN: 978-3-642-31444-5
		Vol. 3 Spatial Analysis and Modelling		Springer/440 pages ISBN: 978-3-642-31309-7
		Vol. 4 Global Environmental Change		Springer/431 pages ISBN: 978-3-642-31336-3
		Vol. 5 Complex Environment		Springer/354 pages ISBN: 978-3-642-31426-1
		Vol. 6 Risk Assessment, Management and Mitigation		Springer/789 pages ISBN: 978-3-642-31318-9
		Vol. 7 Social and Economic Impact and Policies		Springer/333 pages ISBN: 978-3-642-31312-7
WLF3 2014	Beijing, China 531 from 45 countries/regions/UNs	Landslide Science for a Safer Geoenvironment Vol. 1 The International Programme on Landslides (IPL)	Kyoji Sassa, Paolo Canuti, Yueping Yin	Springer/493 pages ISBN: 978-3-319-04998-4
		Vol. 2 Methods of Landslide Studies		Springer/851 pages ISBN: 978-3-319-05049-2
		Vol. 3 Targeted Landslides		Springer/717 pages ISBN: 978-3-319-04995-3
WLF4 2017	Ljubljana, Slovenia 588 from 59 countries/regions/UNs	Advancing Culture of Living with Landslides Vol. 1 ISDR-ICL Sendai Partnerships 2015-2025	Kyoji Sassa, Matjaž Mikoš, Yueping Yin	Springer/585 pages ISBN: 978-319-53500-5

(continued)

WLF	Place/participants	Title	Editors	Publisher/pages
		Vol. 2 Advances in Landslide Science	Matjaž Mikoš, Binod Tiwari, Yueping Yin, Kyoji Sassa	Springer/1197 pages ISBN: 978-319-53497-8
		Vol. 3 Advances in Landslide Technology	Matjaž Mikoš, Željko Arbanas, Yueping Yin, Kyoji Sassa	Springer/621 pages ISBN: 978-3-319-53486-2
		Vol. 4 Diversity of Landslide Forms	Matjaž Mikoš, Nicola Casagli, Yueping Yin, Kyoji Sassa	Springer/707 pages ISBN: 978-3-319-53484-8
		Vol. 5 Landslides in Different Environments	Matjaž Mikoš, Vít Vilímek, Yueping Yin, Kyoji Sassa	Springer/557 pages ISBN: 978-3-319-53482-4
WLF5	2020 (publication) 2021 (Forum)	Understanding and Reducing Landslide Disaster Risk Vol. 1 Sendai Landslide Partnerships and Kyoto Landslide Commitment	Kyoji Sassa, Matjaž Mikoš, Shinji Sassa, Peter T. Bobrowsky, Kaoru Takara, Khang Dang	Springer In Process
		Vol. 2 From mapping to hazard and risk zonation	Fausto Guzzetti, Snježana Mihalić Arbanas, Paola Reichenbach, Kyoji Sassa, Peter T. Bobrowsky, Kaoru Takara	
		Vol. 3 Monitoring and early Warning	Nicola Casagli, Veronica Tofani, Kyoji Sassa, Peter T. Bobrowsky, Kaoru Takara	
		Vol. 4 Testing, modelling and risk assessment	Binod Tiwari, Kyoji Sassa, Peter T. Bobrowsky, Kaoru Takara	
		Vol. 5 Catastrophic landslides and Frontier of Landslide Science	Vit Vilimek, Fawu Wang, Alexander Strom, Kyoji Sassa, Peter T. Bobrowsky, Kaoru Takara	
		Vol. 6 Specific topics in landslide science and applications	Željko Arbanas, Peter T. Bobrowsky, Kazuo Konagai, Kyoji Sassa, Kaoru Takara	

Preface II

Landslides represent one of the major natural hazards that cause loss of human lives and a high number of injuries, significant damage of constructions and infrastructure, high economic loss, and significant consequences in environment. It would be expected that in circumstances of global climate change, overall consequences of landslides will be higher and higher. The response of science society is in significant development of landslide science and its application in different parts from landslide identification and investigation to mitigation and remediation, from landslide observation and monitoring to landslide modeling, as well as in landslide vulnerability, susceptibility, hazard, and risk mapping, with the final aim to contribute to the landslide risk reduction. Although for the most parts of researches, the field of research within the landslide science is very clearly defined including some new methods, and new technical achievements in landslide science enable the development of new landslide research "windows." The Volume 6 entitled Specific topics in landslide science and applications was established as an additional volume of World Landslide Forum 5 that included contributions from several landslide research "windows" that aren't always present as usual parts of landslide conferences. These contributions can be divided into the following four general topics presented as the Volume parts:

- Impact of large ground deformations near seismic faults on critically important civil-infrastructures;
- Recent Progress in the Landslide Initiating Science;
- Earth Observation and Machine Learning;
- General Landslide Studies.

In total, 33 contributions from 22 countries (Australia, Canada, China, Chinese Taipei, Croatia, the Czech Republic, France, Greece, Honduras, Indonesia, Italy, Japan, Nepal, New Zealand, Norway, Russia, Sri Lanka, Switzerland, Thailand, The Netherlands, the United Kingdom, the USA) have been submitted and, after the review process, accepted for publishing in this Volume. This Volume 6, thus, carries 1 Theme Lecture, 6 Keynote Lectures, and 31 research articles prepared by 129 authors in total from all over the world.

The first part, Impact of large ground deformations near seismic faults on critically important civil-infrastructures, encompasses nine articles related to the landslide disasters caused by well-known strong earthquakes in different parts of world during past years (1999 Chi-Chi Earthquake, Taiwan; 2004 Mid-Niigata Prefecture Earthquake, Japan; 2007 Kashiwazaki Earthquake, Japan; 2008 Wenchuan Earthquake, China; 2011 Tohoku Earthquake and Tsunami, Japan; 2015 Gorkha Earthquake, Nepal; 2016 Kumamoto Earthquakes, Japan; 2016 Kaikoiura Earthquake, New Zealand; 2018 Hokkaido Eastern-Iburi Earthquake, Japan). This part starts with the Theme Lecture entitled Recent earthquakes that hit areas covered and/or underlain by pyroclastic matters and their impacts on lifelines by Konagai, K., Tang, A. K. and

Eidinger, J. M. that presented and analyzed consequences of two major earthquakes in Japan last years (2016 Kumamoto Earthquakes; 2018 Hokkaido Eastern-Iburi Earthquake). These earthquakes caused extreme appetences of multiple landslides and ground subsidence and their analyses will enable estimation of landslide runout distances based on simple empirical equation in the future. The other articles describe the experiences gained from different landslide appearances triggered by strong earthquakes and their influence on human lives, infrastructure, and other constructions.

The second part, Recent Progress in the Landslide Initiating Science, contains 13 articles that deal with the analyzing and explanation of different types of landslide initiation process with different triggering conditions. This part starts with two Keynote Lectures: Water exfiltration from bedrock: an effective landslide triggering mechanism by Askarinejad, A. and Springman, S. M. and Controls on landslide size: insights from field survey data by Qiu, H., Cui, Y., Yang, D., Hu, S., Pei, Y., and Ma, S. The first Keynote Lecture explains and discuss a process of landslide initiation caused by exfiltration of water from a bedrock into the slope following the results of centrifuge physical modeling. In the second Keynote Lecture, the group of Chinese researchers presented their experiences in the determination of initiated landslide size based on field survey and terrain topography. The other articles describe different aspects and achievements in landslide initiating analyses based on in-field observations, numerical modeling, as well as physical in situ and in laboratory modeling of landslide initiation processes.

The third part, Earth Observation and Machine Learning, contains one Keynote Lecture and three articles dealing with earth observation big data and their use in landslide behavior prediction. The Keynote Lecture entitled HRT Point-cloud for Catastrophic Landslides in the twenty-first century: Overview of data acquisition and products, and proposals for new acquisition and processing concepts by Gomez, C., Allouis, T., Lissak, C., Hotta, N., Shinohara, Y., Hadmoko, D. S., Vilimek, V., Wassmer, P., Lavigne, F., Setiawan, A., Sartohadi, J., Saputra, A., and Rahardianto, T. present research on present technologies for landslide monitoring, as well as emerging systems, and propose "new" ideas for point-cloud processing and data usage without having to grid or interpolate data. The other three articles are related to landslide displacement patterns predicting and landslide stability predicting using machine learning as well as automated dissemination of landslide monitoring bulletins for early warning applications.

The fourth part, General Landslide Studies, encompassed landslide studies in a wider area of landslide researches mostly related to the risky geology conditions in areas prone to landsliding. This part contains three Keynote Lectures and nine articles on different topics. The first Keynote Lecture entitled Engineering geological appreciation in landslide mapping for a natural gas pipeline project: challenges and risk reduction measures by Marinos, V., Papazachos, K., Stoumpos, G., Papouli, D., Papathanassiou, G., and Stimaratzis, T, presents challenges and risk reduction methods applied during the field investigation and construction of natural gas pipeline project over flysch slopes prone to landsliding in Greece. The second Keynote Lecture entitled Loess stratigraphy and loess landslides in the Chinese loess plateau by Li, T., Haider, M., Shen, W., and Li, P. describes the processes in loess that, because of loose structure, is sensitive to the water content changes and susceptible to failure occurrences caused by rain or engineering activities. The main type of landslides in loess deposits are described and their mechanisms are explained. The last Keynote Lecture An engineering geology model of the Jettan rockslide, northern Norway by Vick, L. M., Berg, J. N., Eggers, M., Hormes, A., Skrede, I., and Blikra, L. H. describes a comprehensive study about the Jettan rockslide, one of three high-risk sites in Troms and Finnmark County in the Scandinavian Mountains, northern Norway and development of gravitational slope deformations and possible metamorphic rock slope collapse. This Keynote Lecture is accompanied with two articles related to similar rockslide in Norway, while the other seven articles in this part of the Volume are related to some other different landslide types and studies all around the world.

Rijeka, Croatia Željko Arbanas
Sidney, Canada Peter T. Bobrowsky
Kyoto, Japan Kazuo Konagai

Acknowledgments The editors of the WLF5 Volume 6 Specific topics in landslide science and applications would like to thank all authors for their contribution, all handling editors who organized the review process of submitted articles, and all reviewers who have reviewed the papers submitted to this Volume and together with editors, ensured adequate quality of accepted and published articles.

Contents

Part I Impact of Large Ground Deformations Near Seismic Faults on Critically Important Civil-Infrastructures

Recent Earthquakes that Hit Areas Covered and/or Underlain by Pyroclastic Matters and Their Impacts on Lifelines 3
Kazuo Konagai, Alex K. Tang, and John M. Eidinger

Landslides in Recent Earthquakes and Damage to Lifelines 19
Kazuo Konagai

Lessons Learned—Landslide Induced Lifelines Disasters from Past Earthquakess 27
Alex K. Tang

Lessons from Recent Geo-Disasters in Hokkaido Under Earthquake 35
Shima Kawamura

Relation Between Horizontal Direction of Crustal Deformation Surveyed on the Control Points and Area Ratio of the Slope Failures Triggered by the 2016 Kumamoto Earthquake (M_j 7.3) 45
Hiroshi P. Sato and Hiroshi Une

Seismic Performance of Buried Pipelines Against Large Ground Deformation of Strike-Slip Faults 55
Farzad Talebi and Junji Kiyono

Reconstruction Strategies for Mw 7.8 Earthquake-Induced Landslide-Affected Settlements in Nepal 61
Tara Nidhi Bhattarai, Dhruba Prasad Sharma, and Lekh Prasad Bhatta

State of Nuclear Power Plant Risk Assessment for Ground Deformation with Seismic Faulting 67
Katsumi Ebisawa, Toshiaki Sakai, Futoshi Tanaka, Ryusuke Haraguchi, Yoshinori Mihara, and Yuji Nikaido

Relationship Between Arias Intensity and the Earthquake-Induced Displacements of Slopes 75
Ching Hung, Chih-Hsuan Liu, and Hsuan-Ho Wang

Part II Recent Progress in the Landslide Initiating Science

Water Exfiltration from Bedrock: A Drastic Landslide Triggering Mechanism 85
Amin Askarinejad and Sarah M. Springman

Controls on Landslide Size: Insights from Field Survey Data 101
Haijun Qiu, Yifei Cui, Dongdong Yang, Sheng Hu, Yanqian Pei, and Shuyue Ma

Geologic and Hydrologic Investigations on Slope Failures Triggered by Extreme Rainfall on Izu Oshima Island, Japan 121
Ikuo Towhata, Takeshi Akima, Satoshi Goto, Shigeru Goto, Junya Tanaka, and Shogo Aoyama

Lessons from Recent Geo-Disasters in Hokkaido Under Heavy Rainfall 131
Tatsuya Ishikawa

Lessons from Geo-Disasters Caused by Heavy Rainfall in Recent Years in Kyushu Island, Japan 141
Noriyuki Yasufuku and Adel Alowiasy

Comparison of Relationship Between Debris-Flow Volume and Peak Discharge in Different Regions 149
Tao Wang and Mingfeng Deng

Investigation of Internal Erosion of Wide Grading Loose Soil—A Micromechanics-Based Study 155
Yifei Cui, Yanzhou Yin, and Chaoxu Guo

Formation Mechanism and Stability of the Instable Block Formed in Xinmo Landsilde 163
Bingli Hu, Lijun Su, Qijun Xie, Fangwei Yu, and Chonglei Zhang

Landslide Field Experiment on a Natural Slope in Futtsu City, Chiba Prefecture 169
Hirotaka Ochiai, Katsuo Sasahara, and Yusuke Koyama

Mechanism of Landslide Initiation in Small-Scale Sandy Slope Triggered by an Artificial Rain 177
Vedran Jagodnik, Josip Peranić, and Željko Arbanas

Experimental Study on the Formation and Propagation of Debris Flows Triggered by Glacial Lake Outburst Floods 185
Huayong Chen, Peng Cui, Xiaoqing Chen, and Jiangang Chen

Quantitative Analysis of Landslide Processes Based on Seismic Signals—A New Method for Monitoring and Early Warning of Landslide Hazards 191
Yan Yan, Yifei Cui, Shuyao Yin, and Xin Tian

Part III Earth Observation and Machine Learning

High-Resolution Point-Cloud for Landslides in the 21st Century: From Data Acquisition to New Processing Concepts 199
C. Gomez, T. Allouis, C. Lissak, N. Hotta, Y. Shinohara, D. S. Hadmoko, V. Vilimek, P. Wassmer, F. Lavigne, A. Setiawan, J. Sartohadi, A. Saputra, and T. Rahardianto

Detecting Change of Patterns in Landslide Displacements Using Machine Learning, an Example Application 215
Giacomo Titti, Matteo Mantovani, and Giulia Bossi

Predicting Rainfall Induced Slope Stability Using Random Forest Regression and Synthetic Data 223
Elahe Jamalinia, Faraz S. Tehrani, Susan C. Steele-Dunne, and Philip J. Vardon

Automatized Dissemination of Landslide Monitoring Bulletins for Early Warning Applications 231
Daniele Giordan, Aleksandra Wrzesniak, Paolo Allasia, and Davide Bertolo

Part IV General Landslide Studies

Engineering Geological Appreciation in Landslide Mapping for a Natural Gas Pipeline Project: Challenges and Risk Reduction Measures 239
Vassilis Marinos, Kostas Papazachos, Georgios Stoumpos, Dimitra Papouli, George Papathanassiou, and Theodoros Stimaratzis

Loess Stratigraphy and Loess Landslides in the Chinese Loess Plateau 269
Tonglu Li, Mumtaz Haider, Wei Shen, and Ping Li

Keynote Lecture: The Jettan Rockslide—An Engineering Geological Overview . . . 289
Louise M. Vick, Jørgen N. Berg, Mark Eggers, Anne Hormes, Ingrid Skrede, and Lars Harald Blikra

Mapping, Hazard and Consequence Analyses for Unstable Rock Slopes in Norway 317
Reginald L. Hermanns, Thierry Oppikofer, Martina Böhme, Ivanna M. Penna, Pierrick Nicolet, and Marie Bredal

Landscape Formation and Large Rock Slope Instabilities in Manndalen, Northern Norway 325
Martina Böhme, Reginald L. Hermanns, and Tom R. Lauknes

Disaster Risk Assessment of the Silk Road 331
Peng Cui, Qiang Zou, Yu Lei, Zhengtao Zhang, and Shengnan Wu

Analyzing the Characteristics of Glacial Debris Flow Activity in Parlung Tsangpo Basin, Tibet 339
Jiao Wang, Qiang Zou, Wen Jin, and Yanju Fu

Rehabilitation of Gully-Dominant Hill Slopes by Using Low-Cost Measures—A Case Study in Nepal 347
Daisuke Higaki, Kishor Kumar Karki, Naoto Koiwa, Mio Takahashi, and Sohan Kumar Ghimire

Site Suitability Analysis for Nature-Based Landslide Risk Mitigation 355
G. A. Chinthaka Ganepola, Udeni Priyantha Nawagamuwa, Anurudda Kumara Karunarathna, Senaka Basnayake, Lilanka Kankanamge, and Dhanushka Jayathilake

Study on the Application of Nature Based Landslide Mitigation in Sri Lanka . . . 361
Udeni P. Nawagamuwa and Markandu Dishan

Slope Stability Around the Northern Part of the Tegucigalpa Basin, Honduras: A Case of Landslide Process at Residential Development Areas 369
Kiyoharu Hirota, Cincy Rosa, and Koichi Hasegawa

Classification of Cryogenic Landslides and Related Phenomena (by Example of the Territory of Russia) 377
Oleg V. Zerkal and Alexander L. Strom

Cutting-Edge Technologies Aiming for Better Outcomes of Landslide Disaster Mitigation 385
Kazuo Konagai

International Consortium on Landslides 407

Part I

Impact of Large Ground Deformations Near Seismic Faults on Critically Important Civil-Infrastructures

Recent Earthquakes that Hit Areas Covered and/or Underlain by Pyroclastic Matters and Their Impacts on Lifelines

Kazuo Konagai, Alex K. Tang, and John M. Eidinger

Abstract

Two major earthquakes highlighted in this review article are the 2016 Kumamoto Earthquakes and the 2018 Hokkaido Eastern-Iburi Earthquake. These two events that hit the southern and northern Japanese islands known as Kyushu and Hokkaido, respectively, have one thing in common from geological and geotechnical viewpoint; the quake-hit areas are covered and/or underlain by volcanic matters. These volcanic matters such as pumice and volcanic ash have crushable nature that can cause large ground deformations, thus resulting in significant service interruption of lifelines and hindering quick recovery of the quake-hit areas. The phenomena to be discussed in this article include a never-seen-before ground subsidence that occurred on a flood plain west of Mt. Aso in Kyushu, and multiple landslides in Hokkaido with the total area of the exposed bare earth reaching 13.4 km^2; the largest area that we've ever recorded since the Meiji era. The observed geometric features of the multiple landslide masses have a striking resemblance to those in a past event; the fact thus inspires a feeling of hope that this resemblance will allow for quick estimation of runout distances of these landslide masses with a simple empirical equation.

Keywords

Earthquake-induced landslides • Volcanic matters • Lifelines • Ground subsidence

Introduction

Japan has more than 100 active volcanoes. Their lavas are in general very sticky because the islands are lined up along the world largest subduction zone where sea-water-rich soils are being dragged down deep into the subducting plate boundaries (see, e.g., National Park Service 2020). Therefore, eruptions of these volcanoes are in general very explosive, and there are many areas east of these volcanoes covered thick with volcanic matters. Takahashi and Shoji (2002) reported that about 18% of the Japanese territory is covered with volcanic ash soils.

Two major earthquakes highlighted in this review article are the 2016 Kumamoto Earthquakes and the 2018 Hokkaido Eastern-Iburi Earthquake (2018 Hokkaido Earthquake, hereafter). These two events have one thing in common; the earthquake-hit areas are covered and/or underlain by volcanic products. Pyroclastic deposits can exist not only on the ground surface but also underground, causing unexpectedly large ground deformations.

The sequence of the 2016 Kumamoto Earthquakes was initiated with a moment magnitude (M_w, hereafter) 6.5 foreshock, and the mainshock of M_w 7.0 occurred 25 h after the foreshock. These earthquakes that hit Japan's southern island of Kyushu have caused extensive damage to a variety of lifeline facilities near ground ruptures that appeared along the known trace of Futagawa fault. Moreover, close to 500 mm of rain fell on some parts along the quake-hit areas on June 20 and 21, causing further extensive damage, highlighting the difficulty to cope with earthquake-flood multi hazards.

K. Konagai (✉)
University of Tokyo, 1-21-4-517 Wakaba, Shinjuku, Tokyo 160-0011, Japan
e-mail: kaz3776k@gmail.com

A. K. Tang
L&T Consultant, 2591 Pollard Drive, Mississauga, Ontario L5C3G9, Canada
e-mail: alexktang@mac.com

J. M. Eidinger
G&E Engineering System Inc., P.O. Box 3592 Olympic Valley, CA 96146-3592, USA
e-mail: john.eidinger@gmail.com

Ž. Arbanas et al. (eds.), *Understanding and Reducing Landslide Disaster Risk*, ICL Contribution to Landslide Disaster Risk Reduction,
https://doi.org/10.1007/978-3-030-60713-5_1

The 2018 Hokkaido Earthquake of M_w6.6 that hit Japan's northern island of Hokkaido can probably be titled as landslide earthquake. According to the Ministry of Land, Infrastructure, Transport and Tourism, Japan (MLIT, hereafter), the total area of the exposed bare earth reached 13.4 km^2; the largest area that we've ever recorded since the Meiji era, which extended from 1868 to 1912 (MLIT 2016). Detached landslide masses have wiped out houses, roads, etc., thus hindering quick recovery of the area.

This review article summarizes some important findings about large ground deformations in these earthquakes and their impacts on lifelines.

The 2016 Kumamoto Earthquake

Ground Zero of the Earthquake and Lifelines

Starting with a magnitude-6.5 foreshock on April 14, 2016, a series of major earthquakes including the magnitude-7.3 main shock on April 16 have hit the central Kumamoto area of Kyushu, southernmost and third largest of the four main islands of Japan, causing deaths, injuries and widespread damage to various facilities. The activity of the fault, whose right-lateral offset appeared in the main event along the previously known section of the Futagawa fault zone, caused extensive damage to roads, bridges, railways, a tunnel and a dam. The observed features of the damage again showed that not only intense shakes but also ground deformations such as landslides, lateral spread of embankments and levees, soil liquefactions etc., which are found within a swath along the fault trace, can be equally or more often responsible for devastations. According to Japan Rail (JR) Kyushu (Fukae 2017), there were about 750 reports of damage to conventional railway lines, of which 73% were caused by landslides, debris flows, rockfalls etc., and majority of these damaged railway sections were found in the vicinity of Tateno notch, the lowest, western portion of the caldera rim of Mt. Aso, formed by continual dislocations of the Futagawa fault over a long geological time (upper Fig. 1). Moreover, the lowest elevation of the notch has progressively been eroded deep by the discharging water from Aso caldera, stretching 25 km north to south and 18 km east to west. This Tateno notch was the ground zero of the Kumamoto Earthquakes in terms of the extensive damage to lifelines, with the fallen Aso Ohashi as the focus of this event. The Aso Ohashi, a 205 m long upper-deck steel arch bridge, which spans Kurokawa River that drains the water from the caldera through the notch, was first considered to have been taken out by a large landslide immediately west of the bridge. However, Chida et al. (2018) reported that the bridge may have largely been compressed by a fault-controlled relative ground displacement between the bridge abutments. The relative displacement reaching almost 2.24 m in the longitudinal direction of the bridge, associated with 0.68 m relative displacement in the bridge's traverse direction, was large enough for its arch ribs to be buckled.

Figure 1 shows a LiDAR image of this landslide west of Aso Ohashi; LiDAR, Laser based altimetry, can penetrate through tree canopy revealing detailed feature of bare earth left behind by past natural hazards, and the LiDAR image of the inner-facing caldera wall shows evidence of not only the most recent one that has hit an important location for traffic, transmission lines and a waterway leading to a penstock, but also scars of past landslides (Konagai et al. 2017). Moreover, cracks are seen immediately behind the fresh scar of this quake-induced landslide showing that there still remain unstable masses perching atop the exposed bare earth. MLIT, thus, has removed a greater part of the unstable soil masses behind the scar, and has taken urgent countermeasure stabilizing the uppermost slope face of about 30,000 m^2 and constructing 7 m high and 300 m long retaining wall near the toe of the slope using un-manned construction machinery to protect reconstruction works for national route No. 57 and JR Hohi railway line (Kyushu Regional Development Bureau, MLIT 2018).

These countermeasures worked out as expected to be sure. However, the sequence of the intense Kumamoto earthquakes was then followed by intense rains in May and June. According to the Water and Disaster Management Bureau, MLIT (2016), the Kumamoto Earthquakes have triggered total 158 landslides in Kumamoto Prefecture. Furthermore, the heavy rains continued in May and June have caused 65 more landslides there. Figure 2 shows locations of quake-induced landslides on the south-facing wall of Tateno Notch as well as those induced by heavy rains in May and June. Landslides on this figure are denoted by polygons as the result of photointerpretation by the Storm, Flood and Landslide Research Division of the National Research Institute for Earth Science and Disaster Resilience NIED (2016). The continuous colour band from blue to red on the background terrain map shows the change in elevation that has occurred over about 3.5 years from July 20, 2013 to February 2, 2017; thus, the most blue and red coloured areas are considered to be the source and depositional areas, respectively, of the landslides that occurred in the 2016 Kumamoto Earthquakes and the continued rainfalls. The figure shows that slurries of some landslides in the post-quake rains have traversed JR Hohi railway line, one of the major arteries connecting east and west of Kyushu island, thus impeding its quick reconstruction. MLIT issued on April 12, 2019 a press release announcement saying that the JR Hohi railway line would reopen in 2020, while the national route No. 57 would be rerouted a little north constructing a 3.7 km long tunnel through the Aso caldera rim, thus avoiding going through the Ground Zero.

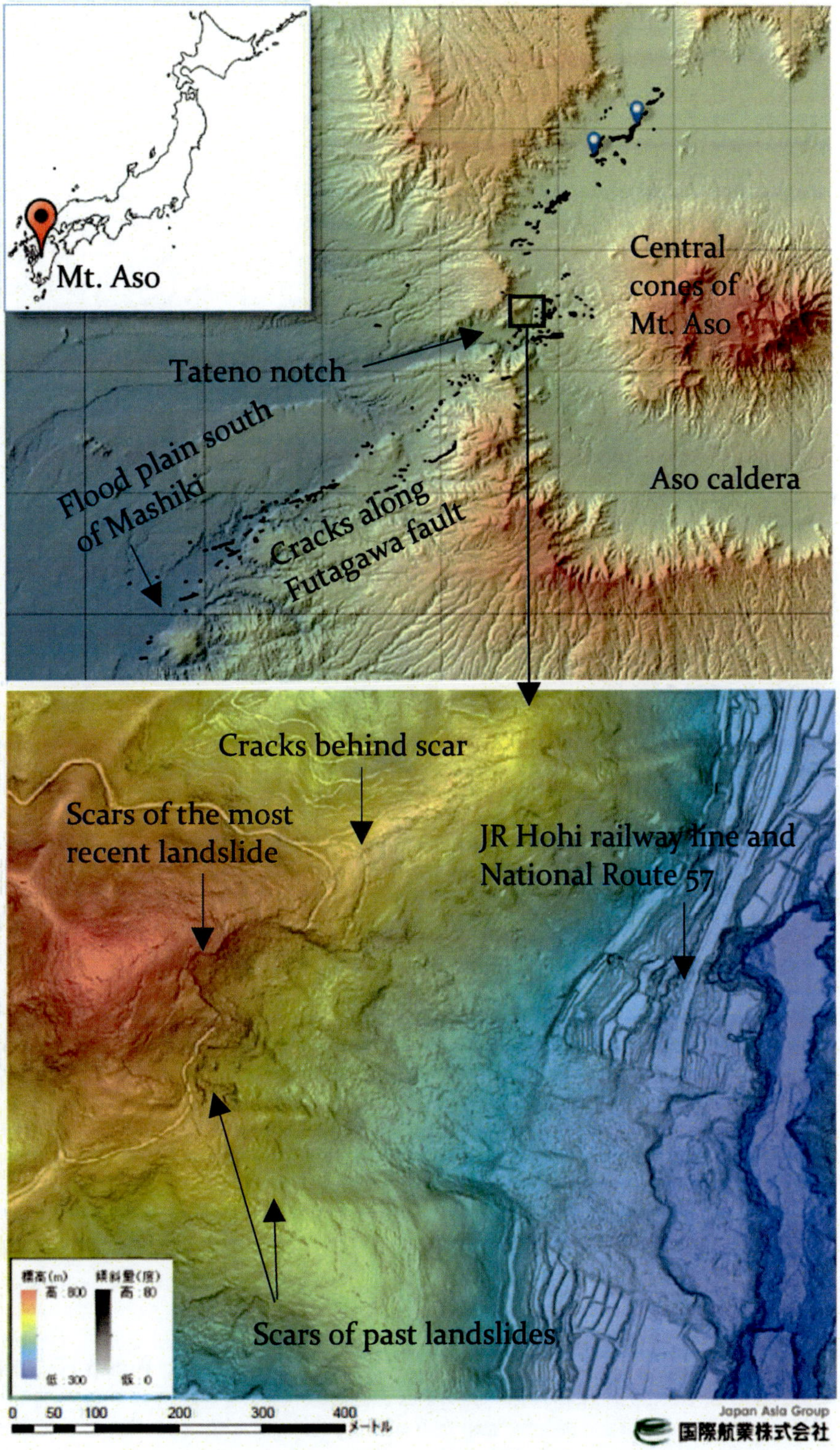

Fig. 1 Scars of past and most recent landslides and cracks appearing behind the scars indicating future risk (LiDAR image from Kokusai Kogyo Co., Ltd. 2016)

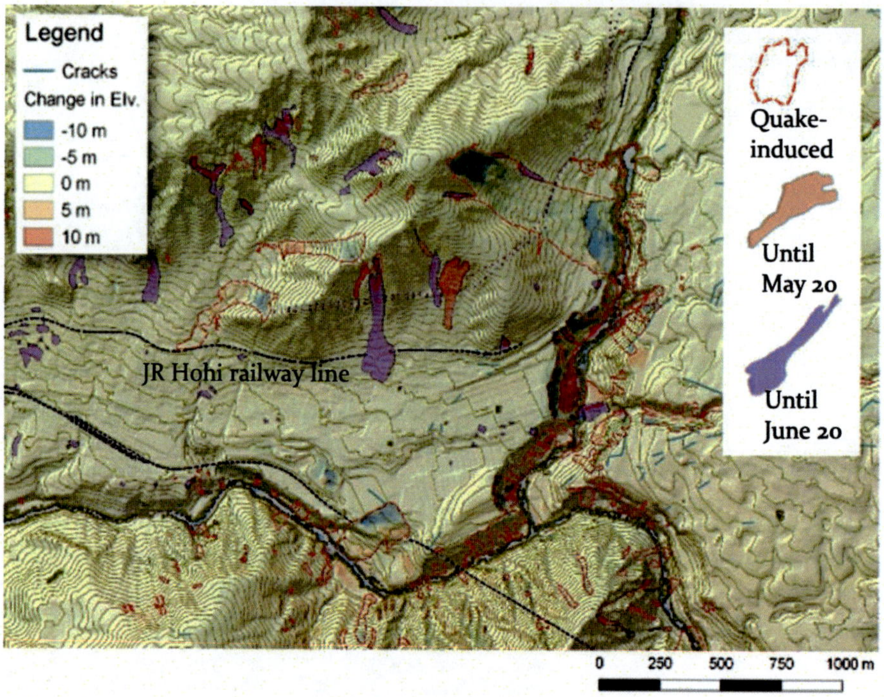

Fig. 2 Locations of quake and rain-induced landslides on the south-facing wall of Tateno Notch (Photo-interpretation by NIED 2016)

Continued Heavy Rainfalls

Not only the inner-facing caldera walls but also wide-spread areas surrounding Mt. Aso are covered thick with pyroclastic flow deposits called Aso-4. The Aso pyroclastic-flow deposits, products of major Pleistocene volcanic eruptions in Japan, are divided into four eruptive units: Aso-1, Aso-2, Aso-3, and Aso-4 in ascending order (Ono et al. 1977; Fujii et al. 2001). Out of these four major eruptions, the fourth eruption (Aso-4) was the largest, with its pyroclastic flow deposits covering the entire Kyushu island and even extending beyond the island. The most distant location of these deposits has been confirmed to be Akiyoshidai in Yamaguchi Prefecture, about 150 km north of Mt. Aso. Even in areas where Aso-4 are not exposed, Aso-4 layer can be found underground beneath fluvial deposits.

Mashiki Town, located on the outer western foothills of Mt. Aso (upper Fig. 1), was another Ground Zero in terms of the damage to houses. Total 3,026 houses were reportedly flattened in Mashiki: this number accounts for 35% of completely collapsed houses in Kumamoto City (Mashiki Town 2017). Just south of the town, there extends a flood plain of Akitsu and Kiyama Rivers with its boundaries (base lines of hills) lying on its NW and SE sides. These boundaries extend along both the Futagawa fault and its bifurcated fault and meet in an acute angle in the north-eastern suburb of the town. Thus, the flood plain lies like a wedge cutting into the foothills of Mt. Aso, and this wedge-shaped flood plain underlays Aso-4 pyroclastic flow deposit beneath its marshy and clayey fluvial soil cover. Except for ground ruptures that appeared along the known trace of Futagawa fault, no clear sign of large ground deformation was seen at first glance on this flood plain. However, when digital terrain models before and after the earthquake are compared, a low-lying area along Kiyama River is found sunken by more than a meter (Fig. 3). Thus, this area was flooded in the continued torrential rain of June 20 and 21 as shown in Fig. 4. The swelling water of Kiyama River overtopped the left riverbank at the location of the red placemark in Fig. 4. The cause of this ground subsidence is not clear yet. It could largely be due to tectonic deformation near the Futagawa fault. Whereas the fault's right-lateral slip was dominant at a shallower depth of its southwestern segment, normal fault slip was more significant at a greater depth of its north-eastern segment (Himematsu and Furuya 2016). In some areas near the confluence of Kiyama and Yagata Rivers, sand ejecta was reportedly found covering roads indicating that liquefaction might have been the cause of some localized ground subsidence. Whatever the cause was, this widespread sunken area has become much more prone to flooding.

Leaning Pump Housings of Clean Water Wells

A phenomenon of note on this flood plain was that many RC housings for pumping facilities of clean water wells were found leaning (Fig. 5). Kumamoto city depend 100% on groundwater from nearly 100 wells, and the main event was the most damaging, resulting in nearly 460,000 customers'

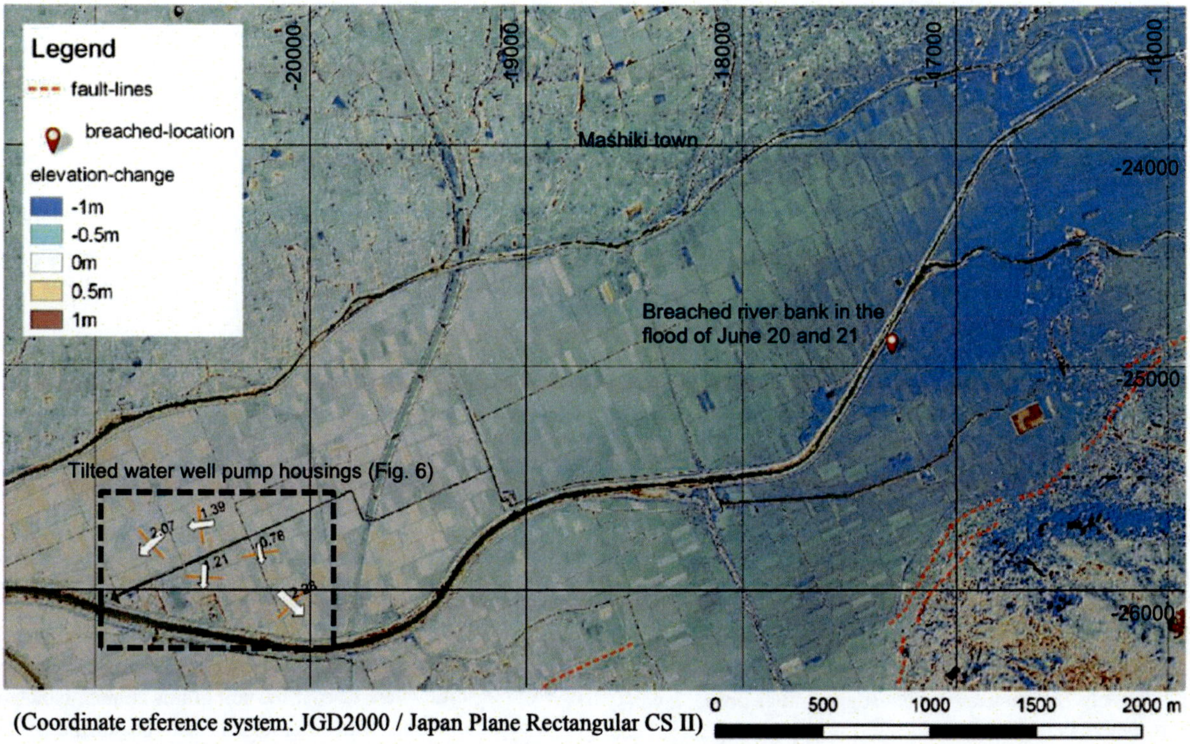

Fig. 3 Elevation change caused by the earthquake and tilted housing for tilting of RC housings of pumps for clean water wells (Konagai et al. 2017). Arrows and their labels show directions and angles of tilt for five pump housings. Red place mark shows the location of river bank breached in the torrential rain of June 20 and 21. DEMs before (Oct. 30, 2009, July 16, 2012) and after (July 25, 2016) the earthquake were provided by the Geospatial Information Authority of Japan

Fig. 4 Inundated flood plain near leaning pump housings of clessan water wells (Photo provided by Asahi Shinbun with Permission Letter No. R/061,689)

Fig. 5 Leaning RC housing of pump for Akita clean water well No. 2. The left near-side column of the housing is about 26 cm lower that the right back-side column. (Photo by Tang A.K., at N32.7666°, E130.7788°)

outages (Tang and Eidinger 2017). One of the primary reasons for the outages was rather artificial due to regulatory requirements for clean water quality. Kumamoto city has set a standard on the allowable turbidity in drinking water. When the turbidity 5 on the Japanese Turbidity Standards (JIS K0101) is reached in any well, its pump stops automatically. If pumps in the tilting housings have stopped due to the increase of turbidity, it may suggest that the soil grain crushing may have occurred in their common aquifer underneath the flood plain due to large strain build-up.

Konagai et al. (2017) measured angles and directions of tilt of these well-pump housings using a total station (Fig. 6). Tilt angles ranged from 1 to 2°, and there was no clear directional regularity observed for the measured housings. However on closer look, it is noted that the steel casing of each well embedded about 170–200 m deep in the ground goes straight up to its housing, which is clamped to the floor is off its exact center, and the steel casing is much closer to one of four corner columns of the housing (Fig. 7). Meanwhile four corner columns of the housing are supported by PC piles, 2 piles for each column, which are all about 20 m long embedded in the uppermost organic clayey fluvial layer (Fig. 7). Now that each housing is found tilting towards the farthest corner from its steel casing, it is considered that the uppermost organic soil layer has sunken as a whole in the earthquake dragging down all PC piles, while the steel casing of each well worked as a strut. Though negative skin frictional forces could have been exerted equally upon all PC piles, the largest distance from the strutting steel casing made the moment from the farthest pile be the largest, and thus caused the housing to tilt in that particular direction.

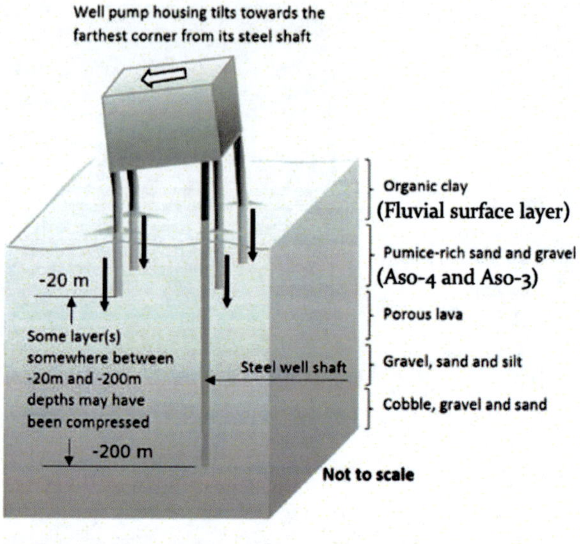

Fig. 7 Schematic illustration of leaning pump housing (Not to scale)

The tilts of housings thus suggest that some layer(s) somewhere between −20 and −200 m depths may have been compressed causing some volume loss. If the soil is incompressible with the presence of confined ground water, this volume loss is to be balanced by creation and/or ejection of volume somewhere else. If the soil grains contain isolated intra-particle voids, the volume loss can be attributed to grain crushing.

Fig. 6 Leaning RC housings of pumps for clean water wells (Konagai et al. 2017). Angles and directions of tilt of five well-pump housings were measured using a total station near Akita Water Distribution Field. Arrows and their labels show directions and angles of tilt in degrees, respectively. (Photo from Google Earth)

Kagami et al. (2019) conducted a series of torsional cyclic loading tests on hollow cylindrical Aso-4 specimens under undrained conditions with the initial effective confining pressure set at 120 kPa. The pumice-rich soil samples were taken from an outcrop of Aso-4, about 8 km east of the inclined well pump housings. This outcrop is exposed on the hanging wall side of the Futagawa Fault, and was considered to be an extension of the same Aso-4 layer underlying the fluvial deposit of the wedge-shaped flood plain. The time history of torsional shear strain applied to the specimens was numerically obtained using the 252 m deep underground strong-motion record from KiK-net Mashiki Station (KMMH16). Immediately after 5 cycles of loading, the excess pore-water pressure ratio reached 1.0 suggesting that the soil sample had liquefied. After the pore water in the specimen was drained, the soil's volume was found to be reduced by 6.5%, suggesting that the grain crushing could have been one of major causes of the ground subsidence. Though this inference sounds very likely, it certainly requires further examination and verification for figuring out possible and rational countermeasures to underground facilities.

The 2018 Hokkaido Earthquake

Multiple Landslides and Their Impact on Lifelines

The major impact of the M_w 6.6 September 6th, 2018 Hokkaido Earthquake was obviously in the form of geotechnical failures. The intense tremor triggered more than 3,300 landslides confirmed over an area of about 20 km × 20 km near Atsuma Town, wiping out homes sparsely distributed along bases of hills (Konagai et al. 2018; Konagai and Nakata 2019). Around 80% of 41 victims in this earthquake were confirmed dead of suffocation. In a week after the quake, some 6,000 households across five municipalities including hard-hit Atsuma and Abira had remained without running water, posing an obstacle for residents to returning to daily life, largely because the newest water purification plant in the northern part of Atsuma town that opened in July, 2018, just two months before the quake, sustained heavy damage from a landslide that occurred immediately behind this plant (Fig. 8).

What was special about this earthquake was that the entire island of Hokkaido was plunged into total darkness, stretching to regions over 300 km from the epicenter of the earthquake. This was largely because the Hokkaido Electric Power Company's coal-fired power plant in Atsuma was heavily damaged by fires that broke out during the earthquake. However, the Japan Broadcasting Corporation (NHK) reported just one year after the earthquake that some transmission line failures in the multiple landslides area were also responsible for the island-wide blackout, though this had already been pointed out by Eidinger and Tang (2019). Landslides resulted in the collapse of two transmission towers on a 66 kV line. The intense shake was responsible for in phase-to-ground faults at a few locations (tension towers) on 275 kV lines. These transmission line faults resulted in dropping of about 600 MW of imported power. In addition to these high voltage transmission lines, low voltage distribution feeders to the last miles were taken out at hundreds of locations by landslide masses, and these masses clogging roads hampered quick recovery works.

Runout Distances

This calamity has left a big question about how far out a landslide mass can travel. Konagai et al. (2018), Konagai and Nakata (2019) examined geometric parameters of total 30 landslide masses to provide a clue as to possible runout distances of them (Table 1). The majority of these landslides were shallow with their sliding surfaces developed in the interiors of surficial deposits of volcanic matters from explosive eruptions of Mt. Shikotsu (about 40,000 years ago), Tarumae (about 20,000 years ago) and Eniwa (about 9,000 years ago).

Konagai and Nakata (2019) assumed that a landslide mass with its initial length L_1 and cross-sectional area A_1 has been decelerated as it traveled over a flat land and stopped completely with its final length of L_2 and cross-sectional area of A_2 immediately when the whole mass left the slope L_1 (Fig. 9). Assuming that the work used up through friction exerted upon the sliding surfaces L_1 and L_2 is nearly equal to the initial potential energy of the landslide mass ignoring an energy dissipation process in the interior of the deforming landslide mass, the runout distance can be given by:

$$X_1 + X_2 \cong \frac{H}{\mu_2} + \left(1 - \frac{\mu_1}{\mu_2}\right) X_1 = aH + bX_1 \qquad (1)$$

where μ_1 and μ_2 are mobilized frictional coefficients on sliding surfaces L_1 and L_2, respectively. Assuming that the variations in μ_1 and μ_2 are substantially small and follow normal distributions, a multiple linear regression analysis for the relationship between the dependent variable H and two independent variables X_1 and X_2 with its intercept set at zero can give us the overall picture of the mobilized frictional coefficients. For the 30 landslides (28 degrees of freedom) listed in Table 1, the average values of $\mu_1 = 0.165$ and $\mu_2 = 0.36$ were obtained with the standard errors of $\sigma_{\mu_1} = 0.058$ and $\sigma_{\mu_2} = 0.069$, respectively. Thus, Eq. (1) is rewritten as:

$$X_1 + X_2 \cong 2.8H + 0.54X_1 \cong 2.8H_1 + 0.54X_1 \qquad (2)$$

Fig. 8 Water purification plant and landslides (Photo from Google Earth, Nov. 9, 2018)

Figure 10 compares the observed and estimated runout distances $X_1 + X_2$ for the examined 30 landslides. Though Eq. (2) helps understand the overall image of devastation, it must be remembered that not a small number of landslides are inevitably on the unsafe (right) side of the prediction line (Eq. 2) drawn on Fig. 10.

Figure 11 plots slope inclinations H_1/L_1 of the chosen 30 landslides against the heights of their top scars H_1. As a whole, the smaller the H_1 values are, the smaller are the inclinations H_1/L_1, and three slopes are found below the $H_1/L_1 = 0.165$ line in this figure.

In the authors' previous report (Konagai et al. 2018), μ_1 was obtained using a small landslide on a very gentle slope shown in Fig. 12 with $H_1 \cong 20$m and $H_1/L_1 \cong 0.2$. This planar landslide mass, after sliding on this gentle slope, hit the opposite wall of the shallow valley and formed a transverse bulge as illustrated in Fig. 13. This bulge was assumed to have developed where wedges of passive soil failure formed one after another at the boundary between the toe part pressed against the opposite valley wall and the slowing tail part with the uniform thickness t as illustrated in Fig. 13. This tail part was gradually shortening until its final length of L_{final} was reached. Equation (3) shows the final equilibrium condition immediately before the final length L_{final} of the tail part is reached:

$$\left(\gamma_t t L_{final} \sin\theta - \gamma_t t L_{final} \mu_1 \cos\theta\right)\cos\theta \geq \frac{1}{2} K_p \gamma_t t^2 \quad (3)$$

where, an unequal sign is necessary because the bulge worked as a surcharge, γ_t = unit weight of the wet landslide mass, and K_p = coefficient of passive earth pressure, which is given by:

$$K_p = \tan^2\left(\frac{\pi}{4} + \frac{\phi}{2}\right)(4) \quad (4)$$

with ϕ = internal friction angle of the landslide mass.

Internal friction angle for the volcanic ash and pumice from Shikotsu quaternary eruption have been examined by several researchers (see, e.g., Kitago et al. 1973; Yagi and Miura 2003). They have reported that values of ϕ for the samples they have taken from Atsuma and Monbetsu areas ranged from 42 to 48° under low effective confining pressures. Assuming that $\phi \cong 45°$, $H = 1.5$ m and $L \cong 30$ m for the Line C3–C4, $\cos\theta \cong 1$ and $\sin\theta \cong \theta$ given that $\theta \cong 0.2$, μ_1 was obtained to be as much as 0.05.

The gentler and the smaller slopes are, the wetter they may have been, because the greater parts of slip surfaces formed in the small and gentle slopes could have been well beneath the seepage lines, given the cumulative precipitation in the epicentral area, at Atsuma station of Automated Meteorological Data Acquisition System, AMeDAS (JMA, 2020) for the period from August 1 to September 6, 2018 (the date of the earthquake), exceeding the average for the same period over a past 30 years (1981–2010) by about 50 mm (Fig. 14).

Strong Resemblance to a Past Case History

The observed geometric features of these landslide masses have a striking resemblance to those on hill slopes in Hachinohe, Aomori Prefecture, covered with pumice-rich volcanic products from an explosive eruption of Mt. Towada (about 13,000 years ago). These multiple landslides were

Table 1 Dimensions of 30 landslide masses examined (Konagai and Nakata 2019)

ID	Location of top scar		L_1 (m)	L_2 (m)	H_1 (m)	H_2 (m)	H (m)	X_1(m)	X_2(m)
	East longitude (degree)	North latitude (degree)							
1	141.8885	42.7776	205.3	108.6	72.1	13.3	85.3	192.2	107.8
2	141.8888	42.7772	156.3	106.5	64.3	2.3	66.6	142.5	106.5
3	141.8898	42.7770	120.7	101.4	40.8	8.8	49.5	113.6	101.0
4	141.8781	42.7636	87.9	14.6	12.3	0.2	12.5	87.1	14.6
5	141.8714	42.7398	70.5	46.5	24.9	-0.1	24.9	65.9	46.5
6	141.9048	42.7451	78.4	93.3	30.8	2.8	33.6	72.0	93.2
7	141.9846	42.7564	159.7	106.3	56.2	8.2	64.5	149.4	106.0
8	141.9803	42.7480	150.4	165.4	69.8	8.7	78.5	133.2	165.1
9	141.9316	42.7627	134.0	121.1	48.9	4.8	53.7	124.8	121.0
10	141.9172	42.7599	71.7	27.3	31.3	4.6	35.9	64.5	26.9
11	141.8976	42.7371	96.2	68.6	34.6	1.5	36.2	89.7	68.6
12	141.8743	42.7149	42.6	11.9	9.7	0.1	9.8	41.5	11.9
13	141.8771	42.7366	105.6	37.0	13.4	0.1	13.5	104.8	37.0
14	141.8779	42.7368	101.1	64.7	14.6	0.2	14.8	100.0	64.7
15	141.9050	42.7461	77.9	131.3	38.7	3.1	41.8	67.6	131.3
16	141.9104	42.7509	115.1	90.2	42.1	1.8	43.8	107.1	90.2
17	141.9633	42.7601	113.6	77.6	65.2	4.4	69.6	93.0	77.5
18	141.9627	42.7602	106.8	68.7	63.6	4.2	67.8	85.8	68.6
19	142.0194	42.7609	144.5	125.4	79.8	14.0	93.8	120.4	124.6
20	141.9661	42.7615	308.8	163.3	93.1	3.5	96.6	294.5	163.3
21	141.9195	42.7400	96.8	46.3	26.9	0.4	27.3	93.0	46.3
22	141.9180	42.7395	106.1	56.3	31.2	0.7	31.8	101.4	56.3
23	141.8976	42.7372	83.7	96.3	39.3	3.7	43.0	73.9	96.3
24	141.9596	42.7614	49.2	32.0	27.1	2.1	29.2	41.0	31.9
25	141.9149	42.7362	65.3	41.8	23.7	-0.1	23.5	60.8	41.8
26	141.9104	42.7509	116.1	85.2	42.8	1.0	43.8	107.9	85.2
27	141.9858	42.7417	189.7	159.8	86.6	4.0	90.6	168.8	159.8
28	141.9802	42.7480	152.0	159.3	71.2	8.3	79.5	134.4	159.1
29	141.9822	42.7442	110.9	85.9	58.3	3.7	62.0	94.4	85.8
30	141.9914	42.7487	99.0	141.5	54.8	4.3	59.0	82.5	141.5

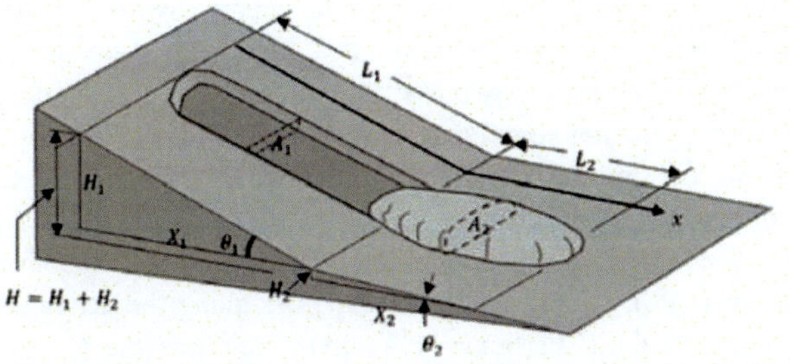

Fig. 9 Schematic view of a planar landslide mass (Konagai et al. 2018; Konagai and Nakata 2019)

caused by the Tokachi Earthquake of 1968. As we saw in the case of the 2018 Hokkaido Earthquake, many victims (33 of the total 46) in this earthquake were also confirmed dead of suffocation (Aomori Prefecture 1969). The earthquake occurred on May 16 at 0:49 UTC (09:49 local time) in the area offshore Aomori and Hokkaido with its moment magnitude put at M_w 8.3. This earthquake was preceded by a heavy rain due to a trough of low pressure (Aomori Prefecture 1969). Figure 15 is the map of 3-days (May 13–15, 1968) rainfall accumulation in Aomori Prefecture (Aomori Prefecture 1969; Kawakami et al. 1969).

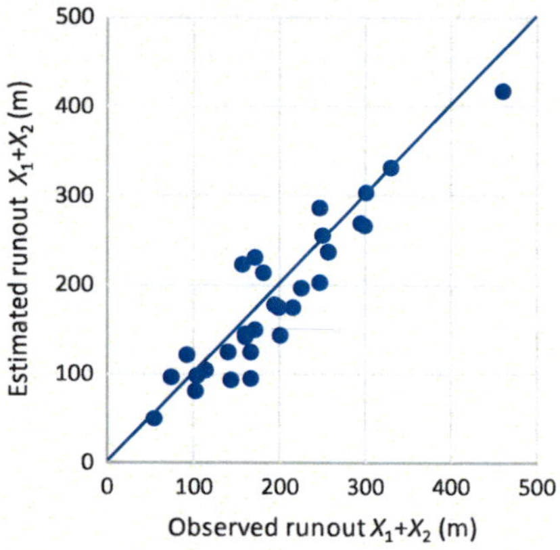

Fig. 10 Comparison of the observed and estimated rounout distances $X_1 + X_2$ (Konagai and Nakata 2019)

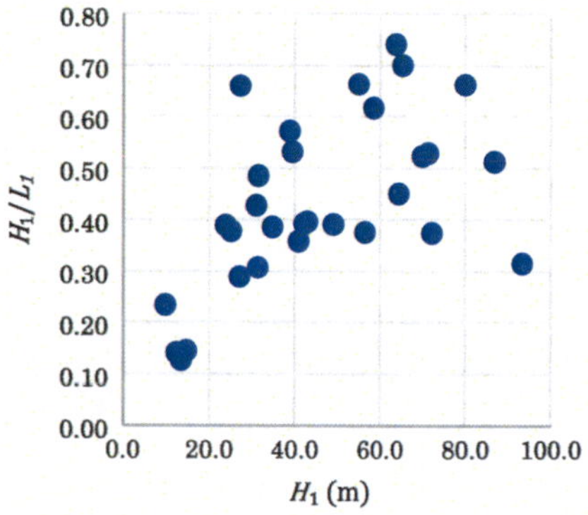

Fig. 11 Slope inclinations H_1/L_1 for different relative heights of top scars (Konagai and Nakata 2019)

It is noted in this figure that the 3-days cumulative rainfall of about 200 mm in the area of the multiple landslides (dashed-line box) was eventually the heaviest in this prefecture. Figure 16a and b show an illustration in Reference (Institute of Geology and Paleontology Tohoku University 1969) and an aerial photo (TO6810Y from Aerial photos retrieving site, GSI) of landslides at Nakazutesu, Gonohe Town, respectively; these landslides and those in Atsuma, Hokkaido (Fig. 16c) look very much alike. Locations of 6 landslides were identified through aerial photographic image interpretation and pinpointed with placemarks on the digital terrain model (Fig. 17) to extract their slope dimensions H_1, H_2, X_1 and X_2 (see Fig. 9); these dimensions were summarized in Table 2. One cannot draw any statistically significant result only with the six examples. Therefore, the parameters in Table 2 were directly substituted in Eq. (2) to compare the actual runout distances of these landslide masses with the estimated ones assuming that the detached soil masses have the same physical features as those of the multiple landslides in Hokkaido (Fig. 18). The slope of 0.9155 (<1.0) of the trendline passing through the origin and these six points in Fig. 18 shows that the actual runout distances are in general slightly longer than the estimated distances, and the coefficient of determination, $R^2 = 0.9728$, of this trendline implies that there is a striking resemblance between the Hachinohe and Hokkaido multiple landslides in terms of not only their appearances but also their runout distances. Kawakami et al. (1969) identified slip surfaces in six borehole logs from different landslide locations (Fig. 19). The identified slip surfaces are all similar lying about 2 m underground with less-dispersed dry density values ranging from 0.82 to 0.99 showing porous and crushable nature of the pumice-rich soil.

Similarities between these two earthquakes are also found in gentler slope failures. Figure 20 shows an illustration of a gentle and shallow landslide that occurred in Towada, Aomori Prefecture in the 1968 Tokachi earthquake (Institute of Geology and Paleontology Tohoku University 1969). This illustration is extremely similar to the landslide shown in Fig. 12 in that both are shallow with bulges traversing their planar landslide masses. Assuming that $\phi \cong 45°$, and substituting its slope dimensions in Eq. (3), μ_1 is obtained to be as much as 0.03.

Mechanical properties of Pyroclastic matters can surely vary from place to place, but the striking similarity between the multiple landslide case-histories from Hokkaido and Hachinohe, Aomori inspires a feeling of hope that the same equation used to estimate their runout distances can be used for the other areas with some necessary adjustment of parameters. Further in-depth studies will be necessary by looking up similar case histories in reports of past earthquakes.

Summary

Both the 2016 Kumamoto Earthquakes and the 2018 Hokkaido Earthquake have shaken the areas covered and/or underlain by volcanic matters. The impacts of these earthquakes were thus obviously in the form of geotechnical failures, yielding significant service interruptions of lifelines. Important findings shown in this article are summarized as follows:

The 2016 Kumamoto earthquakes caused extensive damage to a variety of lifeline facilities along the activated Futagawa fault. Particularly Tateno notch was the ground

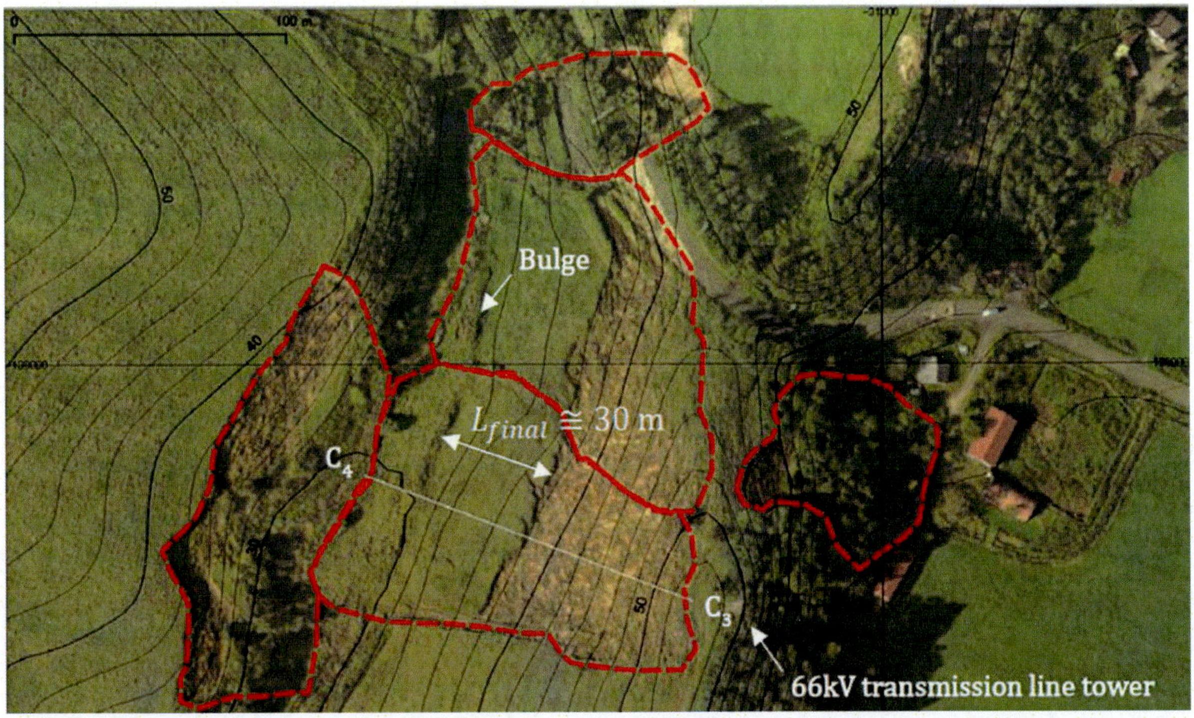

Fig. 12 Coherent landslide mass compressed against the other side wall of valley (Konagai et al. 2018)

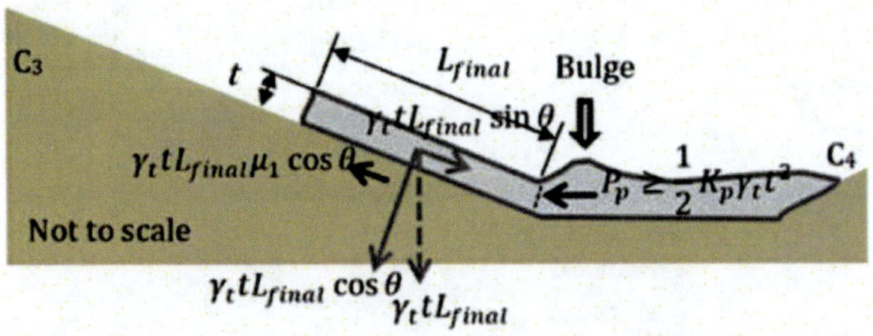

Fig. 13 Cross-section of landslide mass that has stopped moving being compressed against the other side wall of valley (Konagai et al. 2018), Not to scale

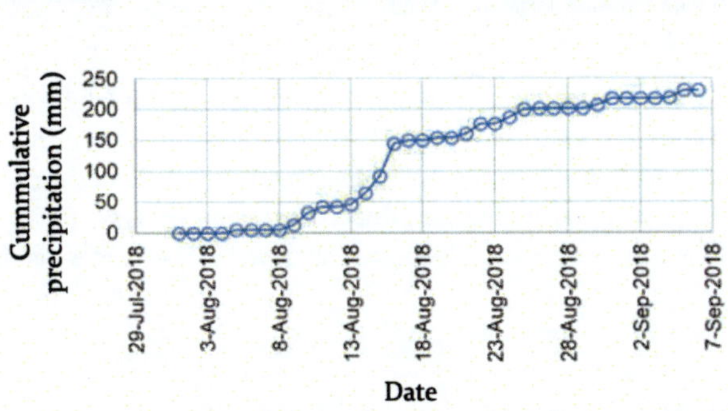

Fig. 14 Cumulative precipitation at Atsuma station of Automated Meteorological Data Acquisition System (AMeDAS) at N42.730, E141.888, for the period from August 1 to Sept. 6, 2018 (Konagai et al. 2018)

zero in terms of severe damage to traffic arteries, with the fallen Aso Ohashi Bridge as the focus of this event. The LiDAR image of the inner-facing caldera wall of this notch showed evidence of not only the most recent landslide but also scars of past landslides suggesting that similar landslides may have occurred over and over on the same caldera wall.

Mashiki Town, located on the outer western foothills of Mt. Aso, was another Ground Zero in terms of damage to houses. Just south of this town, there extends a wedge-shaped flood plain of Akitsu and Kiyama Rivers that underlays Aso-4 pyroclastic flow deposit beneath its marshy and clayey fluvial soil cover. A never-seen-before phenomenon on this flood plain was that many four-legged RC housings for pumping facilities of clean water wells were found tilting. The steel casing of each well embedded about 170–200 m deep in the ground goes straight up to its housing, and is clamped to the floor of the housing eccentrically off its exact center. Meanwhile, four corner columns of the housing are supported by 20 m long PC piles, which were dragged down by the uppermost fluvial soil deposit. Since the 170 to 200 m long steel casing of the well is eccentrically strutting the housing, the housing is considered to have tilted towards the farthest corner of it. Therefore, tilting of these RC housings may have been a reflection of

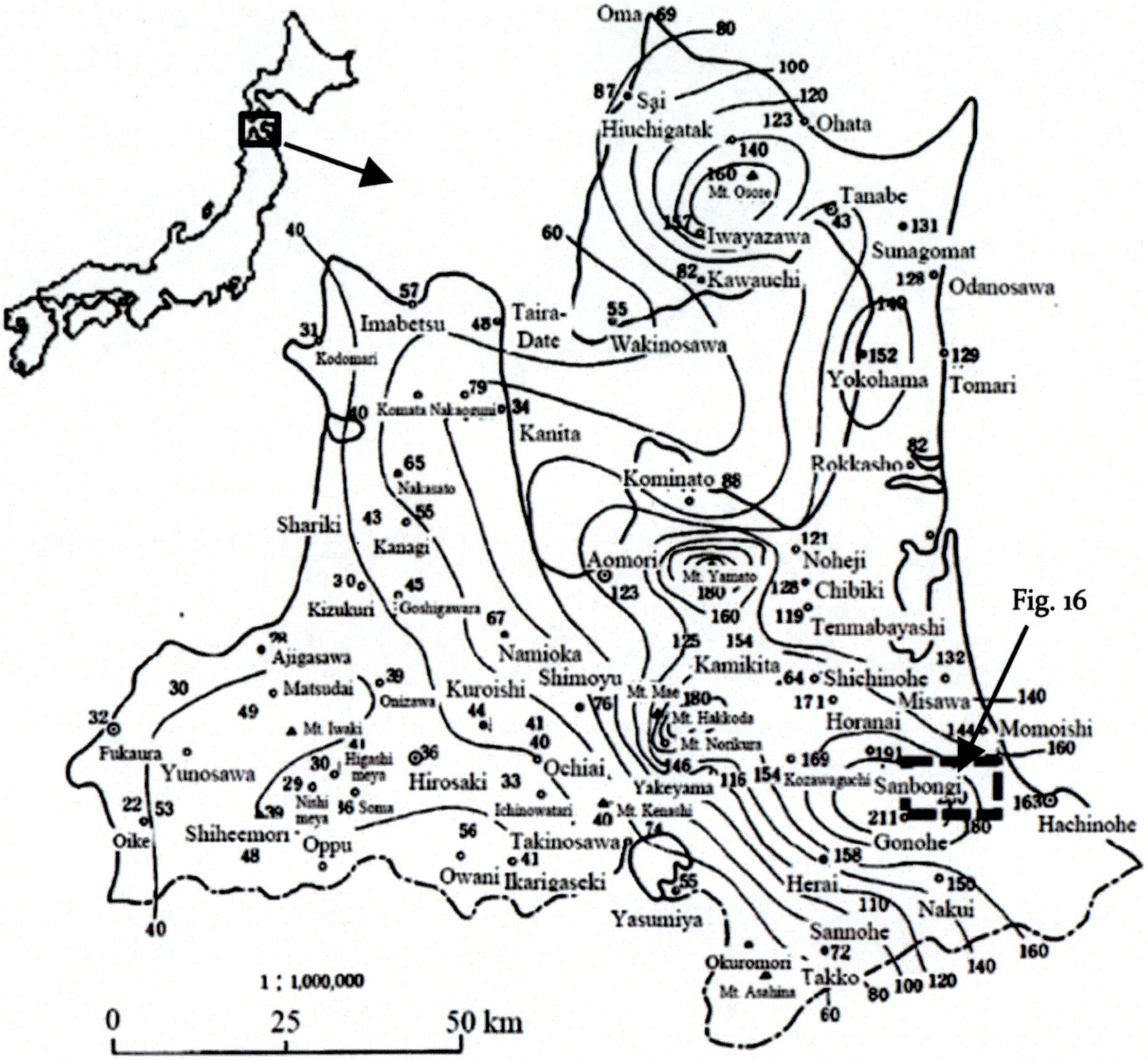

Fig. 15 3-days (May 13–15, 1968) rainfall accumulation in Aomori Prefecture (Aomori Prefecture 1969)

volume loss associated with deformation buildups deep inside aquifers. A series of soil hollow cylinder tests on pyroclastic soil samples suggested that the volume loss can be attributed to grain crushing in the pyroclastic soil layer called Aso-4 (Konagai et al. 2017).

The intense tremor triggered more than 3,300 landslides confirmed over an area of about 20 km × 20 km near Atsuma Town, wiping out homes sparsely distributed along bases of hills. A multiple linear regression analysis for the dimensions of total 30 coherent landslide masses have given

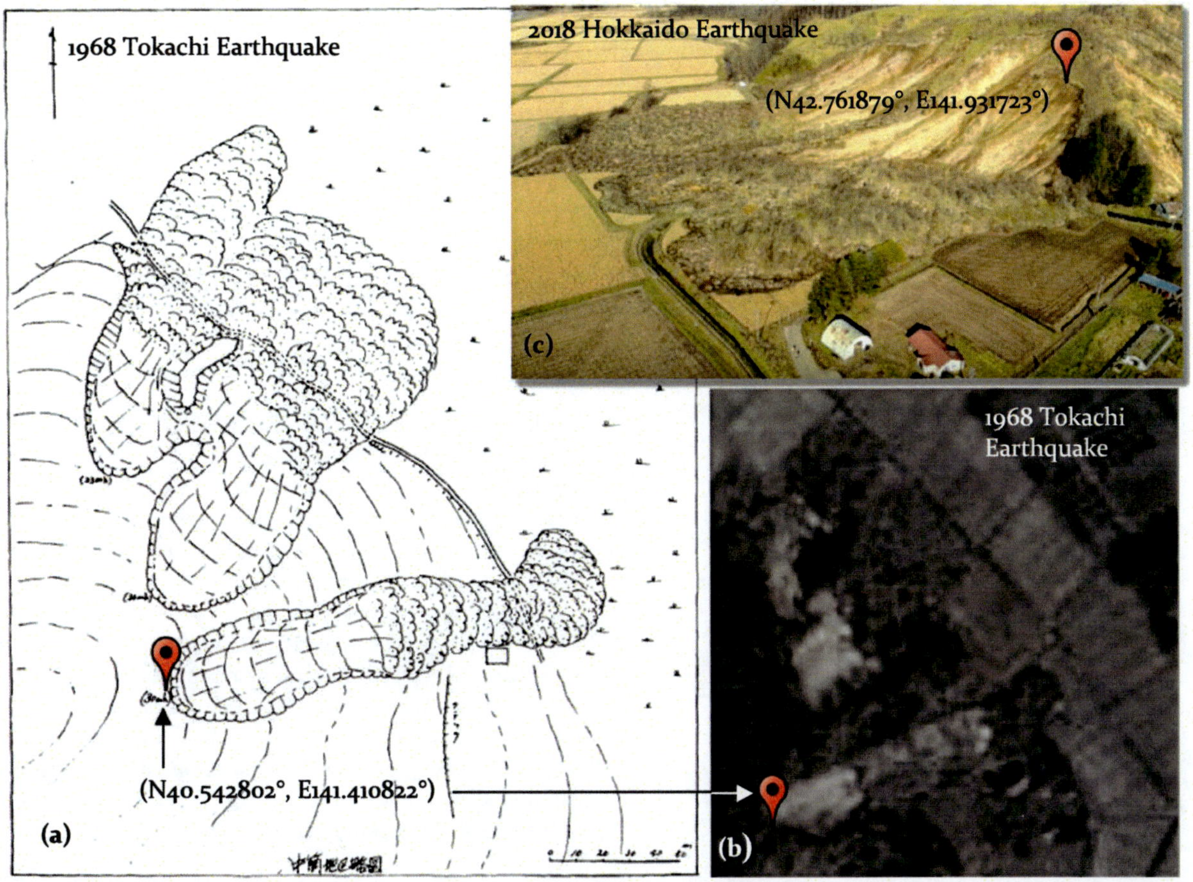

Fig. 16 **a** Illustration in Reference [9] and **b** aerial photo (TO6810Y [10]) of triple landslides at Nakazutsu, Gonohe Town, Aomori (1968 Tokachi Earthquake). These landslides have a striking resemblance to **c** those in Atsuma Town, Hokkaido (2018 Hokkaido Earthquake)

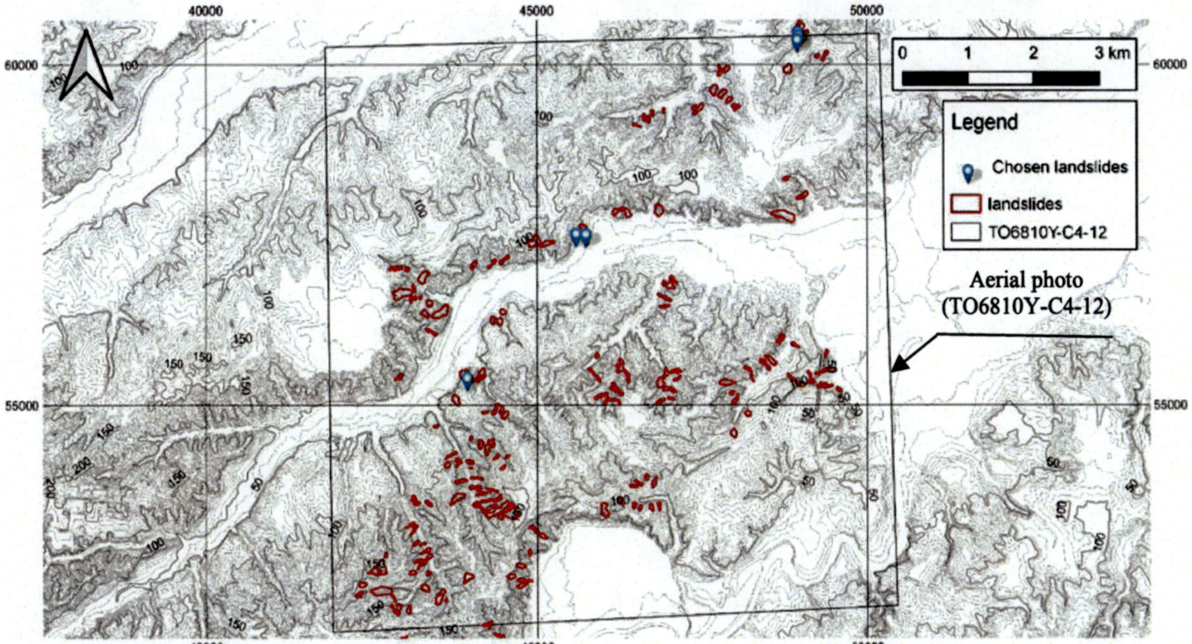

Fig. 17 Photo-interpleted locations of landslides in the multiple landslide area west of Hachinohe City. A slightly inclined parallelogram in the middle of the map is the area covered by a 1–40,000 scale aerial photo, TO6810Y, taken on Oct. 5, 1968 from an altitude of 6,000 feet by the Geospatial Information Authority of Japan (2020)

Table 2 Dimensions of 6 landslide masses shown in Fig. 17 with blue placemarks

Place	Latitude (deg.)	Longitude (deg.)	H_1 (m)	H_2 (m)	H = H1 + H2 (m)	X1 (m)	X2 (m)	Actual $X_1 + X_2$	Estimated $X_1 + X_2$
Toyomanai	40.498328	141.35166	25	0	25	64	50	114	104.56
Nakazutsu 1	40.543854	141.41089	23	0	23	71	62	133	102.74
Nakazutsu 2	40.543356	141.41098	30	0	30	101	53	154	138.54
Nakazutsu 3	40.542802	141.41082	30	0	30	141	35	176	160.14
Kaminarasaki 1	40.517267	141.37122	55	5	60	177	100	277	263.58
Kaminarasaki 2	40.517182	141.37293	33.2	7	40.2	83	87	170	157.38

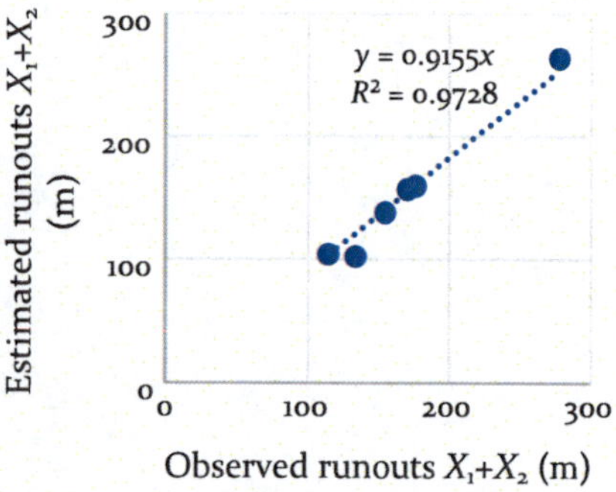

Fig. 18 Comparison of the observed and estimated runouts $X_1 + X_2$ values of valley

us both the average values of mobilized frictional coefficients $\mu_1 = 0.165$ on the slip surfaces and $\mu_2 = 0.36$ on the flat rice fields. These values allow for quick estimation of runout distances of these landslide masses, and thus helps understand the overall image of devastation (Konagai et al. 2018; Konagai and Nakata 2019).

Though these values of μ_1 and μ_2 are in essence exclusively for the landslide-prone epicentral area of the 2018 Hokkaido Earthquake, the striking resemblance between the Hokkaido and Hachinohe multiple landslides in terms of not only their appearances but also their runout distances inspires a feeling of hope that the same equation used to estimate their runout distances can be used for the other areas with some necessary adjustment of parameters.

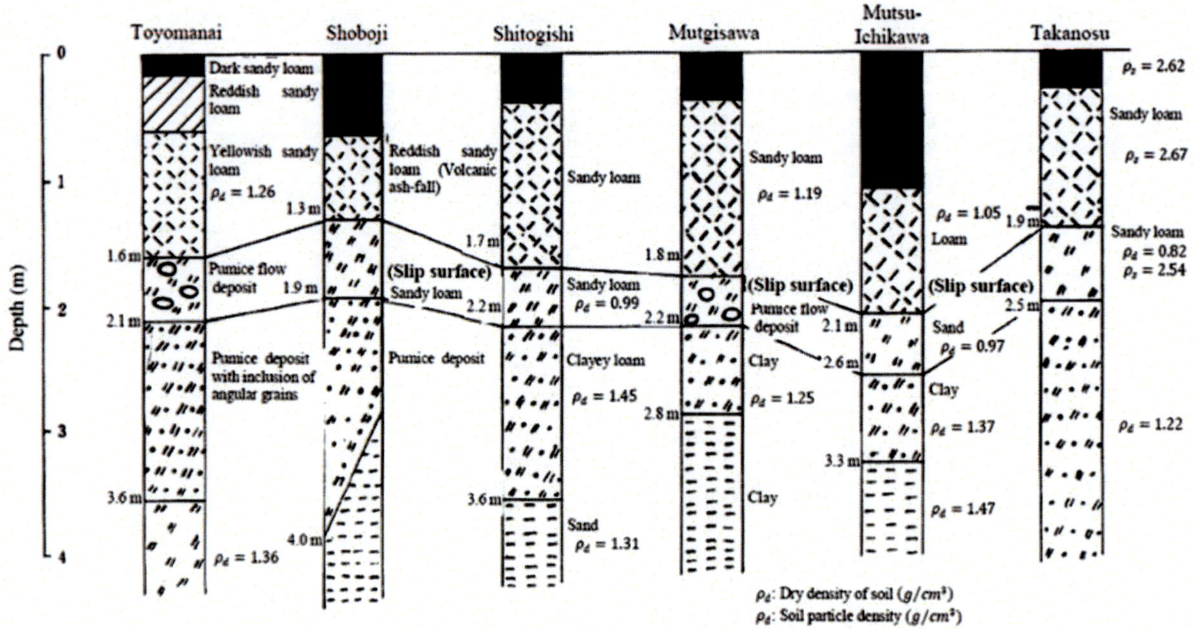

Fig. 19 Soil profiles at 6 landslide locations (Kawakami et al. 1969)

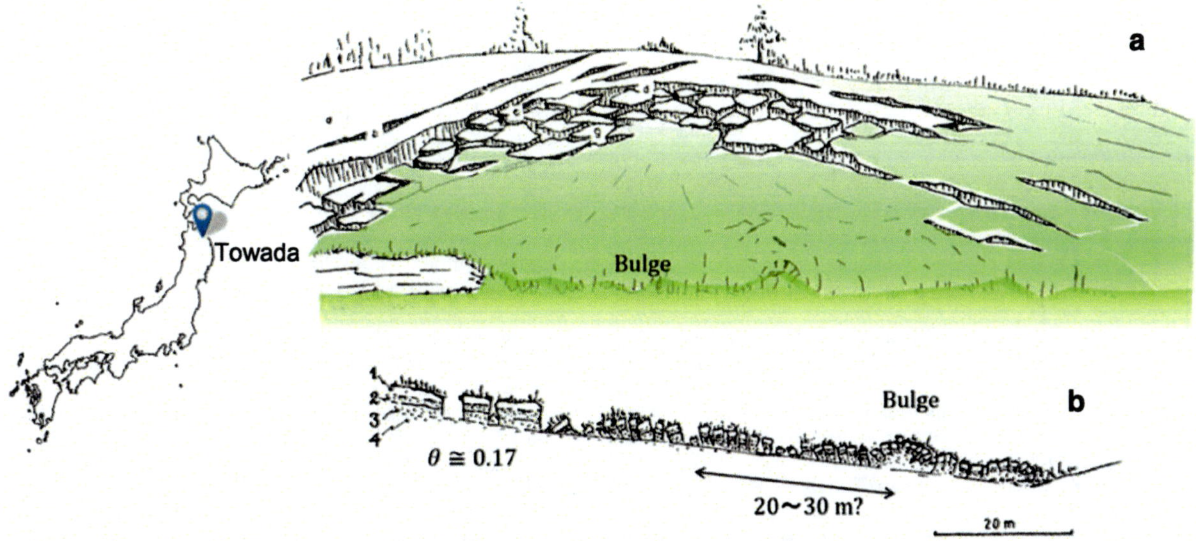

Fig. 20 Illustration of a gentle and shallow landslide that occurred in Towada, Aomori Prefecture in the 1968 Tokachi earthquake (Institute of Geology and Paleontology Tohoku University 1969). Location: N40.618°, E141.234°

Acknowledgements The authors performed the post-earthquake lifelines performance investigations as the joint team members of The Council of Lifeline Earthquake Engineering (TCLEE) composed mainly of lifeline earthquake engineering experts of the American Society of Civil Engineers (ASCE) and earthquake engineering experts of the Japan Society of Civil Engineers (JSCE). TCLEE, originally standing for "Technical Council on Lifeline Earthquake Engineering," was a committee of ASCE. In 2014, ASCE ended its support of TCLEE as a standalone committee; and now is examining lifelines under the general topic of community resilience. The second and third authors have independently elected to continue as The Council on Lifeline Earthquake Engineering (with the same acronym TCLEE) to maintain a sharp focus on the performance of lifelines in earthquakes.Many people from the JSCE provided us with tremendous support and information. The authors would like to express their sincere thanks to Dr. Yukihiro Tsukada, Executive Director of the JSCE, Dr. Takanobu Suzuki, Chairman, Dr. Yoshihisa Maruyama, Secretary General and all members of the Subcommittee of Lifeline Earthquake Engineering, JSCE Committee of Earthquake Engineering. The authors are indebted to Mr. Masataka Shiga, PhD candidate at the University of Tokyo, and Dr. Ms. Alessandra Mayumi Nakata, who have joined these field surveys and helped the authors flying a UAV to collect 3D images of terrains. Dr. Takashi Kiyota, Associate Professor, the University of Tokyo, Dr. Takaaki Ikeda, Professor at the Nagaoka University of Technology, Mr. Takeshi Hara, Mr. Satoshi Suenaga, Kubota Corp., and Mr. Jiro Nakamura, Chubu Electric Power Co., Inc., also joined the field surveys and provided the authors with their generous supports from all aspects. The authors' field surveys were partially supported by the Grant-in-Aid for Scientific Research (A), No. 16H02744.

References

Aomori Prefecture (1969) General report of damage in Aomori Prefecture caused by the 1968 Tokachi Earthquake, Kyodo (in Japanese).

Chida T, Choi JH, Hirakawa Y, Kawasaki T, Watanabe H (2018) The possibility of the collapse of Aso Ohashi considered ground deformation and landslide disaster. J Jpn Soc Civil Eng Ser A1 (Struct Eng Earthq Eng (SE/EE)) 74(4): I_381–I_394. https://doi.org/10.2208/jscejseee.74.I_38

Eidinger J, Tang AK (2019) Hokkaido, Japan, M_w 6.7 Earthquake of September 6, 2018 Lifeline Performance, The Council of Lifeline Earthquake Engineering (TCLEE) No. 4. URL: https://www.geengineeringsystems.com/ewExternalFiles/TCLEE%20No.%204.pdf. Accessed 13 Apr 2020

Fujii J, Nakajima T, Kamata H (2001) Paleomagnetic directions of the Aso pyroclastic-flow and the Aso-4 co-ignimbrite ash-fall deposits in Japan. Earth Planets Space 53: 1137–1150. https://earth-planets-space.springeropen.com/articles/https://doi.org/10.1186/BF03352409. Accessed 13 Apr 2020

Fukae R (2017) Damage to railway systems and their reconstructions. In: The JSCE Conference in commemoration of the first anniversary of the 2016 Kumamoto Earthquake, in Japanese. Kyushu Railway Company. https://committees.jsce.or.jp/eec2/node/96. Accessed 13 Apr 2020

Geographical Information Authority of Japan, GSI (2020) Aerial photos retrieving site. https://mapps.gsi.go.jp/maplibSearch.do#1. Accessed 13 Apr 2020

Himematsu Y, Furuya M (2016) Fault source model for the 2016 Kumamoto earthquake sequence based on ALOS-2/PALSAR-2 pixel-offset data: evidence for dynamic slip partitioning. Earth, Planets and Space, 68(169). https://earth-planets-space.springeropen.com/articles/https://doi.org/10.1186/s40623-016-0545-7. Accessed 13 Apr 2020.

Institute of Geology and Paleontology Tohoku University (1969) Phenomena and disasters associated with earthquakes–Examples from the Tokachi-oki Earthquake of 1968 in eastern Aomori Prefecture, vol 67, p 1–99. https://tohoku.repo.nii.ac.jp/?action=pages_view_main&active_action=repository_view_main_item_detail&item_id=118275&item_no=1&page_id=33&block_id=38. Accessed 13 Apr 2020

Japan Meteorological Agency, JMA (2020) Weather record retrieving site. https://www.data.jma.go.jp/obd/stats/etrn/index.php. Accessed 13 Apr 2020

Kagami Y, Kiyota T, Ikeda T, Muhammad U, Katagiri T (2019) Study on ground subsidence of volcanic soil caused by the Kumamoto

Earthquake, SEISAN KENKYU, vol 71, no 6, pp 1071–1074. https://doi.org/https://doi.org/10.11188/seisankenkyu.71.1071. Accessed 13 Apr 2020

Kawakami F, Asada A, Yanagisawa E, Mori Y (1969) Report on the Tokachi-Oki Earthquake of 1968. Tohoku University, Department of Civil Engineering ((in Japanese))

Kitago S, Takaiwa M, Kamiya M (1973) Geotechnical features of pumice covering the central part of Hokkaido. Bulletin of the Faculty of Engineering, Hokkaido University, vol 68, no 2, pp 223–241. https:// eprints.lib.hokudai.ac.jp/dspace/handle/2115/41162. Accessed 13 Apr 2020

Konagai K, Shiga M, Kiyota T, Ikeda T (2017) Ground deformation built up along seismic fault activated in the 2016 Kumamoto Earthquake. J Jpn Soc Civil Eng Ser A1 (Struct Eng Earthq Eng (SE/EE)) 73(4):I_208–I_215. https://doi.org/https://doi.org/10.2208/jscejseee.73.I_208

Konagai K, Nishiyama S, Ohishi K, Kodama D, Nanno Y (2018) Large ground deformations caused by the 2018 Hokkaido Eastern Iburi Earthquake. JSCE J Disaster Fact-Sheets, FS2018-E-0003, 1–8. http:// committees.jsce.or.jp/disaster/FS2018-E-0003. Accessed 13 Apr 2020

Konagai K, Nakata AM (2019) Runouts of landslide masses detached in the 2018 Hokkaido Eastern Iburi Earthquake. JSCE J Disaster Fact Sheets, FS2019-E-0001, 1–6. https://committees.jsce.or.jp/disaster/FS2019-E-0001. Accessed 13 Apr 2020

Kokusai Kogyo Co., Ltd. (2016) Map of landslides and ground fissures, ELSAMAP. https://www.kkc.co.jp/service/bousai/csr/disaster/201604_kumamoto/index.html. Accessed 13 Apr 2020

Kyushu Regional Development Bureau, MLIT (2018) Outline of stabilization of the slope west of Aso Ohashi Bridge. https://www.qsr.mlit.go.jp/bousai_joho/tecforce/sabo/index.html. Accessed 13 Apr 2020

Mashiki Town (2017) Examination of responses taken by Mashiki Town in the 2016 Kumamoto Earthquakes. https://www.town.mashiki.lg.jp/bousai/kiji0032410/index.html. Accessed 13 Apr 2020

Ministry of Land, Infrastructure, Transport and Tourism, Japan, MLIT (2019) JR Hohi Railway Line will open in 2020, Press Release on April 12, 2019. https://www.mlit.go.jp/common/001285346.pdf. Accessed 13 Apr 2020

National Park Service, U.S. Department of Interior (2020) How magma melts and mixes below ground and influences eruption characteristics. https://www.nps.gov/para/learn/nature/how-magma-melts-and-mixes-below-ground-and-influences-eruption-characteristics.htm. Accessed 13 Apr 2020

Ono K, Matsumoto I, Miyahisa M, Teraoka Y, Kambe N (1977) Geology of the Taketa district, with geological sheet map at 1:50,000, Geological Survey of Japan, p 145

Storm, Flood and Landslide Resarch Division, NIED (2016) Distribution of sediment disasters in the 2016 Kumamoto Earthquakes (as of June 27, 2016). https://www.bosai.go.jp/mizu/dosha.html. Accessed 13 Apr 2020

Takahashi T, Shoji S (2002) Distribution and classification of volcanic ash soils. Global Environ Res 6:83–97. https://www.researchgate.net/publication/228852848_Distribution_and_classification_of_volcanic_ash_soil. Accessed 13 Apr 2020

Tang AK, Eidinger MJ (2017) Kumamoto, Kyushu, Japan, Earthquakes of M_w 6.0 April 14, 2016, M_w 7.0 April 16, 2016-Lifeline performance: The Council of Lifeline Earthquake Engineering (TCLEE), No. 2. https://learningfromearthquakes.org/2016-04-15-kumamoto-japan/11-resources/62-new-report-available-lifeline-performance. Accessed 13 Apr 2020

Yagi K, Miura S (2003) Mechanical property of crushable volcanic soils and its evaluation-Hokkaido volcanic ash grounds. J Jpn Soc Eng Geol 44(3):142–153. https://www.jstage.jst.go.jp/article/jjseg1960/44/3/44_3_142/_article/-char/ja/. Accessed 13 Apr 2020

Water and Disaster Management Bureau, MLIT (2016) Briefing of sediment disasters in the 2016 Kumamoto earthquakes (as of Sept. 14, 2016). https://www.mlit.go.jp/river/sabo/jirei/h28dosha/160914_gaiyou_sokuhou.pdf. Accessed 13 Apr 2020

Landslides in Recent Earthquakes and Damage to Lifelines

2020 Kyoto Japan

Kazuo Konagai

Abstract

This review article summarizes some important findings in recent past earthquakes. One of these findings is that multiple landslides in the epicentral area of the 2004 Chuetsu Earthquake were found along stripes of large co-seismic stress changes, which appeared far away from a surface seismic fault trace. A similar pattern of stress changes may have occurred over and over in the same epicentral area, thus suggesting that we will perhaps be able to be prepared for the next knowing these patterns. Also, it is shown herein that a multiple linear regression analysis for the geometric parameters of many landslides in the epicentral area of the 2018 Hokkaido earthquake can give us the average values of frictional coefficients on both hill slopes and flat depositional areas. These values can help estimate possible runout distances in this area. The observed geometric features of these landslide masses have a striking resemblance to those in the other earthquake-hit area draped with volcanic products, in terms of not only their appearances but also their runout distances.

Keywords

Earthquake-induced landslides • Lifelines • Co-seismic stress changes • Volcanic products

Introduction

In the 2016 Kumamoto Earthquake, Kyushu, Japan, there were about 750 reports of damage to conventional railway lines, of which 73% were caused by landslides, debris flows, rockfalls etc. (Fukae 2017). All were beyond the direct jurisdiction of the railway company. The cascading sequence of accidents at Fukushima Daiichi Nuclear Power Plant in the 2011 Great East Japan Earthquake and Tsunami, leading to the meltdown of its reactors, included a collapse of a tower of a 66 kV transmission line due to a single landslide (TEPCO 2012). Neither the tower nor the slope was an item of the seismic category "S", which requires the most stringent safety margin than the others (NRA 2016).

In February 2013, the Nuclear Regulation Authority, Japan, compiled a draft of new regulatory requirements (NRA 2013). One of the regulatory requirements says that no construction of a critically important facility of "Category S" immediately above a suspected fault is allowed, and relevant organizations need to look back over the past 400,000 years or so to confirm if there was any activity of the fault immediately beneath the planned important facility. However, a fault has its extension over several ten miles or longer, and there often appear a thick swath of large ground deformations such as landslides along the fault. Even though the fault does not traverse the station at all, landslides can take away lifelines leading to "Category S" facilities. However, once NRA considers the fault is active, discussions (even though it was not intended) are liable to be made in an exclusive manner only for facilities within their jurisdictional area, and hardly take one step forward for what-if scenarios. Since an earthquake can cause more than one damage point of an entire system, we need to think from a multidimensional perspective what the actual impact can cascade on our society.

This paper introduces new findings and problems about important infrastructures in recent-past earthquakes that include the 2004 Chuetsu (Mid-Niigata Prefecture) Earthquake, Japan, and the 2018 Hokkaido Eastern Iburi Earthquake, Japan.

K. Konagai (✉)
International Consortium on Landslides, Kyoto, 606-8226, Japan
e-mail: kaz3776k@gmail.com

Ž. Arbanas et al. (eds.), *Understanding and Reducing Landslide Disaster Risk*,
ICL Contribution to Landslide Disaster Risk Reduction,
https://doi.org/10.1007/978-3-030-60713-5_2

Damage to Lifelines Far Off Surface Earthquake Faults

The 2004 Chuetsu Earthquake, Japan

The 2004 Mid-Niigata Prefecture Earthquake (Chuetsu Earthquake, hereafter) jolted low-raised mountainous terrain of Higashiyama mountains (Fig. 1) having an actively folded geological structure (Yoong and Okada 2005). In common with other historical earthquakes in active folding regions, the Chuetsu Earthquake reportedly triggered and/or reactivated thousands of landslides, including 362 landslides with widths of larger than 50 m and 12 large-scale landslides with individual volumes of larger than 1 million cubic meters (Sato et al. 2005). The economic loss due to these landslides was initially estimated at 8 billion US dollars, making this one of the costliest landslide events in history (Kieffer et al. 2006). It was noteworthy that the swath of landslides appeared about 10 km far off the surface fault rupture trace.

Konagai et al. (2009a) obtained Lagrangian components of tectonic displacement from digital elevation models (DEMs, hereafter) of the surface terrain in the epicentral area. The DEMs before and after the earthquake are raster graphic images with pixels arranged in a 2 m × 2 m square grid; each pixel has information of its elevation. Changes in elevation in Eulerian description for DEM were first converted to Lagrangian displacements, and then the moving average method was adopted to obtain the whole picture of tectonic landform changes (Fig. 2).

Given a laterally homogeneous stratified media (Fig. 3) representing the source region, Kazumi and Konagai (2009, 2018) then obtained spatial distribution of fault slips on a multi-segment fault model; this slip distribution allowed the second principal invariant of deviatoric stress tensor, J_2, in the interior of the ground to be obtained as an index for rocks and soils deformability.

Figure 4 shows the spatial distribution of $\sqrt{J_2}$ for the epicentral area of the Chuetsu Earthquake at 5 m below the ground surface. Parallel belts of large $\sqrt{J_2}$ values are observed at an about 4 km regular interval. It is noted that all the existing landslides (white polygons in Fig. 4) that had been identified as of 2000 by the National Research Institute for Earth Science and Disaster Prevention (NIED 2000) are almost precisely clustered along the higher values of $\sqrt{J_2}$. It can be deduced from the concordant behavior of higher $\sqrt{J_2}$ values and landslides existing before the Chuetsu earthquake that a similar pattern of stress distribution may have repeated over and over in the past.

The stripes of $\sqrt{J_2}$ extend further deep in the interior of the ground. The spatial variation of $\sqrt{J_2}$ at the 75 m depth was obtained for the entire target area and shown in Fig. 5 with locations of damaged tunnel sections of the Joetsu line of the Japanese bullet train network called "Jo-etsu Shinkansen". Both the first invariant of the Cauchy stress tensor, I_1, and the second principal invariant of the stress deviator tensor, J_2, were calculated at a regular interval for each tunnel such that the examined points would include the damaged sections. Assuming that the coefficient of lateral pressure before the earthquake was unity, the initial

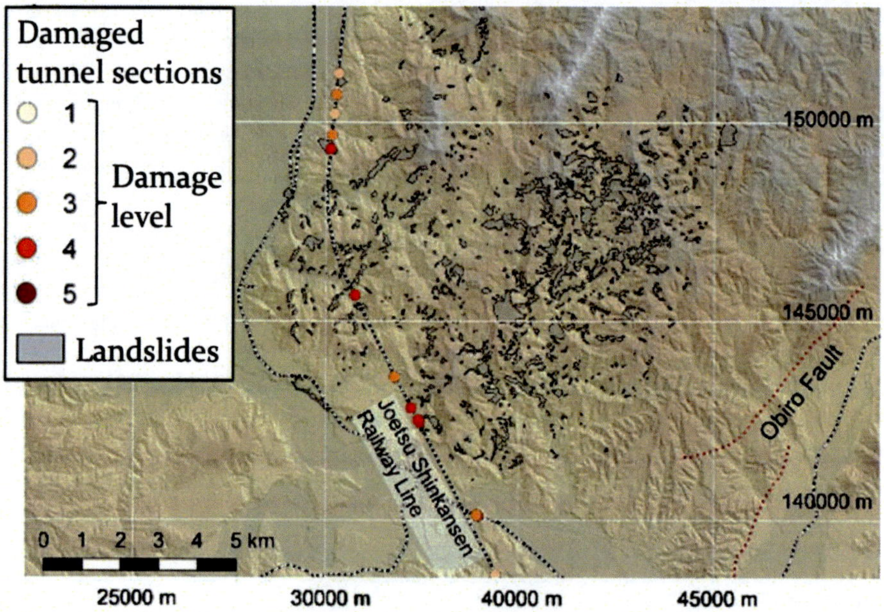

Fig. 1 Multiple landslides in the epicentral area of the 2004 Chuetsu Earthquake. Coloured circles show locations of damaged sections of railway tunnels and their damage levels. Topographical mapping is on the JGD2000/Japan Plane Rectangular Coordinate System VIII

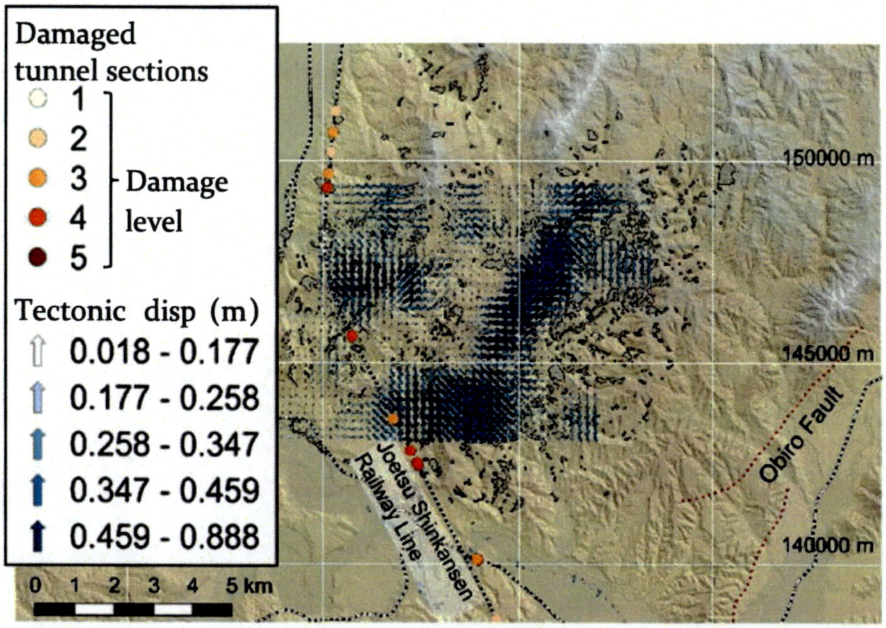

Fig. 2 Lagrangian components of tectonic displacement detected over the epicentral area (Konagai et al. 2009a). A swath of large lateral tectonic displacements appeared about 10 km west of Obiro Fault

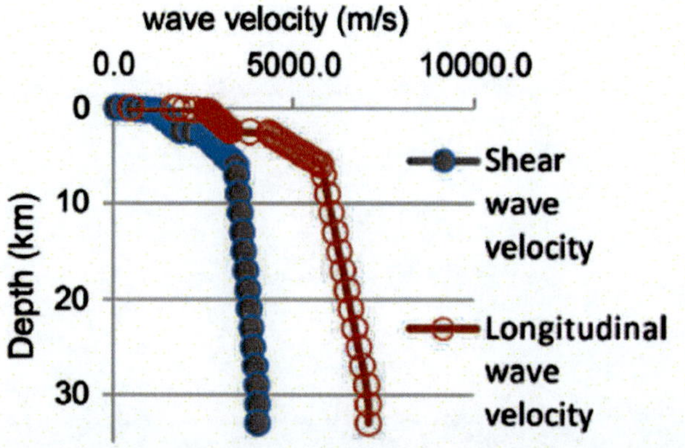

Fig. 3 Stratification assumed for target terrain (Kazumi and Konagai 2018)

confining pressure at each point was set at its overburden pressure.

When the values of I_1 and $\sqrt{J_2}$ are plotted on a scatter diagram (Fig. 6), it is noted that points for the damaged sections (solid circles in Fig. 6) are clustered along the upper bound of the entire cluster of points (Kazmi and Konagai 2018). If the optimum boundary between the clusters of damaged and undamaged points is approximated by a straight line, the line would be considered as a failure criterion, which has the following expression:

$$\sqrt{J_2} = b \cdot I_1 + c \qquad (1)$$

where b and c are the slope of the best-fit line and Y intercept, respectively. It is less likely from Fig. 6 that the criterion line for the sedimentary rocks over the examined stretch of tunnels would intercept the $\sqrt{J_2}$ axis above $\sqrt{J_2} = 0$. Therefore, the optimum b value was obtained to be 0.0939 with c set at 0 (dashed line in Fig. 6). Equation (1) with b and c set at reasonable values may be used to examine the margin of safety of both existing and new tunnels for a given scenario earthquake in the target region. However, the coefficient b, which physically describes the effect of the internal friction angle, is very small, perhaps reflecting the presence of weak and clay-rich discontinuities, such as bedding planes, joints, shear zones, and faults. These unexplained phenomena should be discussed as much more reliable pieces of information are gathered. In any case, the above findings show clearly that stress buildups in the interior of the ground can be equally or even more serious even though these locations are far away from surface fault ruptures, and can affect lifeline performances.

Runout Distances of Landslide Masses

The 2018 Hokkaido Earthquake, Japan

The major impact of the Mw 6.6 September 6th, 2018 Hokkaido Eastern Iburi Earthquake (2018 Hokkaido Earthquake, hereafter) was obviously in the form of geotechnical failures.

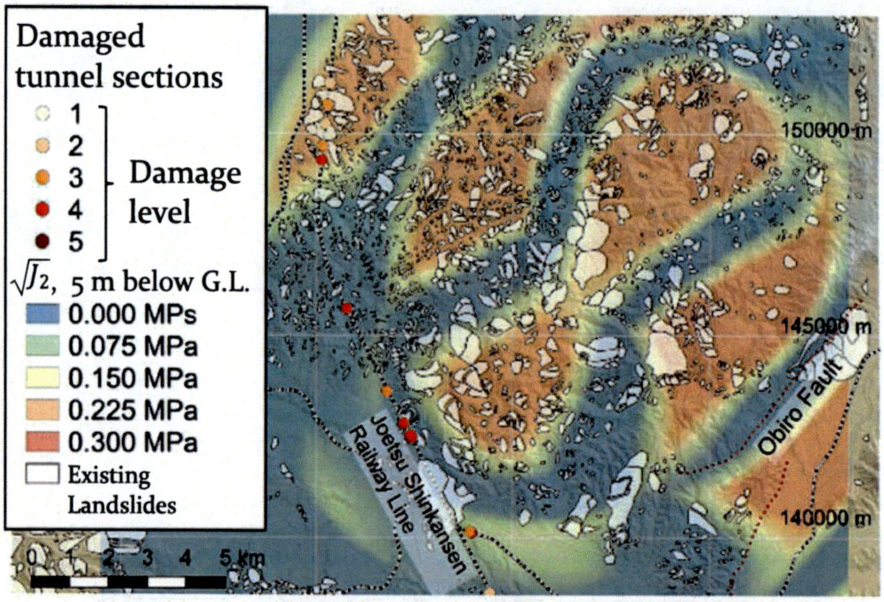

Fig. 4 Distribution of $\sqrt{J_2}$ over the entire target zone at the depth of 5 m from ground surface with locations of existing landslides (NIED 2000)

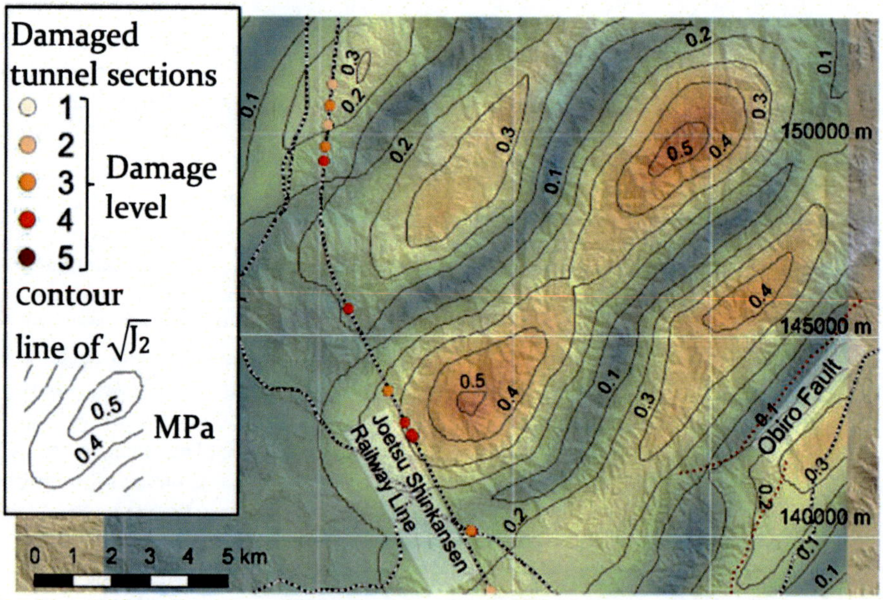

Fig. 5 Distribution of $\sqrt{J_2}$ over the entire target zone at the depth of 75 m from ground surface with locations of damaged railway tunnel sections (Kazumi and Konagai 2018)

The intense tremor triggered more than 3,300 landslides confirmed over an area of about 20 km × 20 km near Atsuma Town, wiping out homes sparsely distributed along bases of hills [1, 2]. Around 80% of 41 victims in this earthquake were confirmed dead of suffocation. In a week after the quake, some 6,000 households across five municipalities including hard-hit Atsuma and Abira had remained without running water, posing an obstacle for residents to returning to daily life, largely because the newest water purification plant in the northern part of Atsuma town that opened in July, 2018, just two months before the quake sustained heavy damage from the landslides that followed the quake.

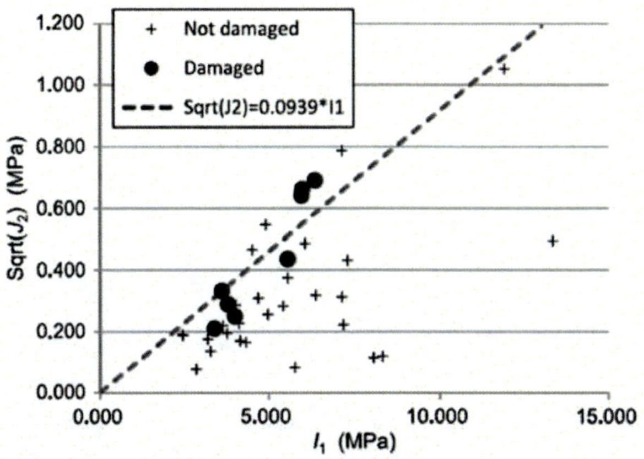

Fig. 6 Scatter diagram of I_1 and $\sqrt{J_2}$ along the tunnels (Kazmi and Konagai 2018)

Multiple Linear Regression Analysis

This calamity has left a big question about how far out a landslide mass can travel. Konagai et al. (2018), Konagai and Nakata (2019) examined geometric parameters of total 30 landslide masses to provide a clue as to possible runout distances of these landslide masses. These landslides were shallow and planar masses of volcanic ash and pumice from explosive eruptions of Mt. Shikotsu (about 40,000 years ago), Tarumae (about 20,000 years ago) and Eniwa (about 9,000 years ago), and the examined masses have deposited over extensive flat rice fields sometimes dotted with farmhouses.

A landslide mass with its initial length L_1 and cross-sectional area A_1 is assumed to have been decelerated as it traveled over a flat land and stopped completely with its final length L_2 and cross-sectional area A_2 immediately when the whole mass left the slope L_1 (Fig. 7). Assuming that the work used up through friction exerted upon the sliding surfaces L_1 and L_2 is nearly equal to the initial potential energy of the landslide mass ignoring an energy dissipation process in the interior of the deforming landslide mass, the runout distance can be given by:

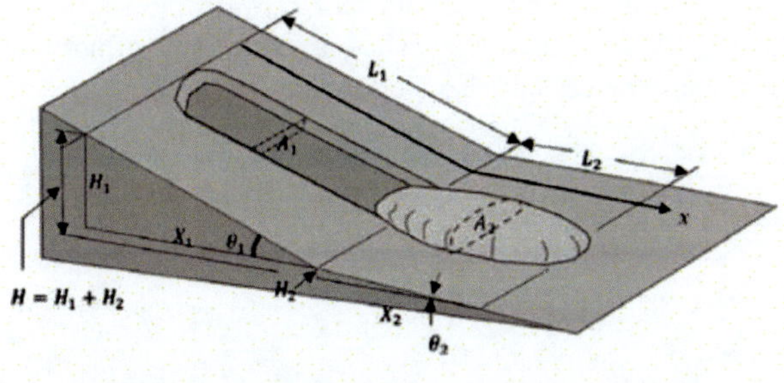

Fig. 7 Schematic view of a planar landslide mass. (Konagai et al. 2018, Konagai and Nakata 2019)

$$X_1 + X_2 \cong \frac{H}{\mu_2} + \left(1 - \frac{\mu_1}{\mu_2}\right)X_1 = aH + bX_1 \qquad (2)$$

where μ_1 and μ_2 are mobilized frictional coefficients on sliding surfaces L_1 and L_2, respectively. Assuming that the variations in μ_1 and μ_2 are substantially small and follow normal distributions, a multiple linear regression analysis for the relationship between the dependent variable H and two independent variables X_1 and X_2 with its intercept set at zero can give us the average values of $\mu_1 = 0.165$ and $\mu_2 = 0.36$ with the standard errors of $\sigma_{\mu_1} = 0.058$ and $\sigma_{\mu_2} = 0.069$, respectively. Thus, Eq. (2) is rewritten as:

$$X_1 + X_2 \cong 2.8H + 0.54X_1 \cong 2.8H_1 + 0.54X_1 \qquad (3)$$

Figure 8 compares the observed and estimated runout distances $X_1 + X_2$ for the examined 30 landslides. Though Eq. (3) helps understand the overall image of devastation, it must be remembered that not a small number of landslides are inevitably on the unsafe (right) side of the prediction line (Eq. 3) drawn on Fig. 8.

Strong Resemblance to a Past Case History

The observed geometric features of these landslide masses have a striking resemblance to those on hill slopes in Hachinohe, Aomori Prefecture (Fig. 9), draped with pumice-rich volcanic products from an explosive eruption of Mt. Towada (about 13,000 years ago). These multiple landslides were caused by the Tokachi-Oki Earthquake of 1968.

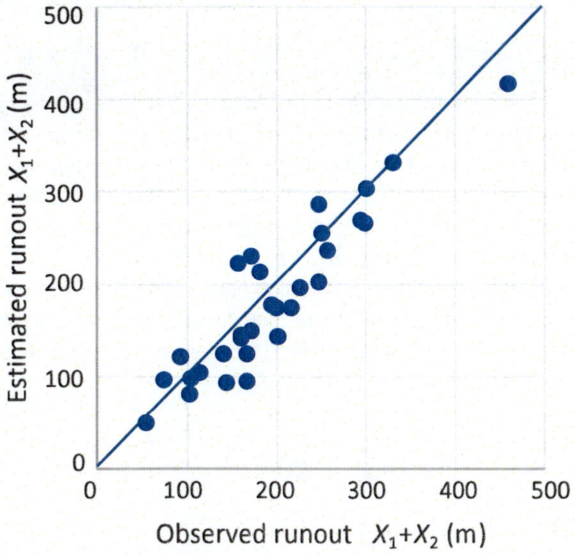

Fig. 8 Comparison of the observed and estimated runouts $X_1 + X_2$ values (Konagai and Nakata 2019)

Fig. 9 Similar landslides detached in **a** the 2018 Hokkaido Earthquake (Credit: Konagai K. and Shiga M.) and **b** the 1968 Tokachi-Oki Earthquake (Copyright: Kyodo News/Amana Images)

Since locations of the landslides described in half-century-old dusty reports are hardly pinpointed in an exact manner on terrain maps, 6 landslides were identified on an aerial photograph, TO6810Y, taken on Oct. 5, 1968, from an altitude of 6,000 feet by the Geospatial Authority of Japan, to extract their slope dimensions H_1, H_2, X_1 and X_2 (see Fig. 7). The extracted parameters were substituted in Eq. (5) to compare the actual runout distances of these landslide masses with the estimated ones assuming that the masses have the same physical features as those in the multiple landslides area in Hokkaido (Fig. 10). The slope of 0.9155 (<1.0) of the trendline passing through the origin and these six points in Fig. 10 shows that the actual runout distances are in general slightly longer than the estimated distances, and the coefficient of determination, $R^2 = 0.9728$, of this trendline implies that there is a striking resemblance between the Hachinohe and Hokkaido multiple landslide areas in terms of not only their appearances but also their runout distances.

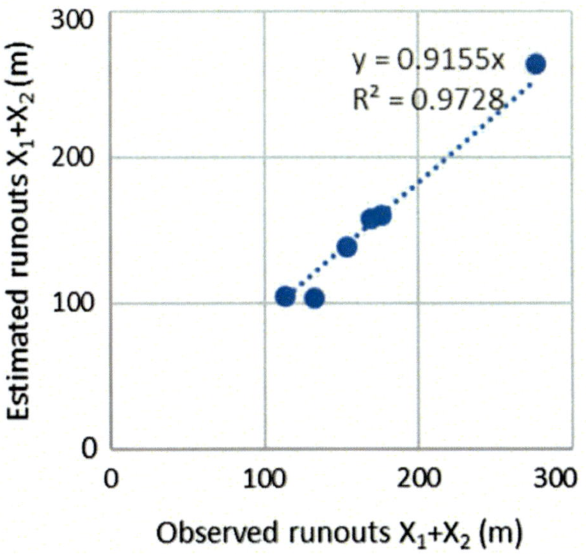

Fig. 10 Comparison of the observed and estimated runouts $X_1 + X_2$ values

Conclusions

Large earthquakes can trigger numerous landslides, which are very often the most devastating secondary effect of an earthquake in terms of all kinds of losses. Sometimes devastating impacts can appear far off surface earthquake faults as was shown in the recent-past earthquakes including the 2004 Chuetsu Earthquake, Japan, where multiple landslides were found clustering mostly along stripes of co-seismic stress changes $\sqrt{J_2}$ about 10 km west or farther beyond of the surface fault rupture trace. Many landslides that had been existing there even before the earthquake are also found within these stripes, suggesting that a similar pattern of stress distribution may have occurred over and over; the fact will help us prepare for the next.

The major impact of the Mw 6.6 September 6th, 2018 Hokkaido Earthquake was obviously in the form of geotechnical failures with more than 3,300 landslides confirmed over its epicentral area of about 20 km × 20 km covered up with volcanic products. A multiple linear regression analysis for the geometric parameters of 30 landslides has given us the average values of frictional coefficients $\mu_1 = 0.165$ and $\mu_2 = 0.36$ on slopes and flat paddy fields, respectively, with the standard errors of $\sigma_{\mu_1} = 0.058$ and $\sigma_{\mu_2} = 0.069$. These values help estimate possible runout distances in this area. The observed geometric features of these landslide masses have a striking resemblance to those on hill slopes draped with volcanic products, in Hachinohe, Aomori Prefecture hit by the 1967 Tokachi Earthquake of M_w 8.3 in terms of not only their appearances but also their runout distances.

Acknowledgements The first half of this paper summarizes some of the outcomes of the Research Project "Detection of Lagrangian displacements of terrains along an active fault and their implementation for land conservation", No 23246087 (Project leader: Kazuo Konagai), Ministry of Education, Culture, Sports, Science and Technology.The following second half reports some findings obtained through a field survey that the author has jointly conducted with the team of lifeline earthquake engineering experts of the American Society of Civil Engineers (ASCE) led by Mr. John Eidinger and Mr. Alex Tang. This survey was partially supported by the Grant-in-Aid for Scientific Research (A), No. 16H02744.

References

Fukae R (2017) Damage to railway systems and their reconstructions. In: The JSCE Conference in commemoration of the first anniversary of the 2016 Kumamoto Earthquake, in Japanese. Kyushu Railway Company. https://committees.jsce.or.jp/eec2/node/96

Keefer DV (2000) Statistical analysis of an earthquake-induced landslide distribution—the 1989 Loma Prieta, California event. Eng Geol 58:231–249

Konagai K, Fujita T, Ikeda T, Takatsu S (2009a) Tectonic deformation buildup in folded mountain terrains in the October 23, 2004, Mid-Niigata earthquake Soil Dyn Earthq Eng 29 2 261 267. https://www.sciencedirect.com/science/article/pii/S0267726108000572

Konagai K, Takatsu S, Kanai T, Fujita T, Ikeda T, Johansson J (2009b) Kizawa tunnel cracked on 23 October 2004 Mid-Niigata earthquake: An example of earthquake-induced damage to tunnels in active-folding zones. Soil Dyn Earthq Eng 29(2):394–403. https://www.sciencedirect.com/science/article/pii/S0267726108000572

Konagai K, Nishiyama S, Ohishi K, Kodama D, Nanno Y (2018) Large ground deformations caused by the 2018 Hokkaido Eastern Iburi Earthquake. JSCE J Disaster FactSheets. FS2018-E-0003, 1–8. https://committees.jsce.or.jp/disaster/FS2018-E-0003

Konagai K, Nakata AM (2019) Runouts of landslide masses detached in the 2018 Hokkaido Eastern Iburi Earthquake. JSCE J Disaster FactSheets. FS2019-E-0001, 1–6. https://committees.jsce.or.jp/disaster/FS2019-E-0001

Kazmi ZA, Konagai K, Sekiguchi K (2009) Extracting earthquake induced Lagrangian ground displacements and their implication for source inversion analysis Soil Dyn Earthq Eng 48:198–208. https://www.sciencedirect.com/science/article/pii/S0267726113000249

Kazmi ZA, Konagai K (2018) Co-seismic stress changes and damage to tunnels in the 23 October 2004 Mid-Niigata Prefecture earthquake. Can Geotech J 55(5):736–748. https://www.nrcresearchpress.com/doi/10.1139/cgj-2016-0658#.XhaQQEf7Rdg

NIED, National Research Institute for Earth Science and Disaster Prevention (2000) Landslide distribution maps (shape files). [October]. https://www.j-shis.bosai.go.jp/map/?lang=en

NRA, Nuclear Regulation Authority (2013) New regulatory requirements for light water nuclear power plants (Earthquakes and tsunamis), Provisional Translation. https://www.nsr.go.jp/data/000067118.pdf

NRA, Nuclear Regulation Authority (2016) Outline of Nuclear Regulation of Japan-Reference documents for the IAEA IRRS Mission. https://www.nsr.go.jp/data/000148578.pdf

Sato HP, Sekiguchi T, Kamiya I, Hasegawa H, Akiyama M, Kojiroi R (2005) Overlaying of landslides distribution on geological, snow depth, and other data for the mid niigata prefecture earthquake in 2004 J Geograph Surv Inst 107 79 90. https://www.gsi.go.jp/common/000024840.pdf

TEPCO, Tokyo Electric power Company (2012) Collapsed fill slope in the Fukushima Daiichi Nuclear Power Plant. Subsequent Report of damage to electric equipment, in Japanese. https://www.tepco.co.jp/cc/press/betu12_j/images/120217c.pdf

Yoong KH, Okada A (2005) Surface deformations associated with the October 2004 Mid-Niigata earthquake: description and discussion. Earth Planets Space 57:1093–1102

Lessons Learned—Landslide Induced Lifelines Disasters from Past Earthquakess

2020 Kyoto Japan

Alex K. Tang

Abstract

Lifelines sustain significant service interruption in medium to large magnitude earthquakes from landslides. There were many examples of long duration service interruptions and recovery to normalcy due to landslide disasters other than earthquake induced, for example heavy rainstorm. This paper intended to bring a focus on landslide impact to lifelines. The attempt is to reinforce the need to encourage more research in order to develop tools to protect lifelines from losses—direct and indirect losses. The cases discussed in this paper were significant earthquake events and rainstorms from 2007 to 2018. The lifelines to be discussed in this paper are telecommunication and transportation. The inherent spatial characteristics of these networks render many segments exposing to landslides, rock falls and mud flows. The paper will not discuss methods of prevention but will identify lesson learned from good geotech engineering.

Keywords

Lifelines • Telecommunication • Transportation • Research

Executive Summary

The strong shaking of medium to large magnitude earthquakes generate landslides and lateral-spread that damage lifelines and interrupt lifeline services. In geotechnical engineering, many landslide/lateral-spread mitigation measures have been developed and implemented. Many of these preventive measures worked well, however field observations showed that some of these methods failed. It is critical to collect and to analyse all the attributes that caused the failures in order to develop a more comprehensive design method and process to mitigate landslides. Preventing loss of lives and services of lifelines are the primary reasons.

For good reasons both telecommunication distribution and transmission networks commonly use the right of way along roads and bridges, therefore landslides/lateral-spreads that damage roads will damage the telecommunication components along that part of the roads.

This paper examines 6 earthquakes between 2007 and 2018 with landslide/lateral-spread damages to Telecommunication and Transportation systems. Good performances observed are also presented.

Telecommunication and Transportation Interruption by Landslides

Telecommunication components presented in this paper are the components of the distribution network. The components of transportation network are roads and railways.

Kashiwazaki Earthquake Japan 2007 ($M_w = 6.7$)

There were five significant landslide/lateral spread locations that impacted both telecommunication and transportation lifelines. One of the landslide sites is still close to traffic in 2018.

1. Landslide at Route 8 between Kashiwazaki and Nagaoka (Lat 37.4139°, Long 138.7129°).

Landslide covered about 80 m of Route 8 road surface. It is located close to the junction of Route 8 and National Road 252. As shown in Fig. 1 the retaining wall collapsed on the road surface due to landslide.

A. K. Tang (✉)
L&T Consultant, 2591 Pollard Drive, Mississauga, ON L5C3G9, Canada
e-mail: alexktang@mac.com

Ž. Arbanas et al. (eds.), *Understanding and Reducing Landslide Disaster Risk*, ICL Contribution to Landslide Disaster Risk Reduction,
https://doi.org/10.1007/978-3-030-60713-5_3

Fig. 1 This landslide covered and damaged a large portion of Route 8. *Source* MLIT

Two landlside reduction techniques were used - retaining structure on the upper part with a reduced slope on the lower part, Fig. 2.

2. Landslide at Ohmigawa JR station (Lat 37.3457°, Long 138.4857°).

This landslide damaged the tracks and the train station, which basically interrupted services between the coastal towns and Kashiwazaki, Fig. 3. The houses at the edge of the landslide along the top of the hill need to be stabilized due to loss of ground. It was evident that landslide was expected at this location. The concrete tunnel shown in Fig. 3 with an * was constructed to prevent possible landslide damaging the track.

3. Lateral Spread on Route 8 (Lat 37.3567°, Long 138.5231°) damaging telecommunication pole.

A lateral spread occurred at this section of Route 8 dislocated telecom poles and damaged the cables, Fig. 4. One lane had to be closed due to ground failure close to the lateral spread.

Fig. 2 MLIT recovery and construction in progress to reduce future landslide

Fig. 3 Two landslides (one much smaller than the other) covered the whole station in front of the concrete tunnel (marked by an asterisk * in this figure) constructed to protect the track from landslide damage. *Source* MLIT

Fig. 4 Lateral spread on Route 8 damaged telecom poles of the distribution system

4. Landslide at Kannon Misaki (Lat 37.4843°, Long 138.6206°).

There were two landslides sites on this road that slopes downhill to the waterfront. Both sites had the same style landslide prevention construction. It was an anchored steel wire net along the top edge of the concrete retaining wall. This landslide damaged both the retaining wall and the steel wire net. Figure 5 shows the landslide and the damaged landslide prevention structure. Lateral spread was also evident as shown in Fig. 5. Part of the retaining wall performed well. This part of the road was still closed to traffic in 2018.

5. Lateral spread damaged telecommunication pole at the incinerator plant area (Lat 37.3935°, Long 138.5857°).

The lateral spread at this site damaged both the road and the utility poles. Figure 6 shows the size of the localized destruction. The utility pole carries both power and telecom links.

Fig. 5 This landslide covered and damaged a large portion of this route as well as the telecom poles on the opposite side of the road

Fig. 6 Lateral spread towards the left damaged utility pole carrying both power and telecom lines

Wenchuan Earthquake China 2008 (M_w = 7.9)

The fact that this earthquake occurred in the mountainous region, there were many extensive landslides but luckily most of them were in the wilderness and not close to any communities, Fig. 7 shows one of the many scenes of massive landslides along the mountainous region. The fact that the earthquake happened during the rainy season in the province causing cascaded damages by debris flows.

1. Landslide at Yingxiu (Lat 31.0416°, Long 103.4767°).

The landslide took out the raised section of this highway connecting to Yingxiu, Fig. 8. A temporary road was paved to allow single lane traffic to access the city.

2. Landslide at road leading to Yingxiu (Lat 31.0259°, Long 103.4699°).

The landslides at this particular location took out two roads serving the Yingxiu area. Figure 9 shows the landslide that took out the roads and a truck was also pushed down the slope. The road below was repaved to allow traffic to reach Yingxiu.

Fig. 7 Massive landslides everywhere in this earthquake

Fig. 8 This portion of the raised highway was collapsed and pushed by landslide to the edge of the river

Fig. 9 This landslide damaged two roads (marked with dotted lines)

3. Major landslide and debris flow at Beichuan (Lat 31.8307°, Long 104.4566°) damaging telecommunication poles.

This location shows damage to telecom poles due to debris flow caused by heavy rainstorm after the landslide that happened above, Fig. 10. In fact a major portion of the town was covered by this debris flow further downstream from this location. The whole town was later relocated to

Fig. 10 Landslide in Beichuan at this location took out the telecom poles of the distribution network

Fig. 12 This section of the highway was blocked. The retaining anchors were disengaged

another part of the province with the same name—Beichuan (Lat 31.6096°, Long 104.4662°). The 'old' Beichuan became the monument of this earthquake.

*4. **Landslide on the road leading to Nanba (Lat 32.1838°, Long 104.8194°).***

Many telecom transmission lines (majority being optical fiber cables) were destroyed by landslides in this earthquake. This particular telecom pole was damaged by landslide and rock falls that followed, Fig. 11.

Tohoku Earthquake and Tsunami, Japan 2011 (M_w = 9.0)

This major event along the Sendai coast was one of the major disasters in Japan. The strong shaking created many landslides and ground deformation kilometres from the coast while the coastal regions were impacted by the tsunami. The following sections provide some highlights of landslide damage outside the tsunami impacted areas.

Fig. 11 At this location temporary telecom link was established, while the damaged link was left as it was damaged by the rocks that came along with landslide

*1. **Landslide on Gosaisho Hwy (Lat 37.0002°, Long 140.7908°).***

This section of the highway was blcoked by landslide. The landslide occurred in one of the major intraplates aftershock of the main event. This M_w 6.6 aftershock on 11 April caused the retaining structures failure resulting in road blocked, Fig. 12. This is a good site for understanding the failure mechanism.

*2. **Landslides on Route 107 Kitakami, Iwate (Lat 39.2916°, Long 141.1596°) damaging telecommunication pole.***

Lateral spread caused this section of the highway closed and later reduced to one lane tarffic, Fig. 13. This site caused both transportation and telecom service interruption. The road was reduced to one lane traffic after the event. By April 2014 this section of the highway was reopened for normal traffic flow.

*3. **Observed good performance at Otsuchi (39.3569°, 141.8953°).***

This location experienced both strong shaking and tsunami damage. A few minor landslides occurred at this

Fig. 13 Both road surface and telecom link were damaged at this lateral spread site

location. The raised train track collapsed while the landslide retaining structure shown in Fig. 14 performed very well.

Kumamoto Earthquake Japan 2016 (M_w = 6.0 and 7.0)

At ground zero of this earthquake was the largest landslide observed in most recent earthquakes in Japan. This landslide damaged a railway line, a bridge, roads, and buried telecommunication cables.

1. Landslides at Aso Bridge area (Lat 32.8836°, Long 130.9888°).

This area was the ground zero of this earthquake. Within 2 km radius from there were more than ten landslide locations that damaged roads, railway and telecom network (also power network). The focus of this event was that the landslide along with ground displacement between the bridge abutments has taken out Aso Ohashi. The landslide damage was more obvious in this case (Fig. 15).

2. Landslide damaged telecom and power poles/towers (Lat 32.8822°, Long 103.9903°).

This landslide site situated on the south-east of the collapsed Aso Ohashi, it took out both telecom and power poles (Fig. 16) interrupting both telecom and power services in this area. The high voltage power tower was standing on three concrete footings. The utility poles close to this tower were damaged, Fig. 17.

Kaikoiura New Zealand Earthquake 2016 (M_w = 7.8)

Landslides caused by this earthquake closed the only major highway and railway connection from Kaokoiura to the northern part of the south island, Fig. 18. Multiple sites of landslides damaged both rail and road interrupted transportation services. The East Coast Link (optical fiber) was severed in several locations due to landslide, lateral spread, faulting, and bridge damage. There is a submarine cable along the east coast called Aqualink, which is owned and operated by Vodafone, NZ and it has a landing repeater site at Kaikoiura. A temporary connection to the cable at the

Fig. 14 This retaining wall performed well

Fig. 15 Multiple landslides destroyed roads, bridge, railway and telecom optical fiber cables. (*Source* insert from MLIT, aerial photo from Geospatial Information Authority of Japan)

Fig. 16 The landslide that damaged the utility poles near Aso Ohashi

Fig. 17 The landslide across from Fig. 16 impacted two high voltage power towers

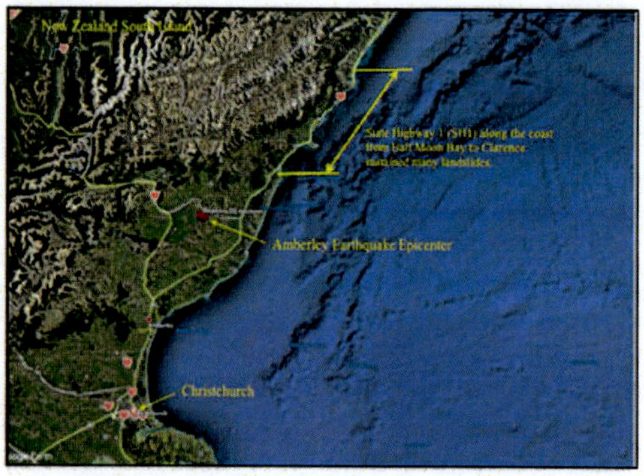

Fig. 18 This aerial view shows the relative location of the epicentre and the landslide zone along Hwy 1 in NZ south Island. (*Source* Google Earth)

Fig. 20 This shows landslide damage to road surface and the train track. (*Source* insert by Mark Mitchell/AFP/Getty Images, aerial photo from Google Earth)

Fig. 19 Aerial view of the bypass around this landslide site close to Half Moon Bay. (*Source* Google Earth)

landing site was agreed by Vodafone and this established a quick communication recovery for other service providers.

1. Landslides around Half Moon Bay (Lat −42.2525°, Long 173.8159°).

A temporary by pass was established by the transportation department to allow traffic after the earthquake, Fig. 19. Since this is the key highway connecting major cities of the south and north of the South Island, therefore temporary detours were necessary to provide emergency and recovery services access to the communities connected by this highway.

2. Landslide around Oahu (Lat −42.2473°, Long 173.8303°).

This landslide took out the train tracks and also blocked the highway. Both trains and cars traffic were disrupted (Fig. 20).

Hokkaido Iburi East Earthquake 2018 (M_w = 6.6)

This earthquake can be titled as landslide earthquake. The total area of landslides was about 13.4 km^2. Figure 21 is a satellite image of a small part of the landslides. Although the landslides occurred in the agricultural region of Hokkaido, the financial impact was significant. As this region provides about 30% of Japan's dairy and agricultural supply. A small NTT East remote telecommunication equipment office was destroyed by landslide. Landslides damaged both telecommunication distribution links and roads to small communities.

Unprecedented power black out had a significant impact to lifeline performance, particularly telecommunication.

1. Landslide damage telecom remote station (Lat 42.7556°, Long 141.9403°).

A small NTT East remote was destroyed by the landslide. This remote site was about 500 m from the foothill but the

Fig. 21 Satellite photo of the landslides in Hokkaido Earthquake impacted area. (*Source* Google Earth)

Fig. 22 The landslide that took out the NTT East remote site. (*Source* insert NTT East, base photo from Google Earth)

landslide reached this location and destroyed the building and the equipment inside, Fig. 22. The damage caused telecom service disruption in this area. A digital node was established quickly to recover service.

*2. **Landslide damage telecom distribution link and block road (Lat 42.7509°, Long 141.8794°).***

There were many examples similar to this shown in Figs. 23 and 24. Luckily the wireless network did not sustain any damage from this earthquake. However, the blackout and prolonged power outage reduced wireless services, the network returned to normal after electric power returned to normal.

Fig. 23 This aerial photo shows the massive landslide that took out the distribution links along this road as well as damaging the road surface blocking traffic. (*Source* Google Earth)

Fig. 24 This shows the downed distribution links, a mix of copper and optical fiber cables

Fig. 25 The landslide retaining structure at this location worked well

*3. **Landslide preventive measures worked (Lat 42.7606°, Long 141.9494°).***

This is one of the few locations of landslide prevention that performed well (Fig. 25).

Conclusion

Landslides/lateral-spreads after medium to large magnitude earthquakes continue to dominate lifelines service interruption and property losses.

There is a need that the lifelines owners have to establish a stronger risk management program with geotechnical engineers and researchers to reduce future losses by studying and analysing the existing vulnerable portions of the system to provide upgrades to prevent landslide damages in future events.

Erosion and corrosion of the landslide prevention structures are also key factors to be considered in the risk management programs. Periodic inspection of the built environment to identify areas to be repaired or upgraded is one of the tasks needed to prevent losses.

With respect to private lifeline service providers there is always a need to balance efficiency and resiliency. The spatial characteristic of the network and the need to reach remote communities, many network paths will be placed in the potential landslide areas. Service providers have to balance the cost associates with resiliency. Many will argue that resiliency is an investment, the opposing argument will be return of investment. The key is that Risk Management has to consider both factors and maintain a good balance.

Acknowledgements The author is grateful to Professor Konagai who continuously encourages and supports my quest for more resilient lifelines to reduce hardship experienced by earthquake impacted communities. During the process of our interface in the past decade, I learned a lot on landslide, lateral spread and other geotech topics. Professor Yue (the University of Hong Kong) provided me with valuable insights in landslide studies and failures.

References

Eidinger J, Tang AK (eds) (2017) TCLEE Report, No. 2 Kumamoto Earthquake Lifelines Performance 2016. (May)

Eidinger J, Tang AK (eds) (2019) TCLEE Report, No. 3 Hokkaido Earthquake Lifelines Performance 2018. (March)

Tang AK, Schiff AJ (eds) (2008) ASCE/TCLEE Monograph No. 31, Kashiwazaki, Japan Earthquake of July 16, 2007 Lifeline Performance

Tang AK (ed) (2010) ASCE/TCLEE Monograph No. 39, Wenchuan, China Earthquake of May 12, 2008 Lifeline Performance. (June)

Tang AK (ed) (2013) ASCE/TCLEE Monograph No. 42, Tohoku, Japan, Earthquake and Tsunami of March 11, 2011 Lifeline Performance. (August, e-book)

Tang AK, Eidinger J (eds) (2013) ASCE/TCLEE Monograph No. 41, Christchurch, New Zealand, Earthquake of September 04, 2009, February 22, 2011, June 13, 2011 and December 23, 2011 Lifeline Performance. (June)

Yue ZQ (2014) The fundamental drawbacks and disastrous consequence of the classical theories of soil mechanics. Published at 15:41 of 2014–10–18 on https://blog.sciencenet.cn/u/QuentinYue as https://blog.sciencenet.cn/blog-240687-836715.html

Yue ZQ, Xu Q (2014) Fundamental drawbacks and disastrous consequences of current geotechnical safety design theories for slopes. Chin J Geotech Eng 36(9):1601–1606

Lessons from Recent Geo-Disasters in Hokkaido Under Earthquake

2020 Kyoto Japan

Shima Kawamura

Abstract

The 2018 Hokkaido Eastern Iburi Earthquake caused a large amount of slope failures and landslides over a wide area. In particular, the enormous damage due to slope failures was centered in Atsuma and Abira town which are to the north of the epicenter. In Atsuma town, a seismic intensity (Japan Meteorological Agency) of seven has been observed for the first time in Hokkaido, Japan. The geology in this region was mainly formed from three kinds of pyroclastic fall deposits (*fa*) erupted from Tarumae volcano, Eniwa volcano, and Shikotsu Caldera. Especially, the serious damage was generated in these tephra stratus. In addition, a large scale slide of rock slopes composed of mudstone also occurred in the Hidaka-Horonai river basin in Horonai district in Atsuma town, and its debris flowed into the river and caused the blockage of the river channel. This paper summarizes earthquake-induced damages on natural slopes and mechanical properties of Tarumae pyroclastic fall deposits.

Keywords

The 2018 Hokkaido Iburi eastern earthquake • Slope failure • Pyroclastic fall deposits

Introduction

The middle and southern parts of Hokkaido were affected by a gigantic earthquake of a moment magnitude Mj = 6.7 occurred at 3:07 AM local time on September 6 2018 (The 2018 Hokkaido Eastern Iburi Earthquake). The details of geo-disasters have been described by many reports (e.g., Kawamura et al. 2019).

In this earthquake, the ground disasters involving a variety of soils such as pyroclastic fall deposits (*fa*) or pyroclastic flow deposits (*fl*) are classified into liquefaction-related one in Sapporo city and failures of natural slopes in Abira and Atsuma town.

The author focuses mainly on geo-disasters of natural slopes formed from pyroclastic fall deposits (volcanic coarse grained soils) and mudstone in the 2018 Hokkaido Eastern Iburi Earthquake. This paper highlights the earthquake-induced slope failure and presents physical and mechanical properties of the pyroclastic fall deposits distributed over this area based on the geo-disasters reports (the 2018 Hokkaido Eastern Iburi Earthquake JGS 2018; Kawamura et al. 2019).

Geology and Topography in Atsuma and Abira Town

The seismic aspects and the distribution of typical pyroclastic fall soils are shown in Fig. 1. The map was drawn based on the data map by Geospatial Information Authority of Japan (2018), Furukawa and Nakagawa (2010), Machida and Arai (2003), the epicenter provided by the Japan Meteorological Agency (2018) and slope failures/sedimentation distribution (red area) provided by Geospatial Information Authority of Japan (2018). In Fig. 1, the red zones (the dotted circle) are the sites where slope failures occurred. As shown in the figure, it is apparent that the locations of the failed slopes in Atsuma and Abira town are within approximately 25 km from the epicenter of earthquake. In this area, pyroclastic fall soils that erupted from volcano groups in western Hokkaido are thickly deposited. Specifically, the deposits are mainly formed from three kinds of pyroclastic fall deposits (*fa*) erupted from Tarumae volcano (the notation is *Ta*) and Eniwa volcano (the notation is *En*) and Shikotsu pumice fall-1 (the notation is *Spfa*). It is estimated that the eruption age of the

S. Kawamura (✉)
Muroran Institute of Technology, 27-1 Mizumoto-cho, Muroran, 0508585, Hokkaido, Japan
e-mail: skawamur@mmm.muroran-it.ac.jp

Ž. Arbanas et al. (eds.), *Understanding and Reducing Landslide Disaster Risk*, ICL Contribution to Landslide Disaster Risk Reduction,
https://doi.org/10.1007/978-3-030-60713-5_4

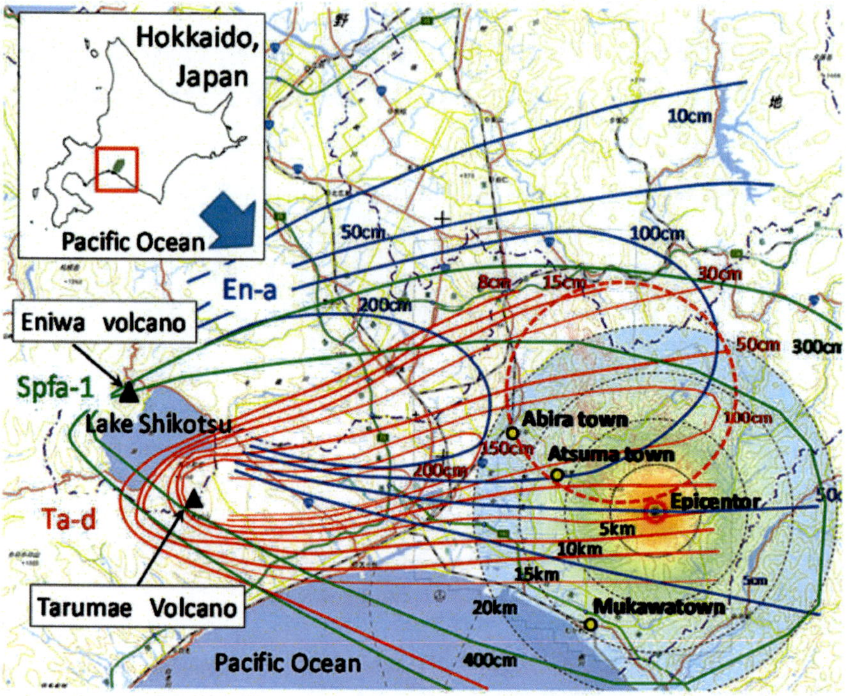

Fig. 1 The seismic aspects and the distribution of typical pyroclastic fall soils

volcanic soils belonging to Shikotsu primary tephra (*Spfa*-1) is 46,000 years, and the eruption ages of the typical deposits of *En-a* and *Ta-d* are 19,000–21,000 years and 9,000 years, respectively. For example, the thickness of *Spfa-1*, *En-a* and *Ta-d* is approximate 300–400 cm (green lines), 10–100 cm (blue lines), 30–100 cm (red lines) in Atsuma town, respectively (Kawamura et al. 2019).

Figure 2 shows the formation process of slopes, tephra deposition, scrapings and slip surfaces from Abira to Atsuma, and to Mukawa town (as an example). As shown in Fig. 2, the aforementioned tephra layers often cover the layer in the Neogene (or the weathered layer in the Neogene) directly. In addition, thin slope sediments may be sandwiched between the Neogene's layer and these tephra layers. The stratum in the Pleistocene-Holocene is well preserved in a flat or a gentle slope. On the other hand, the slopes below the transcendental line often lose the tephra layer due to the solifluction and the formation of valleys (river erosion).

According to the investigation of the disaster sites, the deposits on *Ta_d* were sliding on its boundary plane in Atsuma town. However, several patterns of sliding planes between *Ta_d* and weathering zone in Neogene, loam layer between *En-a* and *Spfa-1* or between *En-a* and weathering layer in the Neogene, *Ta_d* and newer tephra layer are also recognized. In any cases, the effect of seismic loadings on the topology and geotechnical properties just before the earthquake, namely the relationship with geology (such as the response to past earthquakes) and the geological structure should be clarified.

Based on the site investigation, the situation of the geo-disaster in each area will be explained in the following chapters in detail. Figure 3 shows the locations of site investigation on geo-disasters.

Geo-disasters in Atsuma and Abira Town

Figure 4 and 5 show aerial photos after slope disasters in Yoshino district in Atsuma town (the point (a), (b) and (c) in Fig. 5 are the locations in Fig. 6). This region is a valley plain formed by the Atsuma river. The collapsed houses are located under the cliff of river terraces or mountain slope on the right bank of the Atsuma river. The earthquake destabilized the sedimentary soils on the cliff, and the collapsed soils engulfed several houses under the cliff, and then flowed in the paddy field (see Fig. 5). As shown in Fig. 5, a natural slope with a relative height of around 31 m and a slope angle of about 31° collapsed at this site, which was the greatest damage caused by this earthquake. Figure 6a–d explain the local situations in Fig. 5, respectively. According to the topographical maps, valleys cannot be identified in this disaster zones. The slope surface has a relatively uniform slope angle with about 31°. In addition, some abrasion marks due to slope failures were observed for some sites, therefore, it is supported that debris materials from these slope failures flowed at a relatively high speed (see Fig. 6b). In the site investigations, the debris seems to be composed of pyroclastic fall deposits from different eruptions (e.g.,

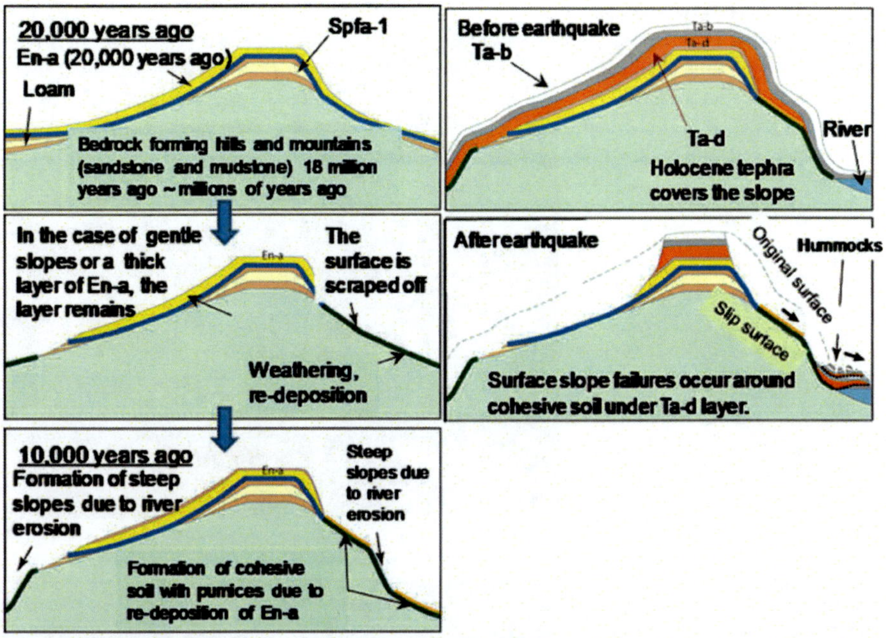

Fig. 2 Formation process of slope, tephra deposition, scraping and slip surface from Abira to Atsuma to Mukawa town (Kawamura et al. 2019)

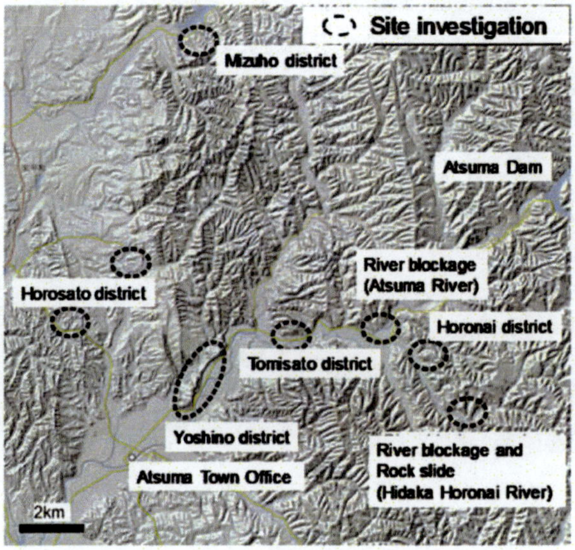

Fig. 3 Locations of site investigation on geo-disasters by the earthquake

sedimentary age, the degree of weathering, see Fig. 6c) and invaded the paddy field (see Fig. 6d). As a result, the surface of paddy field was heightened up to around 1 m. From this fact, it may be said that the failed sediments flowed down with high speed and energy. From the above investigations, it is found that surface slope failures on the Tarumae pyroclastic fall deposits (*Ta–a–Ta–d*), in particular, on *Ta–d* layer occurred frequently in Yoshino district.

In Tomisato district (see Fig. 3), slope failures occurred frequently around the water purification facility. In particular, the hills on the right bank of the Atsuma river collapsed. Figure 7 shows aerial photo after the earthquake. Letter symbols (a) and (b) in the Fig. 7 show the location as described later. The slope failure occurred at the northern slopes on water purification plant. The collapsed soils engulfed houses in the vicinity of the water purification plant facility and flowed down to the river channel. As a result, a river channel blockage occurred on the upstream side of the Narayama Bridge. The slope with a relative height of around 75 m and a slope angle of about 27 degrees at this site failed (see Figs. 7 and 8a). This site is at the point (a) in Fig. 7. In compared with Yoshino district, it is obvious that the sediments fail and flow down like a valley in Tomisato district. Since the depth of the sliding plane was deeper than that of Yoshino district, therefore, it was concluded for this site that the pyroclastic fall deposits thickly deposited in the valley where water flows in are destabilized by strong seismic loadings. Similarly, Fig. 8b is other view from the situation (see the point (b) in Fig. 7). On the slopes located to the west of the water purification plant facility, the collapsed soils flowed widely to the paddy field. In the investigation at the sediment condition around the head of the northern slope, pyroclastic fall deposits and andosol were observed.

In Horonai district located to the east of Tomisato district (see Fig. 3), surface slope failures and a large-scale rock slide and river blockage occurred in the Hidaka-Horonai river. In the regions indicated in Figs. 3 and 9, a couple of surface failures with a long-distance flow were confirmed (see Fig. 10). The slope failures with long-distance flows

Fig. 4 Slope failures in Yoshino district (Provided by Prof. Shimizu, Hokkaido University)

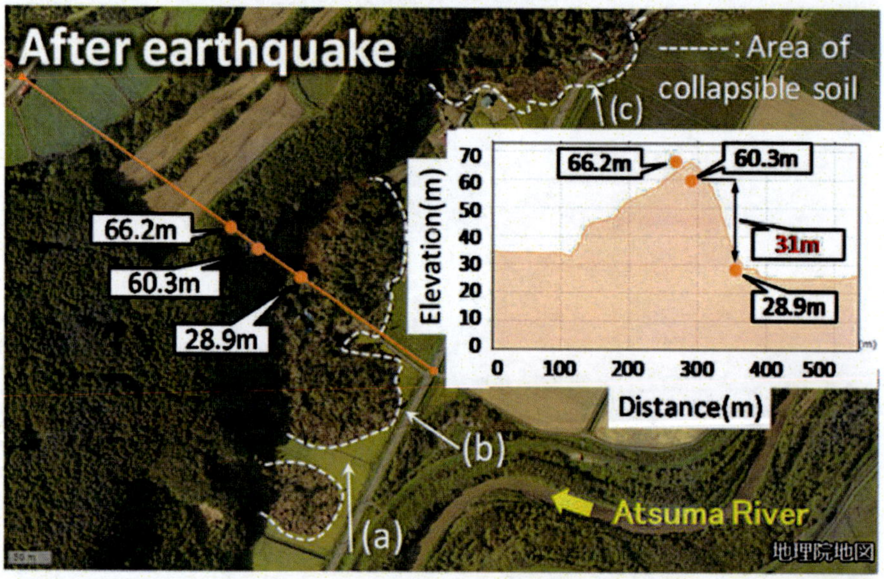

Fig. 5 Slope failures in Yoshino district (based on Geospatial Information Authority of Japan)

concentrated in the valley topography, similar to those in Tomisato district. Figure 11 shows slope angles based on the topographical map (Kawamura et al. 2019). It can be seen from Fig. 11 that the frequency of occurrence of slope failures becomes higher in the order of 20–24°, 25–30°, 35–34°, 15–19°, 35–39°, and 50–54°. It is interesting that slope failures occurred even on relatively gentle slopes. Similarly, Fig. 12 summarizes the number of slope failures in the Hidaka-Horonai river basin (Kawamura et al. 2019). Approximately 80% of the failures occurred in the valley topography. Therefore, it can be also said that topographical features is one of the causes of slope failures. As one of the causes of slope failure with a long-distance flow like debris flow, the influence of Typhoon 21 (September 5, 2018) which passed just before the earthquake has been also pointed out.

As aforementioned above, a large-scale rock sliding was also induced by the earthquake on the site, the valley and the river channel were blocked by its debris (see Fig. 9). It has been reported that the ridge with a width of about 400 m moved over about 500 m and dammed the Hidaka-Horonai river over a relative height of 50 m and a length of about 1,100 m (Ministry of Land, Infrastructure, Transport and Tourism 2018). Figure 13 shows the situation of a natural lake formed by channel blockage in the surveyed field (October 4, 2018) (see the point (a) in Fig. 9).

Surface slope failures have also been confirmed around this site. Figure 14 shows the site situation of surface slope

Fig. 6 Local situations in Yoshino district: **a** at the point (a) in Fig. 5, **b** at the point (b) in Fig. 5, **c** and **d** at the point (c) in Fig. 5

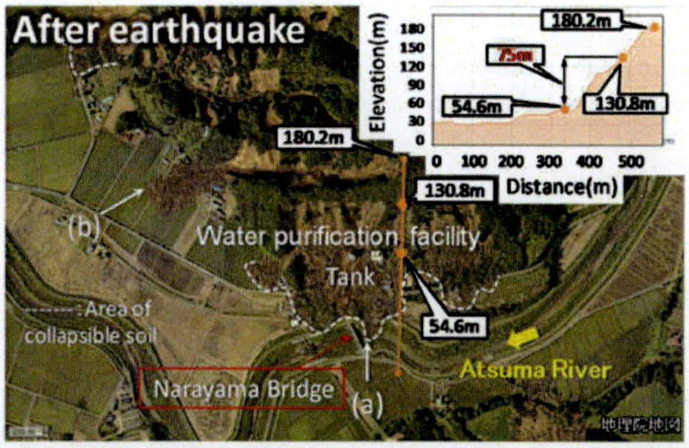

Fig. 7 Slope failure in Tomisato district (based on Geospatial Information Authority of Japan)

failures occurred simultaneously (see the point (b) in Fig. 9), the slope angle was 25–30°, and the depth of slipped layer was about 3 m from the surface. In the site investigation, it was apparent that volcanic soils of *Ta* (the eruption source of Tarumae volcano) were thinly deposited on the mudstone. In the photo, the deposited layers are classified visually into different layers despite belonging to the same *Ta–d* deposits. The visual difference may mean the difference in the degree of weathering. In particular, the coarse-grained soils with reddish-brown in *Ta–d* deposits were well observed mainly in Yoshino and Tomisato districts.

In Abira town, slope failures in gentle slopes and hills occurred in the eastern and south eastern parts adjacent to Atsuma town. An example in Fig. 15 includes slope failures

Fig. 8 Local situations in Tomisato district: **a** at the point (a) in Fig. 7, **b** at the point (b) in Fig. 7

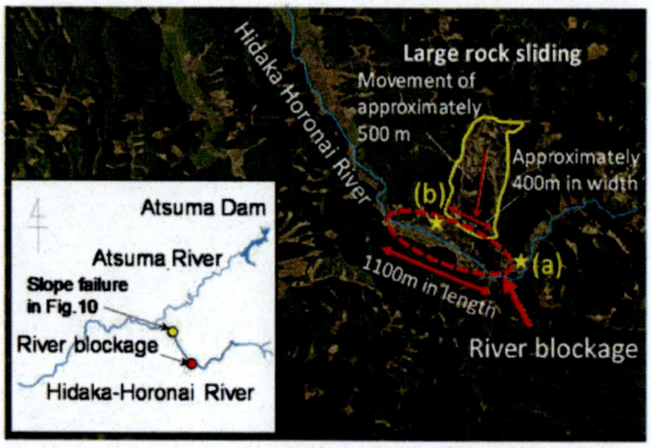

Fig. 9 Location of large-scale rock sliding in Hidaka Horonai River basin (based on Geospatial Information Authority of Japan)

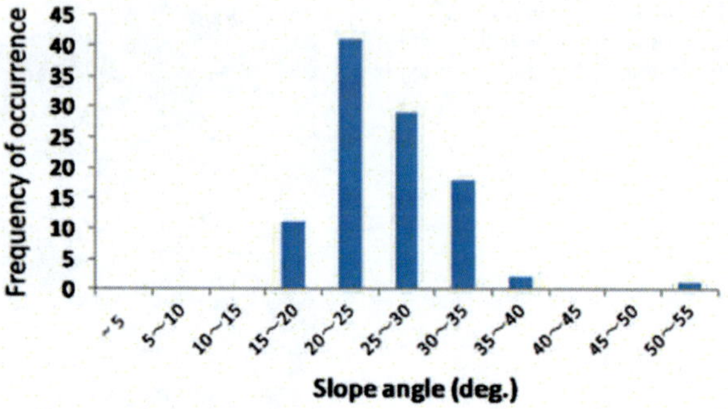

Fig. 11 Frequency of slope failure occurrence on slope angle (Kawamura et al. 2019)

Fig. 12 Classification of slope failures into topography (Hidaka Horonai river basin)

in Mizuho district (see Fig. 3) in Abira town, where a hard mudstone of the Karumai Formation, the Late Pleistocene to Holocene tephra (*Spfa-1*, *En-a*, *Ta–d*, *Ta–c*, *Ta–a*), loam and top soils are covered in this area. In this site, the weathered part of mudstone and the pumice-loam layer below *En-a* or *Spfa-1* have become the main sliding surfaces.

In Horosato district in northern Atsuma town (see Fig. 3), as described above, a large number of slip failures occurred on a gentle slope (see Fig. 16). In the collapsed soils, volcanic soil (*Ta–d*) was observed in layers. This situation of slope failure is similar to that occurred on the gentle slope compose of the Kusa senri gahama volcanic soils by the 2016 Kumamoto earthquake. According to the disaster

Fig. 10 Slope failure with long-distance flows in the Horonai district (from Japan Asia Group, PASCO Corporation.)

Fig. 13 Situation of a natural lake formed by channel blockage in the Hidaka-Horonai river

Fig. 16 Sliding surface at failure site in Horosato district, northern Atsuma town

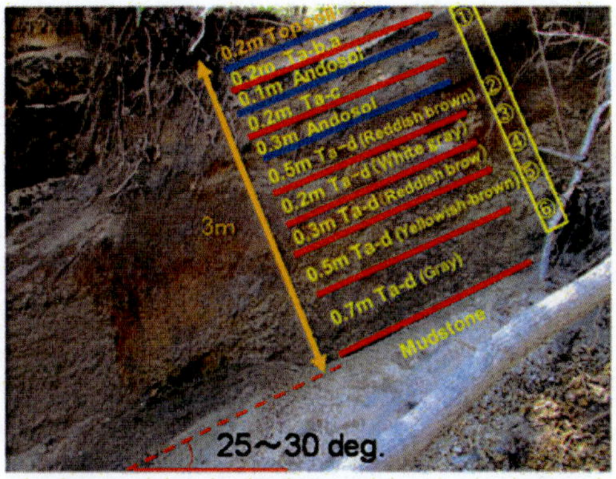

Fig. 14 Surface slope failure in around large rock slide at the point (b) in Fig. 9

Fig. 15 Sliding surface at failure site in Mizuho district in Abira town

report (the research report of the 2016 Kumamoto earthquake disaster, JGS 2017), it was confirmed that the pumice fall layer is crushed on slip surface. Therefore, it is important to evaluate the dynamic stability on slopes with volcanic deposits.

Physical and mechanical characteristics of the volcanic soil samples taken from the outcrop in Fig. 14 were also investigated. The sampling points are indicated by the number from 1 to 6 in Fig. 14. Table 1 is the index properties of *Ta_d*. It is evident that the soil of each layer is mostly composed of coarse-grained soil particles. The density of soil particles Samples 3 to 6 show $\rho_s = 2.49$ to 2.50 g/cm^3. Sample 2 is a low value of 34.4% in comparison with natural water content, but the others show 138.5–197.3%. In addition, according to the results of the ignition loss test, they indicated Li = 5.0–7.1% and almost same as those of volcanic coarse-grained soils in Hokkaido (approximately 10% or less). Moreover, the value of Sample 6 at the boundary with mudstone seems to be high. It may be possible that it has been affected by water for a long time. Therefore, it requires to clarify the effect of weathering on physical properties of soils.

In addition, a series of consolidated drained (CD) and undrained triaxial compression tests (CUB) were performed on the collapsed soils of *Ta–d* in Fig. 14 (Kawamura et al. 2019; Koga et al. 2020). In the tests, the consolidation was performed under effective confining pressure $\sigma'_c = 49$, 98, and 196 kPa. Each specimen was prepared by the multiple sieving pluviation method (MSP method) so as to be the in-situ dry density (0.382 to 0.440 g/cm^3). As a result, cohesion c_d, c' and shear resistance angle ϕ_d, ϕ' were c_d = 33.0 kPa, $\phi_d = 29.0°$, $c' = 8.0$ kPa, and $\phi' = 42.5°$, respectively.

Figure 17 shows the relationship between ϕ_d, ϕ' and effective mean principal stress at failure p'_f comparing with other volcanic soils in Hokkaido (Miura et al. 2003). As shown in the figure, ϕ_d and ϕ' decrease with an increase of

Fig. 17 Dependency of ϕ_d and ϕ' on effective mean principal stress at failure c.

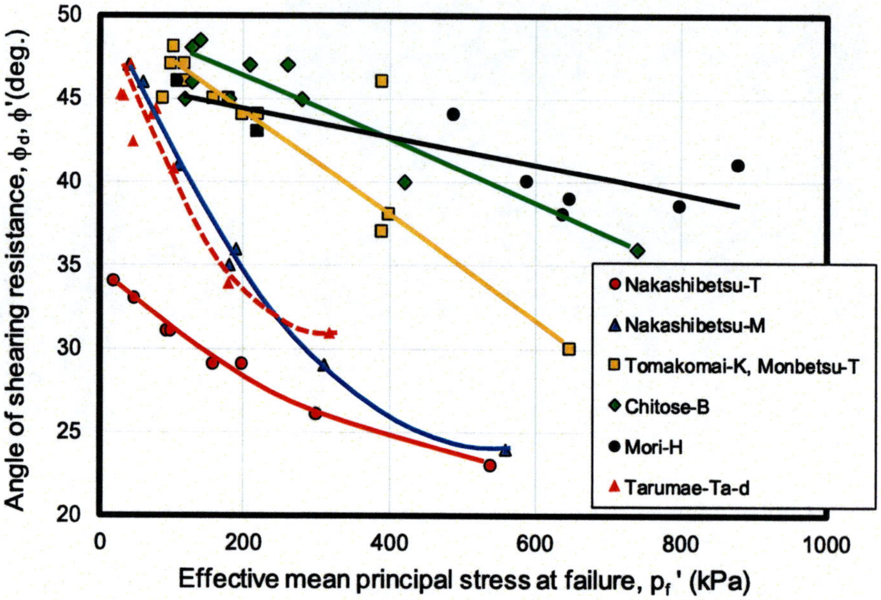

p'_f, similar to those of other volcanic soils. From the figure, it was found that Tarumae pyroclastic fall deposits (*Ta–d*) indicates a significant dependency on the magnitude of effective confining pressure. It has been reported that the reason for this is due to particle breakage (Koga et al. 2020). Therefore, it is important to assess particle brakage (including weathering) for stability of slope formed from pyroclastic fall deposits.

Conclusions

In the site investigation, it was revealed that slope failures caused around pyroclastic fall deposits (*Ta_d*, *En-a*, *Spfa*-1) originating from Tarumae volcano, Eniwa volcano and Shikotsu caldera, particularly the layer of Tarumae pyroclastic fall deposits. For the large-scale rock slide, the sliding surface of the main part of the slope was considered to be the

Table 1 Index properties of sampling soils

	②	③	④	⑤	⑥
Natural water content, w_n (%)	34.4	142.2	197.3	160.7	138.5
Soil density, ρ (g/cm^3)	2.78	2.5	2.49	2.49	2.49
Ignition loss, Li (%)	–	4.99	5.88	4.85	7.10
10% grain size, D_{10} (mm)	0.9	2.8	0.6	1.4	0.6
Mean grain size, D_{50} (mm)	5.1	6.1	4.1	3.9	3.4
Coefficient of uniformity, U_c	6.7	2.5	8.5	3.9	8
Fines content (<75 μm) (%)	1.5	0.3	4.5	1.0	4.9
Liguid limit, w_L (%)	–	–	93.9	–	–
Plastic limit, w_P (%)	–	–	N.P.	–	–

stratified surface of the mudstone-silt layer of the Miocene Karumai Formation. Further studies are required to the dynamic behaviour of natural slopes composed of pyroclastic fall deposits in earthquakes.

Acknowledgments The author wishes to express their sincere gratitude to all those who provided us with valuable information, particularly the Hokkaido Regional Development Bureau, Ministry of Land, Infrastructure, Transport and Tourism and Hokkaido Government, and Dr. S. Kawajiri, Dr. W. Hirose and Dr. T. Watanabe (members of the Geo-hazard Investigation Committee of the 2018 Hokkaido Eastern Iburi Earthquake, JGS). This study was also undertaken with the financial supports of KAKENHI (Grant-in-Aid for Science Research (B) JP17H03318, JSPS and River Center of Hokkaido (General Incorporated Foundation, RIC).

References

Furukawa R, Nakagawa M (2010) Geological map of volcanos 15, Tarumae volcano 1:30,000. Geological Survey of Japan, AIST ((in Japanese))

Kawamura S, Kawajiri S, Hirose W, Watanabe T (2019) Slope failures/landslides over a wide area in the 2018 Hokkaido Eastern Iburi Earthquake. Soils Found 59:2372–2391

Koga T, Kawamura S, Kawamura K (2020) (2020) Physical and mechanical behaviour of volcanic soils in failed natural slopes due to the 2018 Hokkaido Eastern Iburi Earthquake. Annual Technical Report, Hokkaido Branch, JGS. 60:149–155 ((in Japanese))

Machida H, Arai F (2003) Atlas of tephra in and around Japan: revised edition, University of Tokyo Press, 336 p (in Japanese)

Ministry of Land, Infrastructure, Transport and Tourism, Hokkaido Development Bureau, (2018) Response to disasters on the 2018 Hokkaido Eastern Iburi Earthquake (in Japanese)

Miura S, Yagi K, Asonuma T (2003) Deformation-strength evaluation of crushable volcanic soils by laboratory and in-situ testing. Soils Found 43(4):47–57

The Geospatial Information Authority of Japan (2018) Information on the 2018 Hokkaido Eastern Iburi Earthquake. https://www.gsi.go.jp/BOUSAI/H30-hokkaidoiburi-east-earthquake-index.html#10 (in Japanese)

The Japan Meteorological Agency (2018) Information on the 2018 Hokkaido Eastern Iburi Earthquake. https://www.static.jishin.go.jp/resource/monthly/2018/20180906_iburi.pdf (in Japanese)

The Japanese Geotechnical Society (2017) Geo-disaster report of the 2016 Kumamoto Earthquake, The Geo-hazard Investigation Committee of the 2016 Kumamoto Earthquake, JGS (in Japanese)

The Japanese Geotechnical Society, Hokkaido Branch (2018) Geo-disaster report of the 2018 Hokkaido Eastern Iburi Earthquake, The Geo-hazard Investigation Committee of the 2018 Hokkaido Eastern Iburi Earthquake, JGS. https://www.jiban.or.jp/?page_id=9397 (in Japanese)

Relation Between Horizontal Direction of Crustal Deformation Surveyed on the Control Points and Area Ratio of the Slope Failures Triggered by the 2016 Kumamoto Earthquake (M_j 7.3)

2020 Kyoto Japan

Hiroshi P. Sato and Hiroshi Une

Abstract

The 2016 Kumamoto earthquake (M_j 7.3) induced more than 3000 slope failures along the dextral strike-slip epicentral fault, as well as along steep slopes on the caldera wall and the central cones in the caldera. We investigated dominant slope aspect (orientation bias) of the slope failures triggered by the earthquake using horizontal direction and amount of the crustal deformation surveyed by Geospatial Information Authority of Japan. Consequently, we observed that the dominant slope aspect of the failure was correlated with the direction of the crustal deformation in terms of the common component of the direction; i.e., both the dominant slope aspect and deformation direction have an E component along the NW(N) and NW(S) sides and a W component along the SE side of the fault line. This correlation was confirmed at the distances of 2000 and 4000 m from the fault line, but not at the distance of 6000 m. It is thought that the dominant inertial force of the faulting directly affected the slopes and triggered failures near the fault. The peak of the slope failure area ratio was 2.2–3.5% at the distance of 2000 m, where the amount of the crustal deformation was 0.6–0.9 m along the NW(N) and NW(S) sides of the fault line. However, along the SE side, the peak of the slope failure area ratio was 5.8% at the distance of 6000 m, where the amount of crustal deformation was 0.5–0.6 m. Minimal crustal deformation might induce more failures on the central cones, because the slope is covered with loose pyroclastic deposits, which are susceptible to deformation due to the earthquake.

Keywords

Kumamoto • Slope • Failure • Aspect • Earthquake • Control point • Orientation bias

Introduction

Approximately 3460 slope failures were triggered by the 2016 Kumamoto earthquake (Xu et al. 2018), which had a foreshock of M_j 6.5 on 14 April and the main shock of M_j 7.3 on 16 April. The epicenters of the earthquake are depicted in Fig. 1. The foreshock and the main shock of the Kumamoto Earthquake were caused by successive slips first on the Hinagu fault and then on the Futagawa fault, respectively. Both faults are dextral strike-slip epicentral faults (Shirahama et al. 2016). Meunier et al. (2007) and Sato et al. (2007) showed the dominant slope aspect (orientation bias) of the earthquake-induced slope failures. Meunier et al. (2007) ascribed it to obliquely incident S-waves travelling from deep underground to the crustal surface. In addition, Sato et al. (2007) mentioned that many large landslides occurred on the slopes facing S to SW in their study area. This observation is consistent with the horizontal dominant direction of crustal deformation observed on the hanging wall.

In the case of 2016 Kumamoto earthquake, Sato and Une (2017) clarified that the slope failures on the NW side of the Futagawa fault has the dominant slope aspect of E and the SE side has the dominant aspect of W, respectively. In this study, we investigated relation between dominant slope aspect of the earthquake-induced slope failures and the earthquake-induced horizontal direction and amount of

H. P. Sato (✉)
College of Humanities and Sciences, Nihon University, 3-25-40 Sakurajosui, Setagawa Ward, Tokyo, 156-0045, Japan
e-mail: satou.hiroshi37@nihon-u.ac.jp

H. Une
Former Geospatial Information Authority of Japan, 1 Kitasato, Tsukuba, 305-0811, Japan
e-mail: hiroune@nifty.com

Ž. Arbanas et al. (eds.), *Understanding and Reducing Landslide Disaster Risk*, ICL Contribution to Landslide Disaster Risk Reduction,
https://doi.org/10.1007/978-3-030-60713-5_5

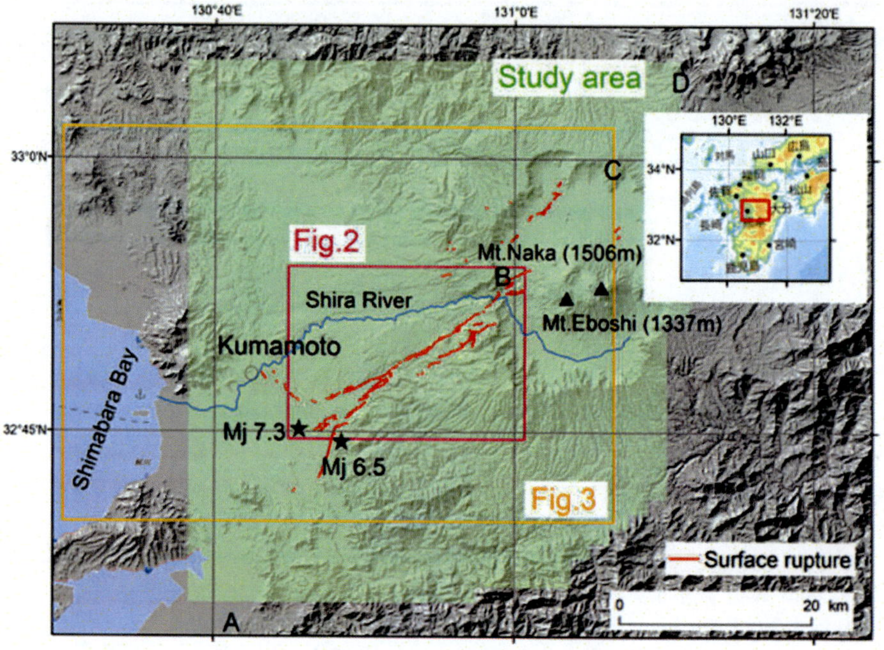

Fig. 1 Study area, hill-shaded map produced from 10 m DEM by GSI. Surface ruptures (Kumahara et al. 2017; Suzuki et al. 2017) and the areas of Figs. 2 and 3 are depicted in this figure

crustal deformation surveyed on the control points by global navigation satellite system (GNSS).

Study Area

The study area is depicted in pale green in Fig. 1. This area is overlapped by the area where the slope failures were interpreted. The inventory map was produced by the National Research Institute for Earth Science and Disaster Resilience (NIED) (2016) and Geospatial Information Authority of Japan (GSI) (2016), as detailed in the next section.

In Fig. 1, Kumamoto City (approximately 12 m in elevation) faces the Shimabara Bay, and the Shira River flows from Aso Caldera into the bay. The volcanos, namely, Mt. Naka and Mt. Eboshi, are seen in this figure as central cones in the caldera.

Data

Slope Failure Inventory Map

The slope failure inventory map was published by NIED (2016) on its website; these locations are depicted as small black polygons in Fig. 2. This map contains the earthquake-induced slope failure areas interpreted using aerial photographs taken on 16, 19, and 20 April 2016 by GSI. The minimum size of the slope failure area is approximately 0.1 ha. We used the polygon data of slope failure areas photo-interpreted by NIED (2016). Each polygon does not distinguish between the source area and dispositional area, and we used it as it is. As some clustered slope failures can often be drawn as a single polygon, it was difficult to separate the polygons into individual slope failures. Therefore, it is meaningless to count the number of polygons to determine the total number of slope failures.

As the provided data contained errors such as multiple overlapped polygons for an individual slope failure site, we cleaned the data to remove these errors. To remove these errors, we interpreted the orthophotos taken before the earthquake using web platforms such as Chiriin-Chizu in Japanese, a literal translation of the GSI Map on web (https://maps.gsi.go.jp/) and Google Earth. If any polygons showed slope failures before the earthquake, these polygons were removed because they were not triggered by the earthquake. Furthermore, if we observed polygons with no corresponding data points of slope failure in the inventory map provided by GSI (2016), these polygons were removed. These steps were taken before performing the analysis.

Figure 2 depicts the surface ruptures due to the earthquake (Kumahara et al. 2017; Suzuki et al. 2017). Along the swath of ruptures, many slope failures appeared. The inset photo in Fig. 2 shows one of these slope failures. The slope failures occurred not only along the surface ruptures, but also on the caldera wall (area A in Fig. 3) and on slopes of the central cones (area B in Fig. 3), which are far from the ruptures lined up along the recognized seismic fault trace.

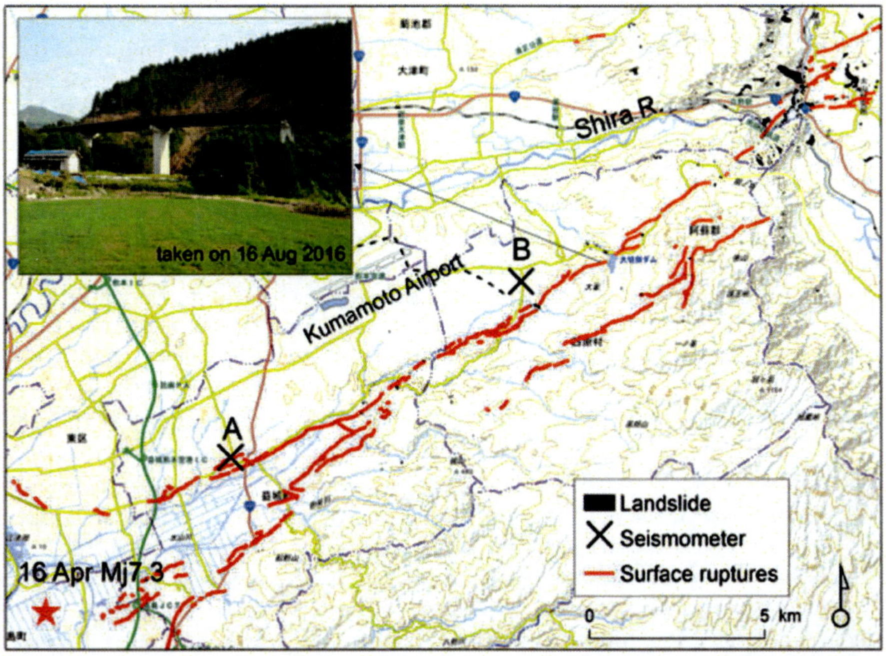

Fig. 2 Slope failures inventory map (NIED 2016) along the surface ruptures (Kumahara et al. 2017; Suzuki et al. 2017). In the figure, A and B is the location of seismometer, mentioned in Fig. 10

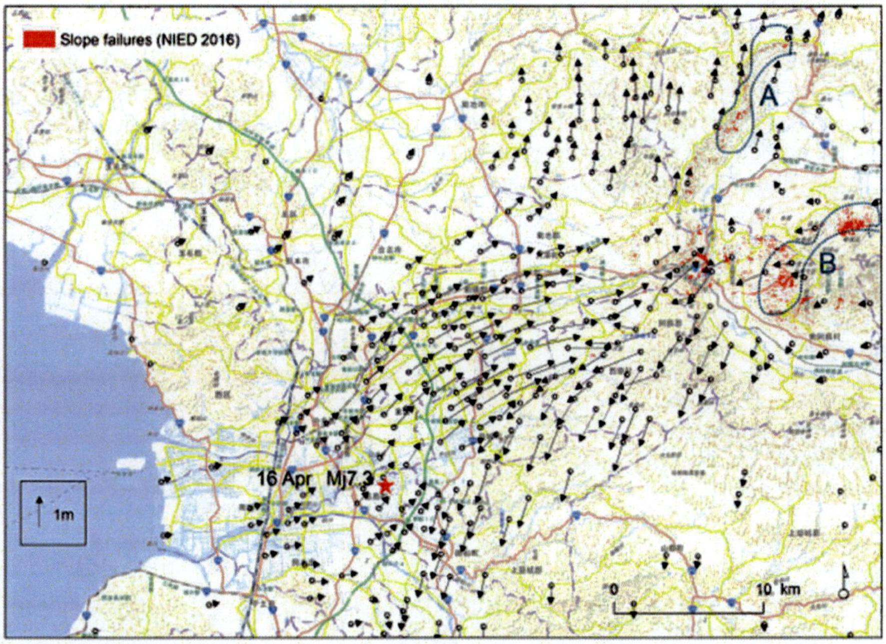

Fig. 3 Horizontal direction of the crustal deformation (Ootaki et al. 2017) and slope failures inventory map (NIED 2016)

Crustal Deformation

Ootaki et al. (2017) calculated the earthquake-induced crustal deformation on the control points shown in Fig. 3. They performed a GNSS survey to calculate the coordinates of these points, and subsequently, compared the changes in the coordinates before and after the earthquake. Although three orthogonal components of displacement are calculated at each point, we used only two horizontal components, namely, its projection on the horizontal plane to investigate the relation between dominant slope aspect of the slope failures and the direction and amount of the detected crustal deformation.

From Fig. 3, it is evident that the directions of crustal deformation differ from NW to SE sides of the fault, i.e., N and NE directions dominate on the NW side while S and SW directions dominate on the SE side. Dominant directions of the crustal deformations were smeared over the whole study area spatially interpolating the values among the given data points, and the locations of landslides were laid upon the map (Fig. 4).

Fault Trace

As depicted in Fig. 2, the surface ruptures do not extend continuously, and the analysis by the geographic information system must separate the study area into NW and SE sides. Therefore, we drew the "fault" line as follows. In Fig. 1, the active Futagawa and Hinagu faults extend between A and B through two epicenters. Although the Hinagu fault was not completely ruptured, Goto and Chida (2018) illustrated the active fault traces for the Hinagu fault through site A.

Along the line that connects sites B and C, using airborne light detection and ranging (LiDAR) to measure data before and after the earthquake, Mukoyama et al. (2016) illustrated two different directions of crustal deformation with line B–C as the boundary: N or NE direction along the NW side and W direction along the SE side.

On the basis of previous studies, as depicted in Fig. 4, we drew the fault line in red from A to D (Fig. 1) through the Hinagu fault, surface ruptures, and line B–C. Line C–D was extended from line B–C by referring to the difference of the direction of the crustal deformation (Ootaki et al. 2017).

Setting the Sides of the Fault Line and Zonation of the Distances from the Fault Line

As depicted in Fig. 4, the fault line divides the target area into its NW and SE sides; the NW side was further divided into the NW(N) and NW(S) sides, as in Fig. 4. The N and NE directions of the deformation were dominant along the NW(N) and NW(S) sides, respectively.

To investigate dominant slope aspect of the slope failures, we must consider the distance from the fault line. Therefore, we divided the NW(N), NW(S) and SE sides into strips according to the distance from the fault line, i.e., 2000, 4000, 6000, and 8000 m, in Fig. 5. For example, a "distance of 2000 m" means that the distance from the fault line is between 0 and 2000 m.

In advance, we overlaid a geological map (https://gbank.gsj.jp/geonavi/) on Fig. 6 (explained in the next section) and investigated the localization of lithological units. The study area is mainly covered with pyroclastic flow deposits and felsic and mafic volcanic rocks. However, we confirmed that the specific lithological unit does not always exist along the fault line or on the specific aspect of slopes.

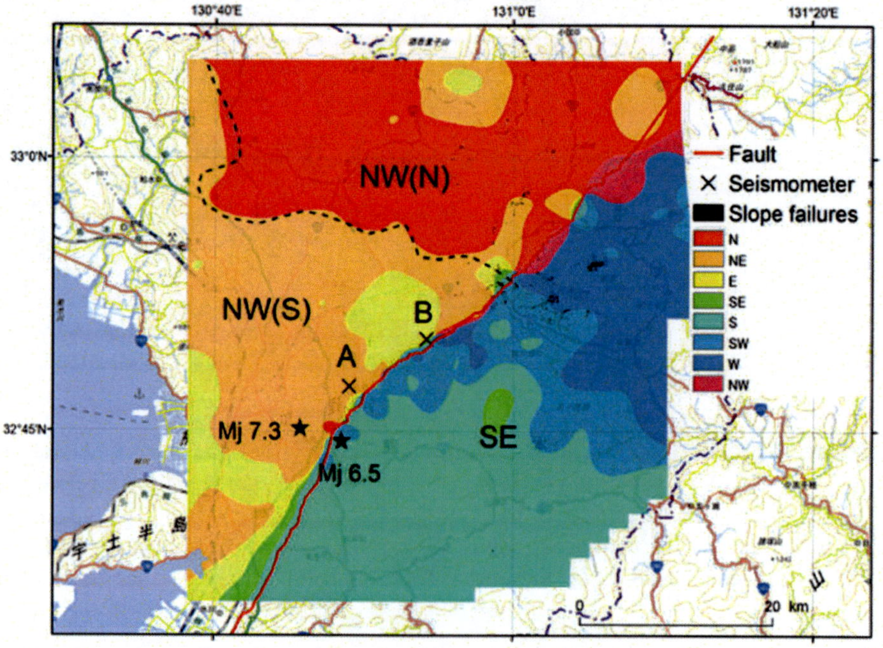

Fig. 4 Interpolation of the horizontal direction of the crustal deformation and division along the NW(N), NW(S), and SE sides of the fault line. In the figure, A and B is the location of seismometer, mentioned in Fig. 10

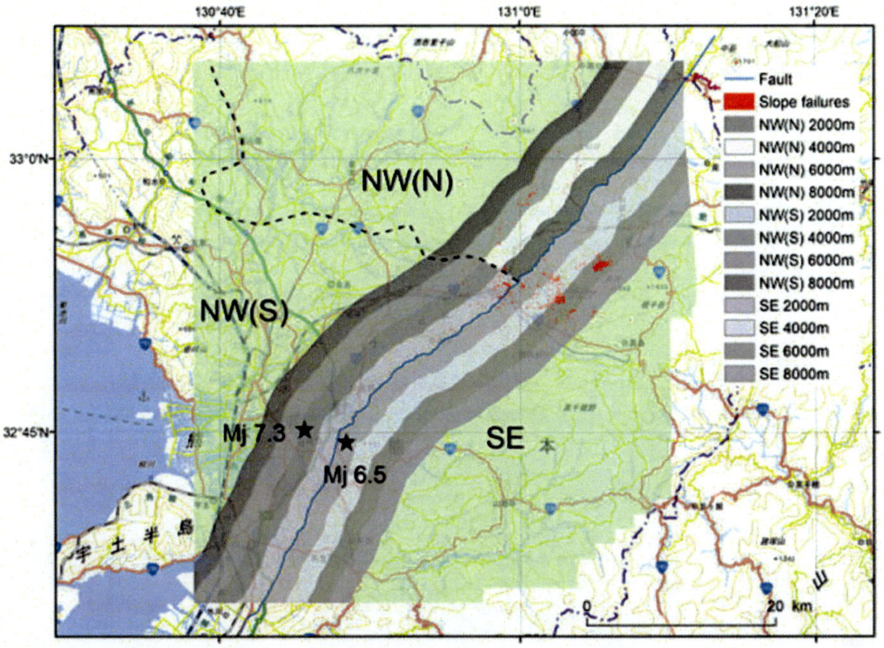

Fig. 5 Zonation of the distance from the fault line

Calculation of the Aspect and Angle Using DEM

The slope aspect and angle, which were calculated using the 10 m digital elevation model (DEM) provided by GSI, are depicted in Fig. 6. We overlaid the slope failures inventory map (NIED 2016) upon the maps showing spatial distribution of slope aspects and angles.

The results and dominant slope aspect of the slope failures are discussed as in the next section.

Results and Discussion

Slope Failure Area Ratio According to the Slope Aspect

In this section, we obtained dominant slope aspect as an area-weighted average of the slope aspects for each stripe (shown in Fig. 5) on each side of the fault.

As depicted in Fig. 7a, the NW(N) side has the dominant slope aspect of E (distance: 2000 m), E and SE (distance: 4000 m), and SE and S (distance: 6000 m). As evident from Fig. 4, the dominant directions of the crustal deformation along the NW(N) side were N and NE. Notably, both the dominant slope aspect of the slope failures and the dominant direction of crustal deformation have an E component, and a correlation was confirmed between both the two.

As depicted in Fig. 7b, the NW(S) side has a dominant slope aspect of both E and SE (distance: 2000 m), and N (distance: 4000 m). The failure area ratio for the distance of 2000 m was the highest among the distances considered, meaning that the area near the fault line along the NW(S) side was susceptible to slope failures.

As evident from Fig. 4, the dominant directions of the crustal deformation along the NW(S) side were NE and E. Notably, both the dominant slope aspect of the slope failures and the dominant direction of crustal deformation have the E component. In addition, a correlation between them was confirmed for the distance of 2000 m.

However, a considerably smaller area ratio for the distance of 4000 m than that for the distance of 2000 m prevents confirmation that both the dominant slope aspect of the slope failures and the dominant direction of the crustal deformation have the common N component.

As depicted in Fig. 7c, the SE side has the dominant slope aspect of W and SW (distance: 2000 m) or NW, W and SW (distance: 4000 m), E and NE (distance: 6000 m), and SE (distance: 8000 m). As evident from Fig. 4, the dominant directions of the crustal deformation along the SE side were W, SW, and S. Notably, both the dominant slope aspect of the slope failures and the dominant direction of crustal deformation have the common W component for the distances of 2000 and 4000 m; moreover, a correlation was confirmed between them. However, the considerably smaller area ratio for the distance of 8000 m than those for the distances of 2000 and 4000 m prevents the confirmation that both the dominant slope aspect of the slope failures and the dominant direction of the crustal deformation have the common S component.

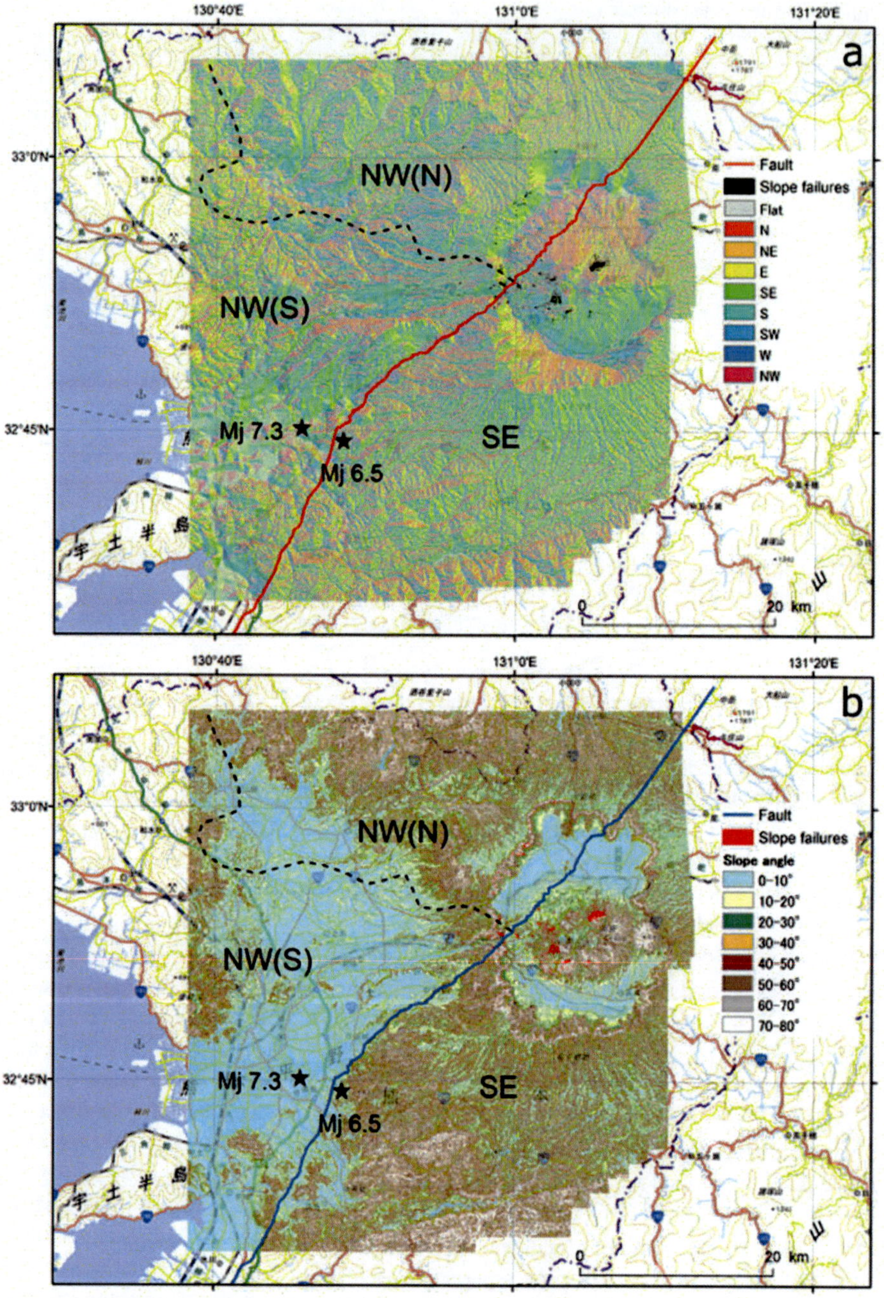

Fig. 6 Slope aspect and angle calculated using the 10 m DEM. **a** slope aspect and **b** slope angle

The above mentioned results for the distance of 2000, 4000 and 6000 m along NW(N), the distance of 2000 m along NW(S) and the distance of 2000 and 4000 m along SE side suggest that the dominant slope aspect of the slope failures are correlated with the direction of the crustal deformation in terms of the common component of the direction. Notably, the E component is along the NW(N) and NW(S) sides and the W component along the SE side. This correlation was not confirmed for the distance of 6000 m (far from the fault line) along SE side. This might be ascribed to the effect of loose pyroclastic deposits on slopes of the central cones (in Area B in Fig. 3).

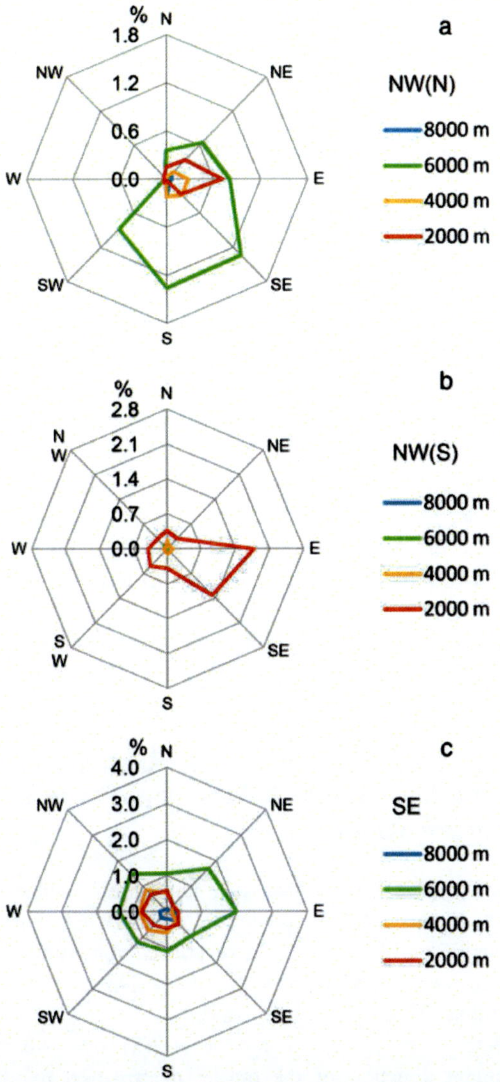

Fig. 7 Slope aspect of the failure area ratio along each side of the fault line and for each distance from the fault line. **a** NW(N), b NW(S), and **c** SE. For the distance of 6000 m in (**b**) has too small value to illustrate the radar chart

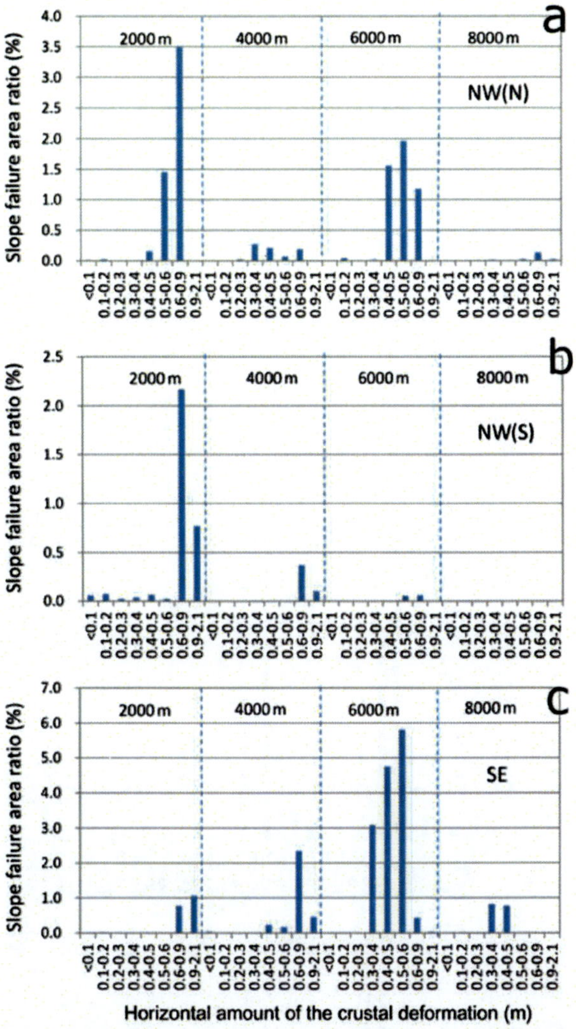

Fig. 8 Horizontal amount of the crustal deformation along each side of the fault line and for each distance from the fault line. **a** NW(N), **b** NW (S), **c** SE

Slope Failure Area Ratio According to the Horizontal Amount of the Crustal Deformation

In this section, we investigate the relation between the slope failure area ratio and crustal deformation amount along each side of the fault line and for each distance from the fault line. The amount was interpolated and calculated using Fig. 3.

The result is depicted in Fig. 8. In Fig. 8a and b, both the NW(N) and NW(S) sides have the highest ratio, i.e., 3.5% and 2.2%, respectively, at 0.6–0.9 m of the deformation amount for the distance of 2000 m. This indicates that the faulting directly affects the slopes and that the crustal deformation induced the failures.

In Fig. 8c, along the SE side, the area ratio recorded was 1.1% at the deformation amount of 0.9–2.1 m for the distance of 2000 m. However, the maximum ratio, i.e., 5.8%, was recorded at the deformation amount of 0.5–0.6 m for the distance of 6000 m where is on slopes of the central cones, covered with loose pyroclastic deposits. Such the slopes might be susceptible to less crustal deformation due to the earthquake.

Slope Failure Area Ratio According to the Slope Angle

We investigated the slope failure area ratio according to the slope angle along each side and for each distance. The result is shown in Fig. 9.

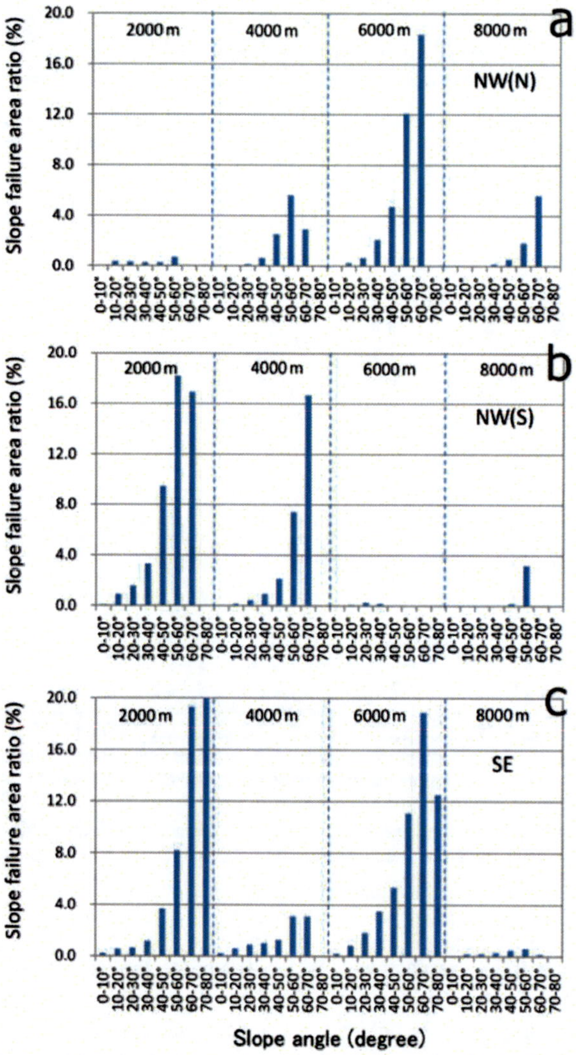

Fig. 9 Slope angle along each side of the fault line and for each distance from the fault line. **a** NW(N), **b** NW(S), **c** SE. In Fig. 9c, the highest ratio is 50.0% in 70–80° for the distance of 2000 m; however, the area in the class is fairly narrow (approximately 242 m^2), and the maximum of the ordinate is set to be 20.0%

For the distance of 2000 m along NW(N) side, Fig. 9a indicates no correlation between the slope angle and area ratio; therefore, for the distance of 2000 m, Figs. 7a and 8a suggest that dominant slope aspect of the slope failures and high peak of the slope failure area ratio were not significantly controlled by the slope steepness for the distance of 2000 m along NW(N) side.

However, in Fig. 9, a good correlation is observed between the slope angle and area ratio for the distance of 2000 m along NW(S) and SE sides, for the distance of 4000 m along NW(N), NW(S) and SE sides, for the distance of 6000 m along NW(N) and SE sides.

Therefore, for these distances along these sides, Figs. 7 and 8 do not always suggest that the dominant slope aspect of the slope failures and higher peak of the area ratio were not significantly controlled by the slope steepness. For example, in Fig. 7a, the distance of 6000 m along NW(N) side has the dominant slope aspect of S and this might be ascribed to the effect of the steep slope on the caldera wall (in Area A in Fig. 3). We must investigate the relation between the slope aspect and slope angle to conclude this suggestion.

Acceleration, Velocity, and Displacement Recorded at the Two Seismographs Located Along NW (S) Side

Seismographic data of the main shock were published by Japan Meteorological Agency (JMA) (2016). Figure 10 shows NS and EW directions of acceleration, velocity, and displacement observed at location A (Mashiki Town) and B (Nishihara Village) in Figs. 2 and 4.

Both A and B are along the NW(S) side, where the dominant direction of the crustal deformation has the E component (Fig. 4). In Fig. 10, maximum displacement amount of E > 45 cm and E > 90 cm was more than that of W at the location A and B, respectively; i.e., the dominant direction of the observed displacement has an E component. Notably, both the dominant direction of the crustal deformation and that of the observed displacement have an E component. As the observed acceleration and velocity in Fig. 10 do not have a dominant E component, it is thought that the dominant direction of the displacement indirectly reflected that of the inertial force of the faulting acting on the mass of slope failures.

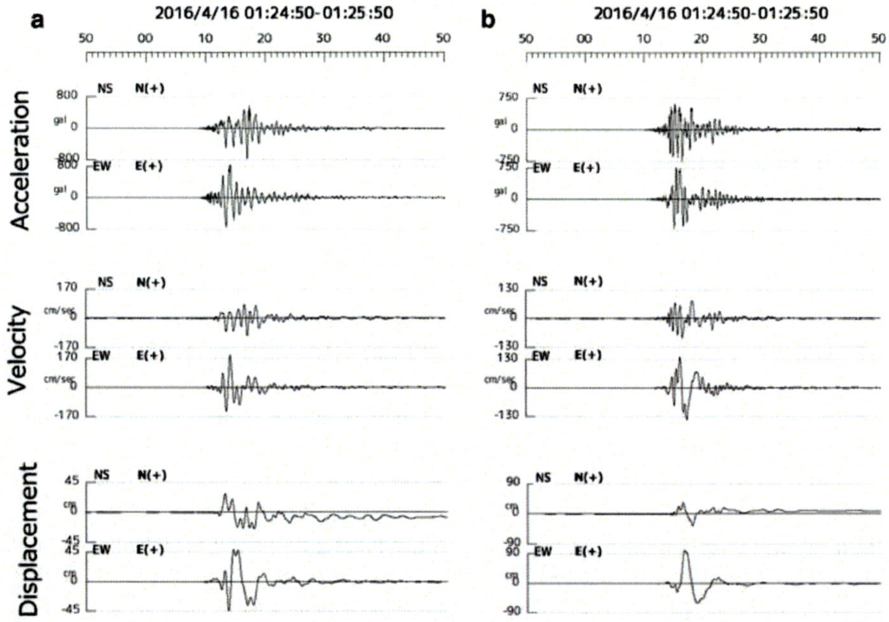

Fig. 10 Acceleration, velocity, and displacement observed at the location A and B in Figs. 2 and 4. Both A and B are along NW(S) side. The data published by JMA (2019) was revised.

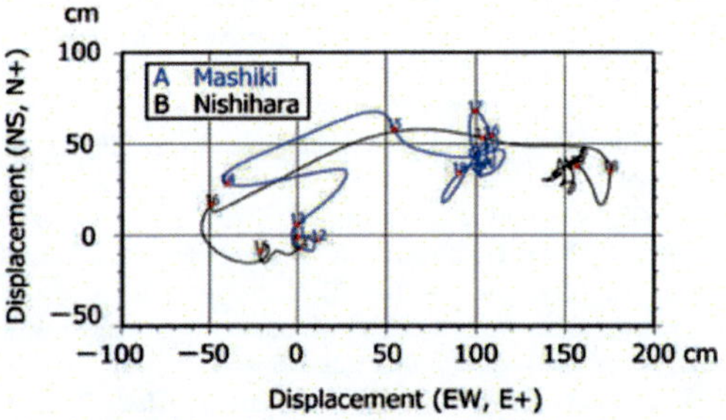

Fig. 11 Horizontal displacement at the location A and B in Fig. 4 (Iwata 2016, revised)

Figure 11 shows horizontal displacement (Iwata 2016, revised) derived from the data in Fig. 10. The number XX stands for the seconds at 01:25:XX (JST) on 16 April, 2016. Notably, the ground moved by 2 m to N75°E between 01:25:16 and 01:25:18 at location B. This fact also supported the existence of an E component to the dominant inertial force along NW(S) side in Fig. 4.

In this study, we observed the correlation of the dominant slope aspect of the slope failure and the dominant direction of the crustal deformation. It is thought that the correlation can be attributed to the dominant inertial force of the faulting.

Conclusions and Final Remarks

As the geological condition was not thought to affect dominant slope aspect of the slope failures near the fault line, we focused on investigating the relation between dominant slope aspect of the slope failures and the direction of the crustal deformation.

Dominant slope aspect of the slope failure was correlated with the direction of the crustal deformation for the distance of 2000, 4000 and 6000 m along the NW(N) side and for the distances of 2000 m along the NW(S) side, as both have the common E component of the aspect and direction, respectively. We observed that the correlation was not affected by slope steepness for the distance of 2000 m along the NW(N) side.

In addition, we observed that the dominant slope aspect of the slope failures was correlated with the direction of the crustal deformation for the distance of 2000 and 4000 m along the SE sides of the fault line, as both have a common W component to the aspect and direction, respectively. However, for the distance of 6000 m along SE side, no correlation was observed and this might be ascribed to the effect of loose pyroclastic deposits on slopes of the central cones. Although the dominant direction of the crustal deformation has E component along NW(N) side, one of the dominant slope aspects was S for the distance of 6000 m along NW(N) side; this might be ascribed to the effect of the steep slope on the caldera wall. The faulting might have directly affected the slopes and triggered the failures near the fault line, but not in the area far from the fault line.

The peak of the slope failure area ratio was 2.2–3.5% for the distance of 2000 m, wherein the amount of crustal deformation was 0.6–0.9 m along the NW(N) and NW(S) sides of the fault line. However, along the SE side, the peak of the area ratio was 5.8% for the distance of 6000 m, where the amount of crustal deformation was 0.5–0.6 m. Less crustal deformation might have induced more failures on slopes of the central cones.

Although seismograph-observed acceleration near the fault line along the NW(S) side is not likely to derive the dominant slope aspect of the slope failures, the dominant direction of the displacement of E observed at the same place showed consistence with the dominant slope aspect of E along NW(S) side. It is thought that dominant direction of the displacement indirectly reflected the dominant direction of inertial force. Therefore, we inferred that dominant slope aspect (orientation bias) of the slope failures is attributed to the dominant direction of inertial force of the faulting.

Acknowledgments The seismographic data in Fig. 10 were measured by Kumamoto Prefecuture. This study was supported by KAKEN 17K01234 "Comprehending InSAR-detected non-tectonic deformation triggered by earthquakes" (Principal researcher: Hiroshi P. SATO) and KAKEN 20K01141 "Interpretation of geomorphological process of landforms induced by non-seismogenic surface deformation revealed by InSAR analysis" (Principal researcher: Hiroshi Une). The authors would like to thank Enago (www.enago.jp) for the English language review. A part of this study has already presented at the 56th Annual Meeting of Japan Landslide Society, Nagano, Japan.

References

Japan Meteorological Agency (JMA) (2016) Municipality-measured strong-motion wave form triggered by the earthquake in Kumamoto region, Kumamoto Pref. https://www.data.jma.go.jp/svd/eqev/data/kyoshin/jishin/1604160125_kumamoto/index2.html (in Japanese). Accessed 24 Apr 2020

Geospatial Information Authority of Japan (GSI) (2016) Distribution map of slope failures. https://www.gsi.go.jp/common/000143473.pdf. Accessed 29 Jan 2020

Goto H, Chida N (2018) Commentary book for 1:25,000 Active Fault Map sheet "Hinagu" and "Yatsushiro, revised", GSI Technical Note D1-No. 914 (in Japanese)

Iwata T (2016) Analysis of strong ground motions at Mashiki town and Nishihara village offices. https://sms.dpri.kyoto-u.ac.jp/topics/masiki-nishihara0428ver2.pdf (in Japanese). Accessed 27 Apr 2020

Kumahara Y, Okada S, Kagohara K, Kaneda H, Goto H, Tsutsumi H (2017) 1:25,000 Active Fault Map "Kumamoto" (revised edition). Technical Report of the Geospatial Information Authority of Japan, D1-No. 868 (in Japanese)

Meunier P, Houvius N, Haines JA (2007) Topographic site effects and the location of earthquake induced landslides. Earth Planetary Sci Lett 275:221–232

Mukoyama S, Sato T, Takami T, Nishimura T (2016) Estimation of ground displacements around Aso-Caldera caused by the 2016 Kumamoto Earthquake, from the Geomorphic Image Analysis of temporal LiDAR DEMs. Japan Society of Engineering Geology. https://www.jseg.or.jp/00-main/pdf/20160822_kumamoto_rep.pdf (in Japanese with English abstract). Accessed 28 Jan 2020

National Research Institute for Earth Science and Disaster Resilience (NIED) (2016) Distribution map of mass movement triggered by the Kumamoto earthquake (revised on 27 Jun 2016). https://www.bosai.go.jp/mizu/dosha.html (in Japanese). Accessed 28 Jan 2020

Ootaki O, Inoue T, Ueda I, Yamashita T, Yamaguchi K, Shirai H, Suzuki A, Mikihara K (2017) Revision of the results of control points after the 2016 Kumamoto Earthquake. J Geospat Inform Author Japan 128:177–187. https://www.gsi.go.jp/common/000147112.pdf (in Japanese). Accessed 28 Jan 2020

Sato HP, Une H (2017) Relation between horizontal direction of crustal deformation and slope failure area ratio revealed by triangulation renewal survey. Proceedings of the Workshop on the Prediction of Landslide Disasters, 7–8 December 2017. Tsukuba, Japan. Technical Note of the NIED, vol 418, pp 79–80. https://dil-opac.bosai.go.jp/publication/nied_tech_note/pdf/n418.pdf (in Japanese). Accessed 31 Jan 2020

Sato HP, Hasegawa H, Fujiwara S, Tobita M, Koarai M, Une H, Iwahashi J (2007) Interpretation of landslide distribution triggered by the 2005 Northern Pakistan earthquake using SPOT5 imagery. Landslides 4:113–122

Shirahama Y, Yoshimi M, Awata Y, Maruyama T, Azuma T, Miyashita Y, Mori H, Imanishi K, Takeda N, Ochi T, Otsubo M, Asahina D, Miyakawa A (2016) Characteristics of the surface ruptures associated with the 2016 Kumamoto earthquake sequence, central Kyushu, Japan. Earth Planets Space 68:191

Suzuki Y, Ishimura D, Kumaki Y, Kumahara Y, Chida N, Nakata T, Nakano T (2017) 1:25,000 Active Fault Map "Aso". Technical Report of the Geospatial Information Authority of Japan, D1-No. 868 (in Japanese)

Xu C, Ma S, Tan Z, Xie C, Toda S, Huang X (2018) Landslides triggered by the 2016 Mj7.3 Kumamoto, Japan, earthquake, Landslides, 15:51–564

Check for updates

2020 Kyoto Japan

Seismic Performance of Buried Pipelines Against Large Ground Deformation of Strike-Slip Faults

Farzad Talebi and Junji Kiyono

Abstract

The response of a pipeline buried in soft soil and hard soil subjected to movement of strike-slip fault at a 90° crossing with and without axial soil–pipe interaction was evaluated. The results of a FEM-based model were compared with those of an existing analytical solution based on the "beam on elastic foundation" theory. The results show that the effect of axial soil–pipe interaction was negligible in soft soil but had a slight effect in hard soil. When the axial soil–pipe interaction is taken into account for the hard soil cases, the bending moment, shear forces and compression stresses of the buried pipeline decrease, and the axial force response of the pipeline increases. For a buried pipeline subjected to movement of a 90° strike-slip fault, the bending moment response was predominantly in the stress field, which indicates that if the 90° strike-slip fault moves, the pipeline is most susceptible to buckling damage.

Keywords

Buried pipeline • Strike-slip fault • FEM analysis • Soil–pipe interaction • Oil and gas pipelines

Introduction

Pipelines are extended across wide areas of a various geographical/geological settings, thus can experience various seismic hazards. Pipelines are generally buried below ground for aesthetic, safety, economic and environmental reasons. When the route of a buried pipeline traverses a geological fault, it is important to assess the performance of the pipeline under seismic activity. Large ground deformation build-up along the fault offset is the main cause of damage to fault-traversing pipelines (Fig. 1). As well as the physical damage to the pipeline, its cascading effects can also cause serious problems for the pipeline network operation (EERI 1999). Moreover, leakage of ecologically harmful stuff such as chemical liquids or gas, natural gas, fuel or liquid waste can cause irreparable environmental damage to our society. The finite element method (FEM) is an extremely versatile numerical tool that can provide in-depth views into the performance of fault-traversing pipelines under various conditions (ASCE 2001). Nonetheless, creating an appropriate soil-and-pipe model on FEM for realizing soil-pipe interactions taking geometrical nonlinearities into account is yet cumbersome. Therefore, the FEM analyses results should be verified by comparing them with available analytical solutions.

Serious damage and rupture to pipelines was reported following earthquakes, for example, the 1971 San Fernando earthquake, the 1995 Kobe earthquake (Nakata and Hasuda 1995), the 1999 Kocaeli earthquake (EERI 1999), the 1999 Chi-Chi earthquake (Uzarski and Arnold 2001) and more recently the 2017 Sarpole-Zahab earthquake (Miyajima et al. 2018). Most of the damage to buried pipelines was caused by ground deformation (e.g. fault movement, landslides, liquefaction), whereas only a few pipelines were damaged by seismic wave propagation (Liang and Sun 2000). The first attempt to evaluate the response of a pipeline crossing a fault was made by Newmark and Hall (1975) based on a simplified analytical model of the pipeline. The most widely used method for evaluation of pipeline response to fault motion when crossing strike-slip and normal faults was proposed by Kennedy et al. (1977) and subsequently adopted by the ASCE in their Guidelines for the Seismic Design of Pipelines (1984). Kennedy et al. extended the work of Newmark and Hall by taking the soil–pipeline interaction into account.

F. Talebi (✉) · J. Kiyono
Graduate School of Engineering, Kyoto University, Katsura Campus, Kyoto, 615-8530, Japan
e-mail: talebi.farzad.57z@kyoto-u.jp

J. Kiyono
e-mail: kiyono.junji.5x@kyoto-u.ac.jp

Ž. Arbanas et al. (eds.), *Understanding and Reducing Landslide Disaster Risk*, ICL Contribution to Landslide Disaster Risk Reduction,
https://doi.org/10.1007/978-3-030-60713-5_6

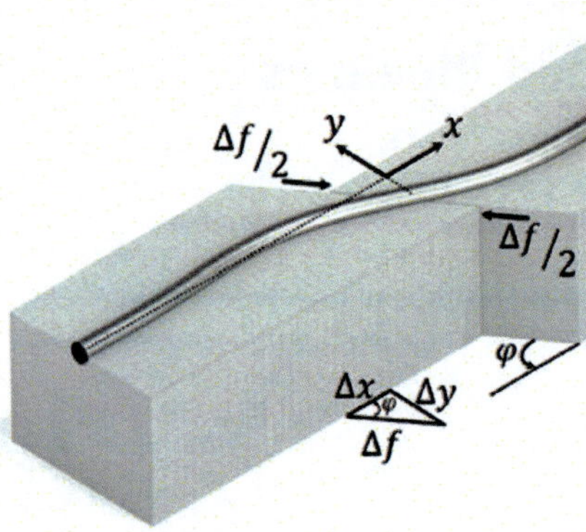

Fig. 1 Pipe crossing an oblique strike-slip fault, showing axes x and y and the fault displacement of Δx, Δy and Δf

However, the bending stiffness of the pipeline at the high-curvature zone was neglected (Karamitros et al. 2007). Wang and Yeh (1985) analyzed the bending stiffness at the high curvature zone of a pipe crossing a strike-slip fault; however, they ignored the effect of the axial forces on the bending stiffness of the pipeline. Karamitros et al. (2007) developed an analytical method for strike-slip faults in which the pipeline is divided into four segments, which are analyzed according to the beam-on-elastic-foundation and elastic-beam theories. They also took the second-order effects into account when calculating the bending strains. However they ignored the effect of the axial forces in furher segments and did a big simplifications in taking axial soil-pipe interaction in to account. Trifonov and Cherniy (2010) extended Karamitros's model for pipes across a normal fault. They removed symmetry conditions about the intersection point and included the transverse displacements when estimating the axial elongation of the pipeline. However, they had similar shortcomings with Karamitros et al. (2007) study.

There is a need to evaluate effects of axial soil-pipe interaction on pipeline performance. In this study, a simple existing analytical method based on the theory of a beam on an elastic foundation was implemented for verification of the FEM model of the pipeline crossing a 90° strike-slip fault. Then, using the FEM based analysis, the effects of the soil stiffness and axial soil–pipe interaction on the performance of the buried pipeline subjected to movement of a 90° strike-slip fault was evaluated to investigate the importance of the axial soil–pipe interaction on the buried pipeline performance.

Analytical Solution

Our model is based on an existing linear analytical solution for a pipeline lying across a strike-slip fault, based on the theory of a beam on an elastic foundation. This analytical solution does not consider the effect of the axial soil–pipe interaction. Therefore, in the boundary conditions only the fault transverse displacements (not the axial displacements) are applied. To ensure reliable results, the analytical solution was conducted only for a pipeline crossing a strike-slip fault at a 90° angle with no axial fault dislocation in the ground deformation.

The differential equation for the elastic response of a pipeline crossing a strike-slip fault at a 90° angle is presented in Eq. (1) with boundary conditions of $w = 0$ for $x \to -\infty, w = \Delta f/2$ for $x = 0$ and $M = 0$ for $x = 0$.

$$-EI\frac{d^4w}{dx^4} - k_t w = 0 \tag{1}$$

$$w(x) = \frac{\Delta_y}{2} e^{\beta x} cos\beta x \tag{2}$$

$$\beta = \sqrt[4]{\frac{k_t}{4EI}} \tag{3}$$

$$M(x) = \Delta_y EI\beta^2 e^{\beta x} sin\beta x \tag{4}$$

$$V(x) = \Delta_y EI\beta^3 e^{\beta x}(sin\beta x + cos\beta x) \tag{5}$$

Here, w is the transverse displacement of the pipeline, M is the bending moment, V is the shear force, E is Young's modulus of the pipeline steel, I is the moment of inertia of the pipeline, Δ_y is the transverse fault displacement and k_t is the elastic constant of the transverse soil spring.

Fig. 2 FEM based displacement contours of pipeline in response to 0.2 m dislocation of strike-slip fault

FEM Results and Verification

To evaluate the response of a pipeline crossing a strike-slip fault, we conducted numerical analyses based on FEM using the ABAQUS code (ABAQUS 2017). We modeled a steel pipeline with an elastic Young's modulus of 200 GPa and Poisson's ratio of 0.3. The pipe is 1 km long with an external diameter of 0.1143 m, thickness of 0.0023 m buried at 0.6 m of sand in two soft and hard soil cases, as shown in Table 1. The transverse soil spring constants in Table 1 were determined based on lateral load test (LLT) results of Hasegawa and Kiyono (2016); the axial spring constants were set based on the Seismic Design Guideline of Japan Gas Association (2016). To improve the accuracy of the FEM results, the meshing and elements size were selected based on the mesh size sensitivity analysis results. The elements on both sides of the fault were symmetrically discretized, with the element size increasing away from the fault line from 0.0125 to 1 m.

Using the FEM-based model presented in section "Analytical Solution", two models of a pipeline crossing a strike-slip fault at a 90° angle with soft and with hard soils were created and analyzed. The results, along with the analytical solutions, are presented in Figs. 3, 4, 5, 6, 7, and 8.

The pipeline responses at the strike-slip fault with a fault angle of 90° were analyzed for the soft and hard soil cases with and without axial soil–pipe interaction.

As shown in Figs. 3, 5, 7, and 8, the FEM analysis responses agrees with that of the analytical solution. The FEM-based transverse displacement response of the pipeline for the soft and hard soils is almost identical to the analytical solution (Fig. 3).

L_c is the first point after the fault line where the pipe deflection reaches zero. Based on Eq. (2) with $x = L_c$ and $w = 0$:

$$L_c = \frac{\pi}{2}\sqrt[4]{\frac{4EI}{k_t}} \tag{6}$$

the L_c of the pipeline in hard soil (1.5 m) is much shorter than in soft soil (4.7 m) because of the soil stiffness.

We analyzed the maximum bending moment of a pipeline crossing a strike-slip fault with a fault angle of $\emptyset = 90°$ for hard and soft soils. When the axial soil–pipe interaction was taken into account, the bending moment was 2% smaller in the soft soil case and 5% smaller in the hard soil than without the axial soil–pipe interaction. For the same pipeline and boundary conditions as above, comparing the FEM models without axial soil–pipe interaction with the analytical models, the maximum bending moment of the pipeline was 1% smaller for FEM in soft soil and 15% smaller in hard soil. The bending moment response of the pipeline in soft soil is in good agreement with the analytical solution. These results indicate a relationship between the root of the soil stiffness ratio and the maximum bending moment of the pipeline crossings a $\emptyset = 90°$ a strike-slip fault. If the maximum bending moment (M_{max}) for a pipeline crossing a strike-slip fault is known for one type of soil (k_t), we can calculate the maximum bending moment $\left(M'_{max}\right)$ of the pipeline in a new soil case with a transverse spring stiffness k'_t.

$$M'_{max} \approx \sqrt{\frac{K'_t}{K_t}} \times M_{max} \tag{7}$$

Thus, we can estimate the maximum bending moment of pipelines in different soils using Eq. (7). The maximum bending moment in cases with hard soil, is approximately 10 times bigger than the maximum bending moment in soft soil (Table 1):

Table 1 Soil–pipe interaction spring properties for different models

Case Name	Solver	Spring type	k (GN/m^3)	Angle
Anal-Soft T. Soil-90°	Analytical	Transverse	0.0018	90°
		Axial	–	
Anal-Hard T. Soil-90°	Analytical	Transverse	0.18	90°
		Axial	–	
FEM-Soft T. Soil-90°	FEM	Transverse	0.0018	90°
		Axial	–	
FEM-Soft T.A. Soil-90°	FEM	Transverse	0.0018	90°
		Axial	0.0009	
FEM-Hard T. Soil-90°	FEM	Transverse	0.18	90°
		Axial	–	
FEM-Hard T.A. Soil-90°	FEM	Transverse	0.18	90°
		Axial	0.09	

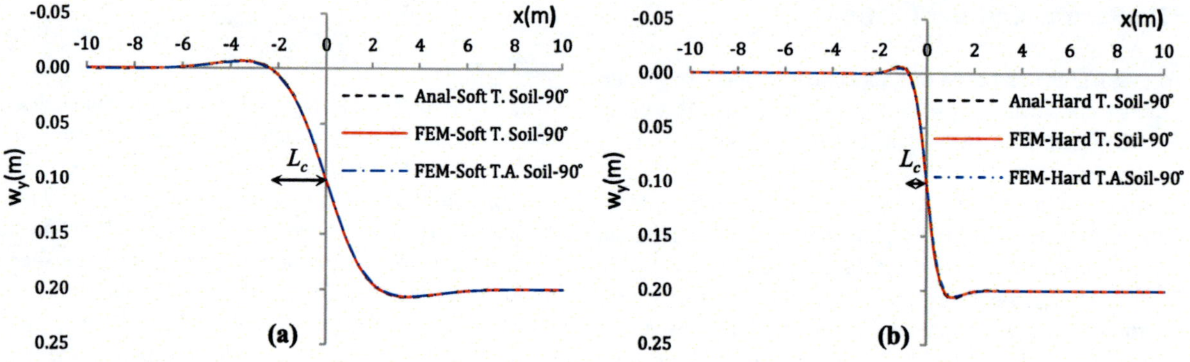

Fig. 3 Transverse displacement in a pipeline crossing a strike-slip fault—analytic solution and FEM solutions with and without axial soil–pipe interaction. **a** Soft soil cases. **b** Hard soil cases

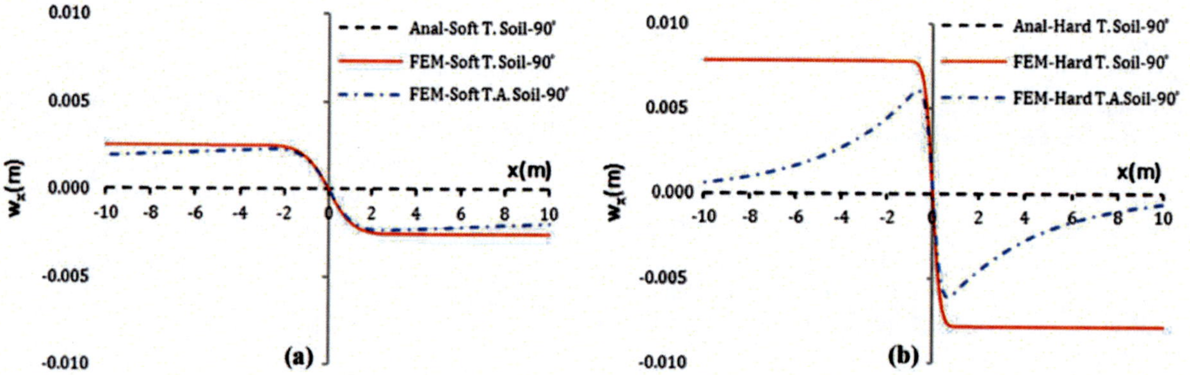

Fig. 4 Axial displacement in a pipeline crossing a strike-slip fault—analytic solution and FEM solutions with and without axial soil–pipe interaction. **a** Soft soil cases. **b** Hard soil cases

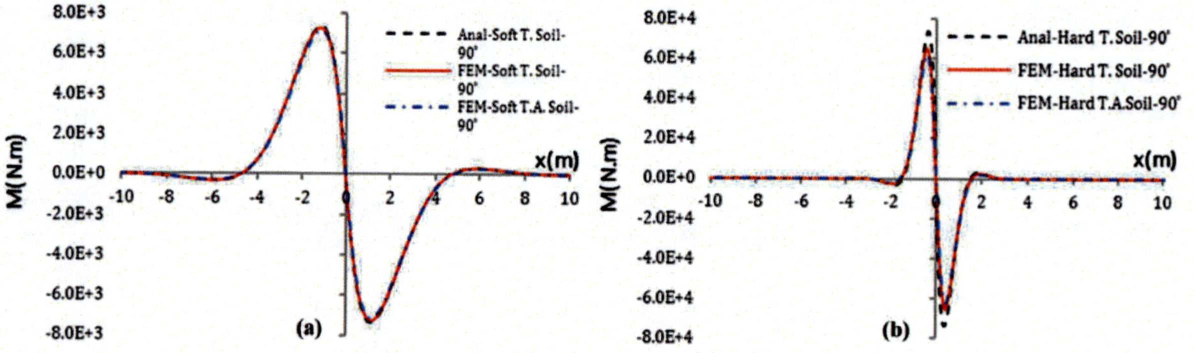

Fig. 5 Bending moment of a pipeline crossing a strike-slip fault—analytic solution and FEM solutions with and without axial soil–pipe interaction. **a** Soft soil cases. **b** Hard soil cases

$$\frac{M'_{Max}}{M_{Max}} \approx \sqrt{\frac{K'_t}{K_t}} = \sqrt{\frac{0.18}{0.0018}} = 10 \tag{8}$$

The maximum bending moment, moment diagram and transverse displacement diagram of the pipeline in soft soil, with and without the axial soil–pipe interaction effect are similar. However, in hard soil the bending moment of the pipeline with axial soil–pipe interaction is 15% larger than that without the soil–pipe interaction.

The FEM results for all the responses except the axial forces and axial displacement responses were in good agreement with the analytical solution results. The differences arise

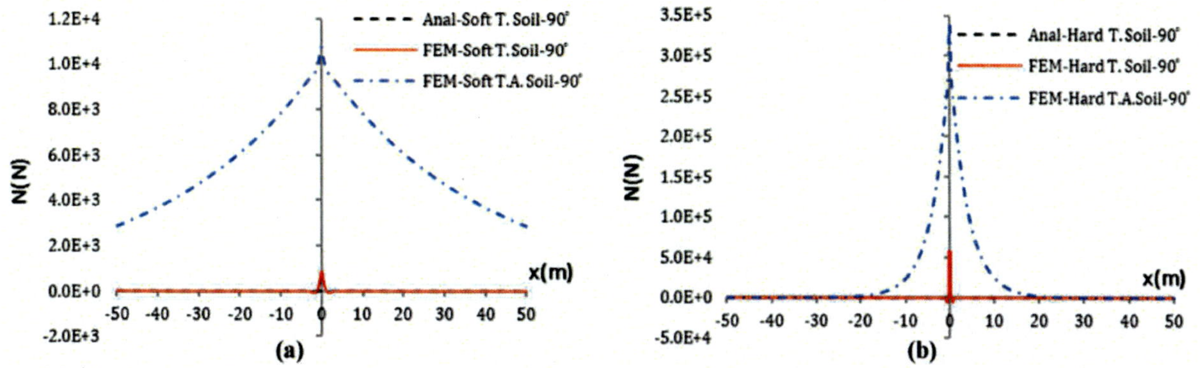

Fig. 6 Axial forces of buried pipeline crossing a strike-slip fault; analytic solution and FEM solutions with and without axial soil–pipe interaction. **a** Soft soil cases. **b** Hard soil cases

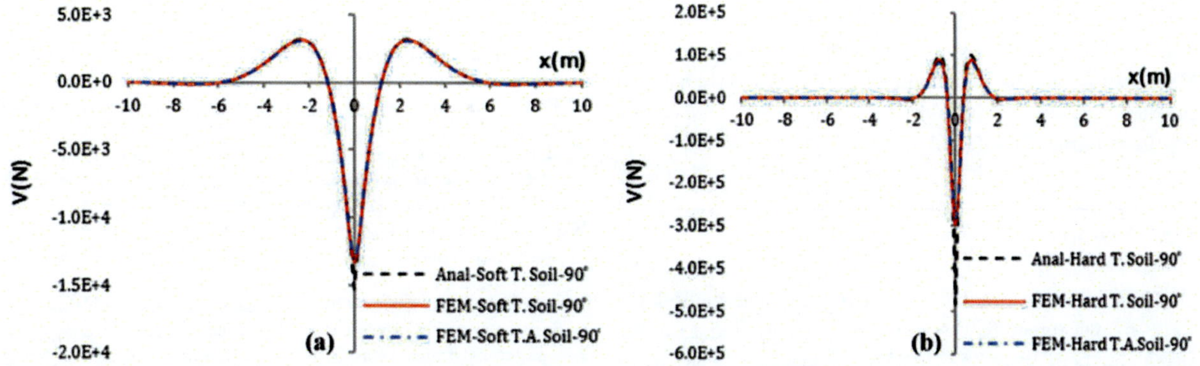

Fig. 7 Shear forces in a pipeline crossing a strike-slip fault; analytic solution and FEM solutions with and without axial soil–pipe interaction. **a** Soft soil cases. **b** Hard soil cases

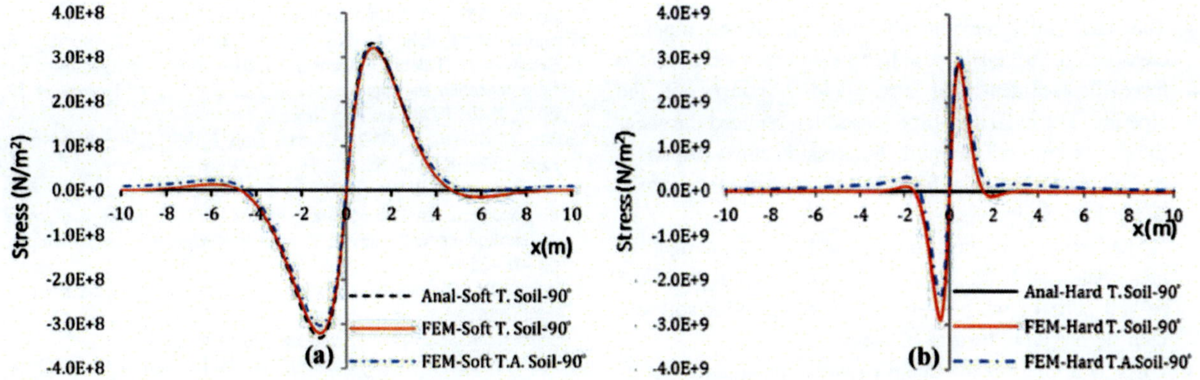

Fig. 8 Maximum stress in a pipeline section crossing a strike-slip fault; analytic solution and FEM solutions with and without axial soil–pipe interaction. **a** Soft soil cases. **b** Hard soil cases

from the lack of axial soil–pipe interaction and axial force terms in the governing equation of the analytical solution. Our results show that neglecting the axial forces for a pipeline subjected to fault movement can lead the analytical method to produce incorrect assessment of the forces acting on the pipe. The analytical results are acceptable only in the case of a 90° strike-slip fault because, in this case, the pipeline's axial forces and axial displacements are negligible because the axial stress of the pipeline is not the dominant stress; rather, the bending stress is the dominant stress. The stress response of the pipeline as the main structural response in the FEM analysis was in good agreement with the analytical solution; thus, the analytical solution can be used as a verification method for the linear FEM model in case of a buried pipeline subjected to 90° strike-slip fault movement, especially in soft soils. The FEM model can be extended with reliable results for other cases.

When modeling a pipeline crossings a strike-slip fault with a fault angle of $\emptyset = 90°$ in soft soil, the axial soil–pipe interaction effect is negligible. In such cases, models without axial soil–pipe interaction are valid.

As it has been remarked that the analytical solution method did not implement the axial soil-pipe interaction. Moreover, with regards to results, the responses of pipeline across the strike-slip fault with 90° in the soft soil for the analytical solution and FEM models with and without axial soil-pipe interaction effect, are almost same. Therefore, the analytical solution is acceptable method for verification of FEM based models with linear materials for problems of pipeline crossing strike-slip faults with 90°, Especially pipelines buried in soft soil.

Conclusions

Using FEM based simulations and an analytical method, the force–displacement fields of a pipeline subjected to a 90° strike-slip fault were studied. The mechanical behavior of the pipeline in four FEM cases and two analytical cases for soft and hard soils was evaluated. Our conclusions are as follows.

- The axial soil–pipe interaction of the buried pipeline crossing the strike-slip fault had a significant effect on the force–displacement and stress-field responses of the pipeline. The axial soil–pipe interaction reduced the shear force, bending moment and compression stress responses and increased the axial force and tensile stress responses of the buried pipeline.
- L_c has an inverse relationship with k and a positive relationship with EI. In another words, the length of the curved zone of the pipeline increases in softer soil and with higher pipe flexural stiffness.
- There is a direct relationship between the square root of the soil stiffness ratio and the M_{Max} ratio of the pipeline.
- The effect of the axial soil–pipe interaction on a pipeline crossing the strike-slip fault with $\emptyset = 90°$ is negligible, especially for pipelines buried in soft soil.
- The force–displacement field responses of the analytical solution and FEM cases of a buried pipeline across a 90° strike-slip fault with and without axial soil–pipe interaction are almost same, especially in soft soil.
- The analytical solution can be used to verify FEM models for cases with linear material only when the buried pipeline is subjected to a 90° strike-slip fault, especially in soft soil.

Acknowledgments This work was supported by Grant-in-Aid for Scientific Research of Ministry of Education, Culture, Sports, Science and Technology, Japan (MEXT Grant), Grant number: 17H01287 (principal investigator: Prof. Junji Kiyono, Kyoto University).

References

ABAQUS/CAE (2017) Dassault Systems Simulia Corp, release 2017 documentation

American Lifelines Alliance-ASCE (2001) Guidelines for the design of buried steel pipe (with addenda at 2005)

ASCE Technical Council on Lifeline Earthquake Engineering (1984) Differential ground movement effects on buried pipelines. Guidelines for the seismic design of oil and gas pipeline systems

EERI (1999) The Izmit (Kocaeli), Turkey Earthquake of August 17, 1999. EERI Special Earthquake Report.

Hasegawa N, Kiyono J (2016) Study of the plastic hinge position of buried steel pipe on fault displacement. J JSCE 72:161–166 (in Japanese)

Japan Gas Association (2016) Seismic design guideline for high-pressure gas pipeline. JGA. Specified pp 206–13 (in Japanese)

Karamitros D, Bouckovalas G, Kouretzis G (2007) Stress analysis of buried steel pipelines at strike-slip fault crossings. Soil Dyn Earthq Eng 27:200–211

Kennedy RP, Chow AM, Williamson RA (1977) Fault movement effects on buried oil pipeline. ASCE Transp Eng J 103:617–633

Liang J, Sun S (2000) Site effects on seismic behavior of pipelines: a review. ASME J Press Vessel Technol 122:469–475

Miyajima M, Fallahi A, Ikemoto T, Samae, M, Karimzadeh S, Setiawan H, Talebi F, Karashi J (2018) Site investigation of the Sarpole-Zahab Earthquake, Mw 7.3 in SW Iran of November 12, 2017. JSCE J Disaster FactSheets

Nakata T, Hasuda K (1995) Active fault I 1995 Hyogoken Nanbu earthquake. Kagaku 65:127–142

Newmark NM, Hall WJ (1975) Pipeline design to resist large fault displacement. In: Proceedings of the U.S. national conference on earthquake engineering, Ann Arbor, University of Michigan, pp 416–25

Trifonov OV, Cherniy VP (2010) A semi-analytical approach to a nonlinear stress– strain analysis of buried steel pipelines crossing active faults. Soil Dyn Earthq Eng 30:1298–1308

Uzarski J, Arnold C, (2001) Chi-Chi, Taiwan, Earthquake of September 21, 1999, Reconnaissance Report. Earthquake Spectra, Professional J EERI Vol. 17, 2001(Suppl. A).

Wang LRL, Yeh YA (1985) A refined seismic analysis and design of buried pipeline for fault movement. Earthq Eng Struct Dynam 13:75–96

Reconstruction Strategies for Mw 7.8 Earthquake-Induced Landslide-Affected Settlements in Nepal

Tara Nidhi Bhattarai, Dhruba Prasad Sharma, and Lekh Prasad Bhatta

Abstract

Nepal was rocked by Mw 7.8 earthquake on April 25, 2015 which damaged lives and physical infrastructures worth millions of dollars. The mainshock and its subsequent aftershocks also triggered thousands of landslides which, at several locations, killed people, damaged highways and public buildings, and destroyed hundreds of settlements severely. Consequently, the government decided to move the landslide-affected villages first to temporary shelters and then relocate to safer sites. As a means of implementing the decision, the National Reconstruction Authority (NRA) devised strategies which include, among others, conducting geological engineering investigation to confirm the state of geo-hazard in each of the reported earthquake-induced landslide affected settlements. Accordingly, 283 settlements were decided to move to safer sites, whereas another 320 settlements were recommended to start reconstruction only after applying control measures to landslides existed in and around the settlements. Several difficulties were also faced during the relocation of those 283 vulnerable settlements (4598 households). But these were resolved by revisiting the vulnerable sites and collecting science-based data by a team of experts and also holding several rounds of professional dialogues with the stakeholders. Finally, after 4 years of rigorous works, majorities of the landslide-affected families have already been well managed as each of them is now having an earthquake-resistant residence constructed in safer locations. Remaining 434 households are being managed with the aim of completing all the administrative processes by the end of June 2020.

Keywords

Gorkha • Earthquake • Landslide • Reconstruction • Strategy

Introduction

On a fine Saturday, a few minutes before noon local time, of April 25, 2015 central Nepal was jolted by Mw 7.8 earthquake. The hypocenter depth of the event was 15 km on a reverse fault dipping at 11° due north (Elliott et al. 2016). Its name was given as Gorkha, Nepal Earthquake considering that the epicentre (Barpak village) of the event was located within the historic district of Gorkha, situated about 76 km northwest of Kathmandu. Nepal has 77 administrative districts. The tremor caused damage across 32 districts out of which 14 districts were considered as "crisis-hit" and the adjoining remaining 18 districts as "severely-hit". Over 8,790 people lost their lives and about 22,300 people got seriously injured (PDNA 2015). Not only the mainshock but also its subsequent aftershocks initiated landslides and rockfalls at several locations. These hazards severely damaged hundreds of settlements distributed across the affected 32 districts of Nepal. Based on the requests received from the district disaster management committees, and also considering the recommendations made by the government-formed rapid geological assessment team, the government declared 475 settlements as the most vulnerable communities. These communities were moved to temporary shelters immediately. At that time, the government had assured them to provide with safer sites to construct a new residence for them at the earliest possible. The decision was

T. N. Bhattarai (✉)
Department of Geology, Tribhuvan University, Kathmandu, 44600, Nepal
e-mail: tnbhattarai222@gmail.com

D. P. Sharma · L. P. Bhatta
Government of Nepal, National Reconstruction Authority, Kathmandu, 44600, Nepal
e-mail: dhrbsharma@gmail.com

L. P. Bhatta
e-mail: lorullekh.lpb@gmail.com

Ž. Arbanas et al. (eds.), *Understanding and Reducing Landslide Disaster Risk*,
ICL Contribution to Landslide Disaster Risk Reduction,
https://doi.org/10.1007/978-3-030-60713-5_7

implemented later by the National Reconstruction Authority (NRA), established nine months after the earthquake to handle all the reconstruction issues.

The main objective of this paper is to highlight the strategies made to implement the above-mentioned government decisions. This paper first provides an overview of Nepal's geological setting, followed by the impact of earthquake-induced landslides on settlements and physical infrastructures. Then the paper briefly describes the government strategies adapted to manage the vulnerable settlements. Finally, the paper discusses how the government strategies were implemented, overcoming social, political, and technical issues raised in due course of time.

Geological Setting

Being located at the convergent boundary of Indian and Eurasian plates, Nepal forms a part of the Himalayan range where reverse thrusting and mountain building process are still going on ever since the collision started 55 million years ago (Searle et al. 1987). During the collision process, in addition to several other geological structures, three main longitudinal thrusts were also developed, namely Main Central Thrust (MCT), Main Boundary Thrust (MBT), and Himalayan Frontal Thrust (HFT), from north to south respectively (Gansser 1964). Being separated several kilometres apart on the surface, these thrusts are interconnected underneath (Zhao et al. 1993) with a major thrust or a decollement, known as Main Himalayan Thrust (MHT). Although the subduction is currently occurring through HFT, its movement has been locked at the ramp (an inclined part of the MHT). The locked part of the ramp is moved only during high magnitude earthquakes that occur in the Himalaya (Bilham et al. 1997). The epicentre of the 2015 Gorkha, Nepal earthquake was also on the MHT. The event ruptured the MHT: about 150 km along the strike (NNW-SSE), and about 60 km along the dip (NE), and with a maximum slip of 5.8 m (Liu et al. 2016). But the rupture did not appear on the surface.

Impact of Earthquake-Induced Landslides

Extensive landslides occurred in hills and mountains owing to the fact that the rupture plane was close to the earth's surface and high magnitude of the event. In addition to damaging physical infrastructures and killing people, more than 69 landslide dams were also formed (Collins and Jibson 2015). Some of these natural dams fully blocked the rivers and in some cases, they temporarily or partially disrupted rivers' natural flows. The main event and its aftershocks produced at least 25,000 coseismic landslides distributed across the central Nepal. The number, area, and volume of the landslides were found to be consistent with other similar cases of earthquake-induced landslides in the world (Roback et al. 2017). Since Nepalese villages, particularly in hills and mountains, are built on slopes of varying inclination, the earthquake severely damaged several settlements. The following cases were observed:

(a) Landslide debris buried villages completely or partly, damaged highways and hydropower infrastructures in its runout route;
(b) Settlements and public infrastructures were slipped down collapsing all the buildings built on it;
(c) Some boulders, resulted from rockfalls, travelled down and damaged residences;
(d) Severe ground ruptures were developed within, upslope, and downslope of villages;
(e) In high elevation areas, snow avalanches buried residential buildings, destroyed tourist trails and killed wild lives;
(f) Unstable rock blocks, likely to fall down anytime, were noticed on the upslope of villages;
(g) Landslides, situated in downslope side of villages, were progressively reactivating threatening settlements in upslope;
(h) Landslides also created natural dams across some rivers. Although it created fears among the locals, they were latter breached naturally causing no harm except intense river bank erosion at the location; and
(i) Although did not happen, there was a fear among the people living on the sides of snow-fed rivers of collapsing of glacial lakes situated in the upstream and causing glacial lake outburst floods (GLOFs) in the downstream due to frequent occurrences of aftershocks.

The above-mentioned conditions were the prominent factors to increase fear among the residents of the affected area. The frequent aftershocks also promoted people to leave such places as early as possible. These were also the basis for the government to evacuate the villagers and keep them in temporary shelters for a short period, also promising them to provide with a safer site for newly constructing their residential buildings in the long run.

Government Strategies

One of the reasons to move vulnerable communities to temporary shelters was that more landslides were expected as soon as the monsoon season begins just after about two months from the date of the mainshock. But, as time passed and the subsequent monsoon did not trigger as many landslides as expected, people started to return to their own land

Table 1 Number of settlements affected by earthquake-induced landslides

	Category I	Category II	Category III
Vulnerable settlements	481	320	283
Types of support required	Awareness raising trainings	Landslide control measures	Settlement relocation sites

to reconstruct their homes. Meanwhile, the National Reconstruction Authority (NRA) was established on December 20, 2015 to handle all the post-earthquake reconstruction issues. Soon after its establishment, the NRA issued a public notice advising people not to start any reconstruction at the 475 vulnerable settlements. It further stated that the government house reconstruction grant (NRs three hundred thousand) would not be provided if someone reconstructs his/her residence on those vulnerable sites. This situation created a tremendous pressure on NRA from the affected families to justify its decision on the basis of scientific information, which was lacking at that time. The following strategies were then adopted by the NRA to handle the situation.

(a) **Collecting information of vulnerable settlements**

As already mentioned above, the decision of declaring vulnerable settlements was based on rapid geological assessment and also request from district disaster management committee. Many villagers wanted to avert the decision and allow them to reconstruct their residential buildings in their native village, receiving the government-declared grant. But there were also increasing public concern about for not evaluating the condition of those settlements which were not visited during the rapid geological assessment due to their remoteness. In addition, several villages in the high mountain were also left behind as the affected family themselves did not report to the concern government office because they had no idea what to do immediately after the earthquake. Considering these facts into account, a notice was issued requesting the general public to contact NRA if their individual house or the whole village is under the threat of earthquake-induced landslides. Consequently, more than 1084 settlements (including those already declared 475 settlements) contacted NRA requesting to conduct a geo-hazard assessment and confirm whether the settlements are safer or not for reconstruction purposes.

(b) **Conducting geological engineering investigation of vulnerable settlements**

A technical committee was formed at NRA to conduct geological engineering investigation of the vulnerable settlements. The committee developed technical guidelines required for the field investigation, and also for classifying the vulnerable settlements into one of the following categories.

Category I: Sites having geo-hazards which can be managed by communities themselves. No financial support is required from the government;

Category II: Sites having geo-hazards which can be managed by applying appropriate engineering countermeasures for which government support is required; and

Category III: Sites having geo-hazards which are extremely difficult to control or economically not feasible. Settlements belonging to this category need to be relocated to a safer site either entirely or in part.

A team comprised of an engineering geologist, watershed management expert, and water-induced disaster management expert was recommended to conduct a field investigation of each vulnerable settlements. Since villages were scattered far away, 32 teams were mobilized to conduct field works in all the affected districts at a time. Their final result is depicted in Table 1.

(c) **Preparing technical guidelines to identify safer sites for settlement relocation**

Since Nepal is a mountainous country, it is not easy to find a suitable place for residential building construction. It is particularly difficult in hills and mountains where majorities of slopes are steep and composed of hard rocks. Soil cover is generally thin, and wherever it is thick, villages are already there. Finding a suitable place to move vulnerable villages was, therefore, a challenge. Nevertheless, technical guidelines were developed to identify a safer site for community relocation. Accordingly, the site was considered to be safer if the natural hillslope is <20°, it is sufficiently away from the runout route of a nearby active rockfall and landslide, it is safe from nearby on-going headward erosion and bank erosion of streams, it has less possibility to be affected by floods, it is not within an active or passive landslide or creep, and it has less possibility to have liquefaction problem or the problem can be addressed while designing the foundation. Engineering geological investigation of the proposed relocation site recommending it as the "safer site" is mandatory to receive the government grant.

(d) **Providing house reconstruction grant**

Nepal government has decided to provide with the financial grant (NRs three hundred thousand) to each of earthquake-affected household. This provision is also for the earthquake-induced landslide-affected households. As a supplement to the total cost of house construction, this amount is intended to encourage people to construct an earthquake-resistant residential building. The grant is provided in three tranches. To receive the grant, all the technical criteria set by the NRA have to be fulfilled, which will be supervised by the government engineers.

(e) **Providing land buying assistance**

In some cases, the impact of the landslide was so widespread that no personal land (safer site) was left to construct a residential building. In such case, the government decided to provide with land buying assistance (NRs two hundred thousand) to the affected families. This assistance is also for landless beneficiaries (those people who are cultivating others' land in the contract). Once they buy the land, these beneficiaries will also receive the house reconstruction grant (NRs. three hundred thousand) to construct earthquake-resistance residence.

Reconstruction Action Plan

Category I Settlements

The NRA encourages the affected communities to start the reconstruction of their residential buildings by themselves in their own land at the same place or nearby to it utilizing the house reconstruction grant. Besides, awareness-raising training related to surface water management and simple bioengineering techniques are also conducted in each of these settlements with the objective of letting them know how to protect the physical environment of their locations.

Category II Settlements

The NRA provides a financial grant to the concern government offices, which, in turn, implements landslide control measures making the vulnerable settlements safer than before. Then the affected families themselves construct earthquake-resistant residential building utilizing the house reconstruction grant.

Category III Settlements

Households belonging to this category are also known as the "geo-hazard beneficiaries". The government has decided to relocate these communities to safer sites. For the purpose, the following approaches were adopted (also see Table 2).

(i) ***Relocation in beneficiaries' own land***

For those geo-hazard beneficiaries who possess safer land in other locations (not within the Category III site), they are provided with house reconstruction grant to construct earthquake resistance residence.

(ii) ***Relocation in integrated settlements***

The NRA has proposed an integrated settlement development scheme for those geo-hazard beneficiaries who do not have safer land to construct their residence. Although the government was ready to move the whole village to a suitable location, villagers' views on it were divided. Some did not like the proposed new location; a few people also mentioned that they do not like to have others' buildings very nearby to that of theirs; further others were in the opinion that they do not like to share facilities like water, toilet, and yard; while other expressed positively to move there. Accordingly, the NRA declared the integrated settlement development schemes if more than ten households agree to move to the proposed location. The villagers are encouraged to find a suitable relocation site by themselves. In addition to the house reconstruction grant, the government will provide another grant to buy a piece of land (NRs. two hundred thousand per household) to construct their residence. The government will also provide a grant to construct physical infrastructures (access road, sewerage management, electricity, school, health post, and drinking water supply) in the settlement (Fig. 1). This grant varies (NRs. three millions to five millions) depending upon the number of households and location (remoteness) of the proposed site.

(iii) ***Relocation in beneficiary-selected site***

For those geo-hazard beneficiaries who do not have safer land and also do not like to move in the integrated settlement, the government will provide with land buying grant (NRs. two hundred thousand per household) to buy safer land at the location of their choice for constructing an earthquake-resistant residence. Obviously, they will also receive the house reconstruction grant.

Table 2 Number of relocated households, belonging to Category III settlements, using different approaches

Approach	Already benefitted	Under progress
Relocation in beneficiaries' own land	573	None
Relocation in Integrated Settlements	1170	434
Relocation in beneficiary-selected sites	2421	

Fig. 1 An intergraded settlement where geo-hazard beneficiaries are also included in Sindhupalchok district

(iv) ***Addressing the relocation problem***

The government tried its best to relocate the vulnerable settlements in safer sites using science, technology, and available resources. But some unexpected hurdles emerged in due course of time that prevented the government to achieve the target fully. The followings were noteworthy:

(a) People do not like to leave their native place even though it is vulnerable (creeping);
(b) Locals want to move the whole village rather than only the effected families;
(c) Political pressure to move even the settlements which were not vulnerable;
(d) Political pressure to consider active flood plain as of safer site for settlement relocation;
(e) Some people already reconstructed their residence in the Category III sites; and
(f) Some school buildings were already reconstructed in vulnerable locations.

The above-mentioned item (a) was partially solved by installing a landslide early warning system. The item (b) was handed over to the local government and negotiation with the affected families is still going on. In case of the item (c) and (d), a detailed engineering geological investigation was conducted by a team of experts who recommended, based on technical findings, not to use the proposed place as the safer site for community relocation. And the case (e) and (f) were unfortunate, and countermeasures are being applied to minimize the impact of likely disaster.

Concluding Remarks

The 2015 Mw 7.8 Gorkha, Nepal earthquake triggered landslides that destroyed several settlements across the most affected 32 districts of Nepal. A geological engineering investigation conducted throughout the vulnerable sites concluded that nearly 283 villages (Category III) are to be relocated to safer sites whereas 320 villages (Category II) need engineering countermeasures to make them safer than before. Nearly 481 villages were identified as normal sites where government's additional support for relocation was not needed. The government managed some of the affected families of category III by relocating them to the integrated settlements. Remaining affected families received a financial grant to buy a piece of safer land at their location of interest.

There were some implementation difficulties, which were solved by using science and engineering-based data along with several rounds of professional dialogues with the stakeholders.

References

Bilham R, Larson K, Freymuller J (1997) GPS measurements of present-day convergence across the Nepal Himalaya. Nature 386:61–64

Collins BD, Jibson RW (2015) Assessment of existing and potential landslide hazards resulting from the April 25, 2015 Gorkha, Nepal earthquake sequence. USGS Open-File Report 2015-1142. https://doi.org/10.3133/ofr20151142

Elliott JR, Jolivet R, González PJ, Avouac JP, Hollingsworth J, Searle MP, Stevens VL (2016) Himalayan megathrust geometry and relation to topography revealed by the Gorkha earthquake. Nat Geosci 9(2):174–180. ISSN 1752-0894

Gansser A (1964) Geology of the Himalayas. Wiley Interscience, Hoboken, NJ, p 289

Government of Nepal (2015) Nepal Earthquake 2015: Post Disaster Need Assessment, Executive Summary. National Planning Commission, Kathmandu, 18p

Liu C, et al. (2016) Rupture processes of the 2015 Mw 7.9 Gorkha earthquake and its Mw7.3 aftershock and their implications on the seismic risk. Tectonophysics. https://doi.org/10.1016/j.tecto.2016.05.034

Roback K, et al. (2017) The size, distribution, and mobility of landslides caused by the 2015 Mw7.8 Gorkha earthquake, Nepal. Geomorphology. https://doi.org/10.1016/j.geomorph.2017.01.030.aa

Searle MP, Windley BP, Coward MP, Cooper DJW, Rex AJ, Red D, Tinh-Dong Li, Xu-Chang X, Jan MQ, Thakur VC, Surendra K (1987) The closing of Tethys and the tectonics of Himalaya. Bull Geol Soc Am 98(6):678–701

Zhao W, Nelson KD, Che J, Quo J, Lu D, Wu C, Liu X (1993) Deep seismic reflection evidence for continental under thrusting beneath southern Tibet. Nature 366:555–559

State of Nuclear Power Plant Risk Assessment for Ground Deformation with Seismic Faulting

Katsumi Ebisawa, Toshiaki Sakai, Futoshi Tanaka, Ryusuke Haraguchi, Yoshinori Mihara, and Yuji Nikaido

Abstract

The authors have developed a concept of fault displacement (FD) probabilistic risk assessment (PRA) methodology, through application to a hypothetical nuclear power plant (NPP). Core damage frequency and its uncertainty of a hypothetical NPP was evaluated using FD hazard, FD fragilities of building and components, and fault tree and event tree models. Important initiating events, accident sequences, components and structure failures, as well as the range of the fault displacement that characterize the core damage risk profile were identified. Through the study, the authors confirmed the feasibility of the methodology. The authors summarized the strategies and issues that can be taken when applying the FD PRA method to an actual NPP.

Keywords

Fault displacement PRA methodology • Concept of methodology application • Hypothetical NPP • Actual NPP

K. Ebisawa (✉) · T. Sakai
Central Research Institute of Electric Power Industry, 1-6-1 Otemachi, Chiyoda-ku, Tokyo, 100-8126, Japan
e-mail: ebisawa@criepi.denken.or.jp

T. Sakai
e-mail: t-sakai@criepi.denken.or.jp

F. Tanaka · R. Haraguchi
Mitsubishi Heavy Industries, Ltd., 1-1-1, Wadasaki-cho, Hyogo-ku, Kobe, 652-8585, Hyogo, Japan
e-mail: f_tanaka@mhi.co.jp

R. Haraguchi
e-mail: ryusuke_haraguchi@mhi.co.jp

Y. Mihara · Y. Nikaido
Kajima Corporation, 6-5-30 Akasaka, Minato-ku, Tokyo, 107-8502, Japan
e-mail: y-mihara@kajima.com

Y. Nikaido
e-mail: nikaidoy@kajima.com

Ž. Arbanas et al. (eds.), *Understanding and Reducing Landslide Disaster Risk*,
ICL Contribution to Landslide Disaster Risk Reduction,
https://doi.org/10.1007/978-3-030-60713-5_8

Introduction

The Chi-Chi Earthquake, a magnitude of 7.6, occurred in Taiwan in Sept. 1999. During this event, a long stretch of the fault offset has traversed a dam and many buildings, etc. destroying them; thus, the event has attracted remarkable attention of researchers. Recently in Japan, concern about the impact of principal and secondary Fault displacements (FDs, hereafter) on nuclear facilities has been growing, and it is currently recognized as an overriding priority of thorough investigation. The current status regarding FD probabilistic risk assessment (PRA, hereafter) methodology in Japan is as follows; the Atomic Energy Society of Japan (AESJ) is conducting two activities (AESJ 2015) and (AESJ 2018). The Japan Society of Civil Engineers (JSCE) published a research report of FD evaluation carrying articles about investigations of Faults and FDs, FD evaluations based on numerical analyses and experiment etc. (JSCE 2015). With regards to international activities, a workshop, with participants from nuclear industry, has recently been held to discuss the latest studies on fault displacement hazards (FDHAWOC 2016).

The authors have an ongoing program to develop fault displacement PRA methodology (Ebisawa et al. 2017), and established its basic framework as part of the project. To confirm the feasibility of the methodology, the core damage frequency (CDF) and its uncertainty were examined applying the developed methodology to a hypothetical NPP. Important initiating events, accident sequences, structures and components such as a pump and an electrical panel, as well as the range of the fault displacement that characterizes the CDF profile were identified. Through the study, the authors confirmed the feasibility of the methodology, and also identified the important source of uncertainties within the methodology (Ebisawa et al. 2018; Haraguchi et al. 2020).

The authors reviewed our papers mentioned above and summarized the strategies and issues that can be taken when applying the FD PRA method to an actual plant, as well as the issues that require further studies, and are sure these contents will be very useful in areas other than nuclear.

This paper describes the outline of the concept of the FD PRA and the CDF profile, and the strategies and issues to apply the FD PRA method to an actual plant.

Concept of FD PRA Methodology

FD PRA Basic Approach

An effective approach for the FD PRA, is to characterize and focus on the credible plant behavior analysis against FD prior to performing detailed CDF assessment, since not all of the impact of FD event on NPP risk may not be significant to justify performing a fully detailed PRA. If the event is deemed to have significant impact on the NPP, a detailed investigation should be performed focusing on aspects with potentially large impact and those that have large uncertainty. To focus on the important accident contributors, screening of events and failure modes are applied at the FD equipment list development stage, initiating event identification stage, and fragility evaluation stage.

Basic Evaluation Process

The FD PRA process consists of accident scenario identification, hazard evaluation, fragility evaluation and accident sequence evaluation as shown in Fig. 1. The overall evaluation process is the same with that of seismic PRA and tsunami PRA.

Application of the Method to the Hypothetical NPP

Hypothetical NPP and FD Hazard of Interest

To confirm the FD PRA methodology, the PRA was performed against a hypothetical pressurized water reactor (PWR) plant consisting of civil structures such as containment vessel (C/V), reactor environmental building (E/B), control building (C/B), turbine building (T/B), waste treatment building (W/B), sea water system (SWS) facility, severe accident management facility (SAF). The plant layout drawing is shown in Fig. 2a. As shown in the figure, piping that cross buildings, such as component cooling water (CCW) piping, SWS piping, and main steam line are also considered. A distributed fault caused by a reverse fault, illustrated in Fig. 2b, has been assumed. To investigate different accident scenarios initiated by FDs, three locations (F1, F2, and F3) of which the surface displacement interacts with the structures are considered. F1 represents a surface FD that crosses the SWS piping trench. F2 represents a surface FD immediately beneath the E/B, C/V and T/B. Finally, F3 is a FD beneath the center of the SAF base mat.

Takao et al. (2014) obtained the distributed FD hazard curve of the relationship between distributed fault displacement and annual exceedance frequency, reviewing the FD database in Japan as shown in Fig. 3. This FD hazard curve is used for the three distributed FDs of interest.

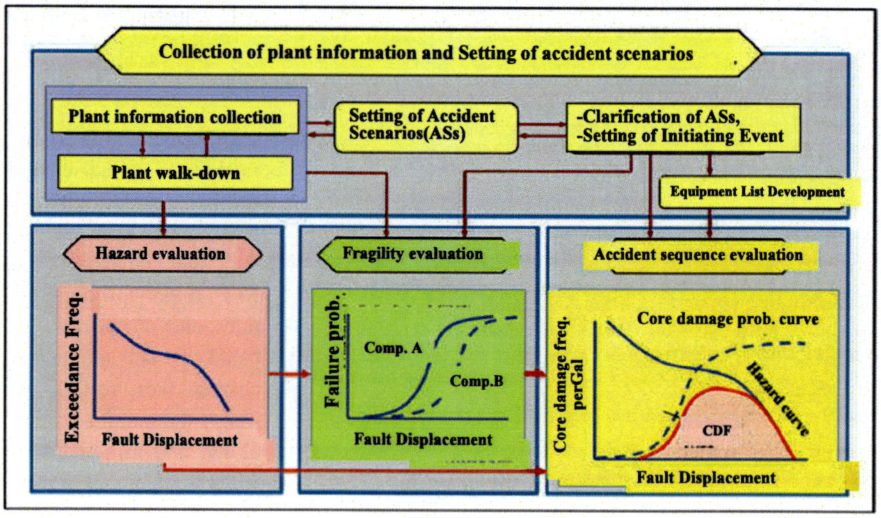

Fig. 1 Evaluation process of FD PRA

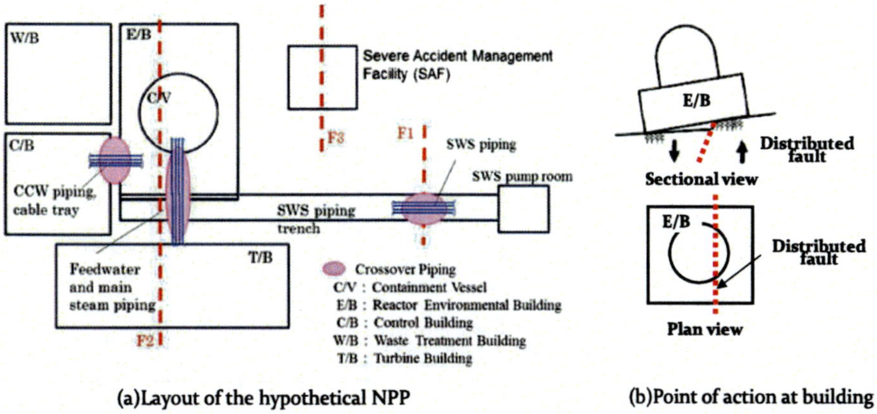

Fig. 2 Hypothetical NPP and distributed FDs assumed in the study

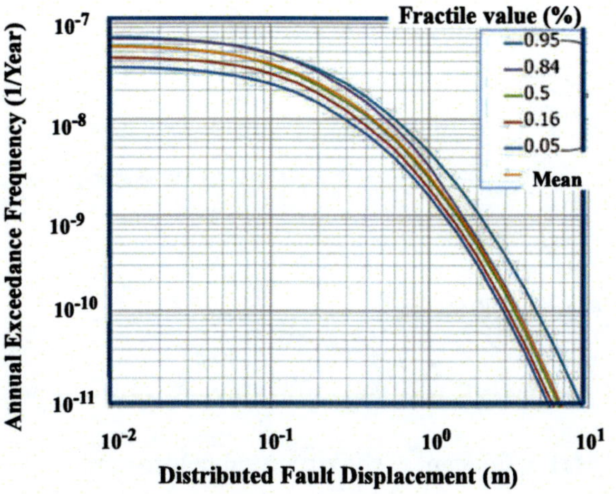

Fig. 3 Distributed FD hazard

Identification of FD Equipment and Initiating Events

A "FD equipment list" is developed to identify the structures, systems and components (SSCs) affected by the FD of interest. This process is important to demonstrate the completeness of the FD induced initiating events, and the failure modes to be considered in the FD PRA.

The development of the FD equipment list starts with the identification of the building and structures affected by the FD. Building and structures directly affected by the FD are identified considering the geometrical condition between the FD and the structure of interest. The initial FD equipment list includes all SSCs within the building directly affected by the FD. Since equipment failure can also be caused by relative displacement of two adjacent structures or buildings, piping and cables that cross the affected buildings are also candidates of vulnerable SSCs to be included in the FD equipment list.

After the initial FD equipment list has been developed, the list will be subjected to screening considering more detail information of the equipment layout. Figure 4 shows the equipment layout in the base floor of the hypothetical plant. Equipment affected by the deformation of the base mat are identified from the component layout at the base floor and the FD location, which is where deformation of the floor is expected to occur. In the case shown in Fig. 4, FD is anticipated to cause displacement to the E/B and T/B causing stress to the crossover piping and cables that extend from the affected buildings. With regards to SAF, which is located above FD F3, portable pumps, emergency mobile power supply unit, and emergency pump vehicle are identified as vulnerable component that can be affected by the FD.

For each component in the FD equipment list, the initiating events triggered by the postulated failure as well as the impact on the mitigation functions are identified.

Fragility Evaluation of Structures and Components

Fragilities of components are evaluated by applying separation of variables method of the seismic PRA (Ebisawa et al. 2018; CEC 1981). The fragility is evaluated using the amount of the FD as a measure and the result is expressed as median capacity displacement A_m, aleatory uncertainty $\beta^R{}_r$ and epistemic uncertainty $\beta^R{}_u$.

In this study, impact on the SSCs function and structure caused by inclination or displacement are considered for the fragility evaluation. Superposition with the load by seismic motion is not considered. Layout of the hypothetical NPP shown in Fig. 4 is considered in the evaluation of failure and fragilities of the components.

Fragility of each component is evaluated based on its dominant failure modes resulting in structural or functional failure. Structural failure is evaluated based on load on the component's member generated by inclination of the floor

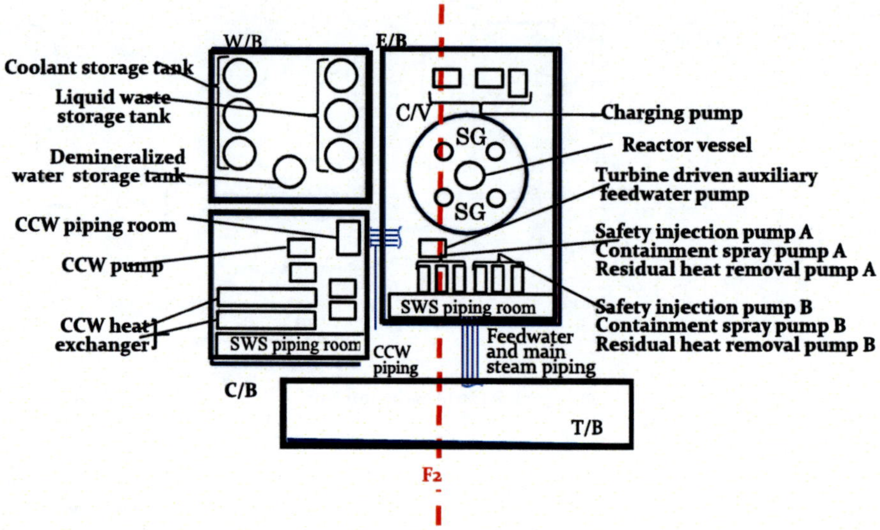

Fig. 4 Conceptual layout drawing of building and equipment

(Fig. 5a). However, actual load generated by the inclination is significantly small compared with the capacity of the component's member and its failure probability is negligible (Ebisawa et al. 2018). Therefore, this failure mode is screened out.

Functional failure of dynamic components is evaluated from a viewpoint of operation continuity under the inclined condition resulting from the FD. Therefore, functional failure of dynamic components is evaluated focusing on the maximum inclination angle of which the lubricant device can maintain its function (Fig. 5b) (Ebisawa et al. 2018).

In the case of failure of crossover piping between buildings, large relative displacement between the buildings may occur in the event of FD (Fig. 5c). Therefore, fragility of such piping is evaluated based on distortion load on the crossover piping generated by the relative displacement. Result of fragility evaluation of main steam piping is shown in Fig. 6, as an example of the fragility of cross over piping.

Accident Sequence and Core Damage Frequency Evaluation

An initiating event tree to address the various FD induced initiating events is developed based on the FD initiating events identified from the FD equipment list. The initiating event tree is shown in Fig. 7. The initiating event tree is developed in a manner similar to the initiating event tree of seismic PRAs. Initiating event with larger impact on plant response is considered in the event headings on the more left-hand side of the event tree. The impact of support

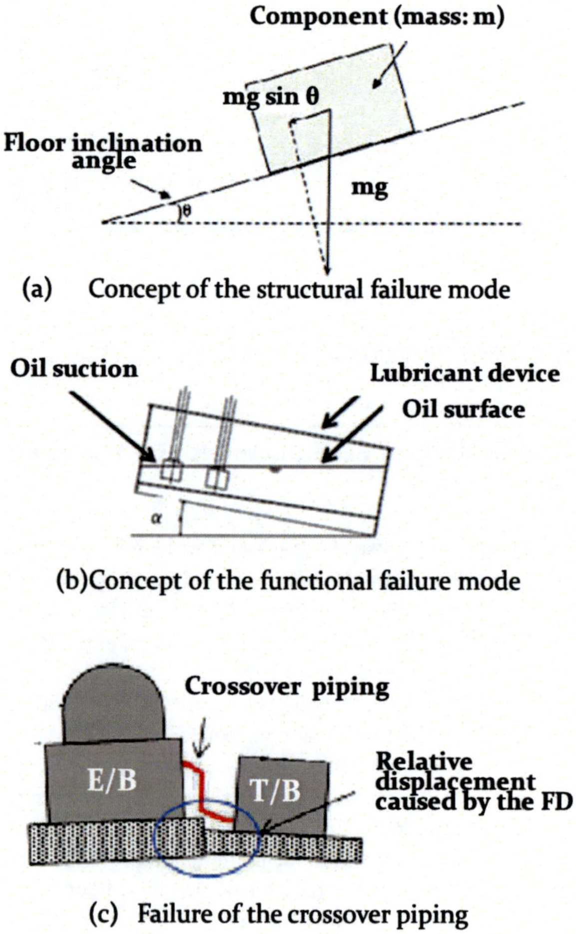

Fig. 5 Failure modes considered in the fragility

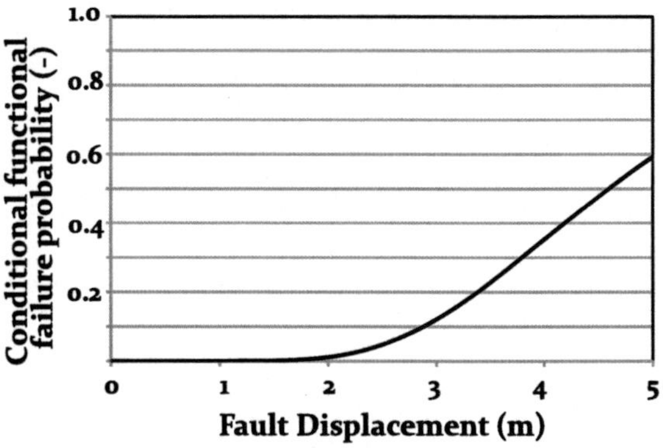

Fig. 6 Example of fragility evaluation result

system and frontline system failures on plant response is addressed in the plant response event tree, which is linked to each of the end states of the initiating event tree.

Since the FD induced support system failures in the accident sequences will be evaluated through the fault trees in the plant response event tree, in the initiating event tree, there are no branches representing support system failures in the sequence where secondary side break has occurred. Fault trees are developed for each event tree headings of the plant response event tree. SSCs considered in the fault trees are the components of the front-line system necessary to achieve the mitigation function, and the components of support systems (i.e., CCW, SWS and Electrical system) that support the frontline systems. Failure modes such as random failures, human errors, FD induced failures and failures due to seismicity are modeled in the fault trees.

Since the multiple distributed FDs at a site may occur simultaneously or independently (only one occur), accident sequence analysis has been conducted for four cases: A case assuming that distributed all postulated FDs occur simultaneously with the same magnitude, and three cases each assuming only one of the three FDs (F1, F2 and F3) occur.

CDF quantification has been carried out with the Risk Spectrum PSA code, using the probabilistic FD hazard curve, event tree and fault tree model as inputs.

Component failure cause by seismicity is considered in the CDF quantification. CDF point estimates of the four cases are shown in Table 1. The CDF for the case assuming that all FDs occur simultaneously, is 3.3×10^{-8} /yr. The dominant initiating event of this case is the "total loss of CCW and SWS". Contribution if other initiating events are a little less but similar. The CDF evaluations for cases with separate FDs show that F2, the event in which the FD takes place underneath the E/B and the TB, has the highest risk contribution. The dominant core damage scenario for F2 involves total loss of CCW and SWS initiating event caused by FD induced failure of CCW piping that cross over E/B and C/B. When F2 involves structural failure of cable trays that also cross the buildings, auxiliary feed water function will be lost along with the mitigation functions necessary to cope with the loss auxiliary feed water function, and the core will be damaged. The large CDF contribution from F2 indicates that the FD induced contribution from F2 indicates that the FD induced failures caused by F2 are the dominant risk contributors when the three FDs actuate simultaneously.

Application of the FD PRA Methodology to an Actual Plant

Depending on the purpose of the FD PRA, the information required for analysis and the problems will vary. When the purpose of the FD PRA is to roughly estimate the CDF in order to grasp the relative risk contribution in comparison with risk from other external hazards, it is possible to cope with it with a simplified evaluation. When the purpose is to evaluate the importance of equipment based on the FD PRA, and to grasp and prioritize the effect of the remodeling work (e.g., structural reinforcement), it is necessary to carry out detailed evaluation. The strategies regarding the detailed

Fig. 7 Example of the initiating event tree

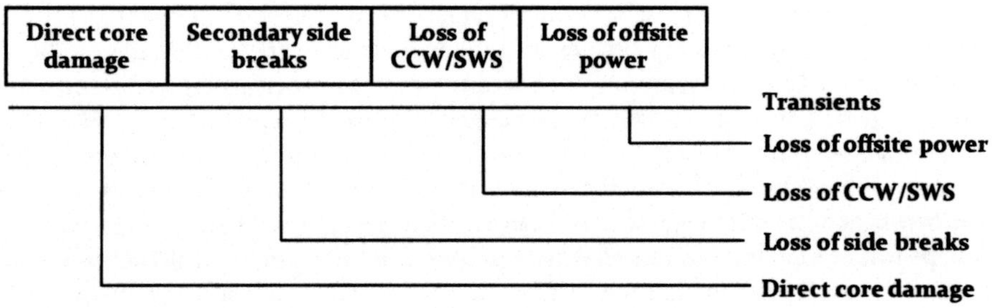

Table 1 CDF contribution per initiating event

Initiating event	Distribute FD of interest			
	F1	F2	F3	F1, F2, F3 simultaneous
Direct core damage	0.0E+00	8.1E-09	0.0E+00	8.1E-09
Secondary side brakes	0.0E+00	5.4E-09	0.0E+00	5.4E-09
Loss of CCW/SWS	3.5E-09	1.6E-08	0.0E+00	1.6E-08
Loss of offsite power	1.2E-10	3.6E-09	1.6E-10	3.6E-09
Total	3.6E-09	3.3E-08	1.6E-10	3.3E-08

All units are in/yr.

evaluation and simplified evaluation are describe in the following sections. Costs required for implementation, which is an important factor in actual operation, is also discussed.

Detailed Evaluation

This section describes the items, information and issues for implementation that need to be considered when carrying out a detailed evaluation of each elements (accident scenario, hazard, fragility, accident sequence) of the FD PRA.

In the accident scenario identification, the item that need to be examined is on how detailed the relationship between assumed fault locations and accident scenarios should be extracted. Necessary information in particular is the detailed information of the SSC's placement and location. Issues are as follows: Combinations of the potential location and type of FD is numerous. In identifying the accident scenarios, it is necessary to consider how to classify the accident scenarios according to the location of the FDs. It is also important to understand the actual situation of each SSC's installation through site plant walkdowns.

In hazard evaluation, the item that need to be examined is the treatment of the seismic approach and displacement approach (Youngs et al. 2003). Necessary information is accumulated information of the ground transfer function related to ground and faults. Issues are as follows: Although the practical implementation of FD hazard using the displacement approach is still under study, the method using the seismic approach has generally been established. In order to develop the displacement approach, it is necessary to obtain detailed ground transfer function data regarding displacement response against a fault.

In fragility evaluation, items that need to be examined are the treatment of capacity and response of target structure and component, and the epistemic uncertainty used in the fragility evaluation. Necessary information is those related to the above items. Issues are as follows: It is necessary to clarify the extent of the building damage when it is judged that the foundation slab and side walls have been damaged. In order to grasp realistic damage limits, it is necessary to analyze damage criteria for dynamic and electrical equipment in particular. Regarding the functionality of dynamic equipment such as pumps, it is necessary to evaluate the functional continuity through actual machine tests. Regarding fragility of electrical equipment such as electric panels, it is necessary to confirm the damage limits through tests that reproduce the effect of inclination caused by FDs. The epistemic uncertainty in the fragility evaluation needs to be quantified by engineering judgment, and the method was proposed by the authors as the fragility SSHAC (Tsutsumi 2019).

In accident sequence evaluation, the item that needs to be examined is the modeling of events specific to FD. Necessary information in particular is specific information such as equipment failure rate and human error rate under FD conditions. Issues are as follows: It is necessary to consider FD effect which may affect the long-term operability of the equipment, when analyzing the achievability of long-term stable cooling. For the human error probability evaluation, it is necessary to analyze in detail the effect of FD event on operator actions, such as degraded accessibility in buildings caused by door deformation and degraded accessibility inside and outside the site due to FDs affecting surrounding roads.

Regarding the cost required for the evaluation, although it is difficult to indicate the cost required for implementing FD PRA, it can be grasped as a sense of the scale of the cost based on seismic PRA, which is carried out commonly in Japan. The costs related to training of personnel are excluded. Since the tasks performed in the FD PRA are similar to that in the seismic PRA, the basic cost for the overall evaluation is considered to be similar. If the equipment to be screened for not being affected by FD can be set more than those of seismic events and the number of accident scenarios are more limited than those of seismic events, the cost may be lower than seismic PRA. The implementation of the fragility SSHAC incurs a cost for the series of processes. The accident sequence evaluation is considered to be equivalent to the seismic PRA, but if the accident scenarios can be limited as described above, the cost will be slightly lower than the seismic PRA.

Simplified Evaluation

This section describes the simplified evaluation regarding accident scenario, hazard, fragility and accident sequence.

- The fault location is roughly set to create an enveloping accident scenario.
- Site and plant walk-down will not be implemented, and information will be collected only through desk top discussions.
- The national FD hazard is used instead of the site specific FD hazard or, the information of the existing similar site is used to develop a simplified hazard.
- The fragility of similar plants is used. For the uncertainty setting, the values set by the engineering judgment are used instead of performing a fragility SSHAC.
- A scope of systems modeled in the accident sequence is limited. Only point estimation such as CDF is performed.

The simplified evaluation is performed within the range of information available at the time of the evaluation. Therefore, there is no particular issue to be found in the evaluation.

Regarding the cost required for the evaluation, costs are examined based on the seismic PRA. The cost of the corresponding elements in the seismic PRA will be reduced due to simplification of the methodology and scope. Therefore, although it depends on the number of items to be simplified and their degree of simplification, it is considered that the FD PRA can be implemented at a cost that is less than half or several percent lower than the seismic PRA.

Conclusion And Future Plans

The results of this study are as follows.

(1) Basic concept of FD PRA methodology has been developed.
(2) CDF and uncertainty evaluation to confirm the feasibility of the methodology have been conducted through application to a hypothetical plant. The analysis provided insights on important initiators, accident sequences and failure modes that have high CDF contribution.
(3) The authors summarized the strategies and issues that can be taken when applying the FD PRA method to an actual NPP.

Acknowledgements This study was conducted as the preparation project regarding nuclear power plant risk assessment and study under the Agency for Natural Resources and Energy, Ministry of Economy, Trade and Industry. The authors hereby express gratitude to the organizations and all who supported the project.

References

Atomic Energy Society of Japan (AESJ) (2018) Report on development risk evaluation methods and measures for fault movement by engineering approach. (In Japanese)

Atomic Energy Society of Japan (AESJ) (2015) Standard for procedure of seismic probabilistic risk assessment for nuclear power plants. (In Japanese)

Commonwealth Edison Co. (CEC) (1981) Zion Probabilistic Safety Study

Ebisawa K et al (2017) Analysis and examination for developing fault displacement PRA methodology. PSA2017

Ebisawa K et al (2018) Improvement of displacement PRA methodology and concept of its application to a hypothetical NPP, PSAM14

Fault Displacement Hazards Analysis Workshop Organizing Committee (FDHAWOC) (2016) Fault displacement hazards analysis workshop. Menlo Park, CA

Haraguchi R et al (2020) Development of fault displacement PRA methodology and its application to a hypothetical NPP. Nucl Eng Des. https://doi.org/10.1016/j.nucengdes.2019.110433

Hideaki T et al (2019) The validation strategy of fault displacement fragility evaluation methodology, SMiRT25.

Japan Society of Civil Engineers (JSCE) (2015) Research report of fault displacement evaluation. (In Japanese)

Takao M et al (2014) Reliability improvement of probabilistic fault displacement hazard analysis. J Jpn Assoc Earthq Eng 14(2). (In Japanese)

Youngs RR et al (2003) A methodology for probabilistic fault displacement hazard analysis (PFDHA). Earthq Spectra 19(1)

Relationship Between Arias Intensity and the Earthquake-Induced Displacements of Slopes

Ching Hung, Chih-Hsuan Liu, and Hsuan-Ho Wang

Abstract

Earthquake-induced landslides have a direct impact on human society. In order to evaluate and mitigate the damage associated slope failure, a well understanding of displacements of slopes induced by earthquakes is pivotal. Acknowledging both the energy-based Arias Intensity as a reliable intensity measure to describe the severity of the earthquake and the finite element analysis as a practical and effective method to conduct seismic hazard analyses. This study is focused on the relationship between the Arias Intensity and the earthquake-induced displacements of slopes. Twenty near-field motions from the 1999 Chi-Chi earthquake and the 1979 Imperial earthquake were scaled to peak acceleration of 70, 100, and 200 gals and applied to the simple models with slope angles of 20°, 30°, and 45°. It was found that the range of the Arias Intensity has effects on the relationships between Arias Intensity and the normalized earthquake-induced slope displacements while the angles of the slopes for the three simple landslide models show a trivial influence on the relationships between Arias Intensity and the normalized earthquake-induced slope displacement, which may be due to the same material properties.

Keywords

Earthquake-induced landslide • Arias intensity • Finite element analysis • Seismic hazard • Empirical relationships

C. Hung · C.-H. Liu (✉) · H.-H. Wang
Department of Civil Engineering, National Cheng Kung University, No.1, University Road, Tainan City, 701, Taiwan
e-mail: N68071504@gs.ncku.edu.tw

C. Hung
e-mail: ChingHung@gs.ncku.edu.tw

H.-H. Wang
e-mail: N66054344@mail.ncku.edu.tw

Introduction

Landslides have a direct impact on human society which can cause economic losses, casualties. The cause of landslides is when the slopes are disturbed by precipitation, earthquakes, erosion, alternation of hydrogeologic regimes, or a combination of these factors (Keefer 1984; Xu et al. 2013; Hung et al. 2018, 2019). Among these factors, earthquakes have been indicated as important in triggering the landslides by a large number of investigations.

Situated at the boundary between the Philippine Sea Plate to the East and Eurasian Plate to the West, seismic activities and subsequent landslides (e.g. Chi-Chi earthquake and Tsaoling landslide) occur frequently in Taiwan due to the collision of the plates. The frequent seismic activities aroused researchers' interest in studying the seismic response, and furthermore, the appropriate countermeasure to mitigate the damage caused by seismic activities or subsequent landslides. To address the hazard, many researchers have utilized seismic hazard analysis, including field investigation, experiment, numerical analysis, etc., and proposed several strong empirical relationships (Joyner and Boore 1981; Campbell and Borzongnia 2003; Jibson 2007; Hsieh and Lee 2011), which can play an important role.

In seismic hazard analysis, one of the ground motion parameters that are most commonly used to represent the earthquake impact on slope stability is Peak Ground Acceleration (PGA). Even with many researchers applying the peak ground acceleration, some studies show different opinions about using PGA to describe the severity of seismic shaking because PGA is a single point in an acceleration-time history and is a raw measure of shaking intensity (Evernden 1975; Jibson 1993). Takazawa and Jennmgs (1980) discovered the effect of duration on the severity of seismic shaking. For earthquakes with short duration, even with higher PGA, the destructive capability is extremely low, and the margin of safety is high. Wilson (1993) indicated that PGA is highly influenced by high-frequency components. From these previous researches, to describe the

Ž. Arbanas et al. (eds.), *Understanding and Reducing Landslide Disaster Risk*, ICL Contribution to Landslide Disaster Risk Reduction, https://doi.org/10.1007/978-3-030-60713-5_9

severity of seismic shaking more comprehensively, except for the PGA, the duration and the frequency should be considered.

Arias intensity (AI) (Arias 1970), an energy-based intensity measure, incorporates duration and frequency. Thus, the estimation of the Arias intensity expected on landslide-prone slope as the impact of an earthquake of defining characteristic is useful for a preliminary delimitation of areas that may be affected by earthquake-induced mass movement (Chousianitis et al. 2014).

Acknowledging finite element analysis as a practical and effective method to investigate earthquake-induced response (e.g., earthquake-induced displacement) (Hung et al. 2017, 2018; Lin et al. 2017), in-depth numerical analyses were conducted to investigate the relationship between Arias Intensity and earthquake-induced displacement. The area that occurred Tsaoling landslide, which is the largest landslide, the most well-known, and best-documented landslide among several large landslides triggering by the 1999 Chi-Chi earthquake, is selected as a study area. A validated Finite element procedure (Hung et al. 2017; Lin et al. 2017) was used in this study. It is hoped that the finding can shed some light on seismic hazard analysis.

Finite Element Procedure

The Finite element procedure was developed and validated in Hung et al. (2017) and Lin et al. (2017), which is based on a commercial Finite element software Plaxis 2D (Plaxis 2016). In this procedure, the analyzed slope is assumed to be in a plane-strain condition. Fifteen-node triangular elements (quadratic in displacement and linear in pressure) were used to model the slope, and a fine mesh was used to divide the system into discrete segments for the study. The geomaterials used in this study were characterized using simple elastoplastic constitutive models obeying the Mohr–Coulomb failure criteria with internal friction angle and cohesion.

For the dynamic analysis, in order to avoid the seismic wave causing unrealistic vibrations in the model, the artificial boundary must be set up on the boundaries. In this paper, viscous boundary conditions are used to ensure the precision of simulation results. The ground motion data were applied (imposing prescribed displacements through table input) at the bottom boundary of the model.

In the finite element analyses, construction simulation of the slope was firstly conducted to reach the initial states of deformation and stress before seismic loading. The ground motion was then activated following the initial phase of stability analysis. Details of the Finite element procedure can be found in Hung et al. (2017) and Lin et al. (2017).

Finite Element Procedure

The objective of this study is to clarify the relationship between intensity measure (Arias intensity) and seismic responses of slope (earthquake-induced displacement). To this end, simple slope models with different slope angles were analyzed. According to Chan et al. (2015), approximately 75% of the landslides in Taiwan occurred on the slopes with angles from 20° to 50° (Table 1). Simple slope models with the angles of 20°, 30°and 45° were considered (Fig. 1). In this study, the effects of material properties were not considered. The material properties of the simple slope models are set to the same. The geomaterial used in the simple slope models are one type of sandstone, which is widely distributed in the area of Tasoling landslide. The parameters are given in Table 2.

In finite element analysis, altogether 20 strong ground motions were employed, 10 of which were taken from the records of the 1979 Imperial Valley, and the other 10 were from those of the 1999 Chi-Chi earthquake. All motions are near-field pulse-like ground motions. To capture the near-field effect, the distance between seismic stations to fault is selected under 30 km. Bhandari et al. (2018) suggested a different distance of 20–50 km around the fault as the near-field range. In order to be comparable (ignore the PGA effect), all the input motions were scaled to have peak accelerations of 70 gal, 100 gal, and 200 gal, respectively. Table 3 shows the information of all the applied seismic records used in this study. A recorded earthquake signal is shown in Fig. 2, showing the dominant frequency and durations of the 1999 Chi-Chi earthquake (CHY080).

Results and Discussion

In the FEA simulation, seven monitoring points were prescribed to measure the displacement varying with time (Fig. 3). Five monitoring points are set up to the surface of the slope (in the sliding zone), and the other three are located on the vertical line extended from the highest points of the slope (outside the sliding zone). To investigate the relationship between Arias intensity and earthquake-induced displacement of slopes, two following impact factors are discussed in this section: the effect of the range of Arias Intensity and the effect of slope angle.

Effect of the Range of Arias Intensity

To understand the effect of the range of Arias Intensity, the seismic records were scaled to PGA = 70, 100, 200 gal, resulting in the Arias intensity to be in the range of 0.01 to

Table 1 Recent landslide in Taiwan (after Chan et al. 2015)

Slope angle	Total	Percentage (%)
<20°	23	12.7
20°–30°	53	29.3
30°–40°	58	32.0
40°–50°	24	13.3
>50°	23	12.7

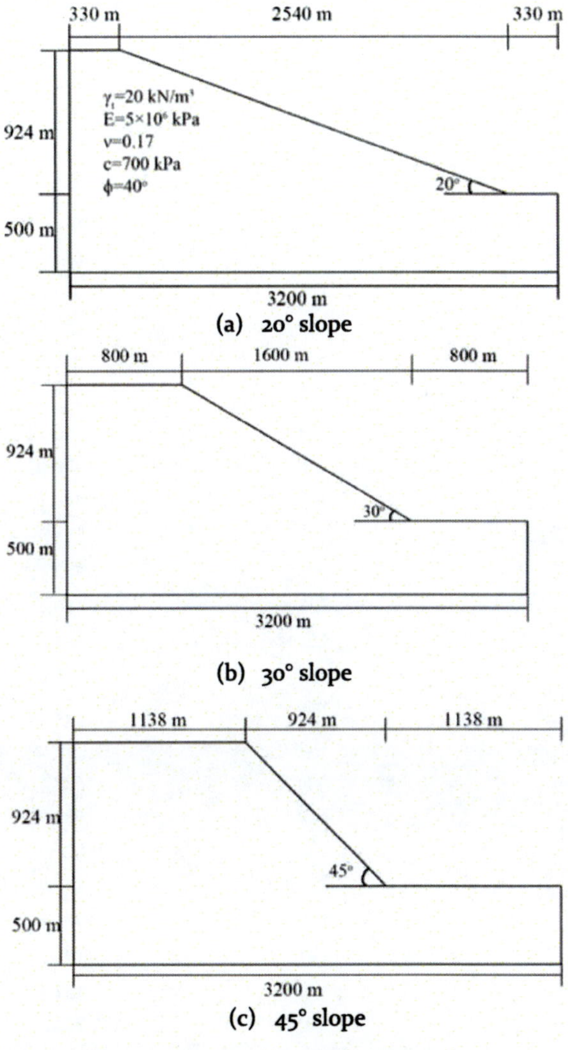

Fig. 1 Analysis model from three different slopes

0.25, 0.01 to 0.4, and 0.3 to 2.0 m/s, respectively. The linear regression method was utilized to present the relationship between Arias intensity and earthquake-induced displacement, which was usually used in the relationship between Arias intensity and displacement (Jibson 1993; Hsieh and Lee 2011; Chousianitis et al. 2014). To make the results easier to compare, the resultant earthquake-induced slope displacements are sorted out by using a normalization method called Min–Max normalization, which normalized the range of the displacement results to the range 0–1. The Min–Max normalization is defined as follows:

$$x* = \frac{x - x_{\min}}{x_{\max} - x_{\min}} \quad (1)$$

The resultant earthquake-induced slope displacements under different range of Arias Intensity are normalized separately, and the results of the relationship between Arias Intensity and normalized slope displacement are shown in Fig. 4. The A, B, C, D, E, F, G, and H in the figures correspond to the monitoring point shown in Fig. 3.

As shown in Fig. 4, the results of five monitoring points near the surface under earthquake show very similar behavior. On the other hand, the results for the other three monitoring points show that the fit-line for the monitoring points in the sliding zone render bigger slopes than the monitoring points outside the sliding zone, but generally, the results are still quite similar, and the main focus of this study is the response in the sliding area. To prevent the flow chart from being too complicated, the displacement of the toe of the slope (monitoring point E), where most houses are located, are used for further discussion. Figure 5 presents the fit-line curve of the monitoring point E. The effect of the range of Arias Intensity on the fit-line curve is obvious. When the range of the Arias Intensity increases, the slope of

Table 2 Geomaterial parameters for simple models (after Lin et al. 2017)

Cohesion (kPa)	Friction angle (Degree)	E (kPa)	γ_t (kN/m^3)	ν
700	40	5×10^6	20	0.17

Table 3 Seismic records used in the analyses

Earthquake	Station	R_{epi} (km)	component	PGA (gal)
1999 Chi-Chi earthquake	CHY024	23.76	UD	141.4
			NS	162.16
			EW	276.34
1999 Chi-Chi earthquake	CHY028	32.3	UD	335.5
			NS	749.9
			EW	624.16
1999 Chi-Chi earthquake	CHY080	10.6	UD	715.87
			NS	841.541
			EW	792.386
1999 Chi-Chi earthquake	TCU067	28.15	UD	230.58
			NS	312.66
			EW	488.86
1999 Chi-Chi earthquake	TCU071	15.07	UD	415.54
			NS	639
			EW	517.82
1999 Chi-Chi earthquake	TCU075	20.06	UD	223.88
			NS	257.32
			EW	325.34
1999 Chi-Chi earthquake	TCU076	15.53	UD	275.38
			NS	420.02
			EW	340.1
1999 Chi-Chi earthquake	TCU078	5.53	UD	171
			NS	302.48
			EW	439.7
1999 Chi-Chi earthquake	TCU079	8.14	UD	382.32
			NS	416.96
			EW	579.78
1999 Chi-Chi earthquake	TCU084	9.23	UD	311.76
			NS	422.82
			EW	989.22
1979 Imperial	UGSG 5054	6	UD	321.1
			S40E	583.9
			S50W	762.9
1979 Imperial	USGS 5053	15	UD	171
			S45W	267.8
			N45W	197.6
1979 Imperial	UGSG 5055	20	UD	205
			S45W	243
			N45W	209.4
1979 Imperial	UGSG 5054	7	UD	347.7
			230	770.4
			140	575.7
1979 Imperial	USGS 5053	17	UD	179.1
			225	272.5
			315	200.4

(continued)

Table 3 (continued)

Earthquake	Station	R_{epi} (km)	component	PGA (gal)
1979 Imperial	USGS 5055	20	UD	223.7
			225	254.1
			315	245.3
1979 Imperial	USGS 0955	27	UD	204.4
			S50W	349.2
			S40E	480.8
1979 Imperial	USGS 0412	27	UD	98.9
			N40W	226.6
			N50E	166.1
1979 Imperial	USGS 0958	27	UP	349.3
			S40E	599.7
			S50W	438.1
1979 Imperial	USGS 5165	26	UP	454.9
			EW	339.3
			NS	473.6

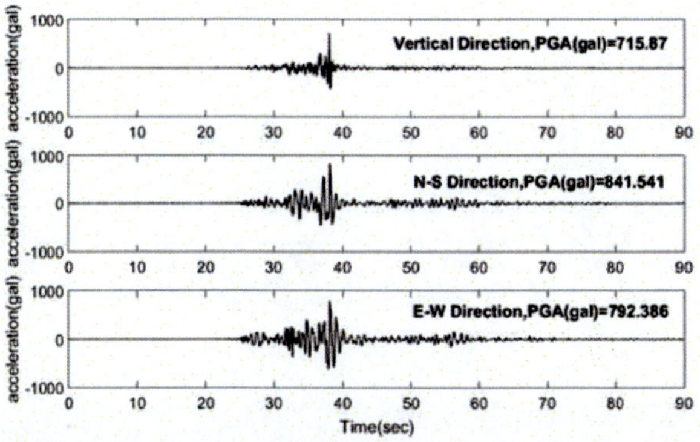

Fig. 2 The 1999 Chi-Chi earthquake from station CHY080

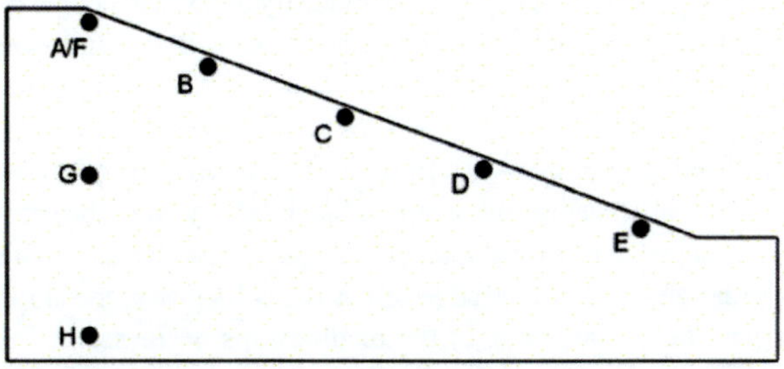

Fig. 3 Locations of the monitoring points

the fit-line curve decreases. In smaller ranges of the Arias intensities, the increase rate is much higher than those in bigger ranges. According to Ravichandran and Arulchelvan (2017), the use of fit-line curves beyond the range of the observed data is called extrapolation and is subjected to a degree of uncertainty since it reflects both the method used to construct the curve and the observed data. Most empirical relationships of Arias Intensity and earthquake-induced displacements of slopes are obtained within a specific range of Arias Intensity. Extending the empirical relationships/fit-line curve beyond the range should be carefully examined.

Effect of Slope Angle

To investigate the effect of angle on the relationship between Arias Intensity and normalized slope displacement, different angles of the same PGA were put in the same figure (See Fig. 6). As shown in Fig. 6, the relationships between Arias Intensity and normalized earthquake-induced displacement show similar results in different angles under the same PGA, a possibility is that same material properties were set to all three simple models. While the original displacement may be different, empirical relationships with normalized

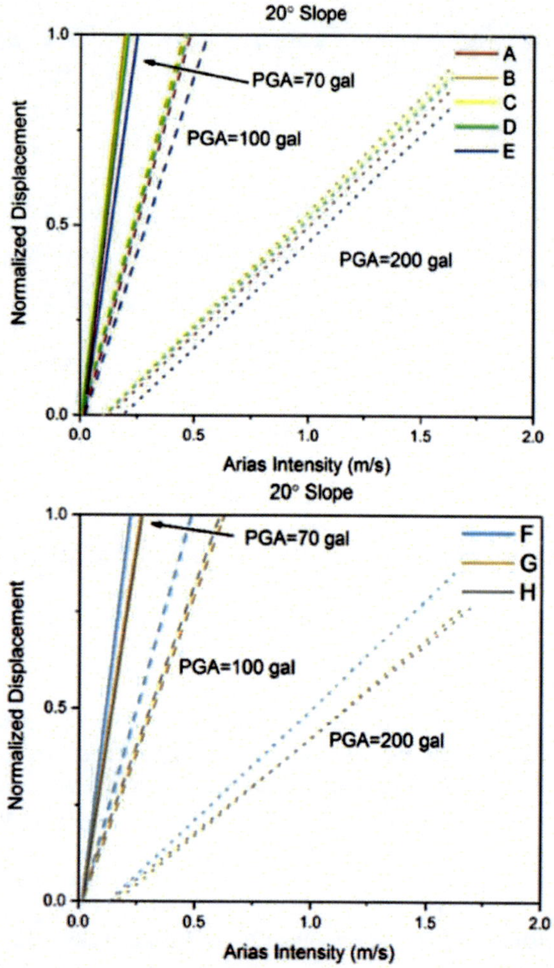

(a) 20° slope

Fig. 4 Relationship between Arias Intensity and normalized displacement for monitoring points (same slope angle with different PGA)

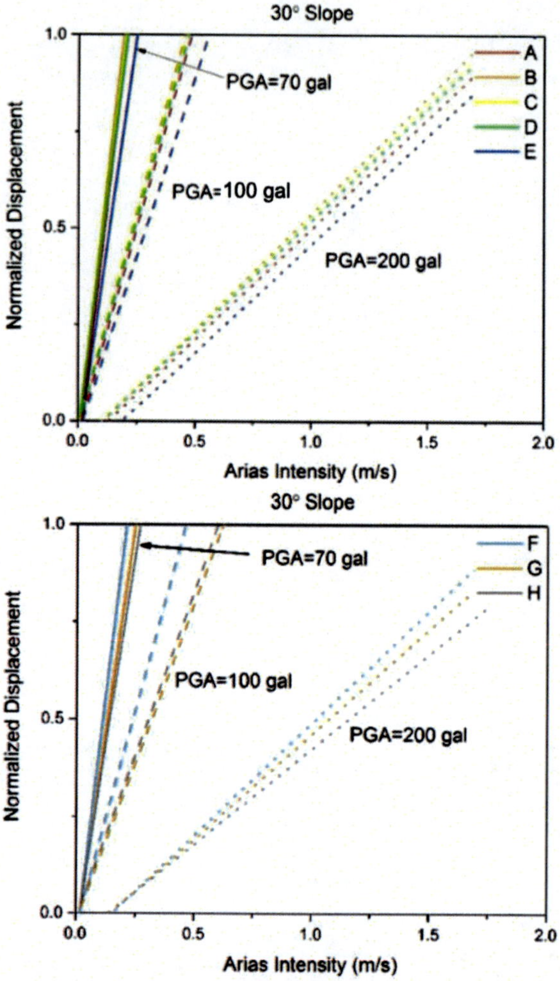

(b) 30°slope

Fig. 4 (continued)

displacements show similar characteristics. This discovery may provide valuable information for future applications on the empirical relationships. Cases with similar properties may utilize the same empirical relationship, which is beneficial for practical projects.

Conclusion

This study is aimed at studying the relationships between Arias Intensity and the earthquake-induced displacements of slopes. In this study, a series of seismic records were utilized, and finite element analysis were performed. The effects of (a) the range of Arias Intensity and (b) the angle of slopes were revealed. The conclusions car be drawn from the results:

1. The range of the Arias Intensity has an obvious effect on the relationships between Arias Intensity and the normalized earthquake-induced slope displacement. With the increase of the range of the Arias Intensity, the slope of the fit-line curve of the relationships decreases.
2. The angle of the slopes for three simple landslide models shows a trivial influence on the relationships between

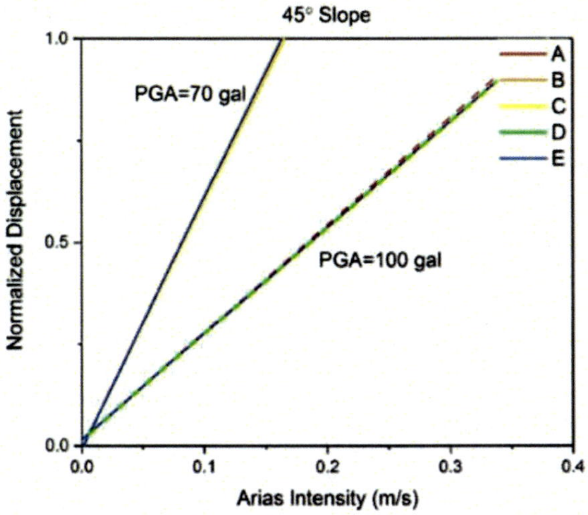

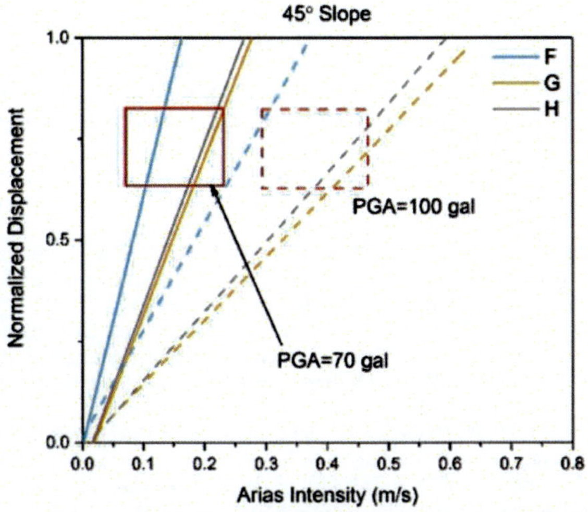

(c) 45°slope

Fig. 4 (continued)

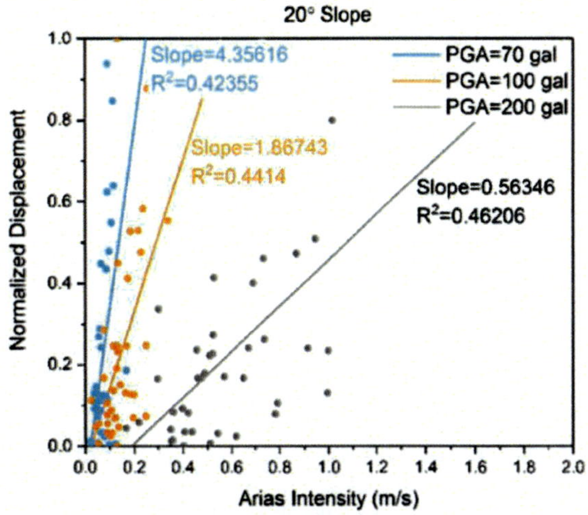

(a) 20°slope

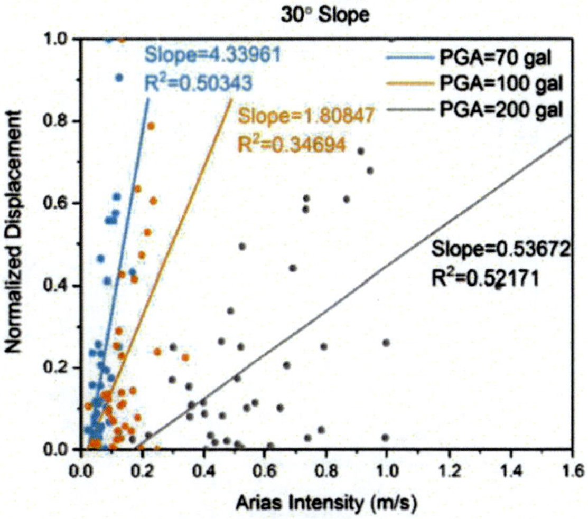

(b) 30°slope

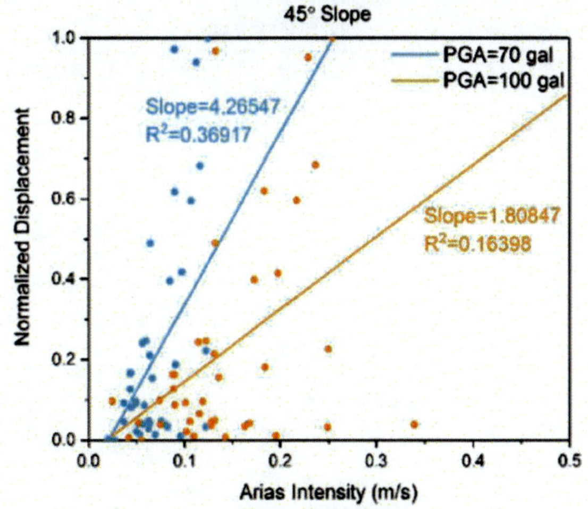

(c) 45°slope

Fig. 5 Relationship between Arias Intensity and displacement for monitoring points E (same slope angle with different PGA)

Arias Intensity and the normalized earthquake-induced displacements of slopes. A possible reason is that all the three simple landslide models have identical material properties.

3. Despite that the best R values fell between a range of 0.5–0.6, which is less satisfactory for the simplified and uniform structure; it is still hoped that the finding can shed some light on seismic hazard analysis.

Acknowledgements The authors gratefully acknowledge financial support from the Ministry of Science and Technology (MOST) in Taiwan: Grants 107-2636-E-006-003. Special thanks go to the Young Scholar Fellowship Program by the Ministry of Science and Technology in Taiwan (The Pilot Directions for MOST Grant for the Columbus Program). The authors appreciate the colleagues in our laboratory for their constructive comments and assistance.

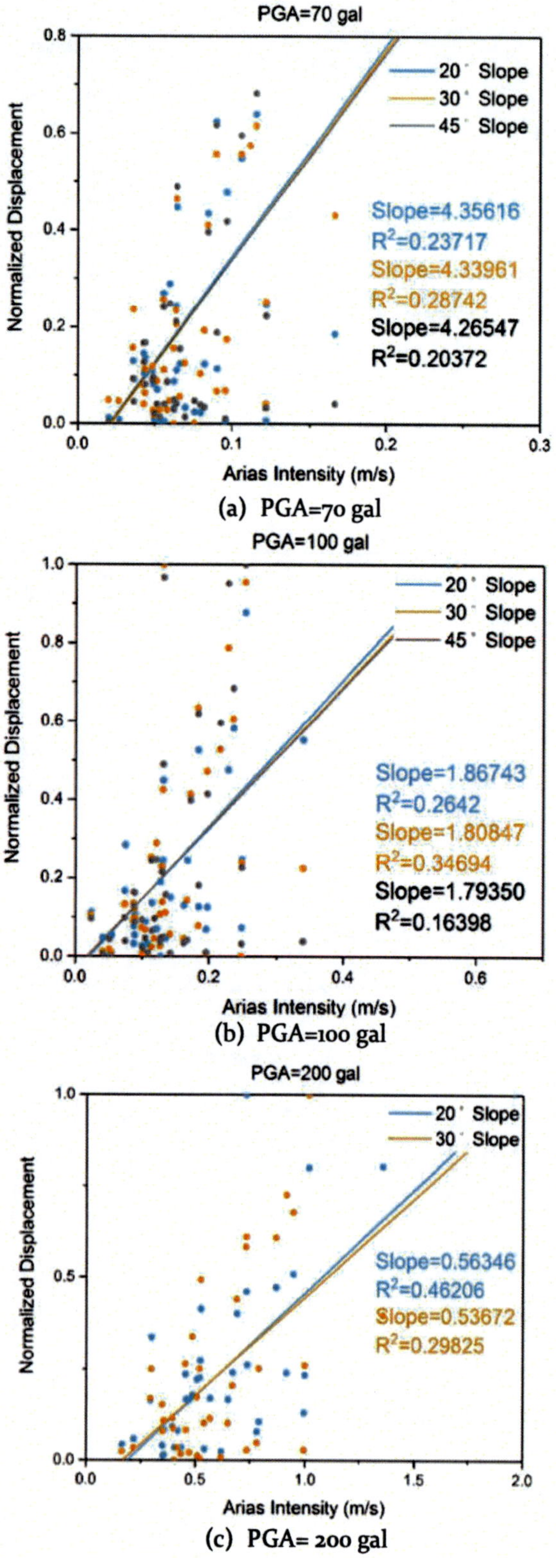

Fig. 6 Relationship between Arias Intensity and normalized displacement (same PGA with different slope angles)

References

Arias A (1970). A measure of earthquake intensity. in: Hansen RJ (Ed), Seismic design for nuclear power plants. Massachusetts Institute of Technology Press, Cambridge, MA, pp 438–483

Bhandari M, Bharti SD, Shrimali MK, Datta TK (2018) The numerical study of base-isolated buildings under near-field and far-field earthquakes. J Earthq Eng 22(6):989–1007

Campbell KW, Bozorgnia Y (2003) Updated near-source ground-motion (attenuation) relations for the horizontal and vertical components of peak ground acceleration and acceleration response spectra. Bull Seismol Soc Am 93(1):314–331

Chousianitis K, Gaudio VD, Kalogeras I, Ganas A (2014) Predictive model of Arias intensity and Newmark displacement for regional scale evaluation of earthquake-induced landslide hazard in Greece. Soil Dyn Earthq Eng 65:11–29

Chan HC, Chang CC, Chen SC, Wei YS, Wang ZB, Lee TS (2015) Investigation and analysis of the characteristic of shallow landslides in mountainous areas of Taiwan. In Taiwanese.

Everden JF (1975) Seismic intensities, "size" of earthquakes, and related phenomena. Bull Seismol Soc Am 65:1287–1315

Hsieh SY, Lee CT (2011) Empirical estimation of the Newmark displacement from the Arias intensity and critical acceleration. Eng Geol 122:34–42

Hung C, Lin GW, Syu HS, Chen CW, Yen HY (2017) Analysis of the Aso-bridge landslide during the 2016 Kumamoto earthquakes in Japan. Bull Eng Geol Env 2017:1–11

Hung C, Liu CH, Chang CM (2018) Numerical investigation of rainfall-induced landslide in mudstone using coupled finite and discrete element analysis. Geofluids.

Hung C, Liu CH, Lin GW, Leshchinsky B (2019) The Aso-Bridge coseismic landslide: a numerical investigation using finite and discrete element methods. Bull Eng Geol Environ

Joyner WB, Boore DM (1981) Peak horizontal acceleration and velocity from strong motion records including records from the 1979 Imperial Valley, California, earthquake. Bull Seismol Soc Am 71(6):2011–2038

Jibson R (1993) Predicting earthquake-induced landslide displacement using Newmark's sliding block analysis. Transp Res Rec 1411:9–17

Jibson RW (2007) Regression models for estimating coseismic landslide displacement. Eng Geol 91(2–4):209–218

Keefer DK (1984) Landslides caused by earthquakes. Geol Soc Am Bull 95(4):406–421

Lin GW, Hung C, Syu HS (2017) Evaluation of an enhanced FS method for finding the initiation time of earthquake-induced landslides. Bull Eng Geol Env 2017:1–9

Plaxis (2016) Reference and material models manual. Plaxis bv, Netherlands

Takizawa H, Jennings PC (1980) Collapse of a model for ductile reinforced concrete frames under extreme earthquake motions. Earthq Eng Struct Dynam 8:117–144

Ravichandran K, Arulchelvan S (2017) the curve fitting model analyzed the survey of crime news awareness in India. 2017 Second international conference on recent trends and challenges in computational models, pp 191–195

Wilson RC (1993) Relation of Arias Intensity to magnitude and distance in California. Department of the interior U.S. geological survey, U.S

Xu WJ, Xu Q, Wang YJ (2013) The mechanism of high-speed motion and damming of the Tangjiashan landslide. Eng Geol 157:8–20

Part II

Recent Progress in the Landslide Initiating Science

Water Exfiltration from Bedrock: A Drastic Landslide Triggering Mechanism

2020 Kyoto Japan

Amin Askarinejad and Sarah M. Springman

Abstract

Landslides triggered through hydraulic perturbations are initiated by a decrease in the effective stresses, and hence the mobilised shear strength of the soil, as a result of the increase in the pore water pressure. This may occur either due to rain infiltration or to exfiltration of water from the bedrock into the slope. Although numerous investigations have been conducted to study the first triggering mechanism (rain infiltration), less attention has been devoted to the latter (exfiltration of water from the bedrock). This paper presents the results of a full-scale landslide triggering experiment, which has been carried out on a natural slope in Northern Switzerland, with emphasis on the observations and measurements regarding exfiltration of water from the bedrock. A series of centrifuge tests was also performed to study the effect of the hydro-geological interactions of the bedrock with the overlying soil mantle on the stability of the slope. Moreover, the behaviour of the full-scale test slope prior to the failure induced by the artificial rainfall event, and/or water exfiltration from the bedrock, was investigated using coupled hydro-mechanical numerical methods. The results of both physical and numerical modelling confirm the importance of the exfiltration from bedrock in triggering of landslides.

Keywords

Water exfiltration from bedrock • Landslides • Geotechnical physical modelling • Hydro-mechanical coupled analysis

A. Askarinejad (✉)
Faculty of Civil Engineering and Geosciences, TU Delft, 2600 GA Delft, PO-box 5048 Delft, The Netherlands
e-mail: A.Askarinejad@tudelft.nl

S. M. Springman
IGT, ETH Zurich, Stefano-Franscini-Platz 5, 8093 Zurich, Switzerland
e-mail: sarah.springman@igt.baug.ethz.ch

Introduction

Hydraulic interactions of soil and bedrock include drainage into, and exfiltration from, the bedrock. Permeable interlayers and bedrock fractures can provide preferential drainage paths for rapid bedrock infiltration processes (Onda et al. 2004) and fractured bedrock layers, which are connected to hydraulic recharge zones, can deliver water to the overlying soil and cause development of high pore pressures at the interface between the soil and bedrock. This process is known as exfiltration (Johnson and Sitar 1990; Everett 1979; Montgomery et al. 2009; McDonnell 2003) and is schematically illustrated in Fig. 1.

Bedrock exfiltration is strongly controlled by precipitation duration and intensity as well as rock joint inclination and permeability. Large storms, in combination with wet antecedent conditions, are highly critical in forcing water to emerge from the bedrock (Uchida et al. 2002). For instance, exfiltration was reported at several locations in Switzerland during the heavy storm events of August 2000 and August 2005 (Rickli et al. 2008).

The ratio of the bedrock hydraulic conductivity to that of the overlying soil layer is generally considered to be very low in slope stability studies. Therefore, less attention has been paid to the hydraulic interactions between the soil and the bedrock, although several studies have indicated that the assumption of an impermeable base might impose major errors in the prediction of pore water pressure distribution in slopes of catchments subjected to hydraulic perturbations (Ebel et al. 2007; Bogaard and Greco 2015).

It has been shown that drainage into the bedrock and exfiltration processes influence the triggering of landslides and determine landslide volume, velocity and runout distance (Onda et al. 2004). For example, continuous emission of significant amounts of water from the exposed bedrock at the source area has been reported for several landslides (e.g. Johnson and Sitar 1990; Cascini 2004; Cascini et al. 2010; Masaoka et al. 2016). Fiebelkorn (1983) measured a rapid

Ž. Arbanas et al. (eds.), *Understanding and Reducing Landslide Disaster Risk*,
ICL Contribution to Landslide Disaster Risk Reduction,
https://doi.org/10.1007/978-3-030-60713-5_10

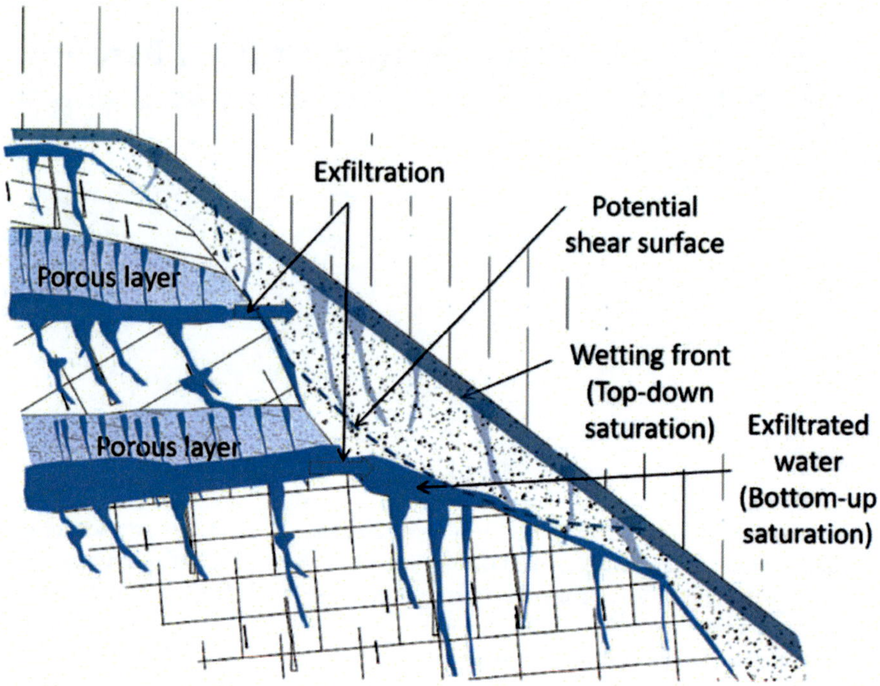

Fig. 1 Schematic diagram of the hydraulic interactions between bedrock and the soil layer

increase in the piezometric response of bedrock to a rainfall event in a debris flow scar in California. Brönnimann et al. (2013) also showed the effect of bedrock exfiltration on the triggering of shallow landslides in Central Switzerland, using comprehensive geophysical and hydro-geological investigations.

Only a few studies have alluded to the importance of the hydro-mechanical interactions of the soil and bedrock on the pore water pressure distribution within the slope and triggering of landslides (e.g. Montgomery et al. 1997; Cascini et al. 2010; Brönnimann 2011; Lanni et al. 2012; Brönnimann et al. 2013; Take et al. 2004; Lucas et al. 2020; Tang et al. 2018; Elia et al. 2017). Therefore, a lack of in-depth knowledge is still apparent in this area due to the large number of variables and conditions.

The influence of water exfiltration from the bedrock on the stability of the unsaturated slopes is investigated in this study using the data from a real scale landslide triggering experiment in Northern Switzerland. Moreover, several hypotheses were tested using coupled hydro-mechanical numerical simulations and the results were compared with the data from in-situ measurements. The details of centrifuge modelling tests on two models subjected to rainfall infiltration and water exfiltration are reported in the last section of the paper.

Full Scale Landslide Triggering Experiment

Test Site

A 38° steep forested slope with an area of approximately 300 m^2 located in Canton Schaffhausen near Ruedlingen village in Switzerland was selected for a landslide triggering experiment. The thickness of the soil on the bedrock was not uniform on the slope and varied between less than 1 m to more than 5 m. The depth and form of the bedrock were determined using Dynamic Probing Light (DPL) (DIN 4094 2003) and Electrical Resistivity Tomography (ERT) (Lehmann et al. 2013; Askarinejad 2013) (Fig. 2).

Several shallow landslides had occurred during and following an extreme rainfall event in May 2002 in this area (Fig. 3). The selected slope was instrumented using time domain reflectometers (TDRs), tensiometers, piezometers, slope inclinometers, and ERT to monitor the hydro-mechanical responses of the soil to both natural and intense artificial rainfall events.

The geological layers in this region were deposited in the foreland basin of the Alps, containing alternate depositions in the Tethys Sea and on land (Tacher and Locher 2008; Sinclair 1997; Platt 1992). Several boreholes, as well as an

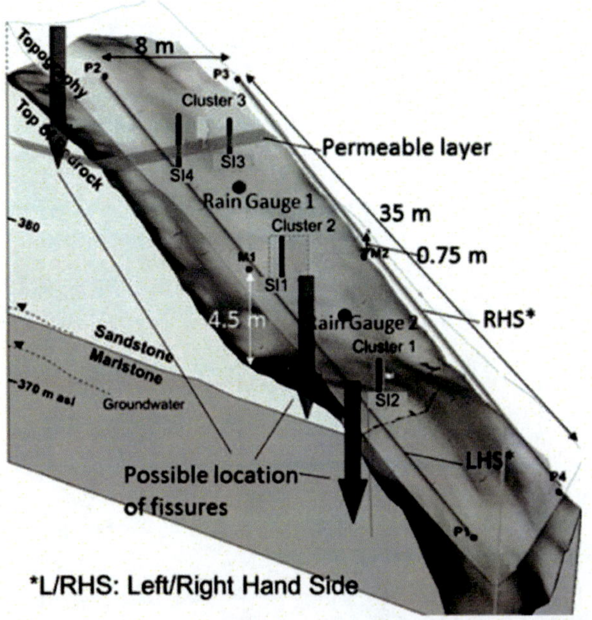

Fig. 2 Shape of the bedrock below extent of experimental area delineated by P_1-M_1-P_1-P_3-M_2-P_4-P_1, with positions of the sensor clusters, Slope Inclinometers (SI, Askarinejad and Springman 2017), and location of the fissures in the bedrock derived from the in-situ hydraulic conductivity measurements and the ERT results (bedrock topography: based on DPL results; after Brönnimann 2011)

outcrop of bedrock revealed horizontal layering of the sedimentary rocks at, or close to, the experimental slope, which consist of fine grained sand-and marlstone (Tacher and Locher 2008). Geological layers derived from the ERT and DPL results, as well as in-situ investigations and the data from geological borehole, are shown in Fig. 4.

Large fissures parallel to the river were mapped (Figs. 2 and 5). These fissures can act potentially as efficient drainage paths for the groundwater (Brönnimann et al. 2009). Moreover, extensive hydro-geological investigations revealed the existence of permeable horizontal layers in the bedrock, which can facilitate exfiltration of water from the bedrock into the overlaying soil mantel. These layers are shown as "Non-lithified sandstone layer" and "permeable layer" in Figs. 4 and 2, respectively. The investigations showed that the permeable horizontal layers can be connected to hydraulic recharge zones (Brönnimann 2011). Exfiltration of water from the bedrock through these permeable layers has already resulted in some local failures in the vicinity of the test site. The extruded water is frozen during winter and the ice stalactites indicate this phenomenon when the air temperature is lower than the freezing point (Fig. 6).

Geotechnical Soil Characteristics

The soil from the selected slope can be categorised as medium to low plasticity silty sand (SM), based on the Unified Soil Classification System (Askarinejad and Springman 2017). Conventional drained and undrained triaxial tests were carried out on reconstituted and undisturbed saturated and unsaturated soil samples by Casini et al. (2013), leading to a critical state friction angle of about 32.5°. Wetting and drying branches of the Water Retention Curve (WRC) of a natural undisturbed sample were determined using the axis translation technique (Hilf 1956) and are shown in Fig. 7.

The saturated hydraulic conductivity of the SM Ruedlingen soil was determined to be between 10^{-6} m/s to 10^{-5} m/s, based on the laboratory measurements (Askarinejad 2013). The unsaturated hydraulic conductivity function for Ruedlingen soil was derived using the Instantaneous Profile Method (IPM) (Daniel 1982) and is shown in Fig. 8. The equation proposed by Averjanov (1950) is used to describe the relative hydraulic conductivity function ($K_r = K_{unsat}/K_{sat}$). The general form of the equation is a power function of effective degree of saturation (S_e):

$$K_r(S_e) = S_e^{\beta} \tag{1.1}$$

The constant β is determined to be equal to 1.7, according to the results presented in Fig. 8.

Landslide Triggering Experiment

Artificial rainfall (using 14 sprinklers installed on the slope) was applied with an average intensity of about 11 mm/hr (Fig. 9); while it was approximately 20 mm/hr on the upper part of the slope and about 7 mm/hr in the lower part. The bedrock on the upper part of the slope was detected to be shallower, where horizontal permeable layers connected to recharge zones were present. The tensiometer installed at the depth of 150 cm in this area registered a quick increase in pore water pressure about 4 h after the start of the rainfall. Similarly, the groundwater table started to develop over the bedrock approximately at the same time, and rose to a height of 1.6 m above bedrock (Fig. 9). This sudden increase can be an indication of the start of exfiltration of water from the permeable layer in the bedrock, as shown in Figs. 2 and 4.

The deformation sensors showed visible movements in the upper right quadrant of the slope, approximately 13 h after the start of the rainfall. A tension crack opened up with

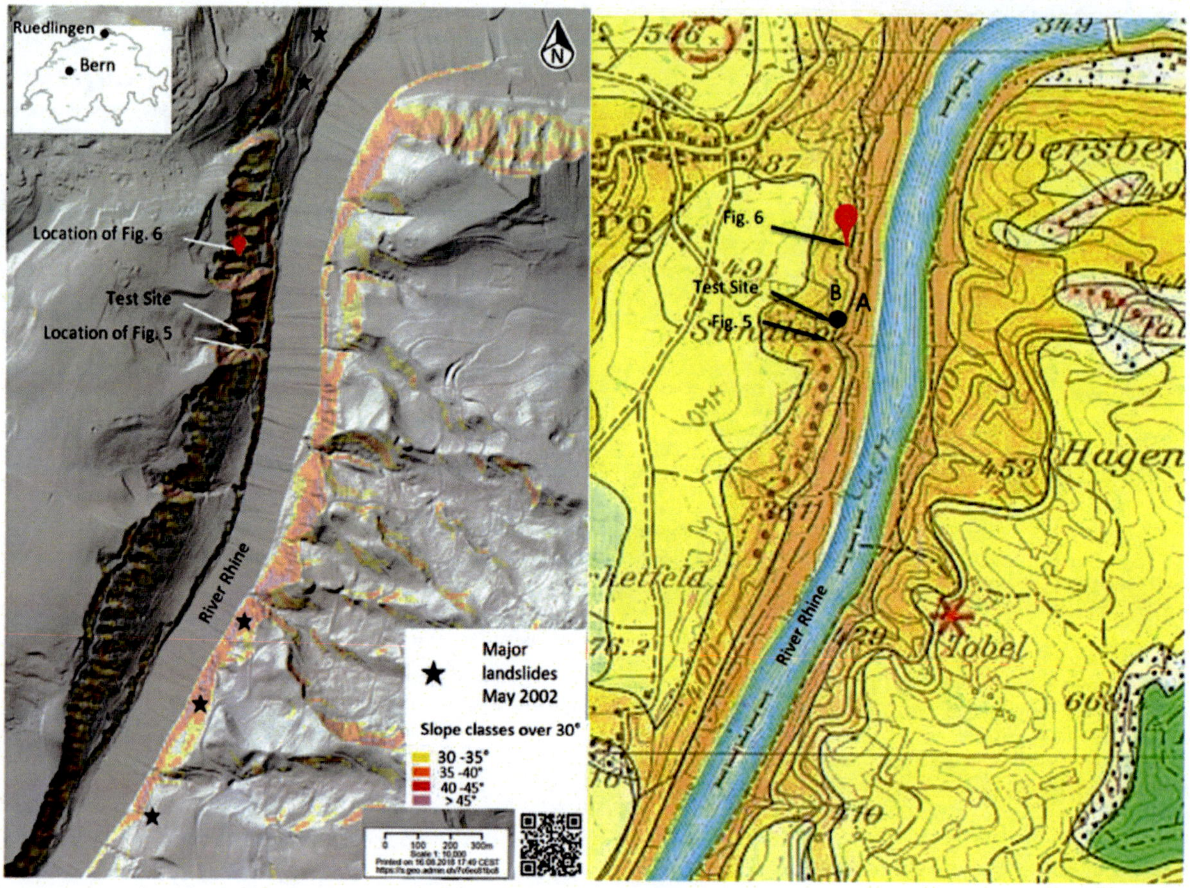

Fig. 3 Left: Relief map of the area in the vicinity of Ruedlingen village, data from www.sf.ch. Right: Geological map of the Ruedlingen area (after Hantke 1967), geological layers A: Lower freshwater Molasse, and B: Upper seawater Molasse, The Swiss Molasse basin is recognized as the 'detritus washed down from the rising Alps and deposited in a peripheral depression related to the Tertiary orogeny (Dickinson 1974)

increasing width in the upper right part and spread towards the left-hand side of the field (Fig. 10). A sliding mass was formed and travelled in a stepwise manner with increasing velocity. The mobilised debris of approximately 130 m^3 moved downslope and was retained by the protection barrier at the bottom of the slope (Fig. 11b). Exfiltration of water from the bedrock at two horizons at the back of the failure scar was detected from the high-resolution photographs (Fig. 11a). The location of these exfiltration zones corresponds well to that of the tensiometer and piezometer, which measured sudden increases in the values of pore water pressure and water table about 4 h after the start of the rainfall (Askarinejad et al. 2018).

It is hypothesised that the occurrence of the water exfiltration from the bedrock might have been one of the main destabilising effects for this slope. Therefore, a coupled hydro-mechanical simulation has been conducted to test the behaviour of the slope for two scenarios of rainfall with, and without, exfiltration from the bedrock. The next section describes the details and results of these simulations.

Coupled Hydro-Mechanical Modelling

Coupled hydro-mechanical simulations were performed using the finite element software "Code Bright" (Olivella et al. 1996). The constitutive modelling of the hydro-mechanical responses of unsaturated soils is performed using the Barcelona Basic Model (Alonso et al. 1990). Moreover, the time dependency of the stresses and strains of the material, i.e. viscosity, is also taken into account for the plastic strains based on the formulation suggested by Perzyna (1966).

A nonlinear formulation, based on the Barcelona Basic Model, is used to define the material behaviour in the elastic region. For the visco-plastic part of the behaviour, the modified yield function, based on the method presented by

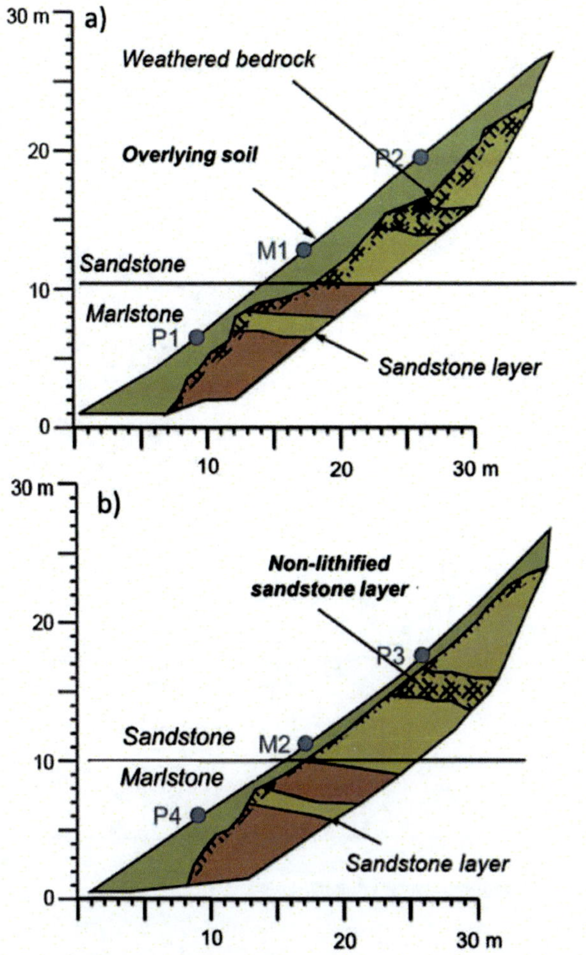

Fig. 4 Geological interpretation of: **a** the left (along the line P_1-M_1-P_2) and **b** the right (along the line P_3-M_2-P_4) hand side of the test site (after Brönnimann 2011)

Desai and Zhang (1987), for unsaturated soils as a function of mean stress (p), deviatoric stress (q) and matric suction (s). The model parameters were defined by comparing the function proposed by Desai and Zhang (1987) with the yield function defined for the Barcelona Basic Model. The details of the equations and parameter selection, as well as the model calibration are discussed by Askarinejad (2013).

Two simulations were carried out to investigate the effect of exfiltration from the bedrock on the responses of the overlying soil layer. The slope is only subjected to a rainfall scenario similar to that of the Ruedlingen experiment in the first model (M_1). Whereas, water exfiltration from the bedrock at two horizontal layers are simulated in the second Model (M_2), in addition to the imposed rainfall. The locations of the water exfiltration from the bedrock are identified from the images after the Ruedlingen landslide (Fig. 11).

Fig. 5 Fissures in the bedrock (photo by Askarinejad)

Geometry of the Models

The geometry of the 2D numerical models, the location of the fissures in the bedrock, and hydraulic boundary conditions are shown in Fig. 12. The lower boundaries of the model are set as "seepage free faces", where the water can drain out from the model and the pressure head is kept constant. The boundary elements defining the bedrock are fixed in all directions.

Applied Rainfall and Exfiltration

Fifteen hours of rainfall, with constant but different intensities, is applied over the lower and upper parts of the slope. This rainfall scenario is a simplified version of the rainfall that was applied to the Ruedlingen slope. Water exfiltration from the porous layers in the bedrock at the upper part of the slope initiated 4 h after the start of the rainfall which is based on the reaction time of the piezometer and the tensiometer installed at 150 cm depth in the upper part of the slope in instrumentation cluster 3 (Fig. 9).

Comparison of the Pore Water Pressure with the Field Measurements

The results of simulation M_1 show that the pore water pressure increases sequentially from the surface downwards (Fig. 13), indicating a "top-down" regime of saturation. Changes in pore water pressure at depth of 30 cm in the upper part of the slope for both simulations with and

Fig. 6 Ice layers showing the location of the water exfiltration from the bedrock (January 2009 photo by Askarinejad)

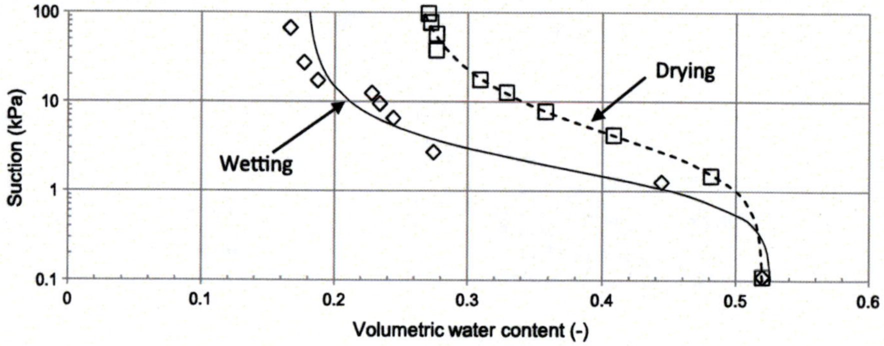

Fig. 7 Water Retention Curve obtained from natural undisturbed samples from the Ruedlingen test site, determined using the Axis Translation Technique (ATT)

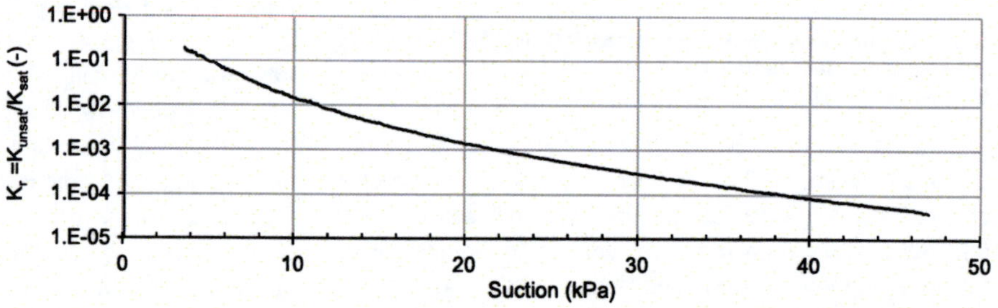

Fig. 8 Hydraulic conductivity function for reconstituted Ruedlingen soil, derived with the Instantaneous Profile Method

without exfiltration are shown in Fig. 14 and compared with field measurements. The results indicate a good agreement between the simulated and measured values with a slight difference which might have been caused by the inhomogeneity of the real soil in the field and existence of macro-pores in the soil structure that cause preferential water paths and also the differences between the Water Retention Curves under restricted flow conditions in the laboratory and those in the field (Askarinejad et al. 2010; Beck 2010).

Moreover, the results show that the exfiltration of water from the bedrock does not have a significant effect on the pore pressure regime at the depth of 30 cm. However, the differences between the models with, and without exfiltration are quite significant at the depth of 150 cm. This comparison confirms that exfiltration could have a major impact on the rapid increase of pore water pressure at deeper points in the soil layer. This response is investigated using advanced physical modelling in the centrifuge. The details of the centrifuge tests are discussed in the next section.

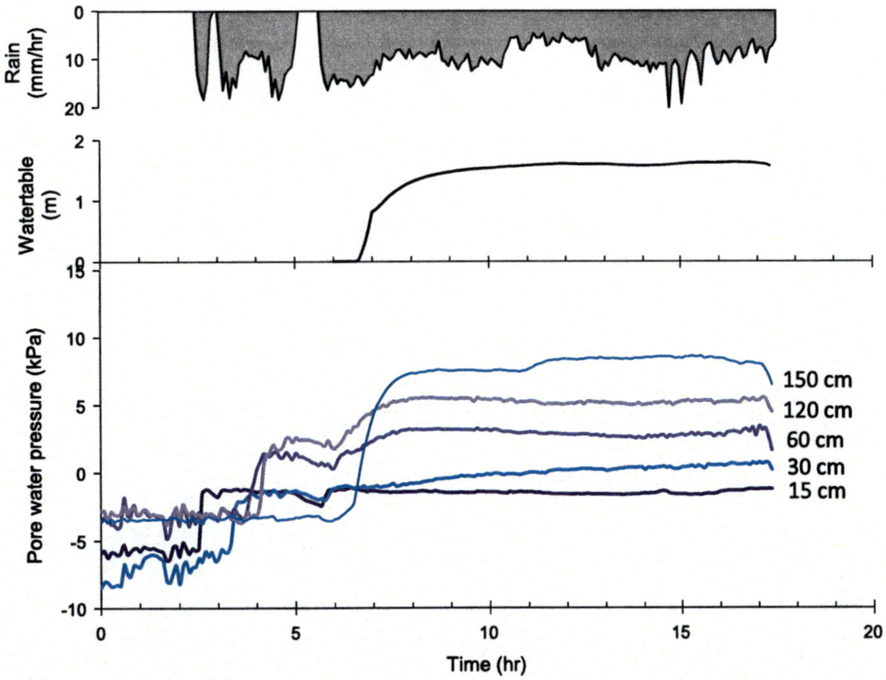

Fig. 9 Average intensity of applied rainfall, development of the groundwater level in the upper part of the slope measured in instrumentation cluster 3 and changes in the pore water pressure at the same place

Centrifuge Modelling

A model slope with geometry of 50 times smaller than the prototype is constructed in geotechnical centrifuge and tested under acceleration field with a magnitude of 50 times the Earth's gravity (50 g). The enhanced acceleration provides a gradient of body stresses within the model similar to the prototype, which ensures similar effective stresses and pore pressures at equivalent depths. A climate chamber and rain simulator was designed and constructed to apply various types of hydraulic perturbations to the model slopes. These hydraulic perturbations include rainfall, and exfiltration of water from the bedrock. The size of the droplets and the Coriolis effect (Schofield 1980; Take and Bolton 2002) have been taken into consideration when simulating the rainfall scenarios. The scaling laws regarding the infiltration process into unsaturated soils are discussed in detail by Askarinejad et al. (2015).

An aluminium triangular frame, with an inclination of 38° provided a rigid underlying layer for the soil slope in the centrifuge rain simulator. The top surface of this "bedrock" was roughened by gluing a thin layer of soil particles onto it. A shallow notch was grooved laterally on the top surface to install filter stones below which there are connections to 4 water pipes. Water could be pumped through these pipes into the slope to simulate the exfiltration of water from the bedrock (Fig. 15). Pore pressure transducers were installed in 4 different places at the interface of the bedrock and the soil mass.

The surface displacements were measured using *inclined Particle Image Velocimetry* (*i*PIV) analysis of the images, captured by photogrammetry cameras at 1 Hz frequency (Askarinejad 2013; Askarinejad et al. 2012, 2014). Druck PDCR81 pore pressure transducers (PPT) were used in these series of tests to measure the pore water pressure at the interface of the soil and bedrock. They were calibrated to monitor both positive and negative values of pore pressure. The PPTs were saturated with a solution of 50% Glycerine and 50% of water to avoid fast desaturation of the sensors (Muraleetharan and Granger 1999).

Two centrifuge tests were performed to study the hydro-mechanical interaction of the soil and bedrock and its effects on the triggering mechanism of landslides due to rainfall (with test name: T_1) and exfiltration (with test name: T_2). The models were prepared using moist tamping to achieve the initial void ratio and water content of e = 1.0 and w = 15%, respectively. These values are similar to the in-situ soil conditions in the Ruedlingen slope. The general setup of the tests is shown in Fig. 15.

The rain intensity in T_1 was 1.27 mm/hr in prototype scale, which is much smaller than that of the Ruedlingen landslide triggering experiment. This choice of lower rain intensity was due to provisions for surface erosion. The variations in the pore water pressure measured by the transducer located close to the toe of the slope and surface

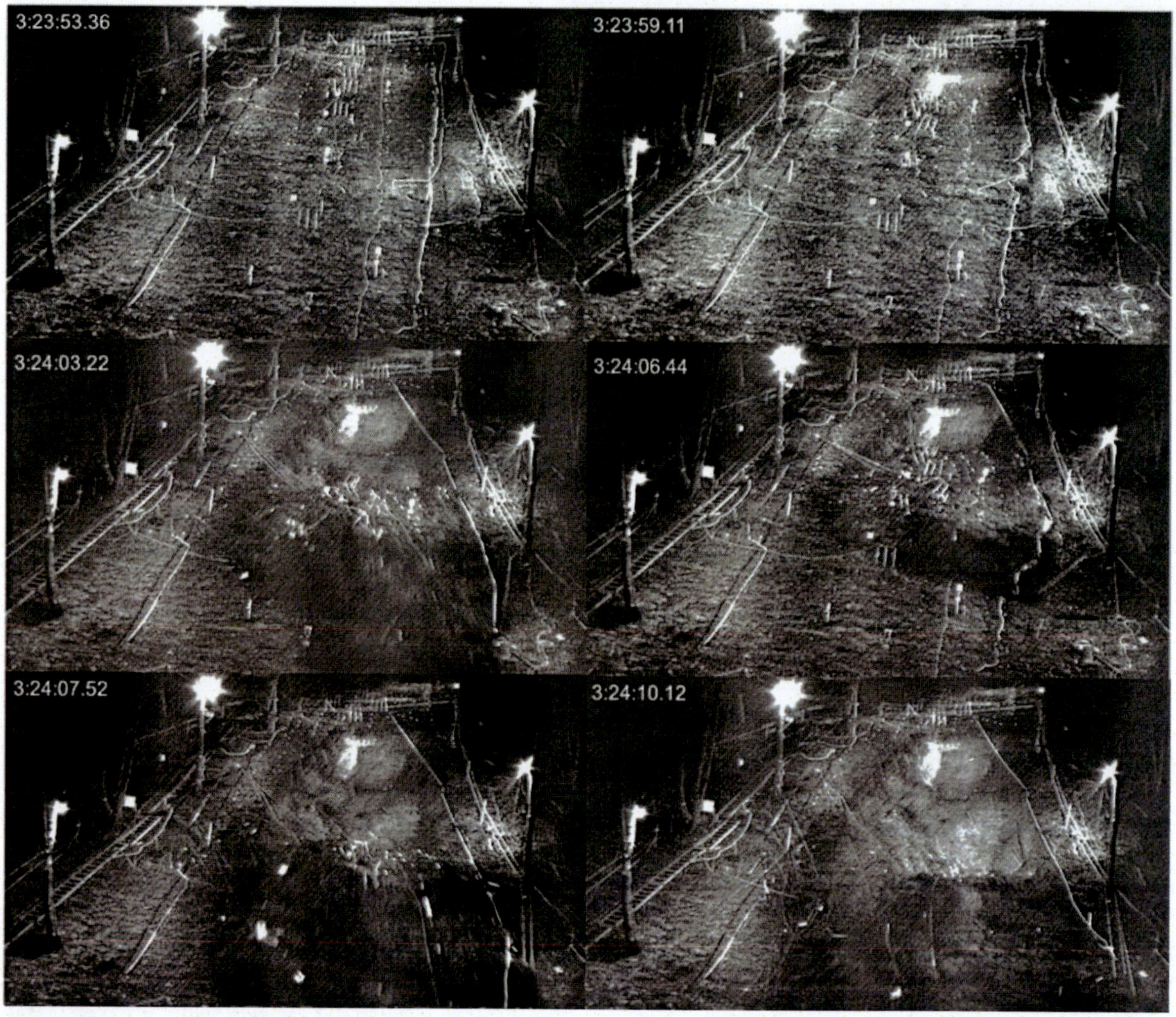

Fig. 10 Sequence of movements of the destabilised soil mass due to artificial rainfall in the Ruedlingen landslide triggering experiment (IGT-ETHZ)

movements at the same location are shown in Fig. 16. The location of the PPT with respect to the surface of the slope is shown in Fig. 17. Surface movements were detected about 30 h (prototype scale) after the start of the rainfall and visible tension cracks developed on the crest of the slope at time 100 h (Fig. 17). The surface movement continued to increase at a constant rate until the next 130 h, when they accelerated about 232 h after the start of the rainfall. The pore water pressure increased at a fairly constant rate, before 150 h. A local compression failure occurred at the toe of the slope, at the position of $iPIV_2$ (Fig. 17) at time 254 h, which corresponds to the time of the sudden increase in the pore pressure. The pore pressure attained a peak of 7.3 kPa at 280 h when a locally fluidised landslide occurred on the left-hand side of the slope (viewed from the crest).

Water was injected through the bedrock to the back of the slope in the model of T_2. The pressure of the water below the porous stone in the groove of the bedrock (Fig. 15) was measured to have an average value of 109 kPa.

The discharge was measured to be approximately 1 m^3/h/m (prototype scale). The pressure drop over the porous stone was calculated to be 0.6 kPa, which is negligible compared to the average exfiltration pressure (109 kPa). Accordingly, the tests replicated the effect of exfiltration of water from interconnected saturated fissures in the bedrock with approximately 11 m of artesian pressure. The slopes failed locally in the vicinity of the exfiltration area with no measurable effect on the upper part of the slope. Sequence of failure is shown in Fig. 18.

The changes in pore pressure in the lower part of the slope (measured by the pore pressure shown in Fig. 15)

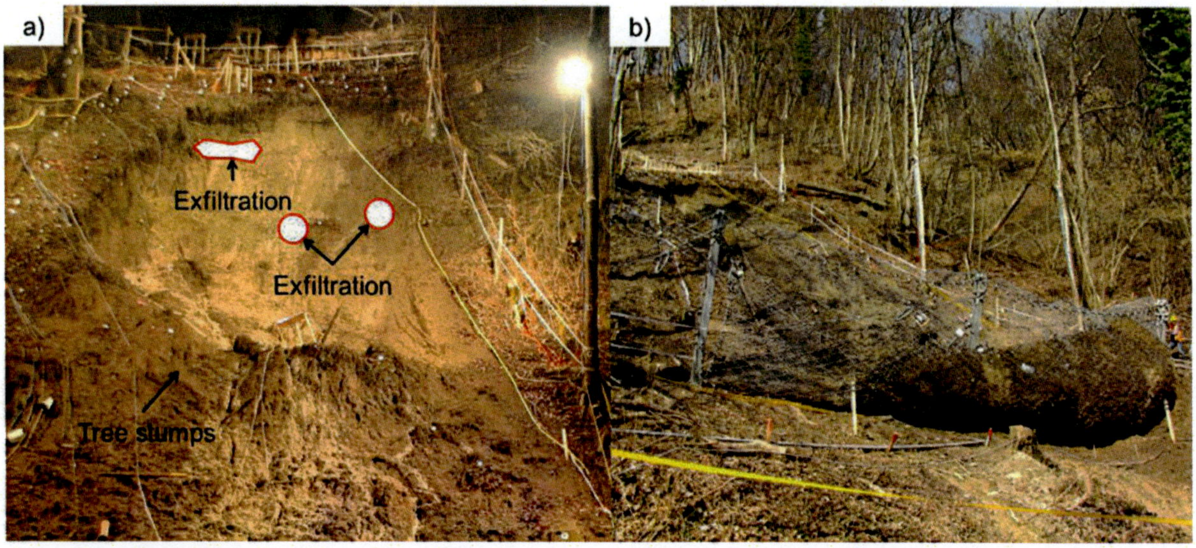

Fig. 11 Post landslide records of observations, **a** locations of the exfiltration from the bedrock after the landslide, **b** protection barrier and retained debris (Photo by Askarinejad)

Fig. 12 Geometry and hydraulic boundary conditions for the 2D numerical model of the Ruedlingen slope

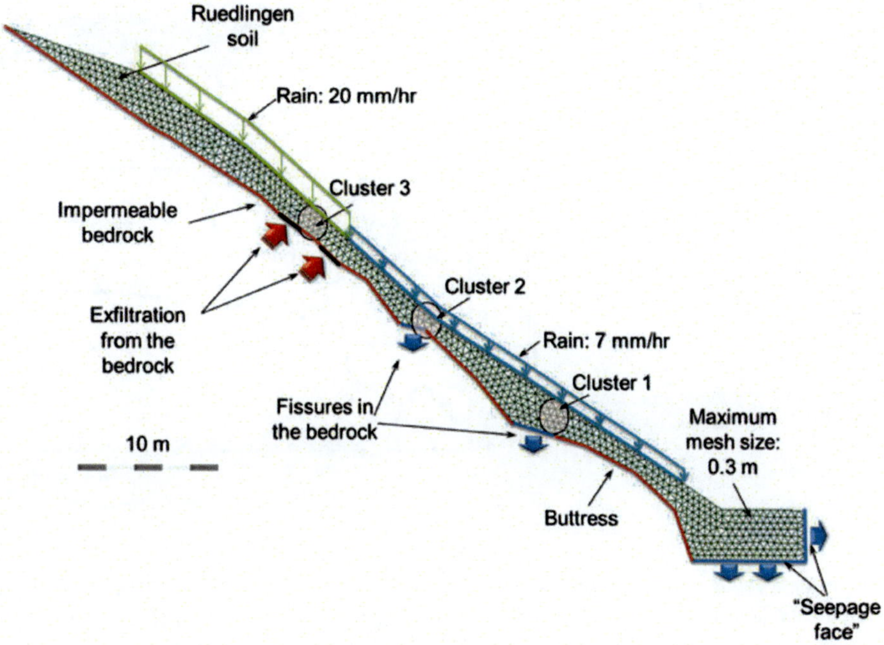

close to the location of the exfiltration are shown in Fig. 19. The pore water pressure increased marginally 0.7 h after the start of exfiltration. The peak pressure in this model occurred 7 h after the start of exfiltration and reached a value of 2 kPa. The results show that the value of the measured pore pressure is higher in tests with rainfall infiltration, but the time to failure is much longer compared to the exfiltration mechanism. The surface displacements were measured using PIV analysis, however, the measurements were only reliable until about 4.2 h after the start of exfiltration.

Summary and Conclusions

The effects of exfiltration on the stability of unsaturated silty sand slopes and the mechanisms leading to the initiation and propagation of the shear deformations, resulting in

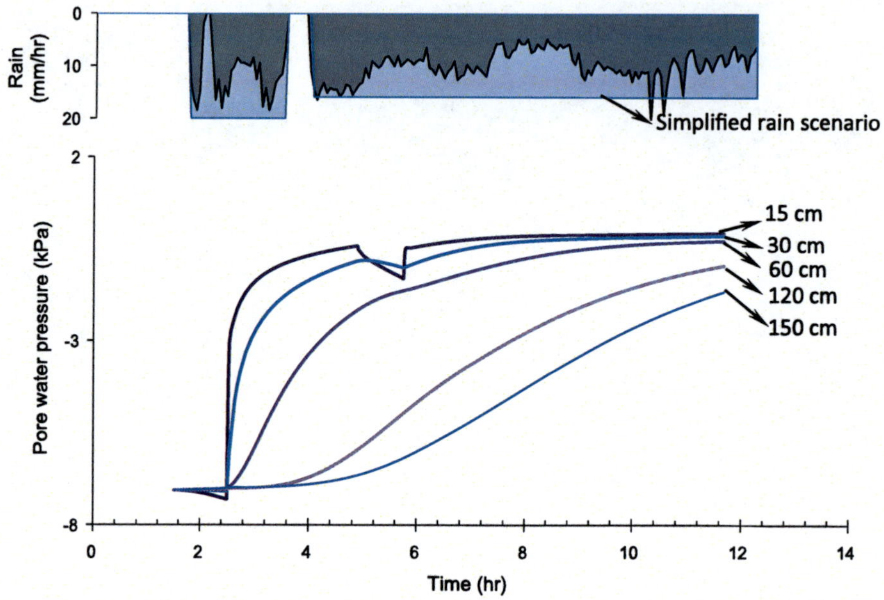

Fig. 13 Pore water pressure development due to rainfall at various depths in the soil profile in the upper part of the slope (simulation without exfiltration)

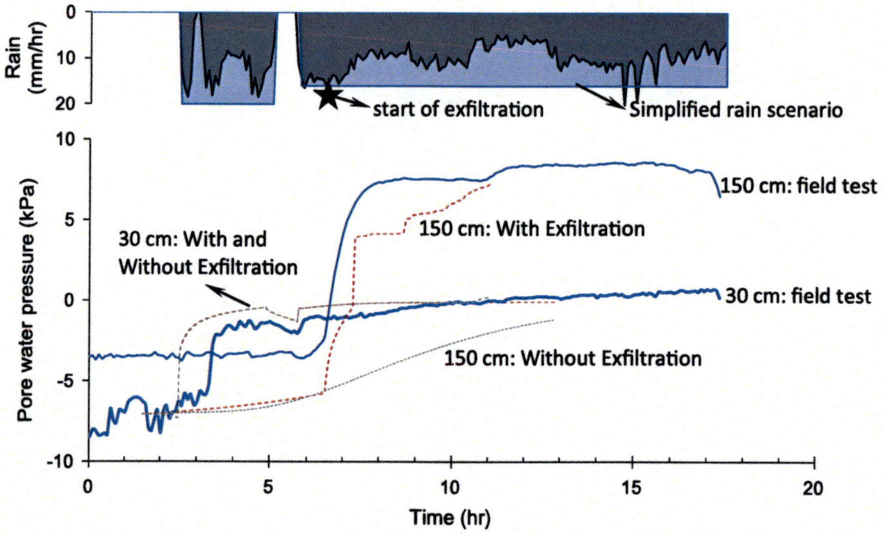

Fig. 14 Pore water pressure at depths of 30–150 cm in the upper part of the slope, based on the field measurements, and numerical simulations, with and without water exfiltration from the bedrock

subsequent rapid mass movements were studied. The measurements and observations from the full-scale landslide triggering experiment illustrated close interaction between the pore pressure development and rate of movements. The sharp rise in the groundwater table measured with piezometers and post failure images indicate exfiltration of water from the bedrock at the back of the failure surface.

The results of 2D coupled numerical simulations suggest the importance of the hydro-geological interactions of the bedrock and the soil mantel. The results revealed that the shallower depth of the bedrock and the presence of a permeable layer in the upper slope might have been factors involved in triggering the landslide in Ruedlingen.

The results of the 2D coupled hydro-mechanical simulations confirm that the evolution of the pore water pressure in the upper slope was strongly affected by the exfiltration of water from the bedrock. However, it should be noted that the effect of the root reinforcement and its corresponding distribution has not been implemented in these simulations and this would contributed largely only root reinforcement over this time frame (Yildiz et al. 2015; Świtała et al. 2017; Sitarenios et al. 2019).

A series of centrifuge tests were conducted to test the effect of the water exfiltration from the bedrock in triggering the landslide. Two similar slopes with bedrock parallel to the surface, one subjected to rainfall and the other one with exfiltration from the bedrock were tested in

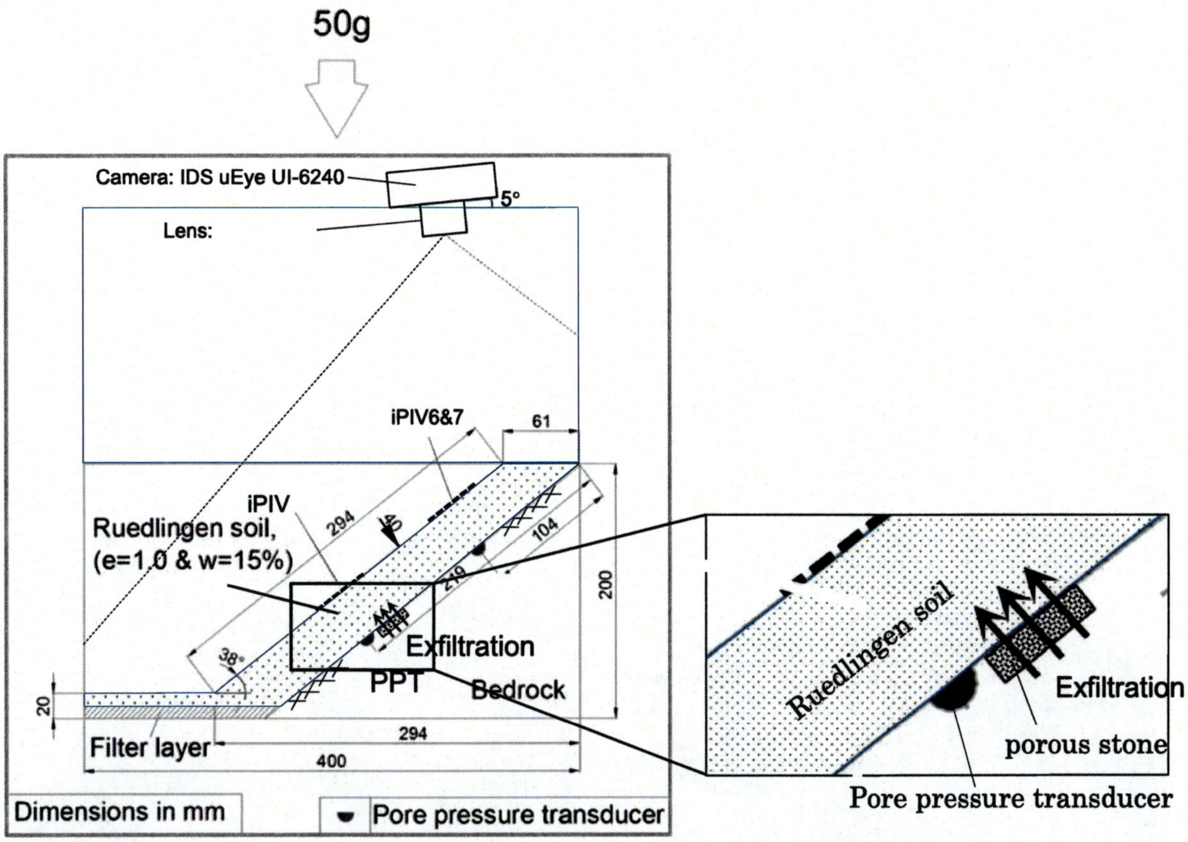

Fig. 15 Cross sectional diagram of centrifuge models, showing the location of the buttress on the bedrock and exfiltration zone

Fig. 16 Pore water pressure above the bedrock, and surface displacement at the same locations in test T_1

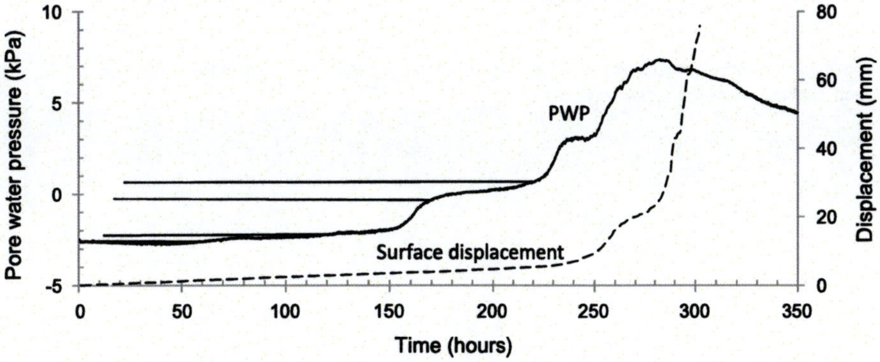

the geotechnical centrifuge. Tension cracks opened up in both sets of these tests and were less prominent in the manifestation of failure through exfiltration. This feature is not possible to represent numerically, however this was less relevant for deep seated changes in pore pressure and downslope soil deformation. The results show that the shape of the failed mass and hence the values of the triggering pore pressure in both models were different. However, the time required to trigger a landslide was much shorter for the slope subjected to water exfiltration from the bedrock (~4.5 h), compared to the model subjected to only rainfall infiltration (~300 h).

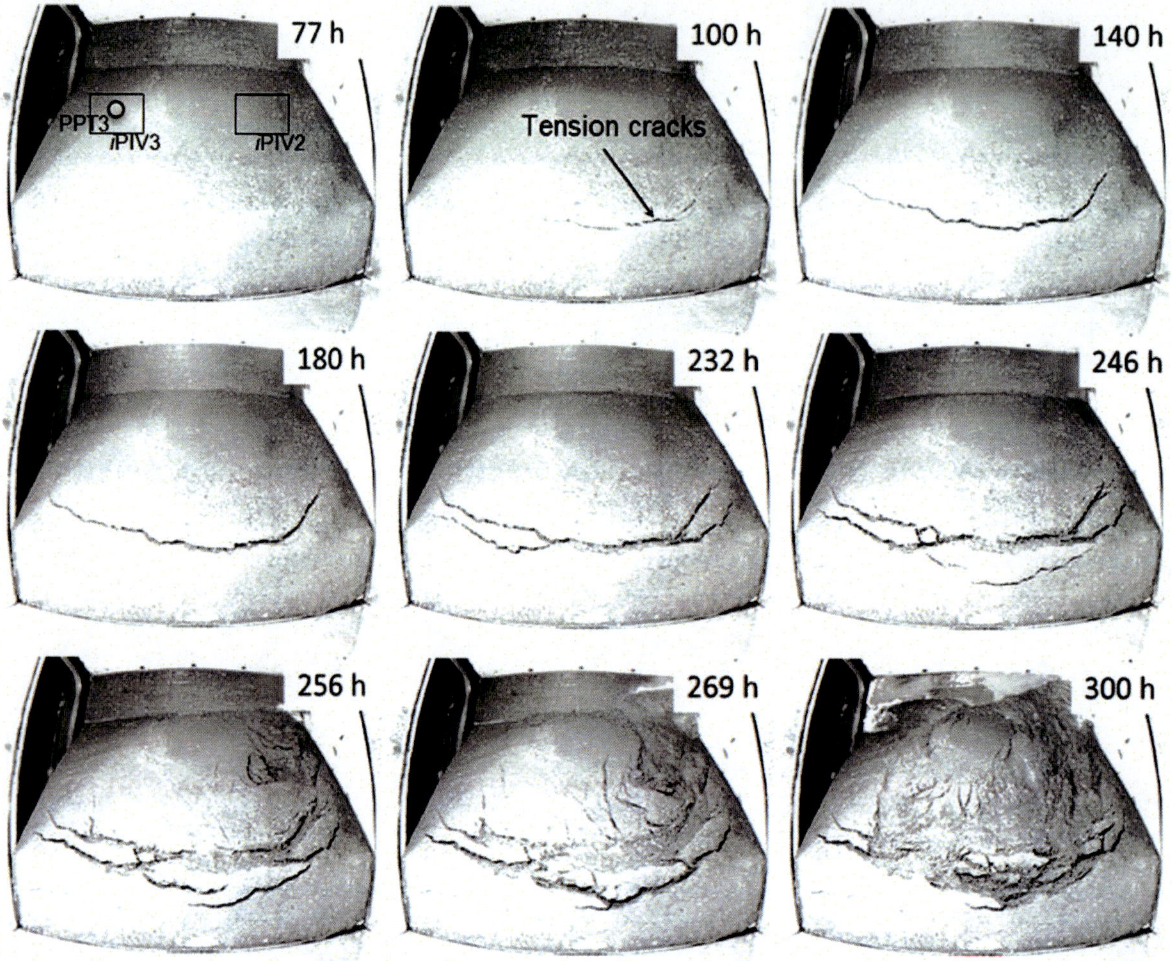

Fig. 17 Sequence of events on the Ruedlingen model slope without buttress in the bedrock in test T_1, with the crest of the slope at the bottom of each picture. Times are at prototype scale

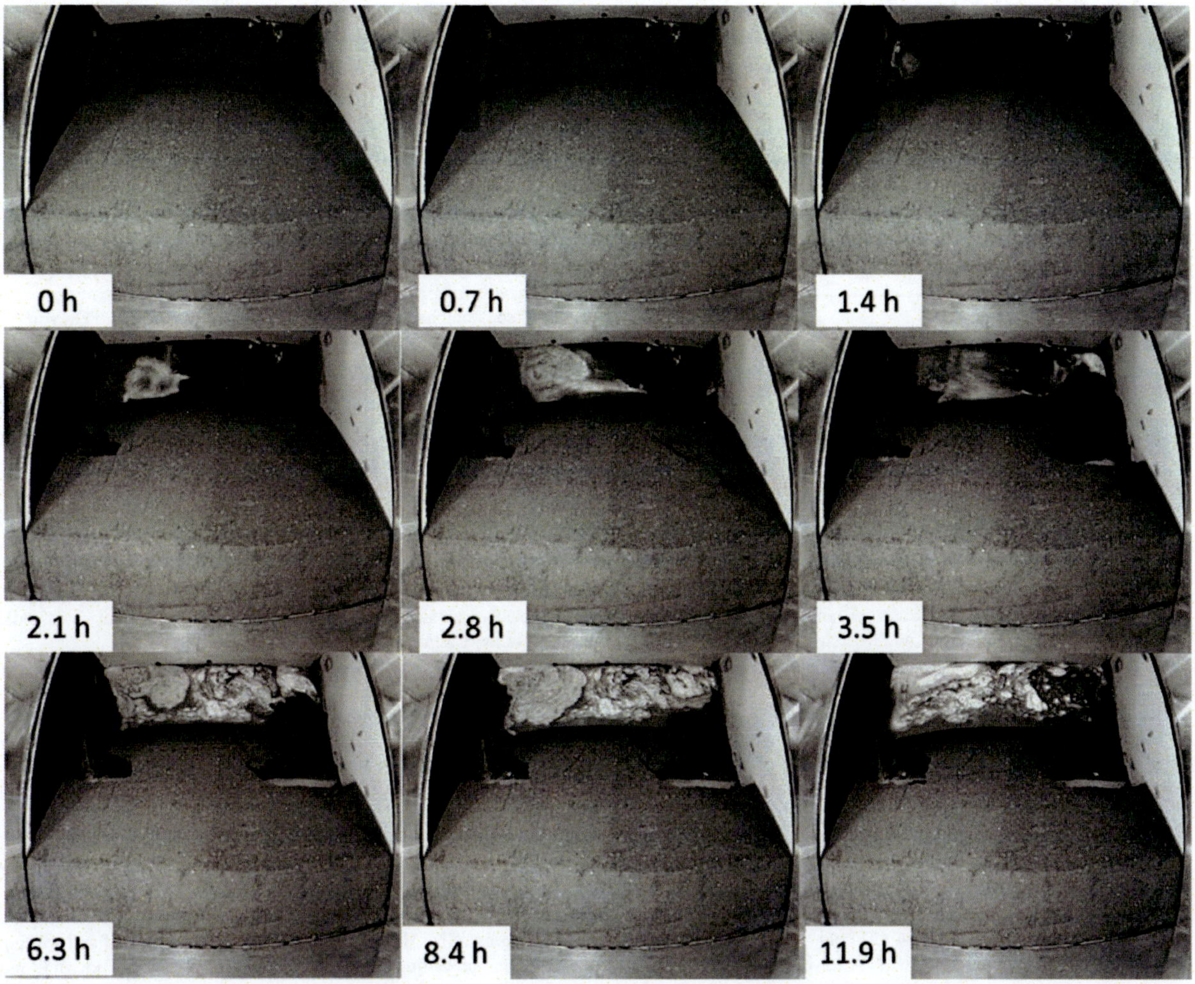

Fig. 18 Sequence of events on the Ruedlingen model slope with exfiltration from the bedrock in test T_2. Times are at prototype scale

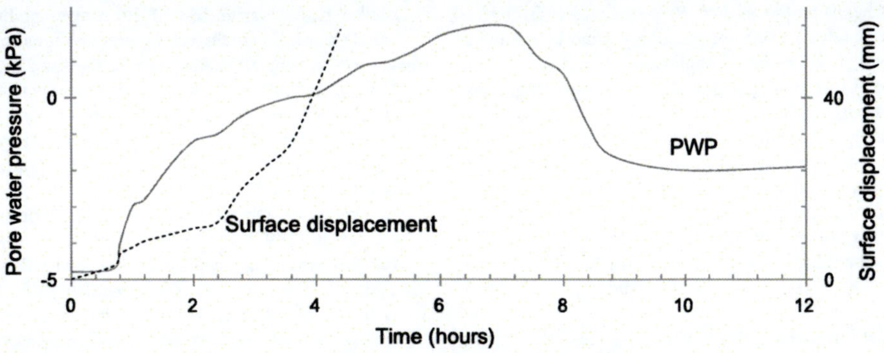

Fig. 19 Pore water pressure at the interface of the soil and bedrock (measured by the pore pressure shown in Fig. 15) and corresponding surface displacement for the centrifuge model with water exfiltration from the bedrock at time T = 0 h for the test T_2

Acknowledgements This research was partially funded by the Competence Centre for Environment and Sustainability (CCES) within the framework of the TRAMM–Project. Additional resources were provided by the ETH Research Fund and EU project SafeLand (EU FP7 grant agreement no. 226479). The authors are very grateful to the Ruedlingen Council, and communities of Ruedlingen and Buchberg. Ms. Cornelia Malecki, Dr. Francesca Casini, Dr. Peter Kienzler, Dr. Jan Laue, Dr. Cornelia Brönnimann, Ernst Bleiker, Markus Iten, Heinz Buschor, and Felix Wietlisbach are thanked for their valuable contributions to this project.

References

Alonso EE, Gens A, Josa A (1990) A constitutive model for partially saturated soils. Géotechnique 40:405–430

Askarinejad A (2013) Failure mechanisms in unsaturated silty sand slopes triggered by rainfall. DSc thesis, ETH Zurich

Askarinejad A, Akca D, Springman SM (2018) Precursors of instability in a natural slope due to rainfall: a full-scale experiment. Landslides 15:1745–1759

Askarinejad A, Beck A, Springman SM (2015) Scaling law of static liquefaction mechanism in geo-centrifuge and corresponding hydro-mechanical characterisation of an unsaturated silty sand having a viscous pore fluid. Can Geotech J 52:1–13

Askarinejad A, Casini F, Kienzler P, Springman SM (2010) Comparison of the in situ and laboratory water retention curves for a silty sand. 5th International conference of unsaturated soils (UNSAT). Barcelona, Spain, pp 423–428

Askarinejad A, Laue J, Springman SM (2014) Effect of bedrock shape and drainage properties on the stability of slopes. International conference of physical modelling in geotechnics, Perth, Australia, pp 1211–1217

Askarinejad A, Laue J, Zweidler A, Iten M, Bleiker E, Buschor H, Springman SM (2012) Physical modelling of rainfall induced landslides under controlled climatic conditions. Eurofuge 2012. Delft, Netherlands, Published on CD only

Askarinejad A, Springman SM (2017) A novel technique to monitor subsurface movements of landslides. Can Geotech J 55:620–630

Averjanov SF (1950) About permeability of subsurface soils in case of incomplete saturation. Engl Collect 7:19–21

Beck A (2010) Measurement of unsaturated hydraulic conductivity with the instantaneous profile method. Semester project ETH.

Bogaard TA, Greco R (2015) Landslide hydrology: from hydrology to pore pressure. Water, Wiley Interdisciplinary Reviews

Brönnimann C (2011) Effects of groundwater on landslide triggering, Nr. 5236. PhD thesis, EPFL

Brönnimann C, Stähli M, Schneider P, Seward L, Springman SM (2013) Bedrock exfiltration as a triggering mechanism for shallow landslides. In press, Water Resources Research

Brönnimann C, Tacher L, Askarinejad A, Kienzler P, Springman SM (2009) Pore water pressure modelling in a rainfall triggered shallow landslide: the sprinkling experiment in Ruedlingen, Canton Schaffhausen. 7th Swiss Geoscience Meeting. Neuchâtel, Switzerland.

Cascini L (2004) The flowslides of May 1998 in the Campania region, Italy: the scientific emergency management. Italian Geotech J 2:11–44

Cascini L, Cuomo S, Pastor M, Sorbino G (2010) Modeling of rainfall-induced shallow landslides of the flow-type. J Geotech Geoenviron Eng ASCE 136:85–98

Casini F, Serri V, Springman SM (2013) Hydromechanical behaviour of a silty sand from a steep slope triggered by artificial rainfall: from unsaturated to saturated conditions. Can Geotech J 50:28–40

Daniel DE (1982) Measurement of hydraulic conductivity of unsaturated soils with thermocouple psychrometers. J Soil Sci Soc America 20:1125–1129

Desai CS, Zhang D (1987) Viscoplastic model for geologic materials with generalized flow rule. Int J Numer Anal Meth Geomech 11:603–620

Dickinson WR (1974) Plate tectonic and sedimentation. Tecton Sediment Soc Petrol Eng Spec Publ 22:1–27

DIN (2003) Subsoil-Field testing-Part 2: borehole dynamic probing DIN 4094–2. Deutsches Institut fuer Normung.

Ebel BA, Loague K, Vanderkwaak JE, Dietrich WE, Montgomery DR, Torres R, Anderson SP (2007) Near-surface hydrologic response for a steep, unchanneled catchment near Coos Bay, Oregon: 2 Physics-Based Simulations. Am J Sci 307:709–748

Elia G, Cotecchia F, Pedone G, Vaunat J, Vardon PJ, Pereira C, Springman SM, Rouainia M, van Esch J, Koda E, Josifovski J, Nocilla A, Askarinejad A, Stirling R, Helm P, Lollino P, Osinski P (2017) Numerical modelling of slope–vegetation–atmosphere interaction: an overview. Q J Eng GeolHydrogeol 50:249–270

Everett A (1979) Secondary permeability as a possible factor in the origin of debris avalanches associated with heavy rainfall. J Hydrol 43:347–354

Fiebelkorn B (1983) A study of possible failure mechanisms of a landslide near Wildcat Canyon Park, Richmond. Masters report, Geotechnical Engineering Department, University of California, Berkeley, California

Hantke R (1967) Geologische Karte des Kantons Zürich und seiner Nachbargebiete. Kommissionsverlag Leemann, Zürich

Hilf JW (1956) An investigation of pore-water pressure in compacted cohesive soils. Ph.D, United State Department of the Interior Bureau of Reclamation, Design and Construction Division

Johnson KA, Sitar N (1990) Hydrologic conditions leading to debris-flow initiation. Can Geotech J 27:789–801

Lanni C, McDonnell J, Hopp L, Rigon R (2012) Simulated effect of soil depth and bedrock topography on near-surface hydrologic response and slope stability. Earth Surf Proc Land 38:145–159

Lehmann P, Gambazzi F, Suski B, Baron L, Askarinejad A, Springman SM, Holliger K, Or D (2013) Evolution of soil wetting patterns preceding a hydrologically induced landslide inferred from electrical resistivity survey and point measurements of volumetric water content and pore water pressure. Water Resour Res 49:7992–8004

Lucas D, Fankhauser K, Maurer H, McArdell B, Grob R, Herzog R, Bleiker E, Springman SM (2020) Slope stability of a scree slope based on integrated characterisation and monitoring. Water 12:447

Masaoka N, Kosugi KI, Yamakawa Y, Tsutsumi D (2016) Processes of bedrock groundwater seepage and their effects on soil water fluxes in a foot slope area. J Hydrol 535:160–172

Mcdonnell JJ (2003) Where does water go when it rains? Moving beyond the variable source area concept of rainfall–runoff response. Hydrol Process 17:1869–1875

Montgomery DR, Dietrich WE, Torres R, Anderson SP, Heffner JT, Loague K (1997) Hydrologic response of a steep, unchanneled valley to natural and applied rainfall. Water Resour Res 33:91–109

Montgomery DR, Schmidt KM, Dietrich WE, McKean J (2009) Instrumental record of debris flow initiation during natural rainfall: implications for modeling slope stability. J Geophys Res Earth Surf 2003–2012:114

Muraleetharan KK, Granger KK (1999) The use of miniature pore pressure transducers in measuring matric suction in unsaturated soils. Geotech Test J 22:226–234

Olivella S, Gens A, Carrera J, Alonso EE (1996) Numerical formulation for simulator (CODE_BRIGHT) for coupled analysis in saline media. Eng Comput 13:87–112

Onda Y, Tsujimura M, Tabuchi H (2004) The role of subsurface water flow paths on hillslope hydrological processes, landslides and landform development in steep mountains of Japan. Hydrol Process 18:637–650

Perzyna P (1966) Fundamental problems in viscoplasticity. Adv Appl Mech 9:244–368

Platt NH (1992) Fresh-water carbonates from the lower freshwater Molasse (Oligocene, western Switzerland): sedimentology and stable isotopes. Sed Geol 78:81–99

Rickli C, Raetzo H, Mcardell BW, Presler J (2008) Hanginstabilitaeten, in Ereignisanalyse Hochwasser 2005 Teil 2-Analyse von Prozessen, Massnahmen und Gefahrengrundlagen. In: Bezzola GR, Hegg C (eds). Bundesamt f€ur Umwelt BAFU, Bern und der Eidgenössischen Forschungsanstalt f€ur Wald, Schnee und Landschaft WSL, Birmensdorf, Switzerland.

Schofield AN (1980) Cambridge geotechnical centrifuge operations. Géotechnique 30:227–268

Sinclair H (1997) Flysch to molasse transition in peripheral foreland basins: the role of the passive margin versus slab breakoff. Geology 25:1123–1126

Sitarenios P, Casini F, Askarinejad A, Springman S (2019) Hydro-mechanical analysis of a surficial landslide triggered by artificial rainfall: the Ruedlingen field experiment. Géotechnique 0:1–14. https://doi.org/10.1680/jgeot.18.P.188

Świtała BM, Askarinejad A, Wu W, Springman SM (2018) Experimental validation of a coupled hydro-mechanical model for vegetated soil. Géotechnique 68(5):375–385.

Tacher L, Locher D (2008) Internal TRAMM report: geological characterization of the Buchberg field site (Ruedlingen). EPFL.

Take WA., Bolton MD (2002) An atmospheric chamber for the investigation of the effect of seasonal moisture changes on clay slopes. In: Phillips R, Guo PJ, Popescu P (eds) Proceedings of the international conference on physical modelling in geotechnics—ICPMG_02. St. Johns, Newfoundland, Canada. 765–770.

Take WA, Bolton MD, Wong PCP, Yeung FJ (2004) Evaluation of landslide triggering mechanisms in model fill slopes. Landslides 1:173–184

Tang A-M, Hughes PN, Dijkstra TA, Askarinejad A, Brenčič M, Cui YJ, Diez JJ, Firgi Ti, Gajewska B, Gentile F, Grossi G, Jommi C, Kehagia Fotini, Koda, Eugeniusz, TerMaat HW, Lenart S, Lourenço S, Oliveira M, Osiński P, Springman SM, Stirling R, Toll DG, Van Beek V (2018) Atmosphere–vegetation–soil interactions impacts on engineered slopes: a review on recent advances. Q J Eng Geol Hydrogeol 51:156–168

Uchida T, Kosugi KI, Mizuyama T (2002) Effects of pipe flow and bedrock groundwater on runoff generation in a steep headwater catchment in Ashiu, central Japan. Water Resour Res 38:24–1–24–14

Yildiz A, Askarinejad A, Graf F, Rickli C, Springman SM (2015) Effect of roots and mycorrhizal fungi on the stability of slopes. XVI European Conference on soil mechanics and geotechnical engineering. Edinburgh, UK, pp 1693–1698. https://doi.org/10.1680/ecsmge.60678

Controls on Landslide Size: Insights from Field Survey Data

Haijun Qiu, Yifei Cui, Dongdong Yang, Sheng Hu, Yanqian Pei, and Shuyue Ma

Abstract

Proper characterization of landslide size distribution is very important for estimating the landslide risk, quantifying the integrated effects of erosion and sediment yield, and determining the magnitude of landslide events. This paper quantitatively addresses the effects of topographic condition and landslide types on landslide size distributions. A detailed landslide inventory based on field work is developed. The landslide spatial distributions are then clustered. There exist some clustered centres in the study area. Statistical analysis showed that the large and very large landslides play an important role in determining the total landslides area in current study area. The double pareto and inverse gamma functions can well fit with the landslide probability distribution and can quantitatively reveal the maximum probability and rollover effect. The probability density of landslide size empirically agrees well with a simple power law relationship above a certain size threshold. Below this size, there exists a rollover effect. The local topographic conditions and landslide type play an important role in landslide size. About 33% of landslides occurred within a slope gradient between 30°–40°. The landslide size decreases with increasing slope gradient, and more frequent small landslides occur on larger slope gradients. About 80% of landslides occurred within a slope height less than 100 m. The landslide frequency decreased sharply with increasing slope height. The local slope height can limit the landslide size. The landslide size increases as slope height increases, and relationship of which can be well fitted by using a power law form. The peak of landslide probability density increases with increasing slope height. Moreover, the landslide size is also controlled by the landslide types and slope morphology. About 62% of the total landslides are retrogressive landslides. Most landslides concentrate on convex slopes. The lowest percentage of landslides occurred on a concave slope. This suggest that the convex slopes are preferentially susceptible to landslides. The probability densities of landslide size are influenced by the landslide types and slope morphology. Many thrust-type landslides were small landslides. More frequent small landslides occur on planar slope. The study here provides a reliable method for landslide hazard mapping.

Keywords

Landslides • Size distribution • Field survey • Topography

H. Qiu · D. Yang · S. Hu · Y. Pei · S. Ma
College of Urban and Environmental Sciences, Northwest University, Xi'an, 710127, China
e-mail: haijunqiu@nwu.edu.cn

D. Yang
e-mail: yangdongdong@stumail.nwu.edu.cn

S. Hu
e-mail: shenghu@nwu.edu.cn

Y. Pei
e-mail: 201620795@stumail.nwu.edu.cn

S. Ma
e-mail: msynwu@stumail.nwu.edu.cn

Y. Cui (✉)
State Key Laboratory of Hydroscience and Engineering, Tsinghua University, Beijing, 10084, China
e-mail: yifeicui@mail.tsinghua.edu.cn

Introduction

Landslides in many parts of the world is the result of various kinds of exogenic and endogenic factors, and often cause destructive impacts on casualties, economic and environmental damage (Crozier 2005; Nadim et al. 2006; Petley 2012; Chen and Cui 2017; Samia et al. 2017; Qiu et al. 2018a; Cui et al. 2019; Ma et al. 2019). Xi'an region in

Ž. Arbanas et al. (eds.), *Understanding and Reducing Landslide Disaster Risk*, ICL Contribution to Landslide Disaster Risk Reduction,
https://doi.org/10.1007/978-3-030-60713-5_11

Shaanxi Province, China has a high exposure to landslide disaster owing to its unique climatic, geologic and geomorphologic characteristics (Zhuang et al. 2015; Qiu et al. 2020a). Especially, the landslide impacts on human beings are continuously increasing duo to the rapid urbanization and economic pressures (Zhuang et al. 2015; Cui et al. 2019; Qiu et al. 2020a).

The size and size distribution of landslides play an important role in determining landslide risk, understanding the processes of landslide dynamics, and estimate erosion and sediment yield (Stark and Hovius 2001; Korup 2005a; Van Den Eeckhaut et al. 2007; Guthrie et al. 2008; Guzzetti et al. 2008; Qiu et al. 2018a). Many authors have researched the landslide size distributions and found that there exists an obvious rollover effect below a certain threshold (Guzzetti et al. 2002; Malamud et al. 2004). The landslide size distribution obeys a power law relationship at large landslide size range (Brunetti et al. 2009; Frattini and Crosta 2013). Some authors have found that the rollover effect of landslide size distribution influenced by material properties, soil moisture, topography (Pelletier et al. 1997; Stark and Hovius 2001; Malamud et al. 2004; Frattini and Crosta 2013). Although the causes of rollover effect have been researches by some authors, how the landslide types influence the landslide size distribution are not clearly understood.

Previous researches have proposed that the landslide size was influenced by material properties, geological and topographic conditions, failure event types, physiographic variables of the landscape (Stark and Hovius 2001; Guzzetti et al. 2008; Frattini and Crosta 2013; Gallen et al. 2015; Qiu et al. 2016; Cui et al. 2017; Guo and Cui 2020). Frattini and Crosta (2013) examined the effects of topography and material properties on the landslide size based on the virtual tiling method. Katz et al. (2014) found that landslide size increases with decreasing slope gradient for a given material strength. However, Qiu et al. (2018a) performed that slope gradient has a significantly control on landslide size based on field observations and simulations, and stated that landslide size increases with decreasing slope gradient. Thus, there exist some controversies regarding on the relationship between slope gradient and landslide size. Moreover, several authors have showed that the contribution of large landslides increase as local relief or slope height increase, and 75% of gaint landslides occur in regions with mean local relief >1000 m (Montgomery and Brandon 2002; Korup et al. 2007; Qiu et al. 2019a). Furthermore, Liu and Koyi (2013) noted that landslide size increases with increasing material strength. Although landslide size has been studied by many researchers, how the topography and landslide types influence the landslide size remain poorly understood. Therefore, it is necessary to address the effects of topography and landslide types on landslide size.

In the present work, the effects of topography and landslide types on the landslide size are examined. We developed a detained landslide inventory based on an intensive field survey, determined the relationship between landslide volume, area, length and width, fit the landslide size distribution by power law, double Pareto, and inverse gamma function, and quantify the response of landslide size and size distribution to slope gradient, slope height, slope morphology and landslide types.

Materials and Methods

Study Area

The study area lies in Shaanxi province of China (Fig. 1). It covers an area of about 10,752 km^2. The elevation in this study area ranges between 345 and 3767 m above sea level, with an average of 1027 m (standard deviation = 663 m) (Zhuang et al. 2015). The slope gradient varies from 00 to 810, with an average value of 160 and a standard deviation of 140. The study area is drained by the Wei River along with its major tributaries (Hu et al. 2017; He et al. 2019). Tectonically, this area is located at the junction of the Erdos, North China, Ganqing and South China blocks (Zhuang et al. 2015). The geological setting is very complex, and include a series of Quaternary thrust faults such as North Qin Mountain fault, Li Mountain fault, and Guanshan fault (Zhuang et al. 2015). This area is characterized by a typical temperate continental monsoon climate with an average monthly and annual precipitation of 53.7 and 585 mm. The maximum daily rainfall, monthly rainfall and annual rainfall are 72.3, 258.8 and 1131.7 mm (Zhuang et al. 2015; Qiu et al. 2020a). Most of the rainfall is concentrated in the rainy season (May to October), which accounts for 77% of the total annual rainfall (Fig. 2). Many landslides occur during this period. About 50% of the annual rainfall occur from July to September. Most of landslides occurred within the same three months, accounting for 77% of the total landslides that occurred in Xi'an (Fig. 3). This suggests that there exists a significant relationship between landslide frequency and mean monthly rainfall. The land use in this area is characterized by croplands, forests, orchards, urban areas and water (Chen et al. 2015).

The study area stretches from the northern Wei River Valley to southern Qinba Mountains. The climate in Wei River Valley is mostly semihumid (Qiu et al. 2020a). Most of this area covered by loess (Qiu et al. 2020a). This region is the major area of agriculture, industry and commerce in Xi'an (Lu et al. 2010). This region consists of six land use types: cropland, forest, grassland, settlement, water and other land according to the land use map provided by the department land and resources of Shaanxi Province (Fig. 4). Cropland (34%) covers the largest area, followed by forest (30%) and grassland (20%). This area is plain characterized

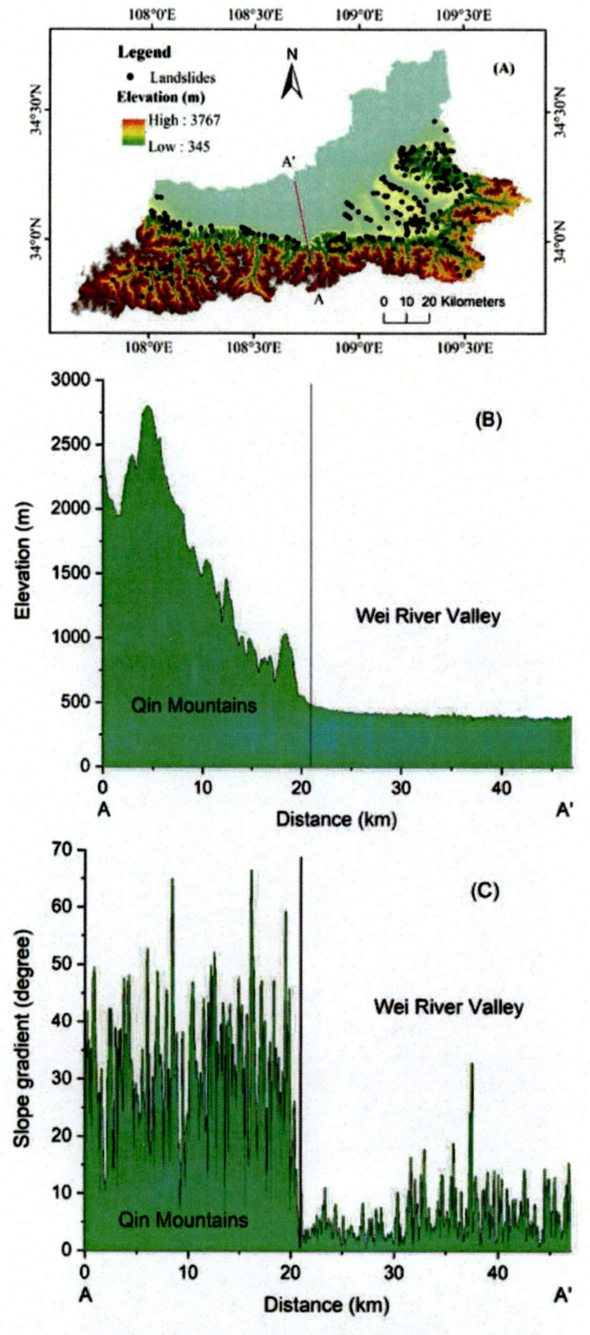

Fig. 1 Location of the study area with elevation (**a**) and landform cross-section (**b**). Black dots show the landslide locations

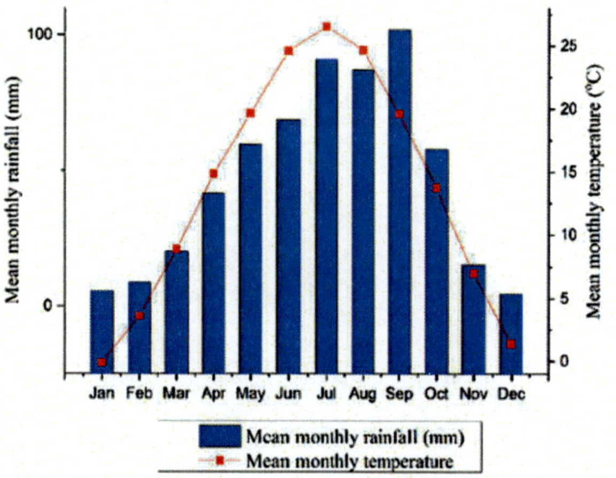

Fig. 2 Mean monthly rainfall and temperature in Xi'an region

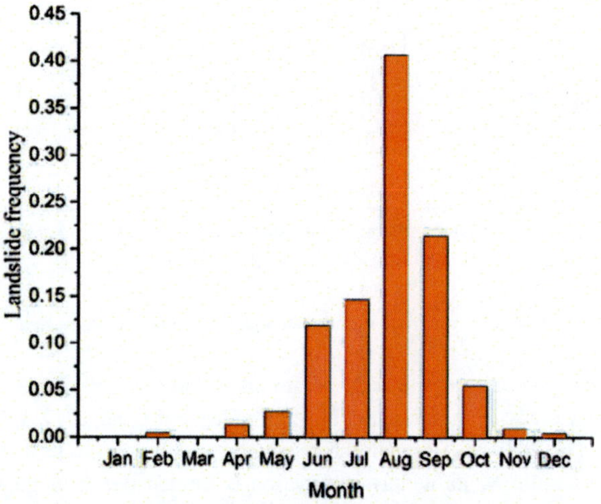

Fig. 3 Landslide frequency distribution within different months in Xi'an

by alluvial deposits. Thus, there is no landslide occurrences in this study area. The Qinba Mountains have a humid subtropical climate (Chen et al. 2015; Wu and Qian 2017). This region lies in an active tectonic area with many folds, faults and thrusts. This area, composed of Proterozoic and Mesozoic granite with highly weathered rock surface, is predominantly characterized by steep and rugged terrain. There are many primarily and deeply eroded V-shaped valleys. The active tectonics, steep topography, intensive rainfall and earthquake triggered considerable landslides, causing both serious damage to infrastructure and injuries to the local people (Qian et al. 2015; Qiu et al. 2020a, b). Moreover, landslides play an important role in the geomorphological evolution (Qiu et al. 2019a; Yang et al. 2019).

Landslide Data

A Landslide inventory can provide detailed landslide information (Guzzetti et al. 1999; Qiu et al. 2017a). It can be developed using many different ways depending on the investigation aim such as the resolution of the remote

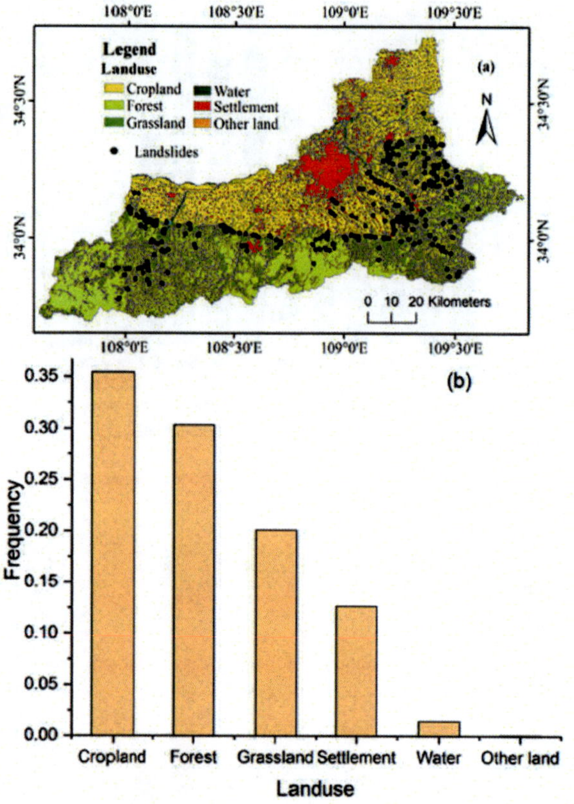

Fig. 4 **a** Land use map. **b** Land use distribution of the study area

sensing images, the scale of the maps and the experience of the study (van Westen et al. 2006; Guzzetti et al. 2012; Qiu et al. 2019a). Although the great advances in remote sensing technology, its widely application is limited due to vegetation canopies, plowing and erosion (Brardinoni and Church 2004; Cheng et al. 2018; Wang et al. 2018). Most importantly, it is very difficult to accurately identify and measure many essential parameters such as ideal slope gradient, slope height, slope morphology by remote sensing technology (Qiu et al. 2019a). Thus, in this work, we conducted a series of intensive field surveys to obtain and map the landslide information in the study area. We collected previously landslide reports and located landslides by Google Earth images before the field surveys (Hu et al. 2019). Figure 5 shows photographs of two typical landslides within the study area. These surveys were aided by recent advances in mobile technologies and unmanned aerial vehicles (UAVs). We obtained high resolution digital elevation models (DEMs) and digital orthophoto maps (DOMs) of some typical landslides in this study area. As a result, 316 landslides were mapped in a GIS platform, which permits efficient data analysis. The information in this inventory includes the landslide locations, area, slope gradient, slope morphology,

Fig. 5 Photographs of typical landslides in the study area

slope height and landslide types. Despite the field work is an efficient tool to obtain some essential parameters, our field work was mainly concentrated in the regions along accessible roads.

Methods

Average Nearest Neighbor

In this work, we used the average nearest neighbor (ANN) tool to analyze the spatial distribution and fractal behavior of landslides (Guthrie and Evans 2004; Qiu et al. 2019b):

$$ANN = \frac{d_o}{d_e} = \frac{\sum_{i=1}^{n} d_i}{n} / \frac{0.5}{\sqrt{n/A}}$$

where.

$$d_o = \frac{\sum_{i=1}^{n} d_i}{n} \text{ and } d_e = \frac{0.5}{\sqrt{n/A}}.$$

Where d_o is the observed mean distance between a landslide and its nearest landslide, d_e is the expected mean distance for landslides, d_i is the distance between landslide i and its nearest landslide, n is the total landslide number, and A is the total area in this study area. If the d_o is less than the average distance for expected mean distance. When ANN < 1, the landslide spatial distribution is clustered. On the contrary, if ANN > 1, the landslide spatial distribution is dispered.

Furthermore, we use the z-score to measure the landslide spatial distribution in this work (Guthrie and Evans 2004; Qiu et al. 2019b):

$$z - score = \frac{d_0 - d_e}{SE}$$

where

$$SE = \frac{0.2614}{\sqrt{n^2/A}}$$

Statistical tests begin by identifying a null hypothesis. Here, the null hypothesis is complete spatial randomness (Qiu et al. 2020b). The z-score indicates whether the null hypothesis is rejected or not. If the z-score is between −1.65 and 1.65, the spatial distribution pattern is random. If the z-score <−1.65, the spatial distribution pattern is clustered. Smaller z-scores shows a more clustered landslide distribution. If the z-score is >1.65, the spatial distribution pattern is dispersed. Larger z-scores shows a more dispersed landslide distribution.

Interdependence Relationship Among the Landslide Parameters

In this paper, in order to establish the interdependence relationships among the landslide parameters such as volume, area, length, width, we used the scaling relationships–power-law equations to fit the empirical data. Power-law equation for landslide is as follows:

$$V_L = \varepsilon_{s1} \times A_{L^{\alpha 1}}$$

$$A_L = \varepsilon_{s2} \times L^{\alpha 2}$$

$$A_L = \varepsilon_{s3} \times W^{\alpha 2}$$

where, V_L is the landslide volume, A_L is the landslide area, L is the landslide length, and W is the landslide width. α1, α2, α3 are the scaling exponents of equations. εs1, εs2, εs3 are constants of equations.

Frequency-Size Distribution of Landslides

Previous researches have shown that the frequency size distribution of landslides obey a power law form at a large landslide size range (Guzzetti et al. 2002; Malamud et al. 2004; Guthrie and Evans 2004; Hungr et al. 2008; Qiu et al. 2018b). In this work, we used power law function to fit the frequency size distribution of landslides at large landslide size. This correlation can be expressed as follow:

$$p = C_1 V_L^{-r1}$$

$$p = C_2 A_L^{-r2}$$

where p is probability density with an area larger than a threshold. C_1, and C_2 are constants depending on the local conditions. r_1 and r_2 are scaling exponent of power law.

Modeling Landslide Size Distribution

In this work, we use the double Pareto and three-parameter inverse gamma probability distribution to fit the lanslide size distribution. The double Pareto model is described as follow:

$$p(A) = \eta \left[\frac{[1 + (m/t)^{-\alpha}]^{\beta/\alpha}}{[1 + (A/t)^{-\alpha}]^{1 + (\beta/\alpha)}} \right] \times (A/t)^{-(\alpha+1)}$$

where $p(A)$ is the probability density function with.

$$\eta = \frac{\beta}{t(1-\delta)} \text{ and } \delta = \pi(c).$$

where m and c are cut-offs with $m = 10^8$ and $c = 1$; π is a uniform sampling function; A is the landslide area in m^2; t can describe the location of rollover in m^2; α and β are exponents of power law scaling related to slope.

The three-parameter inverse gamma probability distribution is described as follow:

$$p(A) = \frac{1}{r\Gamma(\rho)}\left[\frac{r}{A-s}\right]^{\rho+1}\exp\left[-\frac{r}{A-s}\right]$$

where ρ, r and s are three parameters of the gamma function; $\Gamma(\rho)$ is the gamma function of ρ. The parameter ρ controls the power law decays, while r controls the maximum probability density of landslide distribution in m^2; s controls the rate of decay for small landslide areas in m^2.

Results

Spatial Distribution of Landslides

Landslide spatial density shows how many landslides are found in a certain area (Qiu et al. 2019b). In this work, we calculated landslide density using kernel density estimation. As shown in Fig. 6, the landslide spatial distribution exists a heterogeneous and was clustered. There exist some clustered centers. Most of the landslides were concentrated along the Qin Mountains.

The spatial distribution analysis indicates that the observed mean distance between the landslides is 1152 m, while the expected mean distance is 2863 m. The nearest neighbour ratio is 0.51. Thus, the nearest neighbour ratio is less than 1, which indicates that the landslide spatial distribution in this study area is clustered. Moreover, the z-score is –17.34 (Fig. 7). This suggests that the landslide spatial distribution is clustered at a significance level of 0.01. Furthermore, we use cluster and outlier analysis to identify statistically significant hot spots, cold spots, and spatial outlier by using the Anselin Local Moran's I statistic (Fig. 8). The large landslides are clustered in the Bailu tableland. There is no significant cluster on landslide area in most of the study area. Figure 9 shows the spatial distribution of landslide length, width, area and volume.

In addition, the aspect influences the rainfall, soil moisture, vegetation and root (Conforti et al. 2014). We create the aspect map from a DEM using the aspect tool in ArcGIS software. Moreover, we classified the aspect into eight categories plus flat areas (Fig. 10a). The result show that the landslide frequency is very high on NE-, East-, SE-, South- and SW- facing slopes. North-, NW- and West- facing slope were not susceptible to landslides (Fig. 10b).

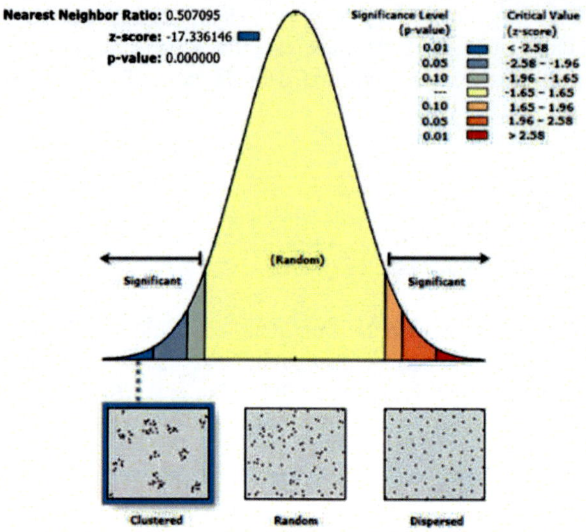

Fig. 7 Spatial distribution pattern based on Average Nearest Neighbour.

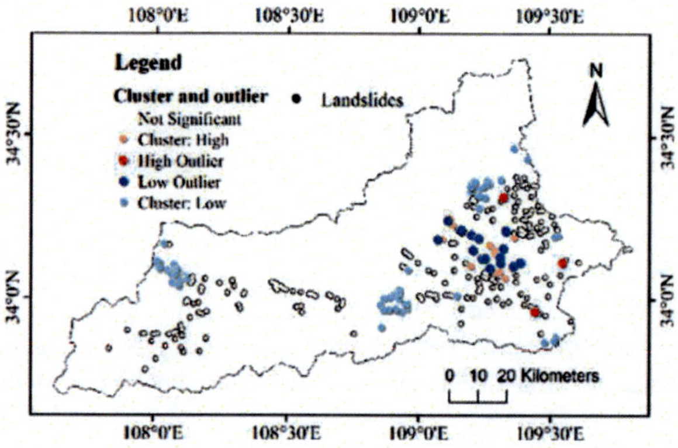

Fig. 8 Spatial cluster and outlier using Anselin Local Morans I statistic

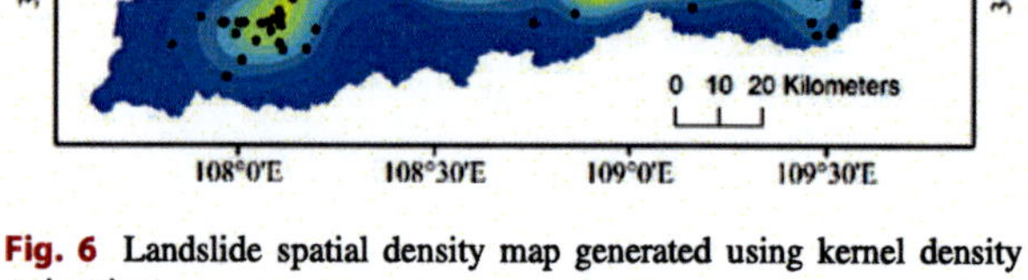

Fig. 6 Landslide spatial density map generated using kernel density estimation

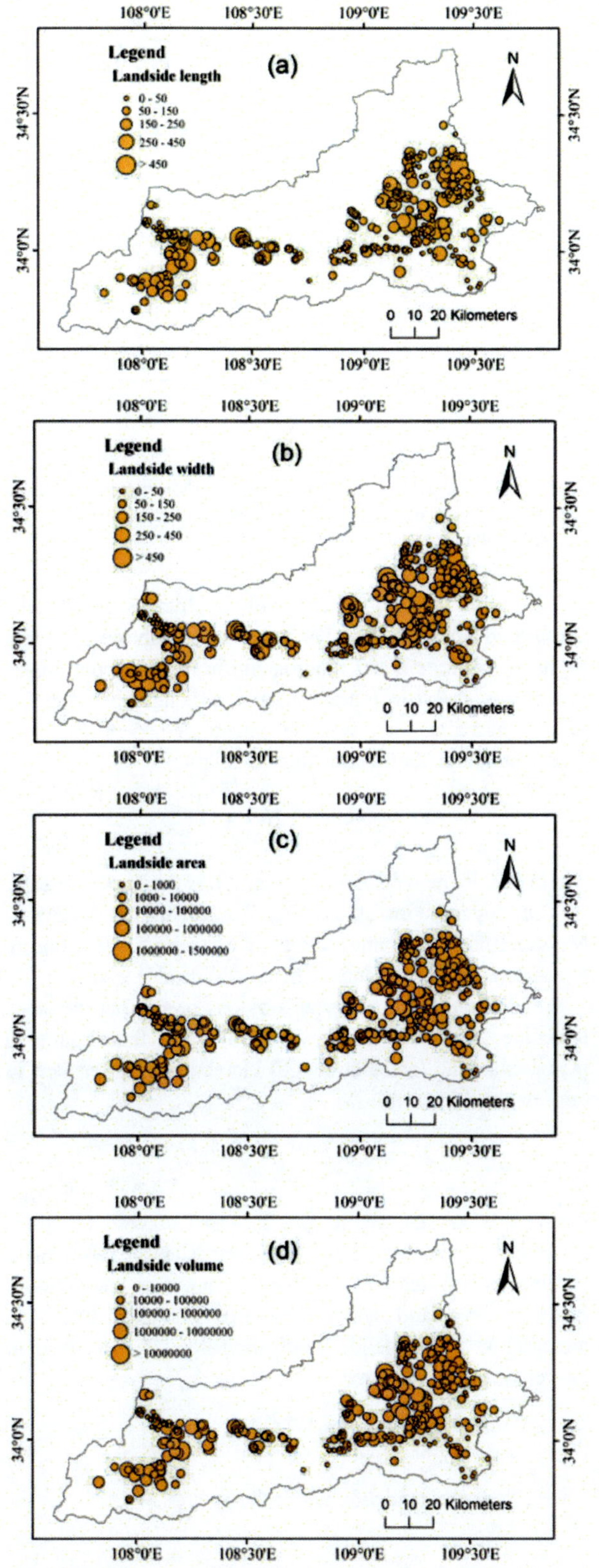

Fig. 9 Spatial distribution of landslide area

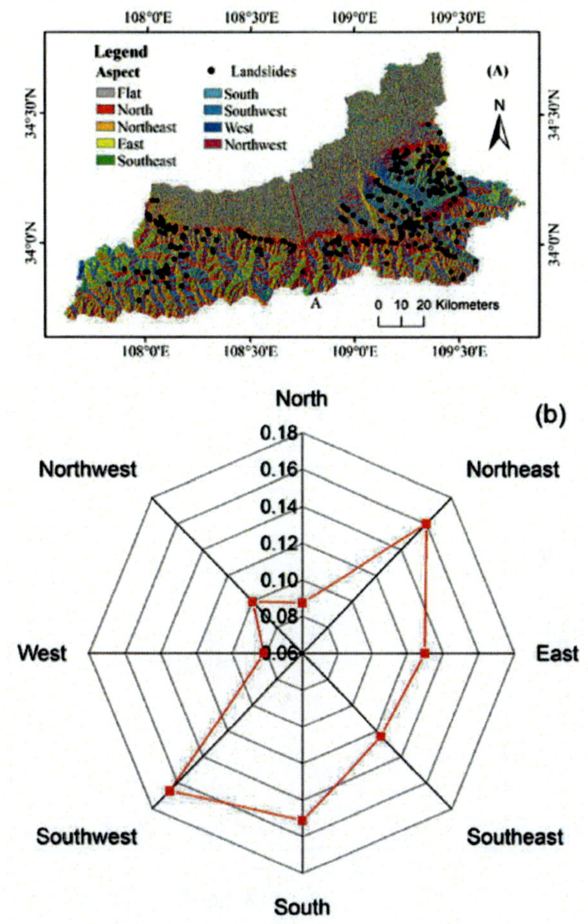

Fig. 10 a Aspect map of the study area. b Landslide frequency distribution within different aspect

Statistics of Loess Slides Sizes and Interdependence Relationship Among the Landslide Parameters

Estimation of landslide volume through measurable area from air photographs is very difficult and more challenging task (Brardinoni et al. 2003; Guzzetti et al. 2009). As shown in Fig. 11, the landslide volumes cover 6 orders of magnitude. The majority of landslides are small scale landslides. The average landslide volume is about 76,1950 m^3 and corresponding coefficient of variation is 6.25. The large and very large landslides play an important role in determining the total landslides area in this study area. The ten largest landslides (3% of total number) account for 74% and the twenty largest landslides (6% of total number) account for 83% of the total landslide volume. Large (>10^6 m^3), medium (10^5–10^6 m^3), and small (<10^5 m^3) landslides account for 8.16, 24.17, and 67.67% of the total landslides, respectively.

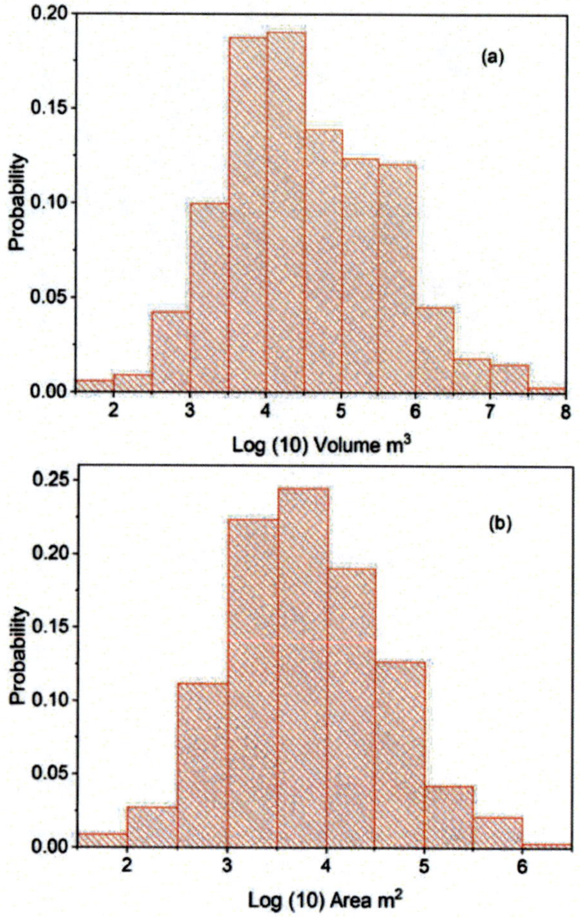

Fig. 11 Probability distribution of landslide volume (a) and area (b)

Moreover, the landslide area varies largely. The smallest landslide observed in this area is 50 m^2, while the largest one is 15 km^2. The ten largest landslides (3% of total number) account for 48% and the twenty largest landslides (6% of total number) account for 62% of the total landslide area. Large (>$10^5 m^3$), medium (10^4–10^5 m^3), and small (<10^4 m^3) landslides account for 6.34, 31.72, and 61.93% of the total landslides, respectively.

The landslide thickness ranges from 0.3 to 115 m, with a mean thickness of 8 m. Most of landslide are shallow. The landslides with thickness <10 m account for 81%. The landslide length ranges from 3 to 1000 m, with a mean length of 109 m. The landslides with length <100 m account for 62%. The landslide width ranges from 6 to 2300 m, with a mean length of 172 m. The landslides with length <100 m account for 32%.

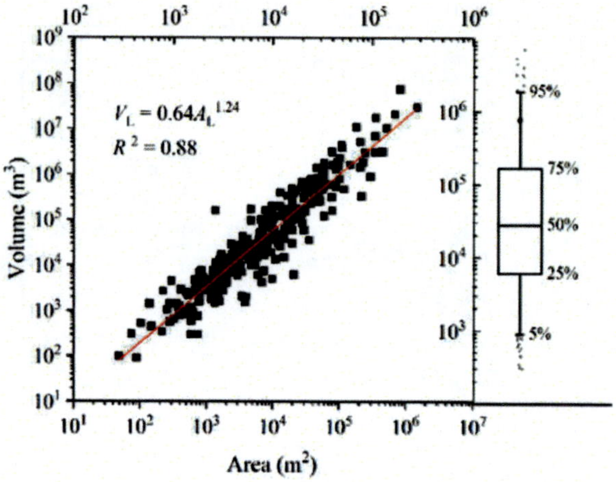

Fig. 12 Empirical relationship between landslide volume and area. The red line shows the regression line. The box plots show statistics of landslide volume

In addition, the landslide volume increases with the increase in landslide area. The present study shows that there exists an obvious linear relationship in log–log coordinates between landslide volume and area, although the landslide volume and area span multiple magnitude orders (Fig. 12). This resulted in the relationship as follow:

$$V_L = 0.64A_L^{1.24}\left(R^2 = 0.88\right)$$

where, VL is the landslide volume. AL is the landslide area.

This suggests that there exists a self-similar behaviour between landslide volume and area. We can use this relationship to estimate landslide volume when landslide area is known.

In addition, there exists an obvious power law between landslide area and length (Fig. 13). The landslide area increases as the landslide length increases. This resulted in the relationship as follow:

$$A_L = 8.62L^{1.58}\left(R^2 = 0.80\right)$$

where, L is the landslide length. This also suggests that there exists a self-similar behaviour between landslide area and length.

Furthermore, there exists an obvious power law between landslide area and width (Fig. 14). The landslide area increases as the landslide width increases. This resulted in the relationship as follow:

$$A_L = 4.03W^{1.61}\left(R^2 = 0.79\right)$$

where, W is the landslide width.

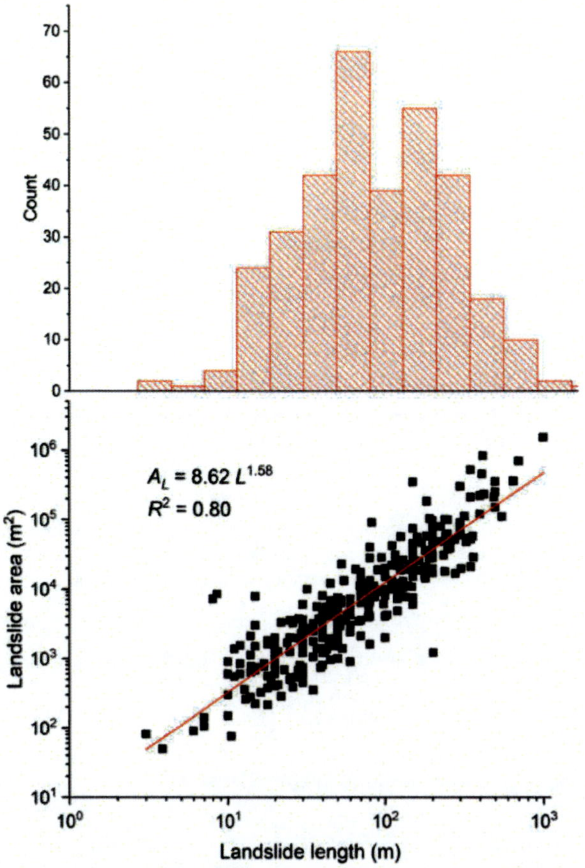

Fig. 13 Empirical relationship between landslide area and length. The red line shows the regression line

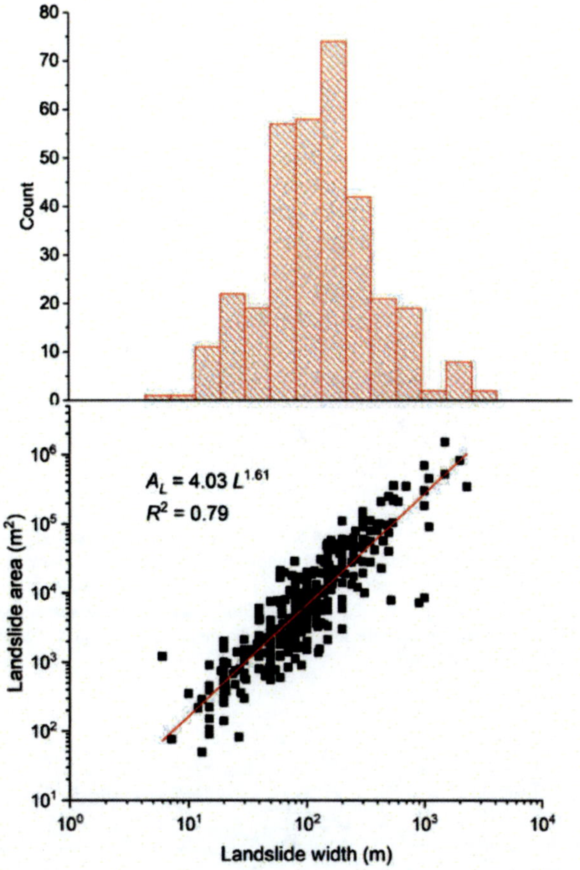

Fig. 14 Empirical relationship between landslide area and width. The red line shows the regression line

Probability Distribution of Landslide Size

The probability density of landslide volume and area was calculated for all the landslides across the entire study area. The probability density of landslide volume and area was ascertained using kernel density estimation, and was plotted on log–log scale plots. The landslide abundance increases with landslide area and peaks at 1000 m^2 before decaying sharply based on a power law. The landslide abundance decreases as landslide volume increases. The probability of landslide volume and area plotted on a log scale shows a very good fit to a double Pareto distribution and inverse gamma distribution. The tails of landslide volume and area distribution have slope of 0.57 and 0.83, respectively. We find the best-fit parameters of inverse gamma and double Pareto distribution by using a maximum likelihood estimation. As illustrated in Figs. 15 and 16, the inverse gamma and double Pareto distribution coincides with the observed probability density distribution of landslides for the entire dataset.

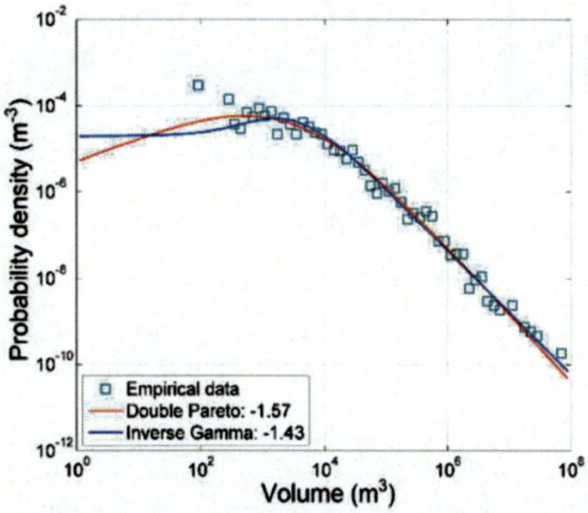

Fig. 15 Probability distribution of landslide volume. Indigo squares show the kernel density estimation of landslide size; Red lines show the maximum likelihood estimation of the Double Pareto function (α = 0.57; β = 1.57; and t = 3088 m^2).); Blue lines show the maximum likelihood estimation of the inverse gamma function ($\rho = 0.43$; $r = 3824$ m^2; and $s = -1022$ m^2)

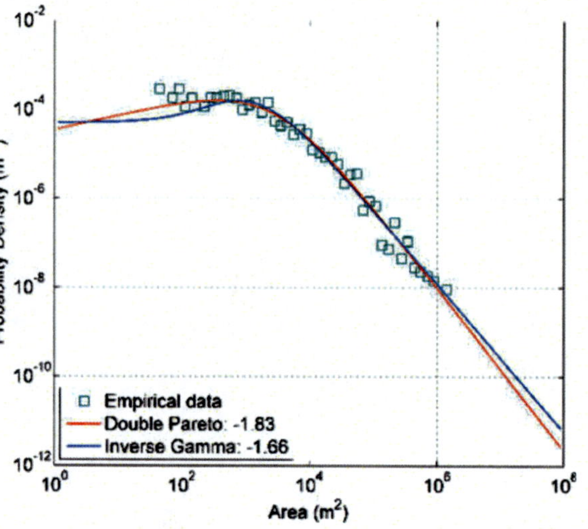

Fig. 16 Probability distribution of landslide area. Indigo squares show the kernel density estimation of landslide size; Red lines show the maximum likelihood estimation of the Double Pareto function (α = 0.83; β = 1.33; and t = 2748 m^2).); Blue lines show the maximum likelihood estimation of the inverse gamma function (ρ = 0.66; r = 2054 m^2; and s = −463 m^2)

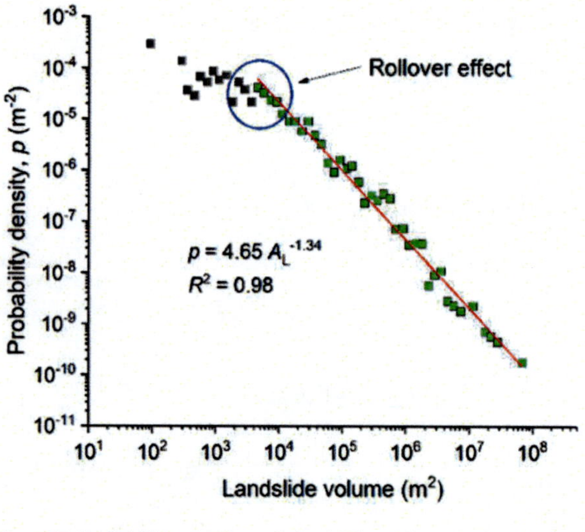

Fig. 17 Probability density of landslide volume using density estimation on log–log axes. The right tail of the landslide distribution scales as a power law function (red line) with a scaling exponent of 1.34

In addition, shown in Figs. 17 and 18, the landslide size distribution curve showed that there exists a clear rollover effect. The larger landslides satisfy a power law relationship with a scaling exponent of 1.34 and 1.47 for landslide volume and area distribution respectively. These resulted in the relationships as follow:

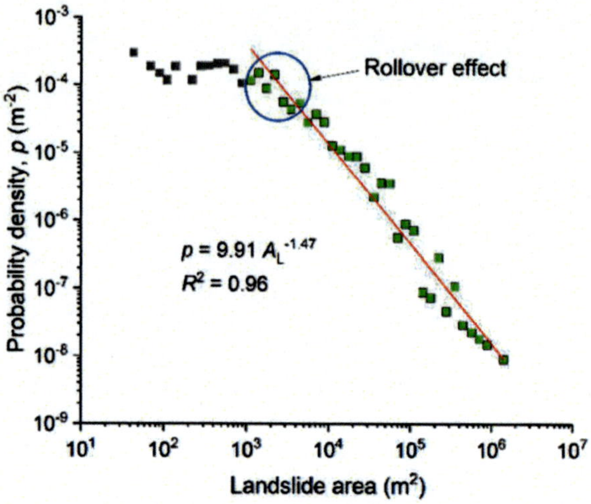

Fig. 18 Probability density of landslide area using density estimation on log–log axes. The right tail of the landslide distribution scales as a power law function (red line) with a scaling exponent of 1.47

$$p = 4.65V_{\mathrm{L}}^{-1.34}\left(R^2 = 0.98\right)$$

$$p = 9.91A_{\mathrm{L}}^{-1.47}\left(R^2 = 0.96\right)$$

where p is probability density.

Relationship Between Landslide Size and Slope Gradient

Slope gradient influences regional hydraulic behaviour, and is usually considered as a basic landslide influencing factor contributing to susceptibility of slope instability (Dai and Lee 2002; Mancini et al. 2010; Qiu et al. 2019c). We calculated the slope gradient based on ArcGIS platform. Figure 19 shows the slope gradient distribution of the study area. The average slope gradient of landslides in this study area was 38.58°. We divided slope gradient into eight categories: <10°, 10°–20°, 20°–30°, 30°–40°, 40°–50°, 50°–60°, 60°–70° and >70°. About 33% of landslides occurred within a slope gradient between 30°–40°. About 24% of landslides occurred within a slope gradient between 40°–50°. Only 6% of landslides occurred within a slope gradient less than 20° (Fig. 20).

Figure 21 shows the probability density of landslide size for different slope gradient. To obtain the relationship between size distribution of landslides and slope gradient, the slope gradient was divided into three classes: < 25°, 25°–50°, > 50°. As shown in Fig. 21, the slope gradient has a significant influence on the landslide size and size

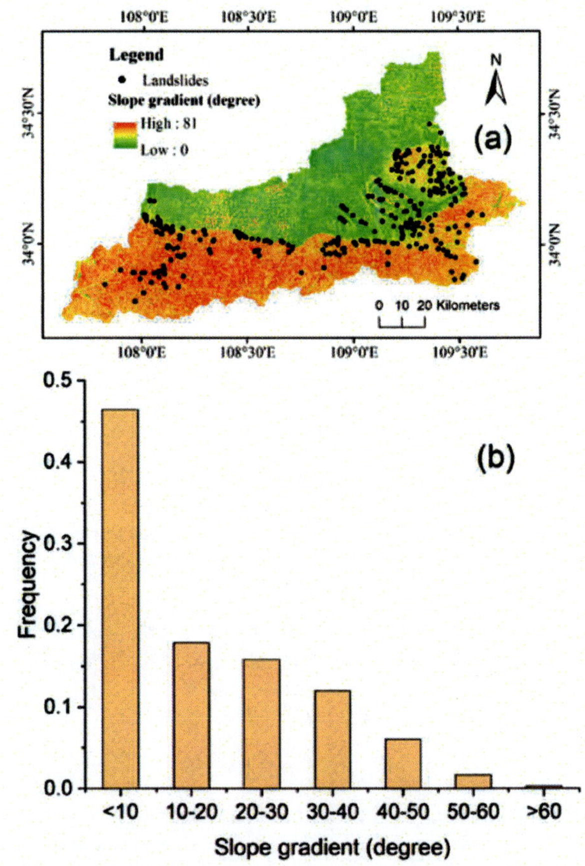

Fig. 19 Slope gradient distribution of the study area.

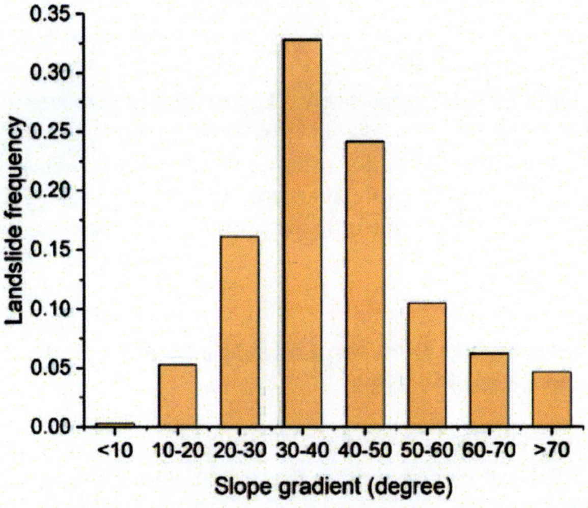

Fig. 20 Landslide distribution across slope gradients in the study area.

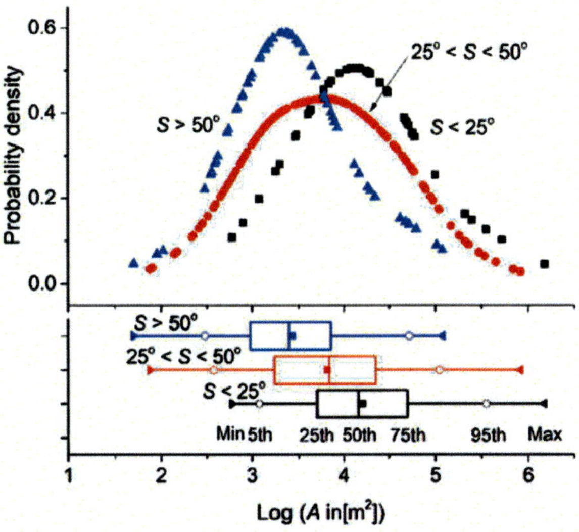

Fig. 21 Relationship between landslide size distribution and slope gradient

distribution. The average landslide areas for slope gradient < 25°°, 25°–50° and > 50° were 80 944 m^2, 33,225 m^2 and 10,015 m^2, respectively. The average landslide area for a slope gradient <25° is about 8 times larger than that for a slope gradient >50°. This is maybe because that it is difficult for steep slopes to support large landslides. Moreover, we found probability density of landslides were also influenced by the slope gradient. Figure 21 shows that more frequent small landslides occur on larger slope gradients.

Relationship Between Landslide Size and Slope Height

We obtained the slope height using SAGA GIS software (Module library: terrain analysis). Figure 22 shows the slope height distribution in the study area. We divided the slope height into seven classes: <50 m, 50–100 m, 100–150 m, 150–200 m, 200–250 m, 250–300 m, >300 m. About 57% of landslides occurred within a slope height less than 50 m. About 80% of landslides occurred within a slope height less than 100 m. Only 6% of landslides occurred within a slope height more than 200 m. As illustrated in Fig. 23, the landslide frequency decreased sharply with increasing slope height.

In addition, as shown in Fig. 24, the landslide size increased with increasing slope height. Although there is little theoretical basis for selecting a functional form, the

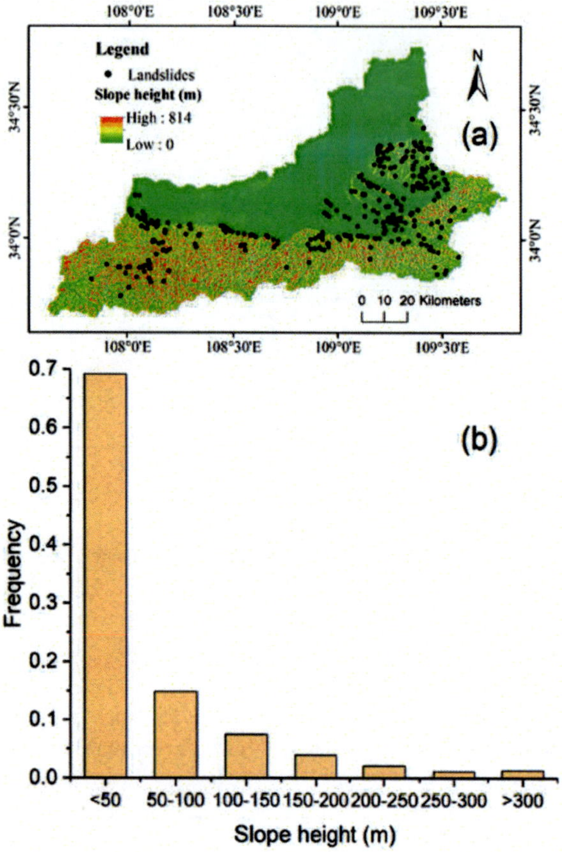

Fig. 22 Slope height distribution of the study area

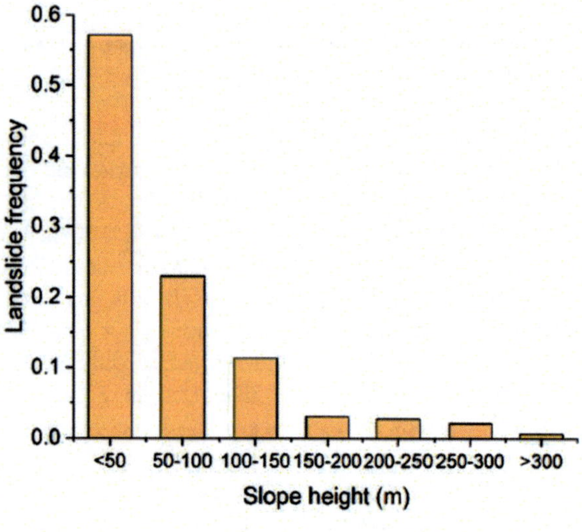

Fig. 23 Landslide distribution across slope height in the study area

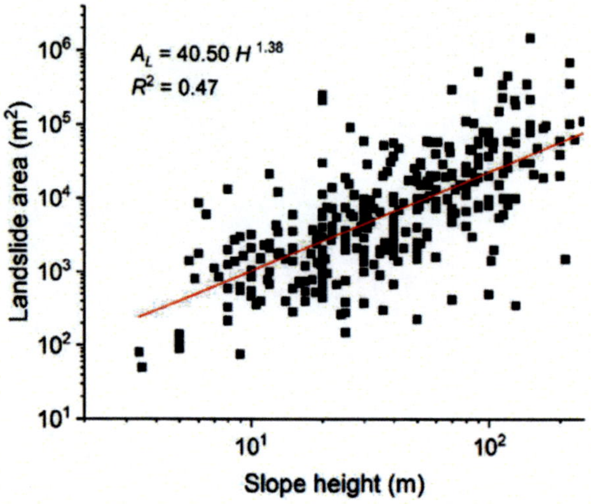

Fig. 24 Relationship between landslide area and slope height. The red solid line is the best fit regression which has a power law form

relationship is well described by a simple power law regression. This resulted in the relationship as follow:

$$V_L = 40.50\,H^{1.38}\left(R^2 = 0.47\right)$$

where, H is the slope height.

Furthermore, we calculated the probability density of landslide size for different slope height. To obtain the relationship between size distribution of landslides and slope height, the slope height was divided into three classes: < 50 m, 50–100 m, >100 m. As shown in Fig. 25, the slope height has a significant influence on the landslide size and size distribution. The average landslide areas for slope height <50 m, 50–100 m and >100 m° were 8,275 m^2, 32,148 m^2 and 118,299 m^2, respectively. The average landslide area for a slope height >100 m is about 14 times larger than that for a slope height <50 m. Moreover, we found probability density of landslides were also influenced by the slope height. Figure 25 shows that more frequent small landslides occur on lows slope heights.

Relationship Between Landslide Size and Landslide Types

We divided landslide into two classes: thrust-type landslide (TTL) and retrogressive landslide (RL). About 62% of the total landslides are retrogressive landslides, and 38 of the total landslides are thrust-type landslides (Fig. 26). The

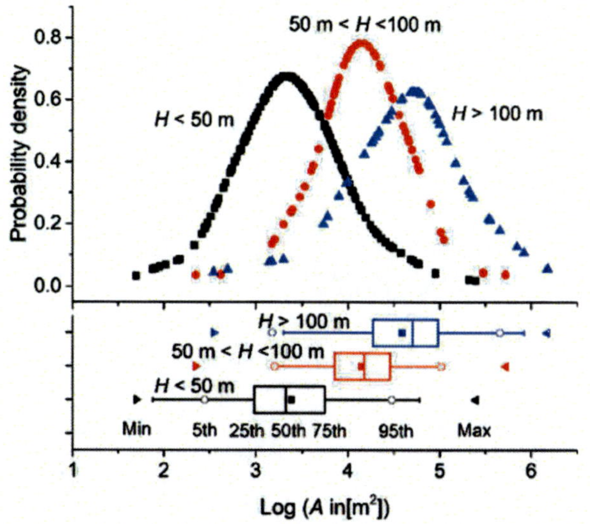

Fig. 25 Relationship between landslide size distribution and slope height

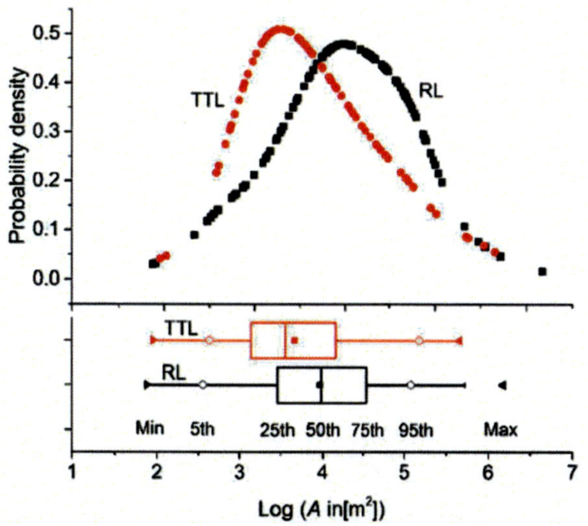

Fig. 27 Landslide size distribution for different landslide types. TTL represent thrust-type landslide. RL represent retrogressive landslide

retrogressive landslides are much higher than thrust-type landslides.

In addition, we addressed the relationship between landslide size and landslide types. Figure 27 shows the probability density of landslide size for different landslide types. As shown in Fig. 27, the landslide type has a significant influence on the landslide size and size distribution. The average area of thrust-type landslide and retrogressive landslide were 26,518 m^2 and 44,265 m^2, respectively. The average retrogressive landslide area is about 1.7 times larger than thrust-type landslide. Moreover, we found probability density of landslides were also influenced by the slope

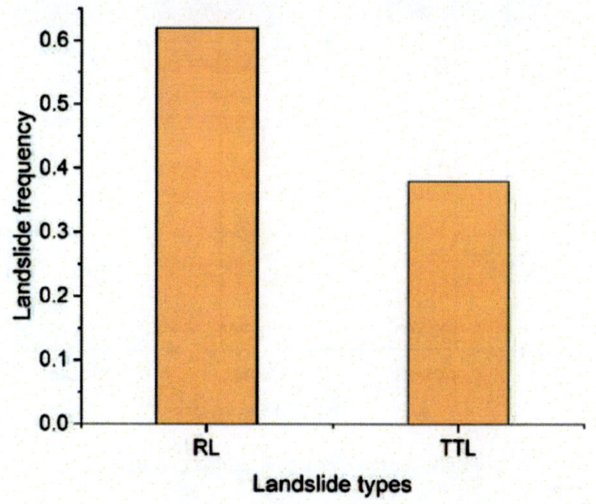

Fig. 26 Landslide distribution across landslide types in the study area

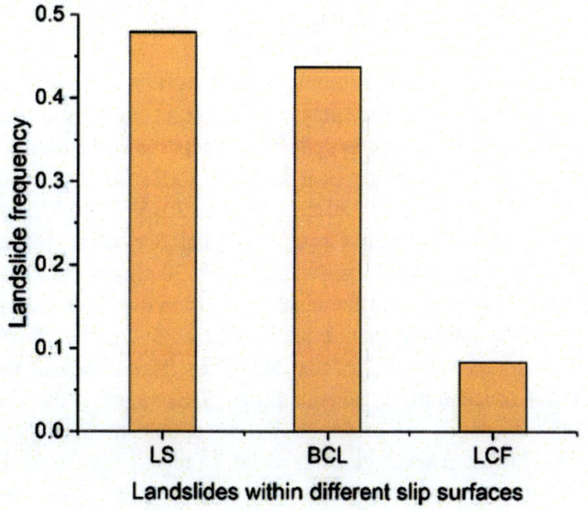

Fig. 28 Landslide distribution within different slope surfaces

gradient. Figure 27 shows that more frequent small landslides occur on thrust-type landslides.

Furthermore, we divided landslide into three classes: landslides in soil (LS), bedrock contact landslide (BCL) and landslides controlled by fissure (LCF). The number percentage of LS, BCL and LCF account for 47.92, 43.75 and 8.33% of the total landslides. The number of landslides in soil are much higher than that of landslides controlled by fissure (Fig. 28). The mean area of LS, BCL and LCF were 52,316, 30,621 and 9,328 m^2. The mean area of LS is about 5.6 times larger than LCF (Fig. 29).

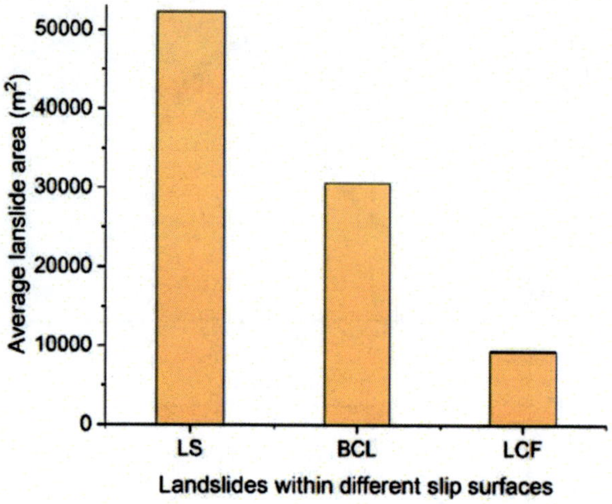

Fig. 29 Average landslide area within different slope surfaces

Relationship Between Landslide Size and Slope Morphology

Slope morphology can influence the landslide susceptibility in several ways, and play an important role in terrain instability. The slope morphology affects the surface runoff or subsurface drainage (Dai and Lee 2002; Conforti et al. 2014; Meinhardt et al. 2015; Qiu et al. 2016). We obtained the profile curvature from a digital elevation model (DEM) using ArcGIS software. Figure 30 shows the slope height distribution in the study area. Moreover, we reclassified the profile curvature into three classes: convex, linear and concave (Fig. 31). Most landslides (about 56% of the total) concentrate on convex slopes. This suggests that the convex slopes are preferentially susceptible to landslides. The lowest percentage of landslides (approximately 12%) occurred on a concave slope. The landslide frequency on the convex slopes is much higher than that on the concave slopes.

Figure 32 shows the probability density of landslide size for different slope morphology. We divided slope morphology into three classes: convex, planar and concave slope. As shown in Fig. 32, the slope morphology has a significant influence on the landslide size and size distribution. The average landslide areas for convex, planar and concave slope were 41,083 m^2, 7,939 m^2 and 29,744 m^2, respectively. Moreover, we found probability density of landslides were also influenced by the slope morphology. Figure 32 shows that more frequent small landslides occur on planar slope.

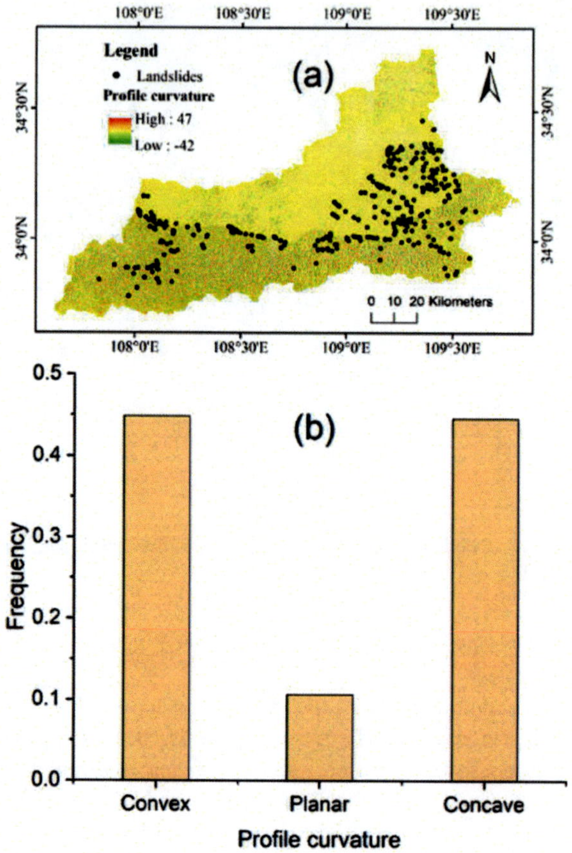

Fig. 30 **a** Profile curvature map. **b** Profile curvature distribution of the study area

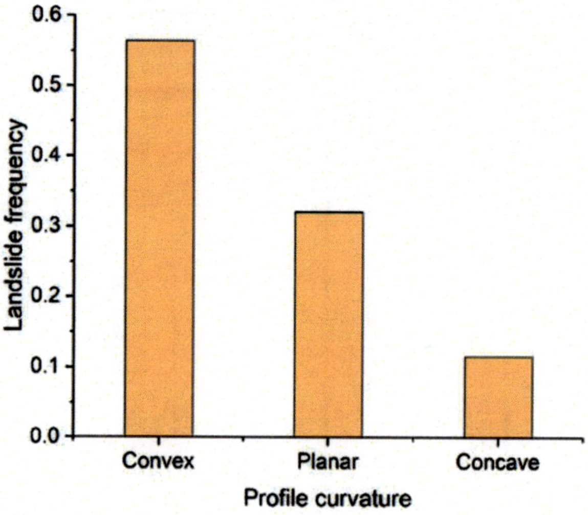

Fig. 31 Landslide distribution across profile curvature in the study area

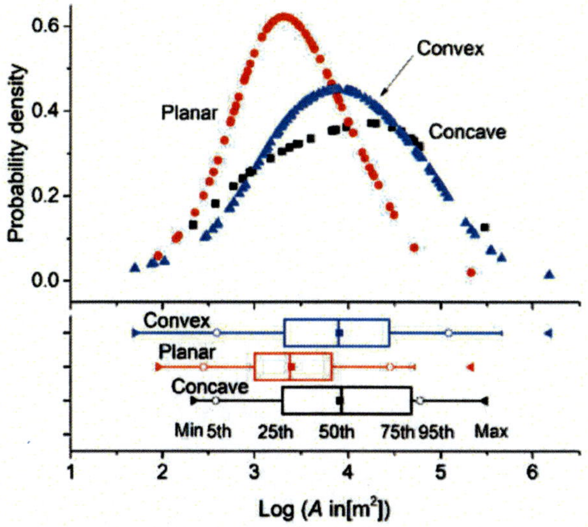

Fig. 32 Relationship between landslide size distribution and slope morphology

Discussion

Interdependent Relationships Among Geometric Parameters of Landslides

Establishment of a quantitative interdependence relationship among geometric parameters of landslides plays an important role in landslide investigation and assessment of a region. These relationships are of great help to use the part of the inventory where the information is most adequate to infer a dependency for the where there is a less information, especially for extremely small or large landslides (Korup 2005b; Guzzetti et al. 2009). It is an extremely challenging task to determine the landslide volume, as it requires detailed information on the surface and sub-surface geometry of the landslides, which is very costly and time consuming (Malamud et al. 2004; Guzzetti et al. 2009). We studied the distribution and interdependence relationship among the landslide parameters such as volume, area, length and width, etc. We found that the landslide area separately spans multiple orders of magnitude, and the large and very large landslides play an important role in determining the total area of landslide. Moreover, the results show that there exists a distinct power law relationship among the landslides parameters for multiple orders of magnitude, although landslides occurred in different tectonic, lithological, topographic and climatic settings, and were triggered by different factors, including intense rainfall, violent earthquakes, rapid snowmelt and anthropogenic forcing (Guzzetti et al. 2009). Especially, there was an empirical relationship between landslide volume and area. This can help us estimate the volume of individual landslide when the landslide area is known (Hovius et al. 1997). In previous studies, many authors have investigated the empirical relationships between landslide volume and area, and showed that the scaling exponent of power law relationship is in between 0.88 and 1.95 with an average value is 1.25 (Table 1). In this present study, the scaling exponent were respectively 1.24, which is very close to the average values published in previous literature (Guzzetti et al. 2009; Larsen et al. 2010). Compared with the empirical relationships between landslide volume and area published in the scientific literatures, together with the new relationship derived from the present work, we can get a conclusion that there exists a similar trend for the relationship linking the landslide volume and area (Innes 1983; Larsen and Torres Sanchez 1998; Martin et al. 2002; Imaizumi and Sidle 2007; Imaizumi et al. 2008). Guzzetti et al. (2009) pointed out that there exists a self-similar behaviour between landslide volume and area, and the relationship is largely and substantially geometrical, which is largely independent of the local physiographical setting and not influenced distinctly by geomorphological or mechanical properties of the failed soils or rocks, or the landslide types. However, compared with previous empirical studies and models, Larsen et al. (2010) found that scaling exponents of the landslide data sets for bedrock landslides ranges from 1.3 to 1.6, whereas the value for soil slides it was from 1.1 to 1.3. Our results are in line with the results of Larsen et al. (2010). In this paper, we not only reveal the power law relationship between loess landslide area and volume, but also derived a quantitative interdependence relationship among other landslide parameters such as area, length and width, etc. These empirical relationships are of great help to regional landslide investigation and assessment, because we can use the part of the inventory where the information is extremely enough to infer a value for the parts of the inventory where there is little information (Korup 2005a, b; Guzzetti et al. 2009).

Landslide Size Distribution and its Rollover Effect

Landslide size distribution is essentially important in evaluating the landslide magnitude (Hungr et al. 1999; Guzzetti et al. 2002; Malamud et al. 2004; Van Den Eeckhaut et al. 2007; Brink et al. 2009), determining landslide hazard (Guzzetti et al. 2005; Valagussa et al. 2019), and quantifying the integrated effects of landslide erosion and sediment (Hovius et al. 2000; Stark and Hovius 2001; Martin et al. 2002; Brardinoni and Church 2004; Guthrie and Evans 2004; Hungr et al. 2008; Chen 2009). Considerable authors have found that many systems (e.g., sand piles, earthquake, landslide) owing to self-organized criticality tend to a critical state and exists sale invariant which can be fitted by a power

Table 1 The scaling exponent of landslides between area and volume. Column 1 list the serial number. Column 2 show the scaling exponent. Column 3 list the number of landslide data. Column 4 show the source of data.

SN	Exp	N	Source
1	1.368	207	Simonett (1967)
2	1.11	29	Rice et al. (1969)
3	1.104	37	Rice and Foggin (1971)
4	1.3852	30	Innes (1983)
5	1.25	45	Whitehouse (1983)
6	0.898	1019	Larsen and Torres-Sánchez (1998)
7	0.88	615	Martin et al. (2002)
8	1.0905	124	Guthrie and Evans (2004)
9	1.95	23	Korup (2005a)
10	1.047	65	Haflidason et al. (2005)
11	1.292	160	Brink et al. (2006)
12	1.31	51	Imaizumi and Sidle (2007)
13	1.4324	539	Guzzetti et al. (2008)
14	1.19	11	Imaizumi et al. (2008)
15	1.45	667	Guzzetti et al. (2009)
16	1.332	2136	Larsen et al. (2010)
17	1.18	275	Qiu et al. (2018a)
18	1.2287	331	Qiu et al. (2018a)
19	1.272	91	Qiu et al. (2019b)
20	1.24	316	This work

law distribution (Bak et al. 1988; Qiu et al. 2018b). However, many observations have found that frequency distribution of some landslide data sets often departs from the power law relationship and exists a so-called rollover effect in the small size range (Stark and Hovius 2001; Dai and Lee 2001; Malamud et al. 2004; Guthrie and Evans 2004; Hungr et al. 2008; Qiu et al. 2018b). In this work, we found that the frequency size distribution of landslides can be well fitted with a power law relationship with scaling exponent 1.34 and 1.47 for landslide volume and area, respectively. These scaling exponents are close to the range of previous published literature (from 0.96 to 2) (Guzzetti et al. 2002). Some authors have shown that the scaling exponents rely on local geological, topographic and geomorphic setting (Van Den Eeckhaut et al., 2007; Qiu et al. 2018a). Some authors attribute the rollover effect of landslide size distribution to incompleteness of landslide inventories because some smaller landslides cannot be inventoried owing to forest canopy in densely forested area (Brardinoni and Church 2004). But, Malamud et al. (2004) examined three well documented landslide inventories and found that the rollover effect of landslide size distribution also exhibits. Furthermore, some authors have indicated that the rollover effect of landslide probability size distribution can be influenced by soil moisture variability, material properties, triggering mechanism, topographic and geological condition (Pelletier et al. 1997; Turcotte et al. 2002; Hungr et al. 2008). Despite the cause of rollover effect on landslide probability size distribution was not clear so far, this work suggests that the slope gradient, slope height, slope morphology and landslide types are essential important influencing factor for rollover effect of landslide size distribution. Thus, there remains much work to be done on comprehensive explanation of rollover effect.

Controls on Landslide Size

Many authors have researched the relationship between landslide size and its controls (Stark and Hovius 2001; Montgomery and Brandon, 2002; Korup et al. 2007; Guzzetti et al. 2008; Frattini and Crosta 2013; Gallen et al. 2015; Qiu et al. 2016) Frattini and Crosta (2013). found that small landslides often occurred on slope with high slope gradient. Katz et al. (2014) reported that landslide size increases as slope gradient increases based on the numerical 2-D discrete element method. But, Chen et al. (2016) found that landslide volume decreases as slope gradient increases based on limit equilibrium simulations. Qiu et al. (2018a) showed that the there is no relationship between landslide size and slope gradient based on simulations. Although we found that the landslide size decreases with increasing slope gradient seen in the field observations. But it is indirect because the dependence of slope gradient on slope length. Furthermore, Qiu et al. (2018a) showed that the slope gradient increases with decreasing slope length. The main control on landslide size is slope length. Larger landslides prefer to occur on longer slopes. Our results in this work found that the mean landslide size decreases with increasing slope gradient based on field observation. But this result is similar to our previous work (Qiu et al. 2018a).

In addition, previous works have shown that landslide size increases with increasing slope height (Montgomery and Brandon, 2002; Korup et al. 2007; Qiu et al. 2017b). Most giant landslides occur in regions with deeply incised valleys and high relief (Montgomery and Brandon, 2002;). This is because local relief limits landslide size, and large landslides rarely occur on areas with low local relief (Korup et al. 2007; Qiu et al. 2016). Our results are similar to finding from previous works (Montgomery and Brandon, 2002; Korup et al. 2007; Qiu et al. 2019a). Furthermore, we found landslide types and slope morphology also control the landslide size.

Conclusion

This paper developed a landslide inventory based on field work. The analysis of this detailed landslide inventory shows the controls on landslide size and size distribution. The landslide spatial distributions are clustered. There exist some clustered centres in the study area. The large and very large landslides play an important role in determining the total landslides area in this study area. The double pareto and inverse gamma functions can well fit with the landslide probability distribution, and can quantitatively reveal the maximum probability and rollover effect. The landslide size distribution well associated with a power law form above a certain landslide size. Below this landslide size, it exists an obvious rollover effect. The results show that local topographic conditions and landslide type play an important role in landslide size. About 33% of landslides occurred within a slope gradient between 30°–40°. Analysis of the landslide size with respect to slope gradient showed that the landslide size decreases with increasing slope gradient. More frequent small landslides occur on larger slope gradients. About 80% of landslides occurred within a slope height less than 100 m. The landslide frequency decreased sharply with increasing slope height. The local slope height can limit the landslide size and their spatial extent. Landslide size increases with increasing slope height, and correlations of which can be fitted using power law form. The peak of landslide probability density increases with increasing slope height. Moreover, about 62% of the total landslides are retrogressive landslides. Most landslides concentrate on convex slopes. The lowest percentage of landslides occurred on a concave slope. The landslide size is also controlled by the landslide types, many thrust-type landslides were small landslides. Furthermore, the probability densities of landslide size are influenced by slope morphology, and more frequent small landslides occur on planar slope. This study is very valuable for landslide hazard mapping.

Acknowledgements This work was funded by the Second Tibetan Plateau Scientific Expedition and Research Program (STEP) (grant no. 2019QZKK0902) International Science & Technology Cooperation Program of China (grant no. 2018YFE0100100), National Natural Science Foundation of China (grant no. 41771539), Strategic Priority Research Program of Chinese Academy of Sciences (grant no. XDA 20030301), and International Partnership Program of Chinese Academy of Sciences (grant no. 131551KYSB20160002).

References

Bak P, Tang C, Wiesenfeld K (1988) Self-organized criticality. Phys Rev A 38(1):364–374

Brardinoni F, Slaymaker O, Hassan MA (2003) Landslide inventory in a rugged forested watershed: a comparison between air-photo and field survey data. Geomorphology 54:179–196

Brardinoni F, Church M (2004) Representing the landslide magnitude–frequency relation: Capilano River Basin, British Columbia. Earth Surf Proc Land 29(1):115–124

Brunetti MT, Guzzetti F, Rossi M (2009) Probability distributions of landslide volumes. Nonlinear Process Geophys 16(2):179–188

Brink US, Geist EL, Andrews BD (2006) Size distribution of submarine landslides and its implication to tsunami hazard in Puerto Rico. Geophys Res Lett 33:L11307

Brink UST, Barkan R, Andrews BD, et al (2009) Size distributions and failure initiation of submarine and subaerial landslides. Earth Planet Sci Lett 287(1–2):0–42

Chen CY (2009) Sedimentary impacts from landslides in the Tachia River Basin, Taiwan. Geomorphology 105:355–365

Chen XZ, Cui Y (2017) The formation of the Wulipo landslide and the resulting debris flow in Dujiangyan City, China. J Mt Sci 14 (6):1100–1112

Cheng D, Cui Y, Su F et al (2018) The characteristics of the Mocoa compound disaster event, Colombia. Landslides 15:1223–1232

Chen XL, Liu CG, Chang ZF et al (2016) The relationship between the slope angle and the landslide size derived from limit equilibrium simulations. Geomorphology 253:547–550

Chen H, Marter-Kenyon J, López-Carr D et al (2015) Land cover and landscape changes in Shaanxi Province during China's grain for green program (2000–2010). Environ Monit Assess 187(10):644

Conforti M, Pascale S, Robustelli G et al (2014) Evaluation of prediction capability of the artificial neural networks for mapping landslide susceptibility in the Turbolo River catchment (northern Calabria, Italy). Catena 113:236–250

Crozier MJ (2005) Multiple-occurrence regional landslide events in New Zealand: hazard management issues. Landslides 2:247–256

Cui Y, Zhou XJ, Guo CX (2017) Experimental study on the moving characteristics of fine grains in wide grading unconsolidated soil under heavy rainfall. J Mt Sci 14(3):417–431

Cui Y, Cheng D, Choi CE et al (2019) The cost of rapid and haphazard urbanization: lessons learned from the Freetown landslide disaster. Landslides 16(6):1167–1176

Dai FC, Lee CF (2001) Frequency-volume relation and prediction of rainfall-induced landslides. Eng Geol 59(3):253–266

Dai FC, Lee CF (2002) Landslide characteristics and slope instability modeling using GIS, Lantau Island, Hong Kong. Geomorphology 42:213–228

Frattini P, Crosta GB (2013) The role of material properties and landscape morphology on landslide size distributions. Earth Planet Sci Lett 361:310–319

Gallen SF, Clark MK, Godt JW (2015) Coseismic landslides reveal near-surface rock strength in a high-relief, tectonically active setting. Geology 43(1):11–14

Guo C, Cui Y (2020) Pore structure characteristics of debris flow source material in the Wenchuan earthquake area. Eng Geol 267:105499

Guthrie RH, Evans SG (2004) Magnitude and frequency of landslides triggered by a storm event, Loughborough Inlet, British Columbia. Nat Hazards Earth Syst Sci 4:475–483

Guthrie RH, Deadman PJ, Cabrera AR et al (2008) Exploring the magnitude–frequency distribution: a cellular automata model for landslides. Landslides 5(1):151–159

Guzzetti F, Carrara A, Cardinali M et al (1999) Landslide hazard evaluation: a review of current techniques and their application in a multi-scale study Central Italy. Geomorphology 31(1–4):181–216

Guzzetti F, Malamud BD, Turcotte DL et al (2002) Power-law correlations of landslide areas in central Italy. Earth Planet Sci Lett 195:169–183

Guzzetti F, Reichenbach P, Cardinali M et al (2005) Landslide hazard assessment in the Staffora basin, northern Italian Apennines. Geomorphology 72:272–299

Guzzetti F, Ardizzone F, Cardinali M et al (2008) Distribution of landslides in the Upper Tiber River basin, central Italy. Geomorphology 96:105–122

Guzzetti F, Ardizzone F, Cardinali M et al (2009) Landslide volumes and landslide mobilization rates in Umbria, central Italy. Earth Planet Sci Lett 279(3):222–229

Guzzetti F, Mondini AC, Cardinali M et al (2012) Landslide inventory maps: New tools for an old problem. Earth Sci Rev 112(1–2):42–66

Haflidason H, Lien R, Sejrup HP et al (2005) The dating and morphometry of the Storrega Slide. Mar Pet Geol 22:187–194

He Y, Qiu H, Song J et al (2019) Quantitative contribution of climate change and human activities to runoff changes in the Bahe River watershed of the Qinling Mountains, China. Sustain Cities Soc 51:101729. https://doi.org/10.1016/j.scs.2019.101729

Hu S, Qiu H, Yang D et al (2017) Evaluation of the applicability of climate forecast system reanalysis weather data for hydrologic simulation: a case study in the Bahe River Basin of the Qinling mountains China. J Geogr Sci 27(5):546–564

Hu S, Qiu H, Pei Y et al (2019) Digital terrain analysis of a landslide on the loess tableland using high-resolution topography data. Landslides 16:617–632

Hovius N, Stark CP, Allen PA (1997) Sediment flux from a mountain belt derived by landslide mapping. Geology 25:231–234

Hovius N, Stark CP, Chu HY et al (2000) Supply and removal of sediment in a landslide-dominated mountain belt: central Range Taiwan. J Geol 108(1):73–89

Hungr O, Evans SG, Hazzard J (1999) Magnitude and frequency of rock falls and rock slides along the main transportation corridors of southwestern British Columbia. Can Geotech J 36(2):224–238

Hungr O, McDougall S, Wise M et al (2008) Magnitude–frequency relationships of debris flow sand debris avalanches in relation to slope relief. Geomorphology 96:355–365

Imaizumi F, Sidle RC (2007) Linkage of sediment supply and transport processes in Miyagawa Dam catchment, Japan. J Geophys Res 112 (F03012). doi:https://doi.org/10.1029/2006JF000495.

Imaizumi F, Sidle RC, Kamei R (2008) Effects of forest harvesting on the occurrence of landslides and debris flows in steep terrain of central Japan. Earth Surf Proc Land 33:827–840

Innes JN (1983) Lichenometric dating of debris-flow deposits in the Scottish Highlands. Earth Surf Proc Land 8:579–588

Katz O, Morgan JK, Aharonov E et al (2014) Controls on the size and geometry of landslides: insights from discrete element numerical simulations. Geomorphology 220:104–113

Korup O (2005a) Distribution of landslides in southwest New Zealand. Landslides 2(1):43–51

Korup O (2005b) Geomorphic imprint of landslides on alpine river systems, southwest New Zealand. Earth Surf Proc Land 30(7):783–800

Korup O, Clague JJ, Hermanns RL et al (2007) Giant landslides, topography, and erosion. Earth Planet Sci Lett 261(3–4):578–589

Klar A, Aharonov E, Kalderon-Asael B, et al (2011) Analytical and observational relations between landslide volume and surface area. J Geophys Res Earth Surf 116(F2).

Larsen MC, Torres-Sánchez AJ (1998) The frequency and distribution of recent landslides in three montane tropical regions of puerto rico. Geomorphology 24(4):309–331

Larsen IJ, Montgomery DR, Korup O (2010) Landslide erosion controlled by hillslope material. Nat Geosci 3:247–251

Liu Z, Koyi HA (2013) Kinematics and internal deformation of granular slopes: insights from discrete element modeling. Landslides 10(2):139–160

Lu X, Li LY, Lei K et al (2010) Water quality assessment of Wei River, China using fuzzy synthetic evaluation. Environmental Earth Sciences. 60(8):1693–1699

Ma S, Qiu H, Hu S et al (2019) Quantitative assessment of landslide susceptibility on the Loess Plateau in China. Phys Geogr. https://doi.org/10.1080/02723646.2019.1674559

Malamud BD, Turcotte DL, Guzzetti F et al (2004) Landslide inventories and their statistical properties. Earth Surf Proc Land 29(6):687–711

Mancini F, Ceppi C, Ritrovato G (2010) GIS and statistical analysis for landslide susceptibility mapping in the Daunia area, Italy. Nat Hazards Earth Syst Sci 10:1851–1864

Martin Y, Rood K, Schwab JW et al (2002) Sediment transfer by shallow landsliding in the Queen Charlotte Islands, British Columbia. Can J Earth Sci 39(2):189–205

Meinhardt M, Fink M, Tünschel H (2015) Landslide susceptibility analysis in central Vietnam based on an incomplete landslide inventory: comparison of a new method to calculate weighting factors by means of bivariate statistics. Geomorphology 234:80–97

Montgomery DR, Brandon MT (2002) Topographic controls on erosion rates in tectonically active mountain ranges. Earth Planet Sci Lett 201(3–4):481–489

Nadim F, Kjekstad O, Peduzzi P et al (2006) Global landslide and avalanche hotspots. Landslides 3:159–173

Pelletier JD, Malamud BD, Blodgett T, Turcotte DL (1997) Scale invariance of soil moisture variability and its implications for the frequency-size distribution of landslides. Eng Geol 48(3):255–268

Petley D (2012) Global patterns of loss of life from landslides. Geology 40(10):927–930

Qian T, Liu S, Li W et al (2015) Early-Middle Jurassic evolution of the northern Yangtze foreland basin: a record of uplift following Triassic continent-continent collision to form the Qinling-Dabieshan orogenic belt. Int Geol Rev 57(3):327–341

Qiu H, Regmi AD, Cui P et al (2016) Size distribution of loess slides in relation to local slope height within different slope morphologies. Catena 145:155–163

Qiu H, Cui P, Regmi AD et al (2017a) Influence of topography and volume on mobility of loess slides within different slip surfaces. Catena 157:180–188

Qiu H, Cui P, Regmi AD et al (2017b) Slope height and slope gradient controls on the loess slide size within different slip surfaces. Phys Geogr 38(4):303–317

Qiu H, Cui P, Regmi AD et al (2018a) The effects of slope length and slope gradient on the size distributions of loess slides: field observations and simulations. Geomorphology 300:69–76

Qiu H, Cui P, Regmi AD et al (2018b) Landslide distribution and size versus relative relief (Shaanxi Province, China). Bull Eng Geol Env 77:1331–1342

Qiu H, Cui Y, Hu S et al (2019a) Size distribution and size of loess slides in response to slope height and slope gradient based on field survey data. Geomat Nat Hazards Risk 10(1):1443–1458

Qiu H, Cui Y, Hu S et al (2019b) Temporal and spatial distributions of landslides in the Qinba Mountains, Shaanxi Province China. Geomat Nat Hazards Risk 10(1):599–621

Qiu H, Cui P, Regmi AD et al (2019c) Loess slide susceptibility assessment using frequency ratio model and artificial neural network. Q J Eng Geol Hydrogeol 52(1):38–45

Qiu H, Cui Y, Pei Y et al (2020a) Temporal patterns of nonseismically triggered landslides in Shaanxi Province China. Catena 187:104356

Qiu H, Hu S, Wang X et al (2020b) Size and spatial distribution of loess slides on the Chinese Loess Plateau. Phys Geogr 41(2):126–144

Rice RM, Corbett ES, Bailey RG (1969) Soil slips related to vegetation, topography, and soil in Southern California. Water Resour Res 5 (3):647–659

Rice RM, Foggin GT (1971) Effect high intensity storms on soil slippage on mountainous watersheds in southern california. Water Resour Res 7(6):1485–1496

Samia J, Temme A, Bregt A et al (2017) Characterization and quantification of path dependency in landslide susceptibility. Geomorphology 292:16–24

Simonett DS (1967) Landslide distribution and earthquakes in the Bewani and Torricelli Mountains, New Guinea. In: Jennings JN, Mabbutt JA (eds) Landform Studies from Australia and New-Guinea. Cambridge University Press, Cambridge, pp 64–84

Stark CP, Hovius N (2001) The characterization of landslide size distributions. Geophys Res Lett 28(6):1091–1094

Turcotte DL, Malamud BD, Guzzetti F et al (2002) Self-organization, the cascade model, and natural hazards. Proc Natl Acad Sci 99 (suppl 1):2530–2537

Valagussa A, Marc O, Frattini P et al (2019) Seismic and geological controls on earthquake-induced landslide size. Earth Planet Sci Lett 506:268–281

Van Den Eeckhaut M, Poesen J, Govers G et al (2007) Characteristics of the size distribution of recent and historical landslides in a populated hilly region. Earth Planet Sci Lett 256:588–603

Van Westen CJ, Van Asch TW, Soeters R (2006) Landslide hazard and risk zonation—why is it still so difficult? Bull Eng Geol Env 65 (2):167–184

Wang J, Jin W, Cui YF et al (2018) Earthquake-triggered landslides affecting a UNESCO Natural Site: the 2017 Jiuzhaigou earthquake in the World National Park China. J Mt Sci 15(7):1412–1428

Whitehouse IE (1983) Distribution of large rock avalanche deposits in the Central Southern Alps, New Zealand. NZ J Geol Geophys 26:271–279

Wu H, Qian H (2017) Innovative trend analysis of annual and seasonal rainfall and extreme values in Shaanxi, China, since the 1950s. Int J Climatol 37(5):2582–2592

Yang D, Qiu H, Pei Y, et al (2019) Spatial and temporal evolution of the infiltration characteristics of a loess landslide. ISPRS Int J Geo-Inf 9:26. https://doi.org/10.3390/ijgi9010026

Zhuang J, Peng J, Iqbal J et al (2015) Identification of landslide spatial distribution and susceptibility assessment in relation to topography in the Xi'an Region, Shaanxi Province China. Front Earth Scis 9 (3):449–462

Geologic and Hydrologic Investigations on Slope Failures Triggered by Extreme Rainfall on Izu Oshima Island, Japan

2020 Kyoto Japan

Ikuo Towhata, Takeshi Akima, Satoshi Goto, Shigeru Goto, Junya Tanaka, and Shogo Aoyama

Abstract

A volcanic slope on Izu Oshima Island in Japan experienced a significant rainfall-induced landslide disaster in October, 2013. Because this slope had been stable for centuries, a special investigation was carried out on the cause. Because of its volcanic origin, the failed slope consisted of layers of ash, sand and lava. While the investigation concerned many disciplines, this paper addresses the post-disaster geotechnical studies. The emphasis was put on the reason why some parts of the slope "did not" fail because the post-disaster construction of infrastructures in the affected area relied on the future stability of the affected mountain slope. In line with this, another focus was on the geohydrological characteristics in the underlying lava layer. We concluded that the lava layer is pervious and allows drainage of infiltrated rain water and that, only during extremely heavy rain, slope failures are triggered in subsurface "valleys" that are carved in lava bedrock. For seven years after the disaster, the remaining parts of the slope have been stable.

Keywords

Rainfall-induced landslide • Volcanic slope • Lava • Seepage • Restoration

Introduction

The volcanic Izu Oshima Island is located in the Pacific Ocean, at 120 km south of Tokyo (Fig. 1), and its size measures 9 km in EW and 15 km in NS directions. The central part of this island is occupied by active Mihara Yama Volcano of 758 m in height (Fig. 2). This island has experienced natural disasters frequently related to volcanic eruptions (Koyama and Hayakawa 1996) and heavy rains. One of the gigantic eruptions occurred in 1338 when lava flowed on the western slope of the mountain (Nakamura 1964; Disaster Prevention Board of Tokyo Metropolitan Government 1990). The mountain slope consists of many layers of lava, loess and ash.

Rainfall Disaster During Typhoon in 2013

The Typhoon No. 26 "Wipha" in 2013 brought 824 mm of rainfall to Izu Oshima Island from October 15th to 16th. The hourly precipitation reached 122.5 mm/h from 2 to 4 AM on October 16th, being most devastating since October 1939 (Fig. 3). Accordingly, the western slope of Mihara Yama Volcano (Fig. 2) failed significantly. The total area of slope failure was 319,000 m^2 with a failed soil volume of 175,000 m^3. 39 people were killed or reportedly missed, while 153 houses were affected (Tokyo Metropolitan Government 2014); see Fig. 4. Figure 3 also shows that the previous worst event was in 1958 when the size of failed slope was smaller and its source was at lower altitude than in 2013 (JMA 1958).

I. Towhata (✉)
Kanto Gakuin University, Yokohama, Japan
e-mail: towhata.ikuo.ikuo@gmail.com

T. Akima · J. Tanaka
Fudo Tetra Corporation, Tokyo, Japan
e-mail: takeshi.akima@fudotetra.co.jp

J. Tanaka
e-mail: junya.tanaka@fudotetra.co.jp

S. Goto
Yamanashi University, Kofu, Japan
e-mail: goto@yamanashi.ac.jp

S. Goto
Waseda University, Tokyo, Japan
e-mail: shigeru.goto2462@gmail.com

S. Aoyama
Kiso-Jiban Consultans, Tsukuba, Japan
e-mail: aoyama.shogo@kiso.co.jp

Ž. Arbanas et al. (eds.), *Understanding and Reducing Landslide Disaster Risk*,
ICL Contribution to Landslide Disaster Risk Reduction,
https://doi.org/10.1007/978-3-030-60713-5_12

Fig. 1 Location of Izu Oshima Island

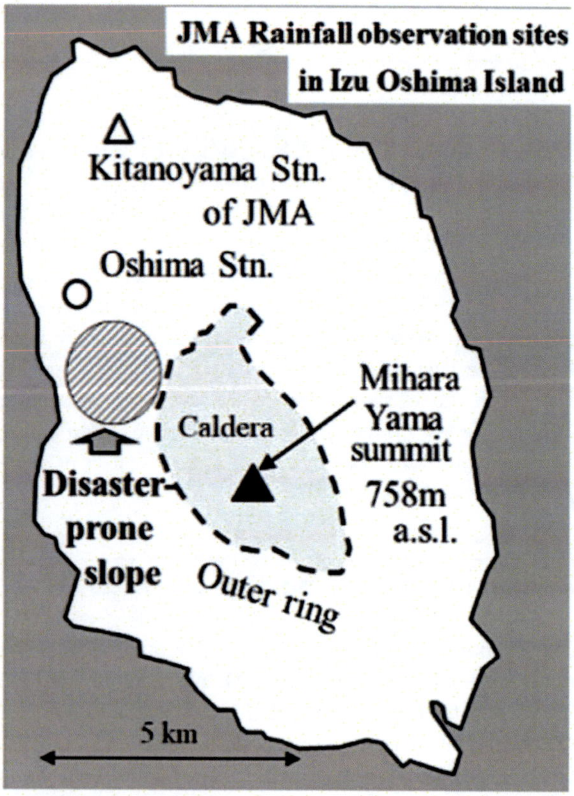

Fig. 2 Brief map of Izu Oshima Island with Mihara Yama Volcano at the center

The failed slope in 2013 lay on volcanic materials that deposited mainly during the eruption in 1338. In spite of many heavy rain events since then, this slope had been stable for more than 650 years. In 2013, shear failure occurred at very shallow soil depth, at the interface between volcanic ash sand and less pervious loess soil underneath (Wang et al. 2019). Afterwards, small failures merged with one another as travelling down (Fig. 4), eroded the slope surface and trees on the way, increased in size and destroyed villages near the coast. Noteworthy are many piping holes in the failed slope (Fig. 5). They suggest that ground water was ejected from water-saturated subsoil during heavy rain and infers the possible effects of artesian pressure on slope stability (Bell and Maud 2000). However, their exact significance in the slope failure is not clearly known yet.

The reconstruction of the community had a question whether or not the remaining parts of the slope were stable because nearly half of the slide materials remained at the top of the affected slope (Fig. 4). To promote the reconstruction, the present study attempts to answer this question from geotechnical engineering viewpoints.

Borehole Investigation

We drilled three bore holes to find out difference in subsoil conditions between failed and intact parts of the slope. Because the island is a national park under strict protection, permission was issued by public sectors and authorities. The location of three bore holes, B, B' and C, is illustrated in Fig. 6, while the numbers 8 to 20 show the sites of more bore holes drilled by the Tokyo Metropolitan Government. All the holes were drilled after the disaster.

B is located below the top scarp of slope failure, while B' was in a intact place beside a slope failure. This arrangement enables to investigate the reason why one part of the mountain slope was stable while other part was not. Figure 7a compares the obtained subsurface profiles. The autobrecciated lava is a material that is formed by quick cooling and volume contraction. Within the range of drilling (down to 10-m depth), the intact subsoil at B' has a thicker lava layer than the failed slope at B. Another point is that B consists of interbedding different materials in which scoria and gravel are more pervious than lava.

Site C is located on a small mound (Fig. 8). At this site, no shear failure happened but the original surface material was eroded and a lava layer was exposed. The bore hole profile at C (Fig. 7b) appears similar to that of the intact site B' (Fig. 7a) where no failure happened.

The collected lava specimen (Fig. 9) shows many cracks. Thus, it was decided to examine the hydraulic property of this material because the underlying lava layer may have affected the degree of saturation of the surface soil during heavy rain. Moreover, many piping holes (Fig. 5) suggested the effects of underground hydrology on slope instability. In this respect, a field permeability test was run on lava inside the bore hole C. The obtained permeability was $k_1 = 1.07 \times$

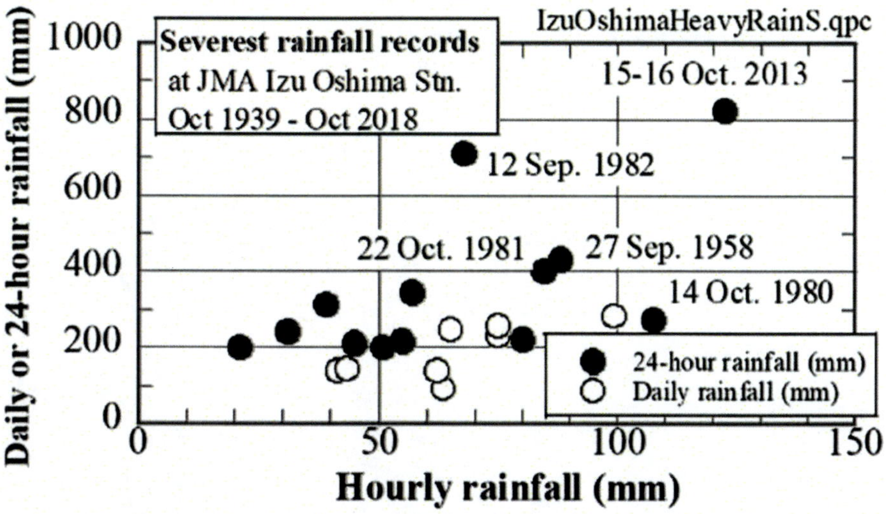

Fig. 3 History of severe rainfall records at Izu Oshima Station of Japanese Meteorological Agency (JMA used to publish daily rainfall, from 0AM to 12PM; recent AMeDAS system publishes 24-h rainfall intensity, starting from any time of the day)

Fig. 4 View of failed slope on the western side of the Mihara Yama Volcano

Fig. 5 Piping holes at the head scarp of the failed slope

10^{-5} m/s for the depth of 6–7 m (thickness, l_1, = 1 m) and $k_2 = 3.20 \times 10^{-6}$ m/s for the depth of 7–8 m (l_2 = 1 m).

Because of financial limitations, permeability was not measured in the range of 8–9 m. Based on the similar extent of breakage between 6–7 m and 8–9 m, the lower part (8–9 m; l_3 = 1 m) is reasonably assigned with a permeability ($k_3 = 1.07 \times 10^{-5}$ m/s) that is higher than k_2. Consequently, the equivalent overall permeability of the three lava layers is determined by

$$k = (l_1 + l_2 + l_3)/\left(\frac{l_1}{k_1} + \frac{l_2}{k_2} + \frac{l_3}{k_3}\right) = 6.01 \times 10^{-6}\,\mathrm{m/s} \tag{1}$$

for vertical water flow. No ground water was encountered down to 20 m during the tests. The obtained permeability of

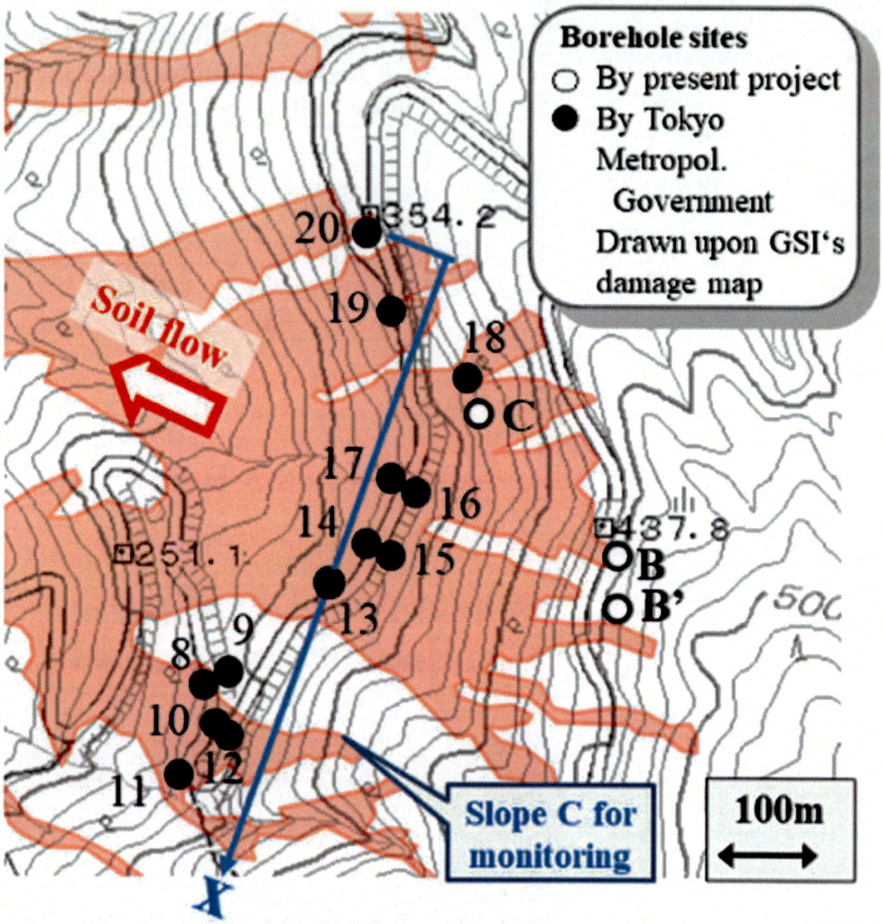

Fig. 6 Area of soil flow and bore hole sites in the western slope of Mihara Yama Volcano (drawn on topographic map by Geological Survey Institute)

lava is less than what Terashima et al. (2014) reported based on laboratory tests on ash (4.2×10^{-5} m/s) and scoria (1.0×10^{-4} m/s). Only loess specimens exhibited low permeability (5.0×10^{-6} and 1.8×10^{-7} m/s). Note that field permeability of soil is often ten times greater than laboratory values because of in-situ cracks and other water channels. Hence, it was concluded that lava is not so permeable as surface materials but not impervious.

Simple Analysis on Slope Stability and Water Flow in Lava

This chapter calculates the factor of safety of the studied slope by referring to different extents of water inundation during heavy rain. Figure 10 illustrates a one-dimensional slope model. The most probable friction angle (φ) for the analysis is 38.5° as Wang et al. (2019) obtained by ring shear tests on labotaroty-reconstituted specimens of tephra which is situated above less permeable loess (aeolian) soil. The unit weights of soil above and below the ground water level were assumed to be $\gamma_1 = 15.0$ kN/m^3 and $\gamma_2 = 19.1$ kN/m^3, respectively.

Pore water pressure, u, in the slip plane (point B in Fig. 10) was assessed by equating the total heads at B and at the water surface (A). Since pore water pressure at A is zero,

$$u \; at \; B = \gamma_w \times \text{(Elevation difference beween A and B)}$$

$$= \gamma_w h cos^2\theta \tag{2}$$

in which γ_w is the unit weight of water. Then, the factor of safety, Fs, against slope failure is derived as

$$Fs = \frac{tan\varphi}{tan\theta}\left\{1 - \frac{\frac{h}{H}}{\left(1 - \frac{h}{H}\right)\frac{\gamma_1}{\gamma_w} + \frac{h}{H}\frac{\gamma_2}{\gamma_w}}\right\} \tag{3}$$

Figure 11 demonstrates the variation of the factor of safety against the level of ground water, h/H, using $\varphi = 38.5°$ and two more hypothetical values (30.5 and 40.5°). It is therein found that the factor of safety becomes less than unity at $h/H = 0.4$ ($h = 40$ cm when $H =$ typically 1 m as was seen near the head scarp). Even for overestimated $\varphi = 40.5°$, $h/H = 0.55$ ($h = 55$ cm for $H = 1$ m) makes the slope

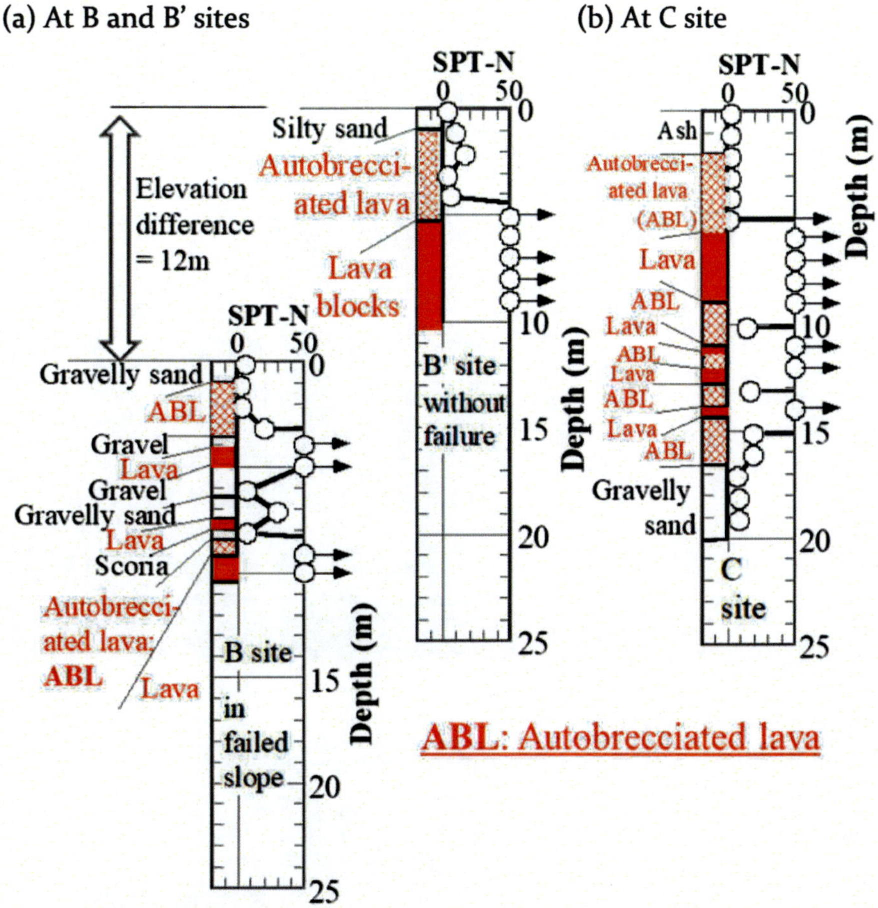

Fig. 7 Bore hole profiles at failled site (B), site without slope failure (B') and eroded site (C)

Fig. 8 Situation around bore hole site C

unstable. Such a shallow water depth is likely to occur frequently under the local rainfall environment (Fig. 12). Despite this, the studied slope had been stable for decades or centuries probably because high ground water table had not occurred in reality.

Loss of Water from Surface Soil into Lava and Deeper Layers

Slope stability analysis often assumes that ground water is retained in the surface soil. This assumption is not necessarily appropriate when the base lava layer is fractured and pervious as in the present case. Figure 13 illustrates a model of water infiltration through lava towards deeper subsoil. Size of layers in this figure (h and l) were determiened by referring to profiles in Fig. 7.

Darcy's law states that the water flux is given by $v = ki = k(1 + h/l)$ where k stands for the permeability of the lava layer

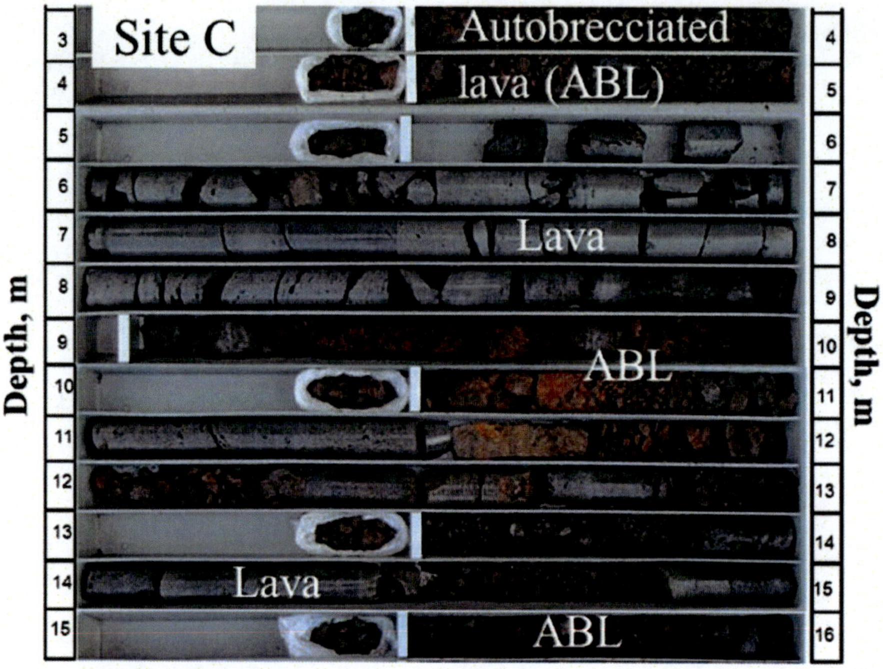

Fig. 9 Appearance of lava collected from depth 6–9 m at Site C

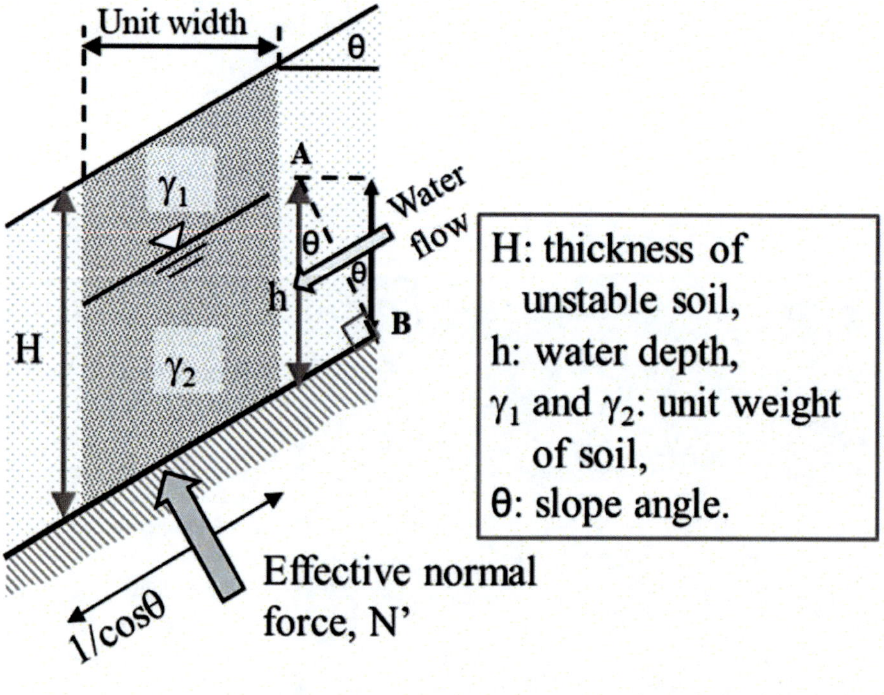

Fig. 10 Infinite slope model for stability analysis

and is equal to 6.01×10^{-6} m/s based on the field tests. Hence, $v = 1.50 \times 10^{-5}$ m/s or 54.1 mm/h. This means that the rainfall intensity has to be continuously greater than 54 mm/h or, 100 mm/h if infiltration into subsoil is 50% of the rainfall, so that the rain water may be retained in the surface soil and may make the soil saturated with water. Such a heavy rain scarecely happens (Figs. 3 and 12). Moreover, the void volume in the surface 6 m is given by $eh/(1 + e)$ where e stands for the void ratio. By hypothesizing e to be 0.7, the void volume in top 6 m is 2471 mm^3/m^2. Hence, infiltration of

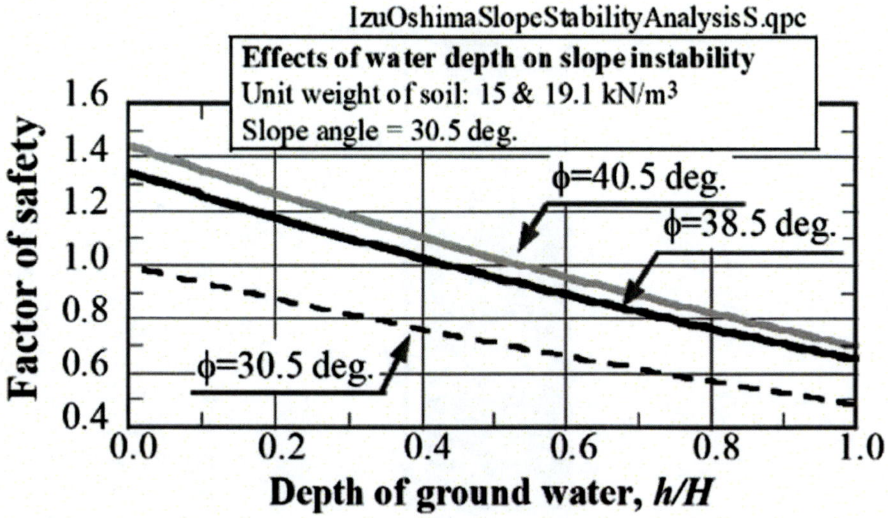

Fig. 11 Effects of ground water depth on slope stability in an infinite slope model

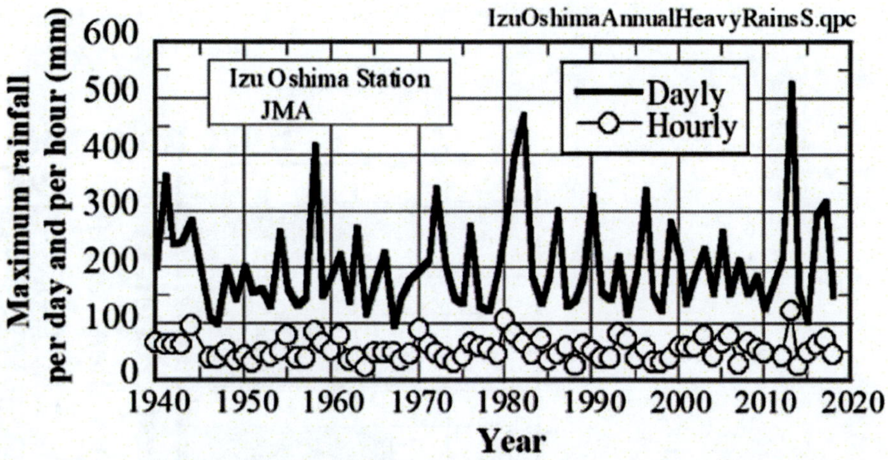

Fig. 12 Maximum daily and hourly rainfall records per year at JMA Oshima Station

400 mm/h should continue for 6 h for full saturation of the surface soil. Figure 12 implies that such an extreme condition has rarely happened. This is probably the reason why the studied slope had been stable without full saturation for many centuries in spite of the rainy climate in this island.

Subsoil Profile Across the Entire Width of the Failed Slope

This chapter discusses a slope stability in the entire width of the affected slope. For this purpose, we borrowed more bore hole data at the middle height of the slope from the Tokyo Metropolitan Government (No. 8–20 in Fig. 6) and interpreted them together with our own data at point C. It is noteworthy that some parts of the slope were affected by the disaster only by surface erosion but not by shear failure; see Fig. 8. To shed light on this, the drainage capacity of lava layers was investigated at those bore hole sites.

Because the depth of the boreholes was limited, the study had to be performed using data only of the shallowest lava layer group. The shallow group was defined arbitrarily as the shallow lava layers that are separated from a deeper lava group by an interval of more than 2 m. At C (Figs. 7b and 9), this shallow lava group was identified between depths of 5 and 9 m. The length of lava parts in the shallow group is denoted by l, and the number of observed cracks per *one* meter in the lava layers by m. Since the greater m means more permeable lava, m is used in place of permeability coefficient that was not measured in the field except at C. In line with Eq. 1, an alternative permeability parameter, k_{eq}, is defined by.

$$k_{eq} = \sum_i l_i / \sum_i \frac{l_i}{m_i} \tag{4}$$

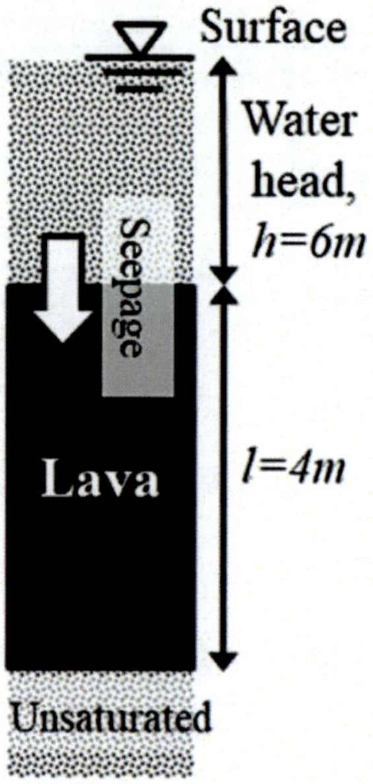

Fig. 13 Simple model for analysis on seepage flow into underlying lava layer

and the total thickness of the shallow lava group, l_{eq}, is determined by

$$l_{eq} = \sum_i l_i \tag{5}$$

Greater l_{eq} and smaller k_{eq} allow the shallow lava group to hold more water above it. Hence, l_{eq}/k_{eq} is called "water-holding index". Two indices are indicated in Fig. 14 in which bore hole sites are classified into those with erosion only and those with shear failure. The sites only with erosion exhibit greater thickness of lava (l_{eq}) and higher water-holding capacity (l_{eq}/k_{eq}). Is it reasonable to conclude that more rain water was held in the surface soil of eroded sites and that the slope "avoided" shear failure? Water storage and lack of failure are contradictory to each other. One possible idea is that the shallow lava reduced the upward flow of ground water from deeper elevations (see piping holes in Fig. 5) and maintained surface stability. Note also that the shear failure site "10" has large numbers of the indices in Fig. 14 similar to erosion sites.

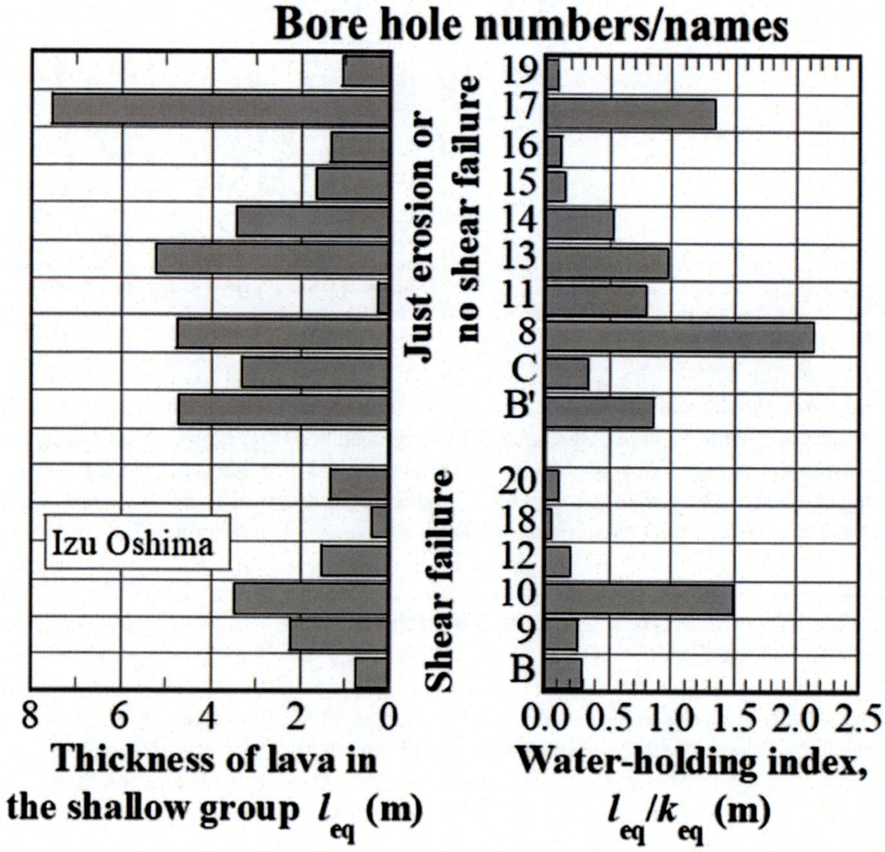

Fig. 14 Thickness and water-holding capacity of shallow lava layer group

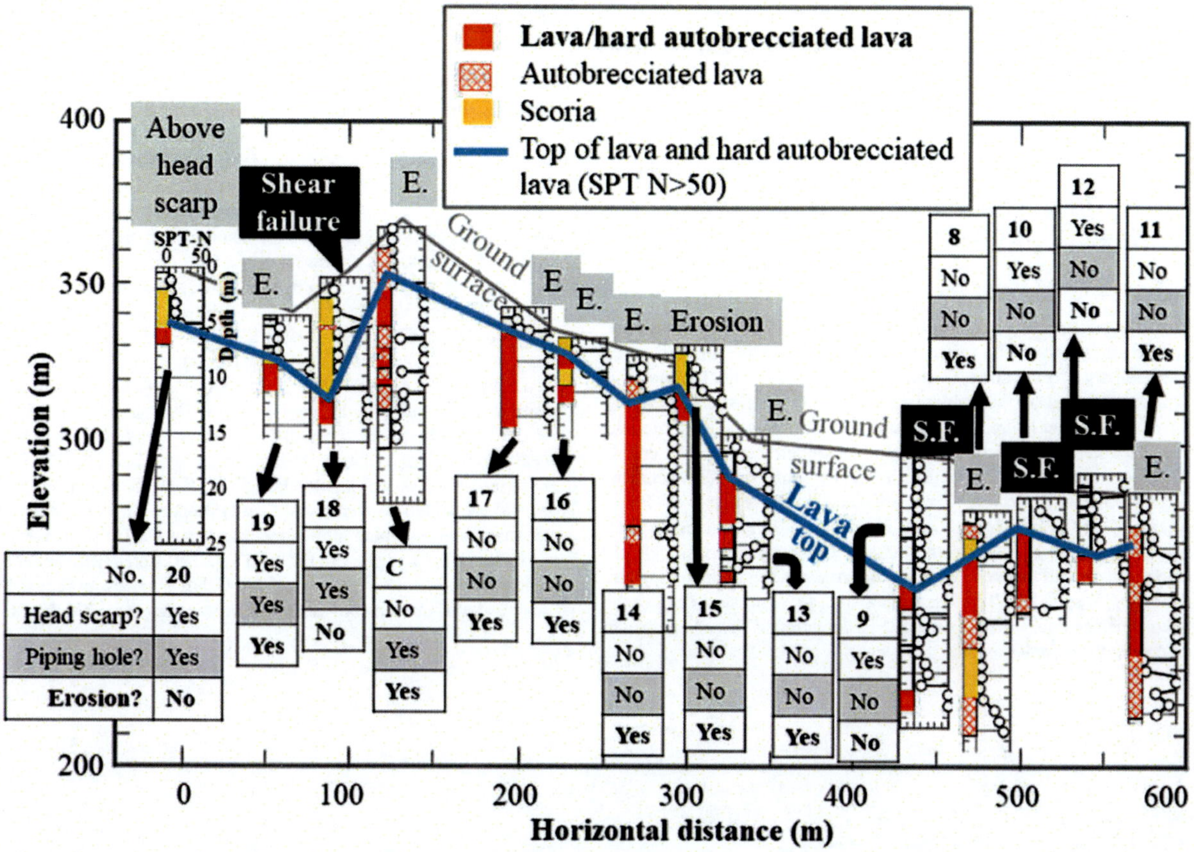

Fig. 15 Cross section of failed slope obtained by interpretation of bore-hole data and type of damage; horizontal distance stands for *X* coordinate in Fig. 6

Figure 15 is a summary of better interpretation of the post-disaster situation at all the bore hole sites. The lava layer in the figure was made by the eruption in AD 1338 as stated previously. The brecciated lava is classified into two kinds in this figure according to the measured SPT-*N*. When *N* is greater than 50, the brecciated lava was considered stable and combined with the intact lava in illustration. Consequently, the top points of lava and hard brecciated lava in bore hole profiles are connected in Fig. 15 to visualize the "subsurface topography".

There is a reasonable correlation between the elevation of subsurface topography and the type of damage. Erosion damage occurred at the subsurface "ridge", while shear failure occurred in the buried "valley". Because the ground water flows laterally into the "valley" and affects the slope stability during heavy rain, it is reasonable that "shear failure" occurred in the "valley" parts (points 12, 8, 9 and 18), while point 10 is probably marginal. Conversely, the subsoil of the eroded parts was intact because water did not remain there (subsurface ridge). Thus, there was a good reason why some parts of the slope remained intact during the heavy rain in 2013 until eroded and is likely to remain stable during future heavy rains. Until 2020, no new failure has occurred in this slope.

Conclusions

This paper addresses the rainfall-induced slope disaster in Izu Oshima Island, Japan, that occurred in October 2013. Efforts were made to find the difference between failed slopes and intact slopes. The major conclusions drawn from this study are as what follows.

(1) Many piping holes were found in landslide source areas. The slope failures were shallow at the beginning but became deeper by erosion.
(2) Two kinds of damage were found in the affected slope; shear failure and erosion. The difference between these two was caused by the subsurface topography of lava layer.

(3) The remaining parts of the mountain slope would be stable in near future. To date (April, 2020), no new slope failure has happened.
(4) The lava layer is pervious to a certain extent against common perception. Rain water is drained into deep parts of the slope and does not easily saturate the surface layer. Therefore. the slope had not failed significantly during heavy rains in the past.

Acknowledgements This study was conducted as an emergency investigation after the disaster with a financial support from the Ministry of Education, Culture, Sports, Science and Technology. It was supported by the Oshima Municipal Government, Ministry of Environment and Geopark of Oshima (Mr. A. Kajiya). The borehole data provided by the Tokyo Metropolitan Government was essential. Collaboration was successfully made by Japanese Geotechnical Society, Japan Society of Civil Engineers, Japan Society of Engineering Geology and Japan Landslide Society. Boring investigations were carried out by the Chisui Kaihatsu Company. Sincere appreciation is dedicaetd to any members of our research group, including Dr. Gonghui Wang, Dr. Taro Uchimura and Dr. Hiroshi Yamaguchi whose names could not be included in the authors' list of this paper due to number limitation.

References

Bell FG, Maud RR (2000) Landslides associated with the colluvial soils overlying the Natal Group in the greater Durban region of Natal, South Africa. Environ Geol 39(9):1029–1038

Disaster Prevention Board of Tokyo Metropolitan Government (1990) Characteristics on volcanic activities in Izu Archipelago, p 40. (in Japanese)

Japan Meteorological Agency (1958) Kanogawa typhoon, Technical Report, 37, p 140. (in Japanese)

Koyama M, Hayakawa Y (1996) Syn- and post-caldera eruptive history of Izu Oshima Volcano based on tephra and loess stratigraphy. J Geogr Tokyo Geogr Soc 105(2):133–162. (in Japanese)

Nakamura K (1964) Volcano-stratigraphic study of Oshima Volcano, Izu. Bull Earthq Res Inst Univ Tokyo 42(4):649–728. (in Japanese)

Terashima T, Matsushi Y, Hattanji T (2014) Hydrological aspects of shallow landslide occurrence in Izu-Oshima Island, triggered by a heavy rainfall of typhoon 26th on October 15–16, 2013. Ann Disaster Prev Res Inst Kyoto Univ 57A:17–24. (in Japanese)

Tokyo Metropolitan Government (2014) Report of committee on disaster mitigation plan after slope disaster in 2013. (in Japanese)

Wang G, Jiang Y, Chang C, Doi I, Kamai T, (2019) Volcaniclastic debris avalanche on Motomachi area of Izu-Oshima, Japan, triggered by severe storm: phenomenon and mechanisms. Eng Geol 251:24–36

Lessons from Recent Geo-Disasters in Hokkaido Under Heavy Rainfall

2020 Kyoto Japan

Tatsuya Ishikawa

Abstract

In the Japanese archipelago, four typhoons continuously hit during mid to late August 2016, which led to significant damage in wide areas of Hokkaido. Considering the social importance of this heavy rainfall-induced geo-disaster, the Japanese Geotechnical Society (JGS) organized "JGS Survey Team for Geotechnical Disasters in Hokkaido, Japan Induced by Heavy Rainfall on August, 2016," comprising experts from both industry and academia of the Hokkaido branch of the JGS. The aim of the survey team was to investigate the phenomena and factors contributing to disaster prevention/mitigation from both short- and mid- to long-term perspectives and provide academic advice to related organizations. This study summarizes the site investigation of a disaster-stricken area along the National Highway Route 274 around the Nissho Pass, as well as the endogenous and exogenous factors for slope failures caused by this historical heavy rainfall.

Keywords

Heavy rainfall • Geo-disaster • Climate change

Introduction

Three typhoons Chanthu (No. 7), Kompasu (No. 11), and Mindulle (No. 9) hit Hokkaido, Japan, in rapid successions during a week from August 17 to August 23, 2016, followed by Typhoon Lionrock (No. 10). These typhoons were responsible for heavy rains triggering severe floods and landslides over wide areas in Hokkaido. Particularly, Typhoon No. 10 was enormously activating a seasonal rain front, and thus having brought a record-breaking heavy rainfall starting from August 29 to 31, which resulted in the most serious geo-disaster ever reported in Hokkaido affecting wide areas of the Hidaka Mountains. Therefore, the Japanese Geotechnical Society (JGS) organized a survey team for the geo-disasters caused by the heavy rainfall in August 2016, mainly comprising experts from the industry and academia of the Hokkaido branch of JGS. The survey team investigated the phenomena and factors contributing to disaster prevention measures from short- and mid- to long-term perspectives and provided academic advice to related government organizations.

Based on the "Geo-disaster report on the August 2016 Hokkaido Heavy Rainfall," written by the JGS survey team (JGS 2017a), this study summarizes the site investigation of the disaster-stricken Hidaka Mountains area along the National Highway Route 274 around Nissho Pass, focusing on both endogenous and exogenous factors for heavy rainfall-induced slope failures. Furthermore, this study highlights future research issues of disaster recovery and disaster prevention/mitigation based on the features of the damage that can be attributed to the heavy rainfall in August 2016 in Hokkaido.

Weather Conditions in Tokachi Region

As shown in Fig. 1, Typhoon Nos. 7, 11, and 9 landed in Hokkaido one after another from August 17 to August 23, 2016. It was for the first time since the Japanese Meteorological Agency started routine observation of weather in 1951 that three typhoons hit Hokkaido in a rapid succession. A week later, Typhoon No. 10, which had been moving north over the Pacific Ocean, made landfall on Iwate Prefecture, maintaining its strong power and causing a record-breaking heavy rainfall in Tokachi region of Hokkaido. Because of this, a remarkable damage was caused to

T. Ishikawa (✉)
Hokkaido University, Kita13 Nishi8, Kita-ku,
Sapporo, 0608628, Hokkaido, Japan
e-mail: t-ishika@eng.hokudai.ac.jp

Ž. Arbanas et al. (eds.), *Understanding and Reducing Landslide Disaster Risk*,
ICL Contribution to Landslide Disaster Risk Reduction,
https://doi.org/10.1007/978-3-030-60713-5_13

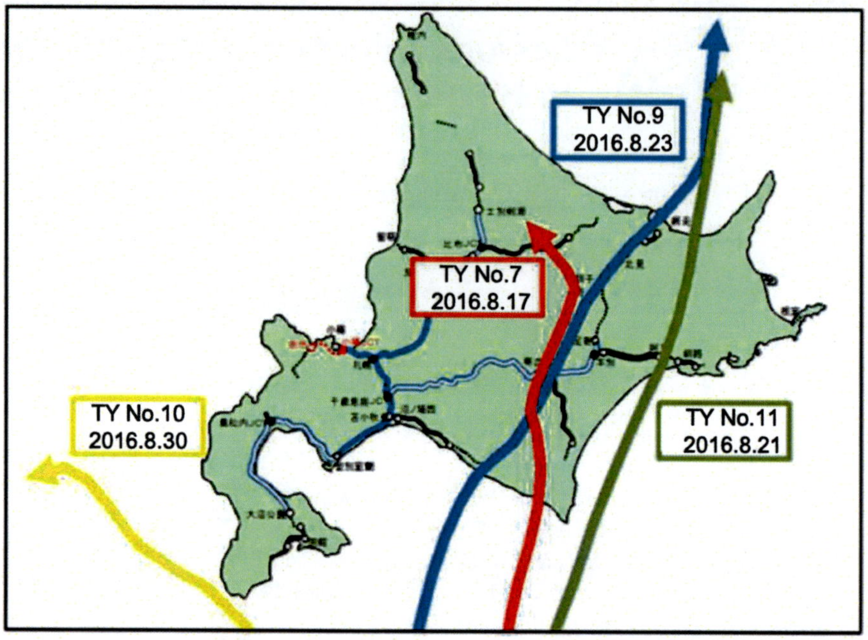

Fig. 1 Tracks of Typhoon Nos. 7, 9, 10, and 11

the region, which was the worst among these typhoons in terms of devastation. The cumulative heavy rainfall (total rainfall) of three days during August 29–31 exceeded 300 mm (i.e., 500 mm) in the Tokachi region around the Hidaka Mountains (Fig. 2). This rainfall caused multiple slope failures in the Hidaka Mountains as well as flood and river bank collapses along the rivers of Tokachi. In particular, extremely heavy rainfall was observed near the Nissho Pass at the National Highway Route 274; this pass was considerably damaged due to a debris flow and the collapse of its embankment. Precipitations observed at various rain gauges have replaced the past highest records. JGS (2017b) reported that this heavy rainfall in areas around the Hidaka Mountains was mainly induced by "orographic rainfall (Fig. 3)." As shown in Fig. 3, Typhoon No. 10 brought a large amount of warm and humid air continuously over and over into the eastern side of the Hidaka Mountains, thereby causing heavy rainfalls from rain clouds that have developed over the mountain slopes.

Figure 4a shows the precipitation that was observed at the National Highway Route 274 Nissho Pass TeleMeter (TM): 55 mm of the maximum hourly rainfall and 488 mm of the cumulative rainfall. Figure 4b shows that this cumulative rainfall is the largest since the start of the observations at the Nissho Pass TM; it is over twice as much as the second-highest rainfall observed in 2006. In two weeks (August 16–31), the total rainfall owing to the four typhoons in 2016 was 929 mm, which equals 73% of the usual annual precipitation at the Nissho Pass, as shown in Table 1. Figure 5a shows the precipitation at the National Highway Route 236 Nozuka Pass TM, in which heavy rainfall ascribed to orographic rainfall was observed. Around the Nozuka Pass, Typhoon No. 10 caused a cumulative rainfall of 713 mm, exceeding that of the Nissho Pass TM. However, no major damage was reported near the Nozuka Pass, although certain smaller-scale landslides were observed (see later Fig. 11) compared with the Nissho Pass; this can be related to the past rainfall history of the site. Comparing the annual precipitation for the past ten years between the Nissho Pass and Nozuka Pass in Table 1, the rainfall of the Nozuka Pass is twice that of the Nissho Pass. Furthermore, as shown in Fig. 5b, there are past experiences of similar rainfall at the Nozuka Pass; this can result in higher resistance to rainfall-induced geo-disasters than the Nissho Pass.

Geo-disasters in the Hidaka Mountains Area

Summary of Damage Along National Highway Route 274

Figure 6 shows the damaged area between Hidaka-cho and Shimizu-cho (42.9 km) on National Highway Route 274. In this area, 10 bridges were noted to be damaged, including falling of bridge, 3 snow shed damages, 6 road pavement severe damages, and 47 cut/embankment collapses (a total of 66 locations). This route passes through the Hidaka Mountains, which is the backbone of Hokkaido. Moreover, on the Hidaka town side, metamorphic and plutonic rocks belonging to the Hidaka metamorphic belt, and non-metamorphic

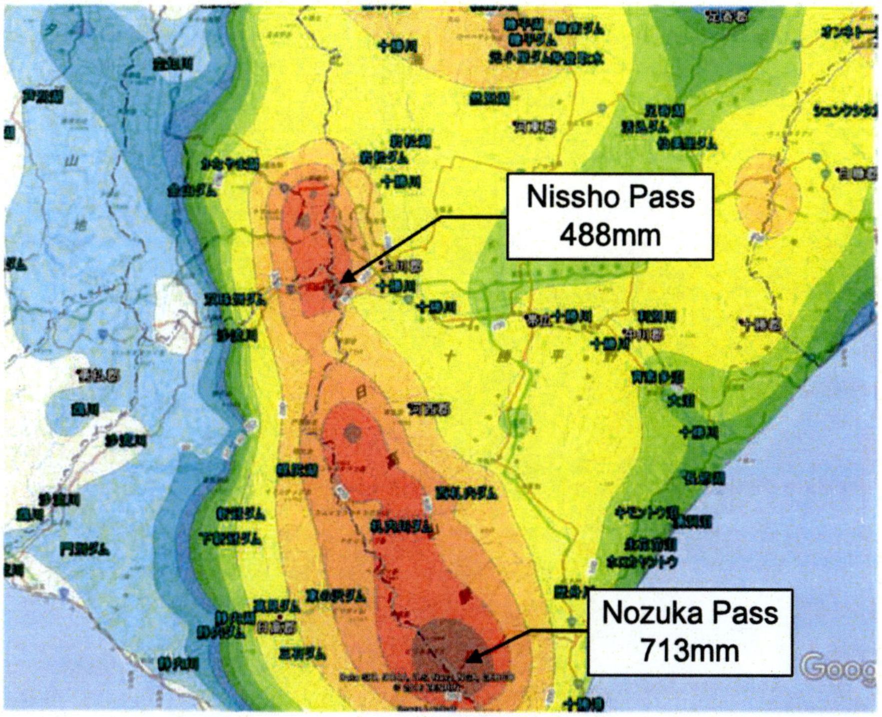

Fig. 2 Rainfall distribution during Typhoon No. 10

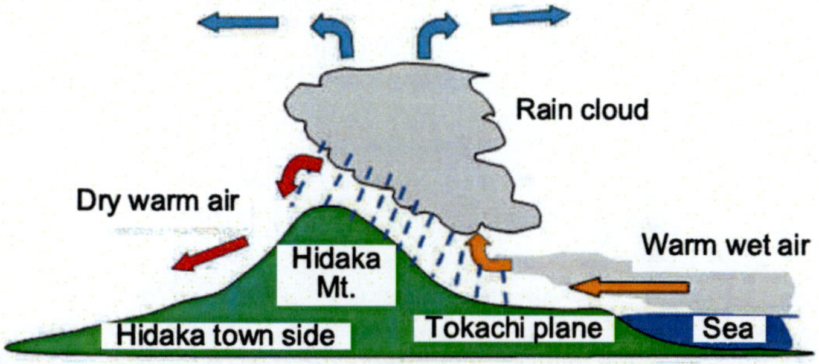

Fig. 3 Orographic rainfall

sandstones and mudstones belonging to the Hidaka and Sorachi-Yezo belts are distributed. However, from the top of the Hidaka Mountains to the Shimizu town side, granite that is more weathered at the shallower parts of subsurface layer owing to climate conditions penetrates the sedimentary rocks in the Hidaka belt. The upper part of the weathered granite is covered by periglacial slope deposits mixed with gravel, which was formed via freeze–thaw action and solifluction. Owing to these considerable damage, National Highway Route 274 was closed for approximately one year from August 30, 2016 (when the geo-disaster struck) to October 29, 2017 (when the restoration was completed).

The following section provides a detailed explanation of the situation and causes of the geo-disaster on Shimizu and Hidaka town sides based on the site investigation of geotechnical aspects.

Geo-disaster at the Shimizu Town Side

A large-scale collapse of embankments and cut slopes was observed near the 7th stage and the entrance of the Nissho Tunnel along National Highway Route 274 (Shimizu town side). Figure 7 shows the embankment collapse and debris flow near the 7th stage. Figures 7a and b demonstrate that the road significantly collapsed at the curved part, and the road body was largely damaged. During the investigation, there were traces of a large amount of surface water and debris flow on the pavement surface flowing down from the higher the 8th stage direction. This indicates that both erosion and scour induced by the surface flow can cause a large-scale collapse at the top of the embankment slope where the road alignment changes from a straight line to a curve. Figures 7c and d show

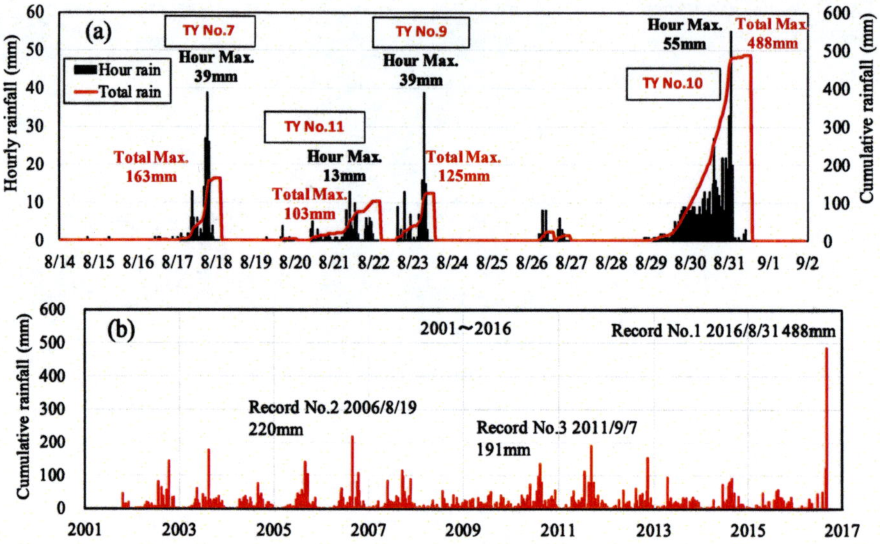

Fig. 4 Precipitation observed at Nissho Pass TM

Table 1 Annual precipitation at Nissho & Nozuka Passes

Nissho Pass TM (mm)				Nozuka Pass TM (mm)			
Year	Rain	Year	Rain	Year	Rain	Year	Rain
2006	1429	2011	1553	2006	2899	2011	3506
2007	1304	2012	1346	2007	2600	2012	2396
2008	989	2013	1291	2008	2120	2013	2969
2009	1490	2014	1239	2009	2965	2014	2642
2010	1748	2015	1053	2010	3562	2015	2735

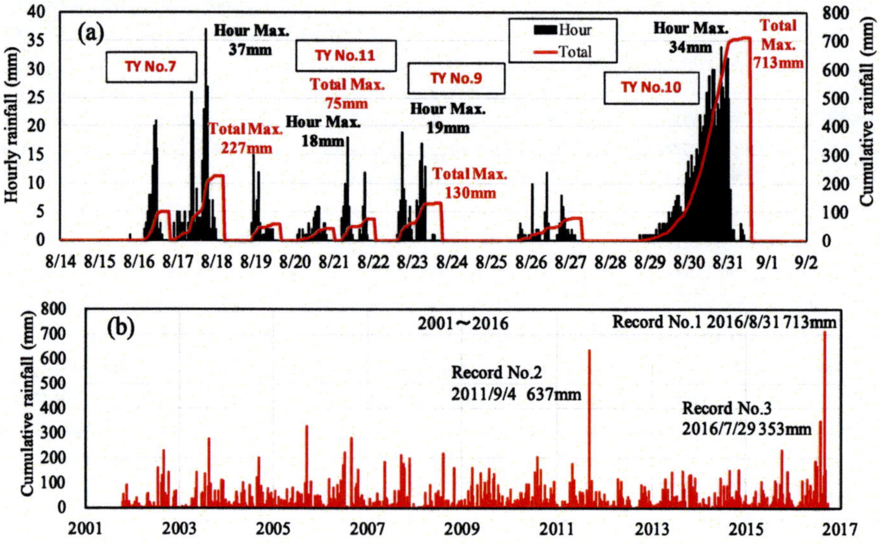

Fig. 5 Precipitation observed at Nozuka Pass TM

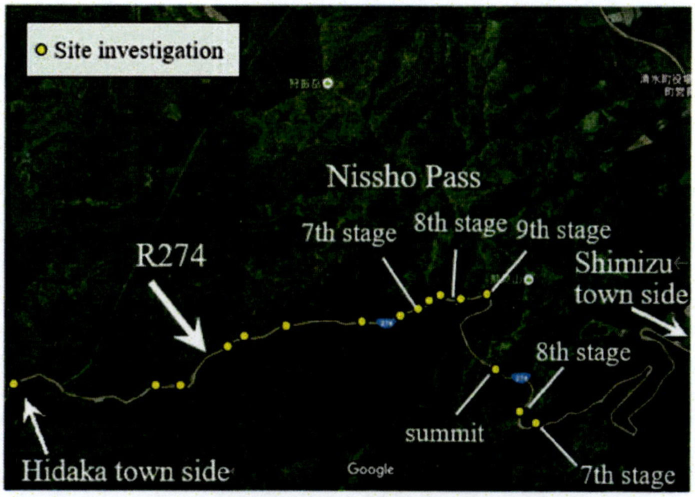

Fig. 6 Damaged spots on National Highway Route 274

the debris flow around the 7th stage (near the Ishiyama tunnel); here, the sediment, stones/rocks, and trees had cascaded on the road surface. The soil comprising the debris flow was "Masado," a fine-grained weathered granite. Figure 8 shows the debris flows and the accompanying collapse of the embankment near the 8th stage.

The features of the geo-disaster from the Shimizu town side to the Nissho Pass along National Highway Route 274 are as follows: (a) The extra heavy rainfall observed at the 7th stage or higher caused a large amount of rainwater to infiltrate into the soil ground that comprised fine-grained weathered granite, thereby weakening the soil ground (Figs. 7a and b). (b) Heavy rainfall caused a debris flow in the catchment area, such as a valley; it combined with the surface water flow to flow down the road surface and accumulate at points where the road alignment changed, thereby eroding and scouring the top of the embankment slope (Figs. 7c and d). (c) A large amount of sediment and driftwood clogged the drainage pipes under the road, reducing the drainage function; this induced further erosion in the surroundings of the pipes and caused the embankment to collapse (Fig. 8).

Geo-disaster at the Hidaka Town Side

Along the National Highway Route 274 (Hidaka town side), river flooding scoured the bridge foundation and damaged the road body at the base of the mountain; the cut and embankment slopes collapsed at the summit. Figures 9a and b show the collapse of the embankment near the ninth stage, which was constructed on sloping grounds. In this case, the shoulder of the road was eroded and scoured by surface water flow on the road surface; a large amount of rainwater infiltrated the embankment from the boundary between the natural ground and embankment. Consequently, as shown in Fig. 9c, the pore water pressure increased, thus reducing the shear strength and leading to a collapse. Figure 9d shows the collapse of the cut slope near the 8th stage, which were ascribed to the following three factors. (a) A water passageway easily formed because the soil ground on the slope was heterogeneous. (b) The hinterland of the slope had a small and shallow catchment topography, where surface water easily accumulated. (c) A large amount of water, resulting from overflowing drain ditches on the berm and crest owing to the exceeding drainage capacity, caused erosion and scour of the slope. However, in the areas lower than the 7th stage, many cases were seen such that the rapid flooding of the Saru river caused erosion and scour of the road embankments (Fig. 10a), backfilling of abutments (Fig. 10b), retaining wall foundations (Fig. 10c), and bridge foundations (Fig. 10d) near the water-colliding front, leading to the collapse.

Lessons from Geo-disasters in Tokachi Region

Issues for Future Study Considering Disaster Situation

The site investigation confirmed that this disaster was caused by the largest rainfall in the history of observations in the last 20 years. Therefore, it is estimated that, because the annual precipitation was relatively less, major geo-disasters rarely hit the region. Furthermore, because the design conditions of drainage work such as drainage ditches and drainage pipes in the earthworks are based on previously experienced rainfall, the drainage capacity is limited to snowy cold regions such as Hokkaido where the annual precipitation is lesser than warm regions. This indicates that, as a future task, the design

Fig. 7 Slope failure and debris flow near the 7th stage

conditions for rainfall must be reconsidered to cope with unprecedented rainfall caused by recent climate change. Therefore, various studies must be conducted on the potential risks of never-experienced geo-disasters associated with climate change to upgrade the conventional design, construction, and maintenance methods for earth structures.

This chapter summarizes certain suggestions for road, railway, and river management in snowy cold regions based on the results of the above disaster investigation. The following four points are deemed as critical geotechnical research items for the future design, maintenance, and disaster prevention/mitigation of earth structures in consideration of the regional characteristics of Hokkaido. (a) Consider the influences of precipitation habituation on the behavior of soil ground. (b) Examine heavy rainfall measures exceeding conventional assumptions. (c) Investigate the possibility of manifesting a new problematic soil/collapse mode caused by climate change. (d) Consider the necessity of treating surface water flow in slope measures during heavy rainfall. The details of each research item are discussed below.

Influences of Precipitation Habituation

On the Pacific side of Western Japan, the annual precipitation is estimated to be at 2000–3000 mm, while in regions with heavy rainfall in Hokkaido (such as southeastern Hidaka Mountains, including the Nozuka Pass), the annual precipitation is 1600–2000 mm. Excluding the mountainous areas, most regions of Hokkaido demonstrate a general annual precipitation of 600–1300 mm, which is only about 33–50% of that of Western Japan. Moreover, in Western Japan, the risk of slope failures increases when the daily precipitation is >200 mm. However, in Hokkaido, the risk is higher when the daily precipitation is >40 mm. This is because when heavy rainfall is less frequently observed in the past, the unstable soil layer thickly covers the slope surface; thus, the collapse easily occurs even during light rainfall. However, if the slope has collapsed before and the

Fig. 8 Slope failure and debris flow near the 8th stage

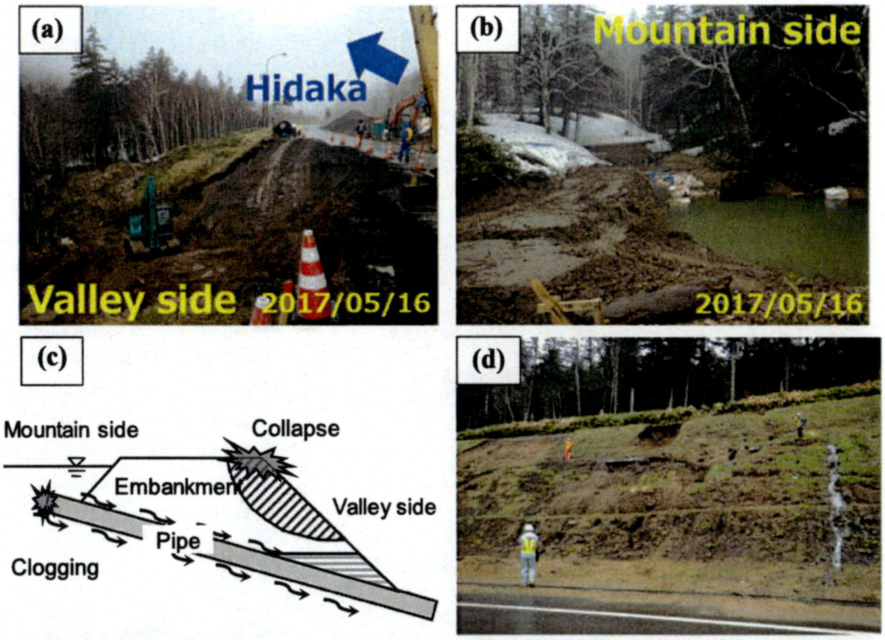

Fig. 9 Slope failures near the 8th and 9th stages

Fig. 10 Geo-disasters caused by river flooding at lower than the 7th stage: **a** embankment, **b** abutment, **c** retaining wall, **d** bridge

collapsed slope has been restored and reinforced, its stability becomes high, which is called "precipitation habituation." For example, Fig. 11 shows the damage to the Nozuka Pass during August 2016, Hokkaido heavy rainfall disaster. However, despite undergoing similar amount of rainfall, the damage at the Nozuka Pass seems lesser than that at the Nissho Pass. This can be attributed to the precipitation habituation of the soil ground. Therefore, it is critical to consider countermeasures using comparative studies with heavy rain areas for climate change in snowy cold regions such as Hokkaido, which is not accustomed to heavy rainfall.

Fig. 11 Surface slope failure at Nozuka Pass

Rainfall Measures Exceeding Conventional Assumptions

In Hokkaido, the effects of typhoons and torrential rainfall were lesser than the rainy warm regions; consequently, most of the soil on the ground is not accustomed to rainfall. Accordingly, compared with heavy rainfall regions in the main island of Japan, the marginal rainfall of slope failures in Hokkaido is lesser, and slope failures are more likely to occur with the same amount of heavy rainfall. According to the site investigation, unexpected rainfall exceeding the drainage capacity of earth structures and accompanying large amounts of seepage and surface water caused blockage and overflow of drainage ditches and drainage pipes at many spots; this resulted in slope failures. These spots were conventionally recognized as necessary countermeasures such as valley-filling embankment. Therefore, it is necessary to seek a comprehensive design, construction, and maintenance of earth structures along with measures to prevent erosion and infiltration by adopting the appropriate drainage treatment as well as measures to increase the strength of earth structures by selecting appropriate embankment materials and compaction methods. In this case, it is necessary to confirm the drainage capacity and validity of existing drainage systems, assuming heavy rainfall in the future. Thus, it is desirable to discuss the necessity of reviewing the design precipitation and installation standards for drainages.

New Problematic Soil/collapse Mode by Climate Change

The site investigation reported that uncommon problematic soil such as weathered granite (Masado) and periglacial sediment, which have not been reported to date, caused geo-disasters around the Hidaka Mountains area. However, three unusual soils, namely, volcanic soils, peat, and serpentine, which were considered problematic soils in Hokkaido, were not reported to be involved in the geo-disasters in this area. The soil ground and soil properties, which have received less interest in geotechnical engineering in Hokkaido to date, have emerged as a cause of this geo-disaster because of unusually heavy rainfall; this heavy rainfall ranked first in the observation history over the past 20 years. If heavy rainfall, such as the geo-disaster investigated here, is observed in other areas with similar topographical and geological compositions, a similar geo-disaster may occur. Therefore, it is estimated that systemizing of new types of geo-disasters specific to Hokkaido, which may occur because of climate change, and evaluation of the potential risk of geo-disasters unexperienced in Hokkaido will have significant implications for disaster prevention/mitigation in future.

Surface Water Flow in Slope Measures During Heavy Rainfall

If the rainfall intensity is higher than the infiltration capacity of the soil ground, the collapse risk of soil ground and earth structures must be evaluated at the wide areas by considering the surface and seepage flow during rainfall. For example, Fig. 12 shows the results of a rainfall-runoff analysis using 3D analysis model of the slopes around the seventh stage of the Nissho Pass, created using a 1 m mesh digital elevation model. This shows that rainwater that cannot infiltrate the soil ground during heavy rainfall flows over the road as surface flow. It tends to flow down the slope from the road toward the embankment slope at slope-failure points. Therefore, when considering countermeasures for torrential rainfall, which cannot be handled by the design precipitation, the flow network of the surface flow should be analyzed by evaluating various information such as geography, topography, geology, soil properties, and incidental structures of roads. Furthermore, it is desirable to eliminate the risk of potential geo-disasters during heavy rainfall and re-examine the design and maintenance methods of road structures, which are the outflow routes of surface flows.

Conclusions

Considering the occurrence of large-scale geo-disasters similar to those reported in this study, it seems that recent disaster countermeasures in Hokkaido have stepped into a new stage. For promoting disaster prevention and mitigation measures of civil engineering structures with long service lifespan against geo-disasters in Hokkaido under this

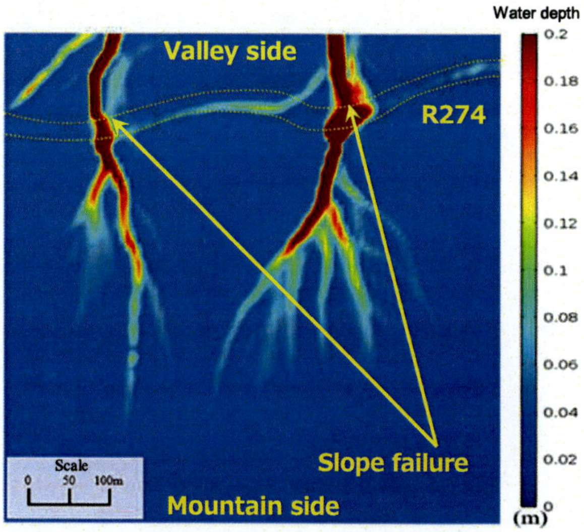

Fig. 12 Distribution of water depth by a rainfall runoff analysis (after Zhu et al. 2018)

frequent disaster stage never experienced before, it is necessary to work on the situation while sharing the sense of crisis throughout society. In particular, it is essential to establish a framework at the earliest, in which the industry, government, and academia can work together on the technical and social issues concerning large-scale geo-disaster prevention discussed in this study. Concrete measures should be drafted and implemented in the future while incorporating the suggestions and recommendations provided in this study. Therefore, there is a requirement to further study the influences of climate change on geo-disasters in snowy cold regions.

Acknowledgments The author sincerely acknowledges all those who provided "JGS Survey Team for Geotechnical Disasters in Hokkaido, Japan Induced by Heavy Rainfall on August, 2016" with valuable information and research supports, particularly the Hokkaido Regional Development Bureau, Ministry of Land, Infrastructure, Transport and Tourism and Hokkaido Government, and Profs. S. Kawabata, Y. Kohata, and S. Kawamura (members of the JGS survey team).

References

The Japanese Geotechnical Society (2017a) Geo-disaster report on the August 2016 Hokkaido Heavy Rainfall, Survey Team for Geotechnical Disasters in Hokkaido, Japan Induced by Heavy Rainfall on August, 2016, JGS. https://www.jiban.or.jp/wp-content/uploads/2017/08/final_report_ver0.12s.pdf (in Japanese)

The Japanese Geotechnical Society (2017b) Proceedings of Symposium on geo-disasters in snowy cold regions associated with climate change. JGS Hokkaido Branch Research Committee on Geo-disaster Risk in Snowy Cold Regions associated with Climate Changes (in Japanese)

Zhu Y, Ishikawa T, and Shimizu Y (2018) Surface flow analysis of Typhoon 10 induced slope failure based on digital elevation modelling. In: Proceedings of the 9th symposium on sediment-related disasters, pp 121–126

Lessons from Geo-Disasters Caused by Heavy Rainfall in Recent Years in Kyushu Island, Japan

2020 Kyoto Japan

Noriyuki Yasufuku and Adel Alowiasy

Abstract

Recently, heavy rainfall events have induced several geo-disasters, floods, sediments, and debris flow at different regions around Japan which in turn have caused severe damage to life and properties. According to several reports from the intergovernmental panel on climate change and other research institutes, the localized torrential rainfall events frequency is expected to increase. Under such circumstances of the anticipated climate change, the increase in the geo-disasters inducing forces such as rainfalls, the deterioration of the social infrastructure, and the decline in the overall geo-disasters prevention capabilities as a result of the changes in social structure due to the reduction in the working power were considered in this study. By carefully comparing and analysing the situation of the repeated geo-disasters and reflecting the obtained results to the geo-disasters mitigation and prevention practice in Kyushu, Japan, developing innovative system and techniques that integrate the academic disciplines, in collaboration with the local residents and government is now strongly needed.

Keywords

Geo-disasters prevention and mitigation • Torrential rainfall • Sediments and debris flow

N. Yasufuku (✉) · A. Alowiasy
Graduate School of Engineering, Kyushu University, Fukuoka, 8190395, Japan
e-mail: yasufuku@civil.kyushu-u.ac.jp

A. Alowiasy
e-mail: adel_owaisi@yahoo.com

Introduction

On the 5th and 6th of July 2017, a heavy rainfall storm has struck Northern Kyushu Island, Japan. The storm affected mainly the Northern part of Fukuoka prefecture (Asakura City) and the Northern part of Oita prefecture (Hita region). The storm which was named by the Japanese Meteorological Agency (JMA) as "Northern Kyushu heavy rainfall in July 2017" has caused severe damage to the mountainous area located between Asakura city and Hita region. Cumulative precipitation of 511.5 mm and 329.5 mm after 12 h were recorded at Asakura meteorological agency observatory and Hita rainfall observation centre respectively. Also, 532 mm cumulative precipitation spanning for 12 h was recorded at the Tsurukawauchi rainfall observatory, which significantly exceeds the cumulative precipitation corresponding to Kyushu Island heavy rainfall events that had occurred in 2009 and 2012 (JGS 2013, 2018). Consequently, several mudflows, debris flows, and landslides simultaneously took place within the affected area.

The slopes located within the affected area are mainly comprised of granodiorite and metamorphic rocks that have experienced severe weathering. Therefore, besides the large-scale collapse of the top parts of the slopes, scouring and collapse of the beds and shores in the middle basin also occurred resulting in generating large amounts of sediments and driftwood that have flooded and accumulated in the downstream region spreading over the private houses and farmlands, causing extensive sediments, driftwoods, and water-induced damages to life and properties.

A Geo-disaster investigation team was formed to conduct in-situ surveys at the affected mountainous area. The survey aimed at investigating the slopes situation and conducting a series of geotechnical tests to define the soil mechanical and hydrological characteristics of the affected area.

The ultimate priority in the case of such large scale geo-disasters is to avoid any human-related damage and fatalities. Several approaches to protect human lives were

Ž. Arbanas et al. (eds.), *Understanding and Reducing Landslide Disaster Risk*, ICL Contribution to Landslide Disaster Risk Reduction,
https://doi.org/10.1007/978-3-030-60713-5_14

developed, such as setting up special infrastructures (Sabo dam "safety system utilizing special facility equipment"). However, considering the current heavy rainfalls, novel approaches based on software measures such as "an evacuation warning system" and "restrictions on land use" to protect human life and to define vulnerable zones are in great need.

Through this paper, a comprehensive summary of the heavy rainfall-induced slopes related Geo-disaster incidents during recent years in Japan is presented. In addition, the learned lessons are reflected aiming at developing a fundamental database for providing appropriate preparations for future similar heavy rainfall-induced geo-disasters and technical issues from a geotechnical point of view. The term "Geo-disaster" is used in this context to describe the slopes disasters such as debris flow including driftwood, slopes failure, landslide, and embankment damages.

Characteristics and Potential of the geo-disasters During Northern Kyushu Heavy Rainfall in July 2017

Based on the aforementioned explanation, Northern Kyushu heavy rainfall in July 2017 event is not only considered as one of the strongest rainfall events based on 1–3 h cumulative precipitation, but also can be considered as one of the strongest heavy rainfall events lasting for a long time spanning from 6 to 12 h. The duration of this heavy rainfall is relatively long-lasting for more than 9 h which is considered an unusual phenomenon. This can be distinctively observed by comparing the cumulative precipitation of this event to the cumulative precipitation record for Hiroshima heavy rainfall in 2014 as illustrated in Fig. 1.

In general, including the recent heavy rainfall events in 2009 and 2012 "Northern Kyushu heavy rainfall events", a distinct frequent occurrence of high precipitation in a relatively short time events were observed. Statistical studies including data for the past 40 years in Japan have revealed that there is a distinct increasing tendency for a more frequent occurrence of heavy rainfall events with hourly precipitation of 50 mm and over 80 mm (JGS 2013, 2018).

According to the report released by the Intergovernmental Panel on Climate Change (IPCC 2013), the global warming will result not only in increasing the heavy rainfall-induced geo-disasters, but will also affect the rainfall concentration, frequency of typhoons and tornados, and wind speed. To cope with the significant increase in the potential of geo-disasters occurrence, effective adaptations and implementations are expected to be in high demand for the coming near future. Under such circumstances, the geotechnical engineering is expected to play a major role.

Types of Heavy Rainfall-Induced Slope Failures and Sediments Movement

The causes leading to slope failures can be generally categorized into predisposition and inducing causes. Generally, the predisposition category takes place due to many reasons including the geological conditions, topographical conditions, and presence of vegetation. On the other hand, in the case of Northern Kyushu heavy rainfall in July 2017, the main cause of the induced slopes failure is the heavy rainfall. For a specific slope to collapse, it is necessary to have sufficient predisposition conditions and a trigger to induce the failure (Iseda et al. 1982). Only when those conditions coexist such phenomenon takes place.

To be more specific, the slope failure can be related to many factors including (1) increase of the sliding force and decrease of the soil strength and resistance due to the increase of the soil degree of saturation, (2) decrease of the

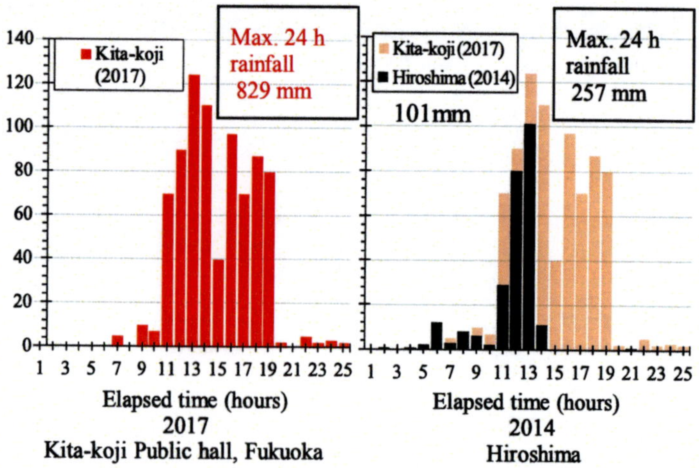

Fig. 1 Comparison of precipitation records (Fukuoka and Hiroshima prefectures in Japan)

effective stress due to the rise of the groundwater table level which results in increasing the pore water pressure, (3) collapsing of a soil layer or a bedrock stratum, (4) generation of an osmotic pressure acting as a slipping force due to seepage flow, (5) scouring, erosion, and transport of sediments during surface runoff (flow), (6) piping phenomenon caused by water flowing in joints and cracks, (7) in the case where the rainfall significantly exceeds the infiltration capacity of the ground, where the water flowing into the ground through the surface layer at a specific rate induces the generation of a hydraulic gradient leading to developing a shearing force that contributes to the sliding force (Cui et al. 2019; JGS 2018; AMeDAS 2018; IPCC 2013; Iseda et al. 1982). It must be noted that such kind of typical relationships change over time due to many reasons such as the progress of weathering of soils and rocks.

Considering Northern Kyushu heavy rainfall in July 2017, the aforementioned 7 factors might have coexisted because of the unprecedented heavy rainfall that lasted for a relatively long time in that area. As a result, several slope failures occurred at the same time within the affected area specifically the zone located at the right bank of Chikugo River in Asakura region. Natural slope failures induced by the coexistence of the aforementioned factors are not well understood yet. Therefore, the development of a comprehensive slope stability analysis and collapsing potential evaluation techniques based on in-situ investigations and a series of laboratory testing is in great need.

Figure 2 summarizes the relationship between the precipitation and the collapsed area ratio of the slope at the right bank of the Chikugo river basin for several periods of time (Jitozono 2017; Committee of Chikugo river 2017). It can be observed that when the precipitation exceeds 100 mm after 1 h, 250 mm after 3 h, 350 mm after 6 h, 400 mm after 12 h, and 450 mm after 24 h, the collapsed area ratio increases rapidly. It is believed that if there is a system that is capable of comprehensively analysing such statistical data and utilizing it in risk assessment reflecting the regional characteristics, it will serve as an effective tool when considering disasters prevention and mitigation countermeasures of a specific area.

Estimating the amount of generated sediments and driftwoods due to extreme heavy rainfall-induced slope failures is important when determining the type and scale of the structure (such as Sabo dam) to be built around rivers' facilities. For that purpose, it is crucial to properly evaluate the scale of the slope failure in relation to the scale of the heavy rainfall event. However, the scale and failure pattern vary significantly depending on many factors such as the geological conditions, topographical conditions, vegetation

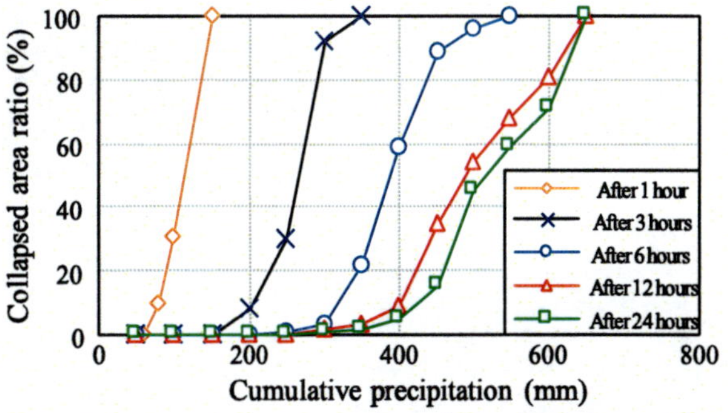

Fig. 2 Relationship between the cumulative precipitation and the collapsed area ratio Committee of Chikugo river (2017)

cover, and rainfall intensity. Until now, an accurate method to estimate the amount of generated sediments does not exist. As a trial to provide a base for such system, the details of the sediment transport were analysed focusing on the heavy rainfall-induced slope failure that has occurred at Akatani river basin located at Asakura area, where the relationship between the cumulative precipitation and the generated amount of sediments is plotted as illustrated in Fig. 3 (Jitozono 2017).

It can be observed that the higher the cumulative precipitation leads to increasing the amount of generated sediments following a power function. Besides, it must be noted that a typical collapse mode occurs once the cumulative precipitation reaches the range of 200 mm to 500 mm. Although this figure doesn't include the actual accurate amount of generated sediments, reflecting the history of such disasters can serve as a basis for predicting the type of failure and the amount of sediments that might be generated under specific cumulative precipitation that might fall on those basins. When considering such heavy rainfall-induced geo-disasters in mountainous areas, enhancing the resiliency of the basin is highly recommended. Therefore, developing a proper evaluation approach as elucidated in Fig. 3 is strongly required, where the system should consider analysing the existing risks and factors to be used in future mitigation and countermeasure plans.

Slope Failures Shapes and Patterns

The geometrical shape, such as the collapsed height and depth of the heavy rainfall-induced collapsed slopes has been analysed by Kasama et al. (2018). The studied area included the basin where slope failures were frequently observed in Asakura city, Fukuoka prefecture (Akatani river, Otoishi river, Shirakidani river, Souzu river, Kita river, and

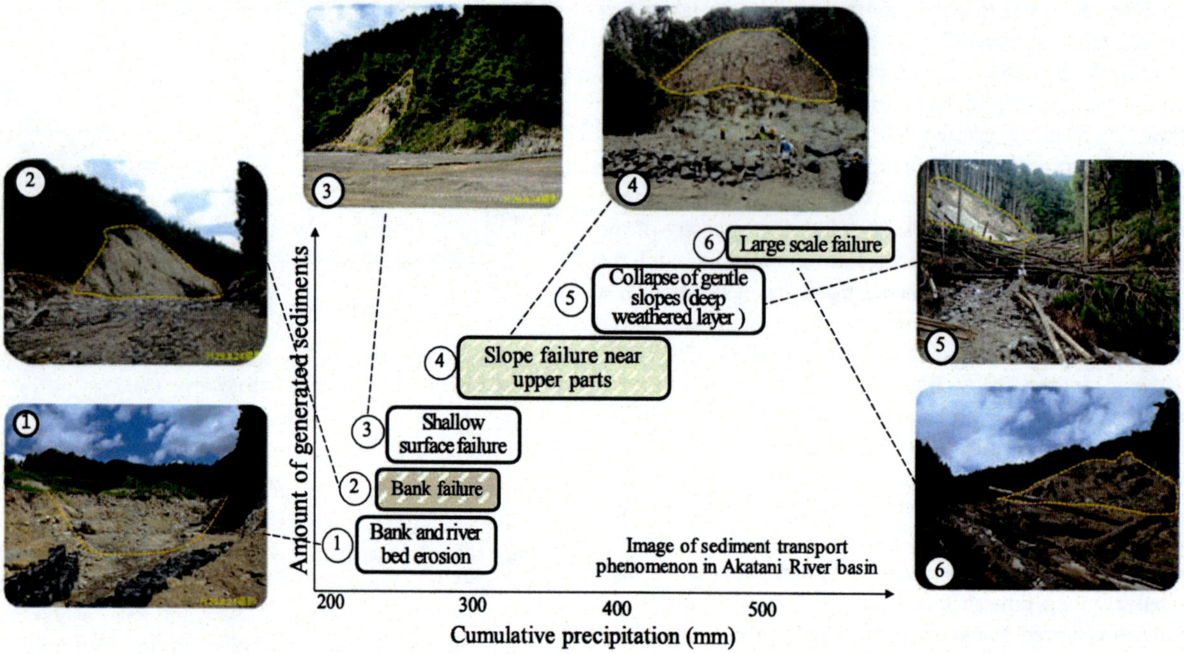

Fig. 3 Relationship between the cumulative precipitation and the amount of generated sediments arranged by the authors after Jitozono (2017)

Naragaya river). A comparison utilizing a laser scanning results of the area profile before and after the collapse was carried out. Some of the results are presented in this study, where a description of the adopted methods and the geological information can be found in Kasama et al. (2018).

Figure 4 shows the frequency distribution and the cumulative frequency of the slopes before and after the failure versus the inclination angle for all the studied basins in Asakura city. In addition, the average cumulative frequency of slope failures due to rainfall events all around Japan is also illustrated for comparison (Koyamauchi et al. 2009). It can be observed that the highest heavy rainfall-induced landslides frequency in Northern Kyushu rainfall disaster 2017 corresponds to an inclination angle of around 40°, while the average failed slopes inclination angle around Japan is about 60°. It must be noted that 80% of the number of slope failures occurred at an inclination angle of 38° or less, which can be considered as a special feature of this region, where the inclination angle is 20° smaller than that of the average angle all around Japan. By conducting cumulative frequency analysis and comparison of Kyushu region to Japan, it is expected that the characteristics of the collapsed slopes subjected to heavy rainfall events can be defined from a statistical perspective and accumulated as a record.

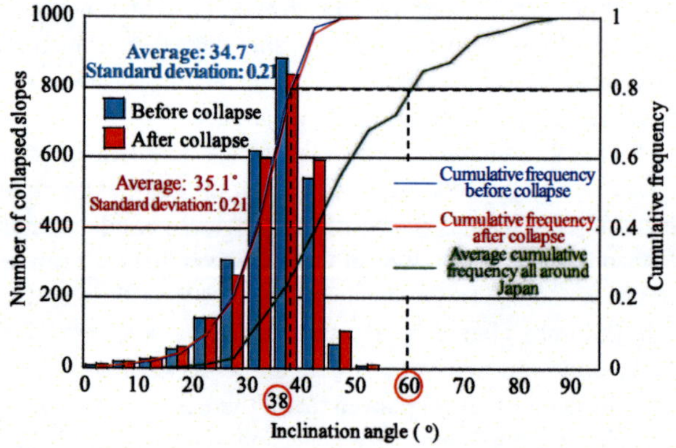

Fig. 4 Slope's angle distribution of collapsed slopes Kasama (2018)

Figure 5 summarizes the average collapsed depth for all the studied basins in Asakura city. It can be seen that the average landslide and slope failure depth in all basins ranges from 0.4 m to 1.3 m. However, several large landslides from deep layers corresponding to an average depth exceeding

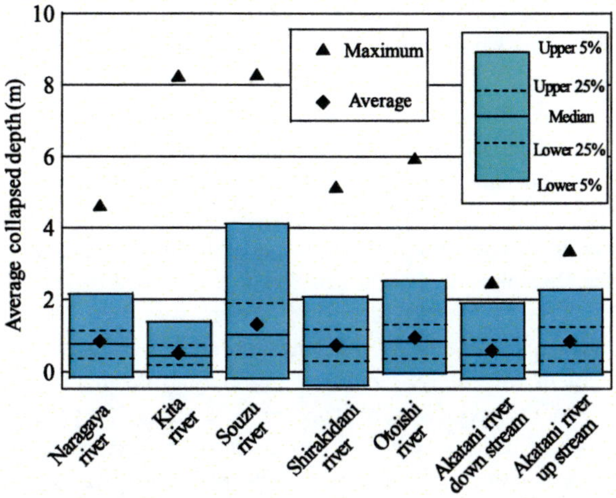

Fig. 5 Distribution of the average collapsed depth Kasama (2018)

8 m have occurred. In the future, integrating various indices reflecting the rainfall analysis and soil volume analysis based on existing statistical slope failure studies, erosion, sedimentation, will serve as a real-time slope risk assessment method. This system is in great need to be established and expected to be used as an efficient tool for preventing and mitigating heavy rainfall-induced landslides.

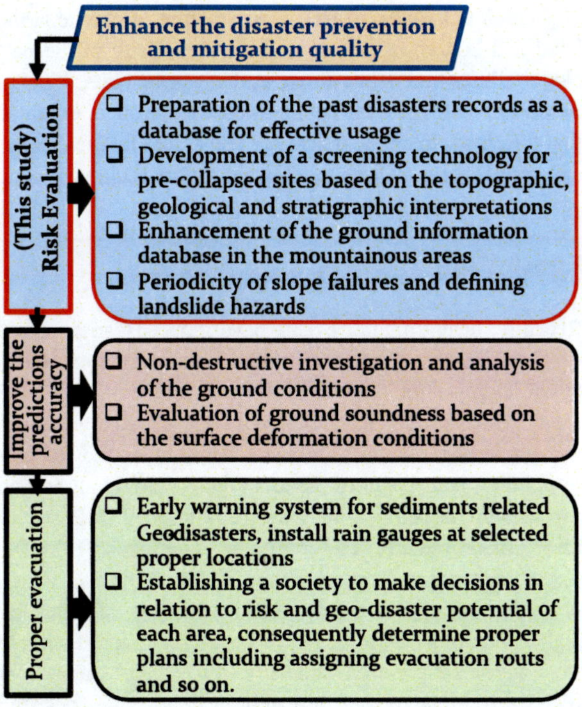

Fig. 6 Geotechnical considerations for slope related geo-disasters mitigation and prevention system

Technical Issues Related to geo-disasters Prevention and Mitigation

Based on the lessons learned from disasters history in Kyushu, some of the issues that should be addressed clearly in geotechnical engineering that are directly linked with the geo-disasters prevention and mitigation for future are summarized, see Fig. 6.

Preparation of the Past Disasters Records as a Database for Effective Usage

Northern Kyushu area has experienced several heavy rainfall-induced geo-disasters that have occurred repeatedly with different scales over the past decades. Although many valuable heavy rainfall-induced geo-disasters are recorded in Northern Kyushu and all around Japan, the records are not efficiently organized to be used in risk management and mitigation. Those records are preserved separately in each department, for example, the administrative office. The records are not effectively used as an integrated geo-disasters information due to the lack of appropriate digital storage format of the data and the lack of proper, accessible, and easy to use system. Considering the current situation, it seems to be essential to create an organized system to collect the geo-disasters history records in a unified format in cooperation with the national and local governments under the leadership of academic and professional engineers specialized in geo-disasters prevention and mitigation. The efforts to create and organize such database and system in an accessible format are expected to be conveyed to the next generation as a reference for future geo-disasters.

Development of a Screening Technology for Pre-collapse Sites Based on the Topographical, Geological, and Strati-Graphical Interpretations

Recently, the accuracy of the laser profile scanning data and the C-X synthetic band radar has remarkably improved. In addition, the image analysis technology using drones has well improved. In the last decade, many organizations have been utilizing the latest technologies in developing topological interpretations and screening techniques based on the analysis of the geological and stratigraphic conditions of the ground. With such technologies, if the vulnerable slopes can be defined by practical indices with high accuracy, it will contribute significantly to the process of improving the quality and efficiency of geo-disasters prevention and

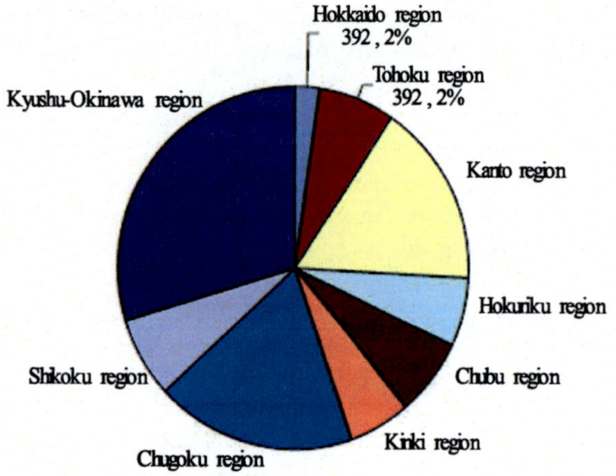

Fig. 7 Slopes geo-disasters frequency since 1967 Ministry of Land, Infrastructure, Transport and Tourism (2013)

mitigation. Such academic and practical integrated approaches are highly recommended.

Enhancement of the Ground Information Database in the Mountainous Areas

Kyushu branch of the Japanese Geotechnical Society has created and published a database including ground and geological information with more than 80,000 boring data in seven prefectures within Kyushu Island, Japan. The database is opened for access to public users through the JGS society. These data are remarkably large in number including both urban and coastal areas. However, to create a database that contributes to the recovery and restoration in the case of a Geo-disaster that occurs in a mountainous region, it is essential to enrich the ground and geological data of the mountainous regions. There are many cases where geotechnical engineers can efficiently respond to the requests of the state and local governments. For example, in the case where engineers collect the data obtained from the ground surveys at the time of a specific Geo-disaster recovery and then establish a system for checking the collected ground and geological data. For recent geo-disasters, it is necessary to construct a suitable ground and geological information database for disasters response that can be effectively used in making decisions to determine a countermeasure after analysing the information for a specific region. In addition, establishing an easy-to-understand Geo-disaster risk analysis method from an academic perspective to objectively explain the reasons why a group of specific decisions can be done as a response for a specific Geo-disaster is strongly needed.

Periodicity of Slope Failures and Landslide Related Disasters

Detailed studies on the periodicity of landslides or slope failures in granite rocks distribution area have been conducted in the 1980s (Shimokawa et al. 1984). Based on the field survey, the relationship between the elapsed years since the occurrence of the previous landslide and the changes in the thickness of the surface soil layer, in addition to the relationship between the rainfall catchment area and the average cycle time of the landslide were summarized. In light of this study, for an average failure depth of about 0.7 m, the periodicity is expected to be about 200 years. In order to estimate the risk and potential of a slope failure corresponding to an elapsed time properly, it is necessary to extensively extend the research to consider the surface soil weathering and restoration rate and slope failure periodicity as functions of time and geology of the ground. In an area comprised mainly of granite, surface soil weathering and restoration at the collapsed site and sediment deposits that have accumulated overtime on the bed become key factors when considering geo-disasters. Therefore, it is not sufficient to specify a landslide disaster based only on an ordinary topographical analysis. In order to reduce the risk of slopes related geo-disasters, it is important to investigate the distribution of sediments on slopes and mountain streams, vegetation at the site, degree of weathering of granite at the head and bottom of slopes, and the periodicity of the past slope failures and debris flows at that relevant site. Finally, based on such practical detailed knowledge, it is expected that the efforts to define landslides and slopes failure disaster susceptible areas will be accelerated by the society.

Conclusions

Recently, heavy rainfall events have induced various types of geo-disasters, floods, sediments and debris flow at different regions not only in Kyushu Island but all around Japan which in turn have caused severe damage to human and properties. Figure 7 illustrates the number and the corresponding percentage of the slope geo-disasters all around Japan in the period ranging from 1967 to 2012 (Ministry of Land, Infrastructure, Transport and Tourism 2013). The slopes related geo-disasters rate in Kyushu-Okinawa region accounts for approximately 31%, which is significantly higher than other regions, where on average more than 390 slope geo-disasters occur annually. According to reports of IPCC (2013) and other research institutes, the localized torrential rainfall events frequency is expected to increase in the future. Under such circumstances of the anticipated climate change, the increase in the geo-disasters inducing forces such as rainfall, the deterioration of the social infrastructure, and the decline in the overall geo-disasters prevention capabilities as a result of the changes in social structures due to the reduction in the working power (where the Japanese society is considered as an elderly society due to the difference in the death to birth ratio) were considered in this study. By carefully comparing and analysing the situation of the repeated geo-disasters and reflecting the obtained results to the geo-disasters mitigation and prevention practice, developing innovative system and techniques that integrate the academic disciplines, in collaboration with the local residents and government is now strongly needed more than ever, see Fig. 6.

References

Committee of Chikugo river right bank basin and Sabo dam restoration technology (Chairman Komatsu T) (2017) Report on Chikugo river right bank basin and Sabo dam restoration technology. (Japanese)

Cui Y, Cheng D, Choi CE, Jin W, Lei Y, Kargel JS (2019) The cost of rapid and haphazard urbanization: lessons learned from the Freetown landslide disaster. Landslides 16(6):1167–1176

Digital Typhoon (2018) AMeDAS torrential rain-past heavy rainfall ranking. http://agora.ex.nii.ac.jp/digitaltyphoon/heavy_rain/[2018.3]. (Japanese)

Intergovernmental Panel on Climate Change [IPCC] (2013) Climate Change 2013, Cambridge University Press, Cambridge, United Kingdom and New York, NY, USA

Iseda T, Ochiai H, Tanahashi Y (1982). Relationship between actual conditions of landslides and rainfall characteristics. Report on a disaster survey caused by heavy rainfall in Nagasaki in 1982 July, Nagasaki University Academic Research Team, 59–71. (Japanese)

Japan Geotechnical Society [JGS] (2013) Report on Geodisasters in 2012 July Northern Kyushu torrential rainfall (Chairman Yasufuku N). (Japanese)

Japan Geotechnical Society [JGS] (2018) Report on "Geodisasters in 2017 July Northern Kyushu torrential rainfall" (Chairman Yasufuku N). (Japanese)

Jitozono T, (2017). Sediment-related disasters in the Ono district of Hita City, textbook of the Sabo Society of Japan seminar. 19–30. (Japanese)

Kasama K (2018) Field study on slope failure analysis and the related deep weathering. Special Grants-in-Aid for Scientific Research of FY2017 (Representative member: Juichiro Akiyama), 17k20140. 3 (3). 127–139. (Japanese)

Koyamauchi N, Tomita Y, Akiyama K, Matsushita S (2009) The actual conditions of the landslide, National Institute of Land and Infrastructure. No. 530. 1–210. (Japanese)

Ministry of Land, Infrastructure, Transport and Tourism (2013). Annual report on disaster management 2012

Shimokawa E, Jitozono T, Hori Y (1984) History of landslides in granite area. Nichirin Kyushu branch annual paper. 37:299–300

Comparison of Relationship Between Debris-Flow Volume and Peak Discharge in Different Regions

2020 Kyoto Japan

Tao Wang and Mingfeng Deng

Abstract

Debris flows frequently occurred in Wenchuan earthquake region from 2008 to 2010, resulting in great damage to localities and being a prolonged threat to reconstruction. Twenty eight events' data including volume and peak discharge of debris–flow were analysed and compared with other debris-flow events in Switzerland, Japan, USA, China (Jiangjia gully), etc. The analysis reveals that there is a strong empirical relationship between debris-flow volume and debris-flow peak discharge in the earthquake region. However, the relationship between debris-flow volume and debris-flow peak discharge in Wenchuan earthquake is different with other regions. The volume of debris-flow with strong influence of earthquake is larger than that with no such influence under same peak discharge of debris-flow.

Keywords

Debris flow • Peak debris-flow discharge • Debris-flow volume • Wenchuan earthquake

Introduction

Debris flows are a phenomenon intermediate between collapse or landslides and fluvial sediment transport (Rickenmann 1999). Debris flows most often occur as a result of intense rainfall or typhoon but there are also other triggering mechanisms such as snowmelt or dam-break failure (Costa 1984; Berti and simony 2005; Breien et al. 2008; Coe et al. 2008; Kean et al. 2011; Chang et al. 2011; Theule et al. 2012). Destabilizing factors (blasting, strong earthquake) are favourable for the formation of debris flow. The 2008 Wenchuan earthquake in SW China in the Sichuan province generated many collapse and landslides, which delivered a lot of loose material. It caused obvious increase of debris flows occurrence in the subsequent years (Tang et al. 2009; Cui et al. 2010; Tang et al. 2011; Xu et al. 2012). Subsequent debris flows shortly after earthquake may have a larger scale than before earthquake for a certain catchment with similar rainfall return period (Ma et al. 2013).

It was reported that debris flows which were initiated from strong sediment entrainment, dam breaching, progressive bulking of runoff, mobilization and transformation of landslides, or a combination of these processes, were one of the main types of geological disasters in the meizoseismal area of Wenchuan earthquake (Tang et al. 2009; Chen et al. 2011; Tang et al. 2011a, b). On 24 of September 2008, debris flows which were induced by rainstorms in Beichuan County inundated large area and killed forty-two people (Xie et al. 2008; Hu et al. 2010; Tang et al. 2011a). From 12 to 14 of August 2010, debris flows were induced in Qingping town of Mianzhu city, Yingxiu town of Wenchuan county and Longchi town of Dujiangyan city (Tang et al. 2011a; Xu et al. 2010). From 9 to 11 of July 2013, debris flows which were triggered by sudden rainstorms along highway from Dujiangyan county to Wenchuan county washed away homes, factories and highways. These towns are belonged to the worst-hit areas of the Wenchuan Earthquake. These events indicate very different relationships between peak debris-flow discharge and debris-flow volume.

Firstly in this paper, the relationship between debris-flow volume and peak debris-flow charge in Wenchuan earthquake area is examined on the basis of 28 debris-flow events. Secondly, data of debris-flow volume and peak debris-flow discharge in Switzerland, Kamikamihori gully (Japan), Jiangjia gully (China) and other areas are obtained

T. Wang (✉) · M. Deng
Key Laboratory of Mountain Hazards and Earth Surface Processes, Institute of Mountain Hazards and Environment, CAS, Chengdu, 610041, China
e-mail: wtwang@imde.ac.cn

M. Deng
e-mail: dmf@imde.ac.cn

Ž. Arbanas et al. (eds.), *Understanding and Reducing Landslide Disaster Risk*, ICL Contribution to Landslide Disaster Risk Reduction,
https://doi.org/10.1007/978-3-030-60713-5_15

and compared with that in the Wenhuan earthquake area. Thirdly, empirical relationship between debris-flow volume and peak debris-flow discharge in Wenchuan earthquake area is compared with other existing relationships. Finally, causes that empirical relationship between debris-flow volume and peak debris-flow discharge in Wenchuan earthquake area is different from other areas are determined.

Study Area and Data Sources

The study area is located in the transitional mountainous belt characterized by rugge mountains and deep valleys (i.e. the terrible Mw 8.0 Wenchuan earthquake zone, Fig. 1). Longmenshan fault belts, the main geology structures in the area, consist of three sub parallel faults: Beichuan-Yingxiu fault (seismic fault) extends through the northern western part of rugged mountains and characterizes by dip-slip reverse faulting (Huang and Li 2009; Wei and Tao 2009); other two faults are Jiangyou-Guanxian fault and Qingchuan- Wenchuan-fault. Each stratum from Proterozoic to Cenozoic group partially outcropped stratigraphic gap existed. The study area is underlain by Granitic rocks, Sinan pyroclastic rocks, Carboniferous limstones, and Trisssic sandstones. Loose quaternary deposits are distributed in the form of terraces and alluvial fans (Xu et al. 2012). The Min River and Fu River are the most important rivers in this region formed by both the tributaries of Yangtze River. The study area belongs to the subtropical monsoon climate zone characterized by high precipitation. The annual average rainfall over a period of 30 year is 800–1600 mm which is concentrated in a rainy season from June to September.

Method

A total of 28 debris-flow events happened in different watersheds along the seismic fault where earthquake strongly disturbed rocs strata are chosen as study cases. These events were triggered by local rainstorms with 10–20 year return period and picked out from clustering-occurrence debris-flow events in Beichuan, Dujiangyan, Mianzhu and Wenchuan (Fig. 1; Table 1). The data of these events including Watershed area, main channel length, slope of main channel, burst-time of debris flow, duration-time of debris flow, density of debris flow, peak debris-flow discharge, debris-flow volume, and loose material volume from scientific papers (Chen et al. 2009; Ma et al. 2011; Geng et al. 2012; Li et al. 2012; Liu et al. 2012; Chang et al. 2012; Yu et al. 2013; Liang et al. 2013; Kong et al. 2014; Yan et al. 2014), and unpublished geological engineering report with local authorities(Table 1). The unpublished geological engineering reports are from local authorities, for example, the emergency investigation report of geohazards in Chenjiaba Town of Beichuan County and Clustering debris flows and landslides in Weijia basin in Western of Becichuan County by Sichuan Insitute of Geological Engineering Investigation.

Empirical Relationship for the Study Cases

From the point of view of the evaluation of a potential hazard, the debris-flow volume, V_d, is one of the most important parameters. In general, a spectrum of possible debris-flow volumes can be expected to occur with different probabilities (Rickenmann 1999). An acceptable recurrence

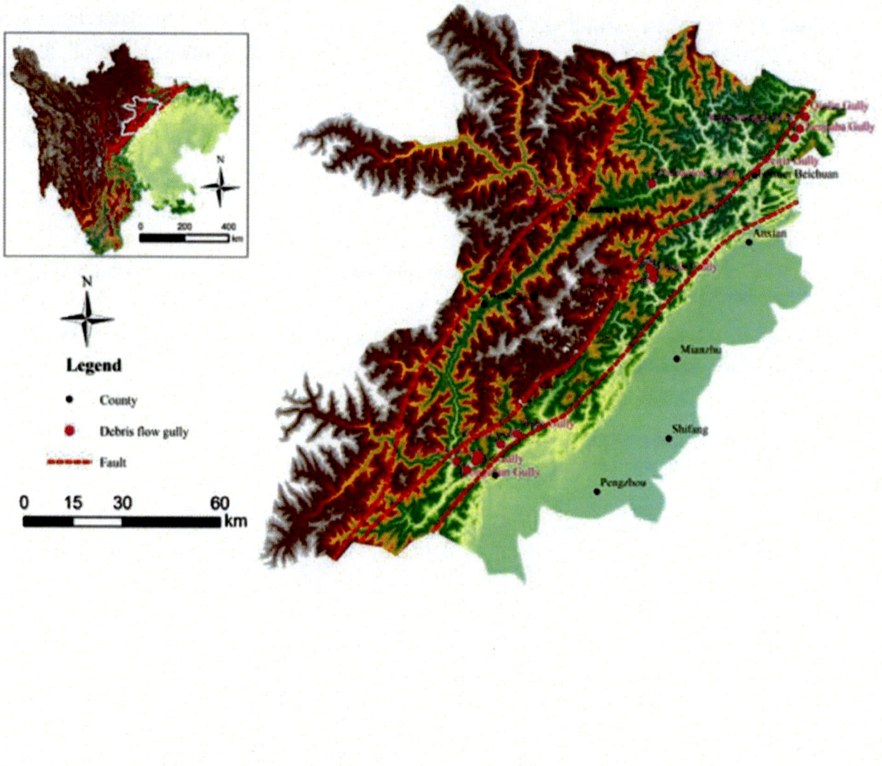

Fig. 1 Location study area and debris flow cases (after earthquake)

Table 1 Debris-flow Volume and peak debris-flow discharge of 28 debris flow events

No.	Gully name	A	L	S	T_1	T_2	ρ	Q_p	V_d	V_L
1	Zhubaotou	25.44	9.64	0.13	18/06/2008	–	2.00–2.20	300.00	10.00	300.00
2	Jianping	3.60	2.80	0.36	13/08/2010	90.00	1.90	80.10	11.50	272.00
3	Chunyashu	0.66	1.56	0.52	13/08/2010	70.00	1.94	150.00	12.60	26.10
4	Bayi	8.50	4.45	0.38	13/08/2010	100.00	1.88	1082.00	116.50	757.60
5	Sanshengong	0.12	0.43	0.67	13/08/2010	40.00	1.80	148.30	9.40	–
6	Xiaogou	5.36	4.80	0.19	17/07/2009	120.00	1.38	157.00	6.60	80.00
7	Fengyongyan	0.30	0.83	0.65	13/08/2010	40.00	2.00	104.00	7.40	–
8	Sunjia	1.54	2.40	0.45	13/08/2010	80.00	2.00	57.00	7.20	77.00
9	Kushui	3.94	2.95	0.28	13/08/2010	20.00	2.10	71.00	1.60	77.10
10	Lengjin	2.69	2.70	0.45	13/08/2010	30.00	1.82	478.60	17.40	197.00
11	Luojia	1.40	2.92	0.33	13/08/2010	135.00	2.12	231.80	30.20	156.00
12	Shuida	0.46	1.30	0.50	13/08/2010	30.00	1.95	515.84	24.14	74.10
13	Zoumaling	5.76	3.39	0.32	13/08/2010	90.00	2.24	866.35	121.64	1614.00
14	Dagan	1.10	2.10	0.37	17/07/2009	30.00	–	176.00	6.30	81.00
15	Wenjia	7.81	3.25	0.47	13/08/2010	150.00	2.22	1530.00	310.00	3000.00
16	Weijia	1.54	2.30	–	24/09/2008	60.00	–	260.00	34.00	510.00
17	Hongchun	5.35	3.60	0.36	14/08/2010	165.00	–	287.00	43.22	384.00
18	Qinglin	23.70	9.90	0.105	24/09/2008	510.00	1.71	697.00	80.50	1372.00
19	Yangchangzi	5.80	3.75	0.21	24/09/2008	200.00	1.49	87.50	4.30	573.30
20	Sanqingcun	2.79	2.65	0.356	24/09/2008	30.00	1.89	150.00	5.00	59.7.00
21	Guxigou	14.08	7.70	0.17	10/07/2013	–	2.00	722.00	40.00	>1000.00
22	Kuagou	1.01	2.73	0.24	08/2008	40.00	1.73	5.32	0.33	4.72
23	Cutou	21.8	7.80	0.26	10/07/2013	180.00	1.85	515.00	50.00	1108.00
24	Liquantai	0.32	0.99	0.67	13/08/2010	40.00	2.00	143.50	9.00	60.00
25	Maojiawan	2.80	2.77	0.51	03/07/2011	45.00	1.75	95.30	5.20	–
26	Mozi	7.30	5.20	0.49	14/07/2008	30.00	1.95	541.00	25.70	>3000.00
27	Miaoer	34.10	9.60	0.11	11/08/2010	–	1.75	520.00	72.12	2663.00
28	Linjiaba	0.05	0.63	0.72	24/09/2008	30.00	1.57	2.83	0.05	4.82

Notes A is the area of basin with a unit of km^2, L is the main channel length of basin with a unit of km, S is the slope of main channel, T_1 is the burst-time of debris flow, T_2 is the duration-time of debris flow with a unit of minute, ρ is the density of debris flow with a unit of g/cm^3, Q_p is the peak discharge of debris flow with a unit of m^3/s, V_d is the debris flow volume with a unit of $\times 10^4$ m^3, V_L is the loose material volume with a unit of $\times 10^4$ m^3

interval with an associated even magnitude has to be defined when designing any protection measures (Rickenmann 1999; Chen et al. 2013).

Previous research has shown that, there is an empirical relationship between debris-flow volume (V_d) and peak debris-flow diachargw (Q_p) (Hungr et al. 1984; Mizuyama et al. 1992; Jitousono et al. 1996; Rickenmann 1999). Mizuyama proposed that the empirical relationships should be different for granular type debris-flow and mudflow type debris-flow. The Empirical relationships for Merapi Volcano (Indonesia) and Sakurajim Volcano (Japan) were presented through analyzing the field data (Jitousono et al. 1996). From the assumption that Froude Scaling must be satisfied for debris-flows of different size, a semi empirical equation was presented (Rickenmann 1999).

By a regression tool in Matrix laboratory, the empirical-statistical relationship between debris-flow volume and peak debris-flow discharge in Wenchuan earthquake area can be presented by this form:

$$Q_p = 0.028V_d^{0.747} \quad (1)$$

Where V_d is debris-flow volume with a unit of cubic meters (m^3), and Q_p is peak debris-flow discharge with a unit of cubic meters per second (m^3/s). The determined coefficient is 0.900.

Twenty eight debris flows are classified into small, medium, large and giant scales by the volume magnitude below 2×10^4 m^3, between 2×10^4 m^3 and 20×10^4 m^3, between 20×10^4 m^3 and 50×10^4 m^3 and over 50×10^4 m^3. There are 3 small, 13 medium, 7 large and 5 giant debris flows. Three lines representing the upper limit, empirical-statistical (Eq. (1)), and lower limit relationships of debris-flow volume and peak debris-flow discharge are plotted in Fig. 2. Equation (1) is the empirical relationship, and has a coefficient 0.028. Coefficients for other two lines are 0.066 and 0.012.

There is a big different debris-flow volume, which range from 0.05×10^4 to 310×10^4 m^3 for the 28 debris-flow events (Table 1). As for the 4 giant debris flow events, watershed area ranges from 5.76 to 23.7 km^2. Debris flows with the volume exceeding 20×10^4 m^3 caused the most serious debris flow hazard. Particularly, on September 24, 2008, about 34×10^4 m^3 materials were dashed out of Weijia watershed and inundated the old Beichuan County; debris flows in Qinglin Watershed blocked downstream Duba river on September 24, 2008; debris flows in Luojia, Wenjia and Zoumaling watersheds blocked Mianyuan river and inundated large parts of Qingping town on August 13, 2010 (Tang et al. 2011a; Chang et al. 2012; Yu et al. 2013; Liang et al. 2013); debris flow in Hongchun watershed blocked Min river and submerged the new Yingxiu town on August 14, 2010 by a volume of 43.22×10^4 m^3 (Li et al. 2012); about 116.5×10^4 m^3 sediments were transferred into Zipingpu reservoir on August 13, 2010, due to debris flow in Bayi watershed (Ma et al. 2011).

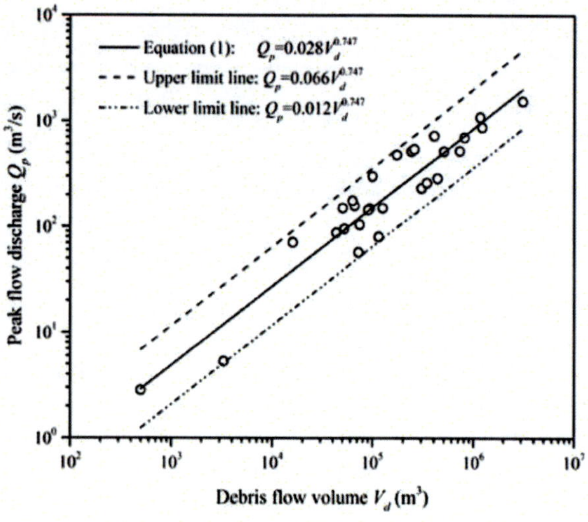

Fig. 2 Peak debris-flow discharge versus debris-flow volume in Wenchuan earthquake area

Among the 28 debris flow events, 57% of them are small to medium scale. Even so, these debris flows can still be destructive because most alluvial-fan and piedmont areas are urbanized and planted. Hydropower stations, highways and lots of recovering projects are suffered greatly. For example, debris flow in Jianping watershed washed away farm tourism facilities, Longxi highway and drinking water supply facilities at the mouth of gully on August 13, 2010; furthermore, about 17.4×10^4 m^3 materials were dashed out of Lengjin watershed and blocked Longxi river. Though many debris flows induced by rainstorm with 10 or 20 years return period, the measured peak debris-flow discharge and debris-flow volume are equivalent to the estimated values induced by a rainstorm of 50 or 100 years return period using traditional methods presented (Cui et al. 2010; Xu et al. 2012; Ma et al. 2013). Meanwhile, the occurrence frequency of debris flows increase significantly.

Comparison of the Empirical Relationship with Other Regions

Debris-flow volume and peak debris-flow discharge data from debris flow events with no influence of strong earthquake in Switzerland (Rickenmann and Zimmermann 1993), Japan (Okuda and Suwa 1981), USA (Pierson 1985), Jiangjia gully (Wang and Zhang 1990), study area before Wenchuan earthquake (Lei et al. 1993; Chen et al. 2003; Liu et al. 2004; Zhou et al. 2012; Deng et al. 2013; Zhong 2014), and other regions are selected to compare with the data in Wenchuan earthquake area (Table 2).

Debris flows in Wenchuan earthquake area exhibit a different relationship between the debris-flow volume and peak debris-flow discharge from other debris flow events with no influence of strong earthquake in Switzerland, Japan, USA, Jiangjia gully, study area before Wenchuan earthquake, and other regions (Fig. 3). It is found that, at same peak debris-flow discharge, debris flow volume in Wenchuan earthquake area is larger than debris flow events in Switzerland, Kamikamihori gully (Japan), Mt. ST. Helens (USA) and study area before earthquake. At same peak debris-flow discharge, debris flow volume in Wenchuan earthquake area is larger than debris flow events in Jiangjia gully in the vast majority of cases. On the whole, debris flow volume in Wenchuan earthquake area is larger than debris flow events in study area before Wenchuan earthquake over all under same debris-flow discharge.

Table 2 Data on debris-flow volume and peak discharge of debris flows in selected regions

Country/Region	N	V_d ($\times 10^4$ m^3)	Q_p (m^3/s)	Source
Swiss Alps (Switzerland)	7	0.40–6.00	120.00–650.00	Rickenmann and Zimmermann (1993)
Kamikamihori valley (Japan)	24	0.02–1.48	13.00–124.00	Okuda and Suwa (1981)
Jiangjia gully (China)	28	0.04–99.90	46.00–3133.00	Wang (1997)
Mt. ST. Helens (USA)	3	0.08–800.00	25.00–6800.00	Pierson (1985)
Study area (before Wenchuan earthquake)	8	2.20–52.00	70.00–3000.00	Several sources[a]
Other field data	38	0.07–9000.00	3.00–655000.00	Sources[b]

[a]Lei et al. (1993), Zhang and Deng (2012), Liu et al. (2004), Li and Zeng (2000), Chen et al. (2003), Zhong (2014), Deng et al. (2013)
[b]The data are gathered from published paper (Rickenmann 1999)

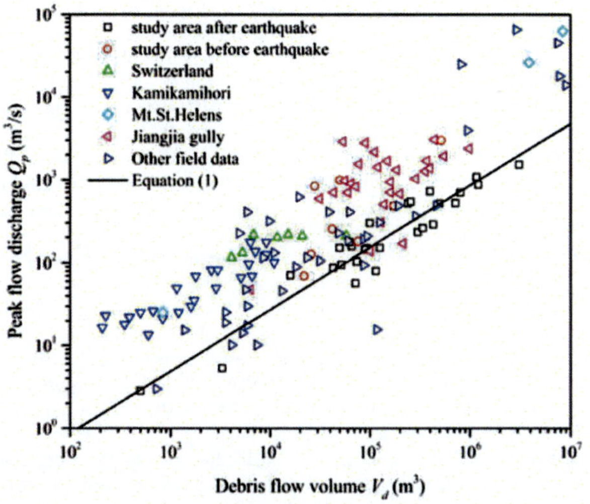

Fig. 3 Peak flow discharge versus debris flow volume in study area (after & before earthquake), Switzerland, kamikamihori gully, Mt. ST. Helens, Jiangjia gully, and other field data

Conclusions

There is a strong power relationship between the debris-flow volume and peak debris-flow discharge in the Wenchuan earthquake area, and the determined coefficient is 0.900.

Debris-flow volume in study area after earthquake is larger than other debris flows with no influence of strong earthquake in Switzerland, Japan, Jiangjia gully (china) and study area before Wenchuan earthquake under same peak debris-flow discharge.

Acknowledgements This study was sponsored the Key Research Program of the Chinese Academy of Sciences (Grant No. KFZD-SW-425) and the National Science Foundation of China (Grant No. 51409243). Furthermore, we would like to thank the anonymous reviewers and editors for their comments.

References

Berti M, Simoni A (2005) Experimental evidences and numerical modelling of debris flow initiated by channel runoff. Landslides 2:171–182

Breien H, Blasio F, Elverhi A (2008) Erosion and morphology of a debris flow caused by a glacial lake outburst flood, Western Norway. Landslides 5:271–280

Chang CW, Lin PS, Tsai CL (2011) Estimation of sediment volume of debris flow caused by extreme rainfall in Taiwan. Eng Geol 123:83–90

Chang M, Tang C, Li WL, Fu R, Zhao XH (2012) Characteristics of giant debris flow hazard of the Zoumaling gully occurred on August 13, 2010 in Qingping area of Mianzhu, Sichuan, China. J Chengdu Univ Technol (Science & Technology Edition) 39(6):629–635 (In Chinese with English abstract)

Chen NS, Hu GS, Deng MF (2011) Impact of earthquake on debris flow-a case study on the Wenchuan earthquake. J Earthq Tsunami 5:493–508

Chen NS, Liu ZG, Xie WY (2003) Survey report on August 28, 2003 Debris Flow In Shimian County, Sichuan Province. J Mt Sci 21 (5):639 (In Chinese with English abstract)

Chen XQ, Cui P, You Y, Li DJ (2013) Layout methods of control works preventing large scale debris flows in Wenchuan earthquake area. J Hydraulic Eng 44(5):586–593 (In Chinese with English abstract)

Chen XZ, You Y, Cui P, Liu JF (2009) Characteristics and risk analysis of Qinglin debris flow induced by "5.12" Wenchuan earthquake in Beichuan county, Sichuan, China. IEEE Int Geosc Remote Sens Sympos 2:408–411

Coe JA, Kinner DA, Godt JW (2008) Initiation conditions for debris flows generated by runoff at chalk Cliffs, central Colorado. Geomorphology 96:270–297

Costa JE (1984) Physical geomorphology of debris flows. Costa JE and Fleischer PJ(eds). Developments and applications of Geomorphology, Springer, Berlin, Germany, pp 268–317

Cui P, Zhuang JQ, Chen XC, Zhang JQ, Zhou XJ (2010) Characteristics and countermeasures of debris flow in Wenchuan Area after the earthquake. Journal of Sichuan University (Engineering edition). 42(5):10–19 (in Chinese with English abstract)

Deng ML, Han B, Zhu X, Guo H (2013) Formation Conditions and Kinetic Characteristics of the Debris Flow in Keji Gully. Yellow River 35(6):90–93 (in Chinese with English abstract)

Geng HH, Jia JC (2012) Features of Sanshengong debris flow hazard in Longchi Town of Dujiangyan city. J Natural Disataers 21(6):136–141 (in Chinese with English abstract)

Hu KH, You Y, Zhuang JQ (2010) Characteristics and countermeasures of debris flows in Beichuan's meizoseismal zrea. Scientia Geographica Sinica 30(4):566–570 (in Chinese with English abstract)

Huang RQ, Li WL (2009) Fault effect analysis of geo-hazard triggered by Wenchun earthquake. J Eng Geol 17(1):19–28 (in Chinese with English abstract)

Hungr O, Morgan GC, Kellerhals R (1984) Quantitative analysis of debris torrent hazards for design of remedial measures. Can Geotech J 21:663–677

Jitousono T, Shimokawa E, Tsuchiya S (1996) Debris flow following the 1994 eruption with pyroclastic flows in Merapi Volcano. Indonesia. Jpn Soc Erosion Control Engng 48:109–116

Kean J, Staley D, Cannon S (2011) In situ measurements of post-fire debris flows in southern California: comparisons of the timing and magnitude of 24 debris-flow events with rainfall and soil moisture conditions. J Geophys Res 116:F04019

Kong YD, Guo XJ, Zou Q, Zeng Q, Yan Y (2014) Giant debris floe hazards in Guxi gully, Wenchuan County of China on July 10, 2013. Mountain Res 32(3):320–324 (in Chinese with English abstract)

Li DH, Xu XN., Hao HB (2012) Formation conditions and the movement characteristics of "8.14" giant debris flow in Yingxiu Town, Wenchuan County, Sichuan Province. Chinese J Geolog Hazard Control 23(3):32–38. (In Chinese with English abstract)

Li ZH, Zeng KF (2000) Primary analysis of debris flow control benefi in Wen County, Gansu Province. Geolog Sci Technol Inform 19 (4):75–78 (in Chinese with English abstract)

Lei XY, Song LS, Shen ZM, Meng WY, Zhang HM (1993) Investigation on the violent debris flow Hazard in Jijiagou, Lueyangcounty of Shanxi province. J Catastrophol 8(2):43–49 (in Chinese with English abstract)

Liang JT, Tang C, Wang J, Chang M (2013) Development Characteristics of small watershed debris flows in meizoseismal area- A case study of Luojiagou gully in Mianzhu city in Sichuan province. J Catastrophol 28(3):100–104 (in Chinese with English abstract)

Liu QH, Tang C, Ma Y (2012) Features of Fengyongyan debris flow hazard on August 13 2010 in Longchi town of Dujiangyan, Sichuan, China. J Chengdu Univ Technol (Sci Technol Edn) 39 (6):636–641 (in Chinese with English abstract)

Liu XL, lyv XJ, Su PC (2004) Characteristics and hazard assessment of debris flow in Chayuan Gully of Wenchuan County in Sichuan. J Natl Disasters 13(1):66–71. (in Chinese with English abstract)

Ma C, Hu KH, Tian M (2013) Comparison of debris-flow volume and activity under different formation conditions. Nat Hazards 67:261–273

Ma Y, Yu B, Wu YF, Zhang JN, Qi X (2011) Research on the disaster of debris flow of bayi Gully, Longchi, Dujiangyan, Sichuan on August 13, 2010. J Sichuan Univ (Eng Sci Edn) 43(S1):92–97 (in Chinese with English abstract)

Mizuyama T, Kobashi S, Ou GQ (1992) Prediction of debris flow peak discharge. Proc. Int. Symp. Interpraevent, Bern,Switzerland, Bd.4, pp 99–108

Okuda S, Suwa H (1981) Some relationships between debris flow motion and microtopography for the Kamikamihori Fan, North Jalan Alp., In Burt TP and Walling DE (eds). Catchment Experiments in Fluvial Geomrpjology. pp 447–464

Pierson TC (1985) Initiation and flow behavior of the 1980 Pine Creek and Muddy River Iahars, Mount St. Helens. Washington Geol Soc Am Bul 96:1056–1069

Rickenmann D (1999) Empirical relationships for debris flows. Nat Hazards 19:47–77

Rickenmann D, Zimmermann M (1993) The 1987 debris flows in Switzerland: documentation and analysis. Geomorphology 8:175–189

Tang C, Zhu J, Li WL, Liang JT (2009) Rainfall-triggered debris flows following the Wenchuan earthquake. Bull Eng Geol Env 68:187–194

Tang C, Van A, TWJ, Chen GQ, Zhao XH, Huang XC (2011a) Catastrophic debris flows on 13August 2010 in the Qingping area, southwestern China: the combined effects of a strong earthquake and subsequent rainstorms. Geomorphology 140:559–576

Tang C, Zhu J, Ding J, Cui XF, Chen L, Zhang JS (2011) Catastrophic debris flows triggered by a14 August 2010 rainfall at the epicenter of the Wenchuan earth-quake. Landslides 8:485–497

Theule JL, Liebault F, Loye A, Laigle D, Jaboyedoff M (2012) Sediment budget monitoring of debris flow and bedload transport in the Manival Torrent, SE France. Nat Hazards Earth Syst Sci 12:731–749

Wang Z, Zhang X (1990) Initiation and laws of motion of debris flow. ASCE, Proc. Hydraulics/hydrology of Arid lands, pp 596–691

Wei YJ, Tao LJ (2009) Characteristics of Wenchuan earthquake and its geological hazard effects. In: Diederichs M, Grasselli G (eds) Proceedings of the 3rd CANUS rock mechanics symposium, Toronto

Xu Q, Zhang S, Li WL, Asch TWJV (2012) The 13 August 2010 catastrophic debris flows after the 2008 earthquake, China. Nat Hazards and Earth Syst Sci 12:201–216

Yan Y, Ge YG, Zhang JQ, Zeng C (2014) Research on the debris flow hazards in Cuntou gully, Wenchuan County on July 10, 2013. J Catastrophol 29(3):229–234 (in Chinese with English abstract)

Yu B, Ma Y, Wu YF (2013) Case study of a giant debris flow in the Wenjia Gully, Sichuan Province China. Natural Hazards 65:835–849

Zhang LL, Deng H (2012) Formation condition and trends of debris flow in Basishi gully. China Water Transport 12(11):185–187 (in Chinese with English abstract)

Zhong S (2014) Research on the disaster of Xiongjia Gully Debris Flow in Sichuan. J Changchun Inst Tech (Nat Sci Ed). 15(1):72–75. (in Chinese with English abstract)

Zhou W, Tang C, Zhou CH (2012) Critical rainfall characteristics for rainfall-induced debris flows in Wenchuan earthquake affected areas. Adv Water Sci 23(5):650–655 (in Chinese with English abstract)

Investigation of Internal Erosion of Wide Grading Loose Soil—A Micromechanics-Based Study

Yifei Cui, Yanzhou Yin, and Chaoxu Guo

Abstract

The Wenchuan earthquake (Mw 8.0) on May 12, 2008 led to abundant loose landslide deposits in the south-western mountainous areas of China. Field observations reveal that this loose soil is ideal source material for debris flows and landslides several years after the earthquake, particularly when mobilized by rainfall. An important slope failure mechanism is rainfall infiltration-induced fines migration within soil slopes. Previous studies of fine particle migration in soil mainly focused on seepage experiments, mid-scale flume tests with rainfall as a boundary condition, and other macro-scale methods. However, these methods have not been able to directly obtain parameters of pore structures, velocity of fine particles, and pore pressure inside soil samples which could be used to quantify the internal erosion process. In this study, the characteristics of wide-grading loose soils (WGLS) pore structure are analyzed quantitatively with serial tomography which uses scanning electron microscopy. The results of statistical analysis of pore size distribution are also presented. Compared with traditional silty soil with bimodal pore size distributions, WGLS corresponds to a second peak with much larger diameter particles. Numerical simulation using the Lattice Boltzmann method (LBM) coupled with the Discrete Element Method (DEM) is used to investigate the jamming probability of fine particle migration through three different samples during seepage. Simulation results indicate that jamming is most prevalent in samples with smaller pores that are dominated by bimodal pore size distributions, which agrees with previous analytical solutions.

Keywords

Shallow failure • Wide-grading loose soil • Micropore structure • LBM-DEM coupling • Fine particle migration

Introduction

Rainfall-induced slope failures pose significant hazards in many parts of the world. Among the types of rain fall-induced slope failure, many post-earthquake landslide deposits are easily remobilized by rainfall and cause subsequent shallow landslides and debris flows (Fan et al. 2018). For example, the large 2010 debris flow in Wenjia Ravine that caused 12 fatalities (Cui et al. 2017), was initiated from loose landslide deposits from the Wenchuan earthquake (Mw 8.0) in 2008. Additionally, slope deposits from the Chi-Chi earthquake were mobilized by heavy rainfall in the typhoon season and resulted in catastrophic debris flows (Shieh et al. 2009). The source material for a debris flow usually contains well-graded particles with a wide range of sizes (clay to boulder) and is collectively called wide-grading loose soils (WGLS) (Guo et al. 2016). Large particles usually form a matrix and the small particles fill the matrix voids and can bond to form larger particles (Guo and Cui 2020). Clearly, a fundamental understanding of the

Y. Cui
State Key Laboratory of Hydroscience and Engineering, Tsinghua University, Beijing, 100084, China
e-mail: yifeicui@mail.tsinghua.edu.cn

Y. Yin
Key Laboratory of Mountain Surface Process and Hazards/Institute of Mountain Hazards and Environment, Chinese Academy of Sciences, Chengdu, 610041, China
e-mail: yinyanzhou@imde.ac.cn

Y. Yin
University of Chinese Academy of Sciences, Beijing, 100049, China

C. Guo (✉)
Fujian Academy of Building Research, Fujian Key Laboratory of Green Building Technology, Fuzhou, 350025, China
e-mail: guochaoxu177@163.com

Ž. Arbanas et al. (eds.), *Understanding and Reducing Landslide Disaster Risk*,
ICL Contribution to Landslide Disaster Risk Reduction,
https://doi.org/10.1007/978-3-030-60713-5_16

initiation mechanism of rainfall-induced failure of loosely deposited soil slopes is required.

Experimental studies of rainfall-induced shallow slope failures have been conducted using flume tests in the past two decades (Iverson et al. 2000; Wang and Sassa 2003; Huang et al. 2009; Cui et al. 2019). More recently, several mid-scale flume models were developed to study the initiation process of slope failures containing WGLS subjected to rainfall infiltration (Guo et al. 2016; Cui et al. 2017; Cui et al. 2019). The experimental results showed that there is a blocking effect during fine particle migration, such that the fine particles accumulate in certain layers of a slope and form a relatively impermeable layer near the slope toe. With increasing pore water pressure in such layers, shear failure and local collapse will occur at the slope toe.

Although these studies offer knowledge on the possible mechanisms associated with infiltration-induced shallow failure, the current understanding of the mechanisms of fine particle migration is limited because of the complexity of WGLS micropore structure. The challenge lies in how to systematically characterize the WGLS pore structure so that it provides a foundation for future numerical simulations of debris flows triggered by fine particle migration.

In this study, WGLS samples were collected from a slope deposit in the Wenchuan earthquake area near Leigu, Sichuan Province, China. Serial tomography (ST) was then used to reconstruct the 3D pore structure of soil using 2D tomographic images from scanning electron microscopy (SEM). A statistical analysis is used to characterize the WGLS pore structure, and a preliminary numerical simulation using the Discrete Element Method (DEM) coupled with the Lattice Boltzmann method (LBM) is used to determine the jamming distribution based on the average fine particle size of a simplified seepage test. This study provides an approach for studying soil micropore structure and is aimed at linking soil micropore structure, fine particle migration during rainfall infiltration and shallow landslide hazards.

Materials and Methods

Serial Tomography Analysis

To study the micropore structure of a typical source material for debris flows, three soil samples were collected along the Subao River basin, near Leigu in Sichuan Province, China. They were selected on the basis of their geomorphological forms and distribution (Table 1). The sampling locations were 130 km from the epicenter of the 2008 Wenchuan earthquake, and these locations experienced seismic intensity of XI during the earthquake (Xu et al. 2014). The particle size distribution was obtained from sieving analysis (Fig. 1) which shows a wide range, from silt to gravel.

Mercury intrusion porosimetry (MIP) is widely used to reconstruct the 3D pore structure of rocks. However, there are two drawbacks: (1) high pressure is needed to intrude mercury or water into rock specimens, which can disturb or distort the microstructure (Gane et al. 2004); (2) the ink bottle effect can occur, resulting in small pores being ignored that are not connected to the sample surface directly or through larger pores (Moro and Böhni 2002). Compared to MIP, the use of ST causes less disturbance to the soil micropore structure during sample preparation (Vogel and Kretzschmar 1996). Soil slices to be analyzed with ST are first scanned using SEM to obtain 2D tomographic images (Vogel 2002; Cui and Jia 2013). These images are then combined using a series of iterations to reconstruct the 3D micropore structure. However, the cutting process of soil slices from a larger specimen can still cause disturbance to the micropore structure if not supported (Sok et al. 2010). Therefore, epoxy impregnation is often used to protect the micropore structure of a sample (Stutzman and Clifton, 1999). This method fills the voids, supports the microstructure, and protects it against shrinkage and cracking; it also enhances the visual contrast between pores and soil particles (Stutzman 2006).

The sample preparation procedures used in this study for SEM scanning followed the recommendations of Stutzman (2006). A mixture of epoxy resin glue, triethanolamine, and acetone was heated to a liquid state. The specimen was then placed in a container and surrounded by the liquid mixture. The low viscosity liquid was drawn into the microstructure by capillary suction. Cutting, grinding and polishing were then carried out to expose a fresh surface. The soil samples were cut into rectangular prisms (50 mm × 50 mm × 40 mm) before they were ground and polished to finer slices (12 mm × 9 mm × 20 mm) for SEM observations. To obtain the 3D soil micropore structure, 50 μm of soil was continuously trimmed using a grinder to obtain slices along the height of a specimen after each scan. The resolution of each scanning slice was 1024 × 768 pixels, with each pixel having an actual length of 12 μm. This procedure was repeated to obtain continuous tomographic images of each specimen, based on a total of 25 serial surfaces separated by a distance of 50 μm.

From digital reconstruction, pore topology can then be used to characterize the number of pores that are completely enclosed by aggregates and their connectivity with each other per unit volume. To grade the soil pore sizes, the numerical technique of Vogel (1997) was adopted.

The standard pore structure elements, also known as a mesh in sieve analysis, were reconstructed using unit

Table 1 Geomorphological forms, descriptions, and wide-graded loose soil distribution near Leigu Town, Sichuan Province, China (Guo and Cui 2020)

ID	Exogenous processes	Formation reason	Distribution
Sample 1	Alluvium (Diluvium)	Formation of the sediment has been eroded and reshaped by water	Alluvial fan distributed at the outlet of valley (groove)
Sample 2	Colluvium (Avalanche deposit)	Collapse of or rock and soil under the force of gravity	Distributed along the slope or below steep cliffs
Sample 3	Taluvium (Slope deposit)	Accumulation of weathered rock and soil along the slope	Widely distributed on the slope

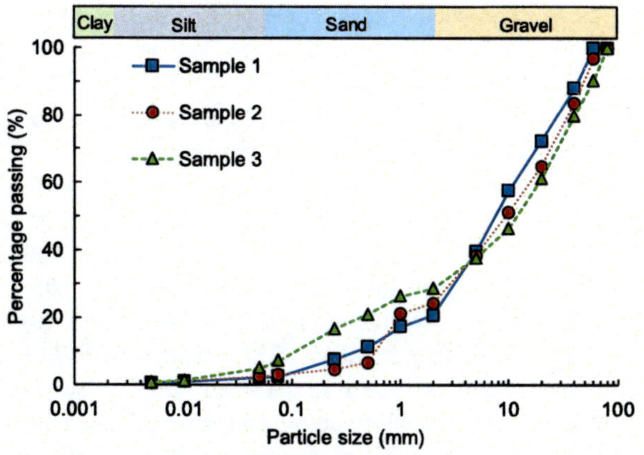

Fig. 1 Measured particle size distributions of wide-graded loose soils (Modified from Guo and Cui 2020)

elements with a resolution of 12 μm in both x and y directions, and 50 μm along the z direction. Structural elements with different diameters (e.g., 2 pixels (24 μm), 6 pixels (48 μm), and 6 pixels (72 μm)) are shown in Fig. 2.

In total, six structural elements (12, 24, 48, 72, 96, and 120 μm) were used in this study for pore size characterization.

Numerical Simulation

The LBM was used to solve the fluid motion process in fluid-structure interactions. This method can efficiently deal with irregular solid boundary problems and is suitable for studying fluid flow in pores (Feng et al. 2010; Leonardi et al. 2016). The basic equation is as follows:

$$f_i(x+e_i\Delta t, t+\Delta t) - f_i(x,t) = -\frac{\Delta t}{\tau}\left(f_i(x,t) - f_i^{eq}(x,t)\right) \tag{1}$$

where f_i is the particle distribution function; e_i is the lattice velocity; Δt is the timestep of the LBM; τ is the relaxation factor and f_i^{eq} is the local equilibrium distribution function computed by Eq. (2):

$$f_i^{eq} = w_i\rho\left[1 + \frac{(e_i\cdot u)}{c_s^2} + \frac{(e_i\cdot u)^2}{2c_s^4} - \frac{(u\cdot u)}{2c_s^2}\right] \tag{2}$$

where w_i is the weighting factor; ρ is the fluid density and c_s is the speed of sound.

The fluid-solid interactions are strongly related to the velocity difference between two materials. Generally, the drag force of a fluid on a particle is calculated by Eq. (3) (Tang et al. 2017):

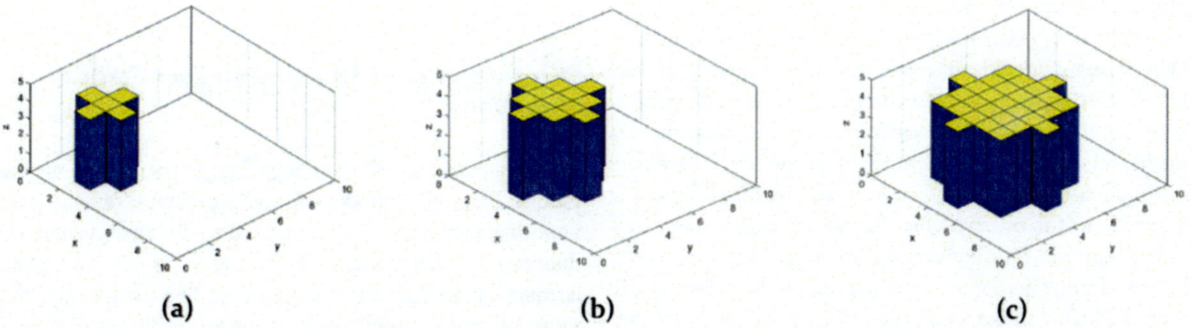

Fig. 2 Pore structural units with different radii: (a) D = 24 μm; (b) D = 48 μm; (c) D = 72 μm (Modified from Guo and Cui 2020)

$$F_d = \frac{\pi r_p^2}{2} C_{d0} \rho_f |u - v|(u - v) \quad (3)$$

where r_p is the radius of the particle; ρ_f is the fluid density; u is the fluid velocity, which is integrated from the velocity distribution on the particle surface from the solid boundary of the raw LBM data; v is the particle velocity calculated from the DEM simulation and C_{d0} is the drag coefficient which is related with the particle Reynolds number. The Reynolds number, Re_p is calculated as follows:

$$\mathrm{Re}_p = \frac{\rho_f |u - v| d_p}{\mu_f} \quad (4)$$

where d_p is the particle diameter and μ_f is the dynamic viscosity of the fluid. The coefficient of drag force is then calculated by different formulae with different values of Re_p, as follows:

$$C_{d0} = \begin{cases} \frac{24}{Re_p} & \mathrm{Re}_p < 1 \\ \frac{24}{Re_p}\left(1 + 0.15 Re_p^{0.687}\right) & 1 \le \mathrm{Re}_p < 1000 \end{cases} \quad (5)$$

In this research, Discrete element method was employed for numerical study on WGLS under steady seepage process (Cui et al. 2016). We simplified the problem into 2D fluid flow by randomly generating pores based on the pore size distribution of each sample in the numerical simulation. This was done to study the interaction between particles and fluid, and subsequent jamming phenomenon of pores.

Results and Interpretation

Micro-Pore Structure Characteristics

The along-plane distribution of pores was obtained by studying and analyzing their spatial distribution. For example, statistical methods were used to analyze the pores of a specific slice in 2D. If a pixel belonged to a pore, then a value of unity was assigned to this pixel index. The pore frequency was then obtained by dividing the index by the total number of pixels occupied by pores in each soil sample. During sampling, the *z* direction was assumed to be normal to the slope surface, whereas the *y* direction was denoted as downslope along the slope surface and the *x* direction was the lateral direction parallel to the slope surface. The planar distribution of pores was then averaged by all slices in the *z* direction. It is observed that Samples 2 and 3 have a relatively uniform distribution (Fig. 3), while the plane distribution of pores in Sample 1 is uneven, with an increase in pore frequency in the *y* direction. This is probably attributable to the depositional process of this particular type of sediment. More specifically, during deposition, the effects of gravity and confining stress will lead to a relatively low distribution of pores near the bottom of the slope, both in terms of pore size and number. This coincides with general observations that bulk density of soil is usually higher at the bottom of a slope (Iverson 2000).

The pore size distribution of the four WGLS samples by volume percent (Fig. 4) was compared with loess data from Wei et al. (2019), prepared with the same method (serial sectioning). It is apparent that the pore size distribution of WGLS is significantly different from the loess, which is mainly composed of fine particles. The loess pore sizes are normally distributed with a single peak (12%) at 13.5 μm and values in the 2–40 μm range. In contrast, the WGLS pore size distribution shows two peaks at 12 μm and 72 μm. The reason for the high percentage of micropores (12 μm) in WGLS is because fine particles are locally concentrated. The diameter of the pores in loess are typically large volume (13.5 μm), significantly smaller than mesopore volume in WGLS (72 μm). Compared with bimodal pore size distributions of silty soil and sand (Oualmakran et al. 2016), WGLS also corresponds to a second peak with larger diameter. The large percentage of pore diameters of 72 μm in Sample 1 may be related to the small sampling area and limited pore structure element (only 6 elements in the current study) used in numerical characterization of pore size distribution of WGLS. Compared with MIP, the ST method is not able to capture the peak pores with diameter <2 μm (Wei et al. 2019). Therefore, the peak pore diameter <2 μm may also be detected in WGLS if MIP is used. However, to cover both large pores and micropores, we used the ST method to examine the pore size distribution of WGLS.

Nonetheless, statistical analysis should be used in future studies to better characterize the pore size distribution with a larger number of SEM slices. In future studies, pore structure of WGLS should be observed at different scales. Indeed, micropores can be characterized using the ST method, while meso- and macropores can be examined using computed tomography (CT) scans with a synchrotron-radiation light source (Baruchel et al. 2006) for high-precision visualization and 3D reconstruction of internal pore structure.

Influence of Pore Structure on Fine Particle Migration

To characterize the effect of pore size distribution on the fine particle migration process in each different sample, each pore diameter (Fig. 4) was replaced by its mean value. The diameter of fine particles was chosen as 25 μm, taking account of the fact that the maximum fine particle size that migrated during experimental flume test is 50 μm (Cui et al. 2014). The fine particle diameters, together with pore size, were then expanded 20 times to improve the calculation

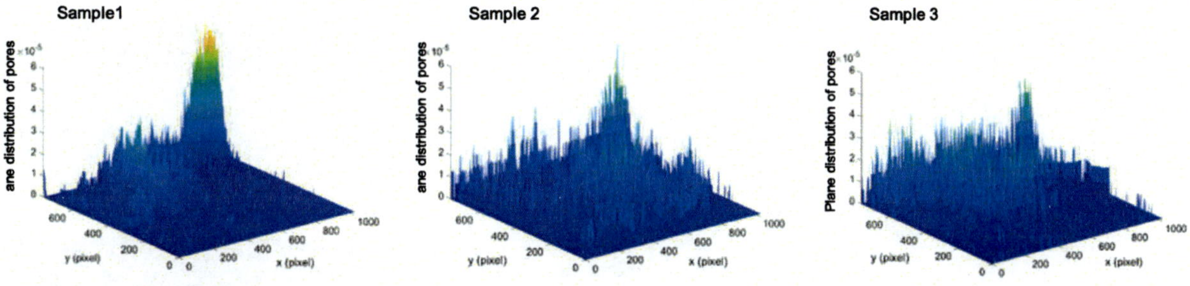

Fig. 3 Probability distribution of pore radius from samples (Modified from Guo and Cui 2020)

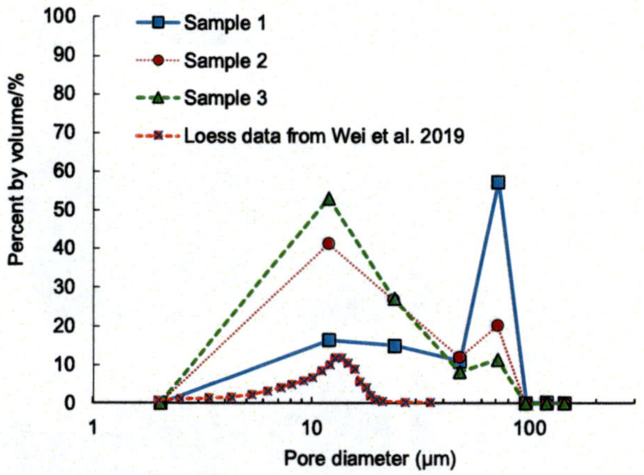

Fig. 4 Pore size distribution of each sample in terms of percentage by volume, compared with loess data (Modified from Guo and Cui 2020, and Wei et al. 2019)

efficiency but maintain the same Reynold's number to minimize the scale effect. When the pore diameter is <25 μm, fine particles cannot pass through the pores, so only the pores with a diameter >25 μm were considered in the pore structure generation. Only one layer of pore structure is simplified in the simulation (Fig. 5a), requiring that the pores are separated by a DEM rigid element with 0.25 mm protrusion and distributed in accordance with the volume percentages of this study.

The results of 3 different samples are shown at two distinct times: 0.08 s and 0.26 s (Fig. 5b–d). It is observed that the fluid initially causes downstream migration of fine particles. Then, there is interaction of particles and pores, resulting in the jamming effect and subsequent arch formation of fine particles that significantly affects the flow field. The blocking effect ratios at 0.08 s are 0.454, 0.468 and 0.538 for the three samples, respectively. These values become 0.0089, 0.0293 and 0.0383 at 0.26 s, respectively. Sample 3 was found to have the highest jamming probability while Sample 1 exhibits the lowest jamming probability, which accords with the pore size distribution in Fig. 4. Sample 1 is dominated by the second peak (72 μm) while sample 3 is dominated by the first peak (12 μm). The same observation is found from analytical jamming analysis using the modified version of Beverloo's law by Cui and Guo (2020).

According to the study by Valdes and Santamarina (2008) on the pore jamming mechanism in dry granular flow conditions, if the ratio of pore size to particle diameter is <3, it is easy to form a stable arch bridge jam. However, in this study the ratio was found to be 1.44 and 2.4 and there were not many jammed pores when t = 0.26 s, which differs from the result of dry conditions. This may be attributed to the drag force along the line of fluid flow from the beginning of the simulation. If multiple particles reach the pores almost simultaneously, clogging is likely to occur (e.g., infinite particles generated in the initial area). Subsequent fluid flow is then conducive to the occurrence of erosion and can also enhance pore jamming. From the analytical solution of Berverloo's law (Beverloo et al., 1961), it can be shown that if the ratio of pore size to particle diameter is <1.4, the number of particles with velocity weight discharged is zero. If the ratio is >1.4, the possibility of pore jamming decreases. The numerical simulation shows the same probability of jamming occurring among the three samples that contain different pore size distributions. Nevertheless, the simplified numerical simulation is not able to present the real case that contains migrated fine particles within the size distribution and different layers of pores along the seepage direction. Therefore, further numerical studies should aim for 3D reconstruction of the pore structure and jamming mechanism to explain the initiation of shallow landslide.

Conclusion

The characteristics of WGLS pore structure have been quantitatively analyzed using an SEM via digitization of soil slices and the ST method. Statistical analysis was performed on pore size distribution. Numerical simulation was then conducted using LBM coupled with DEM to investigate the jamming probability of fine particle migration during seepage. Compared with traditional silty soil with bimodal pore

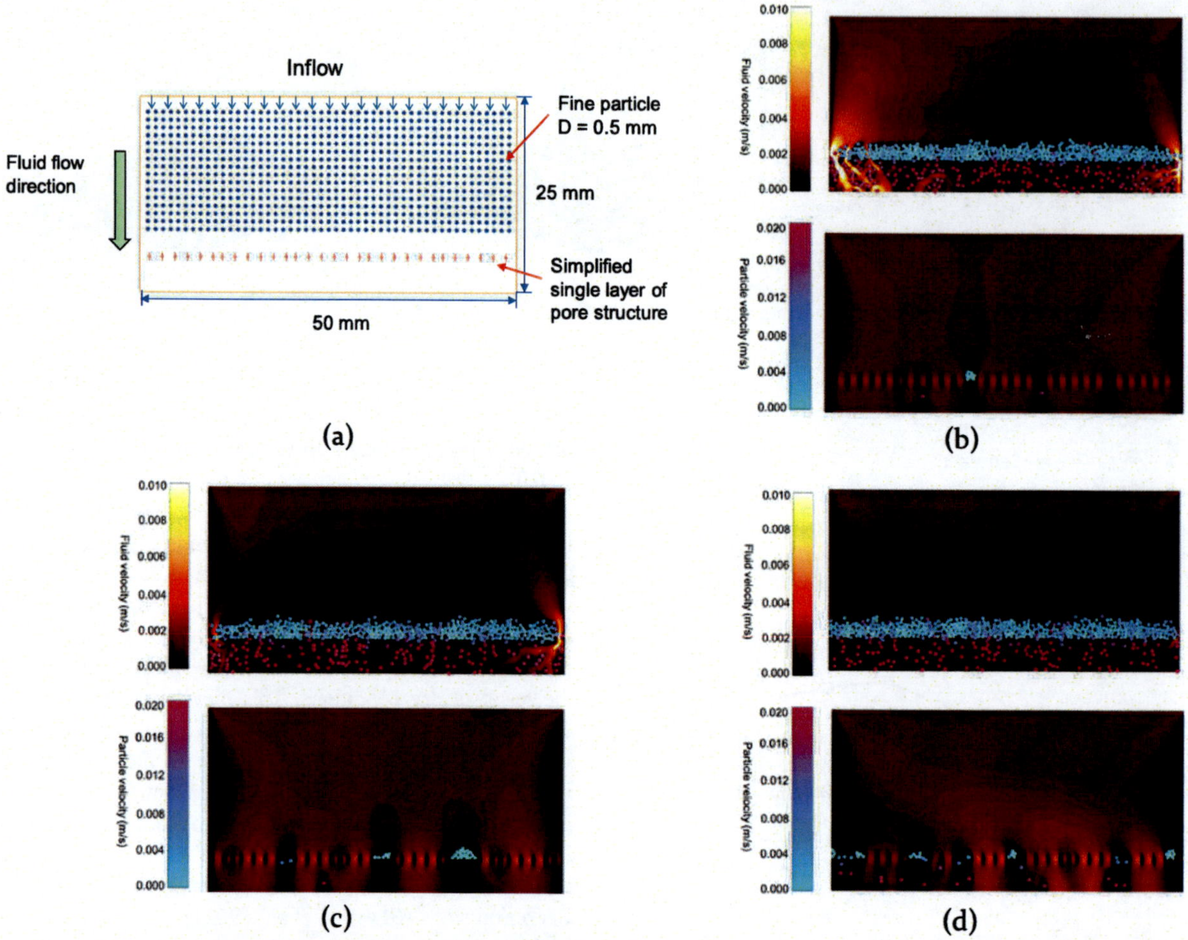

Fig. 5 **a** Numerical model set up with fine particles and single layer pore structure (Sample 1). The simulation results for different samples at 0.08 s and 0.26 s are shown for **b** Sample 1; **c** Sample 2; and **d** Sample 3

size distributions, WGLS shows a pore size distribution corresponding to a second peak with much larger diameter. Simulation results indicate that jamming is most prevalent in samples in which the smaller pore sizes dominate the bimodal pore size distributions. This is in agreement with previously published analytical solutions. This study provides an approach to link soil micropore structure, fine particle migration during rainfall infiltration and the seepage process.

Acknowledgements The study here is funded by the National Natural Science Foundation of China (grant number 42077238 and 41941019), the Research Fund Program of the State Key Laboratory of Hydroscience and Engineering (2020-KY-04), and the Chinese Academy of Sciences (XDA23090401). Dr. Chaoxu Guo acknowledges support from the Fujian province Science Foundation for Young Scientists (grant number 2017J05074) and International (regional) cooperation and exchange projects from National Natural Science Foundation of China (grant number 41861134011). We thank Warwick Hastie, PhD from Liwen Bianji, Edanz Group China (www.liwenbianji.cn/ac), for editing the English text of a draft of this manuscript.

References

Beverloo WA, Leniger HA, Velde JVD (1961) The flow of granular solids through orifices. Chem Eng Sci 15(3–4):260–269

Baruchel J, Buffiere JY, Cloetens P, Michiel MD, Ferrie E, Ludwig W, Maire E, Salvo L (2006) Advances in synchrotron radiation microtomography. Scripta Materialia 55(1):41–46

Cui Y, Zhou X, Guo C (2017) Experimental study on the moving characteristics of fine grains in wide grading unconsolidated soil under heavy rainfall. J Mountain Sci 14(3):417–431

Cui P, Guo C, Zhou J, Hao M, Xu F (2014) The mechanisms behind shallow failures in slopes comprised of landslide deposits. Eng Geol 180:34–44

Cui Y, Nouri A, Chan D, Rahmati E (2016) A new approach to the DEM simulation of sand production. J Petrol Sci Eng 147:56–67

Cui Y, Jiang Y, Guo C (2019) Investigation of the initiation of shallow failure in widely graded loose soil slopes considering interstitial flow and surface runoff. Landslides 16(4):815–828

Cui Z, Jia Y (2013) Analysis of electron microscope images of soil pore structure for the study of land subsidence in centrifuge model tests of high-rise building groups. Eng Geol 164:107–116

Fan X, Juang H, Wasowski J, Huang R, Xu Q, Scaringi G, Van Western CJ, Havenith HB (2018) What we have learned from the 2008 Wenchuan Earthquake and its aftermath: a decade of research and challenges. Eng Geol 241:25–32

Feng YY, Han K, Owen DRJ (2010) Combined three-dimensional lattice Boltzmann method and discrete element method for modeling fluid-particle interactions with experimental assessment. Int J Numer Methods Eng 80:245–299

Gane PAC, Ridgway CJ, Lehtinen E, Valiullin R, Furó I, Schoelkopf J, Paulapuro H, Daicic J (2004) Comparison of NMR cryoporometry, mercury intrusion porosimetry, and DSC thermoporosimetry in characterizing pore size distributions of compressed finely ground calcium carbonate structures. Ind Eng Chem Res 43(24):7920–7927

Guo C, Cui Y (2020) Pore structure characteristics of debris flow source material in the Wenchuan earthquake area. Eng Geol 267:105499

Guo X, Cui P, Li Y, Zhang J, Ma L, Mahoney WB (2016) Spatial features of debris flows and their rainfall thresholds in the Wenchuan earthquake-affected area. Landslides, 1–15

Huang CC, Ju YJ, Hwu LK, Lee JL (2009) Internal soil moisture and piezometric responses to rainfall-induced shallow slope failures. J Hydrol 370:39–51

Iverson RM, Reid ME, Iverson NR, LaHusen RG, Logan M, Mann JE, Brien DL (2000) Acute sensitivity of landslide rates to initial soil porosity. Science 290(5491):513–516

Julio RV, Carlos Santamarina J (2008) Clogging: bridge formation and vibration-based destabilization. Canadian Geotech J 45(2):177–184

Leonardi A, Wittel FK, Mendoza M, Vetter R, Herrmann HJ (2016) Particle–fluid–structure interaction for debris flow impact on flexible barriers. Comput-Aided Civil Infrastruct Eng 31(5):323–333

Moro F, Böhni H (2002) Ink-bottle effect in mercury intrusion porosimetry of cement based materials. J Colloid Interface Sci 246:135–149

Oualmakran M, Mercatoris BCN, Francois N (2016) Pore-size distribution of acompacted silty soil after compaction, saturation, and loading. Can Geotech J 53:1902–1909

Shieh CL, Chen YS, Tsai YJ, Wu JH (2009) Variability in rainfall threshold for debris flow after the Chi-Chi earthquake in Central Taiwan China. Int J Sediment Res 24:177–188

Sok RM, Knackstedt MA, Varslot T, Ghous A, Latham S, Sheppard AP (2010) Pore scale characterization of carbonates at multiple scales: integration of Micro-CT, BSEM, and FIBSEM. Petrophysics 51(6)

Stutzman PE, Clifton JR (1999) Specimen preparation for scanning electron microscopy. In: Proceedings of the 21 st international conference on cement microscopy. international cement microscopy association, Duncanville, TX, pp 10–22

Stutzman PE (2006) Sampling preparation for and examination with the Scanning Electron Microscope. In: Petrographic methods of examining hardened concrete: a petrographic manual. Virginia Transportation Research Council Report No. FHWA-HRT-04-150

Tang Y, Chan DH, Zhu DZ (2017) A coupled discrete element model for the simulation of soil and water flow through an orifice. Int J Numer Analyt Methods Geomech 41(2)

Vogel HJ, Kretzschmar A (1996) Topological characterization of pore space in soil-sample preparation and digital image-processing. Geoderma 73(1–2):23–38

Vogel HJ (1997) Morphological determination of pore connectivity as a function of pore size using serial sections. Eur J Soil Sci 48:365–377

Vogel HJ (2002) Topological characterization of porous media. Morphology of condensed matter. Springer, Berlin, Heidelberg, pp 75–92

Wang G, Sassa K (2003) Pore-pressure generation and movement of rainfall-induced landslides: effects of grain size and fine-particle content. Eng Geol 69(1):109–125

Wei T, Fan W, Yu N, Wei Y (2019) Three-dimensional microstructure characterization of loess based on a serial sectioning technique. Eng Geol 261(1):105265

Xu C, Xu X, Yao X, Dai F (2014) Three nearly complete inventories of landslides triggered by the May 12, 2008 Wenchuan Mw 7.9 earthquake of China and their spatial distribution statistical analysis. Landslides 11:441–461

Formation Mechanism and Stability of the Instable Block Formed in Xinmo Landsilde

2020 Kyoto Japan

Bingli Hu, Lijun Su, Qijun Xie, Fangwei Yu, and Chonglei Zhang

Abstract

Xinmo landslide buried the village and caused many casualties and property damage. After the landslide, a deformed block came into being on the left side of the landslide. Researchers mainly concentrated on the main body and the causes of the landslide without a systematic analysis of this block. This work is to discuss this block based on field investigation and UAV images. In the investigation, four scarps and three cracks were found. The lower two scarps indicated a downslope movement of the foot of the block, while the other two scarps and three cracks showed that main displacement of the block was towards to the free face formed in the landslide. Combined previous researches, it can be inferred that bedrock in the lower two scarps was the locking section, which stopped the movement of the block and the formation of the block can be divided into three stages. Then, two potential slipping direction were concluded and the FOS (Factor of Safety) were calculated with the Finite Element Method (FEM). Results show that the slip is more easily towards the free face which would do less harm. But it cannot be excluded that it would slip down the slope which could do much harm to the village and the road. So, monitoring is recommended on the foot of the block to get an advanced notice.

Keywords

Xinmo village • High position landslide • Formation mechanism • Stability

B. Hu · L. Su (✉) · Q. Xie · F. Yu · C. Zhang
Key Laboratory of Mountain Hazards and Earth Surface Process, Chinese Academy of Sciences; Institute of Mountain Hazards and Environment, Chinese Academy of Sciences, Chengdu, 610041, China
e-mail: sulijun1976@163.com

B. Hu · L. Su
University of Chinese Academy of Sciences, Beijing, 100049, China

L. Su
CAS Center for Excellence in Tibetan Plateau Earth Sciences, Beijing, 100101, China

Introduction

Landslide is one of the common geological disasters in mountainous areas, which poses a serious threat to the safety of lives, and also causes great losses to the infrastructure construction in mountainous areas, which severely restricts the development of those areas. Especially high position landslides possessing higher potential energy can move farther with a short time, which could cause greater danger (Kilburn and Pasuto 2003). But it is difficult to understand the mechanism of the initiation and movement of a high position landslide for lacking sufficient information before and after slipping (Guthrie et al. 2009; Xu et al. 2010; Burgh et al. 2012; Wu et al. 2013; Wang et al. 2017). So more detailed investigations are needed to improve understanding of the mechanism of high position landslides, which is helpful for landslides prevention and reduction.

Xinmo landslide is a high position landslide causing heavy damage to lives and properties (10 deaths, 3 injuries, 73 missing, and 103 houses completely destroyed). Many works have done to understand the mechanism of the landslide (Su et al. 2018; Ouyang et al. 2017; Su et al. 2017; Fan et al. 2017a, b; Wang et al. 2017), e.g. cause analysis, numerical simulation, etc. But fewer were about the western instable block except for Fan (2017) did some displacement monitoring to guide the rescue operation and Ouyang et al. (2017) simulated the possible movement of the instable block with Massfllow. It is still far away to understand the instable block, both the formation mechanism and stability. UAV and field geological survey were used to analyze the formation mechanism and stability of the block, which could

Ž. Arbanas et al. (eds.), *Understanding and Reducing Landslide Disaster Risk*, ICL Contribution to Landslide Disaster Risk Reduction,
https://doi.org/10.1007/978-3-030-60713-5_17

provide help for landslides prevention and reduction of the landslide.

Background

Geological Setting

The study area locates in the transition zone of the Tibetan plateau to the Sichuan basin with the characteristics of complicated geo-structure and terrain. The range of altitude in this area is 2700 m to 5230 m, with the altitude difference ranging from 500 m to 2500 m. "V" shaped valleys are well developed in this area. Ximo village is located in the kernel of an arc-structure composing a serials of south protruding arc-folds, inverted arc-folds and associated faults (Fig. 1). This arc-structure includes Jiaochang arc-structure zone and Shidaguan arc-structure zone (Huang et al. 2002; Xu et al. 1992). Songpinggou fault (supposed to be the causative fault of Diexie earthquake in 1933; Tang et al. 1983; Qian et al. 1999) and Minjiang fault (supposed to be the causative fault of Diexie earthquake in 1933; Huang et al. 2002; Tang et al. 1983) are located on the two sides of Xinmo village. Moreover, this area was also within the intensity zone of VII for the Wenchuan earthquake in 2008 and X for the Diexi earthquake in 1933. Bedrock in the landslide belongs to T2z composing thick layer metamorphic quartz sandstone and thin layer slate (195°∠52°). The slope deposit is loess mixed with gravels.

Hydrometeorology

This area characterizes as the semi-arid valley climate and plateau monsoon climate with an annual average temperature of 11.2 °C. Annual rainfall of this area was 490.7 mm, with accumulated rainfall up to 353.3 mm from May to September (Fig. 2).

Minjiang river and Songpinggou river (branch of Minjiang) are the main rivers in this area. Songpinggou river was dammed by Xinmo landslide. The river bed of Songpingou was also changed by the damming. Bedrock fissure water and reservoir pore water are the main forms of groundwater.

Description of the Instable Block

The shape of the block is inverted trapezoid with an upper width of 250 m, bottom width of 100 m, height of 450 m, and average thickness of 58 m, and the estimated volume reaches 455 × 10^4 m^3 (Xu et al. 2017). To get a better understand and analyze of this block, we climbed from the

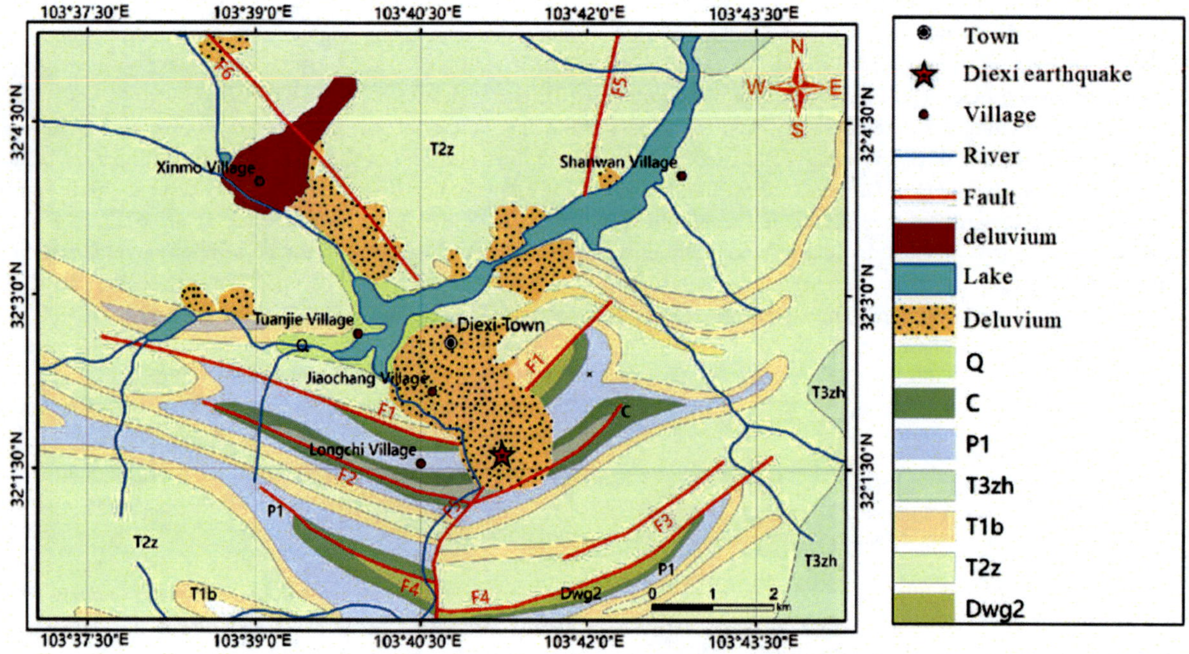

Fig. 1 Geological map of the study area with the location of Xinmo landslide, strata, and active faults. (Refer to Huang (2002); Q, Quaternary deposits; T3zh, Upper Triassic Zhuowo Formation; T2z, Middle Triassic Zagunao Formation; T1b, lower Triassic Bocigou Formation; C, Carboniferous; P1, Lower Permian; Dwg2, Devonian Weiguan Formation. F1, Jiaochang Fault; F2, Guancaigou Fault; F3, Chuanhuagou Fault; F4, Shidaguan Fault; F5, Minjiang Fault; F6, Songpinggou Fault)

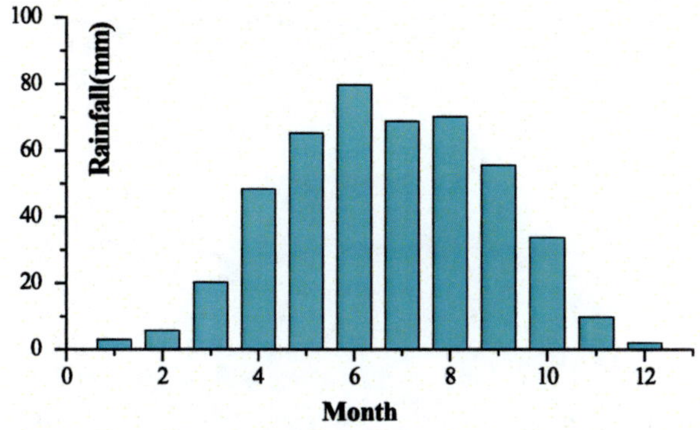

Fig. 2 Graph of average monthly rainfall in Mao County. Data is from Maoxian meteorological bureau from 1983 to 2010

west side to do the investigation. During the climbing, some deformation features were found in the block, namely three serials cracks and four obvious scarps (Fig. 3). These cracks and scarps are depicted in this part.

Cracks

Along the investigation route, three serials cracks were found on the block surface (red lines in Fig. 3), namely C1, C2, C3. These cracks distribute in the block from the inside out. C1 locates at the lower of the block with a length of more than 60 m and a width of about 0.2 m in the direction of 210°. C2 is at the top of C1 with a length of more than 100 m and a width of about 0.3 m in the direction of 220°. C3 is at the inner of the block with over 100 m length and around 0.6 m width with the direction of 230°.

Scarps

Four scarps were also discerned in the investigation (Fig. 3). They are distributing on the foot, right, and top of the block, respectively (we call them S1, S2, S3 and S4 in this article).

S1 is on the foot of the block with bedrock exposed (Fig. 3), which is a small plate composed of gravels and dead trunks. The scarp is about 7 m in height and the plate is about 15 m in weight. Based on the stability in the landslide. This plate can be divided into two parts, stable part and instable part. The stable part locates on the right side with growing trees, while the instable part is on the left side closing to the landslide with dead trunks lying. S2 is a little bit above S1. Bedrock does not expose here, and the deluvium is also loess with gravels. The height of the S2 is about 3 m. Trees below the S2 fell towards slope direction indicating a big disturbance happened in the landslide. But trees up the S2 seem to have a minor disturbance.

S3 is on the top of the block (Fig. 3) with an altitude difference of 20 m and bedrock exposed with many fissures, small cracks and bending deformations. The deluvium of this part consists of big stone blocks and dead tree trunks.

S4 with wedge shape is on the right of the block, which might be a big crack before. In this investigation site of S3, the deposited loess have a 12 m altitude difference and many "saber" trees with a bending direction of 140°.

Features of Deformation

As described in the previous section, the part down S1 caught a seriously destroyed. The area between S1 and S2 had an obvious displacement, but the part a little bit above S2 showed a more stable feature from the phenomenon of trees' growth. So the evidence indicates S1 and S2 are the boundaries of the block from the landslide. Besides, it can be seen from Fig. 3 that there could be a locking section composed of bedrock under S1 and S2 (Fig. 3 light blue area). This locking section stopped the movement of the upside. The locked section should be continuous from the left to the right of the landslide (Fig. 3, both light blue and red zones in the embedded figure). The purple zone in the landslide area could be destroyed in historical earthquakes.

The wedge scarp(S4) also affirms the lower part of the block had a small deformation, while the upper part had a relative big movement. Moreover, the main movement type was sinking, and caused the formation of the scarps instead of cracks. S3 was the trailing edge of the block, and had a maximum displacement. S3, S4 and three serials cracks could demonstrate the deformation direction of the upper part of the block (Fig. 3 white arrow).

Formation of the Instable Block

Based on the analysis of the investigation, we can divide the formation of the block into three stages in the sliding process.

Stage1

When the source of the landslide (bedrock on the top of the hill) began to fall, the upper part of the western block was dragged to move down the slope. S3 formed in this process,

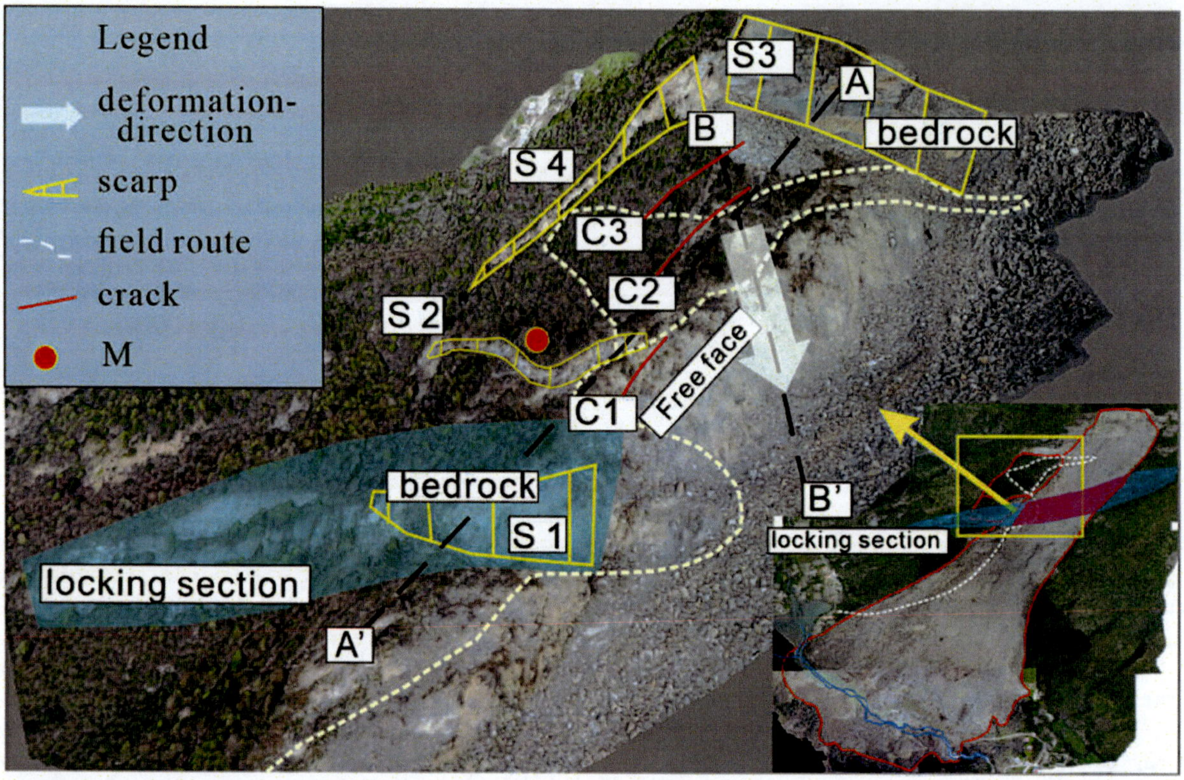

Fig. 3 Instable block and full view of Xinmo landslide(embedded). Yellow lines indicate Scarps; red lines indicate cracks; light blue area, locking section; light purple area, locking section destroyed in historical earthquake; white arrow, deformation direction of the block; white dashed line, investigation route; A-A', potential slip direction down the slope; B-B', potential slip direction to the free face; M, recommended monitoring zone; red lines in embedded image, boundary of Xinmo landslide.(image of instable block is from UAV in April, 2019; embedded is from UAV in June, 2017)

and S4 also began to form in the upper side. Compression energy also began to accumulate due to the movement of the upper part.

Stage2

In this stage, the deluvium of the middle part began sliding, and this block had the same motion trend. The locking section in the block foot stopped this trend and stabilized the block. However, colluvium below S1 lost stability under the impact of the main sliding. Thus, S1 and S2 came into being. At the same time, S4 kept expanding down the slope, but the closer to the locking section, the smaller the vertical displacement of the S4 was. The compression energy of the block continued to grow in this stage.

Stage3

After the slipping of the main part, a free face formed on the left part of the block. The compression energy of the block accumulated in stage2 released in the form of deformation and displacement to the free face. In this stage, cracks in the block surface came into being as well as extending the scarps.

Stability Analysis (More Information of Potential Slip)

As described in the former part, cracks in the block indicate an obvious deformation towards the free face (B-B' in Fig. 3), and B-B' could be inferred as a potential slip

Table 1 Parameters in calculation of FOS

Slip surface	Density (kg/m^3)	Friction	Cohesion	FOS
A-A'	1.95×10^3	30°	35 kPa	1.292
B-B'	1.95×10^3	30°	35 kPa	1.046

surface. Furthermore, a potential slipping down the slope like the eastern part cannot be excluded for causing serious damage to the village down the slope (A-A' in Fig. 3). So, there are 2 potential slip directions based on the investigation: towards to the free face and down the slope (A-A' and B-B' in Fig. 3). Because of the locking section on the foot of the block, the movement down the slope was prevented. It is assumed that the locking section would not be broken down, which means slip surface steer by the locking section and slip to the free face following the deformation direction.

Here UAV images were dealt to get the DEM, and 2-d slip mass as well as slip surface were acquired. Then the Geo-slope software was used to calculate the FOS (Factor of Safety) of the block. The parameters of the deluvium were referred to Engineering Geology Manual (Chang et al. 2007) (Table 1), namely friction is 30°, and cohesion is 35 kPa (Fig. 4). FOS of slipping down is 1.292, while FOS towards to free face is 1.046. The situation of slip towards free face is more easily compared with the situation of slipping down slope. This is also consistent with the deformation analysis.

When slipping towards the free face, it would not do so much harm because it is far away to the road and the ground surface is rugged. But slipping down the slope is not excluded. Ouyang et al. (2017)'s calculation indicated when slipping down the slope, many damages to the village and the road could happened. So monitoring equipment is needed to reduce the risk and the recommended position to detect is the lower part of the block (Fig. 3). Advanced notice can be detected when deluvium slip down the slope.

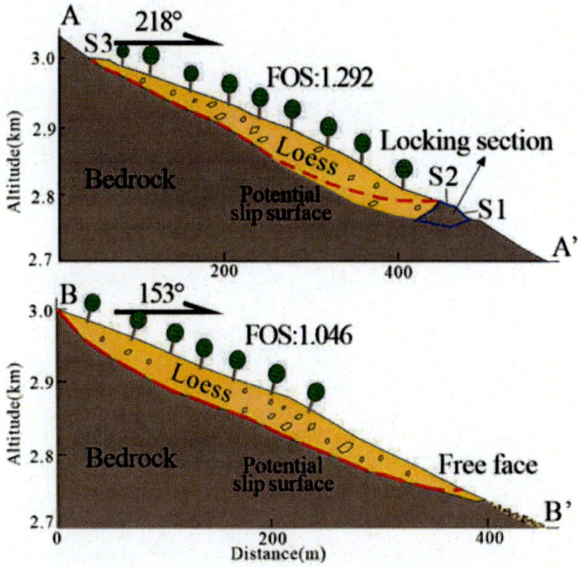

Fig. 4 Two potential slipping surface described in Fig. 3 and the FOS (Factor of Safety)

Conclusion

This work demonstrated the features of the instable part of Xinmo landslide. There are four scarps and three cracks around and in the block. From these features, the formation mechanism of the block was analized. The result shows it is the locking section that prevented the persistent movement of the block and caused the formation of the scarps around the block (some wedge scarps) and also the cracks in the block body. Fan (2017)'s monitoring data can strongly support this conclusion. Further, the potential moving directions of the block were calculated with FEM software Geo-slope, which suggested it is easily to slip towards to the free face than the down slope direction. But, slipping down slop could cause more damages to the road and villages according to the former study. So monitor was recommanded on the foot of the block to get a advance notice.

Acknowledgements This study has been financially supported by the National Natural Science Foundation of China (41790432), the NSFC-ICIMOD joint project (41761144077), the "Belt & Road" international cooperation team project of CAS (Su Lijun).

References

Burght LVD, Stoffel M, Bigler C (2012) Analysis and modeling of tree succession on a recent rockslide deposit. Plant Ecol 213(1):35–46

Chang SP, Zhang SM et al (2007) Engineering geology manual(Fourth Edition), China Architecture & Building Press, Beijing. (9787112216420). 160p. (in Chinese)

Fan JR, Zhang XY, Su FH et al (2017a) Geometrical feature analysis and disaster assessment of the Xinmo landslide based on remote sensing data. J Mt Sci 14(9):1677–1688

Fan XM, Xu Q, Scaringi G et al (2017b) Failuremechanismand kinematics of the deadly June 24th 2017 Xinmo landslide, Maoxian, Sichuan China. Landslides 14:2129–2146

Guthrie RH, Friele P, Allstadt K et al (2012) The 6 august 2010 mount meager rock slide-debris flow, Coast Mountains, British Columbia: characteristics, dynamics, and implications for hazard and risk assessment. Nat Hazards Earth Syst Sci 12:1277–1294

Huang ZZ, Tang RC, Liu SL (2002) A further discussion on the arc structure in Sichuan and the seismogenic structure of Diexi earthquake in 1933. Earthq Res China 02:65–74 (in Chinese)

Kilburn CRJ, Pasuto A (2003) Major risk from rapid, large-volume landslides in Europe. Geomorphology 54(1–2):3–9

Ouyang CJ, Zhao W, He SM et al (2017) Numerical modelingand dynamic analysis of the 2017 Xinmo landslide in Maoxian County China. J Mt Sci 14:1701–1711

Qian H, Zhou RJ, Ma SH, Li XG (1999) Discussions on the 1933 Diexi earthquake with M 7.5. Earthq Res Sichuan 3:11–17 (in Chinese)

Su LJ, Hu KH, Zhang WF et al (2017) Characteristics and triggering mechanism of Xinmo landslide on 24 June 2017 in Sichuan China. J M. Sci 14:1689–1700

Tang RC, Jiang NQ, Liu SL (1983) Recognition of the geological setting and the seismogenic condition for the Diexi magnitude 7.5 earthquake. J Seismol Res 6(3):327–338. (in Chinese)

Wang Y, Zhao B, Li J (2017) Mechanism of the catastrophic June 2017 landslide at Xinmo Village, Songping River Sichuan Province China. Landslides 15:333–345

Wu JH, Chen JH, Lu CW (2013) Investigation of the hsien-du-shan rock avalanche caused by typhoon morakot in 2009 at kaohsiung county, taiwan. Int J Rock Mech Min Sci 60(2):148–159

Xu Q, Fan X, Huang R et al (2010) A catastrophic rockslide debris flow in Wulong, Chongqing, China in 2009: background, characterization, and causes. Landslides 7:75–87

Xu Q, Li WL, Dong XJ et al (2017) The Xinmocun landslide on June 24, 2017 in Maoxian, Sichuan: characteristics and failure mechanism. Chin J Rock Mechan Eng 36(11):2612–2628 (in Chinese)

Xu ZQ, Hou LW, Wang ZX (1992) Mountain building processes of the Songpan-Ganzi Orogeny. Chinese Geology. 12:14–16 (in Chinese)

Check for updates

2020 Kyoto Japan

Landslide Field Experiment on a Natural Slope in Futtsu City, Chiba Prefecture

Hirotaka Ochiai, Katsuo Sasahara, and Yusuke Koyama

Abstract

Climate change has contributed to increasingly frequent heavy rainfall events across Japan, increasing the risk of landslide disasters on steep slopes. Despite the urgency of this issue, few studies have assessed actual changes within a slope in real time during simulated rainfall-induced landslides. This study employed a rainfall simulator to study slope changes and relevant mechanisms in a sandy slope in Futtsu City, Chiba Prefecture. Inclinometers, extensometers, piezometers, and several video cameras were installed along the slope to capture slope movements, groundwater levels, and visual data during a simulated rainfall event leading to a landslide. Changes measured in groundwater flow and soil structure were attributed to the formation of soil blocks. Slope instability was promoted by increased groundwater levels when the discharge capacity was no longer able to accommodate the increasing precipitation rates, resulting in the occurrence of a landslide. The data and video captured in this experiment will be widely disseminated among several cooperating organizations and is also intended to serve as material for disaster-awareness videos to promote public safety.

Keywords

Rainfall simulator • Landslide experiment • Natural slope • Groundwater • Soil blocks

Introduction

According to the Japan Meteorological Agency, in recent years, heavy rain events have become increasingly more frequent with climate change, raising the risk of natural disasters on steep slopes in mountainous areas. Specifically, in urban areas, serious damage from debris flows originating from rainfall-induced landslides is becoming a major problem.

Though many studies have examined the mechanisms of rainfall-induced landslides, studies that assessed changes within the slope during actual landslides are extremely limited. Only three experiments concerning landslides on natural slopes have been conducted in Japan using rainfall simulators including international studies, there are seven in all (Table 1).

During the experiment performed at the Ikuta Ryokuchi Park, Kawasaki City in 1971, debris moved much faster than expected and caused an accident, engulfing 25 people and killing 15 (Oka 1972). Immediately after the accident, the Loamy Slope Landslide Experiment Accident Investigation Committee (1974) was established to analyze detailed records of the experiment, including videos. Regarding the high-speed movement of landslide debris down the slope, it was assumed that the debris slid over a thin layer of water formed on the slope at a maximum speed of 17 m/s, but the detailed mechanism behind this phenomenon was unknown. Subsequently, a large-scale rainfall simulator was built at the National Research Institute for Earth Science and Disaster Resilience (NIED; formerly known as the National Research Center for Disaster Prevention) in Tsukuba City, Ibaraki

H. Ochiai (✉)
Japan Forest Technology Association, 7 Rokubancho, Chiyoda, 102-0085, Japan
e-mail: hirotaka_ochiai@jafta.or.jp

K. Sasahara
Kochi University, 2-6-1 Akebono, Kochi, 783-8520, Japan
e-mail: sasahara@kochi-u.ac.jp

Y. Koyama
NHK (Japan Broadcasting Corporation), 2-2-1 Jinnan, Shibuya, 150-8001, Japan
e-mail: koyama.y-fg@nhk.or.jp

Ž. Arbanas et al. (eds.), *Understanding and Reducing Landslide Disaster Risk*,
ICL Contribution to Landslide Disaster Risk Reduction,
https://doi.org/10.1007/978-3-030-60713-5_18

Table 1 Landslide experiments on natural slopes [modified after Askarinejad et al. (2013)]

Reference	Location	Triggered volume or sprinkled area (m^3/m^2)	Water supply method	Instrumentation	Slope angle (deg.)	Soil type
Oka (1972) Loamy slope landslide experiment accident Investigation committee (1974)	Japan	270 m^3	Rain guns	Extensometer/ Tensiometer/ Piezometer/ Strain probe/ Groundwater level gauge / Flowmeter	-	Loamy sand deposit
Yagi et al. (1985)	Japan	10 x 25 m^2	Rainfall simulator	Extensometer/Piezometer/ Strain meter	-	-
Yamaguchi et al. (1989) Kawabe et al. (1989)	Japan	10 x 30 m^2	Upper trench	Extensometer/Piezometer/ Inclinometer	-	-
Harp et al. (1990)	USA	1.6 x 1.3 m^2	Upper trench	Extensometer/Piezometer	30	Gravelly silty sand
		30 x 5 m^2			43	Weathered disintegrated granite sand
		3.2 x 4.0 m^2			70	Coarse to medium sand with 20% clay
Ochiai et al. (2004)	Japan	30 x 5 m^2	Rainfall simulator	Extensometer/ Tensiometer/Piezometer/ Strain probe/ Photogrammetry	33	Weathered disintegrated granite sand
Teysseire (2005)	Switzerland	55 m^2	Rainfall simulator	TDR/Tensiometer/Moisture point	42	Moraine (Gravelly silty sand)
Springman et al. (2009) Askarinejad et al. (2012) Lehmann et al. (2013)	Switzerland	130 m^3	Rainfall simulator	TDR/Tensiometer/Piezometer/ Soil deformation probe/ERT/ Soil pressure sensors/ Photogrammetry	38	Silty sand

Prefecture, in 1974, and landslide experiments with multiple artificial slopes were conducted using the rainfall simulator with a maximum precipitation intensity of 200 mm/h (e.g., Fukuzono 1985). Iverson and LaHusen (1989) were able to confirm excess pore water pressure during landslide initiation, which is the cause of landslide fluidization, for the first time under the rainfall simulator in 1987. Subsequently, from 2001 to 2004, the APERITIF (Area Prediction of Earthquake and Rain-induced Rapid and Long Travelling Flow Phenomena) project was conducted by Sassa et al. (2005). They generated fluidized landslides on a weathered granite slope in a national forest in Ibaraki Prefecture using a rainfall simulator in 2003, and Ochiai et al. (2004) reported slope changes during the landslide. In addition, in 2009, the TRAMM (Triggering of Rapid Mass Movement in Steep Terrain) project was conducted in Switzerland, which was presented in detail by Springman et al. (2009), Askarinejad et al. (2013), and Lehmann et al. (2013).

The present report outlines a landslide experiment using a rainfall simulator that was conducted on a natural slope in Futtsu City, Chiba Prefecture, on December 12, 2018 at the request of NHK, Nippon Hoso Kyokai (Japan Broadcast Corporation). In preparation for the experiment, NHK located videos of past slope landslide experiments in storage. During the main experiment, the behavior of the slope was monitored and was filmed with multiple cameras. The massive amount of video generated is expected to serve as resource for future studies that will examine landslide mechanisms in more detail.

The experiment represents a collaboration between NHK, The Japan Landslide Society, Japan Forest Technics Co., Ltd., OSASI Technos Inc, and Daiki Rika Kogyo Co., Ltd., with the full cooperation of Futtsu City, Federation of Five Construction Organizations in Futtsu City, the Chiba Prefectural Police, the river engineering department of Chiba Prefecture, and many other entities.

Materials and Methods

Slope Conditions

The experiment site was sparsely forested with hardwoods such as oaks and stone oaks with an east-northeast-facing slope of approximately 40° located in Kokubo, Futtsu City, Chiba Prefecture (Fig. 1). The ground surface was a smooth slope covered with Simon bamboo and ivy, and part of the lower edge of the slope had previously been quarried for sand. Based on the results of a portable penetration test (Japanese Geotechnical Society, 2017) using a number of blows (penetrate 10 cm) with a 5-kg weight dropped from a height of 50 cm, the thickness of the surface soil layer was approximately 1 m. To spray water across an area of the slope (10 × 10 m), we transported water from a large tank (40 m^3) installed at the bottom of the slope to the top of the slope using a fire engine. PVC pipes with a length of 10 m with multiple holes 5 mm in diameter were horizontally placed at three levels, and the water pressure within each pipe was adjusted with a valve. The large tank was continuously filled from a nearby agriculture reservoir using water-supply trucks. A canal in the flat ground at the bottom of the slope removed the surface flow generated by spraying to a sediment basin, and 1-m^3 sandbags were arranged in a line to ensure safety.

Installation of Observation Equipment

To understand the changes in the slope that lead to landslides, we installed inclinometers, extensometers, and piezometers (to a depth of 80 cm) provided by OSASI at three locations along two survey lines (a total of six locations), and buried tensiometers provided by Daiki Rika Kogyo at four depths (20, 50, 80, and 120 cm) for automatic measurements (Figs. 2 and 3).

Video Equipment

Videos were taken by NHK via the installation of multiple cameras— handheld video camera, an 8 K video camera, a drone video camera, a 360° camera, and fixed cameras. When it is possible to apply a timecode, we set the time according to the Internet immediately before beginning the experiment to synchronize the time code on videos.

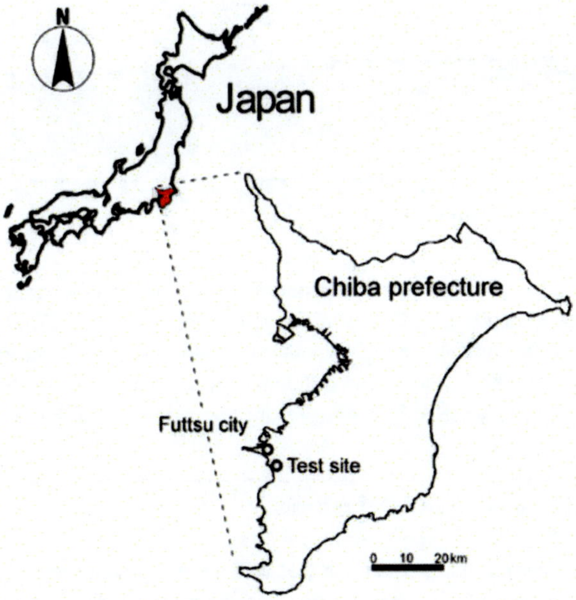

Fig. 1 Location of the experiment site

Experimental Outline

The pre-experiment was conducted in the afternoon of December 11, 2018. With a precipitation intensity of 140 mm/h, water was sprayed for approximately one hour to adjust the distribution of precipitation. After precipitation continued into the evening of December 11 (AMeDAS Kisarazu 29.5 mm), we started spraying with a maximum precipitation intensity of 140 mm/h on 10:53 on December 12, 2018. Subsequently, we increased the precipitation intensity to 188 mm/h at 12:00, 200 mm/h at 13:00, and 240 mm/h at 14:00, ultimately reaching 300 mm/h at 15:00 while observing groundwater level changes and movements of the ground surface. At 15:18 (4 h 25 min after the start of spraying), a landslide initiated immediately below the ridge near the peak, and the surface layer and vegetation slid down the slope surface. The advancing front of collapsed sediment stopped approximately 3 m before reaching the large sandbags. Immediately after the occurrence of the landslide, the slip surface contained much water, reflecting the elevated groundwater levels due to precipitation. The slip surface after the landslide featured many pull-out and broken roots. Many horizontal roots were pulled out and exposed on the surface of the scarp.

After the start of precipitation, water flowed from large pipe-like pores in several locations along the foot of the experimental slope, which changed in volume in response to the changes in the slope (Fig. 4).

Outline of Slope Changes

Changes in the slope along survey line 1 are shown in Fig. 5. As measured by the piezometers, the groundwater level in the upper slope became elevated three minutes after the start of precipitation, which was followed by a rapid increase in the middle and lower slopes. The levels stayed constant until around 13:00, and then gradually increased as the precipitation intensity began increasing at 13:00. After 14:10, groundwater levels in the upper slopes began increasing and in the lower slopes decreasing. The extensometers showed a displacement of the ground surface in the slope direction. The extensometers on the middle and lower slopes reached a constant state when the groundwater level increased, then gradually moved downslope. In conjunction with the drop in groundwater levels after 14:00, all extensometers recorded notable movements. When the precipitation intensity increased to 300 mm/h at 15:00, groundwater levels on the upper slope increased again along with accelerated movements in the middle slope, which led to the landslide initiation. As Fig. 6 shows the records in the slope along survey line 2, groundwater levels decreased in the upper and middle

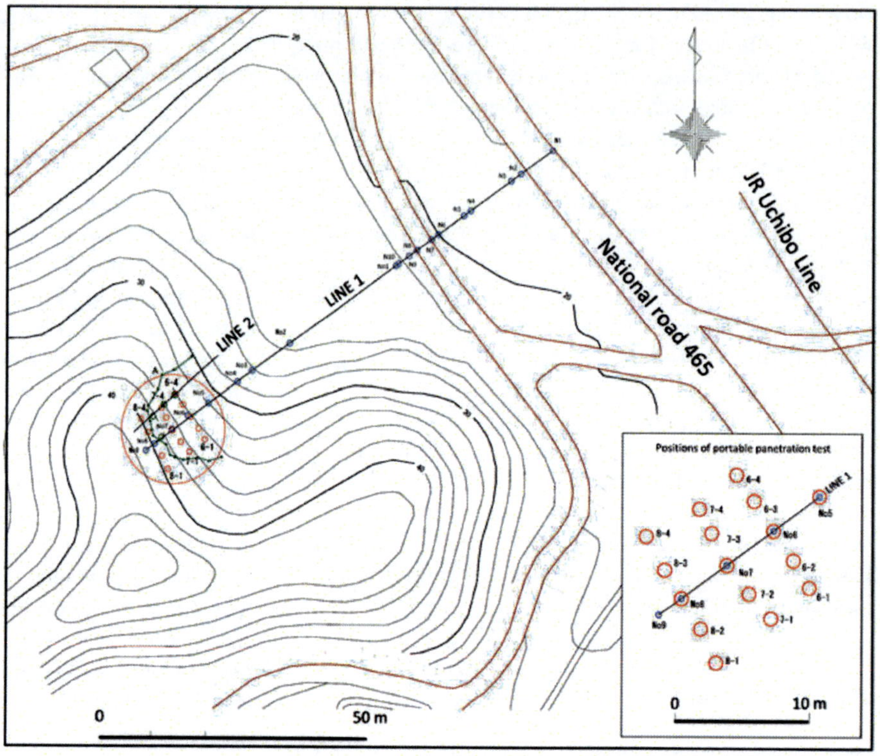

Fig. 2 Map of the experimental slope (Kokubo, Futtsu City, Chiba Prefecture)

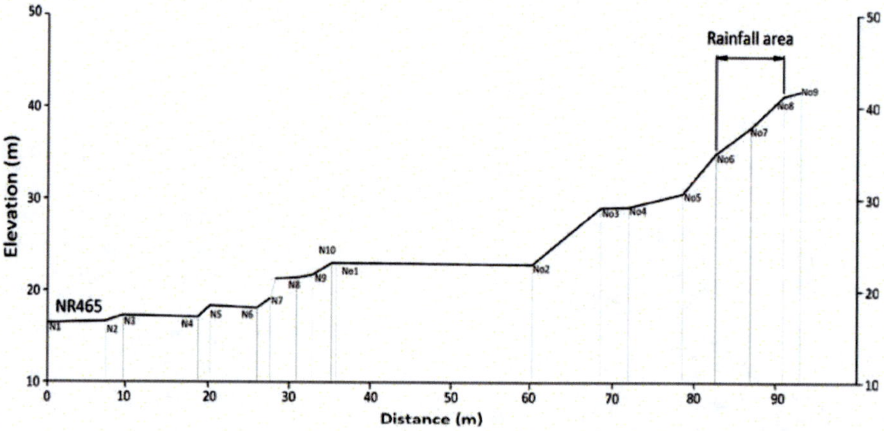

Fig. 3 Profile of the experimental slope (Kokubo, Futtsu City, Chiba Prefecture)

slopes after 14:00. This differences by location may be explained as follows: As the soil structure around piezometers was disturbed due to slope movements, saturated groundwater was discharged, lowering the groundwater level in the soil. Changes measured by the extensometers and in the groundwater levels along the same survey line were not simultaneous, possibly due to the formation of soil blocks and changes the distribution of large pores in the soil that functioned as paths of discharge.

Figure 7 shows the soil moisture in the lower slope along survey line 1. After 14:00, the increase/decrease in water pressure at a depth of 80 cm in the upper part of the slip surface was notable. Subsequent changes in water pressure at a depth of 120 cm displayed similar patterns. These results indicate changes in the groundwater flow structure within the soil surface at different depths. This change in soil structure was likely caused by the formation of soil blocks.

Discussions/Conclusions

Iverson (2015) pointed out that no field experiment is strictly reproducible because of the idiosyncrasies of the natural settings and materials involved, although field experiments

Fig. 4 Drained groundwater from the outlet of the pipe at footslope

can be large enough to avoid scaling problems, and complex enough to mimic nature. However, in order to fully understand the landslide mechnism, that we rarely observe on natural hillslopes, it is necessary to find out peculiar natural phenomena by field experiments and reproduce and confirm it in detail in highly reproducible laboratory experiments.

1. When we sprayed artificial rain over an area of 10 m 10 m on a 40° sandy slope with a precipitation intensity of 140–300 mm/h, a landslide occurred after 4 h and 25 min. The depth of the landslide was approximately 1 m.
2. On the experimental slope, water was discharged from pipes at the slope foot during precipitation prior to the landslide initiation, and the flow rate changed according to the slope changes. However, it was not possible to measure the change in flow rate, so it should be try in the future.
3. The soil formed blocks as the ground surface changed in the process of the landslide initiation. This was assumed

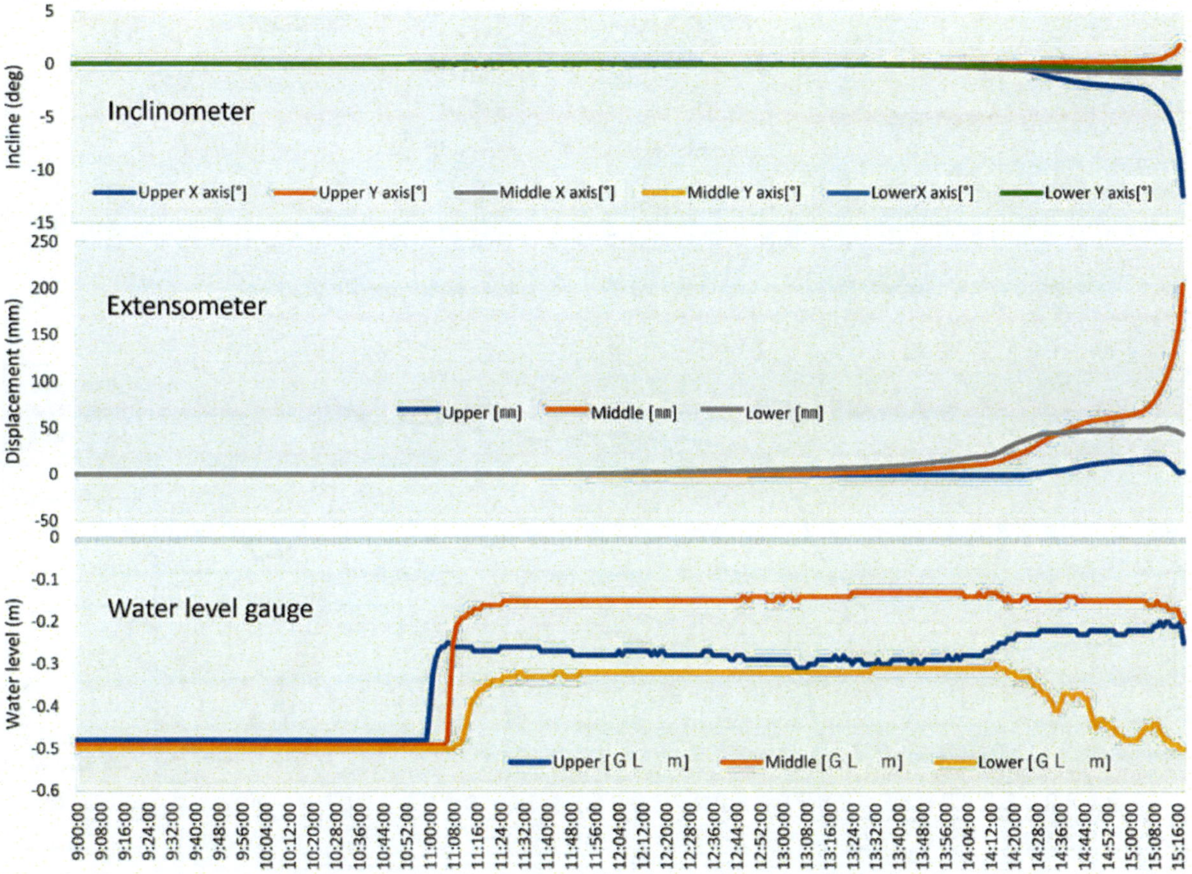

Fig. 5 Inclinometer, extensometer, and Piezometer records of Line 1 on the landslide experiment (2018/12/12)

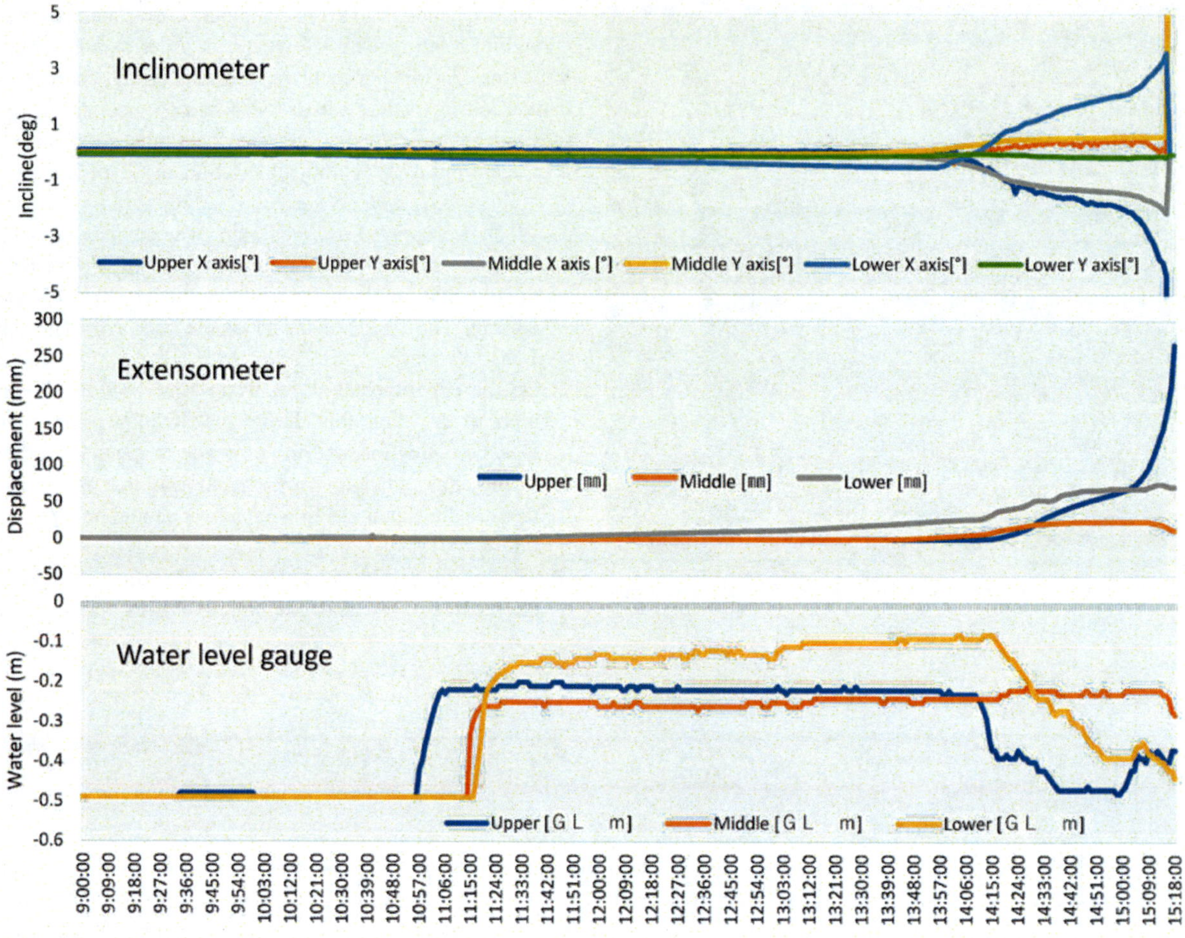

Fig. 6 Inclinometer, extensometer, and Piezometer records of Line 2 on the landslide experiment (2018/12/12)

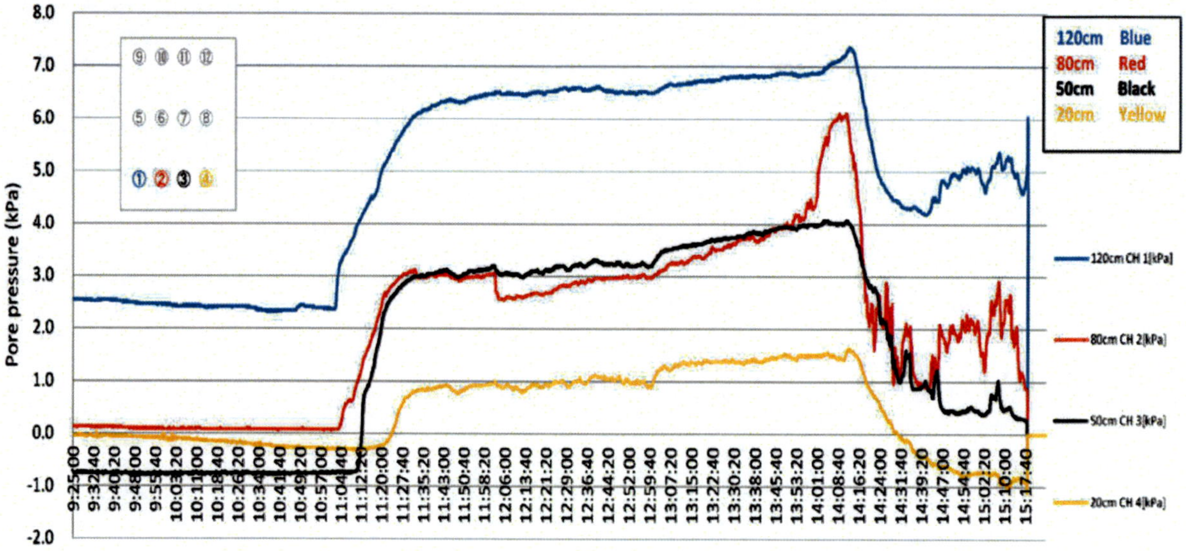

Fig. 7 Pore pressure change of four depth tensiometer records of Line 1 (Upper slope) at the landslide experiment (2018/12/12)

to be the result of the heterogenous distribution of groundwater levels in the soil surface caused by the new discharge system created by the soil blocks. This change in the discharge structure of the soil was also present at different depths in the same location.

4. A decrease in pore water pressure due to discharge slowed movements in the soil layer briefly, but the discharge capacity was insufficient relative to the further increase in precipitation intensity, resulting in another increase in groundwater levels. This in turn invited slope instability.

NHK and partners exchanged memorandums to encourage the wide use of the present data, including via videos to promote disaster prevention awareness among residents, and shared information regarding the experiment. Specific guidelines for the use of the present data will be presented by the Japan Landslide Society.

Acknowledgements Japan Forest Technics Co., Ltd. cooperated in conducting preliminary surveys for the present research. OSASI Technos Inc. and Daiki Rika Kogyo Co., Ltd. provided necessary measuring equipment for the experiment and also cooperated on the actual measurements. This experiment was only possible because of the full support from related organizations, such as Futtsu City, a Federation of Five Construction Firms in Futtsu City, the Chiba Prefectural Police, and the river engineering department of Chiba Prefecture. We would like to thank Editage (www.editage.com) for their writing support on the abstract.

References

Askarinejad A, Casini F, Bischof P, Beck A, Springman SM (2013) Rainfall induced instabilities: a field experiment on a silty sand slope in northern Switzerland. Rivista Italiana di Geotechnica 3:50–71

Fukuzono T (1985) A method to predict the time of slope failure caused by rainfall using the inverse number of velocity of surface displacement [in Japanese]. J Japan Landslide Soc 22(2):8–13

Harp EL, Wells WG, Sarmiento JG (1990) Pore pressure response during failure in soils. Geolog Soc Amer Bull 102:428–438

Iverson RM, LaHusen RG (1989) Dynamic pore-pressure fluctuations in rapidly shearing granular materials. Science 246:769–799

Iverson RM (2015) Scaling and design of landslide and debris-flow experiments. Geomorphology 244:9–20

Japanese Geotechnical Society (2017) JGS 1433–2012 Method for portable dynamic cone penetration test, Japanese Geotechnical Society Standards Geotechnical and Geoenvironmental Investigation Methods, vol 2, 300p

Kawabe H, Shibano H, Nishio K, Yamaguchi I (1990) On the occurrence of the landslide based on the field experiment [in Japanese]. J Jpn Landslide Soc 27(1):1–10

Lehmann P, Gambazzi F, Suski B, Baron L, Askarinejad A, Springman SM, Holliger K, Or D (2013) Evolution of soil wetting patterns preceding a hydrologically induced landslide inferred from electrical resistivity survey and point measurements of volumetric water content and pore water pressure. Water Resour Res 49:7992–8004

Loamy Slope Landslide Experiment Accident Investigation Committee (1974) Survey report of the landslide experiment accident in loamy slope [in Japanese], Prime Minister's Office, 160p

Ochiai H, Okada Y, Furuya G, Okura Y, Matsui T, Sammori T, Terajima T, Sassa K (2004) A fluidized landslide on a natural slope by artificial rainfall. Landslides 1(3):211–219

Oka H (1972) Impacts by the "artificial landslide": re-examine the rage of nature [in Japanese]. Kagaku Asahi 32(1):152–153

Sassa K, Fukuoka H, Sekiguchi T, Moriwaki H, Ochiai H (2005) Achievements of Aerial Prediction of Earthquake and Rain Induced Rapid and Long-Travelling Flow Phenomena (APERITIF) Project, Annuals of Disaster Prevention Research Institute, Kyoto University, No. 48 C, 1–22

Teysseire P (2005) Geotechnical properties of moraine [in German]. Doctoral thesis no. 16322. Institute for Geotechnical Engineering Zürich, ETH Zürich, Zurich, Switzerland. https://doi.org/10.3929/ethz-a-005166289

Yagi N, Yatabe R, Enoki A (1985) Laboratory and field experiments on prediction method of occurring time of slope failure due to rainfall [in Japanese]. J Jpn Landslide Soc 22(2):1–7

Yamaguchi I, Nishio K, Kawabe H, Shibano H, Iida C (1989) Initiation and fluidization of an artificial landslide: Field experiment in Yui, Shizuoka Prefecture, Japan [in Japanese]. Shinrin Kosoku (Areal Survey) 158:3–9

Mechanism of Landslide Initiation in Small-Scale Sandy Slope Triggered by an Artificial Rain

2020 Kyoto Japan

Vedran Jagodnik, Josip Peranić, and Željko Arbanas

Abstract

Physical modelling of landslides using scaled landslide models behavior began in 1970s in Japan at a scaled natural slope physical model. In October 2018, at the Faculty of Civil Engineering University of Rijeka, started a four-year research Project Physical modelling of landslide remediation constructions behavior under static and seismic actions, funded by the Croatian Science Foundation. The main Project's aim is the modelling of landslide remedial constructions' behavior in physical models of scaled landslides in static (rainfall triggered landslides) and seismic conditions (earthquake triggered landslides) and their combination under 1 g conditions. In this manuscript we presents the preliminary results obtained in landslide initiation test of a sandy slope, constructed of 0–1.0 mm the Drava River sand and exposed to an artificial rain typical for local conditions in Croatia by rainfall simulator. The results of landslide development were monitored by observation of volumetric water content and pore water pressure as well as by of surface displacement by *structure from motion* (SfM) surface observation displacement monitoring inside the model displaced mass. In this paper the preliminary results of provided test will be presented related to initiation and development of the observed instability of the sandy slope model and compared with the results of numerical analysis of rainfall infiltration combined with limit equilibrium stability analysis.

Keywords

Physical modelling • Sandy slope • Artificial rainfall • Landslide initiation • Failure mechanism • Soil suction • Infiltration analysis

Introduction

Landslides, as a one of the major geohazards that threatened vulnerable areas in mountains, along river banks, coasts and islands as well as in urban areas, cause severe consequences expressed in lost human lives and economic losses all around the world. An increase in the frequency and/or intensity of heavy rainfall, as well as shift of locations and periodicity of rainfall due to climate change significantly increases landslide risk in landslide prone areas. Urban expansion due to population growth, development of coastal and mountain areas, deforesting, construction of roads and railways increase exposure to the landslide hazard. Strong earthquakes in seismically active areas are the second triggering factor for numerus rapid and long runout landslides with similar consequences caused by rain induced landslides.

The answers of scientists and engineers included in landslide risk reduction were in significant developing of landslide science during last decades. Some of the answers are based on different types of modelling of landslide related researches especially in landslide triggering and landslide process evolution. Landslide modelling for a long time was based on only numerical modelling techniques. Physical modelling of landslide analysing small scale landslide models behaviour was started in 1970s and 1980s in Japan (Oka 1972; Kutara and Ishizuka 1982; Yagi et al. 1985; Yamaguchi et al. 1989) on natural slopes exposed to artificial rainfall. The laboratory experiments of landslide behaviour in a scaled physical model (also called flume test)

V. Jagodnik · J. Peranić · Ž. Arbanas (✉)
Faculty of Civil Engineering, University of Rijeka, Rijeka, 51000, Croatia
e-mail: zeljko.arbanas@gradri.uniri.hr

V. Jagodnik
e-mail: vedran.jagodnik@gradri.uniri.hr

J. Peranić
e-mail: josip.peranic@gradri.uniri.hr

Ž. Arbanas et al. (eds.), *Understanding and Reducing Landslide Disaster Risk*, ICL Contribution to Landslide Disaster Risk Reduction,
https://doi.org/10.1007/978-3-030-60713-5_19

started in 1980s and 1990s in Canada (Hunger and Morgenstern 1984), Japan (Yagi et al. 1985) and Australia (Eckersley 1990) under 1 g conditions. After the initial experiences at in field and laboratory researches, the small scale landslide modelling has found a wide application around the word in different aspects of landslide investigations, analysing different types of landslides (e.g. flows, slides, falls and toppling) as well as different types of material including in landslide movements (rock mass, sandy, silty and clayey material) to the recent days (e.g. Cui et al. 2014; Li et al. 2018; Spolverino et al. 2019; Cui et al. 2019; Cogan and Gratchev 2019).

The research Project Physical modelling of landslide remediation constructions' behaviour under static and seismic actions, funded by the Croatian Science Foundation, started in October 2018 at the Faculty of Civil Engineering, University of Rijeka, Croatia. The main Project aim is modelling behaviour of landslide remedial constructions in physical models of scaled landslides in static and seismic conditions (Arbanas et al. 2019). The manuscript presents the initial results obtained in landslide initiation test of a sandy slope exposed to an artificial rain typical for local conditions in Croatia by rainfall simulator. The results of landslide development were monitored by observation of surface displacement by *structure from motion* (SfM), terrestrial laser scanner (TLS) and a pair of high speed cameras, as well as by pore water pressure and displacement monitoring inside the displaced mass. While analyzes of the observed small scale landslide movements are still ongoing, in this paper we will describe the surface observations of landslide movements and conclusions that followed from movement analyses (Arbanas et al. 2020).

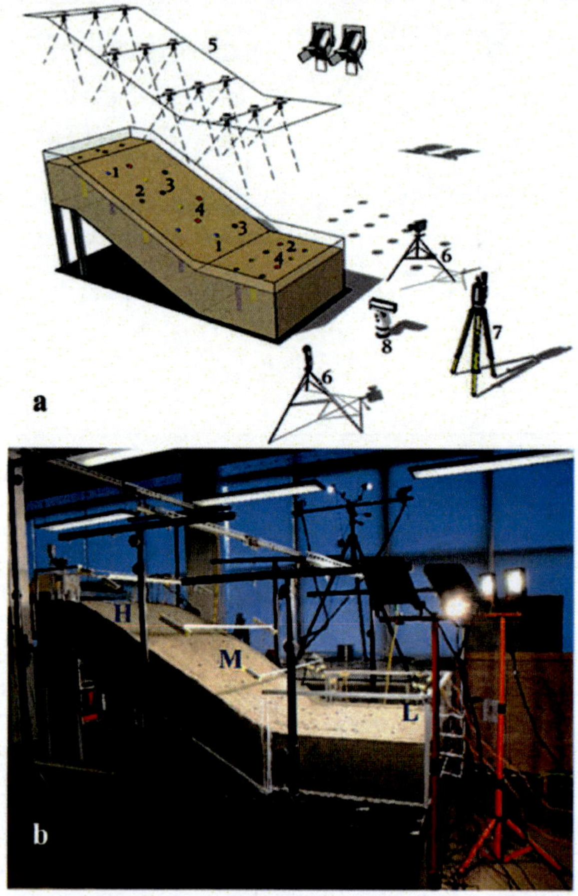

Fig. 1 **a** Schematic view of physical model: 1-tensiometers, 2-pore pressure transducers, 3-strain gauges, 4-accelerometers, 5-rainfall simulator (sprinkler system), 6-high speed cameras, 7-terrestrial laser scanner (TLS), 8-infrared camera; **b** Photo of physical model. Symbols H, M and L define positions of the measurment equipments (Arbanas et al. 2019, 2020)

Physical Model

Physical model of a scaled slope under static actions caused by artificial rainfall, was designed to enable initiation of a landslide caused by controlled artificial rainfall and equipped with adequate photogrammetric equipment and complex sensor network with ability to measure displacements, soil moisture, pore pressures and soil pressures within a landslide body (Fig. 1a). The flume is constructed as a steel base construction with transparent Plexiglas sides in layout dimensions of 2.30 × 1.00 m that lies on the working table (Fig. 1b). It was provided that the maximal height of the soil material in the slope is to 30 cm. An inclination of the flume slope can be adjusted for a type of used soil material from 20° to 45°. To prevent a sliding of soil mass at the contact of the flume base, the geogrid is fixed to the flume base. The flume construction is completely impermeable and the only one drainage from the flume is in the toe of the slope where a series of valves enabled maintaining of a controlled ground water surface in the foot of the slope.

Soil Properties

For the first series of test in the constructed flume, the fine sandy material was chosen to be built-in in the flume slope. Fine graded 0–1.0 mm (Fig. 2) the Drava River Sand (D_{60} = 0.37 mm, D_{10} = 0.19 mm, C_u = 1.947) was built-in in the flume at the inclination of 30° in 6 cm height 5 layers at the, 30 cm in total, at relative density D_r = 0.5 and reached void ratio of e = 0.78. Each layer was compacted by a manual compactor at the water content of w = 2% (Ladd 1978). The specific gravity of sand is G_s = 2.7.

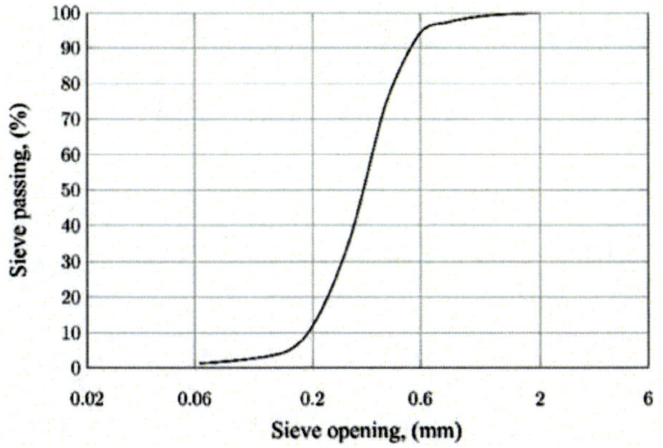

Fig. 2 Grain size distribution curve of the Drava River Sand built-in in the model slope (Arbanas et al. 2020)

Monitoring and Observations

Monitoring system established at the flume test followed the principles used in observation of real landslides; it was consisted of geotechnical monitoring systems and geodetic monitoring systems (Mihalić Arbanas and Arbanas 2015).

Geodetic monitoring is based on innovative photogrammetric equipment for multi-temporal landslide analysis (Zanuta et al. 2006) from stereo image sequences obtained by a pair of high speed cameras (Feng et al. 2016). Terrestrial laser scanner (TLS) as well as *structure from motion* (SfM) photogrammetry surveys enabled determination of reached 3D landslide surface or high resolution digital model of a landslide model before and after sliding very precisely (Bitelli et al. 2004). The TLS Faro Focus HD 130 with the range of 0.6–130 m and resolution of 70 Mpixel as employed for the model slope surface scanning.

Geotechnical monitoring is consisted of a complex network of miniature sensors equivalent to those used in in a field landslides geotechnical monitoring. Built-in sensors have ability to measure displacements MEMS (Ooi and Wang 2014), pore pressure and moisture transducers, soil pressures and contact forces using load cells and strain gauges, as well as soil suction (tensiometers) (De Dios et al. 2010; Ooi et al. 2014; Scaioni et al. 2013; Lu et al. 2015). All sensors used in the test are connected to data loggers for continuous data collection during the test.

Rainfall Simulator

One of the most important issues in physical models to simulate landslide initiation and motion caused by rainfall infiltration is full controlled intensities of artificial rainfall. Although it is possible to use some of the already developed rainfall simulators (e.g. Iserloh et al. 2012, 2013; Lora et al. 2016), there was necessity for developing of a rainfall simulator that can be adjusted to simulate the rainfall intensities characteristic for precipitation events in Croatia. Construction of rainfall simulator includes ordinary elements usually used for watering plants and has an ability to control intensity, flow, and uniformity of wetting as well as the time of rainfall simulation. Rainfall simulator is consisted of a system of tree sprinkler branches with four different axial-flow full-cone nozzles. The minimal intensity that can be achieved with only one nozzle at each branch is 10 mm/hour while ta maximal intensity can reach more than 600 mm/hour using 4 nozzles at each branch. These intensities can enable a wide range of possible rainfall intensities to model different scenarios of real rainfall in Croatia and worldwide and to simulate possible effects of climate change to unexpected rainfall intensities.

Test Results

Description of Test Conducting

After the sandy slope construction and installation of the monitoring equipment, the test was conducted by exposing the slope to the artificial rain from three nozzles, one nozzle in each part of the slope with a uniform artificial rainfall intensity of 58.23 mm/h. The same intensity was maintained during the test 121 min, when the intensity was increased on 121.52 mm/h and after 7 min (adjustment of intensity using water pressure valve) reduced on 81.01 mm/h that as maintained to the end of the test, Fig. 3.

The first movements in the slope were occurred 126 min after the rainfall started, Fig. 4. The landslide initiation was started 11 min after complete saturation of the slope foot (115 min after the rainfall started), and the landslide occurrence was presented in complete loss of soil strength in the slope toe and development of the main scarp. It was noted that after the movement, the ground water started to run out through the sliding mass, Fig. 4. After the landslide initiation, the further instability distribution was retrogressive with developing of new scarps 135, 139 and 148 min after rainfall started.

After developing of the fifth crack, the rapid spread of the landslide to the top of the slope as occurred in next two minutes including an initiation of a debris flow originated from the crest of the slope (Fig. 5). In this moment (150 min after the rainfall started), the rainfall was suspended that affected stopping of landslide activity. While this test was conducted with the task to identify conditions of a landslide initiation and mechanism and type of landslide occurrence through the reached level of landslide development, final collapse of the slope was not enabled (Arbanas et al. 2020).

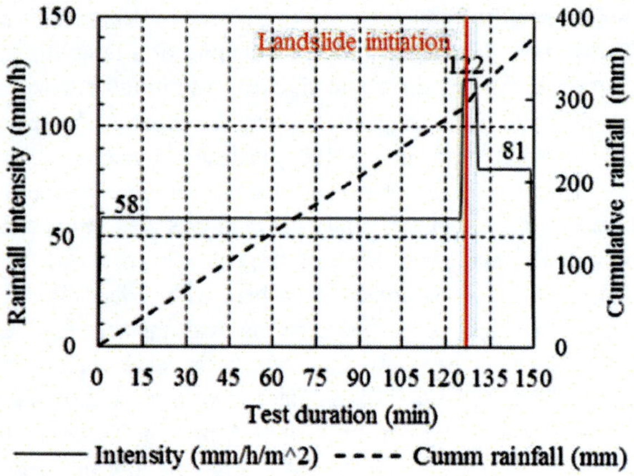

Fig. 3 Diagram of rainfall intensity during the test

Fig. 4 Photo of landslide initiation (126 min after the rainfall started)

Analysis of Measured Data

Analysis of measured data of the volumetric water content during the test in the upper (H), middle (M), and lover (L) part of the slope enables the reconstruction of the landslide development, Fig. 6a–d.

In the first stage of the test the water infiltration in the slope started and the rise of volumetric water content was faster in shallower layers of the slope in all three parts of the slope (Fig. 6a–d). It is clearly noticeable that the upper layers of the slope reached the constant values of volumetric water content around 0.2 m^3/m^3 from 5 to 30 min after rainfall starting in the foot of the slope and from 10 to 40 min in the middle part of the slope. In the top of the slope, the same values of volumetric water content of around 0.2 m^3/m^3 were reached form 5 to 20 min and stayed almost constant to the end of the test. The differences in these volumetric water contents are caused by establishing of a seepage trough the slope, and by increasing of hydraulic conductivity caused by suction decreasing and saturation of the slope from bottom layers to the surface (Arbanas et al. 2020).

Fig. 5 Final landslide development to the top of the slope (150 min after the rainfall started)

An increased inflow and rising of the ground water table in the foot of the slope begun 30 min after rainfall started that is evident at the deepest sensor (L at −24 cm), and this process was continued till the ground water level reached and flooded the surface of the slope foot 117 min after the rainfall started that was maintaining through the test. The forming of the ground water table in the middle of the slope begun 30 min later, Fig. 6a–d. A ground water table was not established at the top of the slope; infiltrated water was drained to the lower part of the slope.

The failure of the slope was occurred immediately after the ground water table reached and the surface of the slope foot when that caused decreasing of soil strength and sliding (Fig. 4) and further retrogressive landslide development to the top of the slope. Finally, in the moment when the slope was completely saturated, the rain intensity was too high for further water infiltration and forming of the debris flow from the top of the slope, Fig. 5 (Arbanas et al. 2020).

Numerical Analysis

To simulate the conducted test as well to check some conditions that are not completely identified during the physical testing, 2D numerical model for rainfall infiltration and limit equilibrium stability analysis was established. Rocscience Slide2 software was employed that enabled stability analysis together with built-in finite element groundwater seepage analysis in the modelled slope (Rocscience 2018). The 2D numerical model, with dimension identical to the physical model, is defined with impermeable boundaries and exposed to the rainfall intensity of 58 $mm/m^2/h$. While the matric suction and permeability of the soil are dependent on volumetric water content in the slope, the relationships between volumetric water content and matric suction as well as permeability were established and presented in Fig. 7.

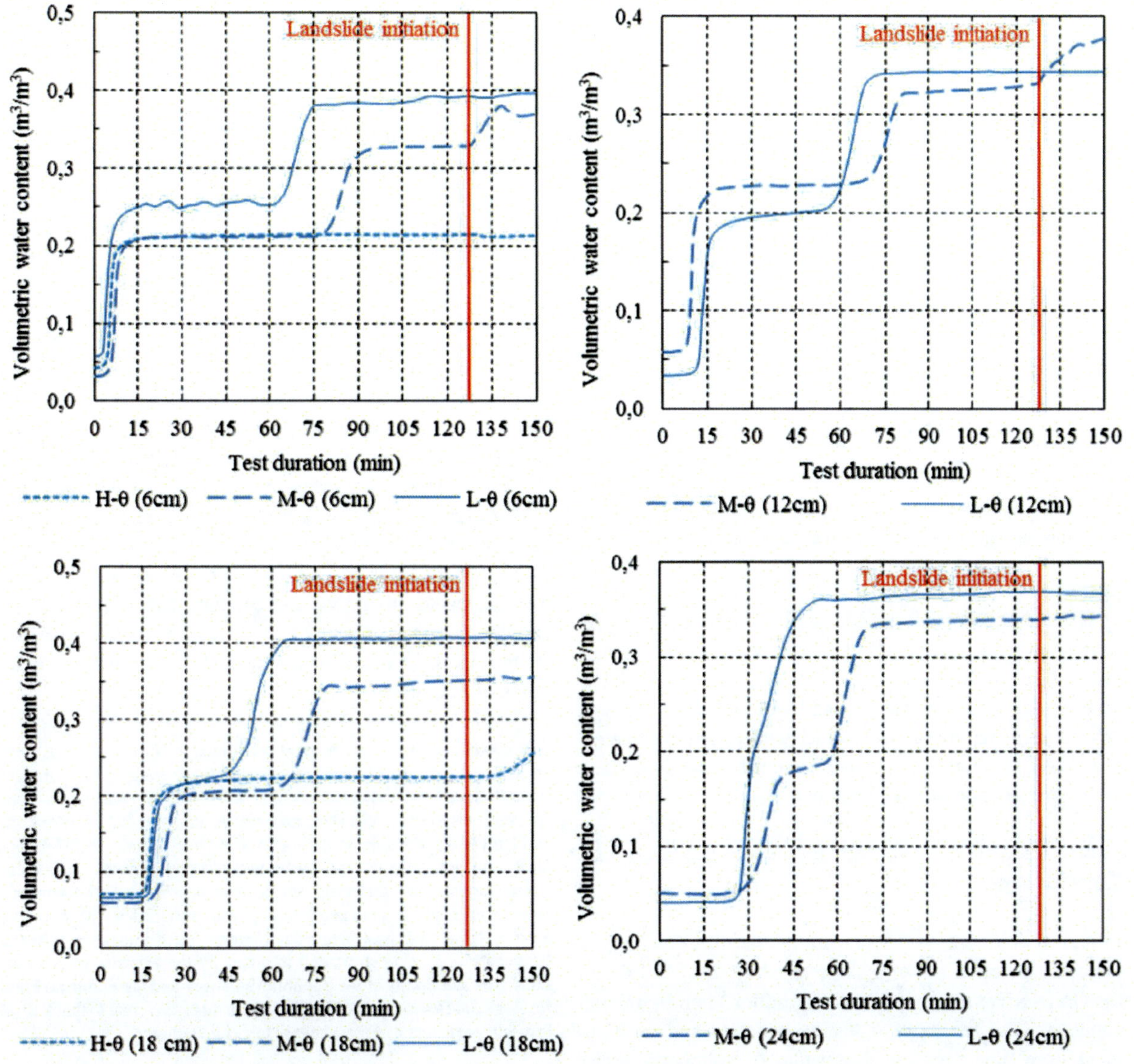

Fig. 6 Diagrams of measured values of volumetric water content in the upper (H), middle (M), and lover (L) part of the slope during the test. Numbers in parentheses designate depth of a sensor

Numerical analysis indicated on very slow infiltration of rainfall in the deeper layers of the slope in the intimal stage of the simulation that was expected related to the high values of matric suction and appropriate low permeability, Fig. 8a, b. With developing of infiltration process and rising of the volumetric water content in the slope, the seepage flow became faster and infiltrated water reached the bottom of the model and the water table was occurred 83 min after the rainfall was started, Fig. 8c, d. After that, the fast rise of the ground water table started, especially in the foot of the slope, and the surface of the toe was reached 98 min after the rainfall was started, Fig. 8e. The failure of the slope was occurred 101 min after the rainfall was started, Fig. 8 e; when the factor of safety (FoS) fell below 1.0. It was indicated that the slope was stable with FoS $\gg$1.0 almost all duration of the test, while the sudden fall of FoS to 1.0 was appear after 90 min following the sudden rise of ground water table, Fig. 8d–f.

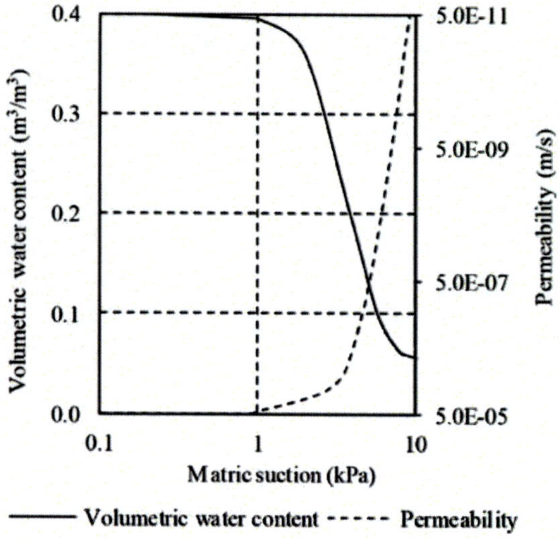

Fig. 7 Diagrams of matric suction and permeability related to changes of volumetric water content in the slope used in numerical analysis

It could be adopted that the results of numerical model have shown relatively good matching with physical testing results (Fig. 6a–d) related to rainfall infiltration. The difference in time of failure was caused by differences in 2D and 3D nature of failure occurrence, as well as not completely confirmed relationships between volumetric water content and matric suction and permeability.

Conclusions

In this manuscript we presented the results obtained in landslide initiation test of a slope constructed of 0–1.0 mm the Drava River sand exposed to an artificial rain by rainfall simulator. The results of landslide initiation and development were monitored by observation of volumetric water content and displacement monitoring within the displaced mass in the model. Under artificial rainfall the model slope built-in in sandy slope started to slide after 126 min, after the ground water table reached the surface of the slope foot 117 min after the rainfall started, and the landslide as developed to the top of the slope.

The landslide initiation in the foot of the model slope, as well as further developing of landslide to the top of the model slope was explained based on results of the volumetric water content and soil suction observation during the test.

Numerical 2D analysis was conducted including stability analysis together with built-in finite element groundwater seepage analysis in the modelled slope and presents good matching with physical testing results. The difference in time of failure was caused by differences in 2D and 3D nature of failure occurrence, as well as not completely confirmed relationships between volumetric water content and matric suction and permeability.

Further analyses of data provided by conducted test as well as results of further test conducted in different slope conditions under different rainfall regimes will give better insights in landslides initiation of sandy slopes exposed to an artificial rainfall.

Acknowledgements The research presented in this paper was supported by Croatian Science Foundation under the Project IP-2018-01-1503 Physical modelling of landslide remediation constructions behaviour under static and seismic actions (ModLandRemSS), and University of Rijeka under the Project Uniri-Tehnic-18-113 Laboratory research of static and cyclic behaviour at landslide activation. These supports are gratefully acknowledged. The part of laboratory equipment used for laboratory testing was provided in the frame of Project Research Infrastructure for Campus based Laboratories at the University of Rijeka, co-funded in a part by the Ministry of Science, Education and Sports of the Republic of Croatia and the European Fund for Regional Development (ERDF). The authors also want to thank to all Projects' members and their help in test conducting.

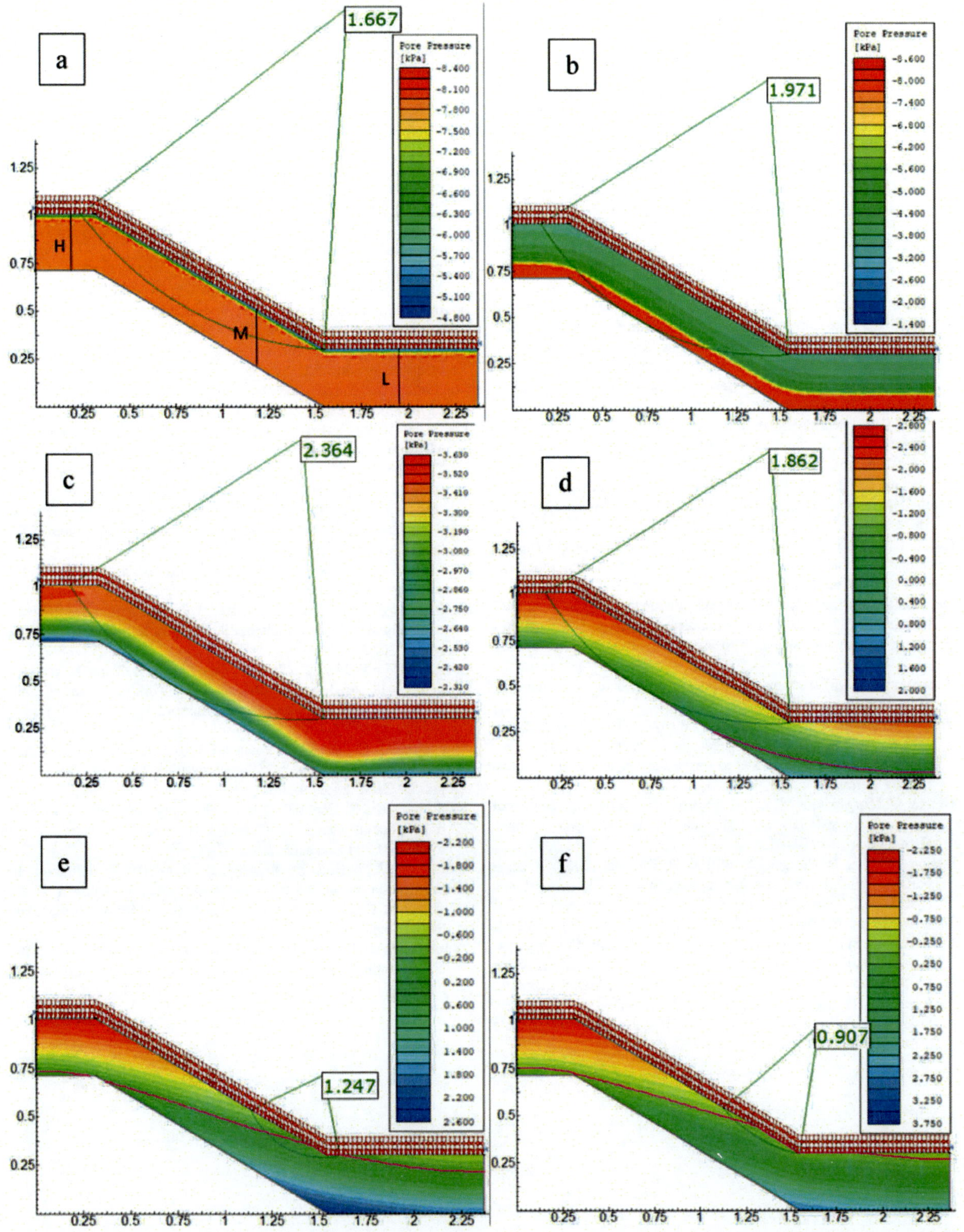

Fig. 8 Infiltration of rainfall in the slope during the test and critical slip surface with related FoS; a 1 min after rainfall starting, b 30 min after rainfall starting, c 60 min after rainfall starting, d 90 min after rainfall starting; e 99 min after rainfall starting f 101 min after rainfall starting when the FoS fell below 1.0

References

Arbanas Ž, Pajalić S, Jagodnik V, Peranić J, Vivoda Prodan M, Đomlija P, Dugonjić Jovančević S, Čeh N (2019) Development of physical model of landslide remedial constructions' behaviour. Proc of 4th Regional Symposium on Landslides in Adriatic Balkan Region, Sarajevo, Bosnia and Herzegovina, 23–25 October 2019. Geotechnical Society of Bosnia and Herzegovina, Sarajevo, pp 103–108

Arbanas Ž, Jagodnik V, Peranić J, Pajalić S, Vivoda Prodan M, Čeh N (2020) Physical Model of Rainfall Induced Landslide in Flume Test: Preliminary Results. Proc. of 4th European Conference on Physical Modelling in Geotechnics, Lulea University of Technology, pp 115–122

Bitelli G, Dubbini M, Zanutta A (2004) Terrestrial laser scanning and digital photogrammetry techniques to monitor landslide bodies. Proc. of XXth ISPRS Congress: Geo-Imagery Bridging Continents, Istanbul, Turkey, 12–23 July 2004. International Archives of the Photogrammetry, Remote Sensing and Spatial Information Sciences, Vol. XXXV, Part 5, pp 246–251

Cogan J, Gratchev I (2019) A study on the effect of rainfall and slope characteristics on landslide initiation by means of flume tests. Landslides 16:2369–2379

Cui P, Guo C, Zhou J, Hao M, Xu F (2014) The mechanisms behind shallow failures in slopes comprised of landslide deposits. Eng Geol 180:34–44

Cui Y, Jiang Y, Guo C (2019) Investigation of the initiation of shallow failure in widely graded loose soil slopes considering interstitial flow and surface runoff. Landslides 16:815–828

De Dios RJ, Enriquez J, Victorino FG, Mendoza EA, Talampas MC, Marciano JJS Jr (2010) A tilt, soil moisture, and pore water pressure sensor system for slope monitoring applications. Science Diliman 21:15–27

Eckersley D (1990) Instrumented laboratory flowslides. Géotechnique 40:489–502

Feng T, Mi H, Scaioni M, Qiao G, Lu P, Wang W, Tong X, Li R (2016) Measurement of Surface Changes in a Scaled-Down Landslide Model Using High-Speed Stereo Image Sequences. Photogrammetric Eng Remote Sens 82(7):547–557

Hunger O, Morgenstern NR (1984) Experiments on the flow behaviour of granular materials at high velocity in an open channel. Géotechnique 34:405–413

Iserloh T, Fister W, Seeger M, Willger H, Ries JB (2012) A small portable rainfall simulator for reproducible experiments on soil erosion. Soil Tillage Res 124:131–137

Iserloh T, Ries JB, Arnaez J, Boix-Fayos C, Butzen V, Cerda A, Echeverria MT, Fernandez-Galvez J, Fister W, Geissler C, Gomez JA, Gomez-Macpherson H, Kuhn NJ, Lazaro R, Leon FJ, Martinez-Mena M, Martinez-Murillo JF, Marzen M, Mingorance MD, Ortigosa L, Peters P, Reguees D, Ruiz-Sinoga JD, Scholten T, Seeger M, Sole-Benet A, Wengel R, Wirtz S (2013) European small portable rainfall simulators: a comparison of rainfall characteristics. CATENA 110:100–112

Kutara K, Ishizuka H (1982) Seepage flow in the embankment and stability of slope during rain (in Japanese). Tsuchi–Kiso Paper No 1330

Ladd RS (1978) Preparing test specimens using under compaction. Geotech Test J 1:16–23

Lora M, Camporese M, Salandin P (2016) Design and performance of a nozzle-type rainfall simulator for landslide triggering experiments. CATENA 140:77–89

Li S, Sun Q, Zhang Z, Luo X (2018) Physical modelling and numerical analysis of slope instability subjected to reservoir impoundment of the Three Gorges. Environ Earth Sci 77:138

Lu P, Wu H, Qiao G, Li W, Scaioni M, Feng T, Liu S, Chen W, Li N, Liu C, Tong X, Hong Y, Li R (2015) Model test study on monitoring dynamic process of slope failure through spatial sensor network. Environ Earth Sci 74(4):3315–3332

Mihalić Arbanas S, Arbanas Ž (2015) Landslides—A guide to researching landslide phenomena and processes. In: Gaurina-Međimurac N (ed) Handbook of research on advancements in environmental engineering. IGI Global, Hershey, pp 474–510

Ooi GL, Wang YH (2014) Applying MEMS Accelerometers to Measure Ground Vibration and Characterize Landslide Initiation Features in Laboratory Flume Test. Proc. of Geo-Congress 2014, Atlanta, Georgia, February 2014, ASCE, pp. 2019–2028

Ooi GL, Wang Y-H, Tan PS, So CF, Leung ML, Li X, Lok KH (2014) An instrumented flume to characterize the initiation features of flow landslides. Geotech Test J 37(5):1–21

Oka H (1972) Impacts by the "artificial landslide": re-examine the rage of nature (in Japanese). Kagaku Asahi 32:152–153

Rocscience (2018) Slide 2D Tutorial. URL: https://www.rocscience.com/help/slide2/#t=tutorials%2FSlide_Tutorials.htm. Accessed 01 March 2020

Scaioni M, Lu P, Feng T, Chen W, Qiao G, Wu H, Tong X, Wang W, Li R (2013) Analysis of spatial sensor network observations during landslide simulation experiments. Europ J Environ Civil Eng 17 (9):802–825

Spolverino G, Capparelli G, Versace P (2019) An instrumented flume for infiltration process modeling Landslide Triggering and Propagation. Geosciences 9:108

Yagi N, Yatabe R, Enoki M (1985) Laboratory and Field Experiments on Prediction Method of Occuring Time of Slope Failure due to Rainfall. J Jpn Landslide Soc 22:1–7_1

Yamaguchi I, Nishio K, Kawabe H et al (1989) Initiation and fluidization of an artificial landslide: Field experiment in Yui Shizuoka Prefecture, Japan (in Japanese). Shinrin Kosoku Areal Surv 158:3–9

Zanuta A, Baldi P, Bitelli G, Cardinalli M, Carrara A (2006) Qualitative and quantitative photogrammetric techniques for multi-temporal landslide analysis. Ann Geophys 49(4/5):1067–1080

Experimental Study on the Formation and Propagation of Debris Flows Triggered by Glacial Lake Outburst Floods

Huayong Chen, Peng Cui, Xiaoqing Chen, and Jiangang Chen

Abstract

In high mountainous areas, debris flows can be triggered by snowmelt, excessive rainfalls, glacier lake outburst floods (GLOFs) or multiple factors combined. Debris flows often pose significant threats to human life and property. In this study, experiments were conducted to investigate the surge waves triggered by glacier avalanches, failure process, GLOFs, debris flow formation and propagation. Additionally, by considering the effect of debris flow scale enlargement, the relationship between the scale enlargement parameter k and the flow density were deduced. The results indicated that the successive surge waves triggered by glacier avalanches have a great effect on the failure of moraine dams. The flow density increased in the flow direction. The maximum flow density in the experiment was approximately 1.49 g/cm^3. The discharge of debris flow at the outlet reached 0.095 m^3/s and the experimental discharge measured at the outlet was approximately 0.1 m^3/s. The error between the calculated peak discharge and the experimental one was approximately 5.26%. The results presented in this paper can provide information for another kind of debris flows triggered by GLOFs in high mountainous areas.

Keywords

Debris flows • GLOFs • Moraine dam • Discharge • Scale enlargement

H. Chen (✉) · P. Cui · X. Chen · J. Chen
Key Laboratory of Mountain Hazards and Earth Surface Process, Chinese Academy of Sciences, Institute of Mountain Hazards and Environment, Chengdu, 610041, China
e-mail: hychen@imde.ac.cn

P. Cui
e-mail: pengcui@imde.ac.cn

X. Chen
e-mail: xqchen@imde.ac.cn

J. Chen
e-mail: chenjg@imde.ac.cn

H. Chen · P. Cui · X. Chen · J. Chen
University of Chinese Academy of Sciences, Beijing, 100049, China

H. Chen · P. Cui · X. Chen · J. Chen
CAS Center for Excellence in Tibetan Plateau Earth Sciences, Chinese Academy of Sciences (CAS), Beijing, 100101, China

Introduction

Debris flows are poorly sorted, fast-moving mixtures of sediment and water that mobilize in the upstream region of a valley, and surge down slopes in response to gravity (Iverson 1997). The debris flow hazards can be triggered by snowmelt (Decaulne et al. 2008, 2010), excessive rainfall (Chen et al. 2014; Zhou et al. 2015; Melo et al. 2018), landslides, rock avalanches or glacial lake outburst floods (GLOFs) (Evans and Clague 1994; Clague and Evans 2000; Breien et al. 2008; Emmer 2016; Frey et al. 2016; Emmer 2018). Especially, in the high mountainous areas, debris flows may be formed by different kinds of geo-hazard chains that proceed as follows: glacier avalanches surge waves moraine dam failure floods debris flows (Cui et al. 2003; Keith and Evans 2015; Wei et al. 2017). Glacial lake outburst floods GLOFs are widely distributed geo-hazards in the Alps, Andes or the Himalayan. When sufficient loose material from the banks or channel bed is available the GLOF can initialize and entrain the materials, which can increase flow density accordingly. The flow transforms from a GLOF to a debris flow, which enlarges the scale of the flow and its capacity to cause damage. Sometimes it can cause a secondary disaster such as debris flow dammed lakes and dam-break floods downstream. Much attention has been paid to the formation, propagation, and discharge prediction or risk evaluation of GLOFs. Little attention has been paid to

Ž. Arbanas et al. (eds.), *Understanding and Reducing Landslide Disaster Risk*, ICL Contribution to Landslide Disaster Risk Reduction,
https://doi.org/10.1007/978-3-030-60713-5_20

the debris flows caused by glacial lake outburst flood triggered by glacier avalanches.

In this paper, experiments were conducted to simulate a debris flow event triggered by GLOFs. Video cameras were employed to record the moraine dam failure and the flood propagation process. The development of the dam breach and the discharge variation during the dam failure process were investigated. The flow densities along the channel were sampled and analyzed. Additionally, the variation of flow discharge along the channel was investigated. The results presented in this paper can provide information for another kind of debris flow triggered by GLOFs in high mountainous areas.

Experimental Design

The dimensions of the tank and blocks were based on the outburst event in the Guangxieco glacial lake, upstream of the Midui gully, in Bomi county, Southeastern Tibet, China (29°27.425′–29°28.032′N, 96°29.905′–96°30.458′E). The experimental setup consisted of a sliding block, an inclined plate with adjustable slopes, a dammed lake, a downstream dam, a channel with a length of 20.0 m, and a debris flow deposition fan. The initial moraine dam was set at a height of 0.6 m (Fig. 1). In order to simulate the moraine dam faiulre triggered by glacier avalanches, the density of the sliding block is 0.9 g/cm^3, which is strictly follow the glacier density under natural condition.

In the field, the particle sizes of the moraine dam change from clay to boulders. In our experiments, the particle sizes of the moraine dam should be relatively small due to the limited model dimensions. The sediment used in this experiment was equivalent to a sample taken from the moraine dam in the Midui gully and excluded particles larger than 20.0 mm. The particle size distribution of sediments for both on-site samples and experimental ones was shown in Fig. 2. In our experiments, the flow discharge in the channel can be obtained by monitoring the flow velocities and flow depths at different cross-sections. Debris flow density can also be measured by monitoring the weight and volume of debris-flow samples taken at each cross-section. Although the experimental setup and sediments used in experiments did not strictly follow the similarity theory, dimensions of our physical model, glacial lake volume, particle sizes and channel width were roughly set based on natural conditions. Therefore, the results obtained in our experiments can reflect the dynamic evolution process and basic mechanism under natural conditions.

Experimental Results and Analysis

Moraine Dam Failure Triggered by Glacier Avalanches

When a volume of glacier plunges into the stilling glacial lake the potential energy of the glacier will be converted into kinetic energy a surge wave will be triggered. The former studies indicated that the value of the initial surge wave was higher than the following surge waves and the wave height decreased when surge waves propagated downstream. When the surge waves have enough energy to reach the moraine dam they will climb up along the dam surface. Here the kinetic energy of the surge wave will convert into potential energy. When the run-up of the surge wave was higher than the crest of the moraine dam, it overflew (Fig. 3b, t = 0.0 s). When the flows overtopped the moraine dam it initialized and entrained particles from the dam (Fig. 3b, t = 8.0 s). The successive surge waves continued to transport the relative particles and the breach developed further, which increased the discharges of dam-break flood (Fig. 3b, t = 60.0 s). During this process a big part of the water was drained from the glacial lake and finally a retained dam formed (Fig. 3b, t = 122.0 s).

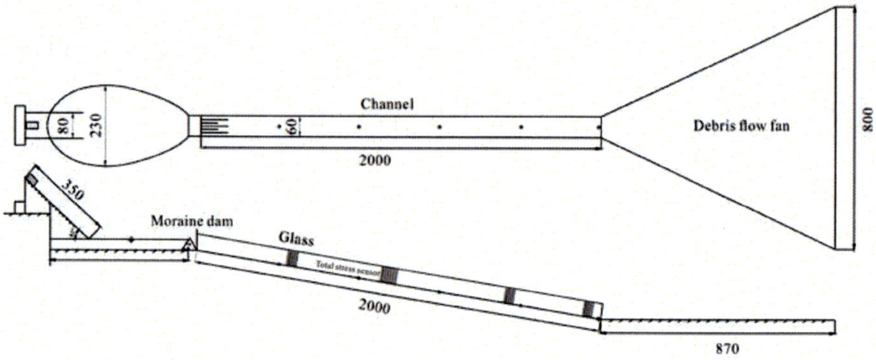

Fig. 1 Sketch of the experimental setup (unit:cm)

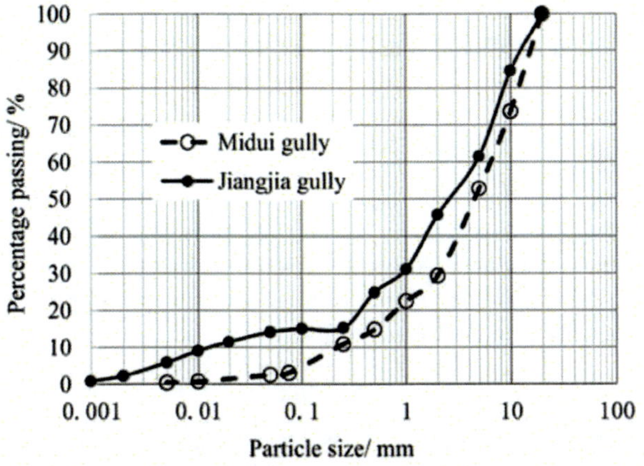

Fig. 2 The particle size distribution of the sediments

Flow Discharge During the Failure Process of the Glacier Dam

It was difficult to calculate the instant discharge at the dam breach due to its rapid development. Fortunately, the discharge could be estimated by variations in water depth and water areas per unit time.

Discharge of the dam breach flow is calculated by

$$\frac{dV}{dt} = Q_1 - Q_2 \tag{1}$$

The difference form can be expressed as follows

$$\frac{\Delta V}{\Delta t} = Q_1 - Q_2 \tag{2}$$

where V is the volume of glacial lake, t is time, Q_1, and Q_2 are the incoming discharge and outlet discharge, respectively, and ΔV is the variation of water volume over time duration Δt.

According to the relationship between the water level, size, and volume, the outlet discharge Q_2 is estimated by

$$Q_2 = \frac{(H_1 - H_2)(A_1 + A_2)}{2(t_2 - t_1)} + Q_1 \tag{3}$$

where H_1,A_1 is the water depth and size of glacier lake at t_1, respectively; H_2,A_2 is the water level and size of glacier lake at t_2, respectively, without discharge $Q_1 = 0.0$ m^3/s. The water area kept constant in the experiment, $A_1 = A_2 = 7.46$ m^2. The water depth of the dammed lake in the experiments was obtained by the Pore Pressure Transducers (PPTs), which were employed to measure the real-time water pressure in the glaical lake. Then the water depth was calculated according to the relationship between the water pressure and water depth ($P = \rho gh$).

From Fig. 4 we found that during the dam failure triggered by the glacier avalanche there was besides one peak discharge ($Q = 0.075$ m^3/s), it had two secondary peak discharges. This was much different from the dam failure, which generally had only one peak discharge. This phenomenon can be explained by the fact that surge waves triggered by the glacier propagate in the glacial lake and will be reflected by the bank. During this process wave energy may dissipate due to water viscosity or boundary friction but the dissipation rate is very low. Therefore, surge waves will persist in the glacial lake for a long time. When a surge wave reaches the dam breach it raises the net water depth at the breach, which strengthens the erosion ability of water flow.

Fig. 3 Surge wave propagation and dam failure process

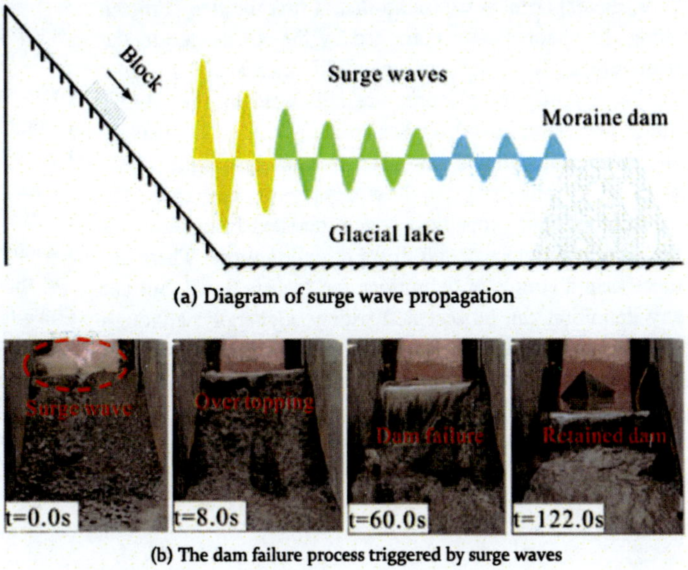

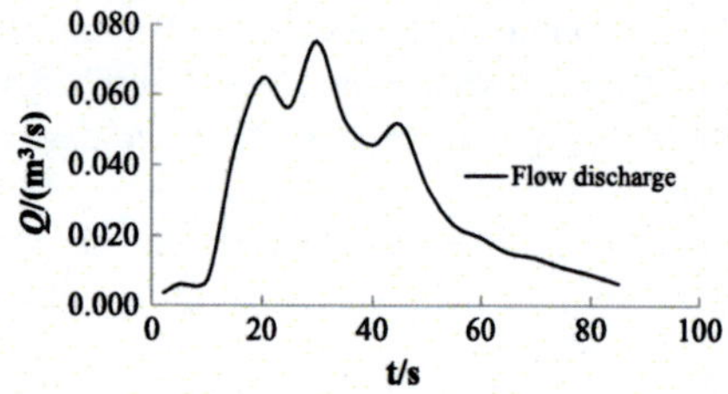

Fig. 4 Flow discharge process without overflow

Then the flow initialized and transported debris from the moraine dam and breach developed fast, which enlarges the flow discharge. When the surge wave retreats from the dam breach the water depth at the breach decreased gradually when the flow discharge decreases. Then the surge wave again reaches the breach and the flow discharge again increases.

GLOF Propagation and Debris-Flow Formation

After the outburst from the glacial lake (GLOFs) the water propagates downstream. If the downstream channel is full of loose materials the GLOFs can initialize and entrain the materials from the gully, forming a flow with higher density and viscosity. When its density reaches a certain value we call it a debris flow. In order to monitor the behavior of GLOFs we designed an experiment roughly based on the Guangxieco glacial lake outburst (upstream of the Midui gully) as shown in Fig. 5. According to the field investigation, we found there were a few landslides distributed along the Midui gully. Additionally, a segment of erodible bed was set to monitor the solid particle supply from the gully bed in Fig. 6.

In the experiment we found that at first the fine particles along the channel were entrained by the floods due to the relatively small flow discharge. With increasing the flow discharge much more debris was initialized by the GLOF. Some flow samples were taken to analyze the flow density and solid concentration along the channel during the whole process. By analyzing the flow density we can know the variation of flow properties, which transited from sediment flows to hyperconcentrated flows or debris flows. There are more than a couple of definitions for "debris flow", but not any definition can be accepted widely. Generally, although we don't have a clear definition for debris flows, we can find some parameters to identify the debris flow. Wu and Tian (1996) considered that when the channel slope is above 1% the two-phase flow is a debris flow rather than a sediment flow and the flow density should be larger than 1.2 g/cm^3. Fei and Su (2004) thought that in debris flows the gravity exerted on the solid phase provided the largest part of energy for the flow movement. The expression was shown as follows:

$$S_v \gamma_s \geq (1 - S_v)\gamma \tag{4}$$

where S_v is the solid concentration in debris flow; γ_s is the desity of solid particles; γ is the desity of water. If the S_v is considered to be 2.7 g/cm^3, then

$$S_v \geq 0.27 \tag{5}$$

From Eq. (5) we know that the critical density of debris flow is about 1.46 g/cm^3. The flow densities at different cross-section were sampled and analyzed. The results indicated that the flow density increased in the flow direction. When the flow arrived at the end of the channel the density was about 1.49 g/cm^3 (Fig. 7), which was larger than both Wu and Tian (1996) and Fei and Su (2004)'s critical value to define a debris flow. According to the field study on the Guangxieco moraine dam failure the density of mixture of water and sediments measure on site was about 1.53 g/cm^3 (Li and You 1992).

Variation of Discharge Along the Channel

The discharge of a GLOF (the liquid phase) along the channel can be calculated by the empirical expression:

$$Q_{L-\max} = \frac{W}{\frac{W}{Q_{b-\max}} + \frac{L}{V_{\max} K_2}} \tag{6}$$

where $Q_{\text{L-max}}$ is the peak discharge at a certain cross-section; L is the distance between the cross-section and the dam site; W is the volume of the water in the glacial lake; $Q_{\text{b-max}}$ is the peak discharge at the moraine dam site; $V_{\max}$ is the maximum average velocity of the flood in the downstream channel, $V_{\max}$ = 3.0 m/s; and K_2 is an empirical parameter, here K_2 = 1.2.

According to Eq. (6), we can evaluate the peak discharge of dam break flood at any cross-section downstream along the channel. But how can we predict the peak discharge of the debris flow after it has entrained the particles from a channel covered by loose solid materials? When the particles are initialized and entrained by the dam-break flood, the discharge of the two phase flow will enlarge. Here we suppose that the discharge of the water is Q_f and the discharge of the granular flow (solid part) is Q_s. Then we have

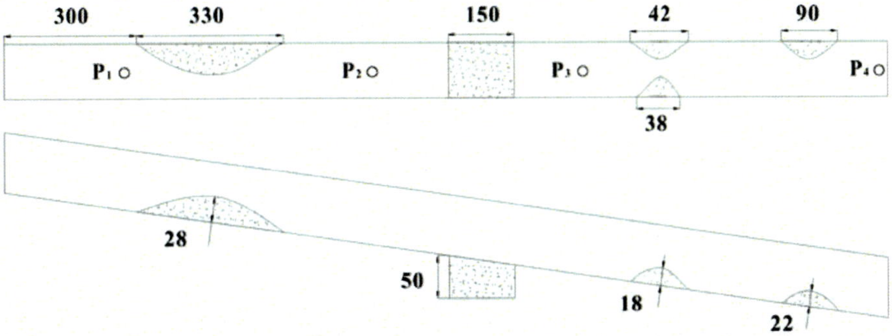

Fig. 5 The distribution of loose materials in modelled gully based on the Midui gully (unit:cm)

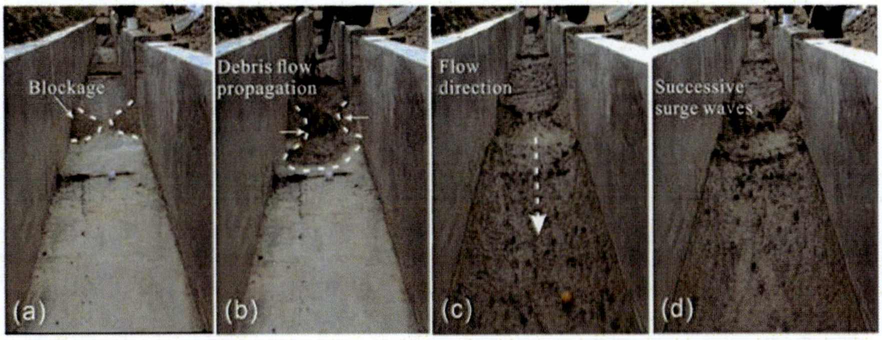

Fig. 6 The propagation of debris flows in the channel

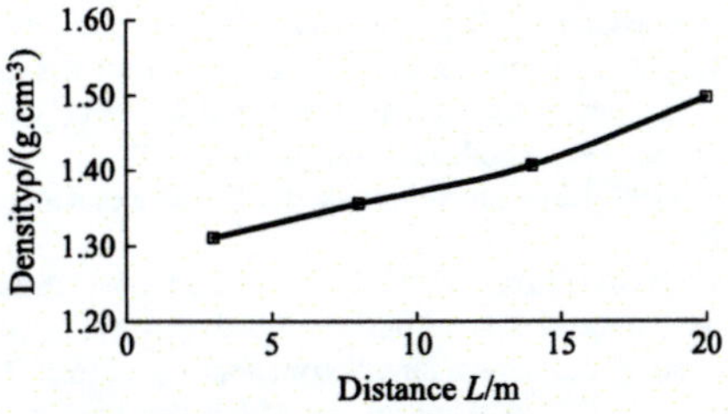

Fig. 7 The variation of flow density along the channel

the enlargement ratio of the debris flow k can be expressed as follows:

$$\frac{Q_d}{Q_f} = \frac{Q_s + Q_f}{Q_f} = k \tag{7}$$

When we used a parameter C to represent the ratio of solid phase and the mixture flow in volume. According to the definition of the solid concentration, C is also a function of k which can be expressed:

$$C = \frac{V_s}{V_d} = \frac{V_s}{V_f + V_s} = \frac{Q_s}{Q_f + Q_s} = 1 - \frac{1}{k} \tag{8}$$

Then the density of the debris flow can be calculated by the following expression:

$$k = \frac{\rho_s - \rho_f}{\rho_s - \rho_d} \tag{9}$$

where k is the scale enlargement parameter; ρ_c is debris flow density; ρ_w is water density; ρ_r is particle density; and C is solid concentration.

Table 1 indicates that the discharge of the GLOF varied slightly along the channel. However the enlargement ratio of the debris flow increased in the direction of flow propagation. The debris flow discharge reached 0.095 m^3/s. Based on experimental data, the maximum flow depth at the outlet of the channel was 0.06 m; the average velocity of debris flow was about 2.8 m/s; and the channel width was about 0.6 m. Then the average peak discharge of the debris flow was about 0.1 m^3/s. The error between the calculated peak discharge and the experimental one was about 5.26% (Table 1).

Table 1 The analysis on the variation of debris flow discharge along the channel

Cross-section	L/(m)	W/(m^3)	$Q_{b\text{-}max}$/(m^3/s)	$Q_{L\text{-}max}$/(m^3/s)	K/(−)	Q_d/(m^3/s)
1	5.0	4.48	0.075	0.073	1.23	0.090
2	10.0	4.48	0.075	0.071	1.27	0.090
3	15.0	4.48	0.075	0.069	1.33	0.093
4	20.0	4.48	0.075	0.067	1.42	0.095

Conclusion

In this paper, experiments were conducted to investigate the surge waves triggered by glacier avalanches, failure process, GLOFs, debris flow formation and propagation based on the natural Guangxieco moraine dam failure, that occurred in July 1988, in Tibet. The following conclusions were drawn from this analysis:

(1) Besides the peak discharge, the profile of discharge had two secondary peak discharges. This was much different from dam failure caused by overtopping without surge waves. The successive surge waves triggered by glacier avalanches have a significant effect on the failure of a moraine dam.
(2) The flow densities at different cross-sections were sampled and analyzed. The results indicated that the flow density increased in the flow direction. The maximum flow density in the experiment was about 1.49 g/cm^3.The flow density measured on site was about 1.53 g/cm^3. In general, the experimental results simulated the formation and propagation of debris flows in the Guangxieco moraine dam failure event.
(3) By considering the effect of debris flow scale enlargement, the relationship between the scale enlargement parameter k and the flow density were deduced. Based on analysis on the GLOFs discharge variation the debris flow discharges at different cross-section were calculated. The discharge of debris flow at the outlet reached 0.095 m^3/s. The experimental discharge measured at the outlet was about 0.10 m^3/s. The error between the calculated peak discharge and the experimental one was about 5.26%.

Acknowledgements The study presented in this paper was supported by the National Natural Science Foundation of China (grant 41520104002, 41771045), Key Research Program of Frontier Sciences, CAS (grant QYZDY-SSW-DQC006), the 135 Strategic Program of the Institute of Mountain Hazards and Environment, CAS, NO. SDS-135-1701, and Foundation of Youth Innovation Promotion Association, Chinese Academy of Sciences (grant no. 2017425).

References

Breien H, Blasio FVD, Elverhøi A, Høeg K (2008) Erosion and morphology of a debris flow caused by a glacial lake outburst flood. Western Norway Landslides 5(3):271–280
Chen HY, Cui P, Zhou GGD, Zhu XH, Tang JB (2014) Experimental study of debris flow caused by domino failures of landslide dams. Int J Sediment Res 29(3):414–422
Clague JJ, Evans SG (2000) A review of catastrophic drainage of moraine–dammed lakes in British Columbia, Quaternary Sci. Rev. 19:1763–1783
Cui P, Ma DT, Chen NS, Jiang ZX (2003) The initiation, motion and mitigation of debris flow caused by glacial lake outburst. Quaternary Sci 6(23):621–628 (In Chinese)
Decaulne A, Porsteinn O, SÃmundsson, Aorsteinn (2008) Debris-flow characteristics in the Gleidarhjalli area, North-western Iceland, Debris-Flow Hazards Mitigation: Mechanics, Prediction, and Assessment
Decaulne A, Ãorsteinn S, Petursson O (2010) Debris flow triggered by rapid snowmelt: a case study in the glei. arhjalli area, northwestern iceland. Geografiska Annaler 87(4):487–500
Emmer A, Vilímek V, Huggel C et al (2016) Limits and challenges to compiling and developing a database of glacial lake outburst floods. Landslides 13:1579–1584
Emmer A (2018) GLOFs in the WOS: bibliometrics, geographies and global trends of research on glacial lake outburst floods. Natural Hazards Earth Syst Sci 18(3):813–827
Evans SG, Clague JJ (1994) Recent climatic change and catastrophic geomorphic processes in mountain environments. Geomorphology 10(1–4):107–128
Fei XJ, Su AP (2004) The movement mechanism of debris flows and hazard prevention. Beijing, tsinghua university press. (In Chinese)
Frey H, Huggel C, Bühler Y et al (2016) A robust debris-flow and GLOF risk management strategy for a data-scarce catchment in Santa Teresa. Peru Landslides 13(6):1493–1507
Iverson RM (1997) The physics of debris flows. Rev Geophys 3(35): 245–296
Keith BD, Evans SG (2015) The 2000 Yigong landslide (Tibetan Plateau), rockslide-dammed lake and outburst flood: Review, remote sensing analysis, and process modelling. Geomorphology 246:377–393
Li DJ, You Y (1992) Bursting of the Midui moraine lake in Bomi, Xizang. Mountain Res 10(2):219–224. (In Chinese)
Melo R, Asch TV, Zêzere JL (2018) Debris flow run-out simulation and analysis using a dynamic model 18(2):1–35
Wei RQ, Zeng QL, Davies T et al (2017) Geo-hazard cascade and mechanism of large debris flows in Tianmo gully, SE Tibetan Plateau and implications to hazard monitoring. Eng Geol 233:172–182
Wu JS, Tian LQ (1996) On the debris flow science. Mountain Res 14 (2):89–95. (In Chinese)
Zhou GG, Cui P, Zhu XH, Tang JB, Chen HY, Sun QC (2015) A preliminary study of the failure mechanisms of cascading landslide dams. Int J Sedim Res 30(3):223–234

Quantitative Analysis of Landslide Processes Based on Seismic Signals—A New Method for Monitoring and Early Warning of Landslide Hazards

Yan Yan, Yifei Cui, Shuyao Yin, and Xin Tian

Abstract

Effective monitoring and early warning methods can greatly reduce the threat of sudden landslide damage to lives and property in mountainous areas. Seismic monitoring is a newly developing remote monitoring method that can make up for the shortage of on-site data obtained from traditional methods. This review article summarizes the authors' recent developments involving the combination of band-pass filter, empirical mode decomposition, and fast and short-time Fourier transform methods to deduce event sequences of actual landslides. Two case studies of landslides in Xinmo and Shuicheng, China are used to follow the methodology development. Through seismic signal processing, seismic signals caused by the two landslides are extracted and successfully denoised, and the time-frequency characteristics of the seismic signal in each stage are analyzed in detail. The results show good correspondence between each seismic signal stage caused by the landslide and the field investigations. This further demonstrates the feasibility of applying monitoring and early warning methods based on seismic signals.

Keywords

Seismic signal • Monitoring and early warning • Signal extraction and identification • Landslides process

Introduction

Major landslides are often called "secondary hazards" triggered by another disaster such as an earthquake or rainstorm. In addition to directly inflicting casualties, landslides can also block transportation lines, destroy buildings, and damage pipeline systems, resulting in severe economic losses. The effective monitoring and early warning of landslides are therefore of great significance to preparation measures for hazards associated with sudden disasters.

Conventional landslide monitoring and early warning require two types of information: (1) degree of deformation regarding surface and interior displacement of a landslide mass, which allows deduction of the creep and deformation mechanism; and (2) hydrological factors, such as pore pressure, precipitation, soil water content, and groundwater level (Wang et al. 2008; Yin et al. 2010). Different models lead to different early warning results. For example, setting too high of a threshold value can lead to failure in issuing a timely alert, whereas setting too low of a threshold can lead to a false-positive alert (Cui et al. 2019).

Landslides can generate ground vibrations, which can be recorded by seismic stations. Recorded signals can vary in amplitude, duration, and frequency components depending on their origin direction and size of the responsible landslide (e.g., volume, dimensions). Processing of such landslide-induced seismic signals provides important information for understanding how the particular landslide event

Y. Yan
Key Laboratory of High-Speed Railway Engineering, School of Civil Engineering, Southwest Jiaotong University, No. 111, North 1st Section of Second Ring Road, Jinniu District, Chengdu, 610031, China
e-mail: yanyan1989@swjtu.edu.cn

Y. Cui (✉)
State Key Laboratory of Hydroscience and Engineering, Tsinghua University, Beijing, 100084, China
e-mail: yifeicui@mail.tsinghua.edu.cn

S. Yin
Key Laboratory of High-Speed Railway Engineering, School of Civil Engineering, Southwest Jiaotong University, Chengdu, 610031, China
e-mail: ysy.swjtu.edu.cn@my.swjtu.edu.cn

X. Tian
Institute of Geophysics, Earthquake Administration China, Beijing, 100081, China
e-mail: tianxin@cea-igp.ac.cn

Ž. Arbanas et al. (eds.), *Understanding and Reducing Landslide Disaster Risk*, ICL Contribution to Landslide Disaster Risk Reduction,
https://doi.org/10.1007/978-3-030-60713-5_21

occurred and how the detached soil mass was transported. Several previous studies have proposed seismic monitoring techniques for landslides. A broadband seismic network can be used to record and analyze landslide-induced ground tremors over a wide frequency range, which makes up for the typical information deficit from on-site observations (Yamada et al. 2012). This review article summarizes the authors' recent attempts to process seismic signals to deduce event sequences of actual landslides that occurred in Xinmo (Yan et al. 2020a) and Shuicheng, China (Yan et al. 2020b). The seismic signal processing includes the extraction and denoising of seismic signals caused by two landslides and time-frequency analysis of signals. Combined with field investigations, the entire landslide process is explained and the movement characteristics of each stage are summarized. Based on the seismic signal reconstruction of the landslide process, this review verifies the feasibility of using seismic signals for landslide monitoring and early warning.

Recent Advances in Seismic Monitoring of Landslides

Among the common methods of landslide monitoring developed over the past two decades, seismic monitoring has received increasing attention because of its potential for remotely monitoring hazards (Moore et al. 2017). Through seismic signal processing, the dynamics of large-scale landslides (e.g., occurrence time, duration) can be quickly obtained, which can be further used for interpreting the landslide movement process (Zhang et al. 2019).

Time-frequency analysis of seismic signals is a common method presently used to study landslide dynamics. The results show that seismic signals related to landslides, including high-frequency and low-frequency landslide components, can reflect the dynamic characteristics of landslides. (Li et al. 2017). However, the scale, velocity, and track of landslides remain difficult owing to the complexity and fast attenuation of high-frequency signal components (Zhang et al. 2019). The common method is therefore to invert signals based on the long-period monitoring signal. The force-time function is first obtained followed by a quantitative analysis of the landslide kinematic parameters. The date and occurrence time of the Todagin Creek landslide were extracted from its seismic signal (Sakals et al. 2012). Based on analysis of the seismic signal amplitude produced by the Baige landslide, Zhang et al. (2019) determined the landslide occurrence time and, combined with the force-time function, spectrogram, and field investigation, discussed the sliding process and movement mode. Fuchs et al. (2018) developed a simple and fully automatic processing chain for the location and classification of rockfalls, landslides, and other hazards. Through comprehensive analysis of the characteristic parameters of seismic signals (e.g., peak degree, maximum amplitude), natural hazards such as landslides can be automatically identified. Kuo et al. (2018) proposed rainfall warning thresholds for different scale landslide events based on the analysis of seismic signals of landslides in Taiwan.

Quantitative Analysis Method Based on Seismic Signals

The quantitative analysis steps of the landslide movement process based on seismic signals is shown in Fig. 1 (Yan et al. 2020b). Effective seismic signals of landslides are extracted using short-time Fourier transform (STFT) and band-pass (BP) filter methods, followed by empirical mode decomposition (EMD) methods for denoising. Fast Fourier transform (FFT), STFT, and power spectral density (PSD) are used to analyze the time-domain and frequency-domain characteristics, compare with the field investigation, reconstruct the entire landslide process, and obtain the movement characteristics of every stage of the landslide. The following is a brief introduction to the principles of these main methods.

BP-filtering is a method of filtering interference signals. Interference and effective signals are usually emitted in different frequency bands. The interference signal is separated by setting the frequency amplitude within the interference signal bandwidth to zero and keeping the effective signal amplitude unchanged. The EMD method focuses on denoising to improve the signal-to-noise ratio. A set of different frequency and amplitude intrinsic mode functions (IMFs) are decomposed from a non-linear and non-stationary signal.

STFT is a method to obtain information regarding the time and frequency distribution of seismic signals. A 128-point length Hanning window is selected as a window function, which is considered static over a relatively short time. Equation 1 shows the transform equation:

$$STFT\{x(n)\}(m,\omega) \equiv X(m,\omega) = \sum_{n=-\infty}^{\infty} x[n]\omega[n-m]e^{-jwn} \tag{1}$$

The power of each frequency unit corresponding to each frequency range component at a certain time can be estimated according to the power spectral density of the seismic signal in the frequency domain. In this study, we integrate the PSD model along F and D to obtain the sum of PSD energy at different frequencies at one time.

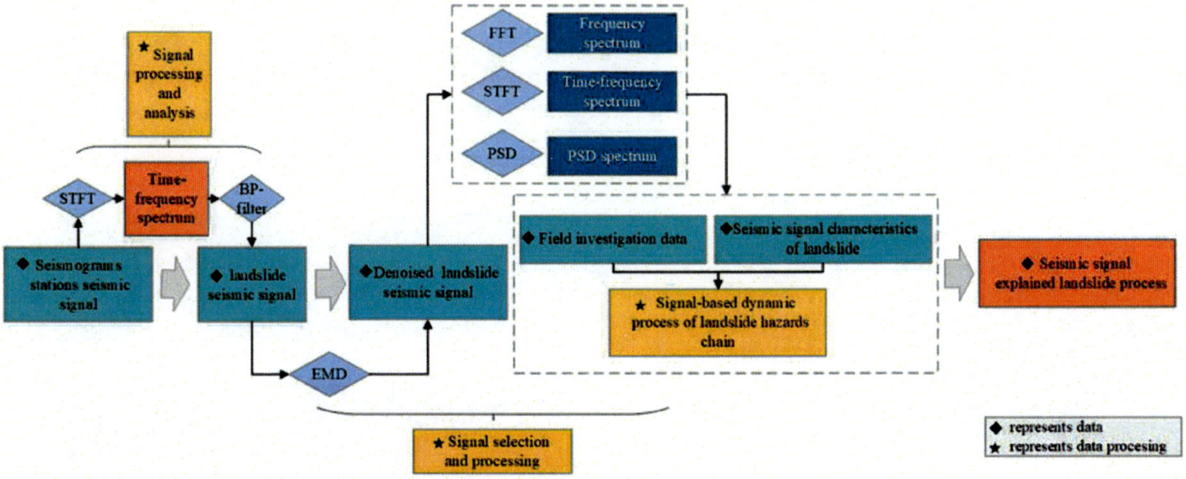

Fig. 1 Flow chart of seismic signal extraction and analysis (Modified after Yan et al. 2020a)

Signal-Based Dynamic Processing of Landslides

Xinmo Landslide

In our recent study of the Xinmo landslide, EMD, STFT, and PSD methods were used to process the seismic signals (Yan et al. 2020a). The EMD method decomposed the signals into 10 IMFs (IMF1–IMF10). The IMF4 of the MXI station and IMF5 of the PWU station are considered the principal components of the seismic signal, as shown in Fig. 2c. From the time-domain perspective, the signals include two parts: the 1st and 2nd main signals. The amplitude and duration of the 2nd main signal are considerably longer than those of the 1st main signal. The vertical seismic signal decays rapidly after the 1st main signal. The time-frequency spectrum in the three seismic signal directions is shown in Fig. 2d. From the frequency-domain perspective, the signal can also be divided into two parts: the 1st and 2nd principal signals. The peak energy frequency of the 1st and 2nd principal signals is ~1 Hz, whereas the 2nd principal signal retains a strong energy frequency up to 4 Hz. The peak frequency of the 2nd principal signal (5 Hz) is slightly higher than that of the 1st principal signal (4 Hz). Because the principal signal magnitude corresponds to the scale of the landslide, the landslide scale corresponding to the 2nd principal signal is larger than that of the 1st principal signal.

By comparing the signal analysis results with the field investigation, the entire physical process of the Xinmo landslide is divided into five stages: stationary, slipping, transition, entrainment-transportation, and deposition, which respectively correspond to intervals A, B, C, D, E in Fig. 2b. The static stage is designated prior to the start of the landslide from 0 to 18 s. At this stage, the velocity of the landslide body remains close to zero and the displacement is relatively small. The sliding stage lasts from about 18 to 28 s, corresponding to the 1st principal signal. The seismic signal in the middle part of the two principal signals corresponds to the transition stage, lasting from about 28 to 68 s. The entrainment-transportation stage lasts from about 68 to 110 s, equivalent to the 2nd principal signal, and the deposition stage occurs from 110 to 150 s, in which the seismic signal appears after the second peak. The amplitude, peak frequency, and frequency range of the signal decrease gradually. The PSD model used to quantitative analysis of the transition stage shows that the energy on the vertical component first decays and maintains a low value for a certain time, then increases gradually. This may be caused by the entrainment process.

Shuicheng Landslide

In our recent study of the Shuicheng landslide, we used BP-filter, EMD, STFT, and PSD methods to pre-analyze 12 station signals with similar signal characteristics (Yan et al. 2020a). Three BP-filters of 0–10, 10–20, and 20–50 Hz were initially used, and the signal was then passed through a 4.5–6-Hz filter to highlight the continuous energy band of 5 Hz. The EMD method was then used to denoise and enhance the effective signal. The results show that the 4.5–6 Hz signal of the XUW station is the clearest. The STFT method was used to obtain the time-frequency spectrum of the XUW stations in the E, N, and Z directions as shown in Fig. 3d–f. The characteristics of the time-frequency spectrum at different times of the landslide are summarized based on the signal analysis and the seismic signal is divided into four stages

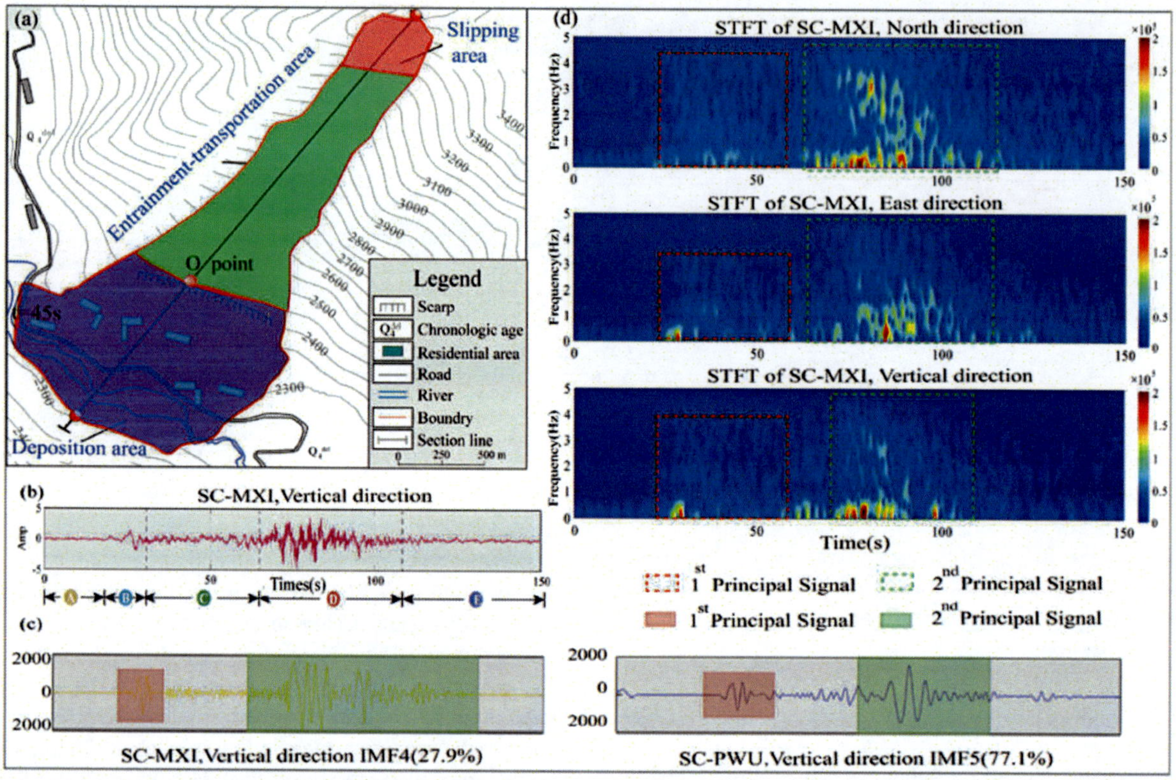

Fig. 2 Contour map of the Xinmo landslide area and signal time-frequency spectra. **a** Contour map of the Xinmo landslide area. **b** Time spectrum of the MXI station in the vertical direction. **c** Vertical seismic signal of IMF4 of station MXI and IMF5 of station PWU. **d** Three direction time-frequency spectra of the MXI station obtained by short-time Fourier transform (Modified after Yan et al. 2020b)

(Fig. 3e, f). The spectrum of the XUW station in the E, N, and Z directions shows that the four peak frequencies form an inverted W-shaped curve. In the first stage, the seismic signal in the E, N, and Z directions is relatively small. In the other stages, the signal is stronger than the first stage and substantially stronger than the background noise.

Through the comprehensive analysis of rainfall, geological survey, and seismic signal, the landslide evolution is determined to consist of four stages: fracture-transition; accelerated initiation; bifurcation-scraping; and deposition (Fig. 3a). These four stages correspond to the four stages of the seismic signal one by one. In the fracture-transition stage, owing to increased soil moisture content, small-scale landslides occurred and local collapse reduced the stability of the landslide area. In the fracture-transition stage, under the action of landslide mass accumulation, the weathering layer in the wide platform and its surroundings gradually destabilized and began to move. Two natural debris flow gullies were distributed on the eastern and western sides of the broad platform. At this stage, the velocity of the main body of the landslide increased and reached the bifurcation of the gullies. In the bifurcation-scraping stage, the landslide divided into two parts and entered the valley on the western and eastern sides of the platform. Owing to the main body of the landslide on the eastern side, the volume and kinetic energy of the landslide in the eastern valley were larger than those on the western side. Some landslide fragments then passed through the western slope, the landslide velocity rapidly decreased, and deposition occurred. In the last (deposition) stage, the landslide body gradually extended to both sides and the constraint of the river valley on the landslide decreased. The entire landslide was ultimately deposited on the bottom of the canyon.

Outlook and Prospect

The method of seismic signal analysis is very promising within the study of landslides and has been well applied to the Xinmo and Shuicheng landslides. This method for remote hazard monitoring is safe, stable, and maintains data integrity. Broadband seismic networks are relatively dense in China and are optimal for establishing rapid monitoring and early warning systems of large-scale landslides based on seismic signals, including determination of the occurrence time, duration time, location, and landslide volume.

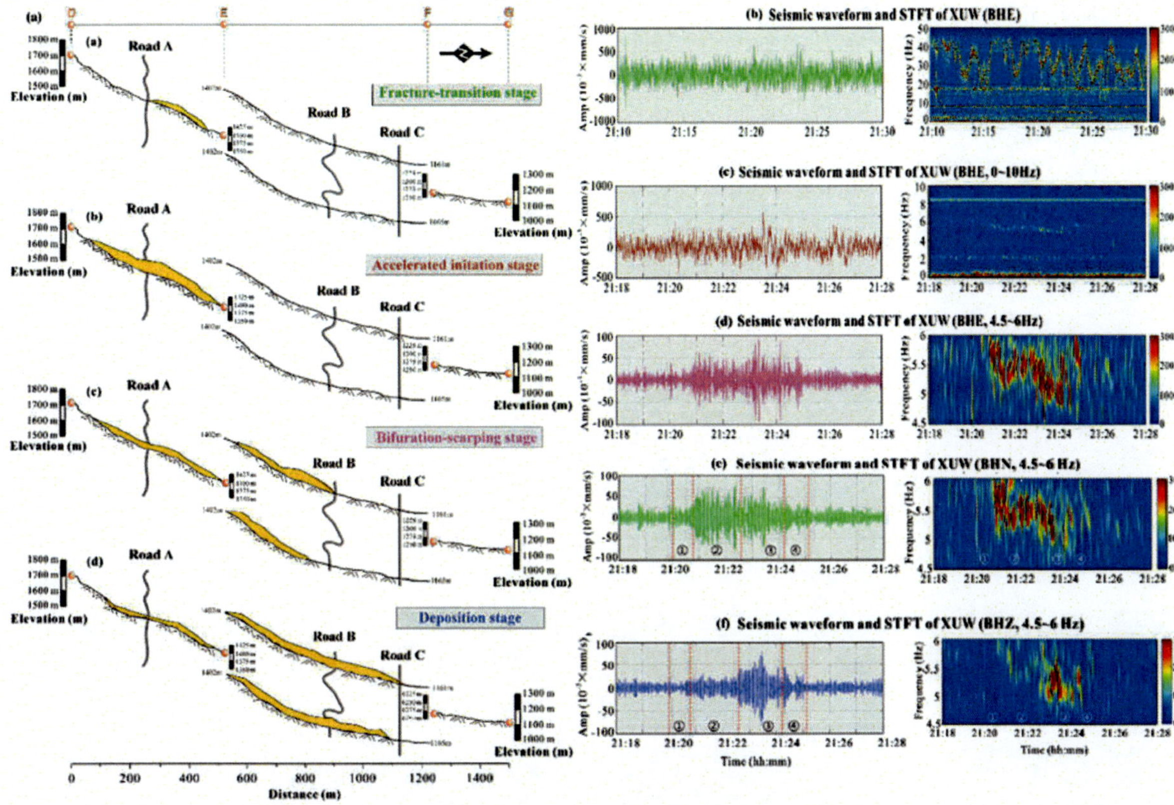

Fig. 3 Schematic diagram of the Shuicheng landslide and signal time-frequency spectra. **a** Schematic diagram of the Shuicheng landslide in stages. **b** Seismic signals of station XUW. **c**–**f** Time-frequency spectra of the XUW station in 0–10-Hz and 4.5–6-Hz frequency bands in the E, N, and Z directions (Modified after Yan et al. 2020a)

However, the study of seismic signals remains relatively new with only a few case studies. We therefore aim to propose a universal monitoring and early warning method in the future based on more samples and in-depth analyses.

The methods presented here can be combined with numerical simulation methods, which can be used to more accurately estimate the location, time, and scale of landslide hazards with the help of seismic network data. The monitoring and early warning of landslide hazards based on seismic signals will then be fully realized. There is no relatively clear distinction between landslide signals and other emergencies (e.g., earthquakes, debris flows, man-made blasting), and the identification of landslide signals is mostly based on the experience and knowledge of involved researchers. Ideally, a description of waveform characteristics of landslide seismic signals and other types of seismic signals can be used to establish a classification standard for landslide signals and achieve their rapid recognition. Such results combined with the rapid assessment of landslide scale will provide a more timely and accurate reference basis for the rapid identification and assessment of landslide hazards to improve the speed and target of rescue measures.

Acknowledgements We thank Esther Posner, PhD, from Liwen Bianji, Edanz Editing China, for editing the English text of a draft of this manuscript.

References

Cui Y, Cheng D, Choi CE et al (2019) The cost of rapid and haphazard urbanization: lessons learned from the Freetown landslide disaster. Landslides 16(6):1167–1176

Fuchs F, Lenhardt W, Bokelmann G et al (2018) Seismic detection of rockslides at regional scale: examples from the Eastern Alps and feasibility of kurtosis-based event location. Earth Surface Dyn 6 (4):955–970

Kuo HL, Lin GW, Chen CW et al (2018) Evaluating critical rainfall conditions for large-scale landslides by detecting event times from seismic records. Nature Hazards Earth Syst 18(11):2877–2891

Li Z, Huang X, Xu Q et al (2017) Dynamics of the Wulong landslide revealed by broadband seismic records. Earth, Planets Space 69(1):27

Moore JR, Pankow KL, Ford SR et al (2017) Dynamics of the Bingham Canyon rock avalanches (Utah, USA) resolved from topographic, seismic, and infrasound data. J Geophys Res: Earth Surf 122 (3):615–640

Sakals ME, Geertsema M, Schwab JW et al (2012) The Todagin Creek landslide of October 3, 2006, Northwest British Columbia Canada. Landslides 9(1):107–111

Wang FW, Zhang YM, Huo ZT et al (2008) Movement of the Shuping landslide in the first four years after the initial impoundment of the Three Gorges Dam Reservoir, China. Landslides 5:321–329

Yamada M, Matsushi Y, Chigira M et al (2012) Seismic recordings of landslides caused by Typhoon Talas (2011) Japan. Geophys Res Lett 39(13):L13301

Yan Y, Cui YF et al (2020a). Seismic signal recognition and interpretation of the 2019 "7.23" Shuicheng landslide by seismogram stations. Landslides, 1–16

Yan Y, Cui YF et al (2020b) Landslide reconstruction using seismic signal characteristics and numerical simulations: Case study of the 2017 "6.24" Xinmo landslide. Eng Geol, 105582

Yin Y, Wang HD, Gao YL et al (2010) Real-time monitoring and early warning of landslides at relocated Wushan Town, the Three Gorges Reservoir China. Landslides 7(3):339–349

Zhang Z, He SM, Liu W et al (2019) Source characteristics and dynamics of the October 2018 Baige landslide revealed by broadband seismograms. Landslides 16(4):777–785

Part III
Earth Observation and Machine Learning

High-Resolution Point-Cloud for Landslides in the 21st Century: From Data Acquisition to New Processing Concepts

2020 Kyoto Japan

C. Gomez, T. Allouis, C. Lissak, N. Hotta, Y. Shinohara, D. S. Hadmoko, V. Vilimek, P. Wassmer, F. Lavigne, A. Setiawan, J. Sartohadi, A. Saputra, and T. Rahardianto

Abstract

The economic downturn at the start of the 21st Century combined with the rise of new economies have helped propelling the need for cheaper, rapid and yet accurate data acquisition methods, including point-cloud producing technologies. As population is ageing in countries like Japan or Taiwan, and as climate change is bound to increase the frequency and number of landslides (at least until sediment stocks are depleted), the need of rapid and low-manpower data acquisition and processing has become increasingly essential. It is within this framework that the present contribution aims (1) to present research on present technologies for landslide monitoring, as well as emerging systems, and (2) to propose "new" ideas for point-cloud processing and data usage without having to grid or interpolate data. The authors thus use a variety of field locations around the pacific and in France using as a method ALS (Airborne Laser Scanning), TLS (Terrestrial Laser Scanning) and SfM-MVS (Structure from Motion —Multiple View Stereophotogrammetry). Then the authors present recent advances in landslide monitoring technology with the YellowScan UAV-ALS. As the resulting mounting amount of data presents constrains to the data processing steps, the authors also present ideas to progress processing: (1) developing a point-cloud signature to avoid gridding, and (2) a conceptual algorithm to process the point-clouds as vectors between a machine and a reflective object, recording the free space and the objects. In this way catastrophic landslides can be characterized from the signature of the data and the distribution of fill and voids without going through the traditional analysis on a Cartesian grid. The authors have worked on those techniques to provide fast and accurate data for landslides hazards and disaster risk.

C. Gomez (✉)
Graduate School of Maritime Sciences, Laboratory of Sediment Hazards and Disaster Risk, Kobe University, Kobe City, Japan
e-mail: christophergomez@bear.kobe-u.ac.jp

C. Gomez · D. S. Hadmoko · A. Setiawan
Faculty of Geography, Gadjah Mada University, Yogyakarta, Indonesia

T. Allouis
Yellow-Scan, Montpellier, France

C. Lissak
Department of Geography, Caen University, Caen, France

N. Hotta
Graduate School of Agricultural and Life Sciences, University of Tokyo, Tokyo, Japan

Y. Shinohara
Department of Forest Environmental Sciences Miyazaki, University of Miyazaki, Miyazaki, Japan

V. Vilimek
Department of Geography, Charles University, Prague, Czech Republic

P. Wassmer · F. Lavigne
Geography Department and CNRS LGP Laboratory, Sorbonne University, Paris, France

J. Sartohadi
Faculty of Agriculture, Gadjah Mada University, Yogyakarta, Indonesia

A. Saputra
Faculty of Geography, University Muhammadiyah Surakarta, Surakarta, Indonesia

T. Rahardianto
Department of Geography, University of Canterbury, Christchurch, New Zealand

Introduction and Conceptual Background

Catastrophic Landslides

In 2013, Ekström and Stark (2013) wrote in *Science* that "Catastrophic landslides involve the acceleration and deceleration of millions of tons of rock and debris in response to the forces of gravity and dissipation". In the research-field of rock avalanches, they were termed

Ž. Arbanas et al. (eds.), *Understanding and Reducing Landslide Disaster Risk*, ICL Contribution to Landslide Disaster Risk Reduction,
https://doi.org/10.1007/978-3-030-60713-5_22

catastrophic landslides, when involving at least 10^7 m^3 in volume (Korup et al. 2007; Schulmeister et al. 2009). Another approach to catastrophic landslide has also involved finding specific processes such as the fragmentation process proposed by Davies and McSaveney (2006). From a hazards and disaster risk perspective, the terminology of "catastrophic landslide" and especially the term catastrophe involves a sudden—somehow unpredictable - event that causes great negative impacts (loss of lives, building and infrastructure destruction).

From an applied sciences perspective, the concept of catastrophic landslide needs to relate to impacts on individuals and societies. Consequently, it can certainly be agreeable to think that catastrophic landslides can be defined as events of "large and rapid-onset", but it seems also appropriate to include smaller-scale mass-movements located nearby and within human settlements and infrastructures, as long as they can generate "catastrophic effects". Indeed, the November 2016 M_w 7.8 earthquake in Kaikoura (New Zealand) triggered less large-landslides than expected (Massey et al. 2018), but those cut the main coastal highway isolating the township of Kaikoura and having catastrophic effects on logistics for the all South Island and on the tourism industry. In the light of this reflection, the present contribution will therefore be interested in landslides that have the potential of generating great harm, regardless of their size and not only those defined by a given size or speed (examples of catastrophic landslides—Fig. 1). Therefore, this excludes slope-confined and small-scale mass movements.

Engineering and scientific interest in landslides has steadily increased since the mid-20th Century, because of technical and conceptual renewal, but also because of the evidenced impacts of climate change on landslides (Gariano and Guzzetti 2006). Considering catastrophic landslides in the recent years, the Wenchuan Earthquake arguably takes the top of the list. In the aftermath of the 2008 Ms 8.0 Wenchuan Earthquake, about 60,000 landslides were triggered (Dai et al. 2011a, b). Transversal studies were accompanied by site-specific work, like at the Niumiangou landslide, that was described with velocities ranging from 19–36 m/s with a drop height of 850 m and a distance travelled of 2.8 km for instance (Cui et al. 2017). In turn, the co-seismic landslides were then the source of secondary hazards with mass movements triggered from the deposits of the first ones…(e.g. Gao et al. (2017) who reported secondary landslides that took the lives of 166 and buried 11 buildings with 1.47 million m^3 of material). On top of those "large spatial-extent events" other ones of lower magnitude have also been reported following the cyclone Tamara for instance, in the Bosna River Basin (Bosnia and Herzegovina in 2014) where more than 50 landslides were generated in four municipalities. The largest reported landslides had surface areas of ~20,000–40,000 m^2 with slip surface of 40 m to ~270 m, which destroyed a few buildings and the local road (Mihalic Arbanas et al. 2017). Finally, the last type of catastrophic landslides concerns multiple smaller-scale events in proximity or within constructed areas, such as the mass movements occurring regularly in Japan, the Philippines, and Indonesia, where the geography, the population dynamic and the local politics have forced people to live within in potentially dangerous areas (e.g. Hadmoko et al. 2017). Those examples are part of a long list of studies considering events of different sizes and velocity in various environments.

Gariano and Guzzetti (2016) have shown that steady increase in landslide research has been notably fuelled by climate-change related studies that have been rapidly increasing in Europe with investigations on this topic over more than 10 sites in the UK, 6 in France, 13 in Italy, 3 in Canada, 5 in Switzerland and 4 in Norway. From this research, it has been recognized that landslides number and frequency is on the rise with climate change (it is not only the effect of a trend). Consequently, to collect data, precise, accurate and in an effective fashion, scientists and engineers have to increasingly rely on the recent developments of geodetic point-cloud technologies, such as ALS (airborne Laser Scanner), TLS (Terrestrial Laser Scanner) and ground-based and airborne photogrammetry (e.g. SfM: Structure from Motion and Multiple View Sterophotogrammetry), to collect high resolution topographic data (HRT) in an effective fashion.

Conceptual Framework for 21st Century Questions: Reconciliation with Epistemology

The present contribution is a presentation of the research advances and some potential directions in geomorphology and geodesy for landslide research, fitting the multidisciplinary framework proposed in the 2006 Action Plan (Sassa 2007) for landslide hazards and mitigation. Indeed, the present chapter can be fathomed from the perspective of an articulation "rhizomic" (from the idea of the French Philosopher Gilles Deleuze) where the development of new technologies (point-cloud technologies and their derivatives) is a response to the temporally-morphing societal needs and scientific challenges, that emerge from the interactions in the multidisciplinary rhizome of ideas. And the tools and methods are not born out of free-air but as the results of both technological, but also "societal" changes. Therefore, we present different technological developments as a product of challenges that needed to be met by scientists and we look at some of the present and forthcoming challenges for the coming generations of scientists and practitioners, concerning catastrophic landslides. In Japan, the economic

Fig. 1 Photograph of different landslides around the Pacific: a GLOF deposit in the Cordillera Blanca in South America, near a local dwelling (1); in Indonesia in the Mts. Menoreh on Java island (2–4), where the landslide took the lives of 28 in April 2017, in the village of Banaran Pulung Ponorogo; in North Kyushu, Japan in 2017, where thousands of landslides and debris flows were triggered within 24 h (5), around Kure-city in Hiroshima in the aftermath of the July 2018 heavy-rainfall events(6–7); and in South Taiwan in the aftermath of typhoon Morokot that occurred in 1999 (the photograph was taken in 2017)

slowdown combined with an ageing population translated into a diminishing man-power and less tax-money available for hazard prevention. In such context, technological and engineering advances, like laser scanners and photogrammetric methods are essential to collect high-accuracy data, easily and at a relatively low-cost.

Point-Cloud Acquisition for Landslide Research and Monitoring

Landslide delineation has been traditionally done using satellite imagery eventually coupled with ground surveys using various range of tools, some of which are parented to laser scanners, such as laser rangefinders (Satangelo et al. 2010). Those surveys yield good accuracy but are very much time-consuming (Satangelo et al. 2010) and they do not fit large-scale events like the one seen in Taiwan or in Japan (Fig. 1), where the number of events need fast methods, and they provide only limited number of observations. TLS-based point-clouds have been used to overcome those difficulties (Jaboyedoff et al. 2012), even if the RMSE (Root Mean Square Error) can still range between 0.8 and 4.2 m under forested cover. The RMSE values are particularly high for slopes > 40 degrees (Razak et al. 2013).

Landside monitoring can be divided between the pre-movement monitoring and post-failure.

Pre-failure measurement typically needs high-resolution in Z to be effective. Therefore, it has seen a recent gain of interest, as technological advances based on radar interferometry (Insar: Colesanti et al. 2003) and LiDAR technologies (Gordon et al. 2001) have allowed accurate measure of vertical deformation (Royan et al. 2014). In France, for instance, TLS-based measurement of topographic change between 2004 and 2007 at Sechilienne Landslide, using a method based on parallel tracks instead of attempting to observe the entire landslide (Kasperski et al. 2010) have provided a mean of following over time slope deformations, even if deformation is only occurring locally. Even in challenging environment with high-atmospheric moisture content, small movement in the range of 3 to 6 cm could detected from TLS with uncertainty in subsequent measurement of 6.3 mm at maximum distance of 950 m. The high-resolution of the measurement allowed the measures of deformation within short-time span: between the 23rd and the 25th September 2014 (Huang et al. 2019). When vegetation is not too much of an obstacle, photogrammetric method has been also used. UAV-based photogrammetry of the Super-Sauze landslide (France) (Niethammer et al. 2012) have shown horizontal displacement between 7 and 55 m in one year, providing daily average of 0.01–0.1 m/day as well as a good record over time of the behaviour of tension cracks notably. Methods have been used on their own, or combined, between UAV photogrammetry and TLS (e.g. Zieher et al. 2018) or combining TLS with satellite imagery (Barbarella et al. 2018), or combining ALS and TLS. For instance, Corsini et al. (2013) combined measures from Riegl® TLS (up to 1300 m distance) and from Leica® ALS (flight about 2500 m above ground), using an iterative procedure that looks for the closest point in each dataset (the ICP: Iterative Closest Point procedure).

Arguably, post-failure measurements have a longer history, because they do not need high-precision nor high-resolution measurements to be surveyed for metrics extraction. Because of the steepness of the slopes however, high-resolution topography data have also benefited of laser technologies sometimes combined with one another to cover the "respective blind-spots) (e.g. along the Yamachiche river —Jaboyedoff et al. 2009). Point-cloud producing technologies have been then used for almost all the existing types of mass movements and landslides, such as rockfalls on lifelines in mountain areas (Abellan et al. 2006) for instance and to attempt measuring displaced volumes for instance (Tseng et al. 2013) (see Corsini et al. 2013 for a list of the different applications of laser technologies for landslide).

Point-Cloud Processing in Landslide Research

Whether one uses point-cloud collected from a laser scanner or from a photogrammetric method, the point-cloud is usually inhomogeneous due to bad returns or lack of visibility, on top of various errors, which are some of the difficulties met when working with such data.

Working without gridding the data, it is possible to regroup the points by voxels of given footprint and volume (Zieher et al. 2018), to which existing data such as the laser intensity, the elevation, the RGB values or any spectral data associated can be computed. For instance, Zieher et al. (2018) computed the geometric curvature c_i at point "i" that represents the deviation of the normal curvature with a defined neighbourhood:

$$c_i = \frac{1}{k}\sum_{j=1}^{k} ||n_i - n_j||$$

Another measure that is related to surface roughness for gridded data is the Index of planarity p_i at a point "i" (e.g. Hansen et al. 2014), defined as follows:

$$p_i = \frac{e_i1 + e_i2}{e_i1}$$

Despite of the development of tools and methods based on point-clouds, most research still relies on the gridding of data over a regular space-scale using different methods (Table 1), which have been made easy to integrate thanks to the interfaces in QGIS® or in ArcGIS®.

Objectives Statement

The socio-economic situation is therefore calling for further improvement of streamlined data acquisition from ALS, TLS and SfM-MVS photogrammetry, while scientific development that have focused on acquisition methods now need the development of new algorithms for data handling that becomes fully 3D and without the need of prior-gridding whenever possible.

The present contribution is therefore two-folds, (1) it first present research of the team at various locations around the world and in various environments, using various point-cloud acquisition technologies from the coast and forested slopes to active volcanoes, and (2) it presents some of conceptual views of point-cloud processing, which could be developed further for landslide monitoring and research.

Short Methodology: Material and Tools

The Material

The present contribution uses (1) a set of UAV-based photograhs collected for SfM-MVS photogrammetry measurement near Asakura-City in 2017 (Japan), (2) it then presents a landslide on the French coast, (3) a set of LiDAR dataset of the Oso landslide acquired between 2003 and 2017 by the USGS for the State of Washington (USA), and then LiDAR and UAV-based photogrammetric data from Unzen Volcano in Japan. Resulting from these datasets and the limitations of each of the methods, the possibilities of UAV-based ALS is explained, before a short section on potential point-cloud direct processing is proposed.

LiDAR Technology

The LiDAR (light detection and ranging) technology aims at measuring the range between an emitter and a target using a laser. The range measurement can be done by either measuring the time it takes for a laser pulse to come back to the instrument (time of flight) or by comparing the change of phase in a semi-continuous signal (phase shift

The range finding technology is generally combined with GNSS (Global Navigation Satellite System) and INS (Inertial Navigation System). The scanner measures distances, the GNSS provides the position at which the measurement is done, and the INS measures in which direction the laser pulse is sent. The combination of those three data sources allow for rendering a 3d model of the survey area in a form of a point cloud. Each point is georeferenced in the global coordinates system. Ground-based LiDAR used for terrestrial landslides monitoring and geodetic survey is usually named TLS (Terrestrial Laser Scanner), while its counterpart is named ALS (Airborne Laser Scanner). The method used in the present contribution follows conventional protocols (e.g. Jaboyedoff et al. 2012)

Table 1 Commonly used gridding algorithm, to transform 3D point-clouds to gridded datasets

Standard representative equations	Name and description
$p(x) = \sum_{i=0}^{n} \frac{q_i(x)p(i)}{\sum_{i=0}^{n} q_i(x)}$ $q_i(x) = \frac{1}{d(x,x_i)^r}$	Inverse Distance Weighting (IDW) [where x is a given point in the population, p(x) is a calculated value and x_i is a point for which the value is known, d is the Euclidian distance btw points and r is the power parameter.](Salekin et al. 2018)
$f(x,y) = \sum_{i=1}^{n} \beta_i f(x_i, y_i)$	Natural Neighbour (NaN) [where for a population i = 1 − n, f(x, y) is the interpolation function at x and y and β is the interpolation weight]. (Sibson 1981)
$\sum_{i,j} Z_{i,j} + A_{i,j} = \sum_{k=0}^{l} x_{k,a} \vert A_{i,j} = \sum_{it=1}^{n} A_n \underline{\underline{\mathrm{def}}} a_n$	ANUDEM or Topo-to-Raster in ArcGIS [where Z is the altitude over a grid i, j and A is a set of topographic attributes (elevation, sink points, streamlines…) over the same grid, calculated from the interpolation of discrete points x, located according to k position quantities, with a measured attributes. The relation between A and a is then iterated n times for A_n approaches a_n.] (Hutchinson et al. 2009, 2011).
$Z_k^* = \sum_{i=1}^{n} \lambda_i Z_i$ $E\{Z_v - Z_k^*\} = 0$ $\sum_{i=1}^{n} \lambda_i = 1$	Standard krigging [where Z_k^* is the krigged value from a set of points i comprised between 1 and n, with a weight λ, and for which the unbiased condition of krigging E must be 0 and the sum of all the weights must be equal to 1] (Chung et al. 2019)

Structure-from-Motion Multiple-View—Stereophotogrammetry (SfM-MVS)

SfM-MVS is a photogrammetric method born out of engineering in the 1970s that reached the fields of geosciences and civil engineering in the late 1990s, thanks to the increase of computing power and the spread of low-cost Unmanned Aerial Vehicles (UAVs). Because of the low- vertical amplitude over widespread areas of the fluvial landforms, SfM-MVS has been developed first for fluvial geomorphology (Westoby et al. 2012, Javernick et al. 2014) before being expended to other fields of Earth surface observations such as tectonic geomorphology (e.g. Saputra et al. 2018), volcanic landforms (e.g. Gomez 2014), large-areas landscape (e.g. Gomez et al. 2015), and landslides (e.g. Saito et al. 2018). For the present contribution, the SfM-MVS method was applied using the Photoscanpro software using photographs taken from DJI® Phantom-pro 4 and Mavick platforms, using on average ~200 to ~600 photographs.

Results

Data Acquisition Methods and Tools

UAV-SfM-MVS Point-Clouds for Landslide Monitoring: Landslides Near Asakura

In the lapse of 24 h between July 5th and 6th 2017, heavy rainfall battered the area of North Kyushu due to the "tsuyu" (the prune rain period) heavy-rainfalls that brought more than 580 mm of rainfall in 24 h, ramping up to 616 mm in 72 h in the vicinities of Asakura-city in North Kyushu. An estimated 1,800 shallow landslides and debris flows have been identified from aerial photography, but for rapid and higher-resolution characterization, UAV-based photographs for photogrammetric analysis of the landslides (Fig. 2).

Advantages: UAV photographs were taken using light-weight commercial UAVs Phantom 4 and Mavick 2 providing rapid coverage (30 min to 2 h per location) of the different landslides. Ground Control Points were collected a

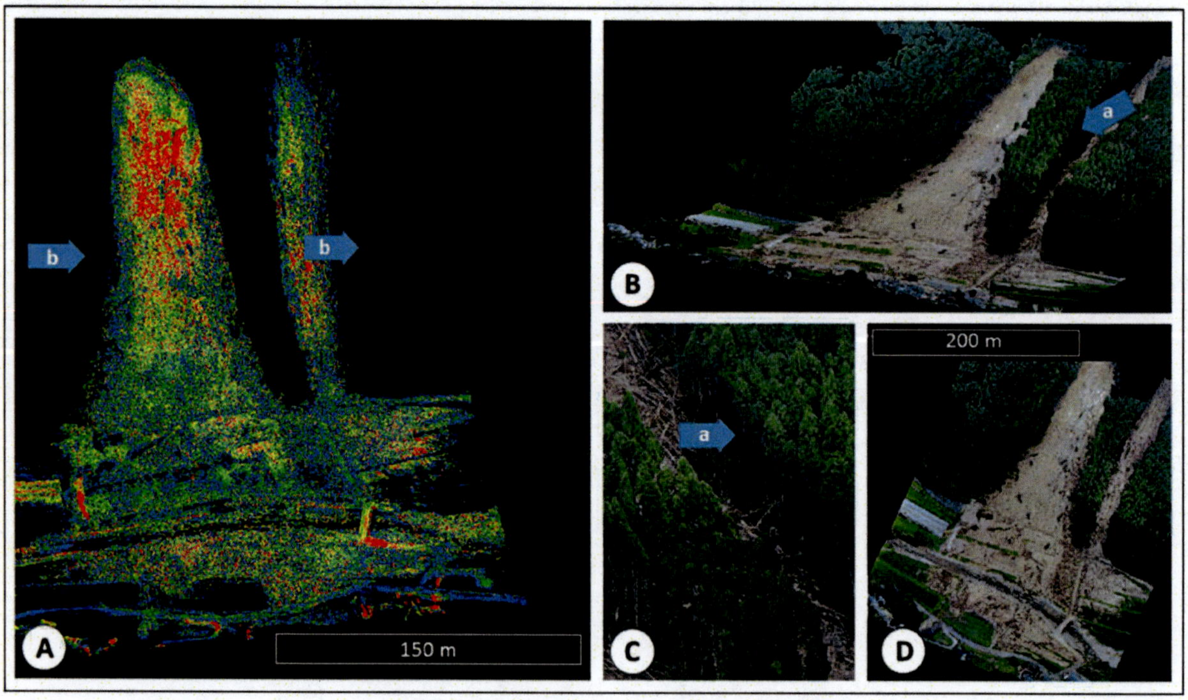

Fig. 2 Shallow landslide and debris-flow track near Asakura. The display is based on a 32 million points dataset using Cloud Compare; with **A** the color intensity using the HSV-scale after segmentation removing vegetation; **B** side view of the debris-flow track and the landslide; **C** Zoom on the debris-flow track; **D** full view of the landslide and the deposit that rode over the terraces and the riverbed. **a** blind-spot zone; **b** no data underneath the vegetation

posteriori and only once on recognizable monuments on and around the road in stable locations.

Disadvantages: The data based on the photogrammetry does not penetrate through the vegetation, and depending on the camera angle, shadow areas appear (a in Fig. 2), especially along subvertical faces, and also under vegetation (b in Fig. 2-A).

TLS and ALS for Coastal Landslide

In coastal areas, TLS-based monitoring of landslides can be limited by various setbacks during acquisition. The prospecting area is limited by the sea, and a combination of ground- with airborne data is therefore essential to characterize the main morphological features and the slope kinematics. The coastal location implies specific slope behaviour, associated with the sea erosion of the toe of the slope. The Villerville landslide (47 ha, Normandy, France) has been considered active since 1982. It is a complex landslide, mainly controlled by hydrogeological processes (Lissak et al. 2014a) and also by sea erosion, which regularly evacuates the landslide deposits. Within the context of climate change and associated sea level rise, the importance of mechanical erosion by the sea is a changing game. To quantify the slope kinematics, several campaigns were lead (3D data acquisition with TLS, ALS...) and permanent monitoring exist (GNSS receivers, fixed benchmarks, inclinometers...). While GNSS receivers provide punctual information of the displacement patterns, approaches based on DEM of Difference is frequently used to spatialize values of slope deformation. In our case, the slope velocity of Villerville landslide is very low and heterogeneous [0.01–0.10 $m*y^{-1}$] and the vegetation covers most of the slope. Consequently, it makes field surveys laborious. Thus, in combination with the TLS data (Fig. 3), 2 ALS surveys (derived DEM: 0.5 m resolution) were operated in 2010 and in 2013 to detect the morphology underneath the vegetation and to control the evolution over the 3-years period. The DoD analysis did not provide significant results.

Advantages: ALS Provide major information of the slope morphological features in highly vegetated area (Lissak et al. 2014b). This tool can be useful in coastal area to get information on landslide kinematics where ground-based LiDAR survey is impossible because of the size of the landslide and because of coastal configuration (limitation by the sea). But in a context of slow moving landslide investigations of ALS must be spaced out in time. TLS surveys, can be an

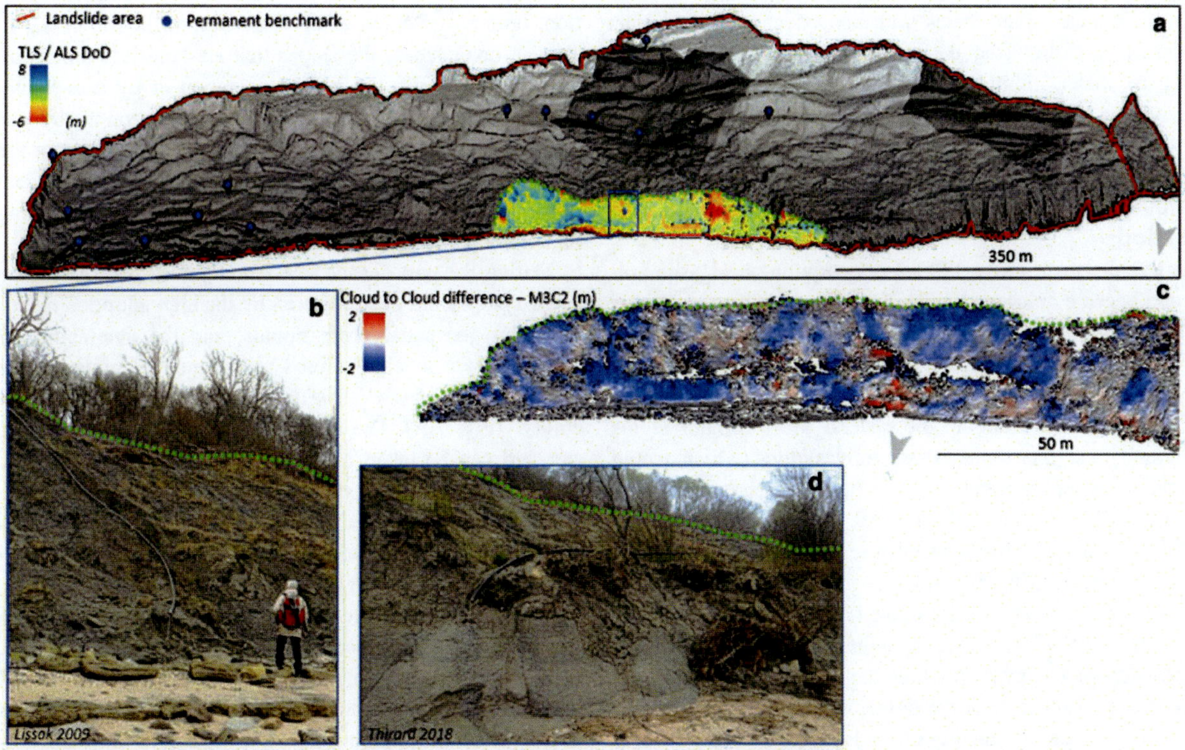

Fig. 3 A Airborne LiDAR DEM (2010) and Terrestrial Laser Scanner (provided by LETG Caen, Normandy Univ.) DEM (2018) combination and the DoD of TLS/ALS at the toe of the landslide, with field photographs (**B** and **D**). In B, the scale is given by the person standing on the right of the photograph, and in D, the apparent bedrock is ∼1.5 m high. **C** Cloud to Cloud difference using the M3C2 plugin in Cloud Compare to quantify the erosion/deposit of clay material at the landslide toe

alternative because it is less expensive than ALS survey and easier to set up for an annual survey and both can be combined (Fig. 3).

Disadvantages: ALS provides high-resolution data but investigation by airplane/helicopter can be expensive and UAV on-board LiDAR sensors are rarely employed and can cover only limited areas. To overcome these difficulties, TLS can be useful. But this tool is mainly efficient for sparsely vegetated areas. Moreover, it allow investigation of limited areas and is very dependent on the configuration of the study site.

LiDAR Survey of the Oso Landslide, in Washington, USA

Repeat LiDAR data has been particularly successful for diachronic data analysis of landslides, especially because the timing of the movement (except the main travel time) can be captured over time, and so even through vegetation, like it is the case for the Oso Landslide in the State of Washington, in USA (Fig. 4).

Advantages: Large areas covered in a densely-vegetated landscape where the human footprint is limited, so that unknown landslides can be detected.

Disadvantages: As vast areas are collected, the processing of the dataset is time-consuming and it can only be set in countries and states with consequent funding. Furthermore, earliest the dataset only provides 1 point per square meter, limiting the possibility to survey detailed evolution of the landslide.

Multi-temporal LiDAR of the Volcanic Dome Movement at Unzen Volcano During Quiescent Period

In the aftermath of the 1990–1995 eruption, the Unzen volcano became quiescent again, if not for the numerous lahars that swept the volcanic flank, so that the volcanic apron was sentenced with a restricted access. Unzen Volcano is however remarkable as the volcanic dome has not fully collapsed during the last eruption and several lobes remain at the summit, and this dome is presently sliding (which would have catastrophic effects and eventually reactivate volcanic activity (Fig. 5). From the LiDAR data over the period of 2007–2016, the dome has shown several signs of movement and instability (Fig. 5), with numerous collapses and gully erosion features around the dome (Fig. 6).

Besides block collapses from the dome itself, the LiDAR also allows the monitoring of the erosion and deposition processes at the base of the dome, which is arguably one of the controls on the stability of the later. Indeed, as the dome slides away from the summit area, it progresses over unconsolidated sediments, which in turn are also being eroded. For this purpose, LiDAR data has shown the head-ward progression of gullies on the volcanic apron (Fig. 6). In existing gullies, the walls' collapse is also being monitored at the year to decadal scale at Unzen Volcano. Both lateral and longitudinal transfers can be recorded.

Advantages: It is only HRT point-cloud technology that allows the monitoring of the dome-sliding combined with the evolution of the supporting unconsolidated material, as the movement is generally at a scale that falls within the error margin of most of the other technologies that can record data over the spatial-scale of the all volcanic dome and apron. Furthermore, because of the danger and the difficulties to reach the area, airborne acquisition is the arguably the best possibility.

Disadvantages: As the LiDAR data has been taken from as early as 1997 at Unzen Volcano, the earliest topographic data have various footprints, with variable resolution and accuracy, which for long-term comparisons bring increased uncertainty on the measurements. Moreover, the steep slopes are not always best suited to traditional ALS.

New Development in Airborne LiDAR: UAV-Based LiDAR Systems

Considering the examples provided in the present contribution, both in USA (State of Washington), in Indonesia, in Taiwan or in Japan, the slopes that are landslide-prone, are all covered by either a dense vegetation or by tree plantations. Consequently, ground or aerial photogrammetry can only provide an image of the canopy structure, but not the ground. Traditional ALS data will often not have sufficient resolution, and TLS is particularly ineffective when working on complex slopes with several ridges.

During the past decade, the development of UAV-borne Lidar systems has been fostered by the high adoption rate of UAV by geospatial professionals, and the availability of compact laser scanners aimed at automotive industry. The big advantage of UAV-based LiDAR is this rapid "deployability" (Fig. 7-A). The compact setup fits in the trunk of a car, and can be taken off a few hundreds of meters away from the area of interested in case of difficult access. The operation is also safer, because no human presence is required on the damaged area. The system compactness and affordability comes with a downside for lower precision than TLS or ALS. Nevertheless, is stays appropriate for mapping terrain slides, at precision below 5 cm. UAV-based LiDAR is providing a measurement density of several tens to several hundred points per square meters, depending on the technology and flight parameters. That is another advantage in order to reach the ground in vegetated areas. Finally, in comparison with photogrammetry, data processing is much faster. It does not require to compute correspondences

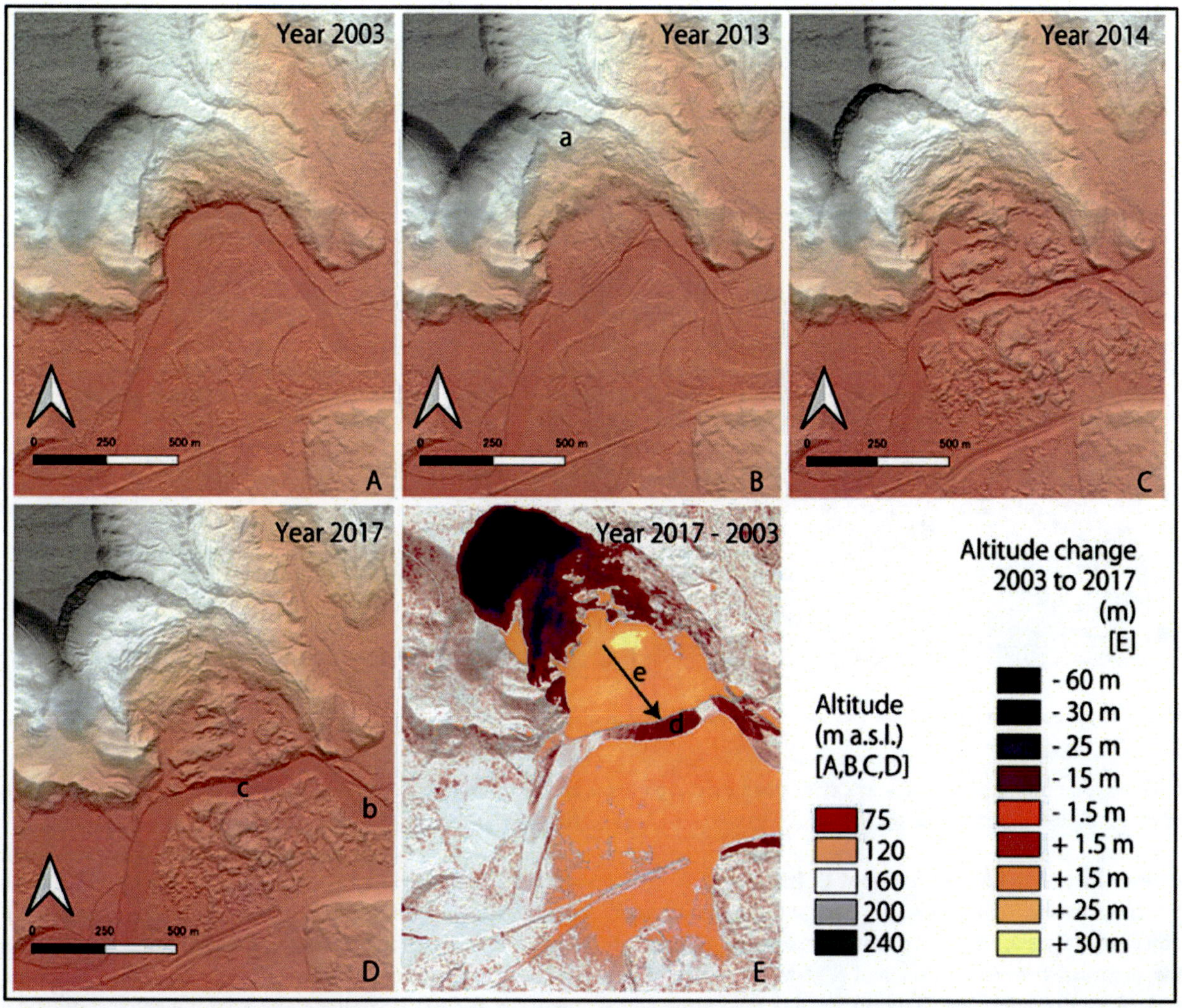

Fig. 4 The Oso landslide in Washington State, USA for the period 2003–2017 investigated using multiple airborne LiDAR dataset. **A** Location where the Oso landslide is to occur on March 22nd, 2014 about 6 km East of Oso Settlement; **B** First sign of movement and deposition at the toe of the slope with erosion of a topographic lump at "a"; **C** LiDAR data post-landslide, where forming a typical landslide staircase slope deposit and the more "hummocky" lower deposit; **D** Modification of the deposit over three years with straightening of the river channel (**c**), and meandering eroding the deposit (**d**); **E** DEM difference between 2017 and 2003 showing the position change of the channel (**d**) and the horizontal forced movement (**e**) as well as the mass moved out of the scarp area and the lower deposit, from which geomorphologic impacts can be calculated (LiDAR data courtesy of the Washington State Government, USA)

between images, or perform bundle adjustment that are time consuming. Direct georeferencing is used instead.

Advantages: UAV-based LiDAR proposes solution that are precise, accurate for virtually any type of environment without the need of any point control in areas that can be otherwise inaccessible and without sufficient line of sight for a TLS or a camera. Furthermore, landslide areas can be dangerous and this technology can bring with virtually no-risk a laser scanner in the heart of the landslide.

Disadvantages: As the technology is new and developing, the price remain relatively prohibitive without specific funding on a research project, thus the democratization of the technology should hopefully reduce the price thanks to increasing volume.

Data Processing Algorithm for Landslides Analysis: Gaps and Propositions

Data processing algorithms exist for classification of point-cloud, cleaning of the dataset, extraction of the error, etc., but in this section, we are concentrating on two new propositions of algorithms providing that we already have surfaces and objects we are interested in, i.e. in landslide

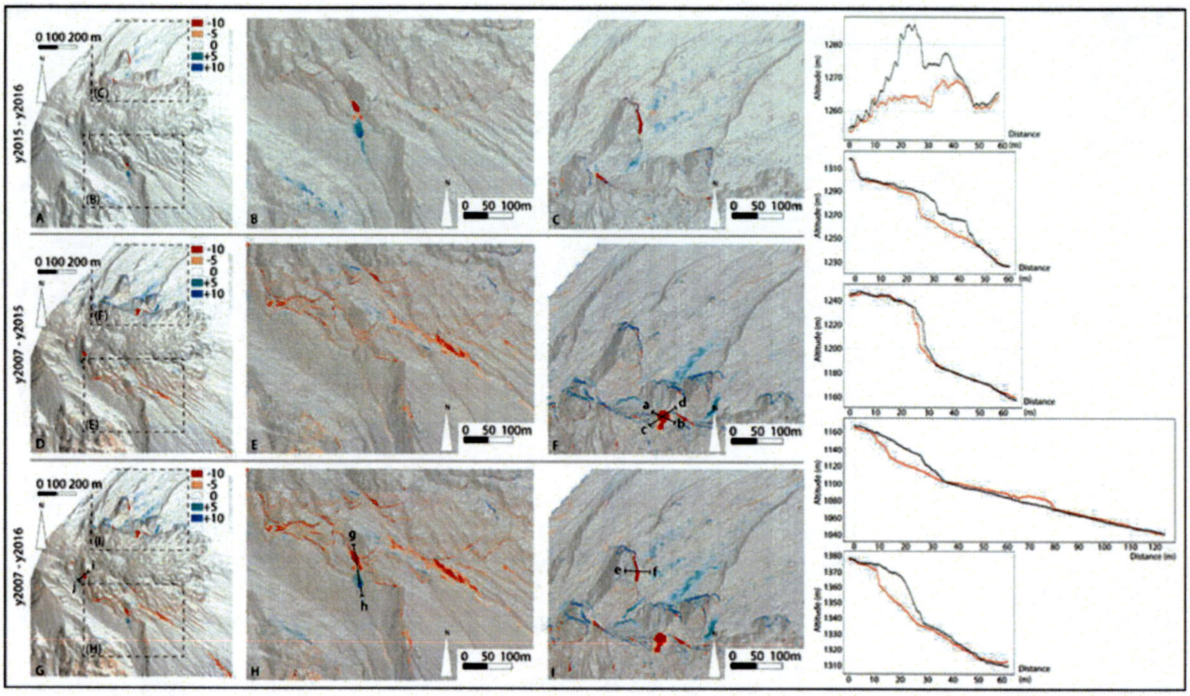

Fig. 5 Dome evolution between 2007 and 2015, 2015 and 2016 and from 2007 to 2016 with transects across the major collapses around the dome

analysis the ground surface(s) - as cavities can exist creating several surfaces at one location.

Points in Point-Clouds Are not Points

Recent advances and efforts in point-cloud classification, object identification and information extraction follows on the progresses in ALS, TLS and SfM-MVS technologies, and we can certainly anticipate to see this field grow at an even faster pace, now that numerous point-cloud data are available online, and that the economic conjuncture is pushing sciences in a corner of lower spending.

If we consider pointclouds in a simple binary distribution space, where the space is either null or with an object, we can fill the '0' space along the pathways of the ray between the recording instrument and the reflective object:

$$\sum_{i=1}^{n} Px_i = \begin{bmatrix} 1\ if\ it\ hits\ an\ object \\ 0\ if\ in free\ space \end{bmatrix}$$

Therefore, pointclouds are characterizing $Px_i = 1$ but mostly overlooking $Px_i = 0$. From this observation, the following question arises: how do we characterize the void-space, knowing that the void-space could be an absence of return for numerous reasons. One way of doing so, is to stop thinking about point-cloud as points, but as end-points of vectors linking the TLS unit (or any other instrument used to produce the pointcloud) to the point, with a known set of shots and return along a vector trajectory, before which the absence of return can be seen as the absence of object (Fig. 8).

It is then possible to group large zones of laser free flight using the weighted network statistical analysis, where the point-cloud including the laser source is computed as an adjacent matrix, where in its simplest form the relation between each connector between a set of points $n_{i,j}$ in the matrix $N = \{n_{i,j}\}$ is computed by a 1 for a connection and a 0 for a no connection. The existence of a connection can be done first using a simple tessellation where all the nearest points are joined together (Fig. 9). From this network, the relations with the networked points can be collected as a point attribute, adding to existing attributes (Fig. 9, subscript a). From this set of data, we can then compute the degree of relation to a nearby point (based on the RGB values, on the return number or intensity and also on statistics of the nearby points, i.e. is there a sudden or gradual change in any of the statistical properties of the nth nearest neighbours), those constructions are named degrees of a node in weighed networks and computed as follow:

$$a'_i = \sum_{j\in\prod(i)} a_{ij}$$

where the degree a'_I is determined as a function of the neighbours it is connected to, based on "i"th linkages. Within this process, we are particularly interested to the "i"th link to the laser unit (point L in Fig. 9). Calculating the

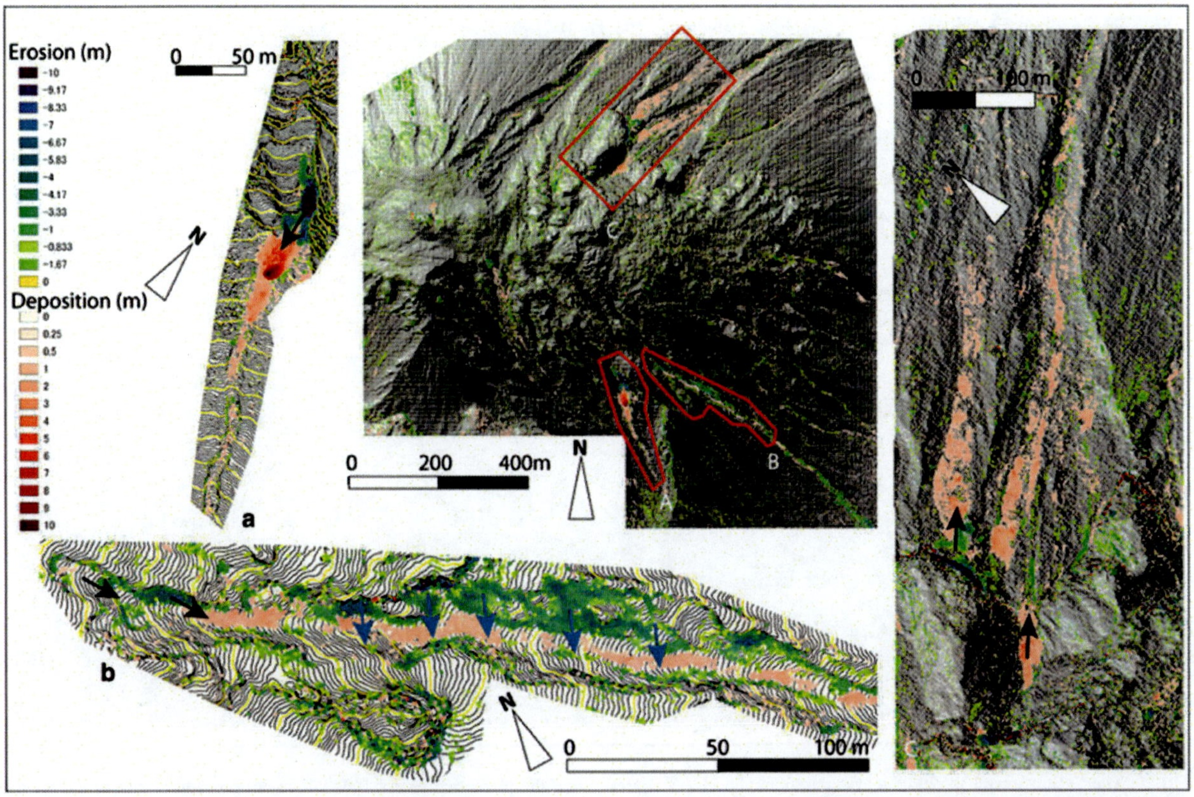

Fig. 6 Gully Erosion and deposition around the dome between 2015 and 2016. The black arrow shows material transfer from the gullies head accompanied by erosion, while the blue arrows show probable gullies' wall erosion and deposition at the bottom of the gully. Although the gully appears to be fully erosional on the longer term (year 2004–2016) the short lived processes are more complex with notably a period of accumulation between 2015 and 2016, showing that the material at the bottom of gully B takes the shape of an erosion feature if considered over the long term, while in fact it is constituted of newly deposited material

sudden length change between L and a point $xn_{(x,\ y,\ z,\ a)}$ between all the connected neighbours of the nth point that are not a connection with P will provide the edge of a volume of free space, allowing to simplify homogeneous volumes of free space between objects and the laser unit, without having to use every single triangle between the connected nearest neighbours and the laser unit (e.g. every laser vectors between the TLS unit and the first return in Fig. 9. Instead, we can use a watershed algorithm for all the points that are not P and belong to the same homogeneous volume to create a single surface that connect to L to form a pyramidal volume of free space. Those new spaces are thus designed with a characteristic path length (P_L) and a clustering coefficient (C_i), so that:

$$P_l = \frac{2}{N(N-1)} \sum_{i,j} d_{i,j}$$

$$C_i = \frac{\sum_{j,k} a_{ij} a_{jk} a_{ki}}{a_i'(a_i' - 1)}, k \neq 0, 1, i, j, k \neq L$$

where I, j, k are the nodes of a triangle between defined nth nearest neighbors others than L. Thus, for each point x, with clustering coefficient C_i it is possible to then use a value LM, so that $x\{C_i\}$ < or > LM will define the boundary between a given free space volume and another. The same network can then be used to compute various forms of variability in the returned signal, RGB values associated, or the eigenvalues of a PCA Brodu and Lague (2012).

Using this method, we have then a set of empty "vectors", allowing us to compute whether some of them are parallel along the vertical plane, so that we can define cavities or secondary surfaces. In the case of rock-formations movement, the deformation of a cavity can then be computed by investigating the change of the shape of the "empty space".

Point-Cloud-Based Multiscale Analysis

Another direction that needs further exploration is the direct use of point-clouds at multiple-scales without using the gridding system, because the latter introduces approximations and interpolated data and generate secondary data that

Fig. 7 The YellowScan® Vx-20 lidar system mounted on UAV. Laser scanner mounted on a DJI® M600 UAV in flight during a test on a slope in Hiroshima prefecture, to "see" through the dense vegetation and the plantation forest; **B** Deployment procedure "out of the car", in Hiroshima; **C** Collected dataset with a display of the return intensity in grey-scale (~22 million points loaded in Cloud Compare)

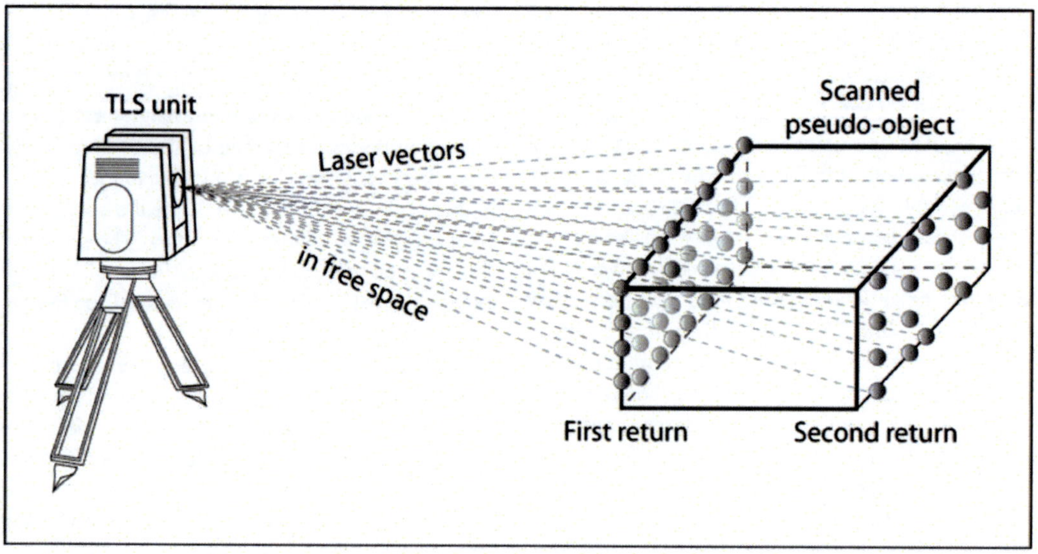

Fig. 8 Defining empty spaces with no return using the portion of space between the first return of the scanned objects and the laser scanner

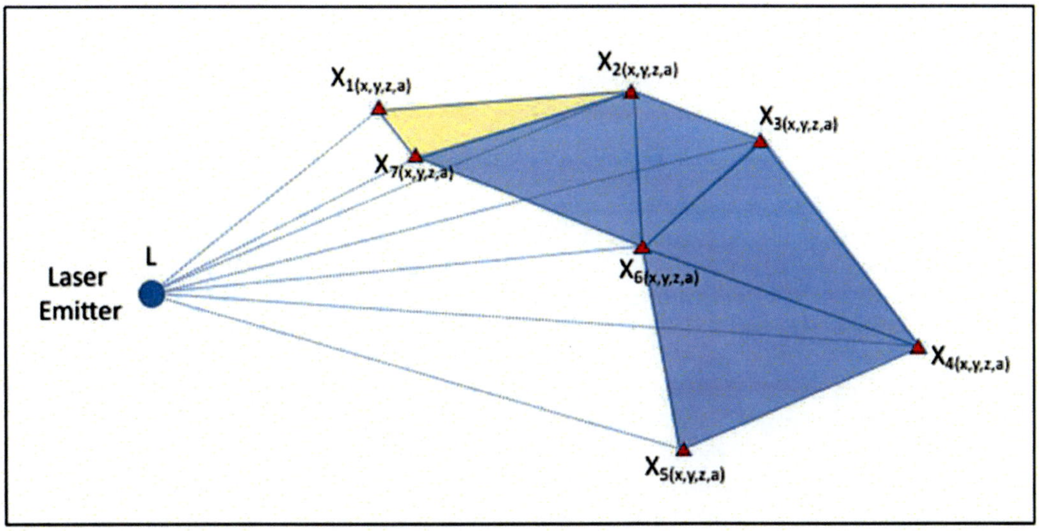

Fig. 9 Tesselation of the point network and attribution of the attributes (subscript a) linked to the distance to the laser emitter and to the nearest neighbour and the potential change in category. The surface in yellow characterized by X_1, X_2, and X_7 has shorter distances to L than X

are more the relict of first generation surface analysis in 2D and 2.5D GIS software. For instance, the variability in altitude, or any other value can be calculated for given areas around a point (e.g. A_1, A_2, A_3… in Fig. 10), or only within given distances, e.g. $A_1 \cup A_2 or A_3 \cap A_1$ allowing to work at given scales. If one does not compute all the variability within A_1 but is not interested at the comparison with any statistical approximation of elements at $>A_2$ distance, it is then possible to work on the point-cloud at a chosen scale, imbricate those scales, and then add the information to the chosen "neighbouring" data to each points from which the calculation is made. This approach, can then in turn be applied to characterize the signature of the pointclouds using classic approaches in multivariate statistics for example for sets of given parameters recorded at each point.

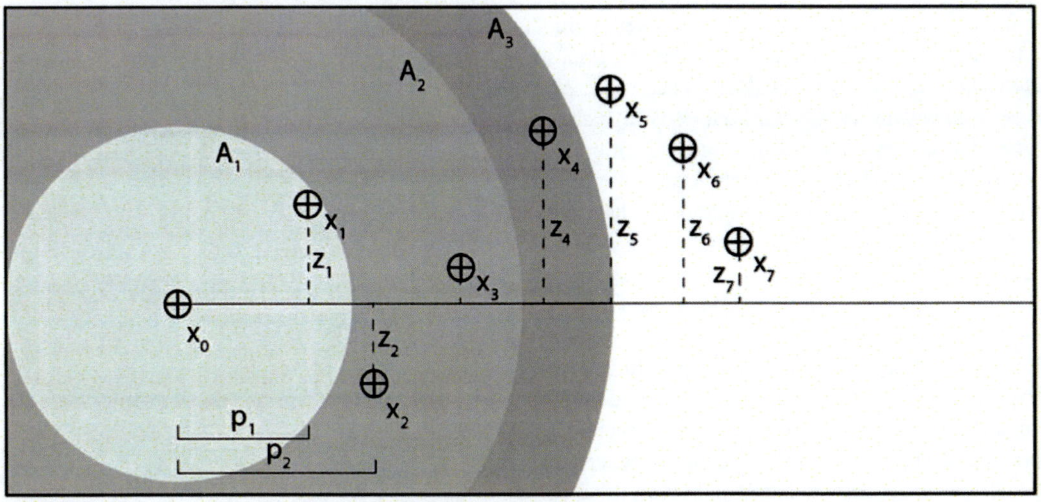

Fig. 10 Point sampling for a random type of value, here Z for all the points X at chosen distances A

Discussion and Conclusion

The acquisition of point-clouds for landslide mapping and monitoring can rely on a combination of airborne and ground-based laser and photogrammetric methods that can be used alone or in combination depending on the funding available, the slope angles, the extent and number of landslides as well as the density of the vegetation cover. The results in this contribution show that ALS can be useful to find landslides in remove areas with dense vegetation cover (in Washington state), while TLS and SfM-MVS provides good results on individual landslides or easily accessible landslides (in Normandy or Asakura for instance). While SfM-MVS is a method that has growth potential in "developing" nations, because it is low-cost, it should however not be seen as a replacement to laser techniques, and eventually not as a competition neither, as it provides different data. Combining the ease of UAV-based SfM-MVS with the reliability of laser technology, ALS on UAV is promising to provide reliable yet rapid and high resolution assessment of landslides, even when lines of sights are not existent or when the field is hardly accessible.

To process the tens to hundred million points generated, it has become necessary to develop more algorithms that allow to work directly from the point-clouds instead of using traditional interpolated dataset. There are several algorithms that already exist, but we propose in this contribution a slightly different vision about point-clouds that should start to be considered for what they are, sets of vectors between an emitter and a receiver, therefore holding information about the "empty space". Furthermore, instead of storing data on grid, those can be added as attribute to every points, allowing to hold multi-scale information at each location, so that vectorization can also be used for data processing improving the computing time when performed in space or over a grid. Indeed, the interpolation methods introduce approximation in space (Table 1), which erase some of the variability existing in the point-cloud (although the ANU-DEM provides elements to constrain potential errors) and its spatial structure. We propose that the signature of point-clouds that is erased in the grid could be used to better define landslide susceptibility.

All those developments are most necessary, to improve the coverage of so-called "developing" or raising nations and also to support data acquisition and computing in economies where funding is shrinking and in countries where the ageing-population limits the number of individuals that can process the data.

References

Abellan A, Vilaplana JM, Martinez J (2006) Application of a long-range Terrestrial Laser Scanner to a detailed rockfall study at all de Nuria (Eastern Pyrenees, Spain). Eng Geol 88:136–148

Barbarella M, Di Benedetto A, Fiani M, Guida D, Lugli A (2018) Use of DEMs derived from TLS and HRSI data for landslide feature recognition. Int J Geo-Inf 7(160):1–22

Brodu N, Lague D (2012) 3D terrestrial lidar data classification of complex natural scenes using a multi-scale dimensionality criterion: applications in geomorphology. ISPRS J Photogramm Remote Sens 68:121–134

Chung SY, Venkatramanan S, Eldin Elzain H, Selvam S, Prasanna MV (2019) Chapter 4—supplement of missing data in groundwater-level variations of peak type using geostatistical methods. In: GIS and geostatistical techniques for groundwater sciences, pp 33–41

Colesanti C, Ferretti A, Prati C, Rocca F (2003) Monitoring landslides and tectonic motions with the permanent scatterers technique. Eng Geol 68:3–14

Corsini A, Castagnetti C, Bertacchini E, Rivola R, Ronchetti F, Capra A (2013) Integrating airborne and multi-temporal long-range terrestrial laser scanning with total station measurements for mapping and monitoring a compound slow moving rock slide. Earth Surf Proc Land 38:1330–1338

Cui S, Wang G, Pei X, Huang R, Kamai T (2017) On the initiation and movement mechanisms of a catastrophic landslide triggered by the 2008 Wenchuan (Ms 8.0) earthquake in the epicentre area. Landslides 14:805–819

Dai FC, Xu C, Yao X, Xu L, Tu XB, Gong QM (2011a) Spatial distribution of landslides triggered by the 2008 Ms 8.0 Wenchuan earthquake, China. J Asian Earth Sci 40:883–895

Dai FC, Tu XB, Xu C, Gong QM, Yao X (2011b) Rock avalanches triggered by oblique thrusting during the 12 May 2008 Ms 8.0 Wenchuan earthquake, China. Geomorphology 132:300–318

Davies TR, McSaveney MJ (2006) Dynamic fragmentation in landslides: application to natural dam stability. Italian J Eng Geol Environ Spec Issue 1:123–126

Ekström G, Stark CP (2013) Simple scaling of Catastrophic Landslide Dynamics. Science 339:1416–1419

Gao Y, Yin Y, Li B, Feng Z, Wang W, Zhang N, Xing A (2017) Characteristics and numerical runout modelling of the heavy rainfall-induced catastrophic landslide-debris flow at Sanxicun, Dujiangyan, China, following the Wenchuan Ms 8.0 earthquake. Landslides 14:1361–1374

Gariano SL, Guzzetti F (2016) Landslides in a changing climate. Earth Sci Rev 162:227–252

Gordon S, Lichti D, Stewart M (2001) Application of a high-resolution, ground-based laser scanner for deformation measurements. In: Proceedings of the 10th international FIG symposium on deformation measurements, pp 22–32

Gomez C (2014) Digital photogrammetry and GIS-based analysis of the bio-geomorphological evolution of Sakurajima Volcano, diachronic analysis from 1947 to 2006. J Volcanol Geoth Res 280:1–13

Gomez C, Hayakawa Y, Obanawa H (2015) A study of Japanese landscapes using structure from motion derived DSMs and DEMs based on historical aerial photographs: new opportunities for vegetation monitoring and diachronic geomorphology. Geomorphology 242:11–20

Hadmoko DS, Lavigne F, Sartohadi J, Gomez C, Daryono D (2017) Spatio-Temporal distribution of landslides in Java and the triggering factors. Forum Geografi 31:1–15

Hansen CD, Chen M, Johnson CR, Kaufman AE, Hagen H (2014) Scientific visualization. Springer-Verlag Pub, 400 p

Huang R, Jiang L, Shen X, Dong Z, Zhou Q, Yang B, Wang H (2019) An efficient method of monitoring slow-moving landslides with long-range terrestrial laser scanning: a case study of the Dashu landslide in the Three Gorges Reservoir Region, China. Landslides 16:839–855

Hutchinson MF, Stein JA, Stein KL, Xu T (2009) Locally adaptive gridding of noisy high resolution topographic data. In: Anderssen RS, Broaddock RD, Newham LTH (eds) 18th world IMACS congress. Modelling and simulation society of Australia and New Zealand and international association for mathematics and computers simulation, pp 2493–2499

Hutchinson MF, Xu T, Stein JA (2011) Recent progress in the ANUDEM elevation gridding procedure. In: Hengel T, Evans IS, Wilson JP, Gould M (eds) Geomorphometry 2011, pp 19–22

Jaboyedoff M, Demers D, Locat J, Locat A, Locat P, Oppikofer T, Robitaille D, Turmel D (2009) Use of terrestrial laser scanning for the characterization of retrogressive landslides in sensitive clay and rotational landslides in river banks. Can Geotech J 46:1379–1390

Jaboyedoff M, Oppikofer T, Abellan A, Derron M-H, Loye A, Metzger R, Pedrazzini A (2012) Use of LiDAR in landslide investigations a review. Nat Hazards 61:5–28

Javernick L, Brasington J, Caruso B (2014) Modeling the topography of shallow braided rivers using Structure-from-Motion photogrammetry. Geomorphology 213:166–182

Kasperski J, Delacourt C, Allemand P, Potherat P, Jaud M, Varrel E (2010) Application of a terrestrial laser scanner (TLS) to the study of the sechilienne landslide (Isere, France). Remote Sens 210:2785–2802

Korup O, Clague JJ, Hermanns RL, Hewitt K, Strom AL, Weidinger JT (2007) Giant landslides, topography, and erosion. Earth Planet Sci Lett 261:578–589

Lissak C, Maquaire O, Davidson R, Malet J-P (2014a) Piezometric thresholds for triggering landslides along the Normandy coast, France. Geomorphologie 20:145–158. https://doi.org/10.4000/geomorphologie.10607

Lissak C, Maquaire O, Malet J-P, Bitri A, Samyn K, Grandjean G, Bourdeau C, Reiffsteck P, Davidson R (2014b) Airborne and ground-based data sources for characterizing the morpho-structure of a coastal landslide. Geomorphology 217:140–151. https://doi.org/10.1016/j.geomorph.2014.04.019

Massey C, Townsend D, Rathje E, Allstadr KE, Lukovic B, Kaneko Y, Bradley B, Wartman J, Jibson RW, Petley DN, Horspool N, Hamling I, Carey J, Cox S, Davidson J, Dellow S, Godt JW, Hodlen C, Jones K, Kaiser A, Little M, Liyndsell B, McColl S, Morgenstern R, Rengers FK, Rhoades D, Rosser B, Strong D, Singeisen C, Villeneuve M (2018) Landslides Triggered by the 14 Nov 2016 M_w 7.8 Kaikoura Earthquake, New Zealand. Bull Seismol Soc Am 108:1630–1648

Mihalic Arbanas S, Scevanj M, Bernat Gazibara S, Krkac M, Begic H, Dzindo A, Zekan S, Arbanas Z (2017) Landslides in the Dinarides and Pannonian Basin—from the largest historical and recent landslides in Croatia to catastrophic landslides caused by Cyclone Tamara (2014) in Bosnia and Herzegovina. Landslides 14:1861–1876

Niethammer U, James MR, Rothmund S, Travelletti J, Joswig M (2012) UAV-based remote sensing of the Super-Sauze landslide: evaluation and results. Eng Geol 128:2–11

Razak KA, Santangelo M, Van Westen CJ, Straatsma MW, de Jong SM (2013) Generating an optimal DTM from airborne laser scanning data for landslide mapping in a tropical forest environment. Geomorphology 190:112–125

Royan MJ, Abellan A, Jaboyedoff M, Vilaplana JM, Calvet J (2014) Spatio-temporal analysis of rockfall pre-failure deformation using terrestrial LiDAR. Landslides 11:697–709

Saito H, Uchiyama S, Hayakawa YS, Obanawa H (2018) Landslides triggered by an earthquake and heavy rainfalls at Aso volcano, Japan, detected by UAS and SfM-MVS photogrammetry. Prog Earth Planet Sci 5(15):1–10

Salekin S, Burgess J, Morgenroth J, Mason E, Meason D (2018) A comparative study of thre non-geostatistical methods for optimising digital elevation model interpolation. ISPRS J Geo-Inf 7–8,300. https://doi.org/10.3390/ijgi7080300

Saputra A, Gomez C, Dekilostidis I, Hadmoko D, Sartohadi J, Setiawan MA (2018) Determining earthquake susceptibility areas Southeast of Yogyakarta, Indonesia—Outcrop analysis from structure from motion (SfM) and geographic information system (GIS). Geosciences 8(132):1–32

Sassa, K. 2007. Landslide science as a new scientific discipline. In: Sassa et al (eds) Progress in landslide science. Springer, pp 3–11

Schulmeister J, Davies TR, Evans DJA, Hyatt OM, Tovar DS (2009) Catastrophic landslides, glacier behaviour and morain formation—a view from an active plate margin. Quat Sci Rev 28:1085–1096

Sibson R (1981) A brief description of natural neighbour interpolation. In: Barnett V (ed) Interpreting multivariate data. Wiley, pp 21–36

Tseng C-M, Lin C-W, Stark CP, Liu J-K, Fei L-Y, Hsieh Y-C (2013) Application of a multi-temporal, LiDAR-derived, digital terrain model in a landslide-volume estimation. Earth Sruf Process Landf 38:1587–1601

USGS (2017) Western Washington 3DEP LiDAR technical data report contract no G16PC00016, 1–35

Westoby MJ, Brasington J, Glasser NF, Hambrey MJ, Reynolds JM (2012) 'Structure-from-Motion' photogrammetry: a low-cost, effective tool for geoscience applications. Geomorphology 179:300–314

Zieher T, Toschi I, Remondino F, Rutzinger M, Kofler CH, Mejia-Aguilar A, Schlogel R (2018) Sensor- and scene-guided integration of TLS and photogrammetric point clouds for landslide monitoring. ISPRS Int Arch Photogram 1243–1250

Detecting Change of Patterns in Landslide Displacements Using Machine Learning, an Example Application

Giacomo Titti, Matteo Mantovani, and Giulia Bossi

Abstract

Machine learning and signal processing can support the definition of landslide alert/alarm systems based on monitoring data. The possibility to rely on a straightforward and automatic procedure to identify hazardous situations could be very useful for risk management and decision makers. In this work, we propose a hierarchical clustering algorithm to identify changes of pattern in the displacements of monitored landslides. Our test site is a large, active Deep-seated Gravitational Slope Deformation (DGSD) in which secondary movements provide sediment for debris flows that threaten downstream settlements. An Automated Total Station (ATS) has been installed in 2012 to measure the three-dimensional displacements of several benchmarks distributed on the source area and to trigger alarms if superficial movements potentially leading to collapses are detected. Results show that the procedure allows to group benchmarks with similar displacement patterns. The unsupervised definition of homogenous areas from a kinematic viewpoint supports an unbiased geomorphological characterization of the large landslide. Moreover, the method allows to trigger alert warnings if some monitored points change displacement pattern. The identification of possible hazardous situation is performed without imposing fixed and arbitrary thresholds and without calibration. The recognition of areas with new types of activity supports the definition of the sediment volumes available for transport for the next debris flow event and assists the definition of reliable risk scenarios.

Keywords

Machine learning • Hierarchical clustering • Automated total station • Landslide monitoring • Rotolon

Introduction

Monitoring data from landslides, both from remote sensing and ground monitoring equipment, has become more and more available. However, the interpretation of this data is most of the times demanded to an expert which infers the landslide mechanism through an expert-based, operator-dependent, empirical and opaque procedure.

Besides, the identification of the possible timing of a landslide collapse is still missing and only few literature examples of accurate and precise forecast of a collapse before occurrence are available. In fact, while there are several robust techniques for landslide detection and for spatial hazard assessment, temporal forecasting of landslide collapse is still difficult and based on few methods that seem to be suitable applied only to specific types of landslide. However, the information about "when" the landslide would detach may be essential for decision makers in order to implement emergency countermeasures and evacuations (Intrieri et al. 2019).

One of the most successful cases of landslide temporal occurrence prediction was performed in Japan for the Otomura landslides (Fujisawa et al. 2010) with the timing closing of a national road and the construction of a safe detour in the days before the collapse. But others successful experiences are found in Switzerland (Loew et al. 2017), and in Italy (Iovine et al. 2006; Carlà et al. 2019). However, these systems tend to provide also false positives (Petley 2014).

G. Titti
DICAM-UNIBO, Department of Civil, Chemical, Environmental and Materials Engineering, Alma Mater Studiorum University of Bologna, Viale Risorgimento 2, 40136 Bologna, Italy
e-mail: giacomo.titti@irpi.cnr.it

G. Titti · M. Mantovani · G. Bossi (✉)
IRPI-CNR, Research Institute for Geo-Hydrological Protection, National Research Council, C.so Stati Uniti, 4, 35127 Padua, Italy
e-mail: giulia.bossi@irpi.cnr.it

M. Mantovani
e-mail: matteo.mantovani@irpi.cnr.it

Ž. Arbanas et al. (eds.), *Understanding and Reducing Landslide Disaster Risk*,
ICL Contribution to Landslide Disaster Risk Reduction,
https://doi.org/10.1007/978-3-030-60713-5_23

The most used methods are based on the concept of tertiary creep (Carey et al. 2007), for example in the simple inverse velocity method of Fukuzono (1985). Recently, these types of methods have been applied to In-SAR data (Mazzanti et al. 2015; Carlà et al. 2019) but limitations intrinsic in the monitoring technique, which are linked to data processing, are still present. Moreover, the Sentinel 1 data return period of 7 days could be a strong constraint for the alarm of not extremely slow landslides.

The development of Machine Learning techniques to detect precursory signals of collapse for both earth and rock slides could be thus extremely valuable especially in case of large amount of data availability.

Method

In this application the Agglomerative Hierarchical Cluster algorithm was used to highlights similarity in displacement patterns in benchmarks of the Rotolon landslide. The method aggregates data into groups or clusters with no preconceptions about data according to a bottom-up approach (Day and Edelsbrunner 1984; Bouguettaya et al. 2015). The cluster analysis is carried out on the distance matrices. Each cell of a distance matrix reports the distance between the two data associated to the cell's row and column. The columns or rows of the distance matrix can be expressed as vectors in a multi-dimensional space (Murtagh and Contreras 2012).

The focal point of the cluster analysis is the concept of distance between those data vectors. One of the most common distance used in cluster analysis is the Euclidean distance (Gülagiz and Sahin 2017). In a N-dimensional vectorial space, the distance is expressed as:

$$d_{i,j} = \sqrt{\sum_{n=1}^{N} \left(x_{i,n} - x_{j,n}\right)^2} \quad (1)$$

where $d_{i,j}$ is the Euclidean distance between the vectors x_i and x_j of dimension N.

To understand the grouping mechanisms that characterized the algorithm selected, it is necessary differentiate hierarchical and non-hierarchical clustering methods. The non-hierarchical methods are based on the separability of the groups formed. The vectors x, which have been assigned to a group during the early stages, may be reassigned to any other group in the next stages. Instead, if the reassignment is not allowed, then, the clustering method is called hierarchical. The avoidance of reassignment represents a potential advantage, since it limits the time-cost for numerical processing and guarantee straightforward results. Thus, a hierarchical clustering method was proposed (Wilks 2011).

The Agglomerative Hierarchical Cluster algorithm is based on the consideration that each observation of the vector x in a N-dimensional space represents a single cluster. Therefore, the first stage is to find from m groups the two closest groups. The process continues until the m-1th stage when the m observations are aggregated into a single group. The clusters have been assigned on the base of the cutoff value selected between the first and the m-1th stage.

The proximity between two clusters may be chosen according to different paradigms. Complete-linkage, or Maximum-distance clustering has been proposed. It based the proximity measure on the maximum distance between each couple of observation members of two different groups (Wilks 2011). Considering C_1 and C_2 two separated clusters of k and h members respectively, the Complete-linkage is based on:

$$d_{C_1,C_2} = max\left(d_{i,j}\right) \text{ with } i = 1,\ldots,k \,\&\, j = 1,\ldots,h \quad (2)$$

The maximum distance ($d_{C1,C2}$) is selected from the $m \times m$ distance matrix which is updated stage by stage. The results are generally expressed by a Dendrogram or Tree-diagram which reports the clustering analysis member by member.

In this work the Agglomerative Hierarchical Cluster algorithm has been applied to evaluate the relational behavior of the monitored benchmarks of the landslide in terms of entity of displacements. The N dimensions of the vectoral space were defined by the monitoring period which is taken into consideration by the analysis. Each dimension is a survey of the benchmarks position and thus a calculation of the cumulative benchmark's displacement.

The most common kind of tree cutting, branch cutting, or branch pruning is based on the static selection of a distance threshold which groups according to the contiguous branches below (Langfelder et al. 2008). In this analysis three different cutoff methods have been proposed. They allow to group benchmarks until the threshold of the smallest maximum displacements distance is larger than two standard deviations beyond the mean value of the distance matrix (Lee and Grauman 2015). The cutoffs are expressed as follows:

$$h_1 = \overline{M_t} + 2 \cdot \sigma_t(M_t) \quad (3a)$$

$$h_2 = \overline{M_i} + 2 \cdot \sigma_i(M_i) \quad (3b)$$

$$h_3 = \overline{M_i} + 2 \cdot \sigma_p\left(M_p\right) \quad (3c)$$

where M_t is the distance matrix based on the entire investigation period, M_i is the distance matrix of the single time-lapse and p is the progressive distance matrix of the p-th stage between 1 to m-1, $\bar{M}$ is the mean value of the relative distance matrix and σ is the standard deviation of the relative period. Therefore, while h_1 and h_2 are static cutoffs, h_3 is dynamic because of related to the progresses of the clustering.

The robustness and efficacy of this method in sections identification of a large landslide that are becoming more active with respect of the benchmarks of the same geomorphological feature, such as main scarp or lateral deposits, was tested. The prospected advantages of using this technique are: (i) the method is robust with respect of spikes due to instrumental or recording errors; (ii) the algorithm does not require calibration; (iii) even though the algorithm is classified as machine learning which is not opaque as other methods such as Artificial Neural Networks. This method allows the user to weight better the outputs and avoids problems with overfitting.

Study Area

In the Rotolon catchment (eastern Italian Alps—45°42′53.9″ N 11°10′20.6″E) a large Deep-seated Gravitational Slope Deformation (DGSD) is located, covering an area of 470.000 m^2 (Fig. 1). The activity of the DGDS causes the collapse of portions of the slope than then evolve into debris flow threatening the settlement (Turcati) located about 4 km downstream. In 2010, 340.000 m^3 of loose debris detached from the frontal part of the DGSD and flowed into the draining channel, damaging a bridge and almost overflooding near the houses.

Since the emergency phases, the landslide has been surveyed, monitored (Frigerio et al. 2014) and broadly studied

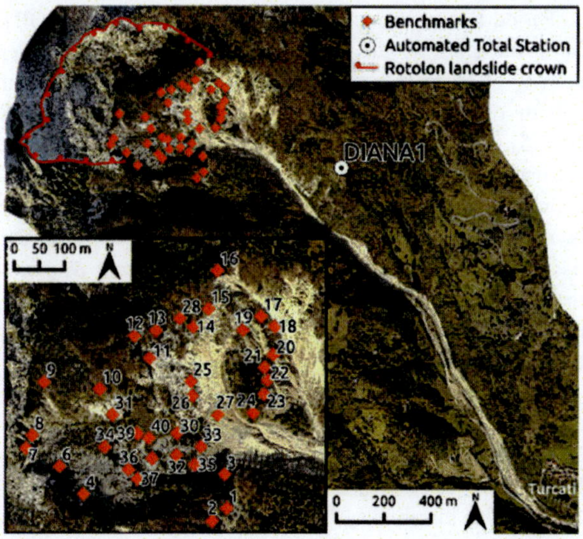

Fig. 1 Study area and monitoring system

by means of multi-temporal LiDAR DTMs and numerical modelling (Bossi et al. 2015a). The superficial displacements of the unstable slope are monitored by a series of wire extensometers and an Automated Total Station (ATS). An Early Warning System, consisting of trip wires, avalanche pendulums and a Decision Support System (DSS), has also been installed to alert the local population in case of major debris flow events (Bossi et al. 2015b).

The landslide kinematic is characterized by constant activity but with different rates associated with different sector of the large source area. For early warning purposes it is extremely important to identify areas with increasing activity with respect to points located in the same geomorphological unit to assess how much material may be involved in the next debris flow event and if a debris flow should be expected soon

Monitoring Data

The topographic station "Diana 1" for the measurement of the superficial displacements of the landslide body was first activated on July 12th 2012. The deployment setting consists of a log cabin built over a concrete slab anchored by micro-piles to the bedrock on the left margin of the mass movement and has an excellent view of the upper part of the landslide. The cabin offers shelter to the topographic monitoring system consisting of an Automated Total Station (ATS) and its management component, equipped with a computer, a modem and a power supply unit.

The topographic monitoring system is formed by 42 passive reflectors, 37 installed over the landslide body and the remaining 5 on stable areas around the slope.

The topographic network operated for two periods: the first one of 18 months between August 2012 and March 2014 and the second one for 17 months between August 2015 and December 2016 (Fig. 2) During the monitoring period some prism went lost due to theft or collapse of parts of the landslide (Table 1).

The ATS has an angular precision of 0.15 mgon and it is equipped with an automatic pointing and collimation system able to identify a prism with an accuracy of ± 7 mm at 3,000 m distance, characteristics that allow to determine a point location from a single reference station with high accuracy. The most distant prism is located at 1,152 m from the topographic station. The ATS measures two angles (horizontal and vertical) and a length (the distance) at each target location. These quantities are affected by systematic errors, attributable to the sensors mechanics, and random errors as result of atmospheric inhomogeneity among the measurements. The former was eliminated by targeting each reflector 4 times (2 straight and 2 reverse readings) each measurement cycle. The latter were reduced through the

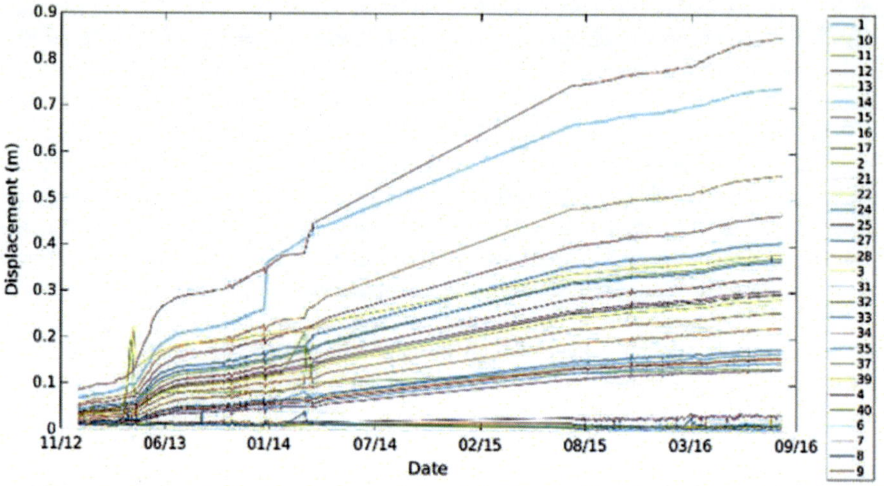

Fig. 2 Cumulated displacements of the 30 benchmarks

Table 1 Prisms no longer visible and the date of the last measurement recorded at their position by the ATS

Prism #	Last measure
14	30/08/2016
17	26/08/2016
18	06/03/2016
19	27/01/2014
20	07/03/2014
26	05/03/2014
30	18/04/2014
36	16/02/2014
38	29/01/2014
R5	18/04/2014

implementation of a calculation model (single station orientation with the application of a scale factor) which considers the 5 benchmarks that surround the landslide body as reference points with fixed coordinates. The system ensures a day and night continuity of the surveys and perform a measurement cycle every 6 h.

Results

The monitoring period considered by the analysis is encompassed between December 2012 and August 2016 and the analysis is based on the data provided by 30 benchmarks in time intervals of 90 days.

Through the analysis of the first dendrogram (Fig. 3A) it is possible to appreciate that the hierarchical clustering identifies automatically groups of similarly moving benchmarks and that these groups represent specific geomorphological features of the DGDS. This occurs even though the approach is blind to the location of all the benchmarks and analyses just their trend of displacement. The method is, thus, geographically validated since clustered benchmarks are located in contiguous areas. For example, as shown in Fig. 3A, benchmarks in cluster 2 are all placed on the main scarp and they all move together (Fig. 4), cluster 1 is in a secondary more active slide with benchmarks 10 and 12 are on the scarp, cluster 3 is formed by points located on rocky outcrops and are almost stable, clusters 4, 5 and 6 are in the most disaggregated part of the source area.

This result could be achieved also with a qualitative analysis, but it would be a procedure operator-dependent, less robust and time consuming, especially for large numbers of benchmarks

Period B was characterized by intense rainfall (Fig. 3B). In dendrogram B the re-grouping highlights the greater activity of adjacent benchmarks 14 and 15 and the similar pattern of the new group formed by points 25 and 28 (Fig. 5). We suggest that regrouping may be used for issue warnings for the operators that may then update risk scenarios on the base on the new volumes of material that may be evolve into a debris flow. This is particularly useful for the monitoring of large landslides with different displacement patterns where the definition of different warning thresholds for different zones can be difficult and, anyway, too much operator-dependent.

The observation period for the hierarchical clustering analysis may of course be changed in function of the monitoring purpose. In Fig. 5 we show that benchmark 14, in the last days of 2013, experienced an abrupt displacement of 10 cm. The hierarchical clustering performed in a month

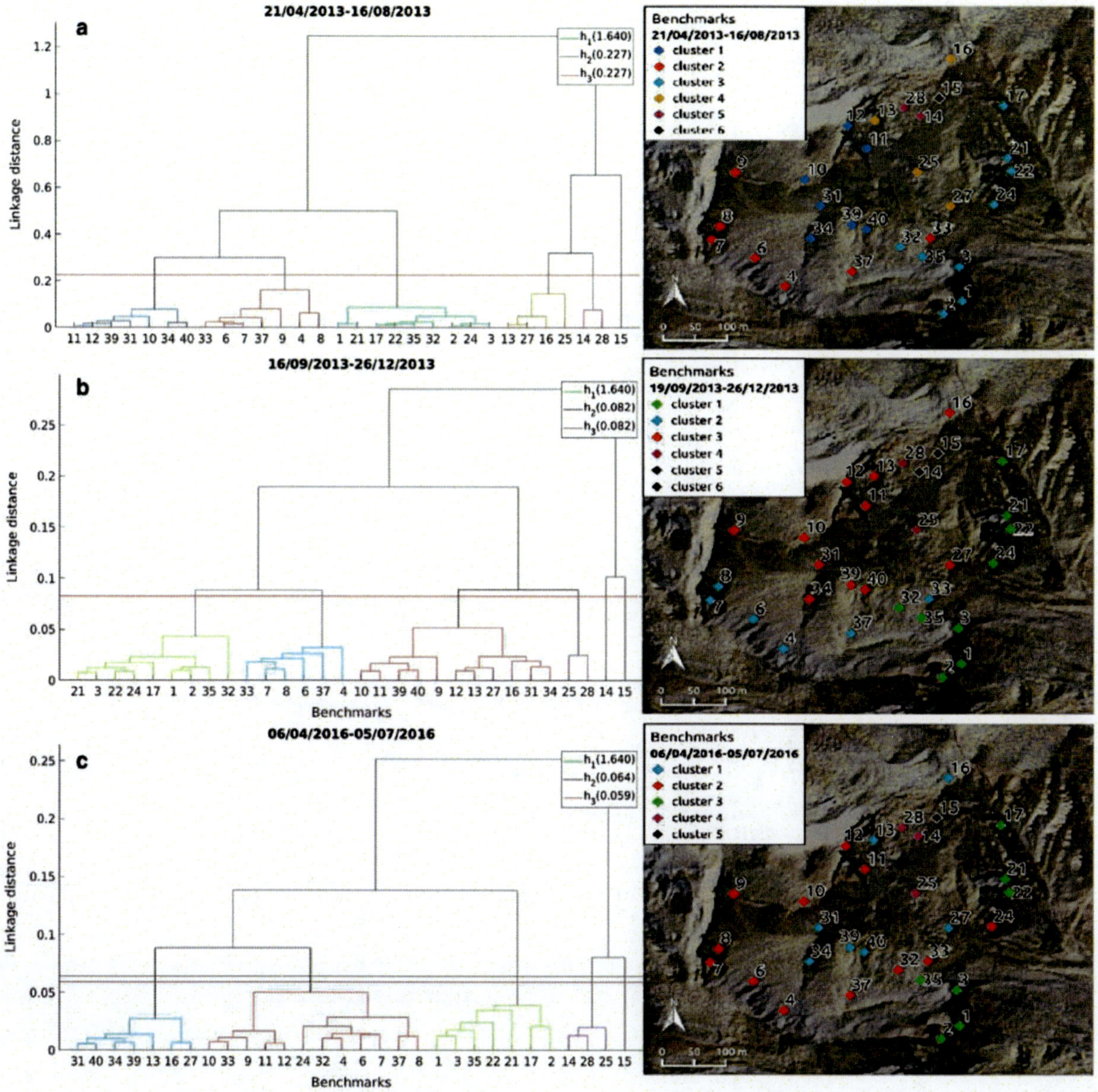

Fig. 3 Dendrograms of the selected periods reporting the clusters based on *h3* cutoff. On the right of each dendrogram the clusters in the source area. The clusters numbering (cluster 1, cluster 2, etc.) is based on the left to right order of the dendrogram clusters

period extracted automatically the peculiarity of benchmark 14 displacement pattern (Fig. 6).

With a fixed threshold and many benchmarks to follow a displacement of 10 cm may be overseen by an operator. However, it is a sign that something is happening in the landslide and an automatic and adaptive detection of these changes may be very useful. In fact, the area below benchmark 14 and 15, a deposit of loose debris, was the main source area for the small debris flows that occurred in the following year.

After this event benchmark 14 did not experienced any new acceleration and re-grouped with benchmark 28 until 2016 (Fig. 3C).

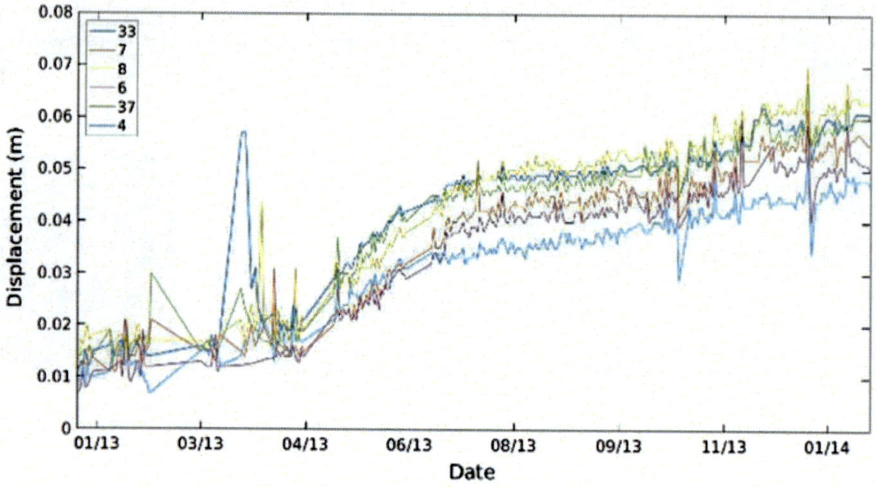

Fig. 4 Temporal displacement pattern of cluster 2 of dendrogram A

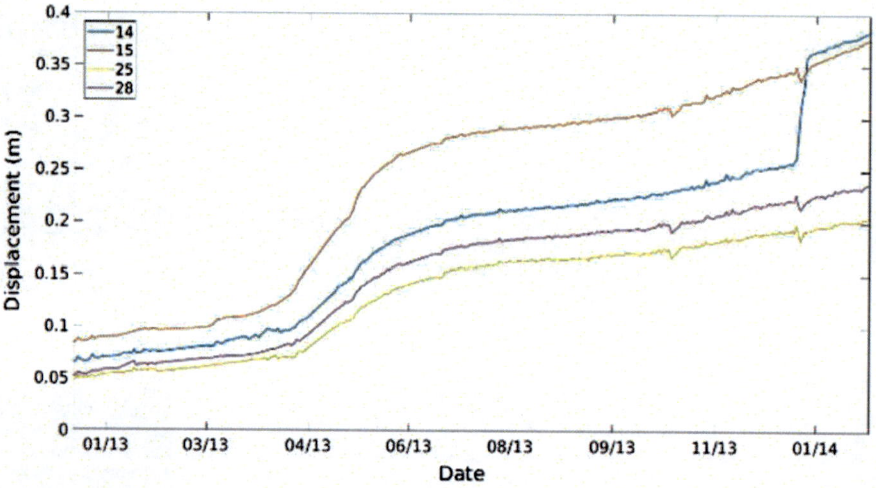

Fig. 5 Cumulative 3D displacement pattern in the year 2013 of benchmarks 14, 15, 25 and 28

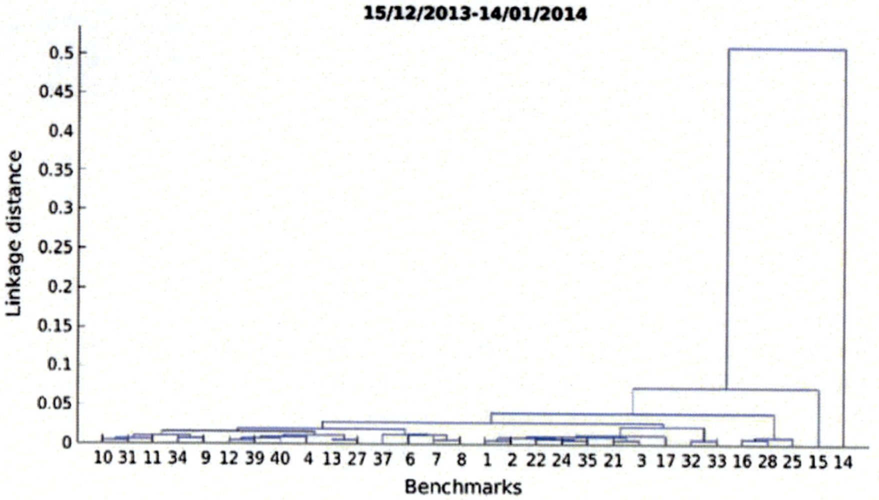

Fig. 6 Dendrogram of a month period in concomitance with the abrupt displacement of benchmark 14

Conclusions

The technique here presented supports the definition of an operator-independent procedure to define areas with similar displacement patterns. The proposed method does not need calibration and it is robust with respect to spikes linked to measurement errors (Fig. 4).

The method allows the issuing of warning when re-grouping occurs, that is when some benchmarks are experiencing displacement not coherent with their previous behavior. Moreover, the possibility to identify through an unsupervised procedure areas with an homogeneous kinematic may guide the production of data-driven geomorphological maps. That would lead to a more precise estimation of the volume of possible new debris flow phenomena and therefore to produce reliable risk scenarios. Besides, it may support a Decision Support System that triggers alarms when abrupt changes in the clustering pattern appear, as indicated by a variation in the style of activity of certain benchmarks.

References

Bossi G, Cavalli M, Crema S et al (2015a) Multi-temporal LiDAR-DTMs as a tool for modelling a complex landslide: a case study in the Rotolon catchment (eastern Italian Alps). Nat Hazards Earth Syst Sci 15:715–722. https://doi.org/10.5194/nhess-15-715-2015

Bossi G, Crema S, Frigerio S et al (2015b) The Rotolon Catchment early-warning system. In: Lollino G, Arattano M, Rinaldi M et al (eds) Engineering geology for society and territory, vol 3. Springer International Publishing, Cham

Bouguettaya A, Yu Q, Liu X et al (2015) Efficient agglomerative hierarchical clustering. Expert Syst Appl 42:2785–2797. https://doi.org/10.1016/j.eswa.2014.09.054

Carey JM, Moore R, Petley D et al (2007) Pre-failure behaviour of slope materials and their significance in the progressive failure of landslides. Landslides Clim Chang Challenges Solut Taylor Fr Gr London, UK 207–215

Carlà T, Nolesini T, Solari L et al (2019) Rockfall forecasting and risk management along a major transportation corridor in the Alps through ground-based radar interferometry. Landslides 16:1425–1435. https://doi.org/10.1007/s10346-019-01190-y

Day WHE, Edelsbrunner H (1984) Efficient algorithms for agglomerative hierarchical clustering methods. J Classif 1:7–24. https://doi.org/10.1007/BF01890115

Frigerio S, Schenato L, Bossi G et al (2014) A web-based platform for automatic and continuous landslide monitoring: the Rotolon (Eastern Italian Alps) case study. Comput Geosci 63:96–105. https://doi.org/10.1016/j.cageo.2013.10.015

Fujisawa K, Marcato G, Nomura Y, Pasuto A (2010) Management of a typhoon-induced landslide in Otomura (Japan). Geomorphology 124(3–4):150–156

Fukuzono T (1985) A new method for predicting the failure time of a slope. In: Proceedings of 4th international conference and field workshop on landslide, 145–150

Gülagiz FK, Sahin S (2017) Comparison of hierarchical and non-hierarchical clustering algorithms

Intrieri E, Carlà T, Gigli G (2019) Forecasting the time of failure of landslides at slope-scale: a literature review. Earth-Sci Rev 193:333–349

Iovine G, Petrucci O, Rizzo V, Tansi C (2006) The March 7th 2005 Cavallerizzo (Cerzeto) landslide in Calabria—Southern Italy. In: Proceedings of the 10th IAEG congress, Nottingham, Great Britain, pp 6–10

Langfelder P, Zhang B, Horvath S (2008) Defining clusters from a hierarchical cluster tree: the Dynamic Tree Cut package for R. Bioinformatics 24:719–720. https://doi.org/10.1093/bioinformatics/btm563

Lee YJ, Grauman K (2015) Predicting important {Objects} for Egocentric {Video} {Summarization}. Int J Comput Vis 114:38–55. https://doi.org/10.1007/s11263-014-0794-5

Loew S, Gschwind S, Gischig V et al (2017) Monitoring and early warning of the 2012 Preonzo catastrophic rockslope failure. Landslides 14:141–154. https://doi.org/10.1007/s10346-016-0701-y

Mazzanti P, Bozzano F, Cipriani I, Prestininzi A (2015) New insights into the temporal prediction of landslides by a terrestrial SAR interferometry monitoring case study. Landslides 12(1):55–68

Murtagh F, Contreras P (2012) Algorithms for hierarchical clustering: an overview. Wiley Interdiscip Rev Data Min Knowl Discov 2:86–97. https://doi.org/10.1002/widm.53

Petley DN (2014) The Mannen rockslide: still standing—the Landslide Blog—AGU Blogosphere. https://blogs.agu.org/landslideblog/2014/10/31/the-mannen-rockslide-2/. Accessed 13 Mar 2020

Wilks DS (2011) Chapter 15—{Cluster} {Analysis}. In: Wilks DS (ed) International {Geophysics}. Academic Press, pp 603–616

Predicting Rainfall Induced Slope Stability Using Random Forest Regression and Synthetic Data

Elahe Jamalinia, Faraz S. Tehrani, Susan C. Steele-Dunne, and Philip J. Vardon

Abstract

Water fluxes in slopes are affected by climatic conditions and vegetation cover, which influence the effective stress and stability. The vegetation cover is the intermediate layer between the atmosphere and the slope surface that alter water balance in the slope through evapotranspiration and leaf interception. This paper studies the data-driven approach for predicting the macro stability of an example grass-covered dike based on actual data and also synthetic data provided by numerical modelling. Two numerical models are integrated in this study. The water balance in the root zone is simulated through a crop model, whereas the hydro-mechanical and safety analysis of the example dike is done using a two-dimensional Finite Element model. The considered period for these analyses is 10 years (3650 daily instances) which will be used to generate a time-series dataset for a secondary dike in the Netherlands. The features included in the dataset are parameters that (i) have a meaningful relationship with the dike Factor of safety (FoS), and (ii) can be observed using satellite remote sensing. The output dataset is used to train a Random Forest regressor as a supervised Machine Learning (ML) algorithm. The results of this proof-of-concept study indicate a strong correlation between the numerically estimated FoS and the ML-predicted one. Therefore, it can be suggested that the utilized parameters can be used in a data-driven predictive tool to identify vulnerable zones along a dike without a need for running expensive numerical simulations.

Keywords

Slope stability • Vegetation • Machine learning

Introduction

The main components of flood protection system in the Netherlands are primary and secondary dikes with the total length of more than 18,000 km. The condition of these engineering structures is assessed based on the infrequent visual inspections usually through ground-based observations. This current method can be systematically augmented by using Earth Observation (EO) data to evaluate the dike condition (Jamalinia et al. 2019a, b; Özer et al. 2018). One crucial aspect of slope stability analysis is the identification of critical points along the slope.

In geotechnical engineering, the analysis and prediction of (in)stabilities is of great importance; however, often little attention is paid to the transient conditions due to vegetated cover and interaction with the environment. This is due to the computational intensity and difficulty in collecting in situ information on the condition of the slope.

Synthetic data driven approaches based on Machine Learning (ML) can be used to develop an efficient estimation of the slope condition and speed up the assessment process, even at the regional scale. In recent years, ML methods have been used in several studies for predicting slope (in)stability (Ada and San 2018; Ghorbanzadeh et al. 2019; Lin et al. 2018; Pourghasemi and Rahmati 2018).

In this research, a Random Forest (RF) approach is used to build and train an ML model on 3650 synthetic data points produced by an integrated crop-geotechnical model on an example geometry. The results show the potential

E. Jamalinia (✉) · S. C. Steele-Dunne · P. J. Vardon
Faculty of Civil Engineering and Geoscience, Delft University of Technology, Stevinweg 1, Delft, 2600, The Netherlands
e-mail: e.jamalinia@tudelft.nl

S. C. Steele-Dunne
e-mail: s.c.steele-dunne@tudelft.nl

P. J. Vardon
e-mail: p.j.vardon@tudelft.nl

F. S. Tehrani
Geo Department, Deltares, Delft, 2600, The Netherlands
e-mail: Faraz.Tehrani@deltares.nl

Ž. Arbanas et al. (eds.), *Understanding and Reducing Landslide Disaster Risk*, ICL Contribution to Landslide Disaster Risk Reduction,
https://doi.org/10.1007/978-3-030-60713-5_24

application of Earth Observation (EO) data for identifying the vulnerable slopes (locations along the slope) without the need for repeating expensive numerical simulations.

Methodology

An integrated crop-geotechnical model (Jamalinia et al. 2020a) is used in this research to calculate the Factor of Safety (FoS) of a dike under realistic climate and vegetation conditions for ten years (daily analysis). The results are used to study the possibility of using ML algorithms to forecast a slope condition based on the observable data from climate, vegetation and slopes.

Numerical Method

In Fig. 1 the geometry of the example dike is shown. This idealised dike is a typical regional Dutch dike (de Vries 2012), which is covered by permanent grass over the surface of the dike with a fixed depth of root zone, 40 cm (shown as green area in Fig. 1). It is assumed that the base boundary is an impermeable layer, while other boundaries of the dike are assumed to be permeable, meaning that the left and right sides have a fixed phreatic surface and the top boundary has a temporal precipitation/evaporation flux applied.

Since standard geotechnical models do not simulate the (dynamic) effects of vegetation, i.e. evapotranspiration and leaf interception, on mass balance and thereby slope stability, the current research utilises an integrated crop-geotechnical model developed by the authors by integrating two existing models (Jamalinia et al. 2020a), although other academic modelling approaches have considered various aspects of the impact of vegetation (Elia et al. 2017). Using this numerical approach enables the study of climatic and vegetation conditions on the stability. The influence of the soil cracking, due to droughts and reducing shear strength, is included in our previous studies (Jamalinia et al. 2019b, 2020a, b). The workflow (Fig. 2) is controlled by Python and explained in detail.

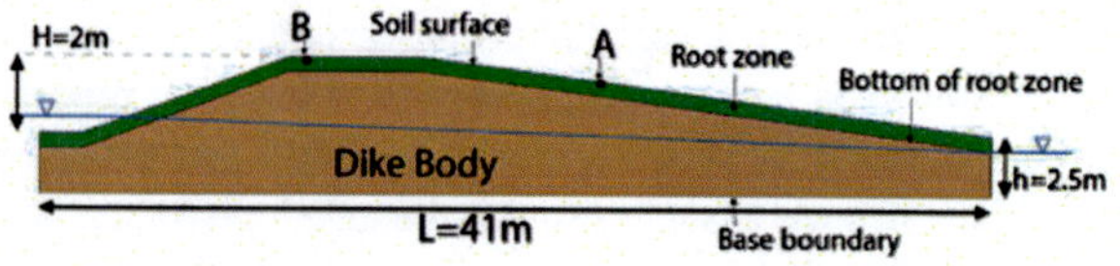

Fig. 1 Geometry representing boundary, root zone layer, and the analysis point

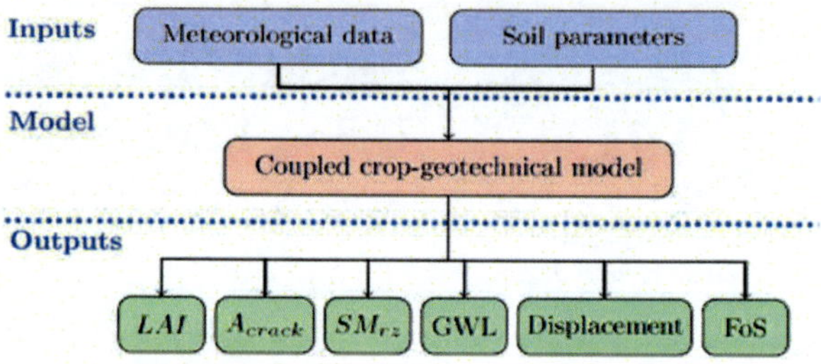

Fig. 2 Flow chart of numerical modeling procedure

The meteorological data (e.g. rain and temperature) and soil parameters are inputs for the integrated crop-geotechnical model. The climate data was obtained from the Royal Netherlands Meteorological Institute (KNMI) at Schiphol Airport station (Amsterdam), which is close (circa 9 km) to the location of the actual dike. The major outputs from the 1D crop model (LINGRA) (Bouman et al. 1996; Rodriguez et al. 1999), shown in Fig. 2 are: Leaf Area Index (LAI), area of leaves divided by the area of ground; crack area (A_{crack}); average soil moisture in the root zone (SM_{rz}). The major outputs from the 2D geotechnical model, Plaxis (2018), are the ground water level (GWL), surface displacement and FoS.

The input parameters for the crop model and the geotechnical model are listed in Tables 1 and 2, respectively.

Data Driven Method

In this study the results of 3650 realisations (simulations) from the integrated crop-geotechnical model, each simulating a period of 10 years from 2009 to 2019, are used in training and testing a RF regressor to predict the safety condition of the example dike. The Random Forest approach is one of the most widely used ensemble learning algorithms. The RF (Breiman 2001) constructs individual Decision Trees (DTs) based on bagging, using bootstrap sampling where samples are taken randomly with replacement from the training set (Qi and Tang 2018). In the DT method the data is divided into smaller subsets and a tree is expanded until the leaf node, where the decision is made about the target value or class in DT regression or DT classification. As the RF method uses the training dataset to create multiple decision trees, the variance of the final model is reduced and then it is less sensitive to over-fitting (Burkov 2019). Each decision tree of the RF predicts an output and RF regression models take the average of all the individual decision tree estimates.

Table 1 Input parameters used for the crop model, modified after Jamalinia et al. (2020a)

Parameters		Value	Unit
Soil	Water content at field capacity (prior to cracking)	0.29	$cm^3 water/cm^3 soil$
	Water content at the wilting point below that wilting starts	0.12	$cm^3 water/cm^3 soil$
	Critical water content below that transpiration is reduced	0.05	$cm^3 water/cm^3 soil$
	Maximum drainage from root zone to lower layers	50	mm/day
Vegetation	Specific Leaf Area: leaf area over leaf mass	0.025	m^2/g
	Remaining LAI after mowing	0.8	$m^2 leaf/m^2 soil$
	Critical leaf area beyond that self-shading occurs	4	$m^2 leaf/m^2 soil$

Table 2 Input parameters for the geotechnical model, modified after Jamalinia et al. (2020a)

Parameters		Value		Unit
		Root zone	Dike body	
Constitutive model (Mohr-Coulomb)	Saturated unit weight	20	12	kN/m^3
	Friction angle (prior to cracking)	23	23	°
	Cohesion (prior to cracking)	2	2	kPa
	Dilatancy angle	0	0	°
	Young's modulus	10	20	MPa
	Poisson's ratio	0.3	0.2	–
	Initial void ratio	0.67	1.2	–
Hydraulic model (van Genuchten*)	Hydraulic conductivity	0.14	0.03	m/day
	Scale parameter (α)	1.47	1.38	m^{-1}
	Fitting parameter (n)	1.97	1.32	–
	Fitting parameter (m)	0.87	−1.24	–

*Hysteresis is not considered

Results

Numerical Simulations

Example temporal inputs and outputs of the numerical simulations are shown in Fig. 3. Time-series of rainfall and temperature (TMP) as climate data is shown in Fig. 3a, b, respectively.

In Fig. 3c–e the variation of crop model outputs over time is shown. Considering the worst-case scenario, it is assumed that cracks do not close during the wet periods, but only expand during unprecedented drier conditions (Fig. 3c). The percentage of the cracked soil area increases in such dry periods and its area remains constant until the next drier period. It is assumed that the cracking happens only in the root zone area with the maximum depth of 40 cm, equal to the root zone depth. The sudden decrease in LAI on 15 June and 15 August annually in Fig. 3d shows mowing events, which were imposed in the crop model based on the mowing schedule of secondary dikes in the Netherlands (Jamalinia et al. 2019a). A higher presence of cracks causes a reduction in the rate of LAI growth after mowing. In the summer of 2018, according to Fig. 3e, the root zone experienced the driest condition during the previous 10 years, and the crack area reached the maximum value during the simulation period. In Aug. 2018, the root zone soil moisture (SM_{rz}) reached its minimum value, and it can be seen that vegetation could not easily re-grow after mowing.

The temporal variation of absolute surface displacement $|U_A|$ at point A (Fig. 1) and FoS are selected as outputs of the FEM model (2D geotechnical model), shown in Fig. 3f, g, respectively. Displacement at point A follows the variation of SM_{rz}, which reflects the response of the $|U_A|$ to the climate and vegetation conditions. The combined effect of rainfall, LAI variations, and crack area influence the water flux into the dike which caused temporal variations of FoS. The maximum crack area in August 2018 and very low LAI

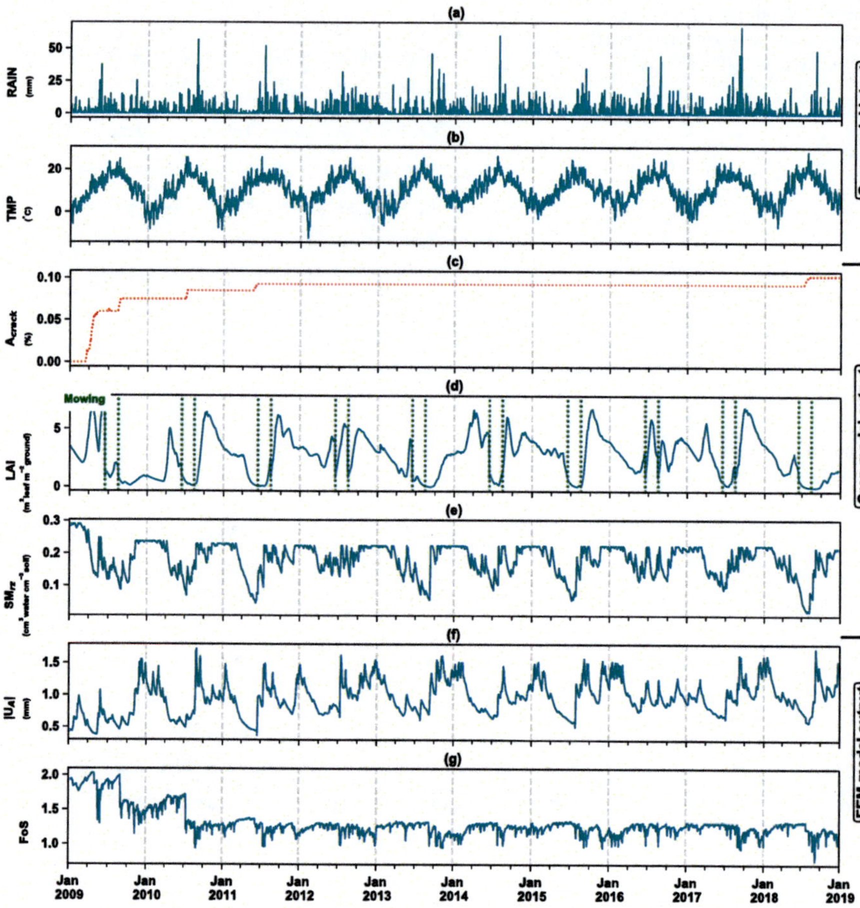

Fig. 3 Time-series of inputs and outputs from the developed integrated models for 10 years

(almost bare soil) caused an increased infiltration due to heavy precipitation event. In addition, the soil had the lowest shear strength due to the maximum crack area, which together led to the minimum FoS.

Feature Selection

There are two criteria to select the features in this study to train the RF regressor: the feature (1) has a strong, meaningful relationship with the FoS; (2) is observable remotely, so experts can monitor these parameters and assess the slope condition based on that feature. Therefore, the features in this study are from (i) climate: rainfall and temperature, (ii) vegetation: LAI, observing anomalies in vegetation could be used as an indicator to distinguish whether a dike is significantly cracked; (iii) slope surface displacement: it can be used as a proxy for both saturation (short term changes) and for accumulation of cracks (long term changes), although long term changes may also indicate subsidence or other processes (Jamalinia et al. 2020a). Using the PSInSAR method Ferretti et al. (2001), it is possible to map surface deformation with millimetre precision.

The lag correlation between pair of key parameters is plotted in Fig. 4. A positive lag means the second term causes the first one. There is a 15 days lag between LAI and Sat_{rz}, which means that root zone saturation affects vegetation growth most after 15 days. There is a strong correlation between saturation at point A (Sat_A) and $|U_A|$, which shows that surface displacement is responsive to the available water in the root zone, which is mentioned in the time-series result as well. Existing correlation in Fig. 4a, b suggest that using

LAI and $|U_A|$ could be good indicators for available water near to the dike surface, and both are reasonably easy to monitor remotely, unlike the SM. The negative correlations between (Fos, SM_{rz}) and (Fos, $|U_A|$) shown in Fig. 4c, d, suggest using $|U_A|$ as an indicator to estimate safety. The cumulative rainfall during the 35 days before an event day, Rain.cu_35 (Fig. 4f), has a stronger correlation with FoS than rainfall on the same day (Fig. 4e). This period has showed the best predicted FoS among other periods (Jamalinia et al. 2020c). Therefore, in the RF analysis a history of rainfall is considered.

Random Forest Regression

The 10-year simulation results from 2009 to 2019 are used to build a predictive model using the mentioned features in previous section. The data set is split to training set (70% of dataset) and testing set (30%), and the number of trees in the RF algorithm set to be 1000.

Here the ability of RF for real time prediction is tested. The features are selected from the same day at which the FoS is calculated, except for rainfall that accumulation during last 35 days is considered. The feature importance values are plotted in Fig. 5 which are derived from the RF regressor as a result of training processes. It turns out that the absolute surface displacement $|U_A|$ has the highest importance (0.52). LAI and cumulative precipitation during the last 35 days (Rain.cu-35) have almost the same feature importance of 0.2, and daily temperature (TMP), has the least effect on FoS and therefore its prediction. As mentioned before, according to time-series results and correlation, vegetation growth and displacement are affected by precipitation, so precipitation impact is embedded in LAI and $|U_A|$.

The predicted FoS from the RF method is plotted against the calculated FoS from the FEM model in the numerical

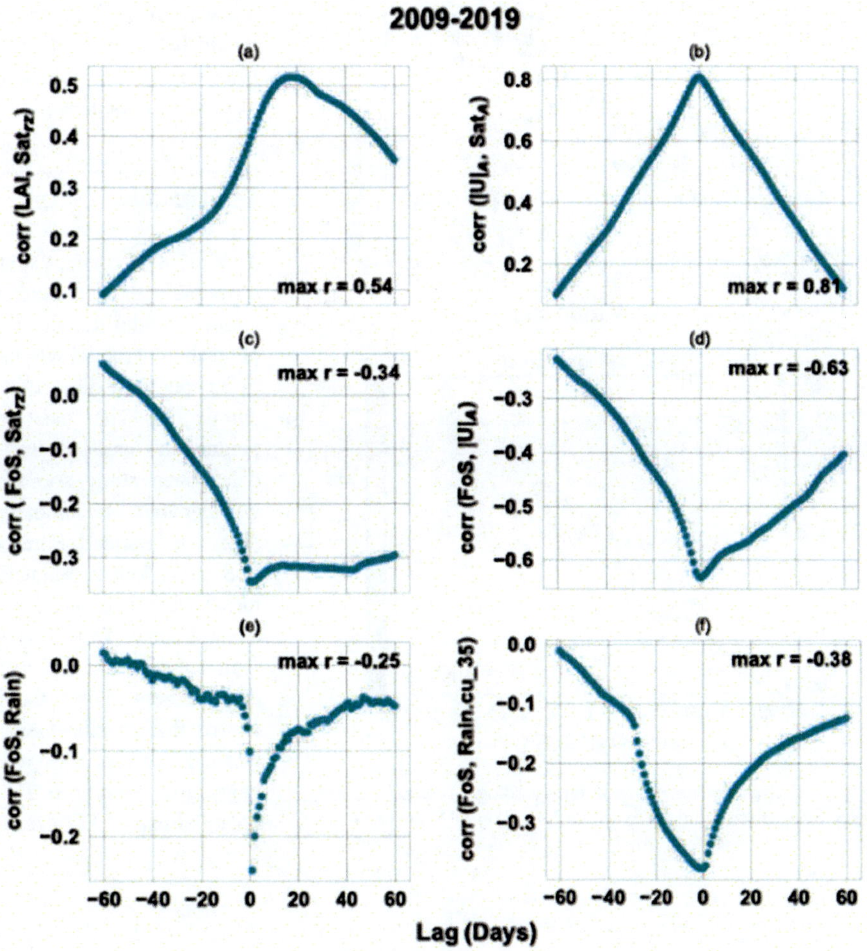

Fig. 4 Lag correlation between pair of key parameters

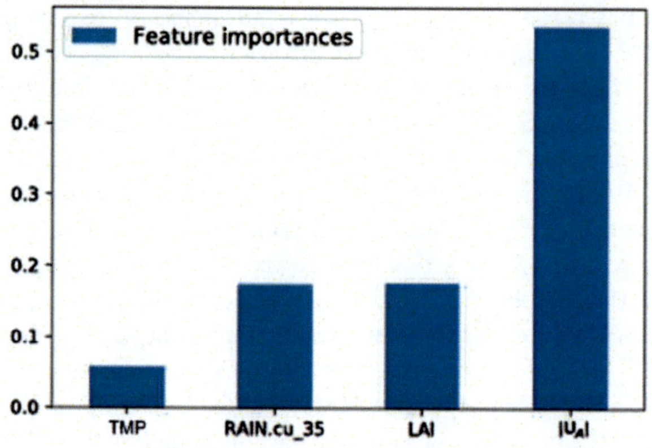

Fig. 5 Feature importance out of RF regression

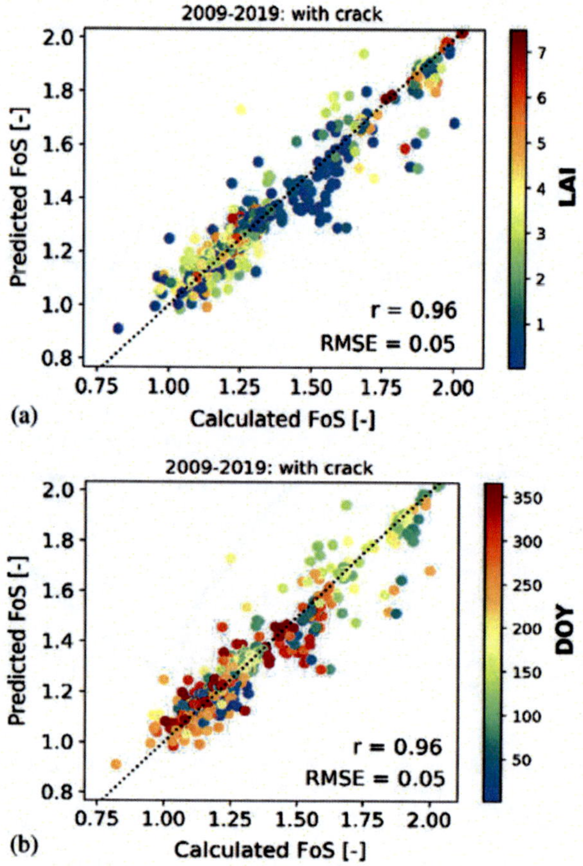

Fig. 6 Correlation between real time predicted FoS and calculated FoS. Scatters are colorized by (a) LAI, (b) Day of Year (DOY)

analysis in Fig. 6. The results show that the RF model yields precise estimation for assessing dike safety only from the observable data. The scatters are colorized by LAI values and day of year (DOY) which suggests that usually at low LAI and winter period (e.g. when LAI is lower than 2 during November to February) outliers occur, where there is not enough energy available for vegetation to grow.

In another analysis, the time-series prediction has been carried out to investigate the ability of the RF method to predict the future FoS from historical data. The training data set has been collected from the first 70% of the dataset and the remaining 30% used as the test set. So, the model is built based on avoiding random train, test split selection. The predicted and calculated FoS are plotted against time in Fig. 7. The temporal trends are well represented, with deviation in the low values and after the cracking event during the summer of 2018. The results demonstrate that RF can be used as a promising method to predict slope condition using observable input data: meteorological data, vegetation and surface displacements. Therefore, doing a numerical analysis for a slope and calculating FoS for a time period would help experts to assess the condition of the slope in future using these observable parameters, without the need to repeat time-consuming simulations.

Conclusion

This proof of concept study investigates the potential use of observable data in predicting slope condition. A one-way coupled model framework composed of a crop model and a geotechnical model was used to calculate the factor of safety of an idealised dike covered with grass for 10-year period simulation. The existing correlation between selected parameters assisted in the feature selection for this data-driven study approach, as well as an assessment of whether they are remotely observable. The supervised ML algorithm, Random Forest (RF), has been used for predicting FoS using key parameters such as: precipitation, temperature, LAI and surface displacement at a selected point on the example dike. The RF algorithm results in a prediction with high accuracy (RMSE = 0.05). Among the features, surface displacement shows the highest feature importance. It is shown that displacement is responsive to the amount of water in the root zone which is affected by the climate and vegetation condition. The results of this study show the potential use of EO data for real time monitoring of slopes

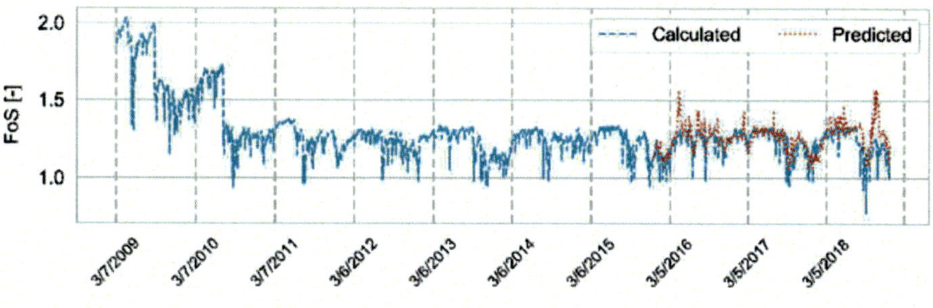

Fig. 7 Time-series prediction of FoS

and detecting the vulnerable locations along the slopes. The results show some deviation, probably due to the strong non-linearities in the physical model, therefore the worth of the RF model is to identify weak areas and allow further detailed investigation.

Acknowledgements This work is part of the research program Reliable Dikes with project number 13864 which is financed by the Netherlands Organisation for Scientific Research (NWO).

References

Ada M, San BT (2018) Comparison of machine-learning techniques for landslide susceptibility mapping using two-level random sampling (2LRS) in Alakir catchment area, Antalya, Turkey. Nat Hazards 90 (1):237–263

Bouman BAM, Schapendonk AHCM, Stol W, van Kraalingen DWG (1996) Description of the growth model LINGRA as implemented in CGMS. In: Quantitative Approaches in System Analysis, vol 7. DLO Research Institute for Agrobiology and Soil, Wageningen, the Netherlands

Breiman L (2001) Random forests. Mach Learn 45(1):5–32. Springer

Burkov A (2019) The hundred—page machine learning. In: (part of title) And add publication details, vol 1. ndriy Burkov Quebec City, Canada

de Vries G (2012) Monitoring droogteonderzoek veenkaden, Report no. 1203255-006-GEO-0001-gbh, Deltares, Delft

Elia G et al (2017) Numerical modelling of slope–vegetation–atmosphere interaction: an overview. Q J Eng Geol Hydrogeol 50:249–270

Ferretti A, Prati C, Rocca F (2001) Permanent scatterers in SAR interferometry. IEEE Trans Geosci Remote Sens 39(1):8–20

Ghorbanzadeh O, Blaschke T, Gholamnia K, Meena SR, Tiede D, Aryal J (2019) Evaluation of different machine learning methods and deep-learning convolutional neural networks for landslide detection. Remote Sens 11(2):196

Jamalinia E, Vardon PJ, Steele-Dunne SC (2019a) Can vegetation indices predict slope (stability) conditions? In: Geophysical Research Abstracts, vol 21, pp 55–64

Jamalinia E, Vardon PJ, Steele-Dunne SC (2019b) The effect of soil–vegetation–atmosphere interaction on slope stability: a numerical study. Environ Geotech ahead of print 1–12

Jamalinia E, Vardon PJ, Steele-Dunne SC (2020a) The impact of evaporation induced cracks and precipitation on temporal slope stability. Comput Geotech 122:103506

Jamalinia E, Vardon P, Steele-Dunne S (2020b) Use of displacement as a proxy for dike safety. Proc Int Assoc Hydrol Sci 382(1):481–485

Jamalinia E, Tehrani FS, Steele-Dunne SC, Vardon PJ (2020c) A data-driven approach for stability forecasting of dikes. in prepara

Lin Y, Zhou K, Li J (2018) Prediction of slope stability using four supervised learning methods. IEEE Access 6:31169–31179

Özer IE, van Leijen FJ, Jonkman SN, Hanssen RF (2018) Applicability of satellite radar imaging to monitor the conditions of levees. J Flood Risk Manag 12(Suppl. 2):e12509

Plaxis BV (2018) PLAXIS reference manual 2018. Delft, Netherlands

Pourghasemi HR, Rahmati O (2018) Prediction of the landslide susceptibility: which algorithm, which precision? Catena 162:177–192

Qi C, Tang X (2018) Slope stability prediction using integrated metaheuristic and machine learning approaches: a comparative study. Comput Ind Eng 118(February):112–122

Rodriguez D, van Oijen M, Schapendonk AHMC (1999) LINGRA-CC: a sink-source model to simulate the impact of climate change and management on grassland productivity. New Phytol v 144(2):359–368

Automatized Dissemination of Landslide Monitoring Bulletins for Early Warning Applications

Daniele Giordan, Aleksandra Wrzesniak, Paolo Allasia, and Davide Bertolo

Abstract

Complex landslides are often monitored by multi-instrumental networks. These networks can be coupled with early warning systems that could reduce human, economic and environmental losses. The large amount of data provided by landslides monitoring networks can create issues related to the data management and processing. The use of different monitoring instruments can create problem in data interoperability and in the definition of multi-source landslide activity maps. In particular during emergencies, when monitoring data are a crucial element to support decision makers and to inform the population about the evolution of the slope instability, the communication of monitoring results should be managed using a dedicated communication strategy. In this paper, we present the developed communication strategy based on the use of a dedicated single page bulletin. This bulletin has been developed and tested during the Mont de La Saxe rockslide emergency and it has supported the landslide management team to inform the population about the evolution of the slope instability.

Keywords

Landslide monitoring • Early warning • Monitoring results dissemination • Risk management

Introduction

Landslides often have a strong impact on the population in many parts of the World (Blaikie et al. 1994; UNISDR 2006; Petley 2012; Alcántara-Ayala 2016). They cause injuries, deaths, economic losses, damages to properties and infrastructures. There are many ways to decrease the impact of landslides (Pecoraro et al. 2019). However, when a landslide threatens highly populated areas, delocalization, landslide stabilization or engineering structures may be hard to implement. The application of monitoring systems could be considered often the only solution (Giordan et al. 2019). Such systems improve the knowledge of the phenomena behaviour, and they can be adopted for early warning applications (Bell et al. 2009; Michoud et al. 2013; Glade and Nadim 2014; Thiebes and Glade 2016). They also have lower economic costs and environmental impact respect to the other solutions (Intrieri et al. 2012, Frodella et al. 2018; Wieczorek and Snyder 2009; Thiebes and Glade 2016). Early warning applications are also considered as essential elements of disaster risk reduction. Thanks to advances in monitoring technology, especially in the real-time transmission of monitoring data, these systems became more complex and precise and, therefore applied worldwide (UNISDR 2007, 2015).

Complex landslides are usually monitored by multi-instrumental networks (e.g., Ayalew et al. 2005, Crosta et al. 2014). In the literature one can find the use of GPS (Notti et al. 2020), RTS (Allasia et al. 2019), GB-InSAR (Lombardi et al. 2017), in-place and robotized inclinometers (respectively, Simeoni and Mongiovì 2007; Allasia et al. 2018), among many others. The monitoring systems provide a large

D. Giordan · A. Wrzesniak (✉) · P. Allasia
National Research Council, Research Institute for Geo-Hydrological Protection, Strada Delle Cacce 73, Turin, 10135, Italy
e-mail: aleksandra.wrzesniak@irpi.cnr.it

D. Giordan
e-mail: giordan@irpi.cnr.it

P. Allasia
e-mail: paolo.allasia@irpi.cnr.it

D. Bertolo
Geological Survey of Aosta Valley Region, Loc. Amérique 33, Quart, 11020, Italy
e-mail: d.bertolo@regione.vda.it

Ž. Arbanas et al. (eds.), *Understanding and Reducing Landslide Disaster Risk*, ICL Contribution to Landslide Disaster Risk Reduction,
https://doi.org/10.1007/978-3-030-60713-5_25

number of accurate measures at a high sampling rate. However, such an amount of data can create issues related to their management and processing (Giordan et al. 2019). This requires automatization procedures and informatics tools for data acquisition, elaboration, validation and dissemination, and for early warning proposes (Allasia et al. 2013). While the attention has focused on technological improvement, (which in fact enabled quick automatic data acquisition, generation of warning and dissemination of monitoring results), the aspect of scientific data communication was omitted.

In this field, the communication aspect plays an essential role because hazard management is an interdisciplinary activity, which requires collaboration among many people (population, politicians, decision-makers, experts, journalists, etc.).

The hazards management, through the application of early warning systems, should reduce human, economic and environmental losses. According to the studies conducted by the United Nations, each early warning system (EWS), to be effective, must be human-centred and should integrate four elements: (i) knowledge of the risks faced; (ii) technical monitoring and warning service; (iii) dissemination of meaningful warnings to those at risk; (iv) public awareness and preparedness to act (UNISDR 2006). The failure of any of these elements can cause the collapse of the whole system. Furthermore, it was noticed that each of those sub-systems is often considered as an individual component, decreasing the EWS efficiency (Garcia and Fearnley 2012; Alcántara-Ayala and Oliver-Smith 2017).

Considering the presented issues and requirements, we propose a human-centred communication concept, suitable for non-expert receivers and helpful in the information flow among all groups involved in hazard management.

In this paper, we present an automatized approach for landslide monitoring results representation and dissemination in the form of a single-page bulletin, made of synthetic and simplified information about the landslide status.

Materials and Methods

Since 2012, we have collaborated with Geological Survey of Aosta Valley to manage the Mont de La Saxe landslide (hereinafter La Saxe). La Saxe is a large and complex slope instability (ca.8 $\times$ 10^6 m^3) located in the region of the Aosta Valley in northern Italy (Crosta et al. 2014). This landslide is the most hazardous in this part of the country, extensively investigated and monitored since 2009. We have applied here our approach for monitoring network management called ADVICE (Allasia et al. 2013). ADVICE (ADVanced dIsplaCement monitoring system for Early warning) is a system for near-real-time and automatic monitoring data acquisition, validation, elaboration and dissemination, and for early warning. Moreover, we experienced two emergencies that occurred in 2013 and 2014 (Manconi and Giordan 2016). This experience helped the technological improvements of our system ADVICE and the development of a scientific data dissemination strategy (Giordan et al. 2019).

As mentioned before, hazard management is a multidisciplinary activity, where the collaboration among all involved (directly and indirectly) people is necessary. In Giordan et al. (2019), we analyzed their role, technical knowledge and needs during emergencies. We determined four groups: GROUP-1, 2, 3 and POPULATION. GROUP-1 is mainly composed of government members such as mayors, council members, politicians. Usually, they are the terminus of the decisional process, and their role is to provide public safety. However, their technical background on natural hazards may be limited, so they need a clear overview of monitoring results and current risk level. They also directly collaborate with population and media, organize educational programs, and they alert immediate risk. For this purpose, they need adequate tools. GROUP-2, mainly composed of engineers, geologists and technicians, supports the GROUP-1. They are, for example, Civil Protection Agency members; they have adequate technical abilities and experience to perform necessary analysis of the landslide evolution. Highly specialized experts (mainly engineers and geologist) in hazard monitoring and EWSs compose the GROUP-3. This group develops monitoring networks and performs analysis, validation and elaboration of the raw monitoring data. The last group is POPULATION, who does not directly participate in emergency management, but they are the final user of early warning applications. The proper data dissemination is fundamental for the correct response of the population at risk. Only well-informed and especially risk-conscious community will act correctly during emergencies.

Many international projects (UNISDR 2007, 2015) and researches (Edwards et al. 2012; Garcia and Fearnley 2012; Wrzesniak and Giordan 2017) already highlighted the problem of the lack of strong links between each sub-system of early warning application. To face this problem, we designed a single-page bulletin, which is easy to read, easy to interpret and easy to disseminate, independently from the reader technical skills. We dedicated this document mainly to ROLE-1 members and POPULATION, but even the experts can utilize it for a quick preliminary overview of the situation.

As the bulletin is dedicated to non-expert people, we used infographics technics to translate numerical data from the monitoring system into graphic visualization. An infographic is a way to represent complex data or information in a visual format for quick delivery and at a glance understandable by the receiver (Smiciklas 2012). The use of this approach is an optimal solution to integrate a variety of information relative to the landslide evolution, which can be used for educative and for early warning proposes. This bulletin is suitable for early-warning applications, thanks to the development of an algorithm for automatic bulletin generation (Wrzesniak and Giordan 2017). The development of an automatized procedure for the production and update of bulletins has been a significant additional value of recent emergencies. The Mont the La Saxe rockslide is one of the most critical, but then the presented bulletin has also been used in different contexts, like the Ponzano landslide (Allasia et al. 2019), and the Planpinciex Glacier (Giordan et al. 2020).

Results

Figure 1 presents an example of two subsequent weekly bulletins for La Saxe landslide from the emergency occurred in March-April 2014. The bulletins correspond to the periods 25/03–01/04/2014 and 01–08/04/2014. We can observe the transition of activity level from moderate to high.

The graphical bulletin layout was designed with a particular focus on communication aspects, which should allow the readers to interpret the results quickly.

In this case, the central part of the bulletin presents the monitoring area and its surrounding at risk. This area is divided into three kinematic sectors, which are countered with coloured dash lines. Such a point of view of the monitored area is an optimal representation because it is an easily recognizable view by the affected population. It shows the elements at risk, such as buildings (civil and public) and infrastructures, making the risk more perceived by habitants. The results from RTS are presented in the form of a deformation map with gradual colour scale and with 3D arrows associated with the prisms' location. The magnitude of the arrows is normalized to make them visible. Such graphical representation of monitoring data is easily accessible for the people without dedicated knowledge, and it provides, in a unique figure, information about distribution, direction and magnitude of the displacement. The behaviour of each sector is shown with an infographic representing a tachometer. Its concept is commonly known so the message can be received at a glance. The movement trend can be defined as: decelerate, constant or accelerate. The trend is analyzed for each sector by comparing the maximum value registered in the considered period and the previous one.

Additionally, the tachometer icon is supported by the numerical value of the maximum displacement in each sector. We highlighted the results of the current period to obtain an intuitive separation of data. General activity of the landslide concludes the bulletin. This information is disseminated in the form of the eye-catching logo coloured according to detected activity level: high, moderate or low. We used gradual colour change (traffic light-type) which increases together with activity level (low = green, moderate = orange, high = red). A short description of the expected evolution of each level is presented in the bottom part of the bulletin.

This bulletin, based on monitoring data, presents the actual status of the landslide. It supports decision-makers in hazard management, and it can be used to disseminate the current landslide status and keep the population informed. To integrate it within the early-warning application, the automatization procedure was necessary. We developed an algorithm for automatic generation of the bulletin, which elaborates monitoring data, plots deformation maps and displacement arrows, defines trend and activity, and associates the corresponding infographic. EWSs can operate at different levels: ordinary level, attention level and alarm level (Intrieri et al. 2012). The authors propose to issue the update with different frequencies: respectively one month, 24 and 12 h. The configurability of our algorithm allows setting the periodicity of one month, one week, 24 and 12 h. Moreover, it can be integrated with threshold-based early warning systems in such a way that it can be issued whenever the threshold level is exceeded. The detailed description of the algorithm development can be found in Wrzesniak and Giordan (2017).

Discussion and Conclusion

Advances in technology increased the capacities of monitoring systems. Modern systems provide a large number of accurate measures at high sampling rates, which are widely combined with early warning applications. Additionally, in the field of hazard management, the aspect of scientific data dissemination must be carefully considered due to the involvement of a variety of people (engineers, geologist, technicians, politicians, council members, population, etc.). The warning messages or update on the risk status must be tailored to the needs of the individual groups.

The population at risk needs to be periodically informed about the hazards and the level of risk, and about how it may be changing. This information should not be technical. Therefore, there is the need for an adequate communication approach, which will educate the public, so the response to the warning will be efficient. Perhaps, the reason that people often ignore warnings is that they do not feel like the warnings are addressed to their values, interests and needs.

The presented bulletin is an optimal tool for both early warning applications (thanks to the development of the algorithm for automatic bulletin generation) and for educational and informative proposes. It facilitates the effectiveness

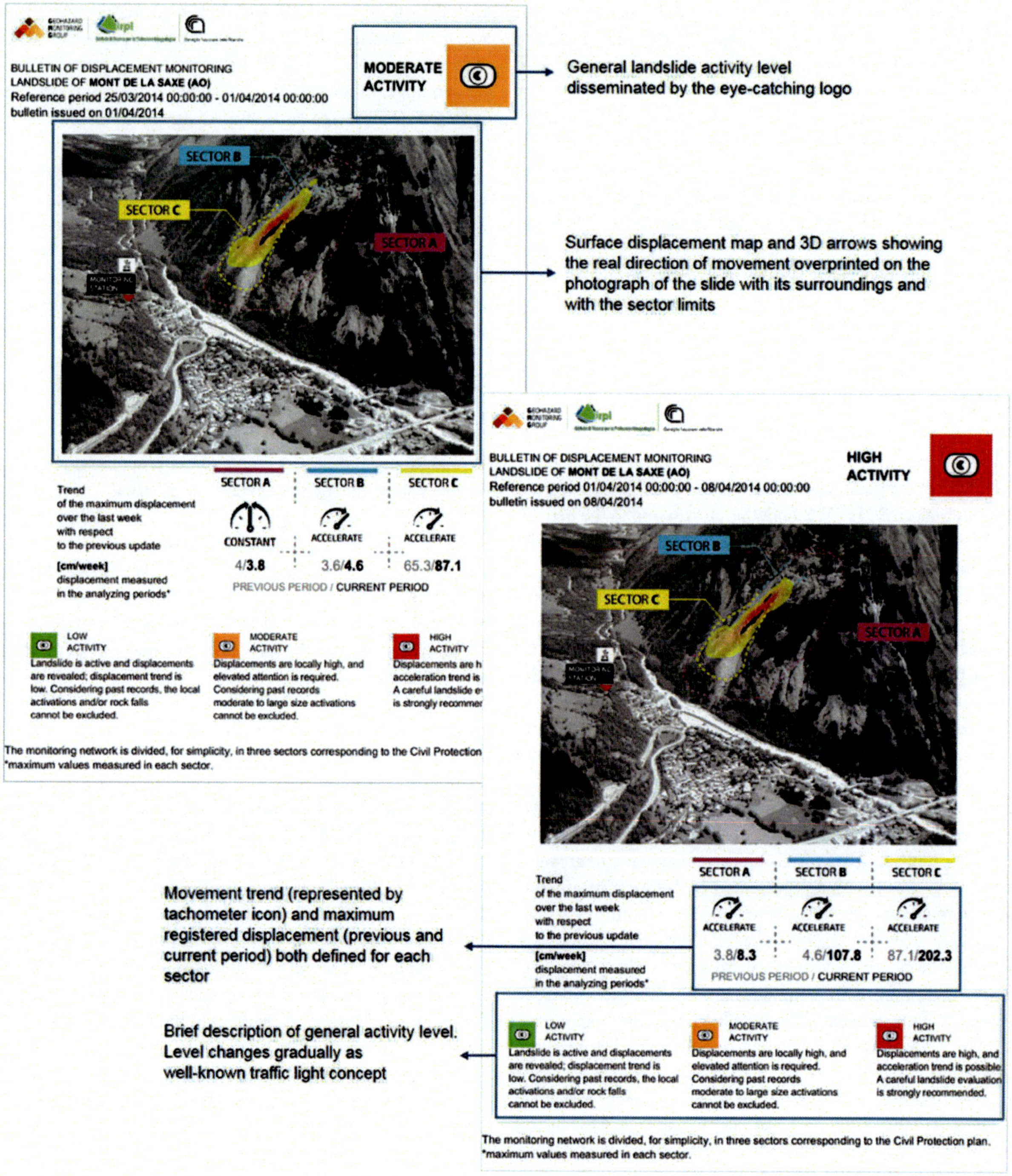

Fig. 1 Example of two subsequent weekly single-page bulletins (generated automatically). The figure modified from Giordan et al. (2019)

of communication path: experts → decision-makers → population; it increases the general awareness, and it supports decision-makers. The bulletin presents complex monitoring data using infographics technics, making it adequate for straightforward communication, especially for non-expert people. An essential confirmation about bulletin usability was provided by the Geological Survey of Aosta Valley, the National Civil Protection Agency and municipal authorities, as we obtained their positive feedback and acceptance to issue the bulletin periodically.

Recently, we have applied a similar bulletin layout for the emergency (October 2019) of Planpincieux glacier (Giordan et al. 2020), located on the southern slopes of the Grandes Jorasses in the Mont Blanc massif of the Alps. This emergency became a media case at the international level, so the communication aspect, especially with journalists, has played a crucial role. The bulletin is public and is published on the website of Montagna Sicura (http://www.fondazionemontagnasicura.org/asset/005–bollettino-planpincieux-04-10-2019-1.pdf). This shows the potential of such a document in other applications related to different naturala hazards.

References

Alcántara-Ayala I (2016) On the multi-dimensions of Integrated Research on Landslide Disaster Risk. In: Aversa S, Cascini L, Picarelli L, Scavia C (eds) Landslides and engineered slopes, vol 1. Experience, Theory and Practice CRC Press, Balkema, Taylor & Francis Group, pp 155–168. ISBN: 978-1-138-02989-7

Alcántara-Ayala I, Oliver-Smith A (2017) The necessity of Early Warning Articulated Systems (EWASs): critical issues beyond response, In: Sudmeier-Rieux K, Fernandez M, Penna I, Jaboyedoff M, Gaillard JC (eds) Linking sustainable development, disaster risk reduction, climate change adaptation and migration. Springer, pp 101–124. ISBN 978-3-319-33878-1

Allasia P, Baldo M, Giordan D, Godone D, Wrzesniak A, Lollino G (2019) Near real time monitoring systems and periodic surveys using a multi sensors UAV: the case of Ponzano landslide. In: Proceedings of the IAEG/AEG annual meeting, 17–21 Sept 2018, vol 1. San Francisco, CA, USA, pp 303–310

Allasia P, Lollino G, Godone D, Giordan D (2018) Deep displacements measured with a robotized inclinometer system. In: Proceedings of the 10th international symposium on field measurements in geomechanics, Rio de Janeiro, Brasil, 16–20 July 2018

Allasia P, Manconi A, Giordan D, Baldo M, Lollino G (2013) ADVICE: a new approach for near-real-time monitoring of surface displacements in landslide hazard scenarios. Sensors 13:8285–8302

Ayalew L, Yamagishi H, Marui H, Kanno T (2005) Landslides in Sado Island of Japan: Part I. Case studies, monitoring techniques and environmental considerations. Eng Geol 81:419–431

Blaikie P, Cannon T, Davis I, Wisner B (1994) At risk: natural hazards, people's vulnerability and disasters. Routledge, New York

Bell R, Glade T, Thiebes B, Jäger S, Krummel H, Janik M, Holland R (2009) Modelling and web processing of early warning. https://homepage.univie.ac.at/thomas.glade/Publications/BellEtAl2009.pdf

Crosta G, Di Prisco C, Frattini P, Frigerio G, Castellanza R, Agliardi F (2014) Chasing a complete understanding of the triggering mechanisms of a large rapidly evolving rockslide. Landslides 11(5):747–764

Edwards SJ, Fearnley CJ, Lowe CJ, Wilkinson E (2012) Disaster risk reduction for natural hazards: putting research into practice. Environ Hazards 11(2):172–176

Frodella W, Ciampalini A, Bardi F, Salvatici F, Di Traglia F, Basile G, Casagli N (2018) A method for assessing and managing landslide residual hazard in urban areas. Landslides 15:183–197

Garcia C, Fearnley CJ (2012) Evaluating critical links in early warning systems for natural hazards. Environ Hazards 11(2):123–137

Giordan D, Dematteis N, Allasia P, Motta E (2020) Classification and kinematics of the Planpincieux Glacier break-offs using photographic time-lapse analysis. J Glaciol 66(256):188–202

Giordan D, Wrzesniak A, Allasia P (2019) The importance of a dedicated monitoring solution and communication strategy for an effective management of complex active landslides in urbanized areas. Sustainability 11:946

Glade T, Nadim F (2014) Early warning systems for natural hazards and risks. Nat Hazards 70:1669–1671

Intrieri E, Gigli G, Mugnai F, Fanti R, Casagli N (2012) Design and implementation of a landslide early warning system. Eng Geol 147:124–136

Lombardi L, Nocentini M, Frodella W, Nolesini T, Bardi F, Intrieri E, Carlà T, Solari L, Dotta G, Ferrigno F (2017) The Calatabiano landslide (southern Italy): preliminary GB-InSAR monitoring data and remote 3D mapping. Landslides 14:685–696

Manconi A, Giordan D (2016) Landslide failure forecast in near-real-time. Geomat Nat Hazards Risk 7(2):639–648

Michoud C, Bazin S, Blikra LH, Derron M-H, Jaboyedoff M (2013) Experiences from site-specific landslide early warning systems. Nat Hazards Earth Syst Sci 13:2659–2673

Notti D, Cina A, Manzino A, Colombo A, Bendea IH, Mollo P, Giordan D (2020) Low-Cost GNSS solution for continuous monitoring of slope instabilities applied to Madonna Del Sasso Sanctuary (NW Italy). Sensors 20(1):289

Pecoraro G, Calvello M, Piciullo L (2019) Monitoring strategies for local landslide early warning systems. Landslides 16:213–231

Petley D (2012) Global patterns of loss of life from landslides. Geology 40(10):927–930

Simeoni L, Mongiovì L (2007) Inclinometer monitoring of the Castelrotto landslide in Italy. J Geotech Geoenviron Eng 133:653–666

Smiciklas M (2012) The power of infographics: using pictures to communicate and connect with your audiences. Que Publishing, Indianapolis (IN)

Thiebes B, Glade T (2016) Landslide early warning systems—fundamental concepts and innovative applications. In: Landslides and engineered slopes: experience, theory and practice: proceedings of the 12th international symposium on landslides, 12–19 June 2016. Napoli, Italy, pp 12–19

UNISDR (2015) Sendai framework for disaster risk reduction 2015–2030. https://www.unisdr.org/files/43291_sendaiframeworkfordrren.pdf

UNISDR (2007) Hyogo framework for action 2005–2015: building the resilience of nations and communities to disasters. https://www.unisdr.org/files/1037_hyogoframeworkforactionenglish.pdf

UNISDR (2006) Global survey of early warning systems. https://www.unisdr.org/files/3612_GlobalSurveyofEarlyWarningSystems.pdf

Wieczorek GF, Snyder JB (2009) Monitoring slope movements. Geological monitoring. Geological Society of America, Boulder, CO, USA, pp 245–271

Wrzesniak A, Giordan D (2017) Development of an algorithm for automatic elaboration, representation and dissemination of landslide monitoring data. Geomat Nat Hazards Risk 8(2):1898–1913. https://doi.org/10.1080/19475705.2017.1392369

Part IV

General Landslide Studies

Engineering Geological Appreciation in Landslide Mapping for a Natural Gas Pipeline Project: Challenges and Risk Reduction Measures

Vassilis Marinos, Kostas Papazachos, Georgios Stoumpos, Dimitra Papouli, George Papathanassiou, and Theodoros Stimaratzis

Abstract

Landslides represent a significant hazard for pipelines because they can generate permanent ground displacement and tend to result in complete failure or significant leaks, major environmental impacts and long periods of service disruption. Hence, landslide-related incidents are regarded as a significant operational risk. The paper mainly focuses on the landslide identification, mostly in the field, along or across a natural gas pipeline, their engineering geological appreciation and challenges in relation to the pipeline and finally, their management to reduce the risk. The case of the Trans Adriatic Pipeline (TAP) and in particular the Albanian section that faced a great number of landslides, is examined here. Whether the "expected" landslide event reaches the Right of Way (RoW) and impacts the pipeline, is influenced by the nature and size of the expected landslide event, controlled by the site geology and geomorphology, the proximity of the existing landslide feature to the pipeline and the position of the pipeline relative to the landslide. When landslides that can threat the pipeline integrity cannot be avoided, more detailed site evaluation is required to support the design and construction of mitigation measures. In this paper, the findings of several field trips are presented, reviewed and assessed. As a result, landslide areas were identified and previously characterized landslide areas were confirmed, delineated or discarded. The most important outcome of the whole process, including not only field trip findings but also desk revisiting of available data, evaluation of ground investigation campaigns, geotechnical monitoring findings as well as a number of discussions between members of the participating bodies, was the avoidance of significant and numerous landslides, the finalization of the preferred route but also the prioritisation of risk reduction measures along the route.

Keywords

Pipeline • Landslide identification • Hazard • Flysch • Risk-reduction measures • GSI

Introduction

Landslides and unstable ground pose a significant threat to buried natural gas pipelines since they can generate permanent ground displacement (PGD) along or across the pipeline alignment. PGD is an important concern since a buried pipeline must deform both along its axis and after bending in order for the movements of the surrounding ground to be accommodated (Nyman et al. 2008).

The pipeline consists of steel tubes with inner diameter generally up to 1.2 m, buried in 1–2 m depth. Pipeline construction is usually done by trench excavation, while trenchless techniques such as boring/jacking, auger bore, HDD (horizontal directional drilling) and micro-tunnelling, may be used in case of river crossings or avoidance of landslides. Such projects also involve a compressor station

V. Marinos (✉) · K. Papazachos · G. Stoumpos · D. Papouli · T. Stimaratzis
School of Geology, Faculty of Sciences, Aristotle University of Thessaloniki, 541 24 Thessaloniki, Greece
e-mail: marinosv@geo.auth.gr

K. Papazachos
e-mail: kpapaza@geo.auth.gr

G. Stoumpos
e-mail: g_stoumpos@yahoo.co.uk

D. Papouli
e-mail: d_papouli@yahoo.gr

T. Stimaratzis
e-mail: thstimar@hotmail.com

G. Papathanassiou
Department of Civil Engineering, Democritus University of Thrace, Thessaloniki, Greece
e-mail: gpapatha@civil.duth.gr

Ž. Arbanas et al. (eds.), *Understanding and Reducing Landslide Disaster Risk*, ICL Contribution to Landslide Disaster Risk Reduction,
https://doi.org/10.1007/978-3-030-60713-5_26

and block valve stations. The first is located along the line to move the product through the pipeline and retain the pressure and the later, located every 50 km aprox, comprises a first line of protection for pipelines for maintenance work or isolate a rupture or leak. Landslides and their interaction with the pipeline integrity must be investigated along and across a zone, called a Right-of-Way (ROW). ROW is a strip of land usually between 18 m and 36 m wide, containing one or more pipelines and it is used for access for inspection, maintenance, testing or in an emergency; Identifies an area where certain activities are prohibited to protect public safety and the integrity of the pipeline.

In accordance with international best practice, routing of a pipeline in mountainous areas is done along ridge crests and spurs, facing morphology challenges like narrow ridges and steep slopes (Fig. 1).

Though, ridge crest alignments are not risk free. An approach for the assessment of the pipeline risk due to landsliding is defined in Fell et al (2005) and Lee and Jones (2014). An initiating trigger event is an incident (or combination of incidents) that causes a landslide event (e.g. heavy rainfall or strong earthquake). A damage case is a level of damage resulting from the impact of a landslide on the pipeline (e.g. exposure, bending and buckling, rupture). Risk is usually expressed as the product of the probability of an event (e.g. pipeline rupture) and its adverse consequences. The assessment of risk along a pipeline route, though, is not in the scope of this paper. Pipeline rupture is not uncommon in incidents caused by landslides. As such, landslide-related incidents often result in leaks that may have severe environmental impact as well as long periods of operational stoppage (e.g. Savigny et al. 2005) with very high costs per day. Regarding pipelines running through mountainous areas, statistics show that landslides are the most common cause of pipeline rupture and as such the most significant operational risk (e.g. Sweeney et al. 2005).

It is generally accepted that avoidance of landslide-prone areas is the most effective hazard reducing option both in terms of cost and time saving. Sweeney (2005) points out that this is due to the fact that the investigation and the subsequent stabilisation of a significant number of landslide areas is not a practical undertaking mainly due to time and cost constraints. Accordingly, in mountain regions, the presence of landslides or the presence of landslide-prone areas is a quite important factor for the finalisation of the pipeline route. Landslide hazard assessment is used to identify "hot spots" along the pipeline route where re-routing or risk reduction measures must be prioritised. In cases where "hot spot" areas cannot be avoided; detailed evaluation of all site conditions is necessary for the support of the design and construction of mitigation measures.

The Trans Adriatic Pipeline (TAP) is a natural gas pipeline. Its' onshore section starts near Kipi on the border of Turkey and Greece, where it connects with the Trans Anatolian Pipeline (TANAP) that comes from the South Caucasus Pipeline (SCPX), continues crossing Northern Greece and Southern Albania and ends near the city of Fier, in the Adriatic Sea.

Fig. 1 Top: In accordance with international best practice, routing of a pipeline in mountainous areas is done along ridge crests and spurs, facing morphology challenges like narrow ridges and steep slopes. Here, an example of Landslide S-AL-4 (now avoided). The rock mass consists of moderately disturbed siltstone with sandstone alternations. A great scale slide is evident in the area that includes not only the weathered zone but the bedrock itself. The crest of the slide reaches very close to the pipeline route that runs on the mountain ridge. Bottom: Overview of landslide area S-AL-4 (Google Earth) in relation to an initial pipeline route (red line) (before this landslide assessment). Such great slide, could primarily threat the pipeline integrity by upslope retrogression in case of re-activation

Across Albania, the pipeline passes through the Pindus (called Albanides in this area) mountain range which presents a rather challenging environment for the construction of the project. The section of the TAP that is analysed for the purposes of this paper spans from 70 km to 131 km approximately (Fig. 2). It is considered as one of the most challenging onshore section of the SCPX-TANAP-TAP pipeline system due to the combination of narrow ridge crests, weak rock masses of flysch sediments, high seismicity, widespread active landslides and unstable side slopes, and very limited access. The route follows a ridge crest alignment, but ridges tend to run NW-SE so it has to cross a number of river valleys. More specifically, the TAP route mainly follows the ridges of the mountains and as such it is typical to run through steep to very steep slopes, which in some areas are delineated by landslides or very steep slopes on both sides.

This paper presents the landslide that were identified along the TAP route in Albania after several focused field-trip investigations and site investigation campaign (from June 2014 to November 2016) by an Aristotle University of Thessaloniki team. As a result of the field investigation, landslide areas were identified, while previously characterized landslide areas were confirmed, delineated or discarded.

These landslides areas were a principal factor for the pipeline route selection process adopted for the mountainous Albania regions. The most important outcome of the whole process, including not only field trip findings but also desk review of available data, evaluation of ground investigation campaigns, as well as a number of discussions between members of the participating bodies, was the avoidance of significant landslides and finalizing of the preferred route for the TAP.

The methodology that was implemented for this process is the main focus of this paper and is presented as a "text-book" approach that may be followed in similar projects.

Fig. 2 The Trans Adriatic Pipeline (TAP) Albania on shore section and the location along the whole project (upper right). The paper focuses on the section (shown in the black rectangle) that spans approximately from 70 to 131 km (KP70-KP131), probably the most challenging onshore section of the SCPX-TANAP-TAP pipeline system. The combination of weak rocks, steep slopes, rainfall and snow melt and strong ground accelerations (earthquakes) creates an environment prone to landsliding

Methodology

The most important scope of the whole process is to identify "hot spots" along the pipeline route, mostly along ridge crests and spurs, for the avoidance of landslide-prone areas, is the hazard reducing option both in terms of cost and time saving. Hence, the finalisation of the route can be established. When re-routing was not possible, risk reduction measures had to be prioritised after detailed evaluation of all site conditions.

The main tools that are used by the landslide experts in order to achieve this goal are: (a) desk revisiting of available data, remote sensing data processing (LiDar, UAV, satellite images) (b) field survey work along the route, (c) evaluation of ground investigation and monitoring campaigns. Since investigation and stabilisation is not the best international practice due to time and cost constraints (Sweeney 2005), the landslide assessment must initially focus on the recognition of landslides and then on the appreciation of their possible failure event scenarios associated with pipeline rupture.

The work was focused on pre-existing landslides since the reactivation of movements, along pre-existing shear surfaces, occurs more frequently than the generation of new-first time landslides. The reactivation of these type of landslides could primarily threat the pipeline integrity. First-time failures, though, can be also characterised by large, rapid displacements. Hence areas characterized by steep terrain, thick weathered mantle and/or weak rock, under high precipitation and seismic motion were also examined and eventually suggested for geotechnical analysis and pipeline verification.

Possible Failure Event Scenarios Associated with Pipeline Rupture

As it was presented in preceding section, the best international practice for pipeline routing is along ridge crests, though the fact that at these areas the landslide risk cannot be neglected. On the other hand, the existence of a landslide area that reach the pipeline will not definitely cause its rupture. In some cases, may only lead to exposure, whilst in other cases the damage may be limited to bending or buckling. The landslide-induced failure types of pipeline, as they have been classified based on industry experience (Nyman et al. 2008; Lee et al. 2009; Young and Lockey 2013), are:

1. **Lateral and vertical displacement**. Pipeline rupture as a result of differential horizontal and/or vertical movement of the landslide main body, upslope retrogression of the

main scarp or failure of the flanks. The potential for pipeline displacement is a function of landslide depth, the behaviour of the materials, the speed of movement and the cumulative displacement that could occur over time.
2. **Spanning**. Pipeline rupture as a result of removal of support along a significant length (e.g. greater than 30 m) due to retreat of the landslide main scarp (upslope retrogression) or failure of the landslide flanks (lateral expansion). The potential for spanning is a function of the vertical displacement of the landslide mass and retreat of an eroding scar across the pipeline alignment.
3. **Loading**. Pipeline rupture due to stresses induced after burial by debris. This failure mode depends on the depth of burial and no measures to face the weight of the material that acts upon the pipeline.
4. **Impact**. If not at the crest and if the pipeline is exposed, impulse due to the momentum of falling boulders may result in pipeline rupture. In general, buried pipelines are less vulnerable to this failure mode. As this is a case of momentum and impulse, the height from whence the boulder originated (i.e. the impact speed) and the mass of the boulder are the defining parameters.

All possible failure event scenarios associated with pipeline rupture are presented in Fig. 3.

Landslide Identification—Site Investigation

Field assessment of problematic areas comprises the in situ detection of landslide-signifying features such as tension cracks, scarps, unusual topography, differences in ground quality, unvegetated and/or steep slopes, surface water ponds and lakes (back tilts) etc. However, desk study is also very important for the detection of existing landslides.

Both at the preparation stage and at field, the effort was to continuously look for key elements that signify the existence of an active landslide or an unstable area. These key elements are presented in the following paragraphs.

Desk Study

The desk study involved three main procedures to prepare an appropriate landslide-prone area inventory.

- Identification of possible old landslide areas from the topographic maps—Sudden changes of slope angle, flat areas or bulge on slopes (square contours).
- Identification of possible old landslide areas from satellite images—unvegetated and/or steep slopes
- Identification of possible old landslide areas from satellite images—surface water ponds and lakes

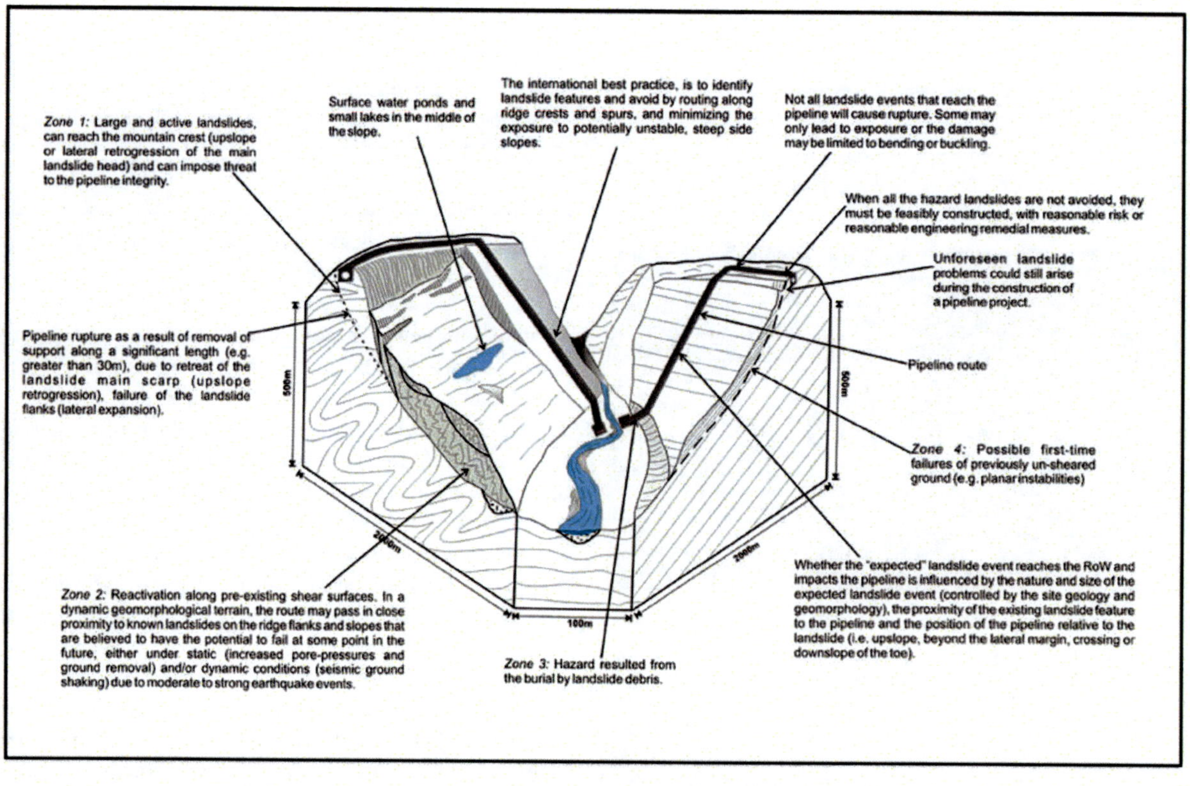

Fig. 3 Possible landslide failure event scenarios associated with pipeline rupture

Engineering Geomorphology—Field Identification

The work was executed mostly on the ridge along the pipeline route following a primary identification of possible landslide areas from the topographic maps and a Google Earth geomorphological study, since LiDar images were not available at the time. Field assessment of problematic areas comprises the in situ detection of landslide-signifying features such as tension cracks, scarps, unusual topography, differences in ground quality, unvegetated and/or steep slopes, surface water ponds and lakes etc. (Fig. 4). The field study involved several procedures and observations, in order to define the actual spatial, geological and geotechnical features of the finally derived landslide areas.

The Engineering Geological Environment—Ground Quality

The geological environment of the investigated area is dominated by flysch formations. Limestones and ophiolites (peridotites) were also identified at the eastern edge but with limited presence. However, the ophiolites are thrusted upon the flysch formations producing heavily folded and sheared rock masses in flysch deposits. Moreover, flysch formations are also disturbed by numerous thrusts due to the mountain building and folding process.

The area receives a high average annual rainfall of about 1500–2000 mm and presents high seismicity with an expected peak ground acceleration (PGA) of 0.4 g and a 10% chance of exceedance in 50 years (1 in 475 year return period).

Ground Quality

The identified landslides were found within the flysch rock mass, so the focus of the engineering geological characteristics is given to them. Flysch sediments that are comprised by rhythmic alternations of sandstone and siltstone layers and associated with orogenesis and paroxysm folding process, creates various rock mass qualities.

In particular, flysch is a weak and complex geological formation consisted of alternations of competent/strong sandstone layers and incompetent, of low generally strength, siltstone/clayey schist beds. The tectonic disturbance has a profound effect on the formation as it transforms the initial structure and can produce tectonic mixtures. Degradation due to shearing and fissuring is also very common. The presence of clayey formations, weathering of silty-clayey members and susceptibility to slaking as well as the presence of groundwater also downgrades the behaviour of the formation.

The primary goal of the field survey was to assess the quality of the rock mass like lithology, rock mass structure, weathering profile, strength, joint characteristics, water presence along the investigated route, based on the Hoek–Brown failure criterion (Hoek et al. 2002; Hoek and Brown 2018). More specifically, for the purposes of the present

Fig. 4 **a** Fresh cracks, **b** Visible irregularities in topography: sudden changes of slope angle, small hills and back-tilts along the slope or flat areas with surface water pond (small lakes) in the middle of the slope, **c** Large openings-tensile cracks parallel and next to the crest indicating movement, **d** Deformed vegetation due to landsliding

work and the characterisation of the rock masses under failure, the extension of the original GSI application charts for heterogeneous and structurally complex rock masses, such as flysch (Marinos 2014; Marinos 2017) was used (Fig. 5). As such, flysch formations are classified in 11 rock mass types, according to the siltstone-sandstone participation and their tectonic disturbance. However, mainly 6 types were mainly met along the route:

Main Flysch Types Along the Route

IV. Moderately disturbed rockmass with sandstone and siltstone in similar amounts
V. Moderately disturbed siltstones with sandstone interlayers
VI. Moderately disturbed siltstones with sparse sandstone interlayers
VII. Strongly disturbed, folded rockmass that retains its structure, with sandstone and siltstone in similar extent
VIII. Strongly disturbed siltstones, folded rockmass, with sandstone interlayers. The structure is retained and deformation—shearing is not strong
X. Tectonically deformed intensively folded/faulted siltstone with broken and deformed sandstone layers, forming almost a chaotic structure

Based on this evaluation, an engineering geological zonation of the investigated TAP route was compiled. This zonation is presented in the following figure (Fig. 6). It is noted that the presented zonation is a direct result of field observations only.

The areas where most identified landslides are encountered (Panarit, Helmës-Staraveckë, Potom) were characterised by the presence of flysch types VIII and X. This coincides with the presence of mainly folded and sheared siltstone flysch, with lower participation of sandstone, and low to very low GSI values (15–25), hence low rock mass strength.

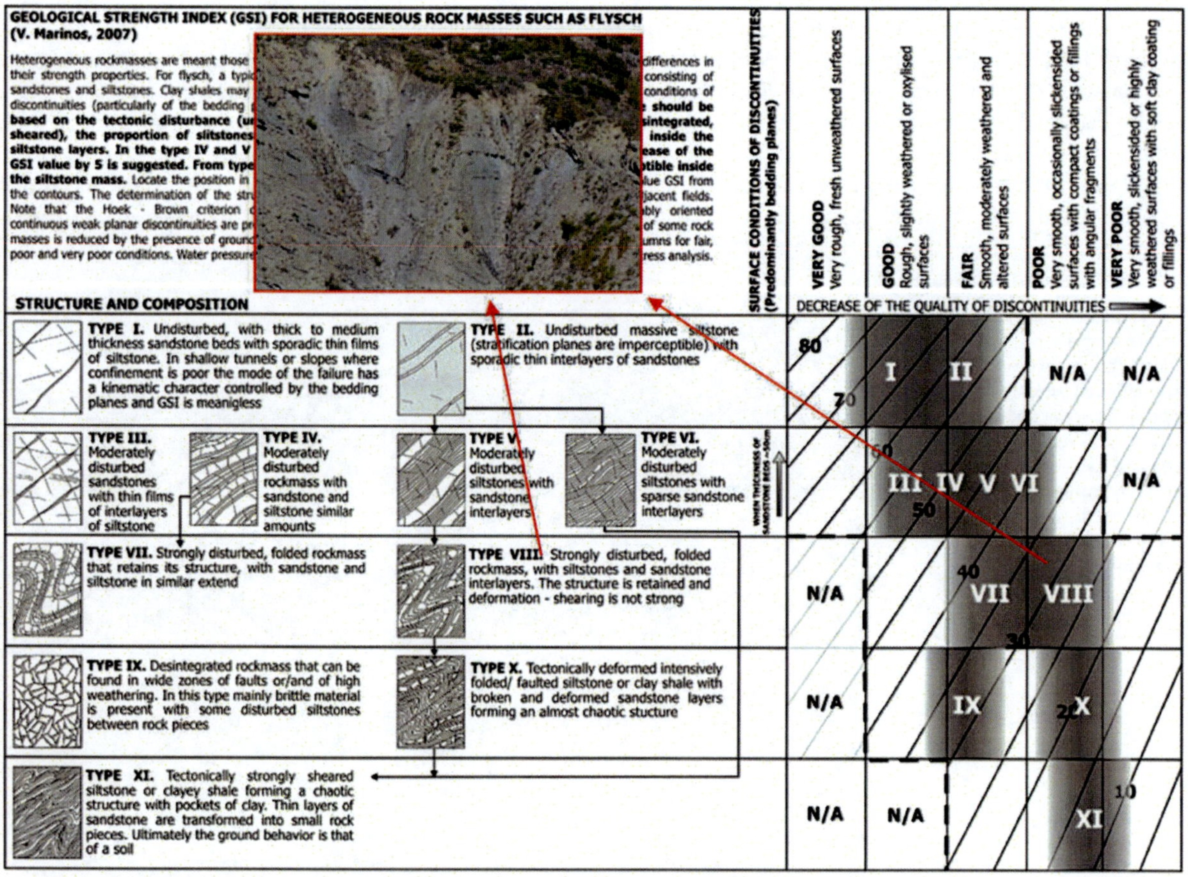

Fig. 5 GSI for heterogeneous rock masses (Marinos 2017). The photo on the chart shows an example of flysch Type VIII—Strongly disturbed siltstones, folded rockmass, with sandstone interlayers

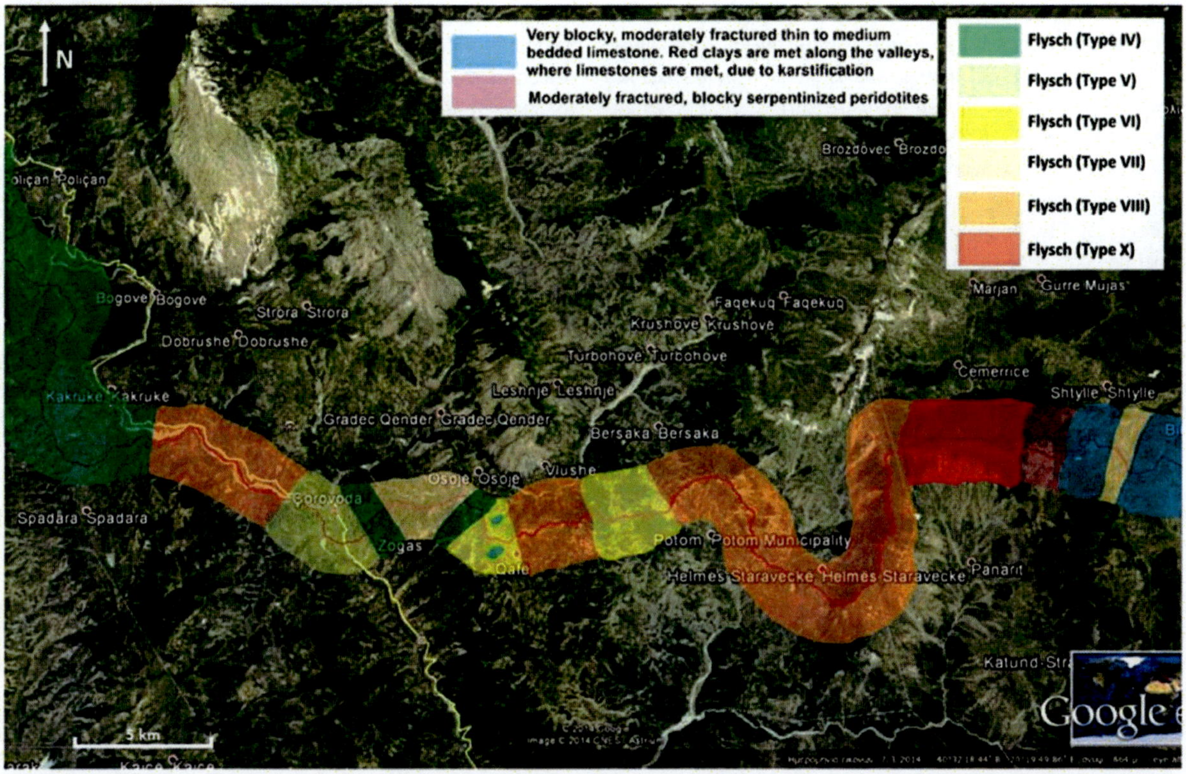

Fig. 6 Zonation of the rock mass quality along the route, based on the GSI chart for heterogeneous and structurally complex rock masses, such as flysch (Marinos 2014, 2017) spanning approximately from ~60 to 130 km (KP70-KP131). Most of the identified landslides along the TAP route are present in flysch types VIII (strongly disturbed siltstones, folded rockmass, with sandstone interlayers) and X (tectonically deformed intensively folded/faulted siltstone with broken and deformed sandstone layers, forming almost a chaotic structure)

Ground Investigation

The execution of sampling boreholes, especially in previously identified areas that might include an unstable mass are of high importance in terms of both confirming and assessing the vertical extent of an existing landslide. The ground investigation campaign consisted the execution of 25 sampling boreholes (maximum depth reached 60 m) in selected pilot areas, though, since there was no time to investigate, in depth, all the identified landslides along or across the pipeline. Examples of borehole cores within the landslide bodies are given in the following paragraphs. Laboratory and in situ tests incorporated unconfined compression tests, point-load tests, 19 shear tests and 12 triaxial tests, as well as 62 Standard Penetration Tests (SPT).

Monitoring was also implemented with the installation of several piezometers and inclinometers. The piezometers were installed since the ground water is a principal factor that causes a landslide event. Generally, the water level was mostly encountered within the landslide materials and the flysch weathering mantle. Water level measurements were taken from December 2014 until June 2015, for a period of seven months and the groundwater level was found to fluctuate from 2 to 4 m depth. Ground water level follows the ground surface and it was encountered within the weathered zone of the flysch rock mass. This is natural, since the permeability of the landslide material or the weathered mantle of siltstones is low to very low. Indeed, in Marinos et al. (2011) the low permeability of its rock mass with very small differences among the types is presented, based on the analysis of a good number of in situ permeability tests in the various litho-types of flysch from the mountains in Greece (continuing, though, to the same mountain chains to Albania). The role of the presence of siltstone interlayers is predominant in all types, even in those types where their participation is very low. Additionally, the history of compression tectonics, from which the flysch formation suffered, led to a homogenization of all types in terms of mean permeability values. This value is of about 5×10^{-7} m/s for the first tens of metres below surface. This overall low permeability and high-water table creates high groundwater pore pressures, after a rainfall or snow melt, that can trigger a pre-existing landslide or create a first-time movement.

Inclinometers were installed in selected areas where either the activity of a landslide was in question or where evidence of fresh and active, and as such with no surface identifiable features, landslides were suspected. For example, inclinometer BH-8B (Fig. 7), with a depth of 59 m, was installed in an area near the identified landslides S-AL-N33 and below identified landslides S-AL-7 and S-AL-8 (shown in Fig. 12). The position of the borehole was directly on the pipeline axis. For the inclinometer BH-8B, measurements were taken from January 2015 until June 2015, for a period of six months. The measurements show the presence of a deep sliding surface at approximately 46–48 m depth. It is noted that this area was avoided via a rerouting.

Landslide Mechanism and Extent Varies with the Terrain Setting and the Rock Mass Quality

Rock slope instabilities in flysch formations are very common due to the heterogeneity, presence of silty-clayey members with low strength and tectonically disturbed structures. The failure mechanisms are different according to the siltstone-sandstone participation, the intensity of tectonic disturbance and the weathering degree. The evaluation of the principal anticipated failure mechanism is essential for the selection of the appropriate geotechnical parameters, used for the stability analysis.

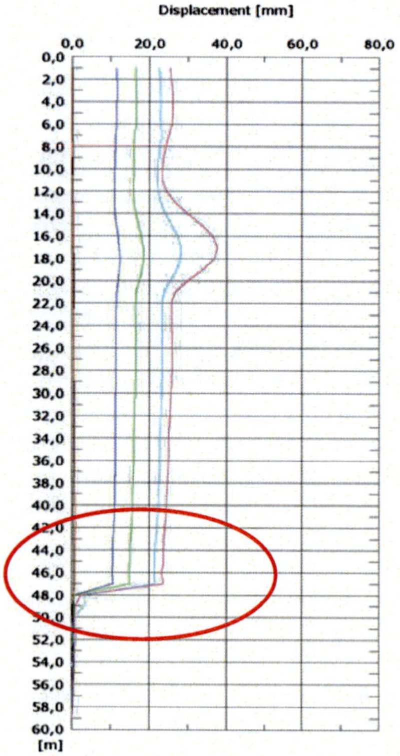

Fig. 7 Displacement of inclinometer BH-8B and depth of observed movement. Such readings, assisted the decision making by avoiding deep seated landslides that threatened the pipeline integrity

A standardization of the qualitative engineering geological characteristics and the assessment of the behaviour in slope stability for flysch rock masses is hereby presented. Its behaviour is controlled by the siltstone-sandstone heterogeneity, the tectonic disturbance and groundwater conditions.

Structurally Controlled Slides

- Controlling factors:
  - potential kinematic favorable conditions
  - shear strength of discontinuities
- Bedding planes the more essential discontinuities:
  - persist in space
  - tectonically sheared due to differential movement during folding.
- Bi-planar slides in a combination of more than one weak persisting discontinuities.

Rotational Landslides

- Controlling factors: lie on the whole rock mass strength and the weathered zone.
- In the first case, the Hoek & Brown failure criterion and the GSI system can be applied for the slope stability analysis or the soil strength characteristics in the weathered zones

In the studied section, highly susceptible areas are mostly underlain by heavily jointed, tectonically sheared and weathered flysch, in proximity with tectonic elements (thrusts and faults), giving rotational slides. Rotational landslides can occur mainly in types VI to XI, where the controlling factors lie on the rock mass strength. In these cases, the Hoek & Brown failure criterion and the GSI system can be applied here for the slope stability analysis.

Planar landslides may occur in types I, III, IV, V. The controlling factors are the potential kinematic favourable conditions and shear strength of discontinuities. The bedding planes here are the more essential discontinuities since they persist in space and are often tectonically sheared due to differential movement during folding. Bi-planar landslides could also develop in a combination of weak persisting discontinuities, although in this case, none of persisting major discontinuity families can provoke alone a landslide; one of these discontinuities is almost always a bedding plane. The bi-planar slide has an almost rotational form, but stability is mainly controlled by the properties of discontinuities not of the strength of the whole rock mass.

Landslide Mapping for Natural Gas Pipelines

Preparation with Landslide Evaluation Sheet Study

The landslide event reaches the RoW is influenced by: (i) Nature and size of the landslide event (controlled by the site ground quality and geomorphology), (ii) Proximity to the pipeline, (iii) Position of the pipeline relative to the landslide. The task of identifying and analysing a significant number of potentially unstable areas along a long pipeline section required the systematic and consistent description and recording of these areas, based on these and other characteristics. This was realized with the use of a landslide presentation and evaluation sheet (example in Fig. 8) that was compiled for this reason

Landslide Field Mapping—Inventory

Three field trips took place along the section of the pipeline from 70 to 131 km. The first field trip focused on the identification of the landslide landslides along the base case route and the two subsequent field trips focused on the investigation of suggested re-routings to avoid landslides but also to get additional detailed observations on specified landslides.

In the first field trip, 27 landslide areas that had been identified at an initial desk study, namely landslide areas S-AL-1 to S-AL-27 (their locations in Figs. 10, 11 and 12) were assessed. Many of these landslides were discarded as non-existent whereas others were differently delineated and their boundaries adjusted. This process pointed towards to the abandonment of the pipeline route from approximately km 77 to km 90 and the identification of an alternative new route.

The second field trip showed that there are much more unstable areas along the route that could threat the pipeline integrity. Many of these landslides (S-AL-N1 to S-AL-N64) were large and showed recent activity. This revealed that many of these sites may fail under static conditions. Most of these landslides were gathered within a specific area, between km 73 and km 90. Hence, a new route to avoid these large and active landslides was suggested.

As a result of this field trip and the subsequent elaboration of all available data, discussions were held between members of all participating bodies. These discussions led to the conclusions that, wherever possible, rerouting options should be examined as a primary mitigation measure in order to avoid areas with large scale and active landslides, since they present a threat to the pipeline integrity. This was the case particularly for the area around the villages of Helmës-Staraveckë, Backë, Potom and Trepollar (km 77 to km 90). Indeed, this led to the need for another extensive field trip that would investigate these re-routing options.

The primary purpose of the third field trip was to investigate the possible alternative reroutings that were examined for the crossing of the area north of the village Backë and through Trepollar (km 77 to km 90) and the area around km 122 to 130. Three local re-routings were also checked (km 99.5 to km 100.3, km 113.3 to km 116 and km 132 to km 134). Additionally, the pipeline route after km 90 was also visited for the re-evaluation process of already identified landslides and for the first-hand assessment of the collected data regarding the main engineering geological factors and presence of other landslides.

The result of this field trip was to conclude to a possibly feasible route for the pipeline and avoid large and active landslides. This route does not avoid all the landslides but could be feasibly constructed, with reasonable risk or reasonable engineering remedial measures.

As another result of the third and final field trip, new landslide landslide areas were identified and previously characterized landslide landslide areas were confirmed, delineated or discarded. The -as accurate as possible- adjustment of the boundaries of each landslide was very helpful for the selection of the optimum route and for the subsequent risk assessment.

The rock mass condition was approached based on in situ engineering geological identification and observation and the lithology of the rock mass was examined as per individual landslide. Qualitative characteristics about the quality of the rock mass (lithology, rock mass structure, weathering and extent of weathering mantle, strength, joint characteristics, water presence) were identified and recorded.

Moreover, considering that structural geometry affects the stability, measurements of key structural elements were taken and each area of interest was classified using the GSI system. Finally, the failure mechanism and the sliding depth for each landslide were assessed. This assessment was aided by the execution of investigative boreholes in several areas.

The identified landslides are related to the all the investigated re-routings but also along the initial base case route. The number of landslides was 83 (ALS-1C to ALS-83A) but not all of them are at present associated with the final pipeline route (Fig. 9).

Trans Adriatic Pipeline - Landslide presentation and evaluation sheet

Identification code	Location	Relative position to pipeline
ALS-60B	108+800	Above the pipeline. 12m from the toe.

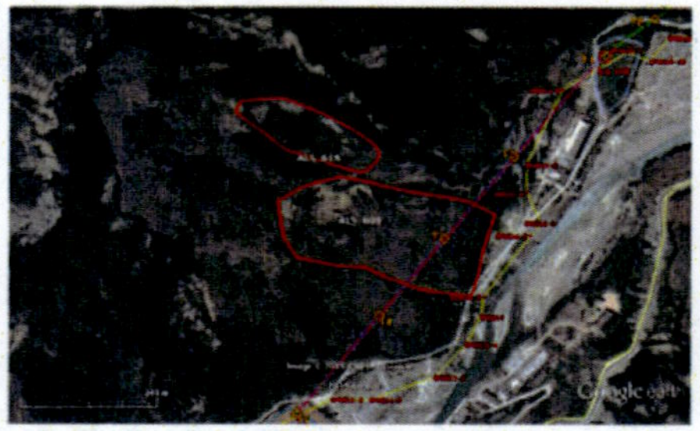

Landslide area
53689 m^2

Landslide max length
465 m

Landslide max breadth
190 m

Geological setting
Weathered and folded flysch rock mass and in superficial soil materials of significant thickness, various origin and variable engineering geological behaviour, namely a) river deposits, b) alluvial deposits, c) landslide materials, d) terrace deposits and e) scree of flysch origin

Terrain
Moderately steep on the south side of the pipe.

Surface water
None on the landslide mass. River adjacent.

Evidence of sliding
Old back-scarp. River bed erosion of fresh material.

Landslide type
Rotational. Deep.

Secondary slides on landslide body
Yes.

Monitoring evidence
Three inclinometers. Inconclusive readings.

Slide surface depth		
Deep		
estimated	measured at monitoring point	found (borehole data)

Landslide activity state	general	pipeline
Dormant / inactive		
Relatively inactive		
Relatively active		
Periodically active	✓	✓
Seasonally active		

Landslide active zones	
Whole landslide	✓
Limited to specific zones	
Connected to secondary units	

Expected reactivation sequence	
Progressive (head loading and downslope expansion)	
Retrogressive (toe unloading and upslope expansion)	
Lateral expansion of secondary units	
Other (state) ***Toe erosion and downslope expansion***	✓

Potential impact on the pipeline upon activation		
low	moderate	high

Likelihood of activation		
low	moderate	high

Risk assessment		
low	moderate	high

Notes - Actions suggested
Re-route or keep the pipeline below the sliding surface by changing the excavation method (e.g. microtunnelling).

Fig. 8 Landslide presentation and evaluation sheet

Fig. 14 View of landslide S-AL-N40 (now avoided). In this case, the pipeline axis (top picture) was situated, where two landslides from both sides of the mountain "are met". The large depth of the sliding material prevented further discussions on routing the pipeline along this area

Fig. 15 Core boxes from BH-3 at S-AL-N40 showing the transition from the landslide material to the bedrock. The depth of the landsliding surface is significant and reaches 30–32 m

Fig. 16 Geotechnical section of S-AL-N40. Landsliding material reaches 30–32 m. The rock mass, below the sliding surface consists of strongly disturbed siltstones, folded rockmass, with sandstone interlayers (type VIII, GSI = 25 − 30). Such sections, were used for geotechnical slope stability analysis under static and dynamic conditions

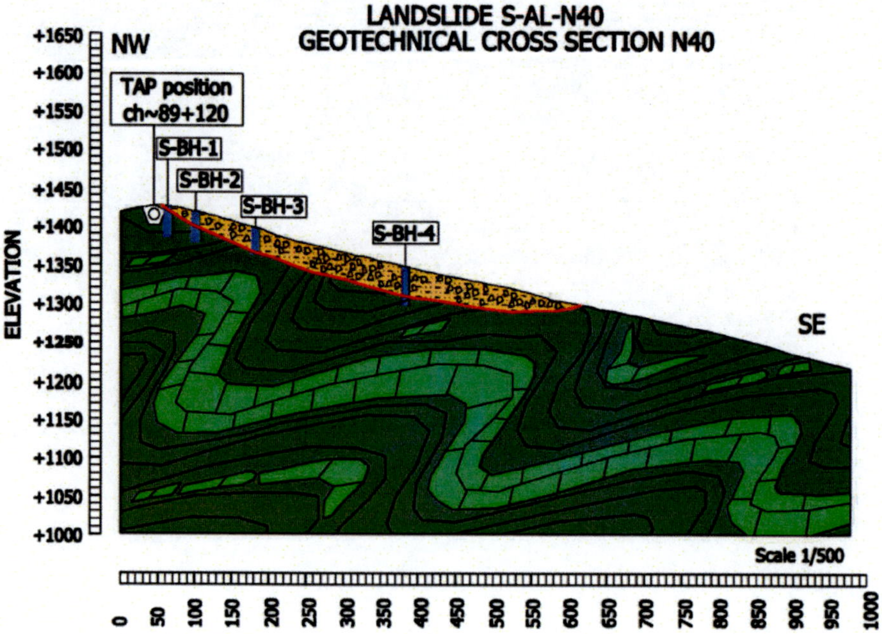

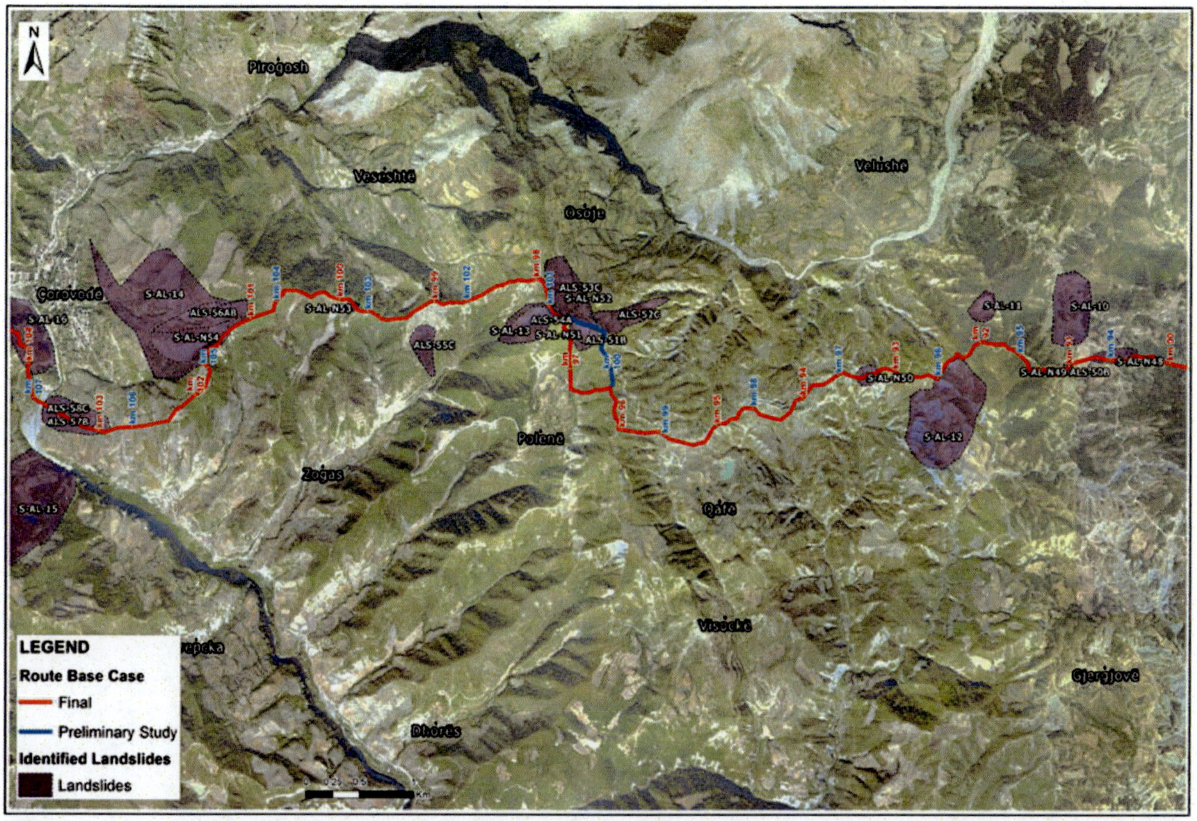

Fig. 17 TAP route—90 to 103 km (Potom to Corovode). In this section, there are not many landslides present, but mainly located around the 101 km (landslides in pink polygons). In this location, some landslides cannot be avoided due to the steep and narrow terrain, hence more detailed site evaluation and the application of remedial measures is required

the application of remedial measures is required. Rock mass, here, consists of moderately disturbed siltstones and sandstones in similar amounts (flysch Type IV in Fig. 5)

Section IV: 103–115 Km (Corovode)

This section of the pipeline runs through slightly weathered, strongly disturbed, folded rockmass that retains its structure, with sandstone and siltstone in similar extent (type VII). Several landslides have been also identified in this area (ALS-60B, ALS-61A, ALS-62A, ALS-63C, ALS-64C, ALS-65B in Fig. 19) but the base case route was mostly followed. Local re-routings and micro-tunnelling, though, were applied to avoid some hazardous areas (detailed example is presented in paragraph 6.4.1). Another route (green line) was also investigated but some "no-go" hazardous zones were found from a focused field survey (Fig. 18).

Case Study of a Landslide—ALS-60B

In the context of this landslide reconnaissance, two landslides, namely ALS-60B and ALS-61A (Figs. 19 and 20), were identified at the area of interest. The pipeline route in this area runs along the Osumi river, at the bottom of the mountain. The main concern was landslide ALS-60B (as coded in a study for landslide hazard assessment along the TAP route), that incorporates a steep slope that inclines towards the north, directly associated with the safe operation of TAP in the event of failure initiation. Thus, an investigation was implemented to study if this is an active landslide and if it reaches to the bottom of the mountain were the pipeline crosses. In that case, and in a future activation of the sliding mass, the pipeline could be buried and highly compressed by thousands cubic meters of debris, resulting to pipeline rupture. The identification of this active landslide was one of the primary reasons why a microtunnel option was considered (now constructed) having gas pipeline to be beneath it.

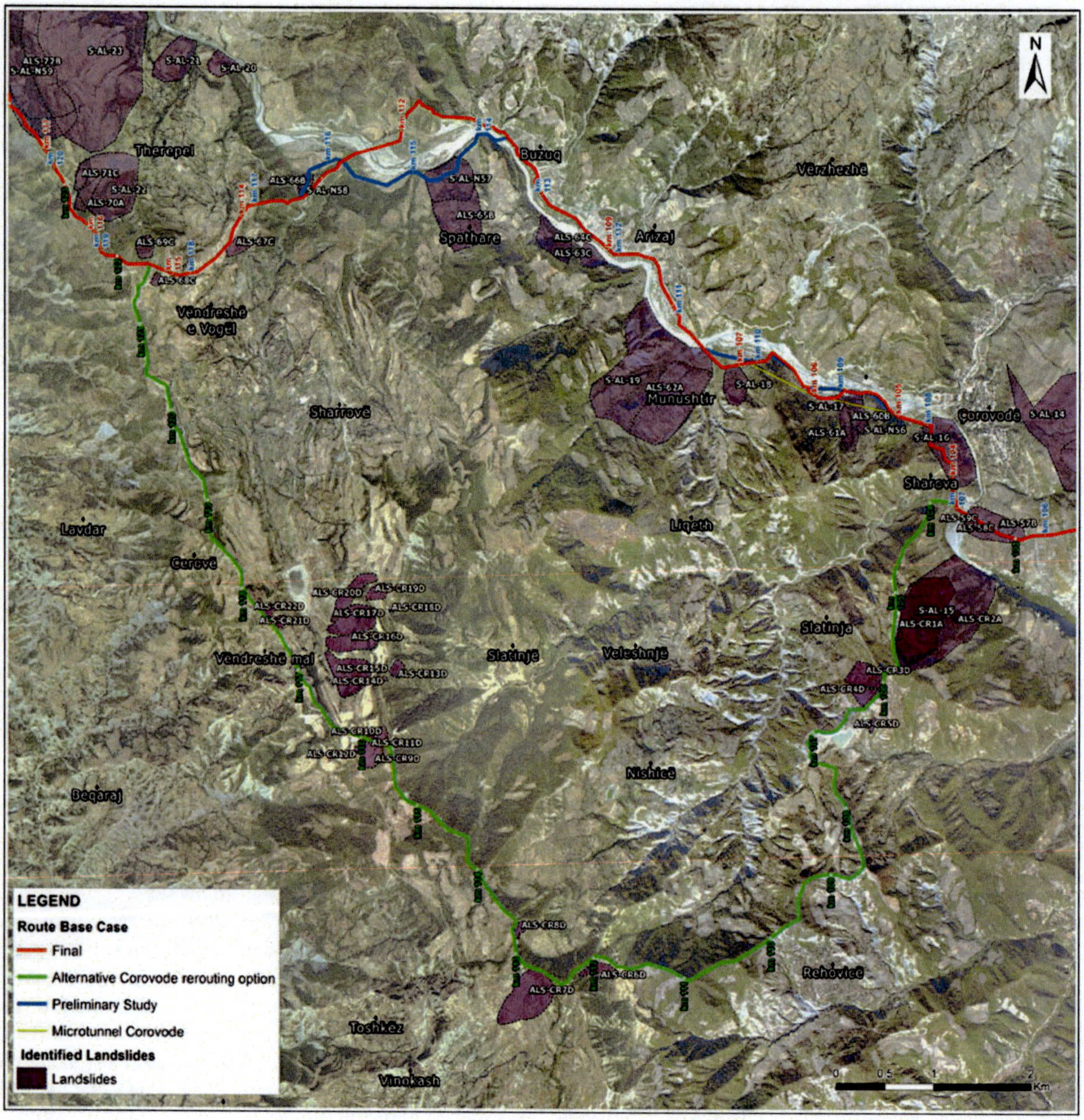

Fig. 18 TAP route—103 to 115 km (Corovode). The red line designates the pipeline route that was investigated and mostly followed. Local re-routings and micro-tunnelling, though, were applied to avoid some hazardous areas. Another route (green line) was also investigated but some "no-go" hazardous zones were found from a focused field survey

The crest of the landslide has almost reached the top of the slope. It is expressed with fresh recent back-scarps with an arch-shape (see Figs. 20 and 21). The failure surface does not follow a specific pre-surface but seems to follow a circular surface. Smaller slides are present within the whole sliding mass, which is expected for a large landslide. As such, one was found next to the river slope. This landslide is a typical form of rotational slide, in a weak siltstone - domained formation, triggered in the past also from the river erosion that undercut the toe of the slope.

The toe of the landslide surely extends to the lower part of the slope, before reaching the river base. This is evident from the recent cracks in this lower part of the slope but also from the findings, identified on the surface till the vertical part of the terrain, next to the river, displaying mixed structureless geomaterial comprised by clayey-silty matrix

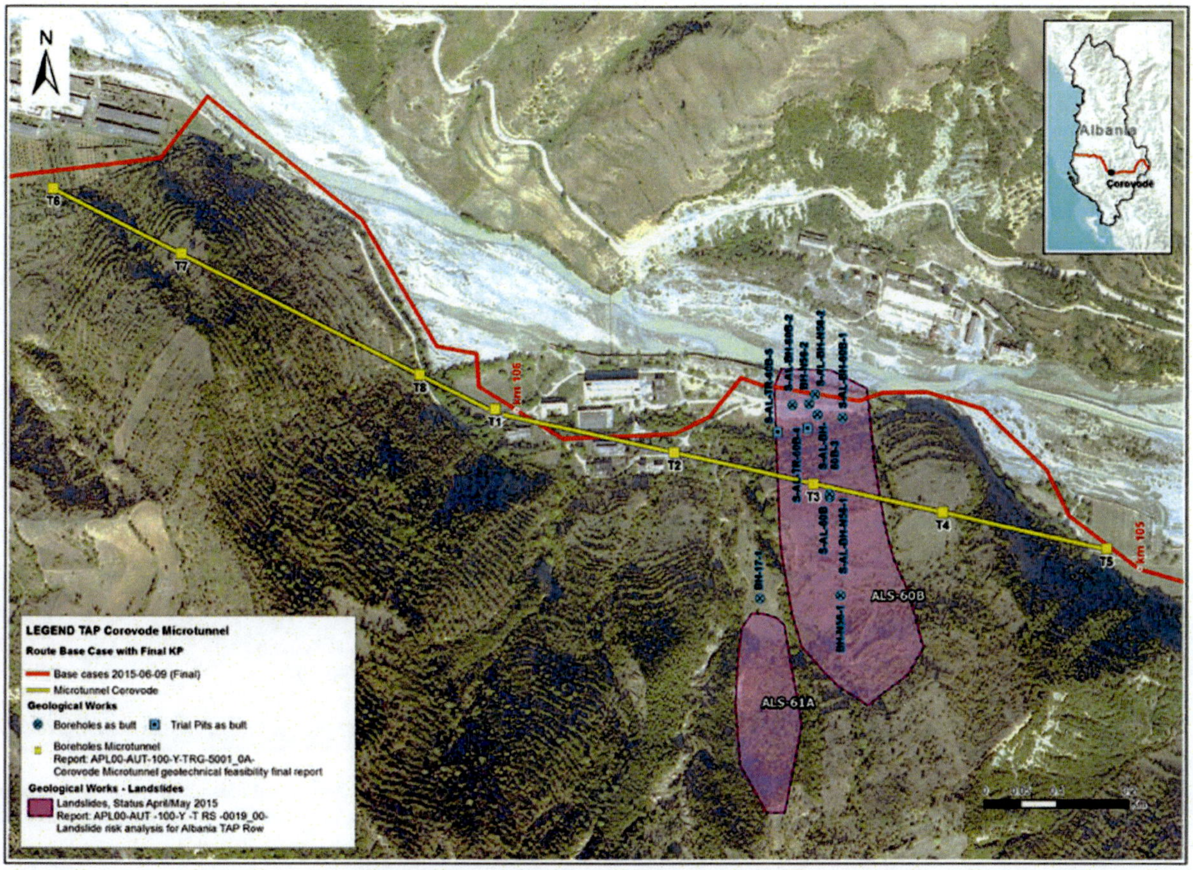

Fig. 19 S-AL-BH-N56-1, S-AL-BH-N56-2 and S-AL-BH-17-1 boreholes location

with several size angular sandstone floaters (see Fig. 22). Some findings of the boreholes and field survey about the limits of the landslide, are described briefly in the following paragraphs.

The lateral boundaries of the landslide ALS-60B are expressed along two streams, one to the east and one to the west. Both these boundaries are depicted by the evidence of outcropping flysch, represented by moderately disturbed siltstones with sandstone alternations. However, shear zones of poorer rock mass can be present in the area (Fig. 23).

Ground Investigation Data

A number of boreholes were drilled to assist building the model of the landslide and answer questions relating to the depth and the extent of the landslide and its relation to the pipeline integrity. Some indicative factual results from the drillings are hereafter presented.

Borehole S-AL-BH-N56-1

This borehole is located along the slope, located on the top of this first section of the slope, of the landslide S-AL-60B. The findings of this borehole showed that landslide geomaterial is present, with a thickness of around 19 m (Fig. 24). It is possible the landslide material covered an old river terrace basin. This possibility is based on the loose sandy material with rounded cobbles that was found around 18-21 m depth in this borehole, while above them heavily disturbed and disintegrated flysch was found. This could be possible if the river was making a meander in this location. In this case, the river was eroding this slope removing weight at the toe, creating favourable conditions for sliding and the flysch material buried the old river deposits. On the other hand, there is a possibility that these cobbles are parts of a loose conglomerate that has also sled and during drilling was completely washed out. In any case, this geomaterial

Fig. 20 View of Landslide S-AL-60B. The pipeline route was based on the bottom of the mountain, along the river. Old slide material might be encountered in the area due to the morphological indications along the slope. The Top left: The back-scarp of landslide ALS-60B. Bottom right: A recent crack on the lower part of the slope before reaching the river base

present the toe of the landslide, since no similar sled material was found in boreholes S-AL-BH-60B-1 and S-AL-BH-60B-3, which are exactly in the toe of the slope.

The results of borehole BH-N56-1 are:

- 0.0–18.7 m: Landslide material.
- 18.7–22.7 m: Possible old river terrace deposits.
- 22.7–50.0 m: Flysch bedrock.

Borehole S-AL-BH-N56-2

This borehole is located almost on the pipeline route, next to the current river level. The findings of this borehole showed that there is a thick layer of soil material, while the flysch bedrock was found around 24 m. This is natural, since this area belongs to the current eroded valley of the river, where flysch products from erosion are deposited. These soil material is of flysch nature (silty-clayey) and it is firm to stiff and not soft with blocks floating in the mass. Indeed, the nature and view of the sled material in borehole N56-1 with the soils in borehole N56-2 are completely different. Again, the findings in boreholes S-AL-BH-60B-1 and S-AL-BH-60B-3, located between the slope and the river certify that there is no continuation in the slide towards the river.

The results of borehole BH-N56-2 are:

- 0.0–9.7 m: River terrace deposits.
- 9.7–24.0 m: Alluvium deposits.
- 24.3–30.0 m: Flysch bedrock.

Engineering Geological Section—Ground Model

Based on the evaluation of all available data from previous studies and the field trips, as well as the elaboration of spatial geological data from the sampling boreholes and in situ observations (superficial layers, extent of weathering mantle, bedrock etc.), the following ground model was compiled (Fig. 25).

Section V: 115–131 Km (Therepelle to Polican)

This section of the pipeline runs through weathered moderately disturbed rockmass with sandstone and siltstone in similar amounts. Many landslides (ALS-66B, ALS-67C,

Fig. 21 The crest of landslide ALS-60B. Top left: Large horizontal and vertical displacements within the landslide S-AL-60B. Several smaller landslides are present within the sliding mass

ALS-68C, ALS-69C, ALS-70A, ALS-71C, ALS-72B, ALS-73B, ALS-74A, ALS-75B, ALS-76B, ALS-77B, ALS-78C, ALS-79C, ALS-80C, ALS-81C ALS-82B and ALS-83A) were identified in this area but only the large instabilities with high risk were finally avoided by a re-routing (red line in Fig. 26).

Interface of Landslides with Pipeline Route—Landslide Evaluation

The assessment of each landslide was based on a critical review of any given landslide and slope stability information that were available for each landslide site, identified along the initial route or reroute.

For this purpose, references to all identified landslides had to be clearly recorded and easy to be accessed by any participant at any phase of pipeline design and construction. It was thus important to create a landslide evaluation "database" in order to have access to information and assess the risk associated with any recorded landslide. Certain criteria were created to classify every landslide, based on: (i) Distance from the pipeline, (ii) Relevant location, (iii) Landslide Activity, (iv) Sliding depth, (v) Mechanism of failure and (vi) Triggering factor.

The following table (Table 1) is an example of a few landslides and the records of each one at a certain phase (after the completion of field works).

Management of Landslides—Pipeline Interaction

Decision Making

As shown in Fig. 27, the process for the identification, assessment and addressing landslides along a pipeline route is dynamic. Depending on the complexity of each project, every assessment step requires a scale of works as such that is described in the previous paragraphs.

In the following figure (Fig. 28), two processes and the all-important two decision-making stages require an appreciation of the risk that each landslide induces upon the pipeline. The hazards associated with this risk must cover

Fig. 22 Landslide material within S-AL-60B (**a** and **b**). The geomaterial is comprised by clayey-silty matrix with large sandstone floaters. This material was identified on the surface till the vertical part of the terrain, next to the river. c: Flysch bedrock in situ (not sled) as found at the lateral slopes, west of the landslide. This is the lateral boundary of the landslide to the west

Fig. 23 Shear zones of poorer rock mass, present in the area of S-AL-60B

Fig. 24 Indicative borehole cores of S-AL-BH-N56-1.Landslide material are present up to 19 m approximately (see detail in top right picture). Possible buried river terrace deposits are present below the sled geomaterial

the aspects of the potential impact of the activation of a landslide upon the pipeline (burial, dynamic loading, undermining etc.), namely rupture or no rupture, the proximity of the pipeline to human activities and/or environmentally significant areas and the cost and scale of repair works that may be required.

The possibility associated with the risk of landslide rupture it is assessed by evaluating the likelihood of activation of a given landslide upon static or dynamic conditions. A credible initiating event can be due to: (i) Increased pore-pressures (heavy rainfall or snow melt), (ii) Seismic ground shaking, (iii) Removal of support by erosion of the landslide or slope toe, (iv) Loading the head of the landslide or slope. This risk assessment is presented in Marinos et al. 2019, while an example is given in Fig. 28.

The first choice in the line of measures in landslide management plan is avoidance (Figs. 27 and 28). Rerouting options, a key process in the flowchart of Fig. 29 should be examined as a primary mitigation measure in order to avoid areas with large scale and active landslides, since they present a threat to the pipeline integrity. Small and large re-routings are typically done for the avoidance of a landslide or a cluster of landslides or even an area that is considered as one with a high landslide potential under either static or dynamic conditions. However, each rerouting and especially a large scale one, creates the necessity for the evaluation of an area that may have not been considered at a previous phase and as such requires a whole new landslide evaluation. Hence, all re-routes to be verified in the field for landslide and constructability issues.

If re-routing is not possible, there are two options: (a) isolate the pipeline from the threat (typically piling is done) and (b) reduce the threat by stabilization of the landslide (Fig. 28). In this case, detailed ground investigation, construction of geotechnical profiles (based on available engineering geological and geotechnical data: mapping and borehole drilling, rock mass and soil properties from laboratory and in situ tests, rock mass classification, groundwater

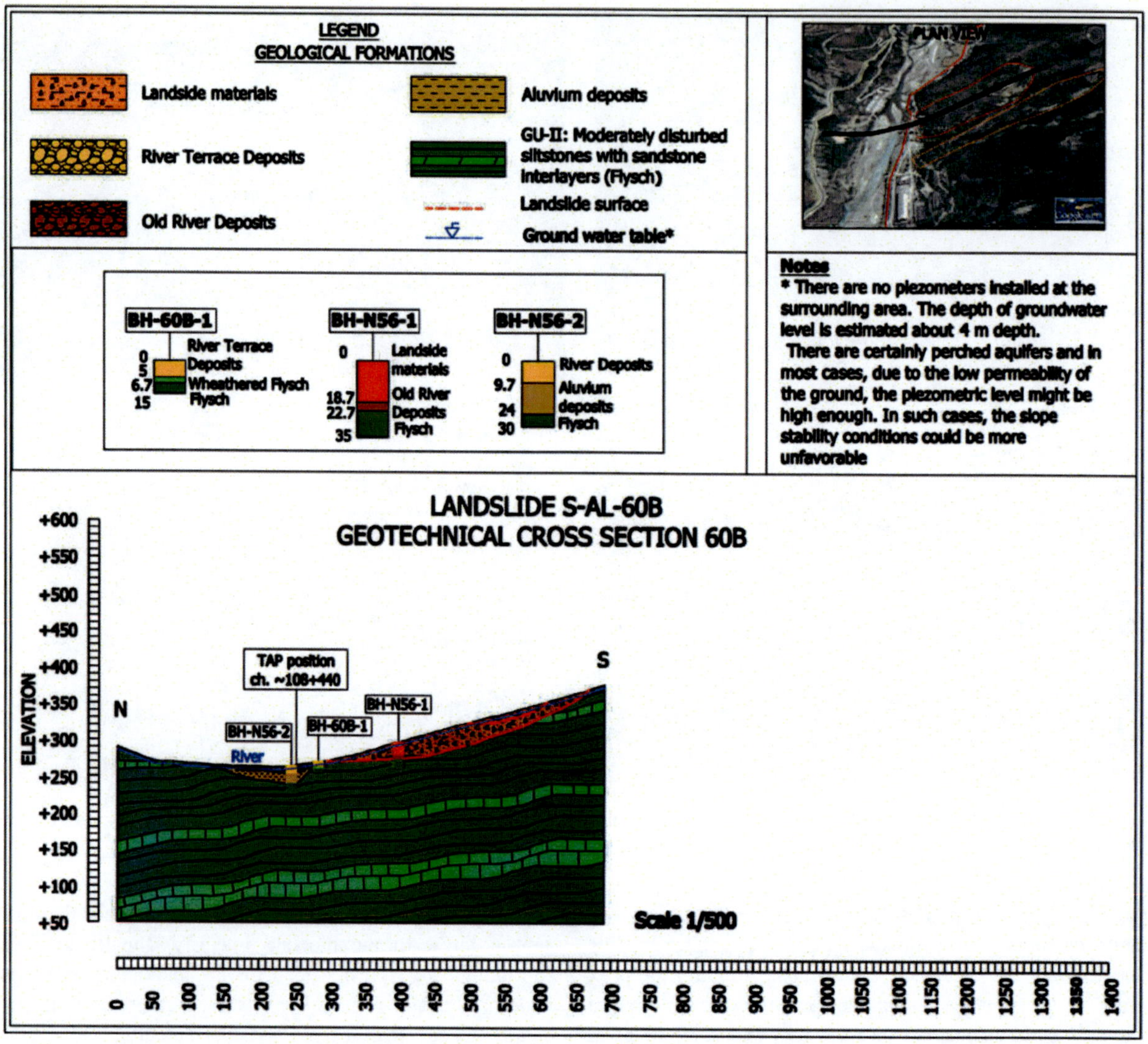

Fig. 25 Ground model of landslide S-AL-60B. Here, three boreholes are projected along the selected section shown in the top right corner. The scarp of the landslide was identified in the field

etc.), geotechnical analysis, pipeline verification and stability analysis—numerical modelling to support the design of engineering solutions are necessary.

The methodology that was followed for the development of geotechnical profiles included the following basic steps for each area:

- The evaluation of all available data from previous studies and field trips (topographical, seismological, seismotectonic, geological and geotechnical)
- The statistical analysis of all available engineering geological and geotechnical data (laboratory tests, rock mass classification), per individual area
- The elaboration of spatial geological data from sampling boreholes and in situ observations (superficial layers, extent of weathering mantle, bedrock etc.)
- The final proposal of geotechnical profiles which include the geometry of the slope, the stratigraphy of the area, the geotechnical properties for each engineering geological formation and seismological info of the site-specific area

As an example, the geotechnical sections of ALS-40B and ALS-60B, landslides that were avoided by the final pipeline route, are shown in Figs. 16 and 25.

As already noted, due to design and site-specific constraints, the selected route may not avoid all identified

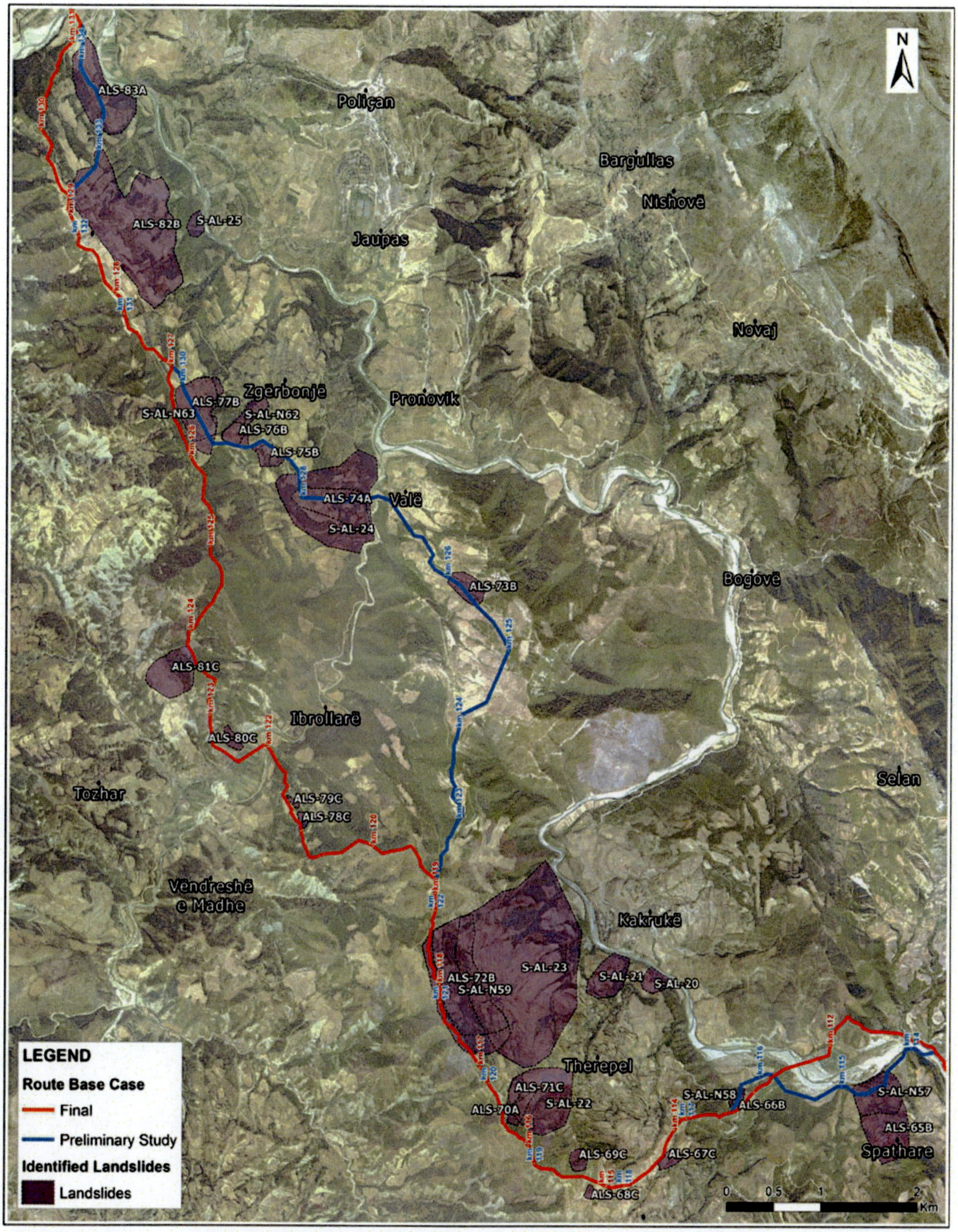

Fig. 26 TAP route—115 km to 130 km (Therepelle to Polican). The blue line designates the base case pipeline route that was partly followed. A re-routing (red line) was decided to avoid some landslide areas. The different colours in the landslide polygons indicate different dates of landslide mapping

Table 1 Example of recorded landslides and their respective attributes and characteristics. All landslides in this section occurred in a tectonically deformed intensively folded/faulted siltstone with broken and deformed sandstone layers, forming almost a chaotic structure with a GSI of 20–25 (Flysch Type X in Fig. 5)

Landslide	Area (m^2)	Distance from pipeline (m)	Measurement of distance from or passing through (for zero distance):	Activity (Active, Unknown)	Sliding surface (deep, medium, shallow)	Failure mode	Remarks	Actions
ALS-1C	1886	10	Flank	Active	Shallow	Rotational	Cracks	Avoid: Move the route 15 m to the NE (discussed in the field)
ALS-7B	6067	28	Crown	Active	Shallow-Medium	Translational	Creeping-cracks	Investigation—Remedial Measures (not considered extensive and deep)
ALS-8B	9513	244	Crown	Active	Shallow-Medium	Translational	Not immediate affect	In an adequate distance from the pipeline
ALS-9B	82025	47	Crown	Active	Medium	Rotational	Creeping-cracks/small lakes	In an adequate distance from the pipeline
ALS-10C	746	6	Crown	Active	Shallow-Medium	Rotational	Creeping-cracks	Investigation—Remedial Measures (not considered extensive and deep)
ALS-11B	37707	204	Crown	Active	Medium	Rotational	Creeping-cracks/small lakes	In an adequate distance from the pipeline
ALS-12B	16076	49	Crown	Active	Medium	Translational	Creeping-cracks/small lakes	In an adequate distance from the pipeline
ALS-13B	31684	23	Crown	Active	Shallow-Medium	Rotational	Creeping-cracks/lakes	Investigation—Remedial Measures (not considered extensive, around 60 m)
ALS-14AB	90259	2	Crown	Active	Shallow-Medium	Rotational	Creeping-cracks/lakes	Investigation—Remedial Measures-Isolate the pipeline from the sliding mass
ALS-15C	3493	2	Crown	Active	Shallow-Medium	Translational	Creeping-cracks	Isolate the pipeline from the sliding mass (from the measures of the ALS-14AB)

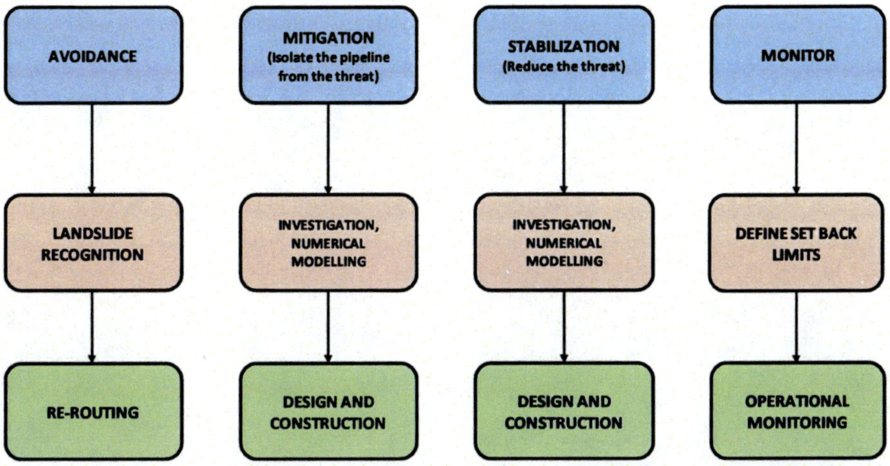

Fig. 27 Landslide Management: Avoidance/Mitigation/Stabilization/Monitor. (after M. Lee, personal communication)

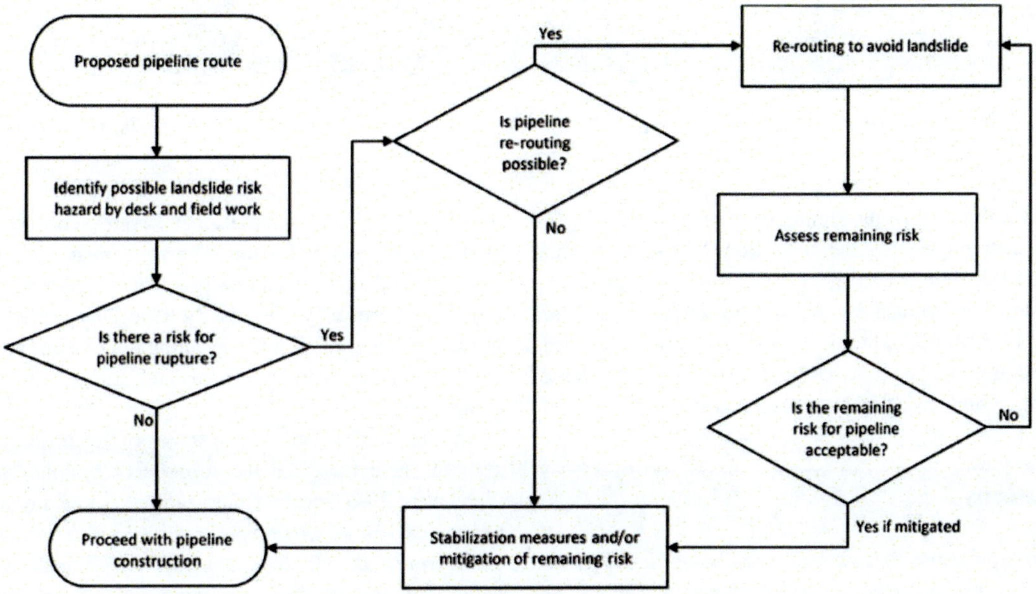

Fig. 28 Decision-making flowchart for pipelines associated with landslides. It can also incorporate various landslides that pose a threat to the pipeline

landslide areas and as such, it may not be possible to eliminate the exposure to all landslide risks. Accordingly, the route may pass in close proximity (<50 m) to either known landslides or potentially unstable areas. The risk profile is expected to be dominated by upslope expansion of existing landslides, resulting in loss of ridge crest and pipeline rupture. However, there remains a possibility that new "first-time" landslides could develop under static and/or seismic conditions, especially on steep ridge flanks areas.

This route should be feasibly constructed, with reasonable risk or reasonable engineering remedial measures. The risk could be significantly reduced through implementation of mitigation measures, such as minor adjustment to the route (to increase the set-back distance from the pre-existing landslide head), the construction of structures to isolate the pipeline from the effects of future landsliding (e.g. pile walls, gabion walls and concrete walls) or reduce the likelihood of future damaging ground movements (e.g. geotechnical slope drainage). Also, erosion control can prevent gully head landsliding and loss of the RoW. Sidecasting, during the construction of the RoW, must be also very cautious since it can lead to man-made instability problems that could impend the pipeline.

When the pipeline runs along the foot of an active and large landslide, the landslide can be avoided by

Fig. 29 Landslide risk assessment results for the pipeline route from approximately 73 to 80 km (see Sect. 6.1 and Fig. 10 for the landslides in this given example) (Marinos et al. 2019)

micro-tunnelling below the sliding mass or by protecting the pipeline with engineering measures from a possible loading from landslide, constructed next to the river. The threat can be furthermore avoided by below the sliding mass (see example of ALS-60B). Monitoring may be required during construction and operation of the pipeline in landslides of moderate or high risk that have not been avoided.

Conclusions

This paper presents: (i) how landslides were identified and engineering geologically appraised along a natural gas pipeline and (ii) the methodology that was followed for addressing the various issues that arise from that identification. These identified landslides were a principal aspect for the pipeline route selection in the Albanian mountain region, probably the most challenging onshore section of the SCPX-TANAP-TAP pipeline system because of the combination of narrow ridge crests, weak rock masses of flysch sediments, high seismicity, widespread active landslides and unstable side slopes, and very limited access.

The most important outcome of the whole process, including not only field trip findings but also desk revisiting of available data, evaluation of ground investigation campaigns as well as a number of discussions between members of the participating bodies,, was the avoidance of significant landslides and finalizing of the preferred route for the TAP, in accordance with international best practice by routing along ridge crests and spurs. The assessment must be based on remote sensing techniques (e.g. LiDar, UAV and satellite images) to primarily identify the landslides, extensive field work evaluation of all findings along the pipeline route, desk study of other available data and ground investigation campaigns.

In total, during the various stages of this process, namely after the field trips and the subsequent evaluation of data, there were 174 landslide delineations (some involving the same area), 66 of which were avoided via re-routings.

However, due to the nature of the difficult geological conditions and its dynamic geomorphological terrain, the route still passes in close proximity to known landslides on the ridge flanks and slopes that are believed to have the potential to fail at some moment in the future, either under static and/or seismic (dynamic) conditions. In such cases, further investigation and monitoring was required to identify the site specific geotechnical conditions and the probable remedial measures.

The landslide related decision making in international oil and gas pipelines is usually risk-based and not driven by codes. A main challenge is to estimate risk and communicate it to higher management. Furthermore, investigation and stabilisation of landslides is not the best international practice due to time and cost constraints (Sweeney 2005)

The results of this work, focused on the engineering geological appraisal, provides a "high level" indication of how landslides, mainly identified in the field, are correlated to possible failure event scenarios associated with pipeline rupture. Hence, hazard and risk is expected to vary through a mountain terrain and RoW and this assessment concludes to identifying "hot spots" along the route, where re-routing must be primarily examined. This process is an important factor and must percid to the finalisation of the route. When re-routing is not possible, such evaluation should be used to prioritise risk reduction measures after detailed estimation of all site conditions. This information can be used to perform numerical analyses for slope stability assessment and the corresponding pipeline verification, in order to provide quantitative support to an expert classification of landslide risk (Marinos et al. 2019). This assessment though, cannot be however a substitute for more detailed site evaluation that would be required to support the design and construction of mitigation measures at the critical slopes according to EN1997 and EN1998.

Acknowledgements We would like to acknowledge E.ON Technologies GmbH (currently Uniper GmbH) S.A. for funding the presented work and, in particular, E. Schmidt, T. Cubaleski and A. Diendorfer for their constructive cooperation. This work was partly inspired by Mark Lee, whom we would like to thank for his insight and advice in developing the adopted hazard assessment approach. We would also like to thank K. Xanthopoulou, for her assistance in editing some of the figures.

References

Fell R, Ho KKS, Lacasse S, Leroi E (2005) A framework for landslide risk assessment and management. In Hungr O, Fell R, Couture R, Eberhardt E (eds) Landslide risk management, Balkema, pp 3–26

Hoek E, Brown ET (2018) The Hoek-Brown failure criterion and GSI—2018 edition. J Rock Mech Geotech Eng 22 pp. https://doi.org/10.1016/j.jrmge.2018.08.001

Hoek E, Carranza-Torres CT, Corkum B (2002) Hoek-Brown failure criterion—2002 edition. In: Bawden HRW, Curran J, Telesnicki M (eds) Proceedings of the North American rock mechanics society (NARMS-TAC 2002). Mining Innovation and Technology, Toronto, Canada, pp 267–273

Lee EM, Jones DKC (2014) Landslide risk assessment, 2nd edn. ICE Publishing

Lee EM, Audibert JME, Hengesh JV, Nyman DJ (2009) Landslide-related ruptures of the Camisea pipeline system, Peru. Quart J Eng Geol Hydrogeol 42:251–259

Marinos V (2014) Tunnel behaviour and support associated with the weak rock masses of flysch. J Rock Mech Geotech Eng 6:227–239

Marinos V (2017) A revised, geotechnical classification GSI system for tectonically disturbed heterogeneous rock masses, such as flysch. Bull Eng Geol Environ. https://doi.org/10.1007/s10064-017-1151-z

Marinos V, Fortsakis P, Prountzopoulos G, Marinos P (2011) Permeability in flysch— distribution decrease with depth and grout curtains under dams. J Mount Sci 8(2):234–238

Marinos V, Stoumpos G, Papazachos C (2019) Landslide hazard and risk assessment for a natural gas pipeline project: the case of the Trans Adriatic Pipeline, Albania section. Geosciences (Switzerland) 9(2). https://doi.org/10.3390/geosciences9020061

Nyman D, Lee EM, Audibert JM (2008) Mitigating landslides for international pipeline projects: challenges and lessons learned. IPC2008-64405. In: Proceedings of IPC 2008: 7th international pipeline conference calgary, Canada

Savigny KW, Porter M, Leir M (2005) Geohazard risk management for the onshore pipeline industry. Bus Brief: Exp Product: Oil and Gas Rev 2:1–3

Sweeney M, Gasca AH, Garcia-Lopez M, Palmer AC (2005) Pipelines and landslides in rugged terrain: a database, historic risks and pipeline vulnerability. In Sweeney M (ed) Terrain and geohazard challenges facing onshore oil and gas pipelines. Thomas Telford, pp 641–660

Sweeney M (ed) (2005) Terrain and geohazard challenges facing onshore oil and gas pipelines. Thomas Telford, London, 758 pp

Young A, Lockey A (2013) The assessment of pipeline integrity in geohazard areas using ILI data. Proceedings of the ASME 2013 international pipeline geotechnical conference IPG2013-1971

Loess Stratigraphy and Loess Landslides in the Chinese Loess Plateau

2020 Kyoto Japan

Tonglu Li, Mumtaz Haider, Wei Shen, and Ping Li

Abstract

Chinese Loess is a dust deposit brought by wind from North-West desert. In the past four decades, the age and the strata of the loess stratigraphy were well defined by dating and correlating with deep sea core isotopic curves. A standard loess stratigraphic profile and the correlation among the different areas are introduced, and typical river terrace with loess deposit is also illustrated. Loess has a loose texture and is sensitive to water, which makes it susceptible to failure during rain or engineering activities. Landslides on the Chinese Loess Plateau are the most common and widespread geological hazard, which results in human deaths and severe injuries. Moreover, it damages infrastructures such as road, tunnels, bridges, railway tracks and engulfs agricultural land every year. Based on the failure modes and motion characters, the loess slope failure can be classified into two main types, shallow slide, and deep-seated landslide. The shallow failure can be further divided into two types of small collapse and shallow flow slide; while the deep-seated failure can be divided into the four types of rapid long run-out slide, rapid flow slide, slow creep slide and quick slump slide. The mechanism of the landslide movement is controlled by the moisture state on the failure surface and the moving path. Based on the Sassa's motion simulation model, the modified models for simulating the catastrophic long run-out loess landslides are proposed which consider the effects of curved moving path, frontal plowing, bed entrainment on the running distance and coverage area of the sliding materials.

T. Li (✉) · M. Haider · P. Li
Department of Geological Engineering, Chang'an University, Xi'an, 710054, Shaanxi, China
e-mail: dcdgx08@chd.edu.cn

M. Haider
e-mail: mumtazhaider@hotmail.com

T. Li · P. Li
Water Cycle and Geological Environment Observation and Research Station for the Chinese Loess Plateau, Ministry of Education, Zhengning, Gansu, 745399, China

W. Shen
Dipartimento di Scienze Biologiche, Geologiche e Ambientali, Università di Bologna, 67-40126, Bologna, Italy
e-mail: shenweichd@qq.com

Keywords

Loess • Stratigraphy • The chinese loess plateau • Landslide • Run-out • Simulation • Failure mechanism

Introduction

Loess is a fine wind-blown dust deposited in the arid to semi-arid regions where favourable weather conditions help in its formation and deposition. Globally, loess covers 10 per cent of the earth surface and is distributed mostly in region spanning 30–55° North latitude and little in 30–40° South latitude, which are monsoon dominated zone. Loess deposits, unlike water-deposited sediments, are not only distributed in river basins; they are also distributed in mountain ranges, the seas and oceans, which strongly prove its wind-blown origin.

Loess in China covers a total area of approximately 631,000 km^2, occupies 4.4% of the national land (Liu 1996), of which, the Chinese Loess Plateau (Fig. 1) has the best developed deposit in the world. The thickness of loess deposits in this area vary from few meters to more than 300 m.

The Chinese Loess Plateau, which has an area of approximately 430,000 km^2, is located in the middle reach of the Yellow River and is bordered by Riyue Mountain to the west, the Maowusu and Tengeri Deserts to the north; Taihang Mountain to the east and Qinlin Mountain to the south (Fig. 1). As monsoon winds carry fine dust from the northern deserts, it is barred by the eastern and southern mountains; this causes the fine material to be deposited on

Ž. Arbanas et al. (eds.), *Understanding and Reducing Landslide Disaster Risk*, ICL Contribution to Landslide Disaster Risk Reduction,
https://doi.org/10.1007/978-3-030-60713-5_27

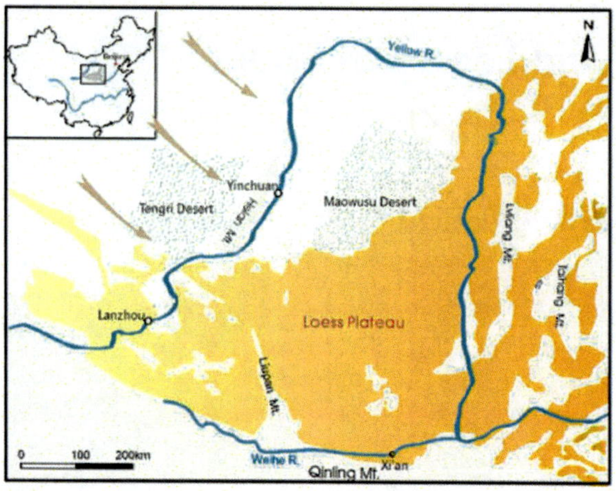

Fig. 1 Loess distribution on the Chinese Loess Plateau

the plateau. Field investigations and laboratory analysis have shown that particle sizes of the loess soil deposit changes from coarse-grained to fine-grained for sampling done from the north western part to south eastern end. Recently, the Loess Plateau has become an important economic zone in the western part of China, because of abundant natural resources such as oil and gas, coal, hydro-energy and it also serves as an important site for agricultural development. The geo-hazards, such as landslides, debris flows, ground collapses have continued to cause havoc to the economic and social development within the area. In China, landslide disaster, apart from earthquake, has claimed the highest number of fatalities in recent years. The physical and mechanical properties of loess are determined by its composition and moisture content. In dry conditions, the loess has a high strength and the slopes could be stable at a steep angle, so we can often see the vertical loess cliffs in the area. However, the strength decreases as the moisture content increases.

In recent 2 decades, loess landslides have been well studied based on the loess stratigrapgy and loess properties. Many typical cases have been investigatied in detail. So this paper aims to review the acievements of the classical loess stratigraphy and sumerizes the up-to-date research results of loess landslide. Therefore, types of the loess landslide are classified based on failure and motion mechanism which could attribute to the moisture content of the loess. The modified models for simulating the catastrophic long run-out loess landslides are proposed based on the mechanism then after. It is of significance to mitigation and further investigation for the landslide in the loess region.

Loess Stratigraphy

During the Quaternary period, the climate became dry and cold and loess started to deposit on the weathered landscape which is covered by Tertiary red clay or earlier rocks. In the early Quaternary period, the Loess Plateau had a series of inner basins and river beds which were filled with lacustrine and alluvial deposits, while the surrounding high lands received the loess sediments, which made up the present loess platforms. Luochuan loess platform is one of the largest loess platforms in northern Shaanxi, with successive loess sediment in the Quaternary. A loess profile in Potou village, Luochuan County, located at the western side of Heimugou gully, has been considered as a standard loess profile on which the Quaternary paleo-climatic and stratigraphic studies have been carried out in detail (Kukla and An 1989). Liu and Zhang (1962) divided the loess strata into three formations based on observed field characteristics. They are: the Wucheng loess, Lishi loess, and Malan loess. Beside these three typical formations, a layer of 2 m or so of Holocene formation covers the upper part of the deposit. Holocene formation consists of a brownish-yellow thin bed in the upper part while a dark brown bed in the lower part, of which the upper is called Holocene loess (L0) and the lower is called black loam (S0).

Due to incessant changes in climate during the Quaternary, the sediments are not uniform in colour, composition and structure on the profile. They occur as alternating beds of brownish-yellow loess and brownish-red paleosol. Loess represents the product of dry-cold climate, while paleosol, as well as the loam, represents that of warm-wet climate. Loess strata on Potou profile is made up of more than thirty loess-paleosol sequences, indicating that the climate must have changed for more than thirty periods during the Quaternary period. The Malan loess is the first layer of typical loess which is a set of uniform loose material corresponding to the last glaciation. Lishi loess includes the first paleosol to the fifteenth loess sequences. Of all the above loess beds, the ninth (L9) and fifteenth (L15) loess beds have relatively higher thickness and coarse particles indicating deposition under harsh weather conditions, while the fifth paleosol (S5) is a thick layer consisting of three sub-paleosols with two thin loess beds among them. So L9, L15 and S5 are considered to be the marked beds for easy identification of the loess stratigraphy in the field. Wuchen loess consists of thin interbeds of paleosol and loess with high content of calcium and stiff of properties, hence it is called as lithoid loess. Below the Wuchen loess underlies Neocene red clay.

Since 1980 a detailed age dating of the loess formation has been recorded through the combined application of radioactive (C14) and paleo-magnetic dating methods, and the correlation of isotope curves along the oceanic bore holes and loess profiles. From the results, loess of the Quaternary are divided into Holocene (Q4), Upper Pleistocene (Q3), Middle Pleistocene (Q2) and Lower Pleistocene (Q1) series. The boundary of the Holocene and Upper Pleistocene is located at the bottom of the black loam, dating at 11,000 years BP. That of the Upper Pleistocene and the Middle Pleistocene is located at the bottom of the first paleosol (S1), with ages dating at 128,000 BP. Also, that of the Middle Pleistocene and the Lower Pleistocene is located in the eighth loess (L_8), with ages dating at 730,000 BP. which corresponds to the B/M boundary of paleo-magnetism. The boundary of the Lower Pleistocene and the Upper Pliocene is located at the bottom of the loess, a little higher than the M/G boundary of paleo-magnetism. The age of the M/G boundary is about 2.47 million years, therefore, the boundary of the Quaternary and Tertiary is considered as 2.50 million years or so. It can be observed that there is a wide difference between the early lithological units and the present chronostratigraphic units. The stratigraphic profile of the Luochuan loess is shown in Fig. 2. The profile also shows that the loess-paleosol sequences from different parts of the Loess Plateau are spatially correlative, as shown in Fig. 3 (Liu 1996).

In the west of the Liupanshan Mountain, 6 terraces can be distinguished along river sides. The highest is terrace VI, on which loess is about 310–505 m thick with 21–23 layers of paleosol, presently the thickest loess deposits in the world. Loess on terrace VI started to sediment at 1.43 Ma BP. Loess on terrace V consists of L0 to S16, 200–400 m in thickness, and dated at 1.23 Ma BP. Loess on terrace IV consists of L0 to S6, 100–200 m in thickness, and dated at about 0.62 Ma BP. Loess on terrace III consists of L0 to S1, about 40–60 m in thickness, and dated at about 0.12 Ma BP. Loess on terrace II is a layer of loessial soil with thickness of 20–35 m and dating at about 0.03 Ma BP. Loess on terrace I consist of present loess L0 and black loam S0, about 2.0 m in thickness, and dated at about 0.01 Ma BP. Figure 4 is a typical section of the Yellow River terrace near Lanzhou city, which represent the loess on six terraces in the west of the Loess Plateau.

In the east of Liupanshan Mountain, there are generally 5 terraces covered by loess, but loess thickness is thinner than that in the west. Loess on terrace V consists of L0 to S1–S16 (Varied in places), about 70–90 m in thickness, and dated at 1.23 Ma BP. Loess on terrace IV consists of L0 to S9, about 40–70 m in thickness, and dated at about 0.80 Ma BP. Loess on terrace III consists of L0 to S6, about 25–45 m in thickness, and dated at about 0.62 Ma BP. Loess on terrace II consists of L0 to S1, 10–17 m, dated at about 0.12 Ma BP. Loess on terrace I consists of present loess L0 and black loam S0, about 2.0 m in thickness, and dated at about 0.01 Ma BP. Figure 5 is a typical section of the Luohe River terrace near Yan'an city, which represent the loess on five terraces in the east of the Loess Plateau region.

The processes of a terrace consist of two episodes: lateral erosion and down-cut. In the tectonic static period, a river has lateral erosion and the riverbed is widened, while during the uplift period, it cuts down vertically to form a new narrow riverbed and the original riverbed and floodplain becomes a terrace. The new riverbed would be the next terrace in the forthcoming static-uplift cycle. Therefore, on the record of river terraces, the Loess Plateau have experienced at least six cycles of uplift which implies the 6 periods of down erosion at the same time. The erosion cases can also be demonstrated by the interaction among some new loess bed with the old ones.

The Properties to Facilitate Loess Landslide

Loess has typical landforms, special structure and textures, as well as special physical and mechanical properties, which favours loess landslides to happen.

Typical Topographies

Loess mainly forms three characteristic types of geomorphic structures. These are loess platform (Fig. 6), loess ridge (Fig. 7) and loess dome (Fig. 8). Other form such as loess Karst is not very common. It could be seen that on the sides of the platforms, ridges and domes are steep slopes which could easily slide in favourable conditions, such as infiltration of rain and irrigation water or during an earthquake jolt. Generally, loess landslides occur on the slopes steeper than 35° and higher than 30 m. Our investigation in the northern Shaanxi Province shows that 80% of loess landslides occurred in slopes with dip angles exceeding 35°. Failure style, for the slopes steeper than 35°, is normally collapse-sliding or purely collapse; the slopes are stable for slope angles lower or close to 35° (Li et al. 2013a).

The sliding is more likely to occur in concaves slopes. The landslides are often continuously distributed on the sides of the loess valleys as the side being the rims of loess platforms, ridges and domes, hence forms the so-called "slide-skirts" or "slide zone". The loess landslides on the southern side of the valley with E–W strike are more than those on the north side because of less sunshine and higher moisture content in the south side.

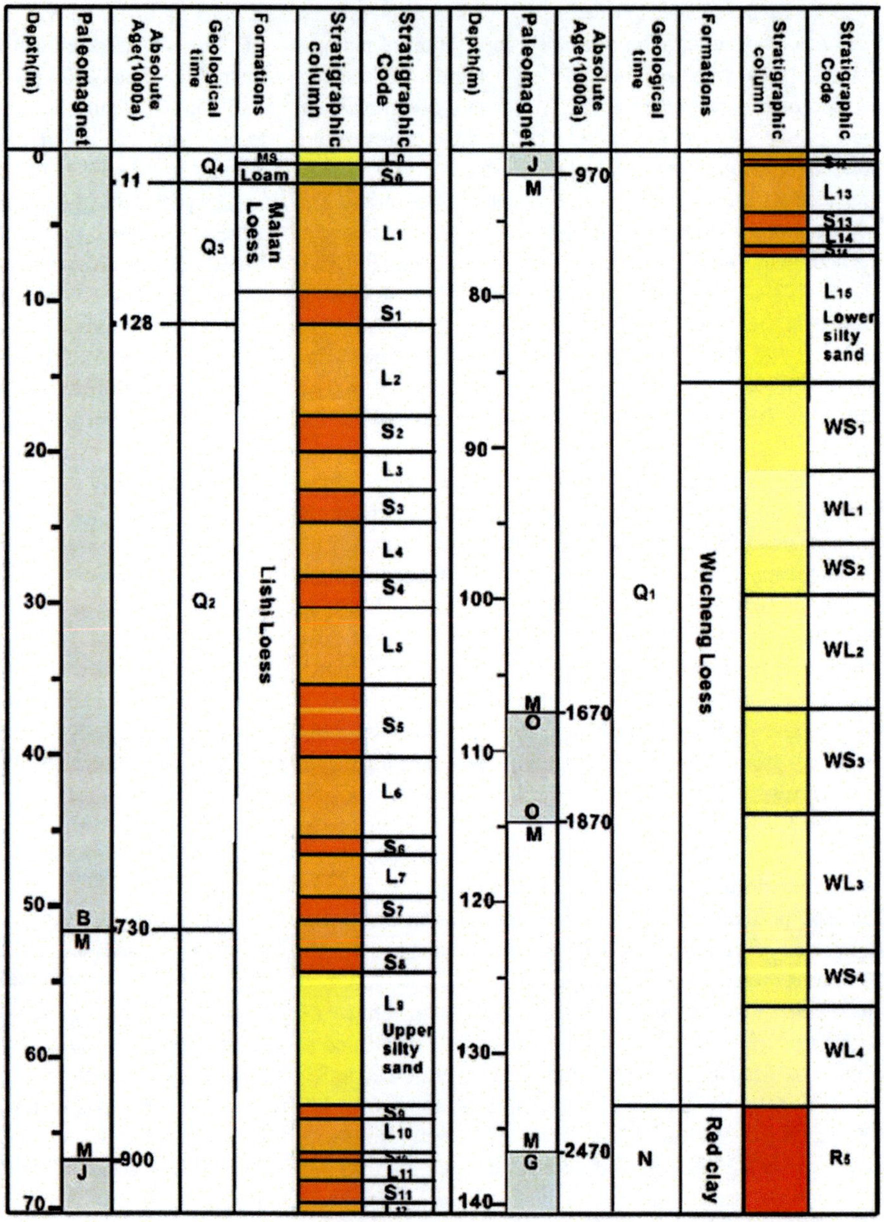

Fig. 2 Stratigraphy of the Luochuan standard loess profile

Joints in Loess

Vertical joint is a unique structural feature in loess, which is due to loess origin by wind-blown processes. Vertical joints in the loess deposits constitute weak planes along which rainfall easily infiltrate into the loess (Fig. 9). The joints are the main erosion path in the loess slopes especially on the margins of the platforms and ridges; the joints generally connect the upper and the lower part of the slopes. As a result, a lot of sinkholes (Fig. 10) form on the slope surface with outlets appears on the slope toe. The process enlarges the joints into wide fissures which allow more water to flow in and form temporary waterfall during rainy periods. Erosion along the joints not only affects the slopes, but also allows more water to flow into the slope which essentially affects the stability of the slope.

Loess Texture

Loess has a porous texture characteristic. The loess in upper Q2 and Q3 formations, L1 to L2 generally have void ratios of 0.8–1.1, that means the ratio of pores to solid particles is about 1:1, so it is a very loose material. As viewing on SEM

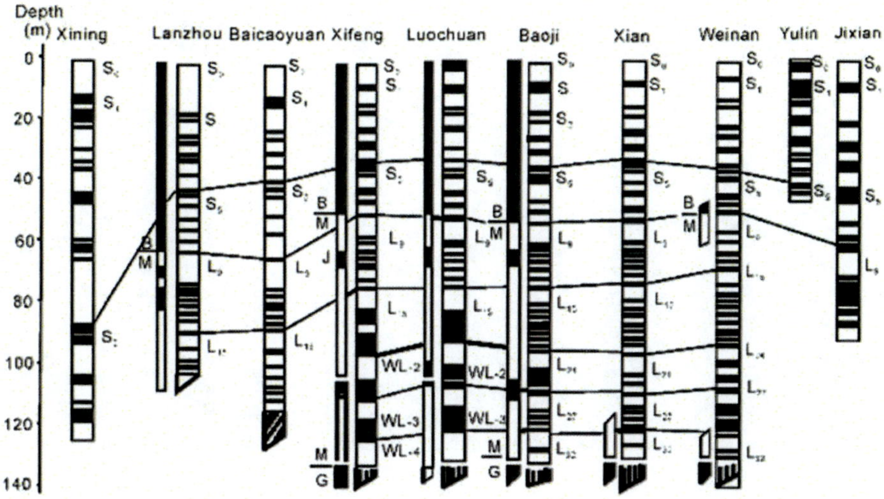

Fig. 3 Correlation of the loess stratigraphy on the Loess Plateau (After Liu 1996)

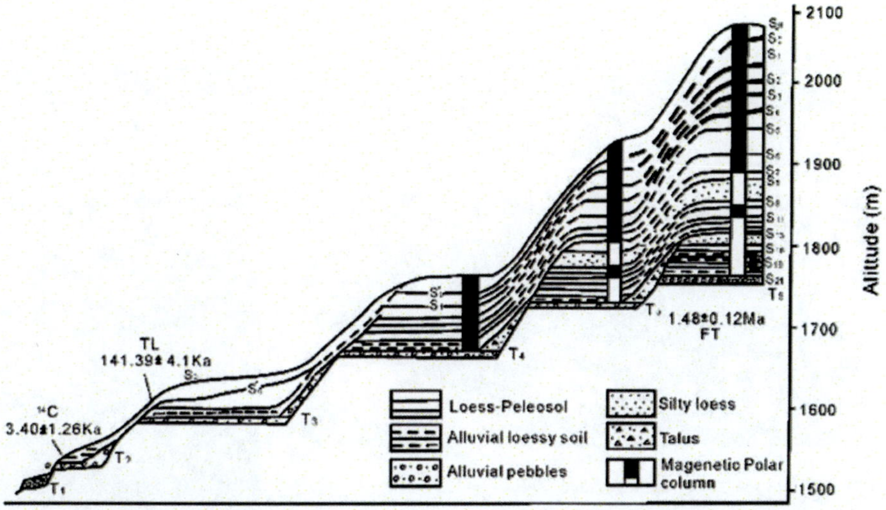

Fig. 4 Loess section on the terraces of the Yellow river (After Zhang 1989)

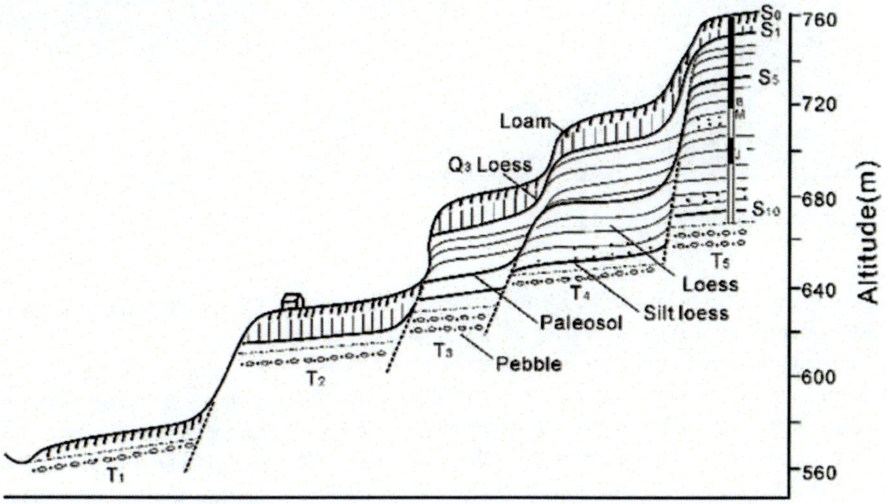

Fig. 5 Loess section on the terraces of the Luohe River

Fig. 6 The topography of loess platform

Fig. 7 The topography of loess ridge

Fig. 8 The topography of loess domes

photos, the micro-holes can be identified clearly (Fig. 11a). The particles are mainly aggregates of silts around the holes contacting at points with weak connection by clay minerals. The connection is water-sensible. As soaked in water, the particles will lose their connection to disintegrate. At last, they become more fine granules and fall in into the holes (Fig. 11b). This typical texture of loess has a special quality

Fig. 9 Vertical joints in loess along which the rain water flow down and plant roots grow in

Fig. 10 Sinkholes on the ground

of collapsibility, which means as the loess has consolidated under given load, if soaked by water, it will create an additional depression.

Types of Loess Landslide

Because of continuous uplifting of the crust and overlapping of the loess deposit, while the down cutting of the river bed, the Loess Plateau has formed the typical topography of high land interpolated by low valleys and gullies. Therefore, slope failures are the most popular geo-disasters in the Chinese

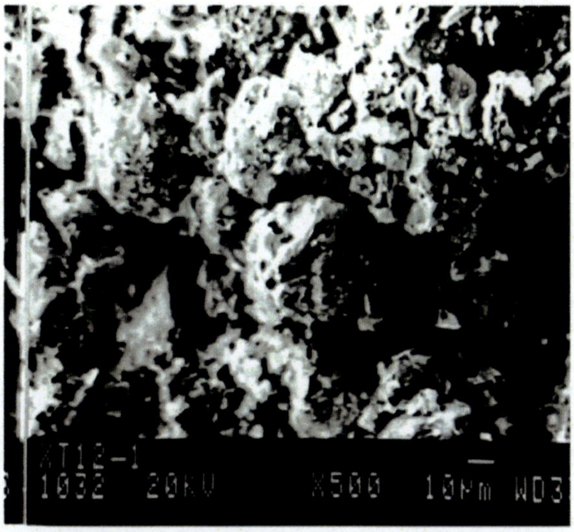

(a) Intact loess specimen

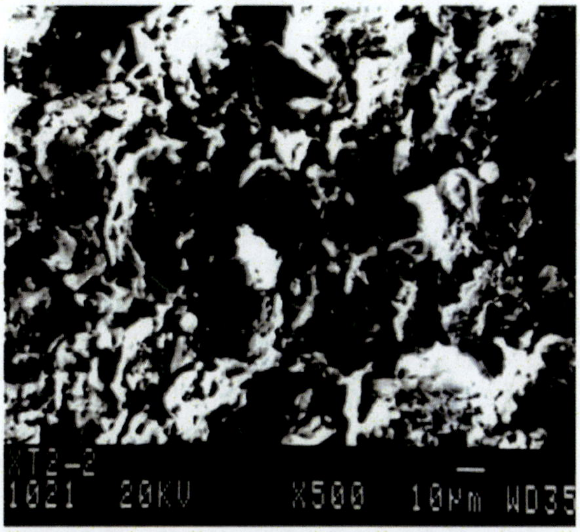

(b) A loess specimen after loaded and soaked

Fig. 11 SEM image of the loess specimens before and after water soaked

Loess Plateau. According the failure mechanism and the motion characters, the loess slope failure can be classified two main types, one is shallow failure, and another is deep failure. The shallow failure can be further divided into the two types of small collapse and shallow flow slide; while the deep failure can be divided into the four types of rapid long run-out slide, rapid flow slide, slow creep slide and quick slump slide.

Small Collapses

Small collapses commonly occurs in the slopes which are steeper than 50°. They are triggered by capillary water rising on the toe of the slope other than water infiltration on the top of slopes. As shown in Fig. 12, for a steep slope, the raindrops hardly fall on the slope surface which always keeps drying. Wetting depth on the top of a slope is no more than 2 m in the extreme rainfall event on the Loess Plateau, so a vertical slope, more than 30 m high in dry, can stay safe while more than 2 m in wet at toe, maybe fail. When rain water falls on the ground off the slope, the capillary water would move in the slope toe. Vertical joints and unloading cracks may cut the steep slope into columns; the strength loss of the wetting toe may cause the column failure. This kind of small collapse is popular in the steep and short cutting slopes along road, back scarps of landslide and sides of erosion gullies. The stress state of the slope is close to uniaxial load. In the worst case, the failure criteria is that the vertical stress exceeds the saturated uniaxial compression strength.

Shallow Flow Slides

Shallow flow slides occur in companying with heavy rains or long-term drizzles. Most of the shallow flow slides are small scale, ranging in volume from tens to hundreds of cubic meters. They generally occur at a depth of 2 m, corresponding to a surface layer of completely saturated loess and on the slopes with angles between 35 and 50°. As mentioned above, for the slopes which are steeper than 50°, they fail as small collapse. Loess has a steady effective friction angle of 30°, so for the slopes which are gentler than 35° are stable even being saturated. A typical case of shallow flow slide occurred in Yan'an area of Northern Shaanxi Province, China in July 2013. In two months intermittently raining, induced 8,135 slope failures, most of which are shallow flow slides, as shown in Fig. 13. The slides destroyed 10,000 living caves and killed 45 people (Wang et al. 2014).

Assessment and prediction of the loess shallow flow slide can adopt the infinite slope model. However, unlike the shallow landslides on impermeable rocks or solid clays, for those on loess, the pore water pressure on the failures surface

Fig. 12 Capillary wetting on the toe of steep loess slope

Fig. 13 Rainfall induced shallow flow-slides in Yan'an area, Shaanxi province, China, in July 2013

should be ignored while the saturated unit weight of the sliding mass should be adopted, because the water in the shallow loess is perched and the underlying loess is unsaturated.

Rapid Long Run-Out Slide

Rapid long run-out slide often occurs on the sides of the loess platforms where the loess slope is high and steep. High river terraces are eroded to leave the wide and flat flood plain and the lower terrace lying between the toe of the slope and the river. The ground water level off the slope is generally very shallow. The main triggering factors are either rise in the groundwater level or surface water erosion. The sliding process can be divided into two phases. In the first phase the sliding mass separates from the sliding bed and slumps down rapidly; while in the second phase it hits on the first terrace at the slope toe, liquefaction of the saturated gravels and silt–clay may reduce the shear resistance which causes the sliding mass to move with high velocity. So this type of landslide has a very high speed and long moving distance (Sassa 1988, 1997).

Saleshan landslide in Gansu province, China is a well-known case which has been mentioned in many literatures. It failed on March 17th, 1983, which destroyed four villages, took 247 lives and 300 domestic animals, filled up two small reservoirs and covered more than 20 million square meters of farmland in a few minutes. The landslide slipped from 300 m high slope with a speed of 30 m/s and moved 1600 m forward on flat ground to the opposite bank of Baxie River. It is a typical rapid long run-out slide. The soil of the slope is composed of the Quaternary loess underlying the Tertiary clay and gravel bed. The red clay is an impermeable bed and is soften by infiltrating rainwater.

Before failure, the slope had been creeping along the flat clay bed for several years which had been proved by the developing fissures and cracks on the slope surface. Because the huge mass suddenly fell down hitting the first terrace of the Baxie River which consists of loose sands and gravels with shallow groundwater. Liquefaction of the loose material on the sliding path is the cause of long run-out, as shown in Fig. 14.

This type of landslide also popularly distributes at the side of a loess platform to the south bank of the Jinghe River, Shaanxi Province, China. The catastrophic one occurred in 1984 which killed 34 people. Those occurring in recent 2 decades are Shutangwang (in 2001), Xiuchidu (in 2002), Dongfeng (in 2003), Taiping (in 2005) and Dabozi (in 2012) landslides. Figure 15 shows a part of the bank that the whole slope is composed of loess, from Holocene to early Pleistocene, with the slope height of 70–90 m and slope angle of 50–70°. The first terrace is composed of 1 m Q4 alluvial silt–clay upper and 2–3 m or more gravel bed down underlying with Q1 solid clay, and the grounder water level is of 0.5–1.0 m in depth. The flowing water of irrigation on the top of the slope is the triggering factor of the slides. Before failing, the top of the slope had developed many fissures by unloading process and water erosion, which allowed a big amount of water flowing in. Rising of the groundwater level and reducing of the loess strength resulted in the failure of the slope at first. As the sliding mass separated from the slope and slumped down quickly, it hit on the first terrace and caused liquefaction of the saturated gravels and silt–clay, which produced a very high speed and a long distance.

In this case, pore water pressure generated by undrained loading and undrained shear in the sliding path is much higher than that inside sliding mass, therefore the shearing takes place in the ground of the path. The high pore pressure causes liquefaction of the loose sands and gravels, and then produces a big difference between the driving force and resisting force. Therefore, the sliding mass could achieve a high acceleration to move as a floating body. The mechanism is shown in Fig. 16 (Li et al. 2013b).

Rapid Flow Slides

Rapid flow slides do not occur often but occasionally occur in areas with suitable topographic conditions. It is characterized by the change of its motion functions in the source area and the path. This type of slide initially starts as a landslide but quickly transforms to a flow slide moving down along a gully to the outlet or a slope to the toe. Rapid flow slide often breaks suddenly and results in disaster to the people living downstream.

A typical case of this type of landslide is the Gaolou flow slide which occurred in Gaolou village, Huaxian County of Shaanxi Province, China on Oct. 6th, 2006 (Li et al. 2013c). It occurred on the top end of a loess gully, which cut in the side of a loess platform. While the flow slide was occurring, most of the displaced material moved down along the gully's

Fig. 14 A rapid long run-out slide, Saleshan landslide occurred in Gansu Province, China in 1983

Fig. 15 A typical topography preferring for the rapid long run-out landslide, the side slope of a loess platform to the south bank of the Jinghe river

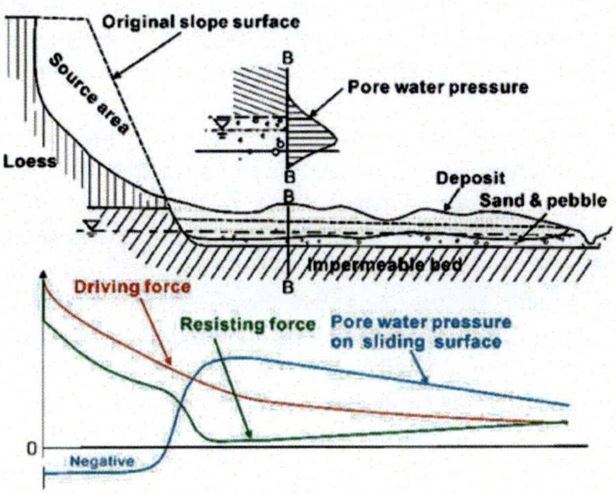

Fig. 16 A profile for the rapid long run-out slide and the variation of the forces with motion to illustrate its mechanism

floor and thereafter accumulated on the outlet of the residential area, where it killed 12 people and destroyed 94 rooms. As shown in Fig. 17, the source area of the slide was an armchair-shape after failure and the back scarp is of 30 m height. The flow path is a U-shaped gully that has a flat bottom and vertical sidewalls transversely. The outside of the gully is an open gentle area on which the village lies and the flow mass is deposited. The landslide descended a vertical distance of 75 m over a horizontal distance of 320 m measured from the top of the back scarp to the outer margin of the deposit. There was a deposition in the source area and the path.

For this type of landslide, the lower portion of the sliding mass is saturated. Especially for the loess in the Upper Pleistocene (Q3) and the top of the Middle Pleistocene (Q2) which has a loose structure with the saturated water content higher than the liquid limit. As the sliding mass starts to move, the structure of the loess near the sliding surface is damaged and the solid changes to liquid at instance, which may result in a fluidization with the sliding mass moving out of the sliding bed with a very high speed. As it moves down, the slope surface in the path may be dry or unsaturated. The pore pressure between the slope surface and sliding mass dissipates which causes a lower or even negative pore pressure. But the pore water pressure in the lower sliding mass could maintain higher value and the sliding mass keeps in liquid state during the moving process, so the friction between sliding mass and the ground could be little higher than that in the fluidized sliding mass. A film of liquid soil sticks on the ground and the sliding occurs in the sliding mass, which conducts a high speed and a long run-out, the sliding mass finally gets to the flat ground and deposit there. The mechanism is illustrated as Fig. 18 (Li et al. 2013b).

In the HeiFangtai loess platform, located in Yongjin County, Gansu province, China, rapid flow slides dominate the sides of the highland. It is the third terrace of the yellow river. The top of the platform is 120 m high relative to the top of the first river terrace and was developed as irrigated farmland in 1970s. The water of the Yellow River is being pumped up for irrigation, hence the water penetrates into the slope to induce a large number of landslides of which, the rapid flow slides are at most in number.

As shown in Fig. 19, the slope is composed of a layer of 40 m Q3 loess, a 4 m Q2 solid clay, a 8 m Q1 gravel in order and the Cretaceous sand-mudstone underneath. Because the solid red clay is impermeable, rainwater penetrating through the loess accumulates on top of the clay and

(a) Source area and moving path of the landslide

(b) Deposit area of the landslide

Fig. 17 A typical rapid flow slide, Gaolu landslide in Huaxian County, Shaanxi Province, China

spills out on the side slope (Fig. 19). Therefore, the groundwater level in the loess is rising with annual irrigation. As the sliding mass slumped down, it quickly changed to a mudflow rushing down along the slope. The flow path in the lower slope is sand-mudstone, which is relatively stiff, part of the mud deposited along the path and finally remains

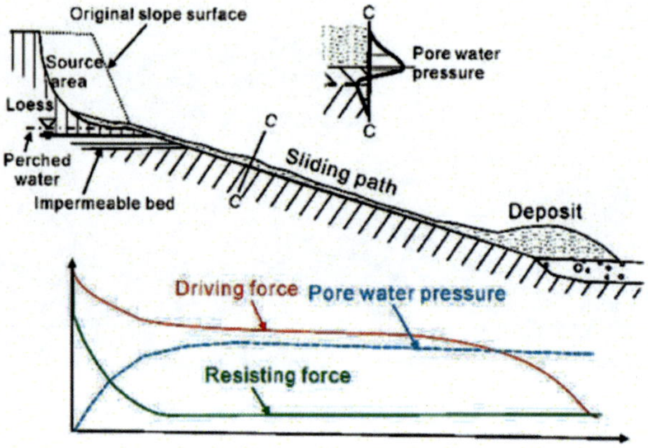

Fig. 18 A profile for the rapid flow slide and the variation of the forces with motion to illustrate its mechanism

deposited on the toe of the slope. The flow slides on the slope are still occurring every year.

Slow Creep Slide

The slow creep slides are characterized by its slow moving rate. This type of landslide is prominent in the Loess Plateau. They generally dominate the sides of riverbanks where the loess coves on the high river terraces, which are composed of gravels, Tertiary red clays and Miocene bedrocks. The lateral erosion or excavations at the foot of slopes often trigger the landslides.

A typical case is the Yanlian landslide which occurred on 21 October, 2010. The landslide is located in Hungling county of Shaanxi province, China, on the right bank of the Luohe River. It moved off and on for 2 days forwarding 20 m at the toe and 8–12 m down at the back scarp. The movement was very slow that there was enough time for working people to evacuate (Fig. 12). The district in which the landslide occurred belongs to the Yan'an oil refinery. The landslide cut off the 14 oil-transfer pipes at the slope toe causing the oil refinery to halt working for a week. It also damaged a coal storehouse, cut off a county road and

Fig. 19 The landscape of Heifangtai loess platform and the flow slides on the side of the loess coved highland

threatened 33 oil tanks near the head scarp. Site investigation and boreholes revealed that the soil of the failed slope is composed of 30 m thick loess, and 1 m of gravel and bedrock underneath. The gravel bed is the deposit of the third Luohe River terrace and the surface of the bedrock is gentle. The head scarp of the landslide is 8–10 m high. It sheared out on the top of the bedrock, so the sliding surface is steep at the back and gentle at the base as Fig. 20 shown. The landslide was induced by water infiltration. The boreholes within and without the landslide revealed that the groundwater is rich in the sliding area but shortage out of the area. Naturally, most of the rainwater runs off rather than penetrating into the ground. During field visit it was found that there is a water heating system working in winter for keeping the crude oil in the transfer pipes and tanks from solidifying. The condensed water of the vapour releasing from pressure-adjusting valves penetrated into ground for more than one decade. The water accumulated on the surface of bedrock caused rising of the groundwater level, which consequently induced the slide.

In general, this kind of landslide often occurs in areas where slopes are composed of Quaternary loess underlain by impermeable clay or bedrock. The failure surfaces are usually on the top of an impermeable bed, so the ground neither generates nor dissipates pore pressure, and the pore water pressure maintains a constant. The shear strength on the sliding surface is almost equal to the shearing stress, thus the landslides have a low speed and move in a long or a short length as illustrated by Fig. 21 (Li et al. 2013c).

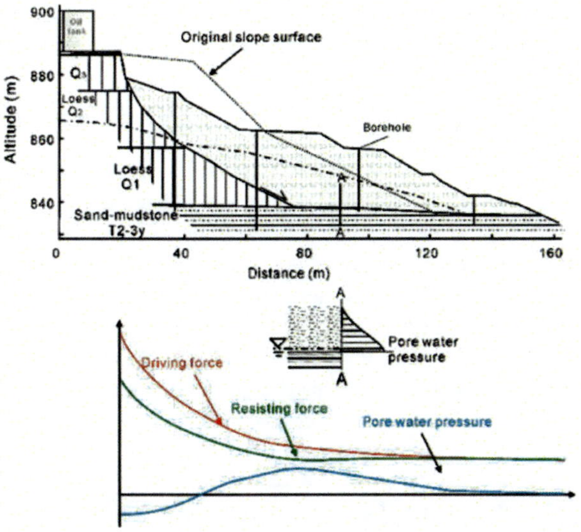

Fig. 21 A profile of the slow creep slide and the variation of the forces with motion to illustrate its mechanism

Quick Slump Slide

The feature of this type of landslide is that it has a short run-out distance, and the vertical displacement is larger than that horizontal. The quick slump slides generally occur on steep loess slopes, with steep sliding surfaces.

A typical case is the Sanmutai landslide in Fuxian county, Shaanxi province,China, which occurred on 6 May, 2006 as shown in Fig. 22. The landslide is located at the side of a loess ridge. The soil of the sliding slope is composed of a layer of 28 m thick loess on the upper and a 5 m thick pebbles at the bottom. The pebbles are the deposit of the second terrace of a stream. Before the slope failed, there were four living caves excavated on the toe of the slope and a row of five houses built in front of the slope in which four families lived. The detached mass slid and stopped at the foot of the slope. It destroyed all the living caves and the surrounding houses and killed two people. There was no rainfall, irrigation channels or springs around the slope. The cause of the slope failure is undercutting and excavations carried out at the slope toe. The slide occurred in the dry loess which has a relative high strength, but shows a big difference between peak and residual strength. Cutting and Caving may have caused the slope to be in a critical state condition.

In this case, the ground water level is far below the toe of the slope, so both the failure surface and the sliding path

Fig. 20 A typical slow creep slide, the Yanlian landslide

Fig. 22 A typical quick slump slide, the Sanmutai landslide in Fuxian county, Shaanxi province occurred on May 6th 2006

were unsaturated, which indicates that no pore water pressure was created during the failure and movement. So it will get to a new equilibrium soon after. Therefore, it has a short run-out distance. The mechanism is shown in Fig. 23 (Li et al. 2013c).

In the Chinese Loess Plateau, casualties made by this type of landslides are recorded every year. A large part of them were due to engineering activities, such as road cut, site levelling and underground excavation. In addition, quick slump slides also occur near the sides of the loess gullies which have been affected by erosion in recent geological time. The gullies are generally far from towns and residential areas, so there are no casualties and serious loss, however, they are the main sources of soil erosion.

Simulation for the Long Run-Out Slide

Sassa (1988) put forward a pioneer depth-averaged model for simulating flow-like landslides. The governing equations of this model, which are deduced in a global Cartesian coordinate, consist of a mass balance equation and two horizontal momentum equations. They can be written as

$$\frac{\partial h}{\partial t}+\frac{\partial Q_x}{\partial x}+\frac{\partial Q_y}{\partial y}=0 \quad (1)$$

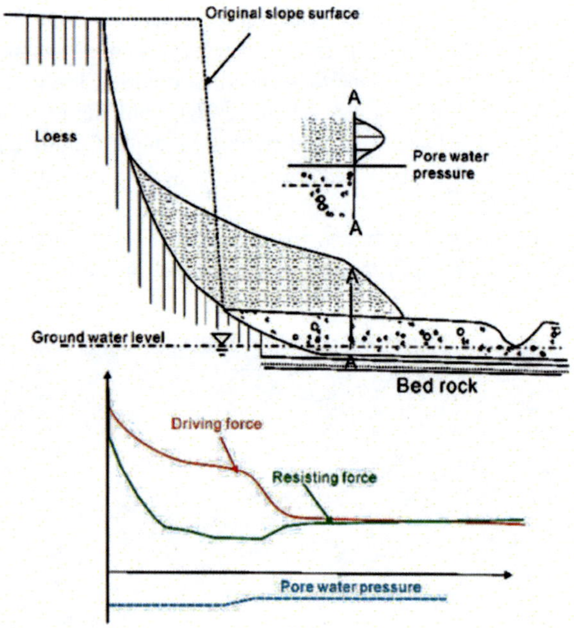

Fig. 23 A profile for the quick slump slide and the variation of the forces with motion to illustrate its mechanism

$$\frac{\partial Q_x}{\partial t}+\frac{\partial Q_x^2/h}{\partial x}+\frac{\partial Q_xQ_y/h}{\partial y}= -kgh\frac{\partial h}{\partial x}+\frac{ght\mathrm{an}\alpha}{\tan^2\alpha+\tan^2\beta+1}-\frac{\tau_bA_bhv_x}{m\sqrt{v_x^2+v_y^2+v_z^2}} \quad (2)$$

$$\frac{\partial Q_y}{\partial t}+\frac{\partial Q_xQ_y/h}{\partial x}+\frac{\partial Q_y^2/h}{\partial y}= -kgh\frac{\partial h}{\partial y}+\frac{ght\mathrm{an}\beta}{\tan^2\alpha+\tan^2\beta+1}-\frac{\tau_bA_bhv_y}{m\sqrt{v_x^2+v_y^2+v_z^2}} \quad (3)$$

where h is the thickness of landslide, $Q_x = v_xh$ and $Q_y = v_yh$ are the discharges in x and y directions, respectively, v_x and v_y are the flow velocities in x and y directions, respectively, k is the lateral pressure coefficients, g is the gravitational acceleration, α and β are the dip angles of sliding surface in x and y directions, respectively, τb is the basal shear stress, Ab is the area of bottom surface in a soil column, and m is the mass of landslide in the soil column.

The above model is very simple, and can be easily solved by finite difference method. This model has been applied in the run-out simulations of landslides worldwide, and the simulation results were generally good. However, the drawback of this model is that it does not consider some common phenomena in flow-like landslide, such as curved topography, bed entrainment, and frontal plowing, which significantly limits its applicability. Therefore, in our recent works, several improvements were incorporated into this model to outcome these deficiencies. These improvements will be discussed in detail in the following sub-sections.

Model for Curved Topography

Some landslides propagate along meandering gullies in which the effect of curved topography can be significant. When a landslide moves on a plain surface, the acceleration of sliding mass in the normal direction of sliding surface is zero. However, a centrifugal (or centripetal) acceleration will exist if the surface is curved. This centrifugal (or centripetal) effect will increase (or decrease) the normal force acting on the sliding mass, which will influence the dynamics of a landslide in turn. Therefore, we modified the momentum equations by adding a centrifugal normal force which can be determined by the velocity of landslide and the curvature of sliding surface (Shen et al. 2018). The modified momentum equations can be written as

$$\frac{\partial Q_x}{\partial t}+\frac{\partial Q_x^2/h}{\partial x}+\frac{\partial Q_xQ_y/h}{\partial y}=-\frac{\partial k_xgh^2/2}{\partial x}+(Ag+B)\frac{gh\tan\alpha}{\tan^2\alpha+\tan^2\beta+1}-\frac{\tau_bA_bhv_x}{m\sqrt{v_x^2+v_y^2+v_z^2}} \tag{4}$$

$$\frac{\partial Q_y}{\partial t}+\frac{\partial Q_xQ_y/h}{\partial x}+\frac{\partial Q_y^2/h}{\partial y}=-\frac{\partial k_ygh^2/2}{\partial y}+(Ag+B)\frac{h\tan\beta}{\tan^2\alpha+\tan^2\beta+1}-\frac{\tau_bA_bhv_y}{m\sqrt{v_x^2+v_y^2+v_z^2}} \tag{5}$$

where A and B are the terms related to the static and centrifugal supporting forces, respectively. The expressions of A and B are given by

$$A=1+\frac{\partial k_xh^2/2}{\partial x}\tan\alpha+\frac{\partial k_yh^2/2}{\partial y}\tan\beta \tag{6}$$

$$B=\frac{C_x}{\cos\alpha}\left(\frac{v_x}{\cos\alpha}\right)^2+\frac{C_y}{\cos\beta}\left(\frac{v_y}{\cos\beta}\right)^2 \tag{7}$$

The improved momentum equations are a generalization of the original momentum equations (Eqs. 2–3). If we take A = 1, B = 0, and lateral pressure coefficient k as a constant, these equations will be exactly same as the original equations.

The Dagou landslide, which is a flow-like landslide with obvious curved topography, was analysed by the improved model. This landslide is located in Tianshui City, Gansu Province, China (Fig. 24). The sliding mass consists of mudstone and a small amount of loess. Figure 25 presents the simulation results of considering and ignoring centrifugal force. It shows that ignoring centrifugal effect will overestimate the run-up of landslide on the slopes and the final run-out distance. This is because centrifugal effect will increase the normal force on the sliding bed and the friction dissipation in this landslide. This modification is possibly able to improve numerical simulation results of those landslides propagating on curved topography.

Model for Bed Entrainment

Bed entrainment is a phenomenon in which volume expansion of a landslide occurs due to entraining loose materials on the flowing path. This phenomenon has long been studied by geologist. Sometimes the landslide volume can expand more than ten times and substantially alter the mobility of a landslide. However, mathematical description for bed entrainment is still a very challenging topic in landslide modelling. The most common method is to define a parameter named bed entrainment rate which reflects the change rate of bed elevation to quantify this phenomenon. Incorporating this parameter into the continuity equation can obtain the following equation

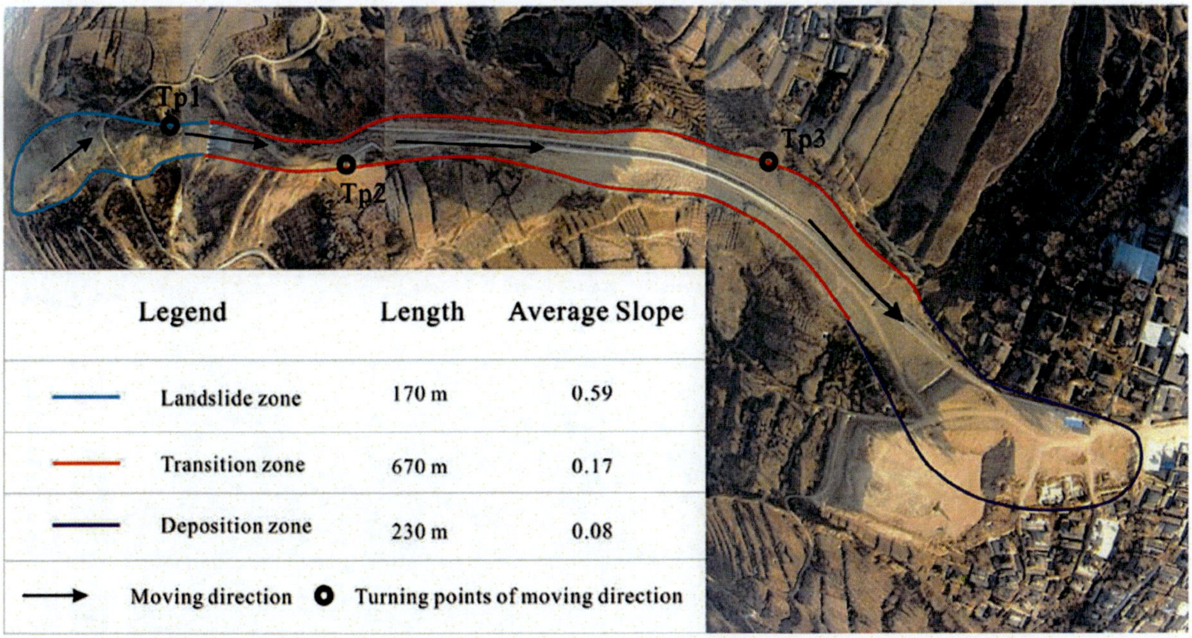

Fig. 24 A Satellite image of Dagou landslide in Tanshui City, Gansu Province, China

$$\frac{\partial h}{\partial t} + \frac{\partial Q_x}{\partial x} + \frac{\partial Q_y}{\partial y} = E_r \quad (8)$$

where E_r is the rate of entrainment, and t is time.

Many entrainment models have been proposed to calculate E_r, while most of them are empirically based which do not have clear physical meaning. Recently, some researchers started to use a physically based model to calculate E_r (Shen et al. 2019a; Ouyang et al. 2015). This entrainment model is based on the momentum conservation of the erodible mass, and the expression of Er is given by

$$E_r = \frac{\tau_b - \tau_e}{\rho_e \sqrt{v_x^2 + v_y^2}} \quad (9)$$

where τb is the basal shear stress in sliding mass, τe is the resistant shear stress in erodible mass, and ρe is the bulk density of the erodible mass.

Equation (9) suggests that bed entrainment will occur when the strength of sliding mass is stronger than that of erodible mass. In fact, bed entrainment will alter the strength of sliding mass, which will change the entrainment rate in turn. To account for this influence, we proposed a linear equation to calculate the rheology change of landslide due to entrainment, and its expression is given by

$$f_a = \frac{f_o V_o + f_b w V_e}{V_o + w V_e} \quad (10)$$

where f_a represents the rheology parameters of sliding mass after entrainment (i.e., ru, c and φ etc.), fo represents the rheology parameters of the sliding mass before entrainment, f_b represents the rheology parameters of the erodible mass, *Vo* is the volume of the original sliding mass, *Ve* is the volume of the entrained mass, and w is a weight coefficient.

This improved model was applied in the simulation of a deep-seated loess landslide (Dabaozi landslide, Fig. 26) in Shaanxi Province of China (Shen et al. 2019b). The sliding mass mainly consisted of loess, while some sand gravel on the first terrace of the Jinghe River was entrained by the landslide. The simulation results are illustrated in Fig. 27. It shows that rheology change plays a dominant role in determining the run-out of a flow-like landslide. Without considering rheology change, the run-out of a flow-like landslide may be significantly underestimated. Moreover, topography change mainly influences the lateral spreading of a flow-like landslide. When the erodible mass is thick, the sliding surface tends to be reshaped by bed entrainment into a concave shape, which can impede the lateral spreading of the sliding mass.

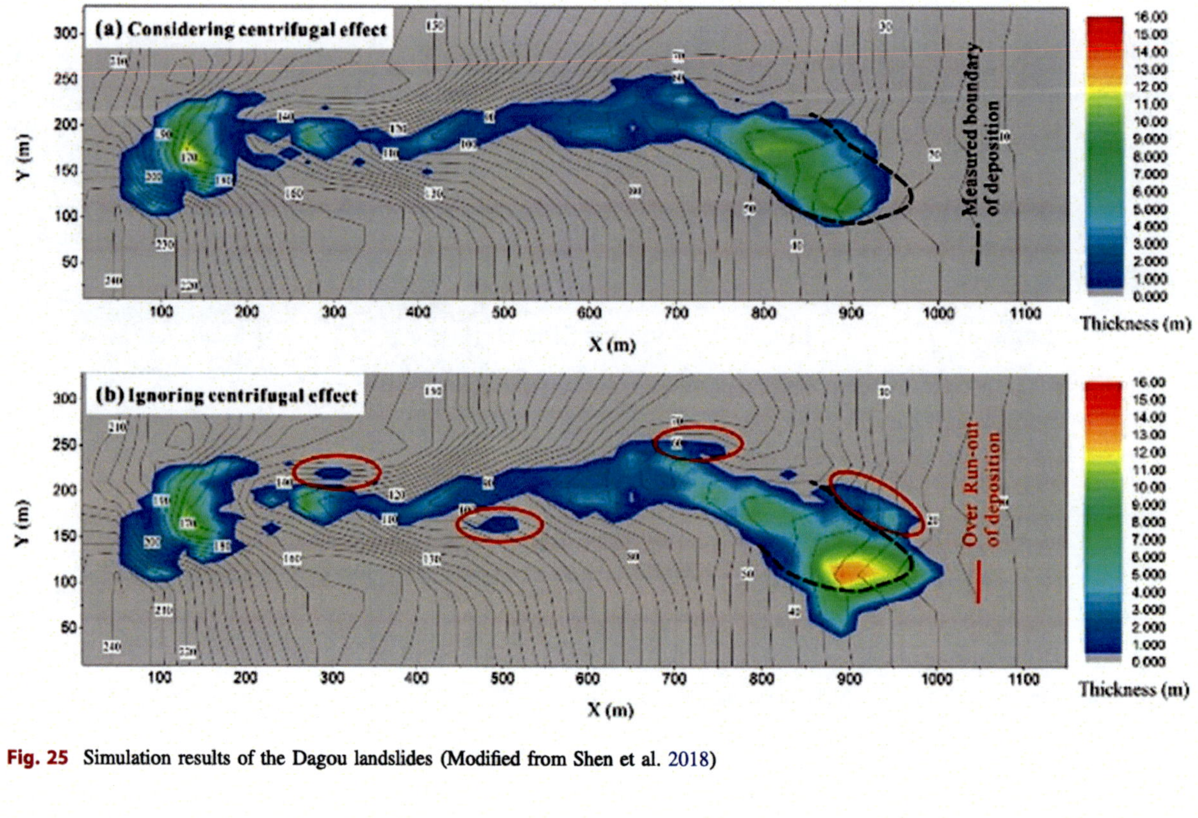

Fig. 25 Simulation results of the Dagou landslides (Modified from Shen et al. 2018)

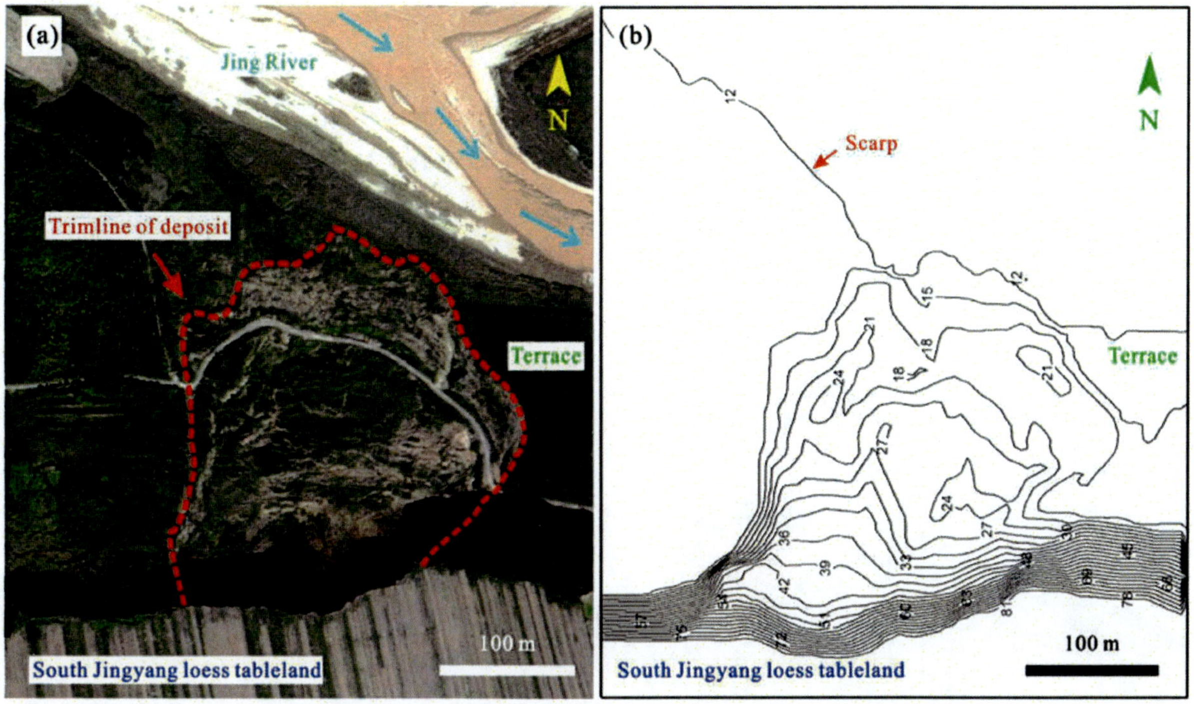

Fig. 26 An satellite image (a) and topography contour of Dabaozi landslide in Jinyang County, Shaaxi Provice, China

Model for Frontal Plowing

Frontal plowing is another common phenomenon in flow-like landslides. Except for entraining the basal material, some landslides may plow into the frontal mass and push it forward. This phenomenon is called frontal plowing, and it is schematically presented by Fig. 28.

Since the landslide and plowing mass basically remain immiscible, they can be considered as two separate materials, and the interaction between them can be simplified as the normal and shear forces on the interactive surface. Therefore, we proposed a two layer model to quantify the interaction between these two materials (Shen et al. 2019b). The governing equations for the two layers can be written as

$$\frac{\partial h_1}{\partial t} + \frac{\partial Q_{x1}}{\partial x} + \frac{\partial Q_{y1}}{\partial y} = 0 \tag{11}$$

$$\frac{\partial Q_{x1}}{\partial t} + \frac{\partial Q_{x1}^2/h_1}{\partial x} + \frac{\partial Q_{x1}Q_{y1}/h_1}{\partial y} = -k_1 g h_1 \frac{\partial h_1}{\partial x} + A_1 \frac{\tan \alpha_1}{G_1} - \frac{D_{x1}h_1}{m_1} S_1 \tag{12}$$

$$\frac{\partial Q_{y1}}{\partial t} + \frac{\partial Q_{y1}^2/h_1}{\partial y} + \frac{\partial Q_{x1}Q_{y1}/h_1}{\partial x} = - k_1 g h_1 \frac{\partial h_1}{\partial y} + A_1 \frac{\tan \beta_1}{G_1} - \frac{D_{y1}h_1}{m_1} S_1 \tag{13}$$

$$\frac{\partial h_2}{\partial t} + \frac{\partial Q_{x2}}{\partial x} + \frac{\partial Q_{y2}}{\partial y} = 0 \tag{14}$$

$$\frac{\partial Q_{x2}}{\partial t} + \frac{\partial Q_{x2}^2/h_2}{\partial x} + \frac{\partial Q_{x2}Q_{y2}/h_2}{\partial y} = -\left(k_1 g \frac{\rho_1}{\rho_2}\left(h_1 \frac{\partial h_2}{\partial x} + h_2 \frac{\partial h_1}{\partial x}\right) + k_2 g h_2 \frac{\partial h_2}{\partial x}\right) + A_2 \frac{\tan \alpha_2}{G_2} - \frac{D_{x2}h_2}{m_2} S_2 - \frac{\rho_1}{\rho_2}\left(A_1 \frac{\tan \alpha_1}{G_1} - \frac{D_{x1}h_1}{m_1} S_1\right) \tag{15}$$

$$\frac{\partial Q_{y2}}{\partial t} + \frac{\partial Q_{y2}^2/h_2}{\partial y} + \frac{\partial Q_{x2}Q_{y2}/h_2}{\partial x} = -\left(k_1 g \frac{\rho_1}{\rho_2}\left(h_1 \frac{\partial h_2}{\partial y} + h_2 \frac{\partial h_1}{\partial y}\right) + k_2 g h_2 \frac{\partial h_2}{\partial y}\right) + A_2 \frac{\tan \beta_2}{G_2} - \frac{D_{y2}h_2}{m_2} S_2 - \frac{\rho_1}{\rho_2}\left(A_1 \frac{\tan \beta_1}{G_1} - \frac{D_{y1}h_1}{m_1} S_1\right) \tag{16}$$

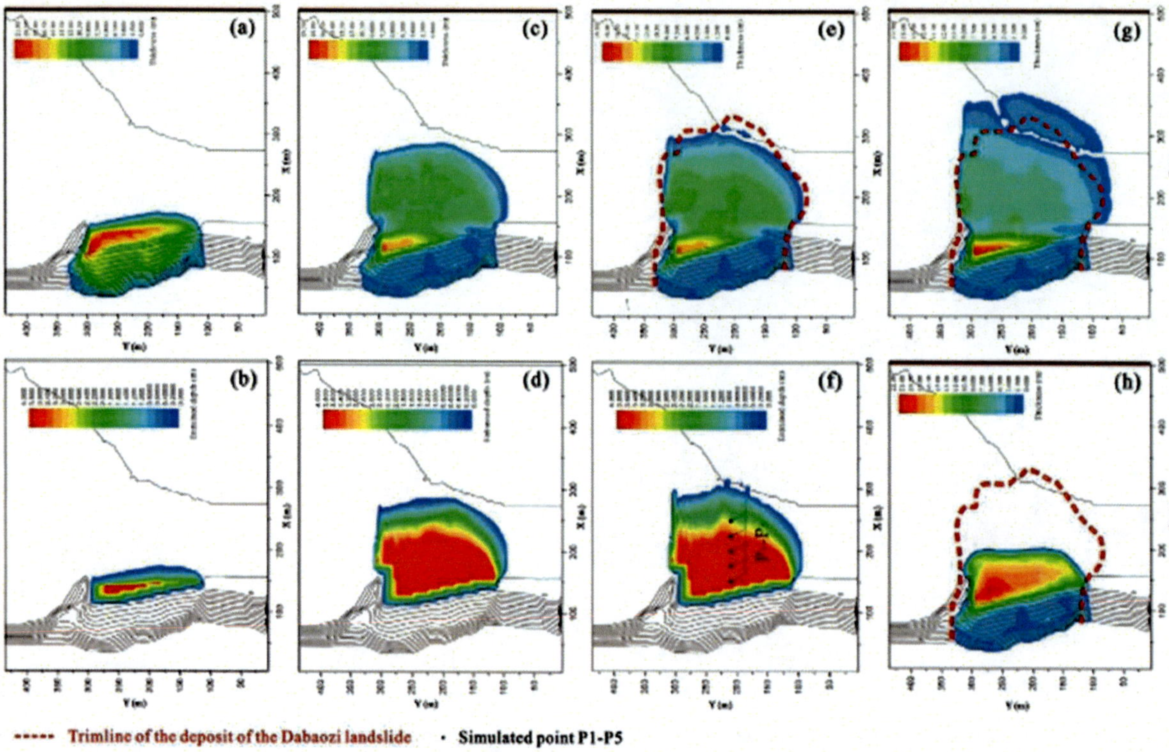

Fig. 27 Simulated thickness distribution and entrainment depth of the Dabaozi landslide at different times and simulation situations (S1: neglecting rheology change; S2: considering rheology and topography change; S3: neglecting topography change). a. thickness, and b. entrainment depth at t = 4.6 s in S2. c. thickness, and d. entrainment depth at t = 16.1 s in S2. e. Thickness (final deposit), and f. entrainment depth at t = 28.0 s in S2. g. final thickness in S3, and h. final thickness in S1 (From Shen et al. 2019a)

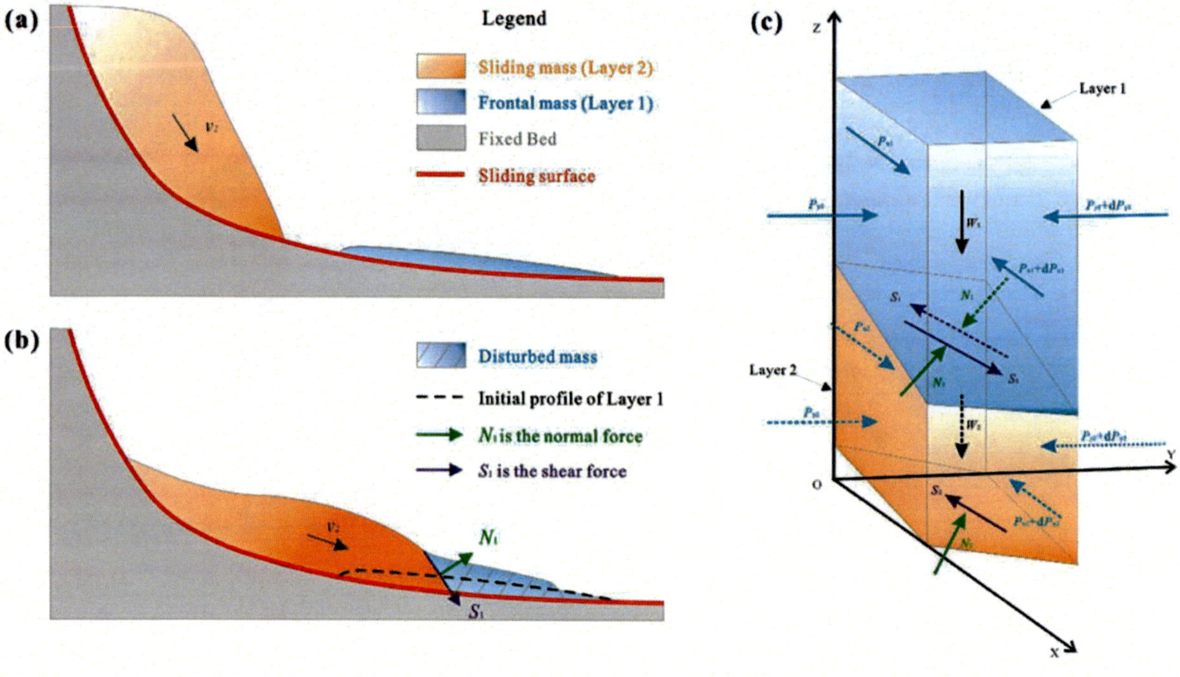

Fig. 28 Schematic diagram of frontal plowing phenomenon in flow-like landslides (Modified from Shen et al. 2019b)

where the expressions of A1, Dx1, Dy1, G1, A2, Dx2, Dy2, and G2 are given as

$$A_1 = k_1 g h_1 \frac{\partial h_1}{\partial x} \tan \alpha_1 + k_1 g h_1 \frac{\partial h_1}{\partial y} \tan \beta_1 + g h_1 \quad (17)$$

$$D_{x1} = \frac{v_{x1} - v_{x2}}{\sqrt{(v_{x1} - v_{x2})^2 + (v_{y1} - v_{y2})^2 + v_{z1}^2}} \quad (18)$$

$$D_{y1} = \frac{v_{y1} - v_{y2}}{\sqrt{(v_{x1} - v_{x2})^2 + (v_{y1} - v_{y2})^2 + v_{z1}^2}} \quad (19)$$

$$G_1 = \tan^2 \alpha_1 + \tan^2 \beta_1 + 1 \quad (20)$$

$$A_2 = \left(k_1 g \frac{\rho_1}{\rho_2} \left(h_1 \frac{\partial h_2}{\partial x} + h_2 \frac{\partial h_1}{\partial x} \right) + k_2 g h_2 \frac{\partial h_2}{\partial x} \right) \tan \alpha_2 + \left(k_1 g \frac{\rho_1}{\rho_2} \left(h_1 \frac{\partial h_2}{\partial y} + h_2 \frac{\partial h_1}{\partial y} \right) + k_2 g h_2 \frac{\partial h_2}{\partial y} \right) \tan \beta_2 + g h_2 \quad (21)$$

$$D_{x2} = \frac{v_{x2}}{\sqrt{v_{x2}^2 + v_{y2}^2 + v_{z2}^2}} \quad (22)$$

$$D_{y2} = \frac{v_{y2}}{\sqrt{v_{x2}^2 + v_{y2}^2 + v_{z2}^2}} \quad (23)$$

$$G_2 = \tan^2 \alpha_2 + \tan^2 \beta_2 + 1 \quad (24)$$

This two-layer model was implemented in simulating the run-out process of a flow-like landslide (Ximiaodian Landslide) in Shaanxi Province, China. Obvious frontal phenomenon was observed in filed investigation of this landslide, which is illustrated by Fig. 29.

The simulated final topography is shown in Fig. 30. It shows that the result of the two-layer model agrees best with the measured result, while other two situations which do not consider frontal plowing significantly overestimate the propagation area. As it can be seen in Fig. 30a, the loess deposit has an average thickness of 20 m, and the thickest part of the deposit is about 24 m, forming a bulb-shaped ridge between the loess deposit and the gravel deposit. The simulated results of the two-layer model (Fig. 30b) somehow reflect these features, and the bulb-shaped ridge is also captured well. However, in the other two situations, the landslide propagate too far because lacking resistance from the frontal gravel mass, so the final shape of the deposit becomes too flat.

The simulated plowing process in profile I-I' (Fig. 29) at three times are presented in Fig. 31. Before frontal plowing occurs (t = 2.00 s), the shape of the loess mass is relatively smooth. In the plowing process (t = 7.47 s and 12.45 s), the velocity of the frontal part of the loess mass decreases and forms a ridge under the resistance of the gravel, and the shape of the loess mass in the middle part becomes fluctuant under the impact of the frontal resistance, forming a valley in the rear part. The frontal gravel accumulates at the ridge at 7.47 s, and departs from the ridge at 12.45 s. Figure 31b shows that frontal plowing has significant influence on the velocity of both the loess mass and gravel mass. The velocity of the loess mass fluctuates and decelerates quickly due to the plowing resistance, which may contribute to the formation of the frontal ridge. The velocity of the gravel mass is about the same of the loess mass at 7.47 s (corresponding to the second stage of the plowing process), while its velocity is larger than that of the loess mass at 12.45 s (corresponding to the third stage of the plowing process). It indicates that the energy is transferred by the loess mass to the gravel mass through this interactive process.

A measured topography, b result of situation 1 by the two-layer model, c result of situation 2 in which entrainment is not considered, and d result of situation 3 in which both loess mass and gravel are in the sliding mass layer (layer 2) (From Shen et al. 2019b).

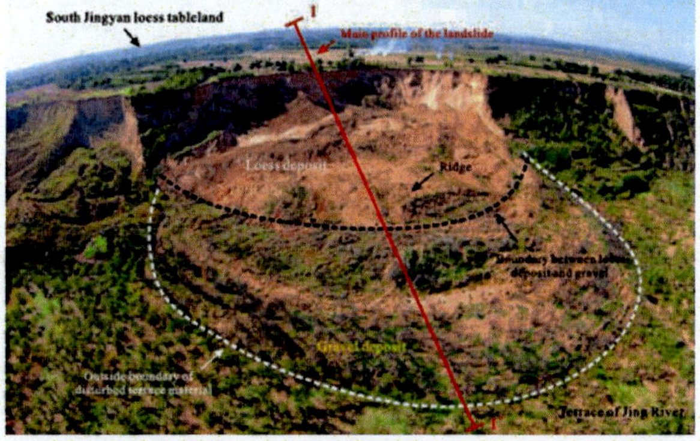

Fig. 29 Panorama view of the Ximiaodian landslide. The terrace material of the Jing River was pushed forward by this loess landslide for about 100 m (From Shen et al. 2019b)

Summary

Loess is a wind blown deposit which is best developed in the Chinese Loess Plateau. The loess highlands of the Plateau have a continuous stratigraphy in the whole Quaternary which reflects the climatic change in the geological record of the deposits, while loess on the river terraces reflect the intermittent uplift of the Plateau and consequent lateral and downward erosion.

Loess landslide occurred extensively on the Loess Plateau due to the typical topography, the special structure and texture, as well as the water sensibille properties.

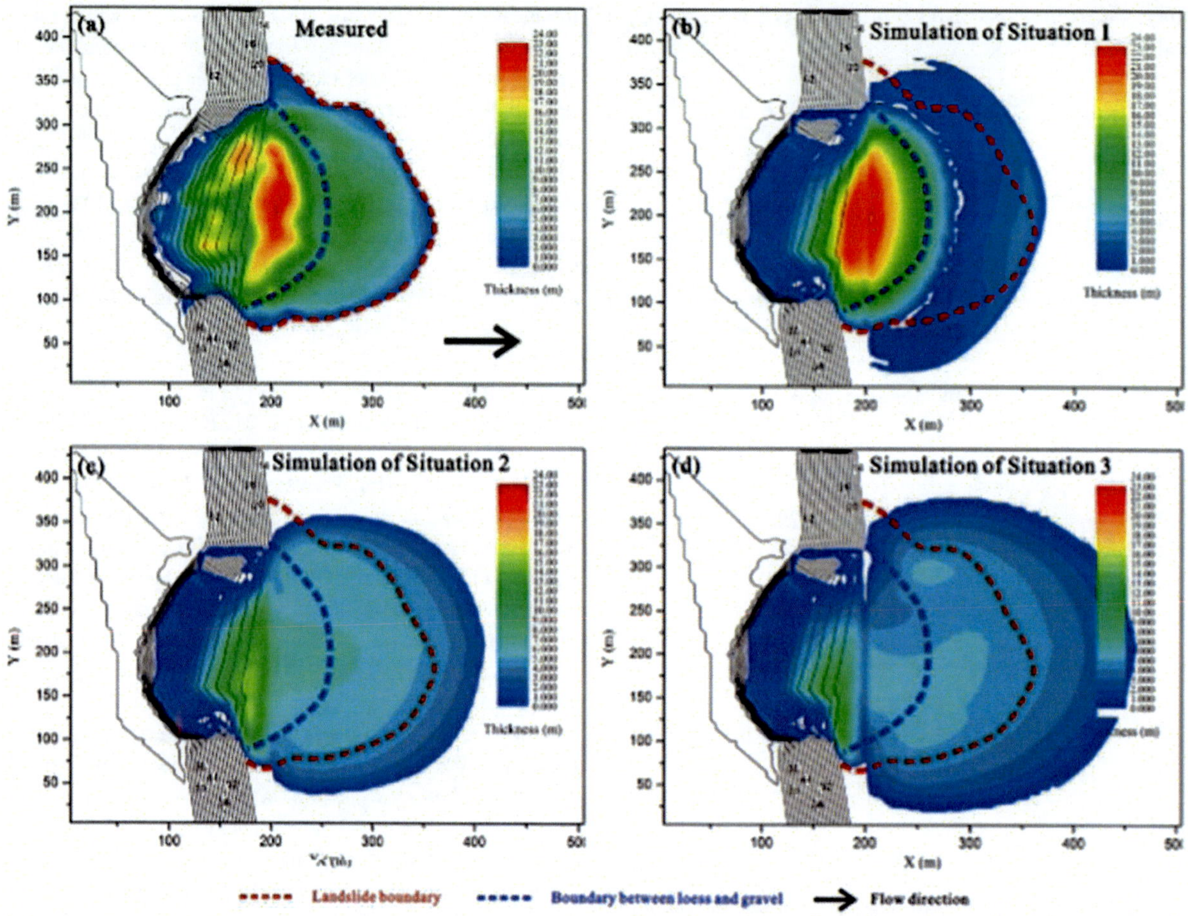

Fig. 30 Comparison of the simulated final topography of the three situations with the measured topography

Loess landslides are the most developed geological disaster in the Chinese Loess Pateau. Based on the failure modes and motion characters, the loess slope failure can be classified into two main types, those are shallow failure and deep-seated failure. The shallow failure has two types of small collapses and shallow flow slide; while the deep-seated failure has the four types of rapid long run-out slide, rapid flow slide, slow creep slide and quick slump slide. The mechanism of the landslide movement is controlled by the moisture state on the failure surface and the moving path. The small collapses caused by capillary wetting on the foot of steep to vertical slopes. The shallow flow slides occurs in raining days because of rain water infiltration. The slow creep slide is that the sliding mass slides on an impermeable bed. In the rapid long run-out slide the sliding mass slides on a saturated loose ground. The quick slump slide is that the less saturated sliding mass slides on a less saturated or dry ground. The rapid flow slide is that the saturated sliding mass slides on an unsaturated or dry ground.

Furthermore, modified finite difference models for different ground conditions are proposed to simulate the propagation of the catastrophic long run-out flow slides. Based model was founded by Prof. Sassa. Our Modifications of the model are conducted by calculating the lateral pressure coefficient k in the sliding mass and the entrainment and centrifugal effect during the transport process. And also two-layer model is proposed to simulate frontal plowing. A series of case studies are conducted to document that the models are suitable to apply in predicting the run distance and coverage area of the long run-out flow slides.

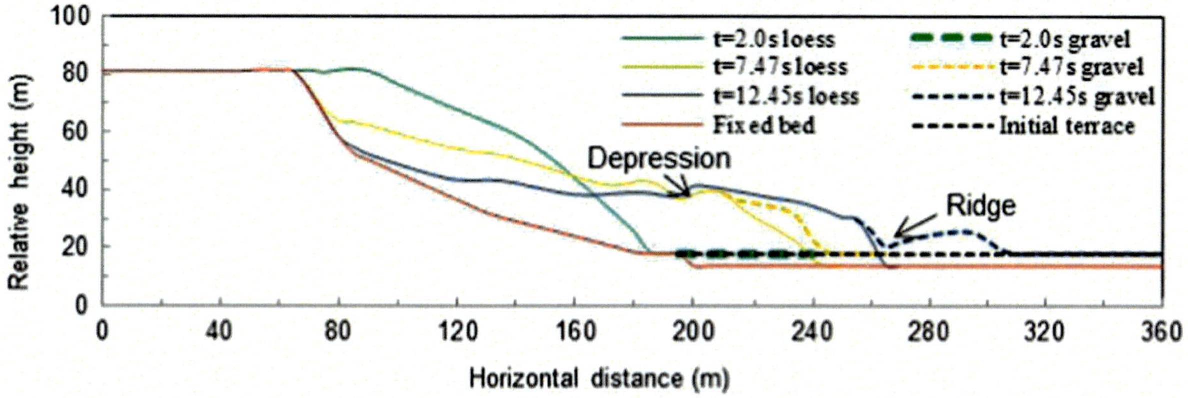

a. Topography of the loess sliding mass and terrace plowing mass in profile I-I'

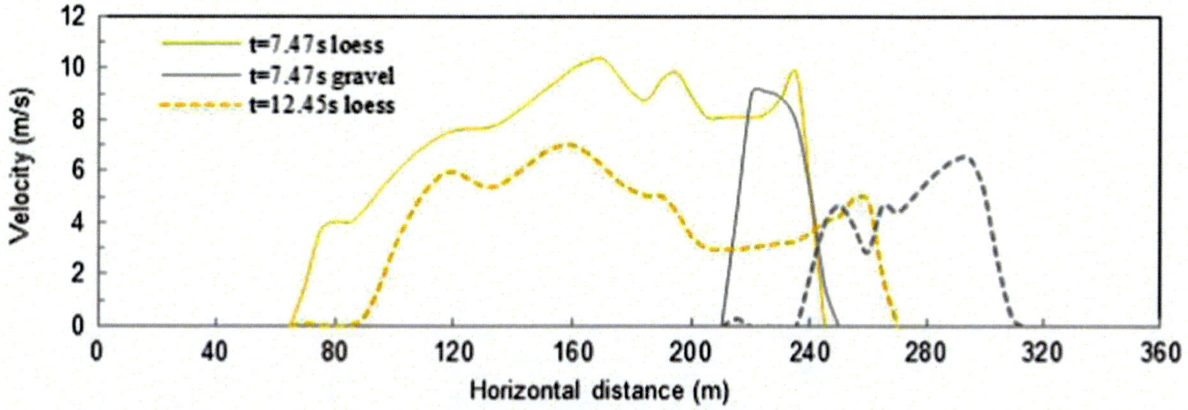

b. Velocity of the loess sliding mass and terrace plowing mass in profile I-I'

Fig. 31 Simulated interaction process between loess landslide and plowing mass (From Shen et al. 2019b)

Acknowledgments The authors acknowledge the supports from the National Natural Science Foundation of China (No.41790442, No. 41772278, No. 41877242), , the National Key R&D Program of China (2017YFC1501302).

References

G Kukla Z An 1989 Loess stratigraphy in Central China Palaeogeogy Palaeoclimatol Palaeoecol 72 203 225

Liu D, Zhang Z (1962) Loess of China. Acta Geol Sin 42(1):1–14 (in Chinese)

Liu D (1996) Geological environments in China and global change. Keynote speech on the 30th IGC, Beijing, pp 1–9.

Li T, Wang C, Li P (2013a) Loess deposit and loess landslides on the Chinese loess plateau. In: Wang F (ed) Progress of geo-disaster mitigation technology in Asia. Springer, Berlin, pp 235–261. ISBN: 978-3-642-291074-4

Li T, Wang P, Xi Y (2013b) The mechanisms for initiation and motion of Chinese loess landslides. In: Wang F (ed) Progress of geo-disaster mitigation technology in Asia. Springer, Berlin, pp 105–122. ISBN: 978-3-642-291074-4

Li T, Zhao J, Li P, Wang F (2013c) Failure and motion mechanisms of a rapid loess flowslide triggered by irrigation in the Guanzhong irrigation area, Shaanxi, China. In: Wang F (ed) Progress of geo-disaster mitigation technology in Asia. Springer, Berlin, pp 421–433. ISBN: 978-3-642-291074-4

C Ouyang S He C Tang 2015 Numerical analysis of dynamics of debris flow over erodible beds in Wenchuan earthquake-induced area Eng Geol 194 62 72

Sassa K (1997) Landslide hazard assessment in cultural heritage, Lishan, Xian. In: Sassa K (ed) Proceedings of international symposium on landslide hazard assessment, Xi'an 13–16 July, pp 1–24

Sassa K (1988) Special lecture: geotechnical model for the motion of slides. In: Proceedings of the 5th international symposium on landslides, vol 1, pp 37–56

W Shen T Li P Li J Guo 2018 A modified finite difference model for the modeling of flowslides Landslides 15 1577 1593
Shen W, Li T, Li P, Shen Y, Lei Y, Guo J (2019a) The influence of the bed entrainment-induced rheology and topography changes on the propagation of flow-like landslides: a numerical investigation. Bull Eng Geol Environ 78(7):4771–4785
Shen W, Li T, Li P, Berti M, Shen Y, Guo J (2019b) A two-layer numerical model for simulating the frontal plowing phenomenon of flow-like landslides. Eng Geol 259:105168
G Wang T Li X Xing Y Zou 2014 Research on loess flow-slides induced by rainfall in July 2013 in Yan'an, NW China Environ Earth Sci 73 12 7933 7944

Keynote Lecture: The Jettan Rockslide—An Engineering Geological Overview

2020 Kyoto Japan

Louise M. Vick, Jørgen N. Berg, Mark Eggers, Anne Hormes, Ingrid Skrede, and Lars Harald Blikra

Abstract

A recognised cluster of gravitational slope deformations in northern Norway is the result of a protracted geological history. A combination of nappe metamorphosis, strain localisation, basin extension, glacial loading and unloading and exposure to weathering has resulting in a significantly weakened rock mass. We present the engineering geology of one such site, the Jettan rockslide, compiled from existing studies and multiple field campaigns. We show that these geological processes manifest in the Jettan rockslide as mylonitic foliation, pervasive jointing and faulting, and ductile shear zones degraded by weathering. These structural factors accommodate the deformation at Jettan and contribute to the overall failure path. Channelised fluid flow through vertical joints and subhorizonal shear zones interplay with permafrost processes to reduce the strength of the presented failure path. These factors work together to form a failure hazard, which threatens to trigger a displacement wave, a scenario validated by the Holocene record and past performance.

Keywords

Rockslide • Rock slope failure • Northern Norway

L. M. Vick (✉) · J. N. Berg · A. Hormes
Department of Geosciences, UiT The Arctic University of Norway, Dramsveien 201, 9009 Tromsø, Norway
e-mail: louise.m.vick@uit.no

A. Hormes
e-mail: anne.hormes@uit.no

M. Eggers
Pells Sullivan Meynink, G3, 56 Delhi Road, North Ryde, NSW 2113, Australia
e-mail: Mark.Eggers@psm.com.au

I. Skrede · L. H. Blikra
Norwegian Water Resources and Energy Directorate, Vestre Rosten 81, N–7075 Tiller, Norway

Introduction

The occurrence of foliated metamorphic rock in steep eroded bedrock slopes has resulted in a cluster of deforming rock slopes in the Troms section of the Scandinavian Mountains, northern Norway. The boundaries of these instabilities are largely controlled by their inherited bedrock structure: foliation, ductile shears, tectonic and exfoliation joints, faults and other less regular structure such as boudinage, lenses, veins and folds (Braathen et al. 2004b; Vick et al. 2020b). These rockslides are identifiable by slope deformations that often extend from near the ridge crest to fjord or valley level. Slope deformations are in the order of 10′s of mm/year, and rarely exceed the creeping state. However past slope performance in Norway is illustrated by examples such as the Tafjord displacement wave disaster (Blikra et al. 2006), where 40 people died, and the Pollfjellet rockslide in Lygen, in 1819, which took 14 lives. Events like Tafjord and Pollfjellet show that catastrophic failure can occur several times a century (Hermanns and Longva 2012). This frequency of catastrophic events is also exemplified in the geological record. Fjord studies show turbidites catalysed by repeated rock avalanche impact through the Holocene (Forwick and Vorren 2002; Hegstad 2014), and dating of subaerial deposits reveal extremely large events recurring in similar time frames (Blikra et al. 2006; Fenton et al. 2011).

If slope deformations indicate instability in the vicinity of populations and infrastructure, a risk analysis must be carried out to decide if the slope need further investigations or an early warning system. The first step of the risk analysis is the analysis of the natural hazard, and the final risk class is assigned one consequences of a failure scenario are assessed (Hermanns et al. 2012). If considered high risk, permanent monitoring and early warning systems are installed. The Jettan rockslide (Fig. 1) is one of three high-risk sites in Troms and Finnmark County (NVE 2020). It is located on the east flank of the Storfjord, with neighbouring slope deformations along the ridgeline. It moves up to 50 mm a^{-1}.

Ž. Arbanas et al. (eds.), *Understanding and Reducing Landslide Disaster Risk*, ICL Contribution to Landslide Disaster Risk Reduction,
https://doi.org/10.1007/978-3-030-60713-5_28

The most realistic failure scenario is a displacement volume up to 6 Mm^3. The annual probability of a catastrophic failure is estimated at between 1/100 and 1/1000, with a displacement wave representing major consequences to villages lining the shores of Storfjord (Glimsdal and Løvholt 2018; NVE 2020). Jettan rockslide has been continuously monitored from 2007, with installations or campaigns of GNSS, laser, extensometers, ground-based radar, satellite corner reflectors, borehole instrumentation measuring displacement, piezometers, temperature, DMS (differential monitoring of stability), tiltmeters, a meteorological station and web cameras. The slide has been the subject of studies, investigations and theses since 1999 (Braathen et al. 2004a; Rønning et al. 2008; Blikra et al. 2009; Blikra and Longva 2009; Nordvik et al. 2010; Elvebakk 2013; Ganerød 2013; Skrede 2013; Blikra and Christiansen 2014; Elvebakk 2014; Ganerød 2014; Eriksen et al. 2017a; Eckerstorfer et al. 2018; Berg 2020).

While extensive, the literature is mostly bit-part studies highlighting specific investigative campaigns, or mechanical behaviour over defined periods. In this contribution, we will synthesise the existing literature, and build on the past understanding with additional results to present an overview of the engineering geological model of the slide. This leads to observations and comments on the mechanisms and preparatory factors causing movement.

Overview and Setting

Location and Climate

Jettan is a part of Nordnesfjellet Mountain, in Troms and Finnmark County (Fig. 2). The unstable area is located between zero and 800 m asl on the W-facing slope of Nordnesfjellet. Jettan neighbours two other RSDs, Midtre Nordnes and Indre Nordnes, forming a complexly deforming mountain ridge. The mountain ridge is bound by the Storfjord on the western side, and Kåfjorden and Manndalen valley on the eastern side. The fjords merge north of Nordnesfjellet, where it is traced northward to the continental shelf. While the Lyngen Alps to the west stand out with the highest mountains in the area, including some peaks over 1800 m, Nordnesfjellet is less dramatic with a plateau

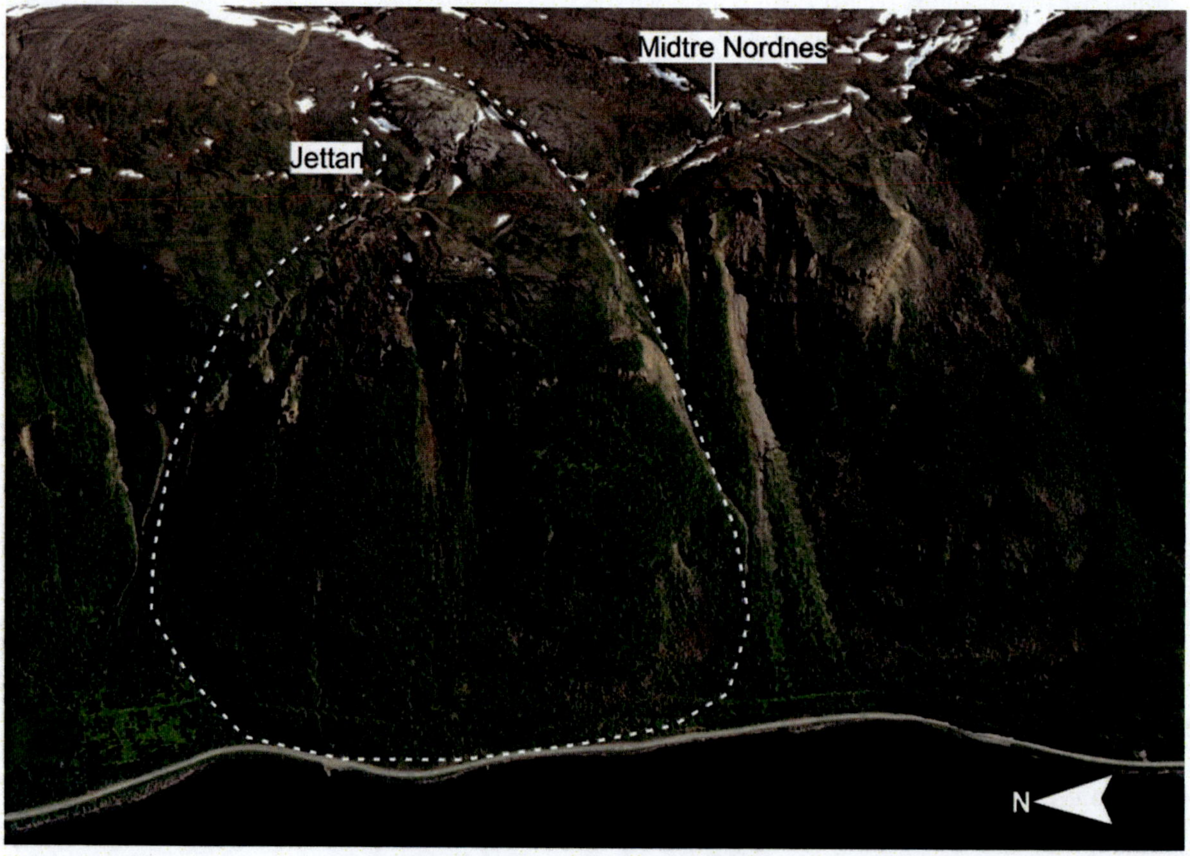

Fig. 1 View SE of Jettan and neighbouring unstable slope Midtre Nordnes (Google Earth)

mountaintop at 900 m asl. Until recently the main road connecting Troms and Finnmark (E6) passed below the rockslide, but a tunnel has now been constructed going through Nordnesfjellet. This has resulted in less traffic in the impact hazard zone.

The county has a mild and humid coastal climate characterized by large amounts of precipitation, and a relatively dry, cold inland climate. Jettan lies at the transition from a steep coastal landscape to flatter plateau-like morphology inland. The average Celsius air temperature is positive in May/June and usually dips under zero from October. The coldest period is in January to March, whereas the warmest is in July. The average air temperature is around 10 °C in the summer and approximately −10 °C in the winter. The ground temperatures react slower to changes, therefore the ground gets above and below zero around a month later than the air temperatures. The average ground temperature in the summer is below 10 °C and in the winter the average is rarely below −5 °C. The wind is strongest in the winter and typically blowing from the south/southeast. The site often experiences higher wind than other areas of the county, as wind is channelled through the fjord system. Summer is the season with the greatest amount of precipitation (May to September), and the annual precipitation is usually over 300 mm. The snow cover is in general greatest in April (eKlima 2019). The slope is in the rain shadow of the Lyngen Alps, so it receives less rain that slopes west of the mountain chain.

Regional Geology

The bedrock of Troms was formed through multiple phases of allochthonus thrusting and Caledonian mountain building followed by rifting and opening of the continental margin. The Caledonian nappes were thrusted, stacked and imbricated to the NW, with foliation, folds and thrusts generally dipping to the W and NW (Gee et al. 2008; Augland et al. 2014). The study area is located in the Reisa Nappe Complex, comprising the Vaddas-, Kåfjord- and Normannvik Nappe (Faber et al. 2019).

After the Caledonian orogeny, an extensional phase began, leading to the opening of the Norwegian and Barents Seas. Bergh et al. (2007) suggest the rifting happened in several stages leading to normal faults and other tectonic structures which strike predominantly NW-SE, NNE-SSW and NE-SW (Indrevær et al. 2013). Generally joints align with the fault orientations, with localized variations (Vick et al. 2020b).

Quaternary History

Glaciers have a big impact on the landforms seen in Norway and in the study area in Troms. U-shaped valleys and fjords have formed through glacial erosion, and isostatic uplift from rebound has exposed steep rock slopes (Corner 2005). The Last Glacial Maximum (LGM) experienced its peak

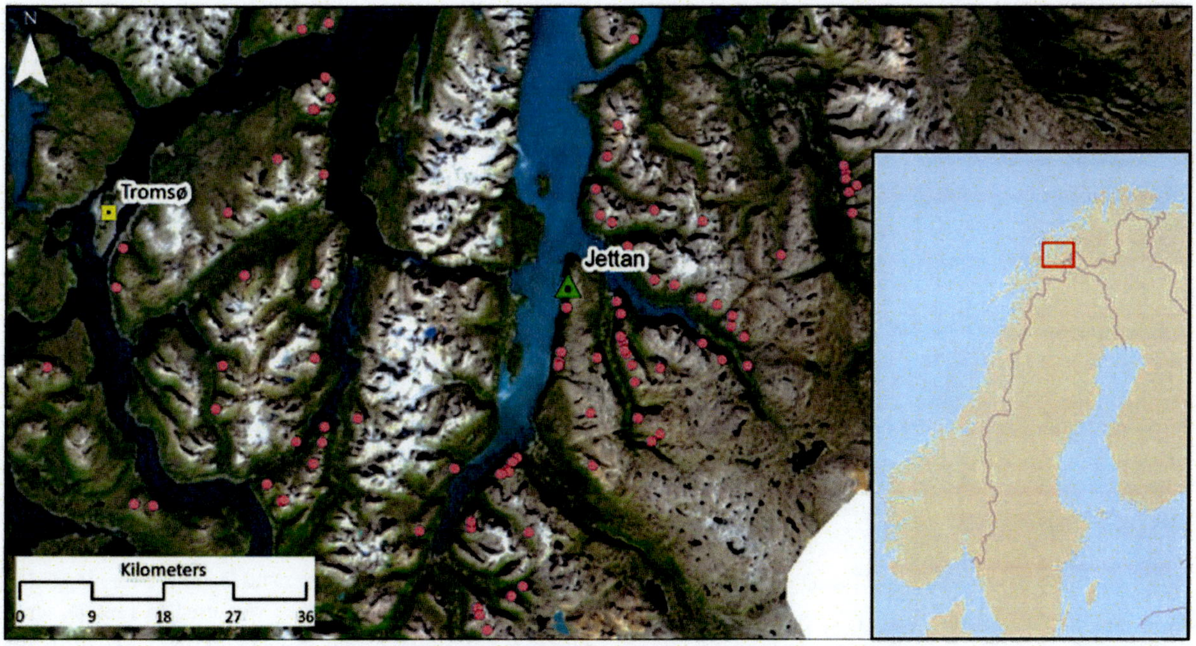

Fig. 2 Location map of the Troms area, showing mapped rock slope deformations (RSDs) as pink dots (from NGU (2019))

between 29 and 21 ka when Barents Sea Ice sheet and Fennoscandinan ice sheet merged (Hughes et al. 2016; Patton et al. 2017) and a rapid deglaciation occured during Bølling interstadial (15.3–14.1 ka) (Stokes et al. 2014). A re-advance during the Younger Dryas (12.9–11.7 ka) resulted in the formation of the most dominant moraine in Troms, the Tromsø-Lyngen moraine, which shows marked terminal and lateral moraines. In the Storfjord-Lyngenfjord area, the Tromsø-Lyngen moraine is seen north of Jettan, at Spåkenes (Fig. 3). From 9.9 to 9.8 ± 150 years ka, the ice had retreated south of Jettan to Ørnes in Lyngen, called the Ørnes event (Corner 1980). At the same time, an ice-front accumulation was deposited at Holmen in Kåfjordalen (Corner 1980).

At Nordnesfjellet marginal moraines are mapped on both slope aspects, with a more dominant moraine morphology on the eastern side (Fig. 3). The moraine ridges are mapped at 700 m asl in the south and 500 m asl in the northern outer end of the peninsula (NGU 2020).

Fig. 3 The major ice-front accumulations signalling deglaciation fluctuations. The Tromsø-Lyngen moraine is mapped at Spåkenes, the Ørnes event is mapped at Ørnes and Holmen. Basemap from NGU (2020)

Hegstad (2014) reconstructs the local post-glacial mass-transport in the Storfjord by looking at sedimentary cores, swath bathymetry and high-resolution seismic data. They note sediment lobes at the base of Nordnesfjellet, and several mass-transport deposits (MTDs) in the fjord adjacent to Nordnesfjellet. The MTDs contain debris flow deposits, turbidites and rafted blocks, representing both rock avalanches and glacial sediment deposits. Frequent mass wasting was observed above glaciomarine sediments dated at the upper boundary to 10.8 ka cal BP, implying high failure rate from Nordnesfjellet shortly after the Ørnes event (Fig. 3). The period of frequent mass wasting falls within a time of enhanced postglacial uplift with the occurrence of major earthquakes (order of magnitude ≥ 6.0). Therefore, earthquake activity has most probably been a major trigger. This period lasted from 11.7 to 10.5 ka cal BP (Dehls et al. 2000).

Core 19 is located close to the run-out area of Jettan URS (Fig. 4). Sedimentation in the core stratigraphy is interrupted by several MTDs including both turbidites and debris flows. The lowermost MTD boundary was dated to 10.3 ka cal BP, and the next to 9.9 ka cal BP (Hegstad 2014). An upper boundary above this was not dated, but correlates to a similar horizon in Core 17, dated to 2.9 ka cal BP. The most notable MTD, dated between 3.0 and 2.9 ka cal BP, is found beneath Indre Nordnes, 2.4 km south of Jettan and attributed to a rock avalanche event (Fig. 4). Many MTDs suggested to represent small, local events are found within all the sediment cores. Hegstad (2014) infer a minimum frequency of smaller events within the study area, estimated to be 2.3 per 1000 years over the last 3 ka, with a marked decrease earlier than this.

Blikra and Christiansen (2014) conducted cosmogenic surface exposure dating of the Jettan backscarp, resulting in an age of 6.5 ± 0.3 ka. This indicates that movement along the fault was reactivated 7–6 ka ago.

Methods

Desktop mapping was conducted prior to field work based on Troms 2016 0.25 m orthophoto series (Kartverket 2019a), and DEMs from Troms 2014 0.5 and Troms 2015 0.25 m lidar projects (Kartverket 2019b). Lithological and geomorphological mapping was over several summer and autumn field campaigns 2017–2019. Geomorphological features including ground cover and features were mapped in the field. A Mavic 2 Pro drone was used to capture aerial imagery and video, to aid in mapping morphological features. For 3D modelling Leapfrog Works by Seequent was used. The model was build using 0.5 m resolution DEM,

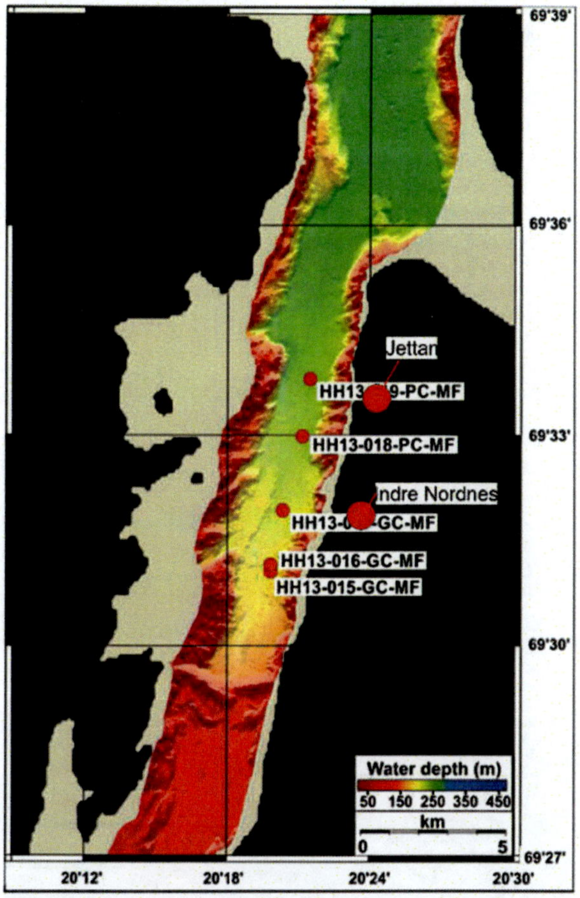

Fig. 4 Bathymetry of the Lyngenford showing core locations, Jettan and Indre Nordnes. Core 19, below Jettan, is the northern most core (HH13-019-PC-MF). From Hegstad (2014)

borehole inputs and surface mapping shapefiles. Borehole interval files specifying lithologies, crushed zones, core loss and faults were created from geological and borehole geophysics reports (Elvebakk 2013, 2014; Ganerød 2013, 2014). Foliation was added as planar structural data.

Lithology, Structure and the Rock Mass

Lithology

Two different nappes are mapped at Jettan: the Kåfjord nappe forms the lower portion of the slope, comprising garnet mica schist with kyanite and staurolite. The boundary between the Kåfjord and overlying Normannvik Nappe is mapped at ca. 200 m asl, and within the Normannvik Nappe packages of blastomylonite, calcite silica gneiss, garnet mica schist and dolomite marble are mapped as present (Fig. 5) (Boyd and Minsaas 1983). Results from this study give an updated overview of the bedrock units, which we term mica schist, calcite marble and dolomite marble. Lithologies are described herewith.

Mica Schist

This unit is a light to dark grey banded rock, which transitions between mylonite, schist and gneiss at outcrop scale (Fig. 6). The layers show a poor gneissosity in the form of alternating of dark and lighter layers. The rock contains different amounts of mica (biotite and muscovite), quartz and garnet. The mica schist is an envelope term which encompasses a highly variable rock. The differences in composition are attributed to (1) a varying garnet content, and (2) changes in grain size from pelitic to psammitic. It often contains, or is adjacent to, quartz veining. Iron coating was found in some outcrops where the rock was garnet rich (Fig. 7). In some places, a thick calcite coating was also observed to cover exposed joint surfaces, most likely a dissolution of calcite layers from above transported by groundwater (Fig. 8).

Calcite Marble

Calcite marble is a light yellowish brown coarse grained laminated marble with a smoothly weathered surface. The marble contains biotite and quartz, but is primarily composed of calcite (Fig. 9). The brown colour of the rock is a

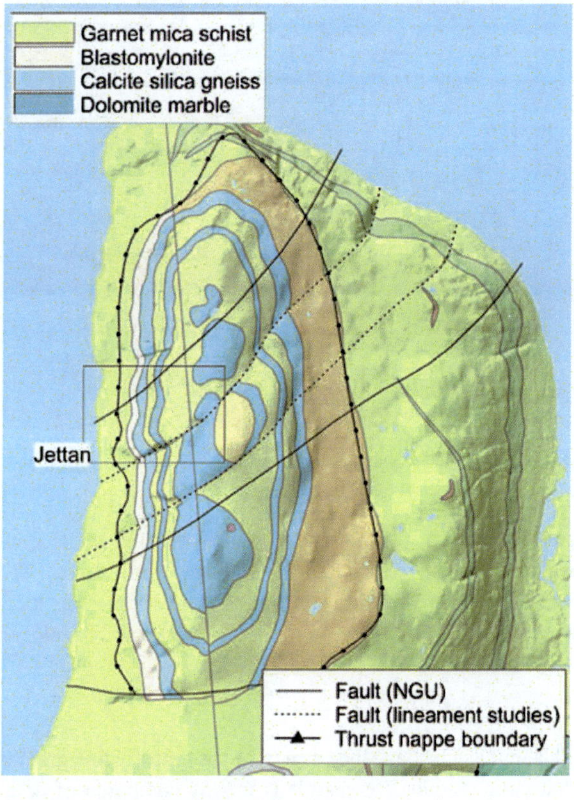

Fig. 5 Lithology, normal and thrust fault surface traces across the Nordnesfjellet peninsula (Boyd and Minsaas 1983)

Fig. 6 Mica schist exposure, showing alternating light and dark bands

Fig. 7 Iron staining on joint planes of the mica schist

consequence of weathering rather than a result of its mineralogy. Crystals are more or less equigranular (1 mm). The intensiveness of the foliation varies from location to location, where in some places it is more distinctly developed than in others.

Dolomite Marble

This unit can be described as a light grey mostly fine grained dolomitic marble. The marble is mainly composed of dolomite but also features calcite, quartz, tremolite and diopside (Fig. 10). The marble does not fizz under application of acid. Crystals are unequal in size. The intensiveness of the foliation varies from not foliated to poorly developed. Large tremolite crystals are found along the joint planes of the dolomite marble, and can measure up to 3 cm long (Fig. 10 inset). The dolomite marble is only found as a cap or dome at the apex of the rockslide, and exposed as a column in the lower reaches of the slope.

Fig. 8 Thick calcite coating on joint planes of the mica schist

Fig. 9 Calcite marble exposure

Fig. 10 Dolomite marble in outcrop. Note interbedded boundary between the calcite and dolomite marbles. Inset: tremolite crystals exposed on a marble joint plane

Outcrop Composition

In reality, each of the lithologies occur as variably thick interbedded units. There are four outcrop composition modes noted. (1) The pelitic mica schist, psammitic mica schist, sheared mica schist, calcite marble and oftentimes thin quartz veins appear as a recurring (unordered) sequence of thinly interbedded units (Fig. 11). (2) massive mica gneiss (Fig. 12); (3) massive calcite schist (Fig. 12); and (4) massive dolomite marble overlying calcite schist (e.g. Figure 10). For the purposes of mapping, modes 1 and 2 are both termed 'mica schist' on the lithology map (Fig. 13), which has been updated from Skrede (2013) following mapping campaigns as part of this study.

Fig. 11 Outcrop composition showing interbedded nature of the lithologies

Fig. 12 Alternate outcrop composition. Left: mica schist occurs as massive. Right: Calcite marble occurs as massive

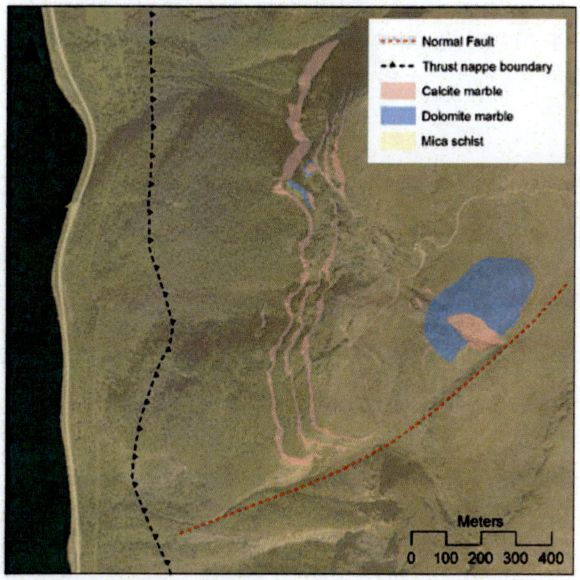

Fig. 13 Lithology map of Jettan, modified in this contribution from Skrede (2013)

Sheared and Weathered Zones

Within the mica schist host rock, layers of sheared and altered material of the same rock type are found (Fig. 14). The layers are preferentially weathered, and in some cases appear as very weak yellow banded and foliated material (Fig. 14a and d), while in others as a heavily deformed but lithified ductile shear zone (Fig. 14b). The shear zones often contain thin lenses of soil (Fig. 14c). The contact between soil and rock is either sharp or gradational, from mica schist with a weak foliation, to mylonite, to thinly laminated unconsolidated soil. The lenses are foliation-parallel and generally accompany quartz veins. The soil lens generally sits above or below the vein, or staggered en-echelon through it. The soil can be described as: light greyish yellow, moist, fine sandy SILT with minor clay, low plasticity thinly laminated undulating fabric, easily indented by thumb. Some gravel up to 100 mm in diameter. Ring shear testing of soil horizons showed a residual shear strength of $c' = 0$ kPa, $\phi' = 17\text{–}18.5°$ (Vick et al. 2020a).

Fluid circulation is controlled by foliation planes, with seepage observed on the top or bottom plane of the soil lenses (Figs. 14a and 15). Some fluid transport is observed on steep discontinuities, with discolouration indicating that groundwater appears to flow vertically along joints and horizontally across the weakness zones.

Vick et al. (2020a) sampled the mica schist for thin section analysis, in and outside of a weakness zone. NF1 was taken at a comparatively lesser affected portion of the shearing, while NF3 is synonymous with its center (Fig. 16). The analysis shows the 'unaffected' schist comprises foliation formed by bands of quartz and biotite, displaying C′-type shear bands (Fig. 16a–c). The NF3 sample features

Fig. 15 Groundwater seeps are found where the lithological 'weakness zones' provide a horizon boundary within the mica schist

Fig. 14 Lithological weak zones

sharper and wavy grain boundaries- the biotite and quartz grains appear flattened and elongated. NF3 additionally exhibits cracks due to the increased grade of weathering of the sample. This weathering is also responsible for the higher opaque content, and is seen in dirty or messy quartz-rich zones (Fig. 16d–f). High metamorphic grade minerals in both samples (e.g. kyanite) indicates that the shearing is inherited, rather than landslide-generated.

Lineaments and Structure

Across the Nordnesfjellet peninsular several clear lineaments striking NE-SW are mapped (Fig. 5) through a combination of published geology maps (Boyd and Minsaas 1983) and linement studies (Andresen 2018). These are interpreted as normal faults, as they slightly offset the lithology (e.g. Fig. 5) and align with regional structural trends (Indrevær and Bergh 2014).

Five joint sets have been identified at Jettan (Table 1). The most dominant joint set (J1) is oriented NE-SW dipping 55–90° towards NW, and because of its steep dip it is also dipping towards SE. Joint set two (J2) is oriented ESE-WNW dipping 50–90° towards N and 60–90° towards S. The third joint set (J3) is oriented NW-SE dipping 70–90° towards SW and NE (Nystad 2014). Additionally there are two minor joint sets oriented NNE-SSW and NNW-SSE (Skrede 2013).

Foliation at Jettan is measured to be sub-horizontal with a dip rarely above 20–25° (Fig. 8b). The dip direction varies from dipping towards SW and NW in general, but with the small dip angle only small variations could lead to a different dip direction (Nystad 2014; Skrede 2013).

The characteristics of these joint sets vary. Some are extremely persistent, for example in Fig. 17 >10 m long exposures of J1 cut vertically through a mica schist outcrop. In general the joints in mica schist are stepped/planar with a constant dip angle and a step across some foliation intersections. In some places however listric joints are observed. In the marble, joints are undulating (Tables 2 and 3).

Engineering Geology of the Rock Mass

Rock mass characteristics have been summarised from mapping in Table 3 along with an estimate of the Geological Strength Index (GSI; Marinos and Hoek 2000). For the purposes of presenting the defect conditions and rock strength characteristics, we have divided the rockslide into seven behavioural rock mass units: massive dolomite marble, bedded dolomite marble, blocky dolomite marble, calcite marble, schist, shear zone, and disintegrated shear zone. On the higher end of the index, the massive dolomite marble with 90 corresponding to the clean and rough joint surfaces and low frequency of defects. On the lower end of the spectrum is the sheared zone, with a GSI of 15–25. Other units fall between these, with the exception of the disintegrated shear zone, which is effectively residual soil. Correspondingly, weathering grade increases between the massive dolomite marble (fresh to slightly weathered) and the disintegrated shear zone (completely weathered).

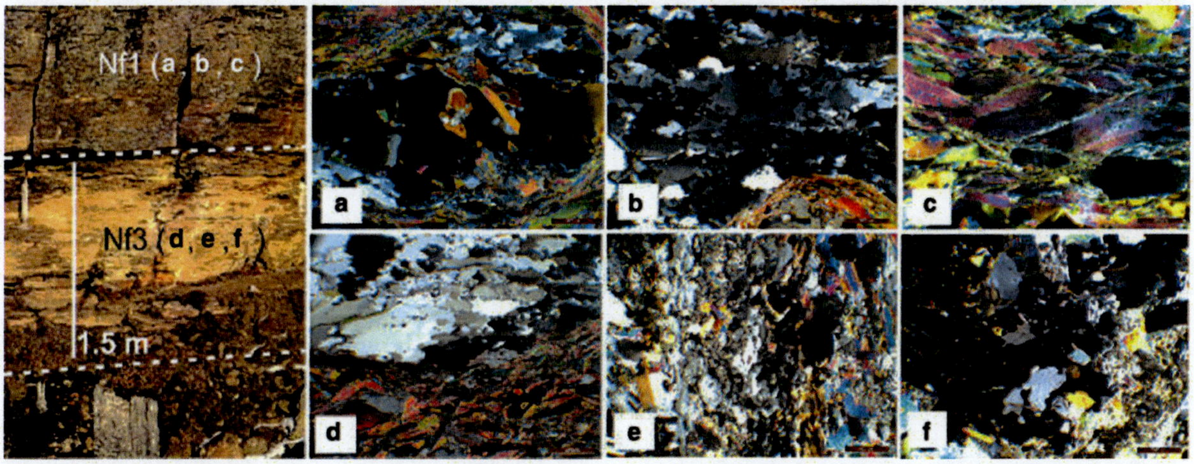

Fig. 16 Thin section analysis of sample materials NF1 and NF3. **a** Matrix in the form of biotite and quartz being deflected around the garnet. **b** Foliation formed by bands of quartz and biotite. **c** C′-type shear bands. **d** Sharper and wavier grain boundaries in quartz and biotite. **e** Dirty-messy quartz rich zones. **f** Garnet seemingly affected by the increased weathering of the rock, "eaten up" appearance. All photos were taken under crossed polarizer

Table 1 Average orientation of the Jettan joint sets

Joint set	Dip	Dip direction
J1	55–90	NW/SE (~302) (Nystad 2014)
J2	50–90	N/S (~195) (Nystad 2014)
J3	70–90	SW/NE (~40) (Nystad 2014)
J4	>60	WNW (Skrede 2013)
J5	>60	ESE (Skrede 2013)
Foliation	<25°	SW-NW (Nystad 2014)

Fig. 17 Highly persistent joint planes in mica schist

Nystad (2014) performed laboratory strength testing of borehole core to evaluate density, E-modulus, Poisson's ratio and unconfined compressive strength (UCS) (Table 2). They tested mica schist, garnet-rich mica schist and calcite marble. The garnet-rich mica schist density is slightly higher (2.9–3.0 g/m^3) compared to the standard mica schist densities (2.7 g/m^3), implying the garnet increases the density. The marble density is standard for what is typical in Norway for this rock, between 2.7 and 2.8 g/cm^3. The Young's modulus values for the rocks at Jettan show good correlation with lithology, with the calcite marble showing the highest values (<53.4 GPa) while the garnet-rich mica schist shows the lowest (>29.6 GPa). Correspondingly the results show the marble can accommodate more deformation with a poisson's ratio between 0.2 and 0.23, while the ratio for the schist samples is very low, ranging between 0 and 0.07. The mica schist shows, on average, the highest UCS (221 MPa), while the calcite marble shows the lowest (97.2 MPa). Interestingly, the presence of the garnet does not increase the compressive strength of the mica schist rock.

Nystad (2015) also performed a tilt table test, giving a friction angle of 34° for the garnet-rich mica schist, 31° for the mica schist and 31° for the marble. The mica schist and the marble correlates well with typical values for these materials, while the garnet-mica schist result suggest a slightly higher shear strength than expected.

Free swelling tests of borehole clay were also conducted from samples at 46.5 m and 53 m bgl in borehole 1. These showed free swelling of 145 (weight percent < 20 μm = 18.8) and 110 (weight percent < 20 μm = 13.7) respectively. Free swelling clays in Norway are generally active between 100 and 200%.

Detailed Geomorphology and Morphostructure

For the purposes of describing the geomorphology of such a large area, we have divided the terrain into five different domains, based solely on geomorphic signature.

The domains include (1) the rockslide apex, (2) the marble cap, (3) the plateau, (4) the active area and (5) the scarp-terrace-scarp area (Fig. 19) These domains have been assigned on the basis of extensive field mapping, as well as aerial photographs (Fig. 19), a slope map (Fig. 20), and a hillshade map (Fig. 18). The geomorphological map is presented in Fig. 22.

Domain 1. Rockslide Apex

The V-shaped intersection of the NE-SW normal fault and the conjugate backscarp creates an apex boundary to the slide, bordering everything within the unstable area of Jettan, apart from the marble cap. A clear dislocation of the rockslide downwards on this structural intersection is evidenced

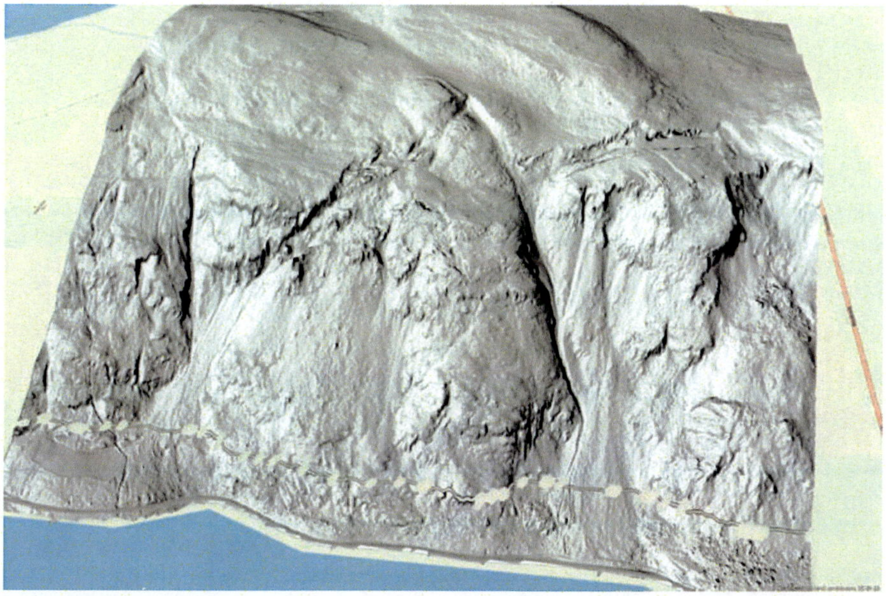

Fig. 18 3D scene of Jettan

Table 2 Samples tested for density, E-modulus, Poisson's ration and UCS from Nystad (2015)

Sample lithology	Density [g/cm^3]	E-modulus [GPa]	Poisson's ratio	UCS [MPa]
Garnet mica schist	3.0	29.6	0	146
Garnet mica schist	2.9	31.7	0.06	206
Garnet mica schist	3.0	42.9	0.07	107
Mica schist	2.7	40.3	0.1	240
Mica schist	2.7	36.5	0.09	197
Mica schist	2.7	39.2	0.07	228
Calcite marble	2.8	38.5	0.2	90
Calcite marble	2.8	51.6	0.23	88
Calcite marble	2.8			97
Calcite marble	2.7	53.4	0.21	107
Calcite marble	2.7			104

by the 30 m vertical offset of plateau (Fig. 23). Evidence of sliding is seen by striae in coating on the exposed fault plane (Fig. 24). The fault scarp continues along strike within Domain 1, where it meets with the backscarp to Midtre Nordnes. Here a group of dislocated and rafted columns, the Four Gangsters, have separated from the in situ scarp material, and balance precariously above the gully. The Four Gangsters are formed by the highly pervasive J1 surfaces presented in Fig. 17. The bedrock in this area contains frequent weak horizons within the mica schist, and very thick coatings of a white precipitate (untested).

In the centre of the domain a mound of blocky dolomite marble protrudes from the ground surface. It has been interpreted as a rock glacier based on movement rates alone (e.g. Blikra and Christiansen 2014), however a definite structural fabric to the rock mass is observed in the field suggesting that it has or is moving cohesively. A second smaller rock glacier has been interpreted at the backscarp intersection. The rock glacier displaces a lobate and flow morphology (Eriksen et al. 2017a). At the NW corner of the domain two 20 m wide linear morphological depressions follow approximately parallel with the backscarp. These are extensional and imply local movement towards the SW. At the western edge of the domain a change in slope to the plateau marks the boundary. Minor scarps have formed around here along J1 and J3, and are fringed by talus.

Domain 2. Marble Cap

The marble forms a rounded cap, or dome, above the backscarps (Fig. 26). It is comprised of massive dolomite

Table 3 Rock mass unit descriptions

Rock mass unit	Lithology	Description	Typical intact strength	Weathering	Degree of fracturing	Defect spacing	GSI (Marinos and Hoek 2000)	Defect persistence	Roughness (NGI 2015)
Massive dolomite marble (Fig. 10)	Dolomite marble	Whitish-grey massive reactive marble with infrequent joints. Joints contain large crystals	190 MPa (Xie et al. 2011)	Fresh to slightly weathered- no discolouration	Massive with widely spaced defects	Bedding >1 m thick. Defects 0.5–2 m spacing	90	>2 m	Undulating rough
Bedded dolomite marble	Dolomite marble	Whitish-grey bedded reactive marble with infrequent joints. Joints contain large crystals	190 MPa (Xie et al. 2011)	Fresh to slightly weathered- no discolouration, some decomposition	Moderately thickly bedded and widely spaced defects	Bedding 0.2–0.6 m thick. Defects 0.5–2 m spacing	70	>2 m	Undulating rough
Blocky/transported dolomite marble	Dolomite marble	Large blocks (>1 m diameter) of above, transported via rockfall, rock glacier and rockslide activity	190 MPa (Xie et al. 2011)	N.A	N.A	N.A	N.A	N.A	N.A
Calcite marble (Fig. 9)	Calcite marble	White-yellow bedded coarse grained. Non-reactive	88–107 MPa (Nystad 2015)	Fresh to slightly weathered, some discolouration	Thinly laminated to thickly bedded	Bedding ranges from laminated to >2 m	85	0.5–2 m	Undulating smooth
Schist (Fig. 6)	Mica schist	Dark blueish grey foliated gneiss	107–240 MPa (Nystad 2014)	Fresh to slightly weathered, some oxidization and calcite coating on defect surfaces	Moderately widely spaced joints. Very thinly spaced foliation	Joints 0.2–0.6 m. Foliation 6–20 mm	55	> 8 m	Stepped smooth and planar smooth
Shear zone (Fig. 14c)	Mica schist	Brown-yellow and grey heavily deformed and fragmented shear zone containing pockets of soil	N.A	Moderately to highly weathered	Very to extremely closely spaced	<20–60 mm	15–25	N.A	N.A
Disintegrated shear zone (Fig. 14d)	Mica schist	Brown-yellow decomposed shear zone	N.A	Completely weathered	N.A	N.A	N.A-residual soil	N.A	N.A

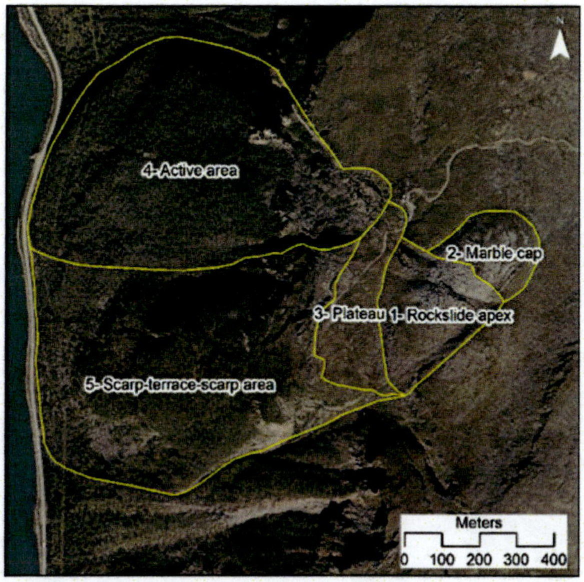

Fig. 19 Geomorphic domains at Jettan

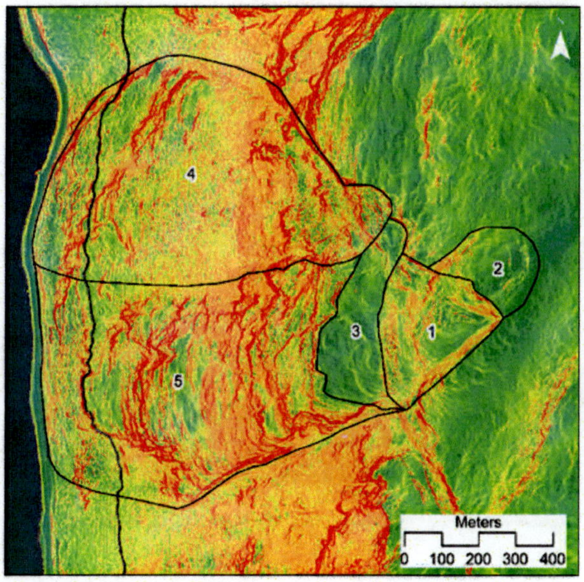

Fig. 20 Slope map of Jettan

marble, which has clearly been more weather resistant than the underlying reactive and softer calcite marble. The dome is geologically isolated due to fault offset (Fig. 5). The rock is heavily fractured. In the center of the dome, tension cracks follow J1 and 3 surfaces, and are several meters deep. Fractures widen to the SE, closer to the fault trace. The larger fractures form morphological depressions in the terrain interpreted to be grabens, ranging up to 10 m in width and 2 m deep (Fig. 27). The axes of the depressions curve to the

Fig. 21 Aerial image of Domain 1

north to border the dome and taper out. These fractures and depressions imply the dome is opening/moving en masse to the W (Fig. 28).

Domain 3. Plateau

The plateau is morphologically striking in its smooth unbroken surface. The pleatau sits between 600 and 670 m asl, with the highest end in the north. The domain is draped in a thick unsorted, broad ranging size of sediments (mostly till). The north and center of the plateau is cross cut by morphological depressions and sink holes. These are formed by till draped over open fractures. The depressions are generally oriented NW-SE sand have small ridges between them, indicating an displacement to the SW. They are up to 3 m deep. The sinkholes often appear in connection with the depressions, usually where fractures intersect. Scarps mark the transition to the lower domains. Water has been observed to pond in the southern section of the domain.

Domain 4. Active area

We have termed domain 4 the active area as it has shown the most activity in recent times (historic and current day). It is the most deformed surface area of the slope.

The upper section is characterised by a mass of dislocated and disintegrated rafted blocks. At the rear of the domain,

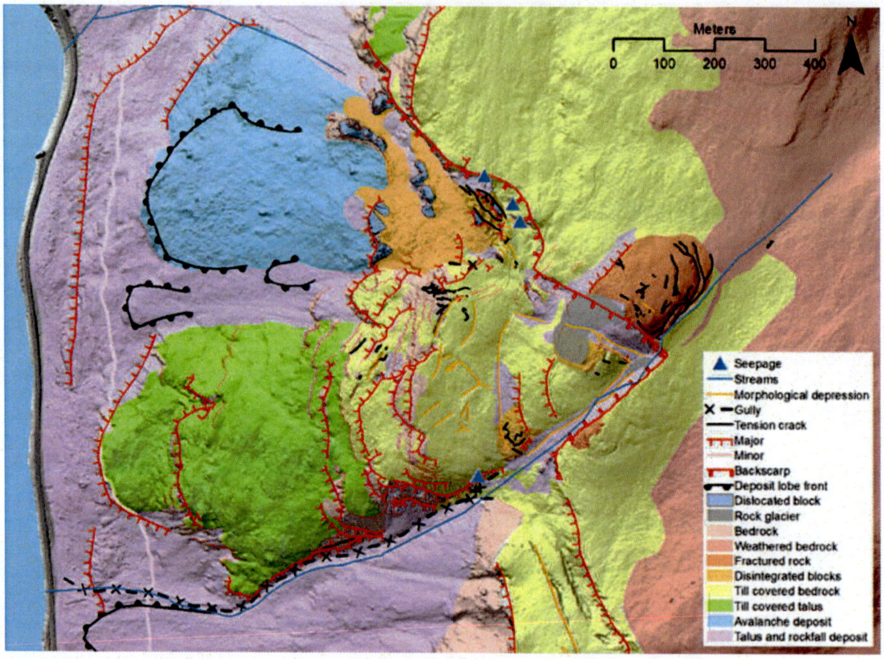

Fig. 22 Geomorphological map of Jettan. Basemap 1 m DEM hillshade

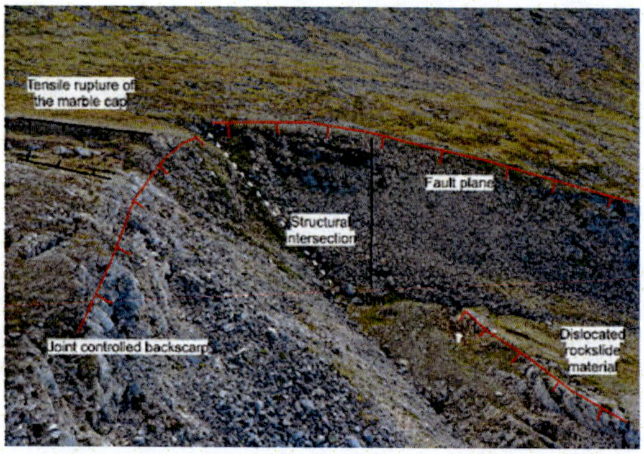

Fig. 23 Rockslide apex created by the intersection of the backscarps

Fig. 24 Evidence of sliding on the exposed fault surface

the NW-SE trending backscarp delineates the limit of the unstable area. The overall strike of the backscarp is disrupted by a small retrogression, or The Bite (Figs. 31 and 32). The 'bite' contains large displaced blocks measuring up to 30 m long, between 10–12 m wide and 10 m high. Deep open cracks can be seen on the upper paleosurface of the blocks. Many of the blocks are precariously perched, some on narrow lenses of weak sheared material (Fig. 33). The Borehole 3 collar is located on one such block. The ravines between blocks strike NW-SE, indicating that The Bite is displacing towards the SW.

Further to the NW the backscarp becomes higher, reaching 100 m at the highest point. Below this sit several tall columns. The largest of these, Jettankallen, is 60 m higher than the surrounding slope. The upper southern boundary is framed by a scarp which follows J1, and is parallel to a series of the same joints which have formed depressions in the landscape above (Fig. 36). The Bite, Jettankallen, and this scarp together triangulate an area of disintegrated blocks (Figs. 34 and 35).

Below this triangle, the ground surface becomes continuously vegetated. High resolution lidar hillshade images show that much of the ground surface is dominated by a lobe of avalanche material (Fig. 37). It seems likely that an avalanche has occurred, sourced from the aforementioned triangle. This has debuttressed The Bite, allowing for

Fig. 25 The four gangsters sit precariously above the gully dividing Jettan and Midtre Nordnes

Fig. 27 Looking NE down the axis of a morphological depression

Fig. 26 Aerial image of the marble cap

Fig. 28 Drone image showing the cap separating from the mica schist near the fault plane, opening towards the W

expansion between the joints of J1, forming dislocated blocks. Incidentally the upper section of domain 4 has been identified as the fastest moving surface area of the RSD (Eriksen et al. 2017a).

Several seepage points have been observed in the backscarp of domain 4 (Figs. 22 and 38).

Domain 5. Scarp-Terrace-Scarp Area

Domain 4 and 5 could be considered mirror twins, where in domain 4 a failure has already occurred; in domain 5 the terrain remains intact. Much of the rock mass in domain 5 does not display signs of deformation, and the only indication that it has moved from its original position is the offset of material above it in domain 1. The NE edge of the domain follows a very large open and instrumented fracture (Fig. 40). It strikes NNE-SSW and follows the J1 and 3 orthogonal joint sets. Ice is found inside fracture all year round. The crack is at least 20 m deep and decreases in width with depth.

Below this, and wrapping around the entire slope within the domain is a scarp-terrace-scarp formation. The scarps are formed by zigzagging intersections of J1, 3, 4 and 5. Some show a smoothed surface with striae. The terrace steps generally follow foliation and are often covered in talus and a thin till cover. The terraces become progressively wider downslope.

At the S edge of the domain a 150 m high NE-SW-oriented wall marks the outer limit of the rockslide.

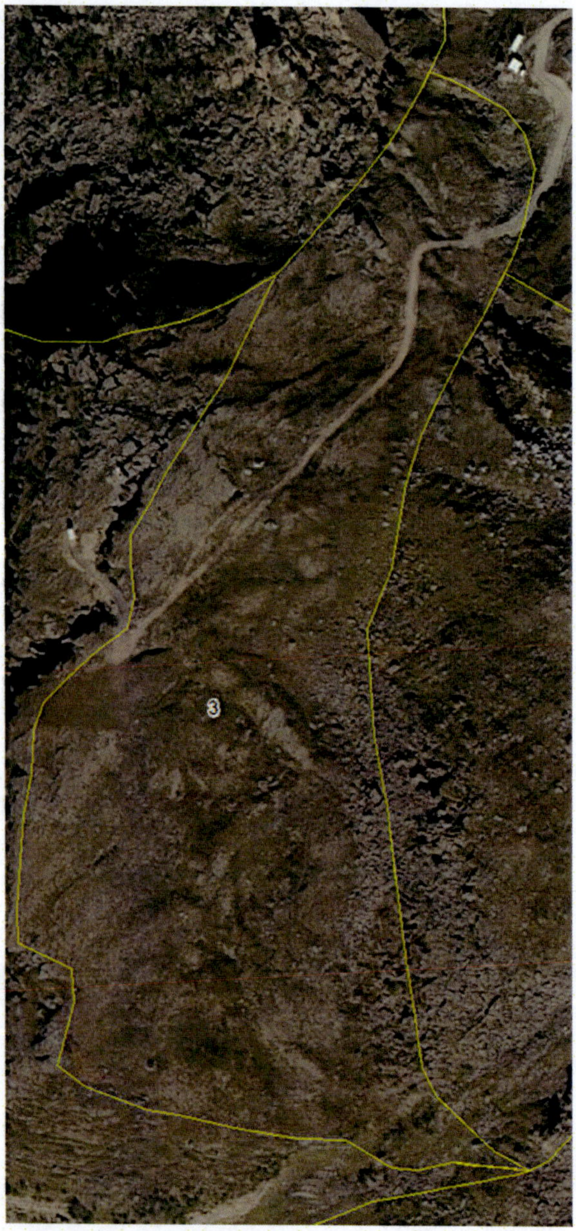

Fig. 29 Aerial image of the plateau in domain 3

Fig. 30 Morphological depressions and sinkholes with draped till over open fractures

Fig. 31 Aerial image of upper portion of domain 4. Note the retrogressive 'bite' interrupting the general strike of the backscarp

It is aligned with a gully which follows the fault trace. At the time of mapping no water was seen in the valley, but imbrication of loose material and erosional features give signs of seasonal drainage. The southern section of the large wall branches into several sets of large scarps, bending to strike NW-SE. Beneath the wall, rock fall deposits fill the valley. Seepage was found in the upper eastern part of the wall, leaking out from a weaker layer in the mica schist (Fig. 15). Calcite coating was observed on several joint surfaces in the large wall, likely sourced from calcite marble layers in the top portion of the wall. Rockfall deposits fringe the scarps in the lowermost parts of domain 5.

Movement

Rates and Direction of Movement

A number of studies have investigated the movement rate and movement direction at Jettan. The most novel presents

Fig. 32 Drone image of The Bite looking SSW. Note the container housing the collar of borehole 3

Fig. 33 A dislocated block within The Bite, which narrows at the base

Fig. 34 Disintegrated blocks below The Bite

Fig. 35 Jettankallen

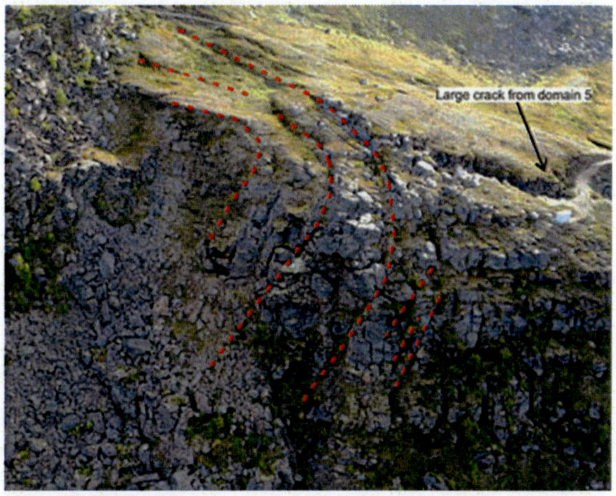

Fig. 36 NW-striking structures as they appear framing the 'historic avalanche' in domain 4

3D surface displacement from interpolated satellite- and ground-based InSAR in connection with structures and geomorphology at the URS (Eriksen et al. 2017b), and interpreted surface displacement using 2D InSAR (Eriksen et al. 2017a).

Three areas with particularly high movement rates stand out (Fig. 41 areas 1, 2 and 3). Area 1 has a movement rate of up to 50 mm a^{-1} and is the most rapidly moving area. 3D displacement vectors strike towards WNW with a shallow plunge angle. Eriksen et al. (2017a) suggests a displacement going into the slope in the eastern part (upper most), and a displacement going out-of-the-slope in the western part (lower most) along listric and planar fractures. Area 2 shows a homogeneous displacement pattern with a steeper dip than Area 1, and a movement rate up to 35 mm a^{-1} towards WNW. Area 3 has been interpreted to be a rock glacier by Blikra and Christiansen (2014).

GNSS are only instrumented in the northern portion of the rockslide. These show a movement of material inwards

Fig. 37 Hillshade image showing The Bite in the upper area, with dislocated blocks and avalanche deposits which reach to the lower slope. Based on a 0.5 m DEM

Fig. 38 Seepage points are mapped in the backscarp where groundwater exits at weak horizons

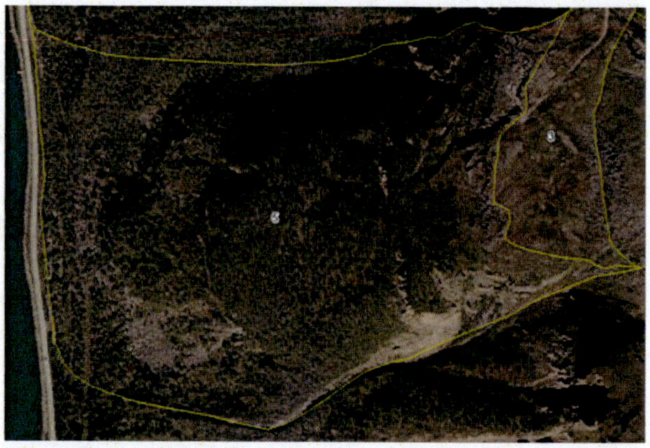

Fig. 39 Aerial image of domain 5

towards the avalanche area, with the largest displacement occurring on a block in The Bite.

Seasonal Variations

Nordvik et al. (2010) assessed a seasonal pattern displaying the largest rates of deformation in the autumn, interpreted as an effect of freezing of groundwater leading to fracture expansion. They also showed that the deformation was affected by the ambient temperature, directly on instruments or heating/cooling of bedrock. A later exceptional study by Blikra and Christiansen (2014) reviewed seasonal controls on deformation at Jettan, and in particular looked at permafrost as a driver. They concluded the topography, the accumulation of snow and the complex joint patterns at Jettan creates a unique setting that leads to localized permafrost at or below the permafrost limit. The study also revealed big seasonal variations in deformation; acceleration in movements from May (springtime), reduction of movement in December (winter) and even a greater reduction of deformation again in February (late winter) (Fig. 42).

Fig. 40 Looking N through a large open fracture in domain 5

Below we describe the seasonal deformation model as presented by Blikra and Christiansen (2014).

Spring to Autumn, High Deformation

Deformation increases when the snow begins to melt, however a deceleration is not seen when the snowmelt is heavily reduced or has stopped, indicating an additional driving mechanism than pore pressure increases derived directly from snowmelt. It is interpreted meltwater infiltrates fractures and raises the permafrost ice temperature from −0.5 °C to just below 0 °C, destabilising the ice. Pore pressure increases and the effective shear strength is reduced leading to higher deformation. While air in the upper fractures can reach 8–10 °C in summer, low temperatures remain in the deeper parts of the fractures, maintaining the permafrost. In October the air cools, lowering the ice temperature in the fractures and stabilising the permafrost, leading to lower rates of deformation.

Late Autumn to Early Winter, Medium Deformation

Accumulation of cold air in the fractures causes a gradual reduction of deformation from early winter. This accumulation of cold air increases the strength of the ice, and the reduction in ice melt water leads reduced pore pressures. When and how long this period of medium deformation happens depends on the extent of the snow accumulation in

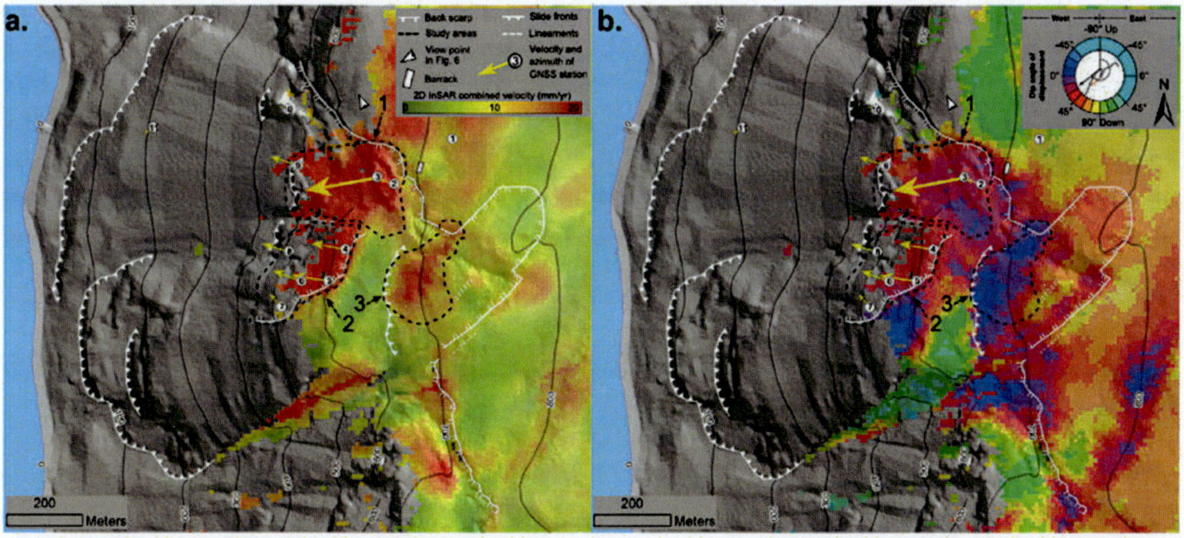

Fig. 41 Movement rates and kinematics from Eriksen et al. (2017b). **a** Combined velocity from 2D InSAR data. **b** 2D InSAR dip of displacement. Yellow arrows show movement vectors from dGNSS

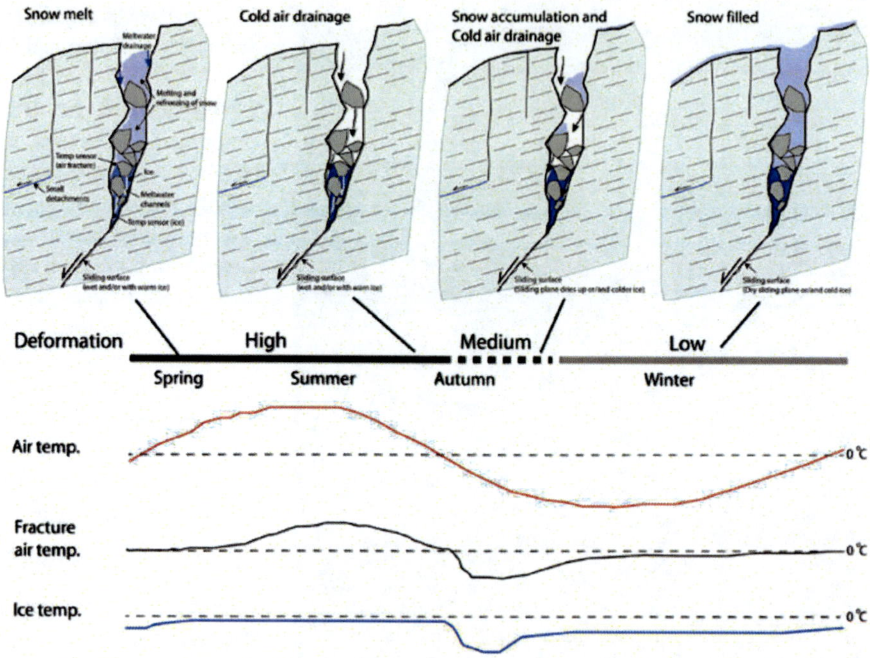

Fig. 42 A model of seasonal deformation, by Blikra and Christiansen (2014)

the fracture and the characteristics of the cold-air drainage (Fig. 42).

When the snow cover occurs early, closing the fracture, cold-air-accumulation in the lower part of the fracture is stopped. The relatively warmer ice temperature in these years means it is less stable, allowing a continued reduced medium rate of deformation in winter. When the snow cover arrives late, cold-air-accumulation lasts for a longer period, reducing the ice temperature and deformation rate. The increased effective shear strength in the fracture could lead to a deformation rate of almost zero (Fig. 42).

Early Winter to Spring, Low Deformation

This period lasts for around 3–5 months and is characterized by a relatively stable minimum rate of deformation. The snow covering the main fractures prevents influence from the environment outside, making a stable temperature just below 0 °C inside the fracture. In the beginning to the middle of May, when the air temperature outside of the fracture goes above 0 °C, the snow cover melts and the cycle restarts.

3D Model and Failure Scenarios

Previous Failure Scenarios

Previous failure analyses have focused on the most realistic hazard scenarios, using volumes calculated from areas within the most active portion of the slide. Henderson and Blikra (2007) constructed potential failure scenarios based on mapping and monitoring data. They concluded with three possible failure scenarios, 1a, 1b, and 2- a combination of 1a and b (Fig. 43). Their respective volumes were calculated to be 7, 4 and 11 Mm^3.

Following surface geophysics campaigns (Rønning et al. 2008) the volume was recalculated to be between 5 and 9 Mm^3 for scenario 1a, and 12 and 22 Mm^3 for scenario 2, using depths between 70 and 120 m (Blikra et al. 2009) (Fig. 44). The larger volumes are a result of the increased depth to the basal sliding zone.

3D Rockslide Model

The purpose of our model is not to redefine the official failure scenarios, but to bring together lithology and geotechnical data for a greater 3D understanding of the rockslide. This work is in the preliminary stages and the 3D model is a work in progress intended to provide the basis for future studies.

Previous work brings together borehole geology, downhole geophysics and DMS column data to assess likely sliding surfaces (Eriksen et al. 2017a). Borehole 1 shows a highly fractured and crushed zone at 45 m. DMS data shows that much of the deformation takes place at this horizon, to the W. In borehole 3, DMS data shows a high amount of deformation between 77 and 87 m depth, which aligns with a crushed zone at 80. Data from borehole 2 shows displacement toward NW at 40–50 m.

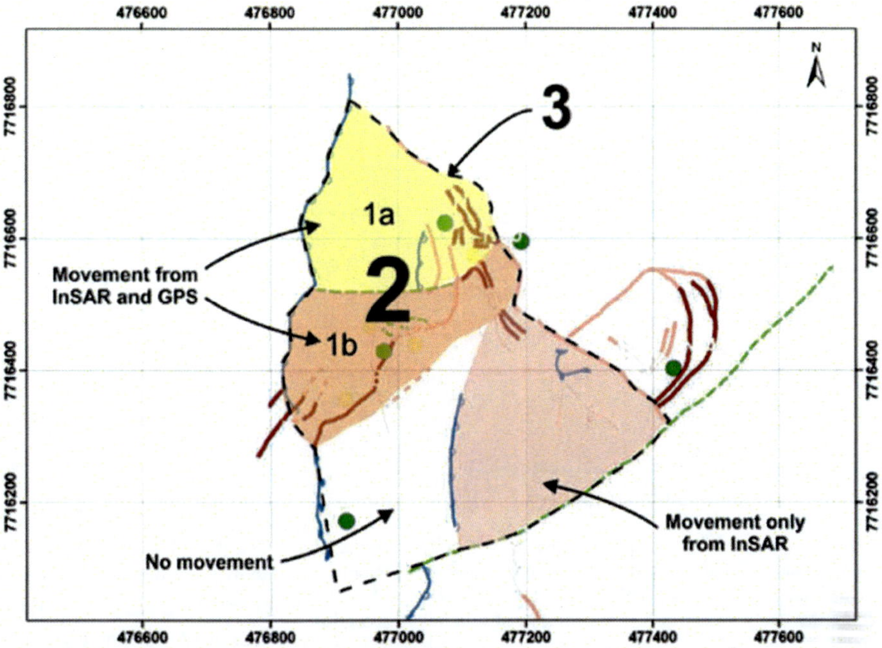

Fig. 43 Failure scenario 1a, 1b and 2 location at Jettan. From Henderson and Blikra (2007)

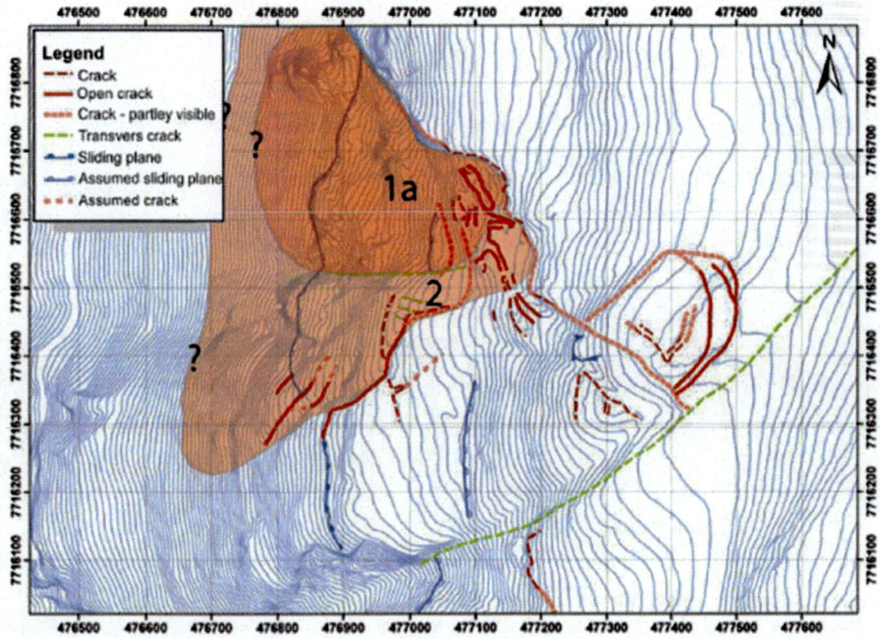

Fig. 44 Failure scenario 1a and 2. From Blikra et al. (2009)

Four sliding surfaces have been interpreted based on borehole crush zones, structural orientations and geomorphology (Fig. 45). WE believe the failure surface of the slide follows a step path of joints, foliation and rock mass breakthrough fractures, and there are likely multiple tiers of failure surfaces. We show the most relevant of these surfaces on two cross sections (locations: Fig. 46, cross section A: Fig. 47, cross-section B: Fig. 48). We display these together with the volume surfaces for volume calculations. Volumes are calculated for 4 scenarios using smooth surfaces for modelling purposes only (Fig. 45 and Table 4). We have also include volumes for the entire instability area and for just The Bite.

Vick et al. (2020b) hypothesise that a heavily segmented sliding surface, with different moving parts is likely the result of both the weakness of the bedrock from fault

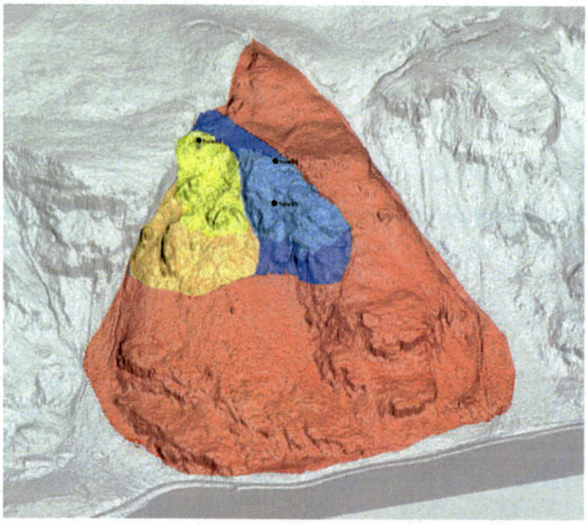

Fig. 45 Various failure scenarios based on borehole interpretations. Light blue = failure scenario 1, yellow = failure scenario 2, gold = failure scenario 3, dark blue = failure scenario 4. Red = the entire instability area, which is not considered as a failure scenario

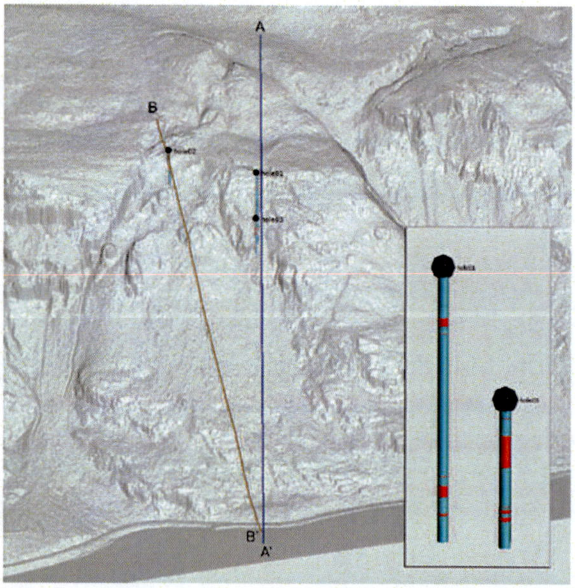

Fig. 46 Boreholes showing 'crushed zones' intervals (pink intervals), cross section A (blue line), and cross section B (brown line)

activity, and the presence of multiple types of structures presenting many potential rupture planes for deformation. They published a geotechnical model of rockslides (based on examples including Jettan) to describe the interplay of brittle structures in the rockslide deformation (Fig. 49). As the steep backscarps cannot geometrically daylight in the slope surface, the basal sliding surface is formed along a series of low angle planes connected by steep joint controlled structures. At Jettan the low angle structures are likely to be foliation. Pinpointing the exact transfer zone from the steeper surfaces to the shallow surfaces is difficult. Foliation is generally mapped at a low angle at Jettan. However with the shear strength of the weakness zones controlled by foliation approaching residual strength as a result of the various processes discussed in the manuscript, with gravitational forcing and high pore pressures it is possible to achieve sliding at angles lower than the friction angle (Glastonbury and Fell 2010, Fig. 49).

Risk Assessment

The hazard class is assigned based on many characteristics, including stage of development of the rockslide. The Jettan hazard scores high, between 6.25 and a 9.50 (NGU 2019). A volume of 6 Mm^3 was used for the most recent displacement wave modelling (NGI 2018). A 6 Mm3 avalanche would trigger a wave which would impact many small settlements around the fjord. Modelling shows a maximum height of the run-up at Lyngseidet (village opposite) to be 10 m in a 1/1000-year event, and 13 m in a 1/5000-year event (Glimsdal and Løvholt 2018). The consequences of this places the risk class at high (Fig. 50).

Discussion

The geological history of the area, both the regional scale Caledonian mountain building and district scale Quaternary glaciation, have created a physical environment which is susceptible to failure. The specific geological events that lead to failure potential on this slope are:

1. The Caledonian allochthon/orogeny events resulted in the formation of a rock fabric producing an anisotropy in the rock mass which is predisposed to sliding (mica foliation planes gently dipping towards the fjord).
2. Lithological variation in the nappe rock mass has concentrated shear strain to specific areas of the rock mass resulting in the formation of ductile shear zones parallel to this anisotropy. These zones of weakness have later been altered by weathering to reduce the overall shear strength.
3. Opening of the Norwegian Sea resulted in the formation of brittle tectonic structures such as the normal faults and joint swarms, which accommodate the deformation at Jettan and contribute to the overall failure path.
4. The LGM caused significant bedrock erosion leading to oversteepened slopes. Ice debuttressing of the slope after glacial retreat removed slope support and facilitated relaxation of the rock mass from stress relief. The pre-existing shear zones accommodated most of this

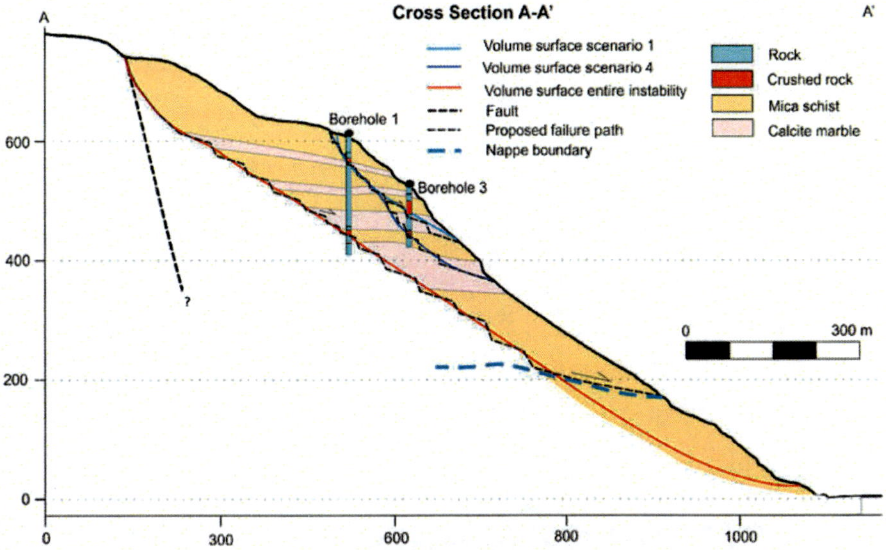

Fig. 47 Cross section A-A′ showing interpolated sliding surfaces and lithology from Leapfrog model

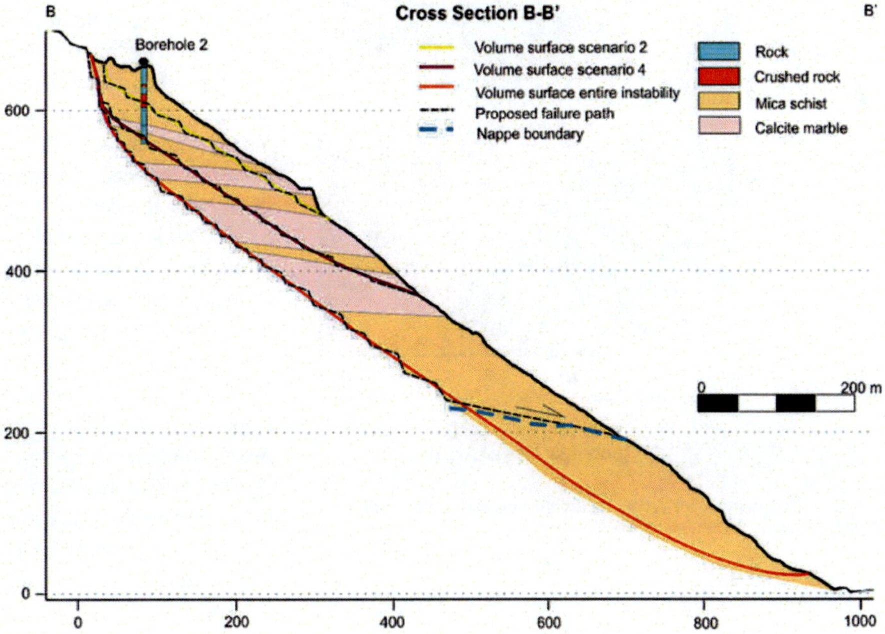

Fig. 48 Cross section B-B′ showing interpolated sliding surfaces and lithology from Leapfrog model

Table 4 Volumes calculated for various failure scenarios

Failure scenario	Volume (Mm^3)
1	1.6
2	1.0
3	4.6
4	8.8
Entire instability	79.1
The Bite	0.1

strain, further degrading the condition of these zones of weakness.

5. The lithological variation and zones of weakness, exacerbated by multiple cycles of glaciation and deglaciation and slope stress changes, plays a major control on the flow of groundwater in the rock mass. Seasonal recharge of groundwater from snowmelt and permafrost is controlled by permeability differences across these lithology boundaries, forming seasonal perched groundwater systems within the rock mass complex. These perched systems are likely controlled by the zones of weakness

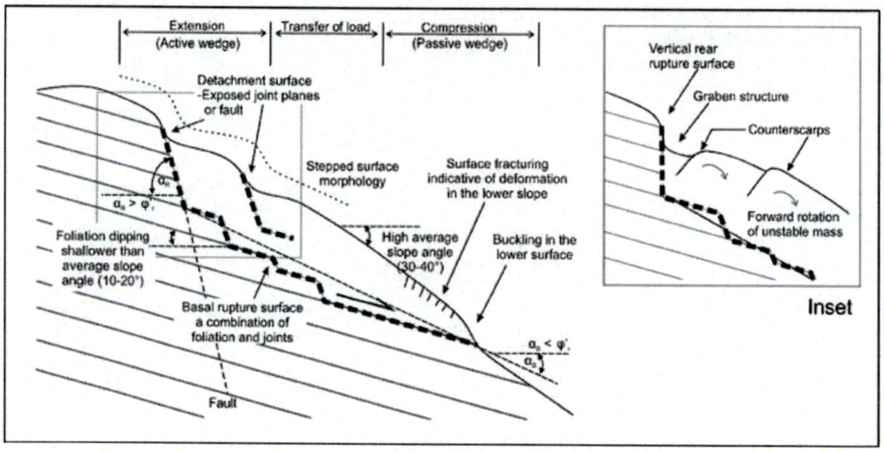

Fig. 49 Geotechnical schematic from Vick et al. (2020b) of rockslides where sliding surfaces are controlled by joints, faults and foliation

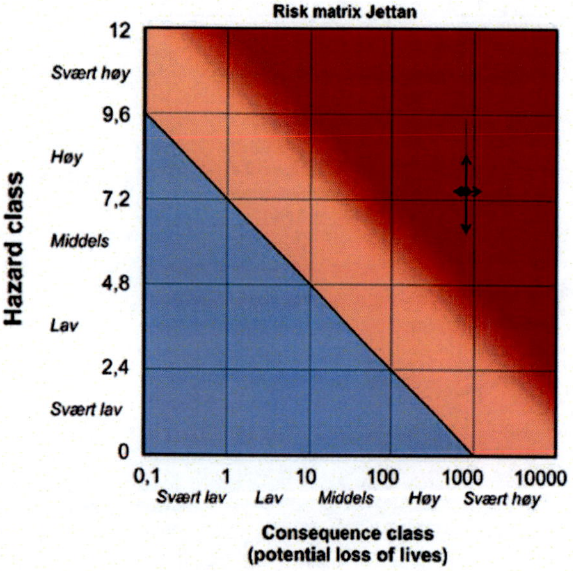

Fig. 50 Risk matrix for Jettan including nominal annual probability (NVE 2020)

resulting in a seasonally higher pore pressures contributing to the periods of increased movements.

6. Significant mass wasting appears to have occurred around 10.8 ka cal BP immediately following the Ørnes event as part of the deglaciation of Jettan (Corner 1980; Hegstad 2014) (Fig. 3). These ancient slope movement events exposed the modern-day slope, contributing to the morphostructural features observed, adding to the predisposition of the current slope to future failure.
7. An avalanche that appears to have occurred in the northern area of the slide (Figs. 22 and 37) has resulted in partial removal of toe support to the upper areas leading to extension in this slope and significant acceleration of movement (Fig. 41).

Mica schist is commonly known for contributing to slope instability due to the anisotropy resulting from the platy alignment of mica minerals in the foliation bands. At Jettan, destabilisation of the rock mass is amplified by the presence of sporadic mylonite (a heavily sheared metamorphic rock), and the foliation-parallel weakness zones. While the intact strength of the mica schist is measured to be high, the Young's modulus and Poisson's ratio were correspondingly low. The low capacity of this rock to accommodate strain is likely why the pervasive joint sets have formed in this rock mass. The joints control the backscarp and scarp formation throughout the site and define the columns at the Four Gangsters and Jettankallen. They are also likely responsible for much of the vertical groundwater flow through the site. Iron staining was noted on many joint surfaces, and it accounts for the presence of thick calcite coating lower in the slope, a good distance from the closest mapped calcite marble outcrop.

The calcite marble conversely has a higher Young's modulus and Poisson's ratio, and displays fewer joint sets. The rock has a lower intact strength, and has a more eroded and weathered appearance giving a clear source for the precipitate coating much of the mica schist. Dissolution is often responsible for slope destabilisation in other parts of the world (e.g. Ostermann et al. (2016)), however no calcite marble was mapped lower in the slope, so it is unlikely a factor here.

No outcrop of the nappe boundary plane was observed during mapping. The boundary is shown on the 1:50,000 scale geology map (Boyd and Minsaas 1983), however, it was unlikely to have been identified in the field at the site due to the slope being covered by avalanche debris. The nappe boundary is likely expressed by a change in lithology and/or metamorphic grade and as such represents a major structural feature in the overall slope. While nappe boundaries are known to control mountain slope deformation in some alpine areas of the world (e.g. (Ambrosi and Crosta 2006)), there is no current evidence to say this is the case at Jettan. Given the boundary may form a major stiffness change in the overall slope rock mass, there remains the possibility of some rock mass damage along this boundary. If a damage zone is present it may have been exacerbated by recurring phases of stress changes acting over two time scales: 1. Repeated cycles of glacial ice loading and unloading of the rock mass during the Quaternary, and 2. Seasonal groundwater recharge during interglacial periods causing hydromechanical 'pumping' of the rock mass. Accordingly, the role of the nappe boundary in the larger scale instability remains a focus for future research.

The ductile weakness zones present in the mica schist are likely related to nappe metamorphosis and thrusting, resulting in a controlling nappe fingerprint on the rockslide. The weakness zones within the mica schist show evidence of strain localisation indicating ultramylonitic ductile shear zones. The shear zones are localised, differing in character in relation to the host rocks as quartz and biotite grains are flatter in appearance, and quartz grains display more evidence of deformation. While some ductile shear zones in other areas have been known to form synchronously with landslide development, the high grade mineral assemblages of these shear zones imply they are inherited. The weakness zones have preferentially formed an element of strength degradation because:

1. The intense shear fabric forms a boundary between the host rock and the shear zones. The fluid flow pathway for groundwater created by this boundary has forced a preferential weathering of the sheared material.
2. The preferential weathering has contributed to the development of soil horizons within the zones. The shear strength of the material in these horizons measured at Jettan is low, with a friction angle (ϕ') between 17 and 18.5° (Vick et al. 2020a). This suggests it is approaching residual shear strengths.

The indicated shear strength of the weakness zones at Jettan is equal to or exceeds the foliation dip. However seasonal high pore pressures acting along the foliation controlled weakness zones, and within the steep joint and fault controlled boundary defects will result in a lower effective shear strength. Furthermore, it is common for compound slides to overcome a low friction angle at the basal sliding surface due to the effect of the load force of the rockslide under gravity (Glastonbury and Fell 2010; Vick et al. 2020b). We therefore assume that if these weakness layers are present at depth within the rockslide, they have likely been utilised as the basal sliding surface, and furthermore may have controlled the location of the basal sliding surface development.

Basal sliding surfaces likely propagated across foliation surfaces, and it is unknown whether a fully formed rupture surface exists. Many authors speculate that a fully pervasive and formed rupture surface is an imminent precursor for failure (e.g. (Hermanns and Longva 2012), so it is unlikely that it is yet formed. However we see some evidence of strain at the toe from rockfall deposits at the base of domain 5.

Movement of the slide has a permafrost control. Blikra and Christiansen (2014) present a model which highlights the control fracture ice plays on the slide. While positive temperatures predictably control the onset of deformation by snowmelt infiltration, the autumn cycle is less predictable with a complex interplay of cold air, ice temperatures and stability. The subtlety of this interaction is perhaps an indication of the heightened sensitivity of the rockslide stability condition to pore pressure changes. This may point to an enhanced susceptibility to long-term climate change.

9. Conclusions

As series of geological events have resulted in structural and mechanical changes to the slope at Nordnesfjellet. Within this context, we review lithological and structural data, borehole information, geomorphology and movement to develop a holistic model of the Jettan rockslide. We determine that the overall rock mass is characterised by anisotropic lithology, zones of weakness and very pervasive joints, which together with seasonal fluctuations act to form sliding planes and lower the overall strength of the slide.

Acknowledgements We would like to thank Sophie Bainbridge, Stephan Höpfl, Martin Mikkelsen and Linn Asplin for field assistance and constructive discussion. We would also like to acknowledge the NVE Manndalen office for field and data access and problem solving.

References

Ambrosi C, Crosta GB (2006) Large sackung along major tectonic features in the Central Italian Alps. Eng Geol 83:183–200. https://doi.org/10.1016/j.enggeo.2005.06.031

Andresen M (2018) Regional structural analysis of rock slope failure types, mechanisms and controlling bedrock structures in Kåfjorden, Troms. Dissertation, UiT The Arctic University of Norway

Augland LE, Andresen A, Gasser D, Steltenpohl MG (2014) Early Ordovician to Silurian evolution of exotic terranes in the scandinavian caledonides of the Ofoten-Troms area-terrane characterization and correlation based on new U-Pb zircon ages and Lu-Hf isotopic data. Geol Soc Spec Publ 390:655–678. https://doi.org/10.1144/SP390.19

Berg JN (2020) Engineering geology of the Jettan rockslide. UiT The Arctic University of Norway, Kåfjorden

Bergh SG, Eig K, Kløvjan OS et al (2007) The Lofoten-Vesterålen continental margin: a multiphase Mesozoic-Palaeogene rifted shelf as shown by offshore-onshore brittle fault-fracture analysis. nor J Geol 87:29–58

Blikra LH, Christiansen HH (2014) A field-based model of permafrost-controlled rockslide deformation in northern Norway. Geomorphology 208:34–49. https://doi.org/10.1016/j.geomorph.2013.11.014

Blikra LH, Henderson IHC, Nordvik T (2009) [Rockslide hazard at Nordnesfjellet in the Lyngenfjord, Troms] Faren for fjellskred fra Nordnesfjellet i Lyngenfjorden, Troms. NGU Report 2009.026

Blikra LH, Longva O (2009) Gravitational-slope features in Troms: indications of paleoseismic activity? NGU Report 2000.001

Blikra LH, Longva O, Braathen A et al (2006) Rock slope failures in norwegian fjord areas: examples, spatial distribution and temporal pattern. Landslides from Massive Rock Slope Fail 475–496. https://doi.org/10.1007/978-1-4020-4037-5_26

Boyd R, Minsaas O (1983) Bedrock geology lyngen [Berggrunnskart Lyngen] 16343 1:50 000

Braathen A, Blikra LH, Berg SS, Karlsen F (2004a) Rock-slope failures in Norway; type, geometry, deformation mechanisms and stability. Nor J Geol 84:67–88

Braathen A, Blikra LH, Berg SS, Karlsen F (2004b) Rock-slope failures in Norway; type, geometry, deformation mechanisms and stability. Nor J Geol 84:67–88

Corner GD (2005) Atlantic coast and fjords. In: Seppälä M (ed) The physical geography of fennoscandia. Oxford University Press, p 468

Corner GD (1980) Preboreal deglaciation chronology and marine limits of the Lyngen-Storfjord area, Troms, North Norway. Boreas 9:239–249

Dehls JF, Olesen O, Olsen L, Blikra LH (2000) Neotectonic faulting in northern Norway; the Stuoragurra and Nordmannvikdalen post-glacial faults. Quat Sci Rev 19:1447–1460. https://doi.org/10.1016/S0277-3791(00)00073-1

Eckerstorfer M, Eriksen HØ, Rouyet L et al (2018) Comparison of geomorphological field mapping and 2D-InSAR mapping of periglacial landscape activity at Nordnesfjellet, northern Norway. Earth Surf Process Landforms 43:2147–2156. https://doi.org/10.1002/esp.4380

eKlima (2019) Free access to weather- and climate data from Norwegian Meteorological Institute from historical data to real time observations. https://eklima.met.no. Accessed 19 Mar 2020

Elvebakk H (2013) [Borehole logging with optical televiewer, Jettan 1, Nordnesfjellet, Kåfjord Municipality, Troms] Borehullslogging med optisk televiewer, Jettan 1, Nordnesfjellet, Kåfjord kommune, Troms. NGU Report 2013.020

Elvebakk H (2014) [Borehole logging with optical televiewer, Bh 2 and Bh 3, Jettan, Nordnesfjellet,Kåfjord Municipality, Troms] Borehullslogging med optisk televiewer, Bh 2 og Bh 3, Jettan, Nordnesfjellet, Kåfjord kommune, Troms, NGU Report 2014.016

Eriksen HØ, Bergh SG, Larsen Y et al (2017a) Relating 3D surface displacement from satellite- and ground-based InSAR to structures and geomorphology of the Jettan rockslide, northern Norway. Nor J Geol 97:233–253

Eriksen HØ, Lauknes TR, Larsen Y et al (2017b) Visualizing and interpreting surface displacement patterns on unstable slopes using multi-geometry satellite SAR interferometry (2D InSAR). Remote Sens Environ 191:297–312. https://doi.org/10.1016/j.rse.2016.12.024

Faber C, Stünitz H, Gasser D et al (2019) Anticlockwise metamorphic pressure-temperature paths and nappe stacking in the Reisa Nappe Complex in the Scandinavian Caledonides, northern Norway: Evidence for weakening of lower continental crust before and during continental collision. Solid Earth 10:117–148. https://doi.org/10.5194/se-10-117-2019

Fenton CR, Hermanns RL, Blikra LH et al (2011) Regional 10Be production rate calibration for the past 12ka deduced from the radiocarbon-dated Grøtlandsura and Russenes rock avalanches at 69° N, Norway. Quat Geochronol 6:437–452. https://doi.org/10.1016/j.quageo.2011.04.005

Forwick M, Vorren TO (2002) Deglaciation history and post-glacial mass movements in Balsfjord, northern Norway. Polar Res 21:259–266. https://doi.org/10.3402/polar.v21i2.6486

Ganerød GV (2014) Geological logging of drill cores from borehole BH 02-13 and BH 03-13 at Jettan, Nordnes mountain in Troms county, Northern Norway. NGU Report 2014.005

Ganerød GV (2013) Geological logging of drill core from borehole NN-01-12 at Jettan, Nordnes mountain in Troms county , Northern Norway. Report no. 2013.042

Gee DG, Fossen H, Henriksen N, Higgins AK (2008) From the early paleozoic platforms of baltica and laurentia to the caledonide orogen of scandinavia and greenland. Episodes 31:44–51

Glastonbury J, Fell R (2010) Geotechnical characteristics of large rapid rock slides. Can Geotech J 47:116–132. https://doi.org/10.1139/T09-080

Glimsdal S, Løvholt F (2018) Displacement wave from rock slope failure in Lyngen. Propogration assessment using probabilistic methods [In Norwegian: Flodbølger etter fjellskred i Lyngen. Oppskyllingsberegninger ved hjelp av probabilistiske metoder]

Hegstad SK (2014) Post-glacial sedimentary processes and slope instabilities off Nordnesfjellet, Lyngenfjorden , northern Norway. Dissertation, UiT The Arctic University of Norway

Hermanns RL, Longva O (2012) Rapid rock-slope failures. In: Clague JJ, Stead D (eds) Landslides: types, mechanisms and modelling. Cambridge University Press, pp 59–70

Hermanns RL, Oppikofer T, Anda E et al (2012) Recommended hazard and risk classification for large unstable rock slopes in Norway

Hughes ALC, Gyllencreutz R, Lohne ØS et al (2016) The last Eurasian ice sheets-a chronological database and time-slice reconstruction, DATED-1. Boreas 45:1–45. https://doi.org/10.1111/bor.12142

Indrevær K, Bergh SG (2014) Linking onshore-offshore basement rock architecture and brittle faults on the submerged strandflat along the SW Barents Sea margin, using high-resolution (5 × 5 m) bathymetry data. Nor J Geol 94:1–34

Indrevær K, Bergh SG, Koehl JB et al (2013) Post-Caledonian brittle fault zones on the hyperextended SW Barents Sea margin: new insights into onshore and offshore margin architecture. Nor J Geol 93:167–188

Kartverket (2019a) Norge i bilder. www.norgeibilder.no. Accessed 17 Jun 2019

Kartverket (2019b) Høydedata. www.hoydedata.no/LaserInnsyn/. Accessed 17 Jun 2019

Marinos P, Hoek E (2000) GSI: A geologically friendly tool for rock mass strength estimation. Proc GeoEng2000 Conf, pp 1422–1442

NGU (2019) Unstable rock slope inventory map. www.ngu.no. Accessed 29 Oct 2019

NGU (2020) The national soils database (Nasjonal løsmassedatabase). https://geo.ngu.no/kart/losmasse/. Accessed 25 May 2020

Nordvik T, Blikra LH, Nyrnes E, Derron MH (2010) Statistical analysis of seasonal displacements at the Nordnes rockslide, northern Norway. Eng Geol 114:228–237. https://doi.org/10.1016/j.enggeo.2010.04.019

NVE (2020) Rock avalanche monitoring. www.nve.no. Accessed 25 May 2020

Nystad TM (2014) [Stability analysis and modelling of the unstable rock slope Jettan at Nordnesfjellet, Troms] Stabilitetsanalyse og modellering av det ustabile fjellpartiet Jettan på Nordnesfjellet, Troms. Dissertation, Norwegian University of Science and Technology

Ostermann M, Koltai G, Spötl C, Cheng H (2016) Deep-seated gravitational slope deformations in the Vinschgau (northern Italy) and their associations with springs and speleothems. Geophys Res Abstr 18

Patton H, Hubbard A, Andreassen K et al (2017) Deglaciation of the Eurasian ice sheet complex. Quat Sci Rev 169:148–172. https://doi.org/10.1016/j.quascirev.2017.05.019

Rønning JS, Dalsegg E, Heincke BH et al (2008) [Geophysical measurements at Nordnesfjellet in the summer of 2007, Kåfjord Municipality, Troms] Geofysiske målinger på Nordnesfjellet sommeren 2007, Kåfjord kommune, Troms. NGU Report 2008.024. 33

Skrede I (2013) [Jettan, Nordnesfjellet, Kåfjord, Troms- inner geometry and structure, kinematics and controllign factors of an unstable mountain slope based on structural analysis, geomorphology and monitoring data] Jettan, Nordnesfjellet, Kåfjord, Troms- indre geometri. Dissertation, UiT The Arctic University of Norway

Stokes CR, Corner GD, Winsborrow MCM et al (2014) Asynchronous response of marine-terminating outlet glaciers during deglaciation of the Fennoscandian ice sheet. Geology 42:455–458. https://doi.org/10.1130/G35299.1

Vick LM, Bergh SG, Höpfl S, Percival J (2020a) The role of lithological weakness zones in rockslides in northern Norway. In: Accepted to: Eurock2020. Trondheim, Norway, pp 14–19

Vick LM, Böhme M, Rouyet L et al (2020b) Structurally controlled rock slope deformation in northern Norway. Landslides.https://doi.org/10.1007/s10346-020-01421-7

Mapping, Hazard and Consequence Analyses for Unstable Rock Slopes in Norway

Reginald L. Hermanns, Thierry Oppikofer, Martina Böhme, Ivanna M. Penna, Pierrick Nicolet, and Marie Bredal

Abstract

Systematic mapping for unstable rock slopes that can cause catastrophic failures has been carried out in Norway for 15 years. In this time a systematic mapping approach was developed that includes a hazard and a consequence analysis. The first weighs morphological signs of rock slope deformation, structurally based stability considerations, the state of slope activity and past events to come up with a hazard score. The second includes the analysis of the volume, the potential run-out area and if applicable the assessment of related displacement waves in case that the failure hits a water body or the assessment of the landslide dam in case that the failure would dam a valley. The goal of the consequence assessment is to assess the potential loss of life for possible failure scenarios. The hazard and consequence analyses are used by the Norwegian Water and energy Directorate to define high risk sites and for hazard zoning in relation to the Norwegian building codes. So far, 523 unstable rock slopes have been found, of which 110 were hazard and risk classified. 48 of the slopes have hazard zones that restrict building activities and 7 sites have been defined as high risk sites that are today under continuous surveillance.

Keywords

Rock slope failure • Displacement wave • Landslide dam • Risk matrix • Mapping approach

The Challenge of Mapping the Hazard from Rock Slope Instabilities in Norway

Although Norway is only the 68th largest country on Earth it is certainly one of the countries with the highest number of large unstable rock slopes. This is mainly due to the fjord landscape with fjords penetrating all along the coastline deep into the mountainous continent. Norway is the country with the seventh longest coastline and relief contrasts are similar as if the eastern European Alps would be flooded up to 2000 m altitude by water. This kind of fjord landscape is formed by continental ice sheets that expanded during last glaciations into the surrounding oceans and eroded deep valleys (Vorren and Mangerud 2008). Similar landscapes exist elsewhere on Earth but are characterised by sparse population density often separated by hundreds of kilometres of uninhabited land. In contrast is the entire coastline of Norway populated and has important economic development. Historically did secondary effects such as displacement waves from rock slope failures and landslide damming of narrow valleys cause much more deaths and economic consequences than direct impact of rock slope failures on settlements (Furseth 2006; Blikra et al. 2006a, b). It is thus necessary for an effective hazard mapping to give a high weight to secondary effects.

This contribution describes the result of systematic mapping for slopes that might fail catastrophically (using

R. L. Hermanns (✉) · T. Oppikofer · M. Böhme · I. M. Penna · P. Nicolet · M. Bredal
Geological Survey of Norway, Geohazard and Earth Observation Team, Leiv Eirikssons vei 39, 7491 Trondheim, Norway
e-mail: reginald.hermanns@ngu.no

T. Oppikofer
e-mail: thierry.oppkofer@ngu.no

M. Böhme
e-mail: martina.bohme@ngu.no

I. M. Penna
e-mail: ivanna.penna@ngu.no

P. Nicolet
e-mail: pierrick.nicolet@ngu.no

M. Bredal
e-mail: marie.bredal@ngu.no

Ž. Arbanas et al. (eds.), *Understanding and Reducing Landslide Disaster Risk*,
ICL Contribution to Landslide Disaster Risk Reduction,
https://doi.org/10.1007/978-3-030-60713-5_29

here the definition given in Hermanns and Longva 2012 as slopes that might reach an area larger than the shadow angle of rock falls including their secondary effects) over the country of Norway. This project was initiated 15 years ago in only one county "Møre og Romsdal" and expanded from there stepwise into other counties. Mapping was completed in Møre og Romsdal in 2019 (Oppikofer et al. 2021), however in more than half of the counties of Norway systematic mapping has still not started.

In order to protect society from disastrous effects of rock slope failures, effective risk management includes, besides hazard mapping, multiple other aspects such as monitoring, early warning, risk communication, or even mitigation measures such as draining (Lacasse et al. 2008). However, these measures are very costly, especially if carried out over decades so they should be applied only for high risk sites. These measures, although essential in risk management, are not part of this publication and the focus is the hazard mapping only. In the following we describe the development of the project over its 15 years life span and the mapping approach that was developed. We will complete with an overview of the knowledge on unstable rock slopes in Norway today and highlight on outstanding gaps of knowledge. We finalize this contribution with giving reference to the online databases where updates become continuously available to the public. Our approach can be followed in similar geographic settings; however, it was developed for the Norwegian fjord situation and Norway's geology. Adaptation to other local geologic conditions and country specific legislations would be necessary to apply this methodology elsewhere.

Early Hazard and Risk Classifications and the Need of a Systematic Approach

In the beginning of the mapping activity the work focussed on well-known unstable rock slopes and 5 unstable rock slopes were quickly classified as "high-risk" objects (Blikra et al. 2006a; Dahle et al. 2008; Blikra et al. 2009). The geological data used for defining the hazard and risk varied from site to site and thus also the standards for the hazard classification. This went parallel with the installation of municipality-run early warning centres that requested financial support from the Norwegian government. This situation led to complex communications between government, municipalities and involved companies (Hermanns et al. 2011) and called for an objective standardized approach to define the hazard and potential consequences from unstable rock slope failures.

The Systematic Hazard Classification System

The classification system was developed using a broad Norwegian and international expertise and 23 persons from European and Norwegian universities as well as different governmental institutions and consulting companies participated in its development (Hermanns et al. 2012). In addition, earlier approaches to define long term slope stability were taken as guide (e.g. Glastonbury and Fell 2008).

The classification is based on an analysis of 32 historic non-seismic catastrophic rock slope failures from Norway and around the world showing that in all cases pre-failure slope deformation occurred prior to the event (Hermanns et al. 2012). The system is thus developed for aseismic failure conditions and is valid for a country like Norway with low frequency of high magnitude events, where no earthquake magnitude ≥ 6 event was ever recorded (Bungum et al. 2000). Furthermore, the classification system is built for the special geographic and geological conditions in mainland Norway that is dominated by crystalline rock and does not present large rock slopes with weak sedimentary or volcanic rocks.

Experiences with multiple failures from the same unstable rock slope over hundreds or thousands of years in Norway and elsewhere indicate the necessity to distinguish between different failure scenarios with different likelihood and magnitude of failure (e.g. Hermanns et al. 2006). These failure scenarios are defined based on different rates or degrees of rock slope deformation, and different structural conditions (Hermanns et al. 2012). The classification includes three criteria describing the morphological development of the slope deformation, two criteria describing the structural feasibility of failure, three criteria describing the displacement rates and other signs of activity and one criteria describing past events on the same slope section. The hazard classification does not include geotechnical aspects as these parameters cannot be obtained easily by geological mapping and are often only obtained by coring into the sliding surface. In the hazard classification the scores for each criterion are added up, resulting in a hazard score with uncertainties for each scenario.

The hazard classification system was designed to assess the hazard relatively in 5 classes (very low to very high likelihood of failure), as the expert panel agreed that with today's knowledge it is impossible to give the likelihood in annual probability. This had the disadvantage that the relative probability classes could not be used within the Norwegian building code (Byggeteknisk forskrift, TEK 17, 2017). It is defined in security classes (S1-S3) representing annual probability of occurrence of 1/100, 1/1000, and

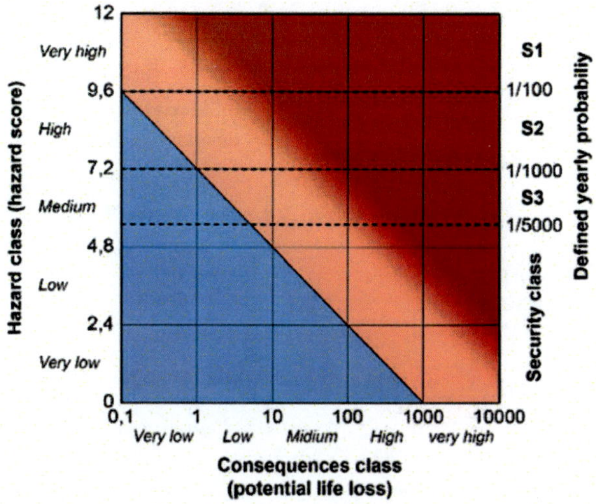

Fig. 1 Hazard and risk matrix for unstable rock slopes in Norway following the risk classification after Hermanns et al. (2012) and the definition of security classes and annual probabilities after Blikra et al. (2016)

1/5000 for landslides. In order to transpose the hazard classes to the security classes used for quantitative hazard zonation, yearly probabilities were defined (Fig. 1; Blikra et al. 2016). This is not based on scientific knowledge but on following considerations: (1) Historically 2–3 large rock slopes failures have occurred per century, the number of unstable slopes falling into the 1/100 a^{-1} class should not be much higher than that. (2) The 1/100 a^{-1} limit was thus placed between very high and high hazard classes, the 1/1000 a^{-1} limit between high and medium hazard classes, and the 1/5000 a^{-1} limit was arithmetically determined using a logarithmic scale. (3) Unstable slopes that do not show any active deformation today should not lie above the 1/5000 line.

The finalization of the hazard and risk classification of unstable rock slopes in Møre og Romsdal offers a good opportunity to estimate the validity of these consideration by looking at the spread of the hazard scores among the three security classes (Oppikofer et al. 2021). This is historically the most rock slope failure-prone county of Norway (Furseth 2006). Of the 78 classified sites, one falls in the highest security class, six in the second highest security class and 17 in security class S3 thus fitting well with historical observations (Oppikofer et al. 2021).

The hazard classification system was recently compared to two alternative methodologies using different failure scenarios of the Åknes rock slope instability in Norway as an example (Oppikofer et al. 2018). Obtained results were comparable in scale but the method by Jaboyedoff et al. (2012), focusing on the strain of the rock mass and state of activity only, resulted in much less geologic understanding of the instability itself. The method by Brideau et al. (2017), based on a combination of a regional analysis of previous failures and activity of the slope, resulted in an inversion of the hazard levels from larger to smaller scenarios as the size of the instability goes into the analysis as factor.

The Consequence Analysis and Mapping Approach

The consequence analysis is very tightly bound into the mapping approach as the workflow is iterative (Oppikofer et al. 2016). The idea is to spend minimal effort on sites with no consequences or low risk and focus on sites that pose a threat to society. As a consequence, all sites get first evaluated with fast techniques that are based on remote sensing data only, and more sophisticated methods including field investigations are applied if necessary.

The detection method of unstable slopes has changed over time depending on data availability. It was first based on aerial photos and helicopter surveys, supplemented locally by available high-resolution digital elevation models from airborne LiDAR data. From 2012 onward, systematic mapping was also carried out using satellite InSAR in areas of interest with data collected on demand (e.g. Booth et al. 2014). With the increasing coverage of airborne LiDAR data, these data became since 2014 the primary source for mapping if available. In 2018, satellite-based InSAR data became a second tool that can be applied nation-wide since SENTINEL-1 data are today processed with a full cover of the country (https://insar.ngu.no/). However, this tool does not work on vegetated slopes and detects only vertical displacement or the E-W component of the horizontal movement. A test with known unstable slopes with displacement measured by other methods indicated that ~25% of unstable slopes are not covered by InSAR data and ~7% do not show any deformation rates although displacement is measured with other techniques. How the availability of remote sensing data influences the possibility of detecting unstable rock slopes is shown in Fig. 2a

The consequence analysis is organized in five working steps (Fig. 3) including volume calculation (step 1), run-out analysis (step 2) for all sites and all defined scenarios, followed by either a displacement wave analysis (step 3) for sites where a run-out analysis indicates that a rock slope failure would hit a water body or a landslide dam analysis (step 4) if the run-out analysis suggest that the landslide might form a dam in the valley. The last step (5) is the assessment of potential life loss in a rock slope failure event combining the areas defined in steps 2–4 (Oppikofer et al. 2016). For each of these five steps methods with different degree of sophistication are available dependent on the hazard and risk level and related acceptance of uncertainties.

Fig. 2 Unstable rock slopes detected in Norway

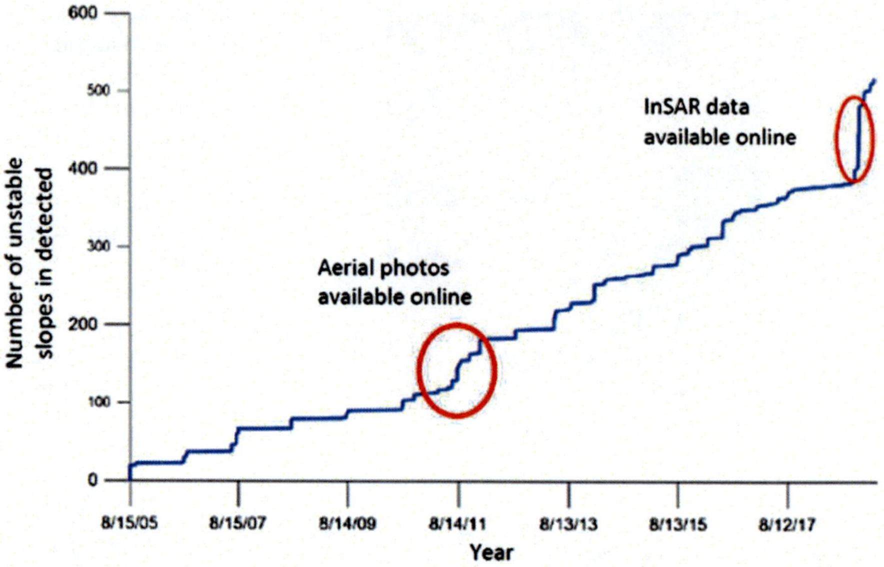

Fig. 3 The five stepped consequence analysis with different levels of sophistication related to the degree of uncertainty that is acceptable for each site (after Oppikofer et al. 2016)

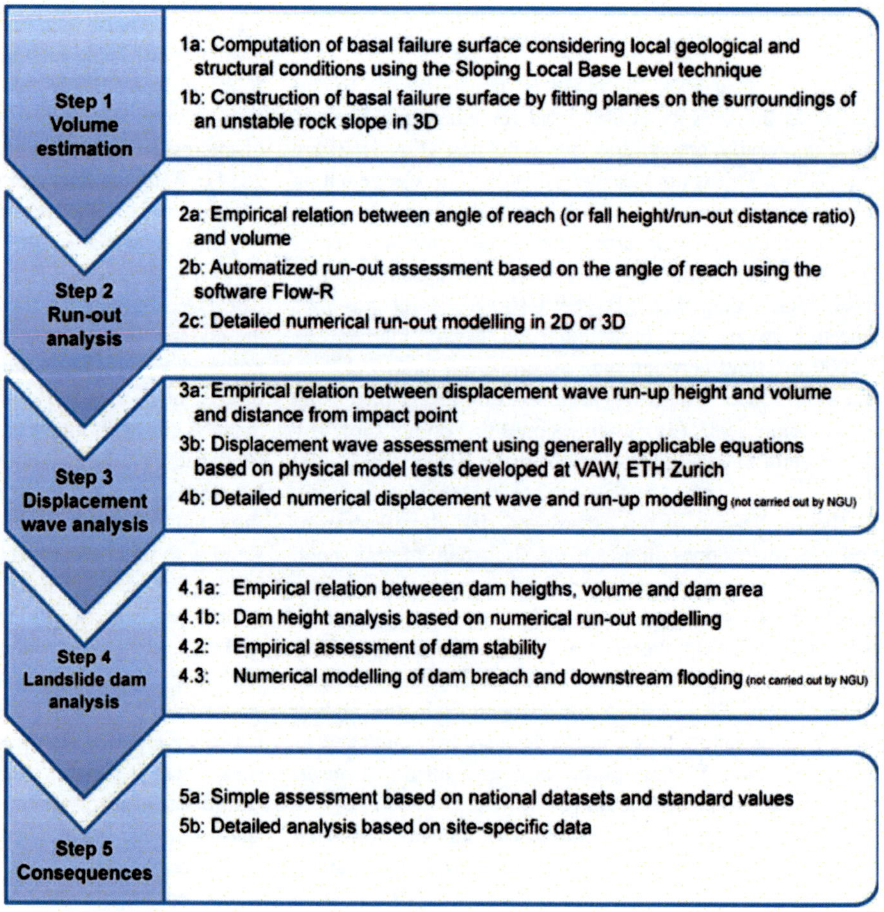

For all working steps our consequence analysis makes use of existing tools, that have been developed elsewhere and that are described in Oppikofer et al. (2016) and in the literature referenced therein. However, two of the tools were specifically developed for our mapping effort as tools with a good temporal work input/low uncertainty output relation did not exist before. These focus on the assessment of the displacement wave analysis and the dam height and dam stability analysis (steps 3a and 4.1a/4.2 respectively). Both methods are based on empirical data and the results are rather conservative in respect of more sophisticated methods. Mapping results are thus seen on a susceptibility map level rather than on a hazard map level (Oppikofer et al. 2019; Oppikofer et al., 2020). The base for the empirical relation between displacement-wave run-up height and volume and distance from impact point is based on a global data set (Roberts et al. 2014). In contrast, the empirical assessment of dam height and dam stability is based mainly on a national Norwegian data set extended by selected published data (Oppikofer 2016, Oppikofer et al. 2020) and on the concept of the Dimensionless Blocking Index (DBI) developed by Ermini and Casagli (2003). The flooding area upriver of potential rockslide dams and the potential downstream flooding area in case of a dam breach do not count in the security classes for landslides but in those for flooding as people can get evacuated out of the hazard zone on time (e.g. Hermanns et al. 2013).

We specially want to highlight that the Geological Survey of Norway and the authors of this paper do not carry out detailed numerical displacement wave and run-up models. This work is carried out by researchers with the specialized expertise (e.g. Glimsdal and Løvolt 2018) within the general landslide hazard mapping program, whose responsibility lies with the Norwegian Water and Energy Directorate (NVE).

For the assessment of potential loss of life, national cadastre datasets (Kartverket 2015) are used (step 5a), following the principals outlined in Oppikofer et al. (2016). For medium and high-risk sites more detailed data on the usage of buildings can be applied.

Mapping Status and Information to the Public

After 15 years of systematic mapping, 523 unstable rock slopes that might lead to catastrophic failures have been registered in Norway. It is important to note here that only ~30% of the mountainous terrain was mapped systematically with aerial photos and/or digital elevation models. Entire Norway has been mapped using satellite-based InSAR, but that does not allow to discover unstable rock slopes with forest cover or unfavourable direction of movement (see above). In addition, 568 suspect rock slopes were investigated but it was concluded that collapses would be too small to fall into our definition of catastrophic collapse (rock fall sizes) or that the slopes are stable. Of the 523 unstable rock slopes mapping, hazard and consequence analysis has been completed on 110 sites of which 48 have hazard zones falling into either security class S1, S2 or S3 (Fig. 4). Seven of them have been classified as high-risk sites, using the definition of Blikra et al. (2016) as they fall in security class S1 or S2 and may impact settlements. Two out of the seven have been classified as high risk sites before our systematic approach was established and fall today in security class S3. All high-risk sites get followed up in Norway by 24/7 monitoring and early-warning practises are in place (www.nve.no/flaum-og-skred/fjellskredovervaking/) carried out by NVE.

We do not show an inventory map of this overview here, as the database is public and gets continuously updated (Oppikofer et al. 2015; Böhme et al. 2021). The data base is connected to two online map services. One (https://geo.ngu.

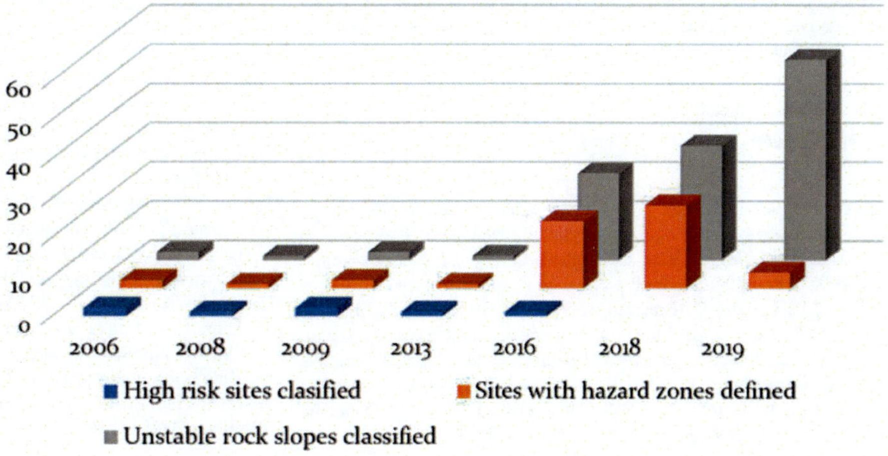

Fig. 4 Overview of high-risk sites classified, hazard zones defined, and hazard and risk classification finalized

no/kart/ustabilefjellparti_mobil/) contains all related geological information while the other (https://atlas.nve.no/) is displaying information relevant for land-use planning. Factsheets with key information on unstable rock slopes can be downloaded for each unstable rock slope as individual reports and all information becomes publicly available after it is quality checked.

Future Challenges and Possible Developments

The hazard classification developed that far does not include any possibility to account for global climatic change. First systematic analysis of the temporal distribution of rock slope failures in Norway suggests that the activity of catastrophic failures has been much higher in the late Pleistocene and early Holocene, directly following deglaciation (Böhme et al. 2015; Hermanns et al. 2017) and was less influenced by the climate variability experienced in the Holocene. However, latest research indicates that some catastrophic failures concentrated on slopes that lie today at the permafrost boundary at the end of warm phases within the Holocene (Hilger et al. 2018). Furthermore, slopes that deformed over several thousand years in similar climatic settings show recent acceleration of movement (Böhme et al. 2019). Actually, ~11% of the unstable rock slopes in our data base lie within today's permafrost terrain (Magnin et al. 2019) and thus suggest that an addition to our classification system might become necessary at least for climate critical slopes like those lying within or at the edge of permafrost environments.

Acknowledgements Mapping of unstable rock slopes is financed in Norway since 2009 through the Water and Energy Directorate (NVE). Discussion with colleagues from NVE was essential to set up the mapping approach and they authored or co-authored multiple papers referenced herein. The list of authorship to this contribution is restricted and listed are only those that today contribute to mapping of unstable rock slopes and participated in the writing of the paper. Multiple colleagues contributed over the years in this effort and got credit as authors in earlier publications and reports referenced herein.

References

Blikra LH, Anda E, Høst J, Longva O (2006a) Åknes/Tafjord-prosjektet: Sannsynlighet og risiko knyttet til fjellskred og flodbølger fra Åknes og Hegguraksla: Geological Survey of Norway report 2006.039. (0800–3416) 20 p

Blikra LH, Longva O, Braathen A, Anda E, Dehls JF, Stalsberg K (2006b) Rock slope failures in Norwegian Fjord areas: examples, spatial distribution and temporal pattern. Landslides from massive rock slope failure. In: Evans SG, Scarascia Mugnozza G, Strom A, Hermanns RL, (eds) NATO Science Series, IV. Vol 49: Dodrecht, Netherlands, Springer, (1–4020–4036–9) 475–496

Blikra LH, Henderson I, Nordvik T (2009) Faren for fjellskred fra Nordnesfjellet i Lyngenfjorden, Troms. Geological Survey of Norway report 2009.026. 29 p

Blikra LH, Majala G, Anda E, Hallvard B, Eikenæs O, Helgås G, Oppikofer T, Hermanns RL, Böhme M (2016) Fare- og risikoklassifisering av ustabile fjellparti. Norwegian Water and Energy Directorate (1501-2832) 47 p

Böhme M, Oppikofer T, Longva O, Jaboyedoff M, Hermanns R, Derron M-H (2015) Analyses of past and present rock slope instabilities in a fjord valley: implications for hazard estimations. Geomorphology 248:464–474

Böhme M, Hermanns R, Gosse J, Hilger P, Lauknes T, Dehls J (2019) Comparison of monitoring data with paleo–slip rates: cosmogenic nuclide dating detects acceleration of a rockslide. Geology 47 (4):339–342

Böhme M, Hermanns RL, Oppikofer T, Dreiås M (2021) The online map service for unstable rock slopes and related hazard zones in Norway. Bergen, in press

Booth AM, Dehls J, Eiken T, Fischer L, Hermanns RL, Oppikofer T (2014) Integrating diverse geologic and geodetic observations to determine failure mechanisms and deformation rates across a large bedrock landslide complex: the Osmundneset landslide, Sogn og Fjordane Norway. Landslides 12(4):745–756

Brideau M-A, Jakob M, McDougall S (2017) Methodology to estimate the rock avalanche frequency for a specific slope. Abstracts of the 4th Slope Tectonics Conference, Kyoto, Japan, p 29

Bungum H, Lindholm CD, Dahle A, Woo G, Nadim F, Holme JK, Gudmestad OT, Hagberg T, Karthigeyan K (2000) New seismic zoning maps for Norway, the North Sea, and the United Kingdom. Seismol Res Lett 71(6):687–697

Byggeteknisk forskrift, TEK 17, 2017, https://dibk.no/byggereglene/byggteknisk-forskrift-tek17/

Dahle H, Anda E, Saintot A, Saetre S (2008) Faren for fjellskred fra fjellet Mannen i Romsdalen: geological survey of Norway report 2008.087. (0800–3416) 21 p

Ermini L, Casagli N (2003) Prediction of the behaviour of landslide dams using a geomorphological dimensionless index. Earth Surf Process Landforms 28(1):31–47

Furseth A (2006) Skredulykker i Norge, Tun Forlag Oslo. (82–529–3043–3) 207 p

Glastonbury J, Fell R (2008) Geotechnical characteristics of large slow, very slow, and extremely slow landslides. Can Geotech J 45 (7):984–1005

Glimsdal S, Løvolt F (2018) Flodbølger etter fjellskred i Lyngen, NGI rapport 20170309–01-R, 29 p

Hermanns RL, Blikra LH, Naumann M, Nilsen B, Panthi KS, D, Longva, O, (2006) Examples of multiple rock-slope collapses from Köfels (Ötz valley, Austria) and western Norway. Eng Geol 83(1–3): 94–108

Hermanns RL, Bunkholt HSS, Böhme M, Fischer L, Oppikofer T, Rønning JS, Eiken T (2011) Foreløpig fare- og risikovurdering av ustabile fjellparti ved Joasete-Furekamben-Ramnanosi, Aurland kommune: geological survey of Norway report 2011.025. (0800–3416) 46 p

Hermanns R, Oppikofer T, Anda E, Blikra L, Böhme M, Bunkholt H, Crosta G, Dahle H, Devoli G, Fischer L (2012) Recommended hazard and risk classification system for large unstable rock slopes in Norway. NGU rapport 2012.029. (0800–3416) 53 p

Hermanns R, Dahle H, Bjerke P, Crosta G, Anda E, Blikra L, Saintot A, Longva O (2013) Rockslide Dams in Møre og Romsdal County, Norway. In: Margottini C, Canuti P, Sassa K (eds) Landslide science and practice. Springer, Berlin Heidelberg, pp 3–12

Hermanns RL, Schleier M, Böhme M, Blikra LH, Gosse J, Ivy-Ochs S, Hilger P (2017) Rock-avalanche activity in W and S Norway peaks after the retreat of the Scandinavian ice sheet. World Landslide Forum, Ljubljana, Springer, pp 331–338

Hilger P, Hermanns RL, Gosse JC, Jacobs B, Etzelmüller B, Krautblatter M (2018) Multiple rock-slope failures from Mannen in Romsdal Valley, Norway, revealed from Quaternary geological mapping and 10Be exposure dating. The Holocene 28(12):1841–1854

Jaboyedoff M, Derron M, Pedrazzini A, Blikra L, Crosta G, Froese C, Hermanns R, Oppikofer T, Böhme M, Stead D (2012) Fast assessment of susceptibility of massive rock instabilities. Landslides and engineering slopes: protecting society through improved understanding. In: 11th international and 2nd North American symposium on landslides and engineered slopes. Banff, Canada pp 3–8

Kartverket (2015) Føringsinstruks for matrikkelen—9.3 Bygningstyper (basert på NS 3457)—fullversjon. Norwegian Mapping Authority. Available: http://kartverket.no/Eiendom-og areal/Matrikkelen/veiledning-for-lokal matrikkelmyndighet/foringsinstruks-for-matrikkelen/9-Vedlegg/93-Bygningstyper-basert-pa-NS-3457—fullversjon-/

Lacasse S, Eidsvik U, Nadim F, Hoeg K, Blikra LH (2008) Event tree analysis of Åknes rock slide hazard. In: 4th Canadian conference on geohazards, Laval, pp 20–24

Magnin F, Etzelmüller B, Westermann S, Isaksen K, Hilger P, Hermanns RL (2019) Permafrost distribution in steep rock slopes in Norway: measurements, statistical modelling and implications for geomorphological processes. Earth Surf Dyn 7(4):1019–1040

Oppikofer T, Nordahl B, Bunkholt H, Nicolaisen M, Jarna A, Iversen S, Hermanns RL, Böhme M, Molina FXY (2015) Database and online map service on unstable rock slopes in Norway—from data perpetuation to public information. Geomorphology 249:69–81

Oppikofer T, Böhme M, Nicolet P, Penna I, Hermanns R (2016) Metodikk for konsekvensanalyse av fjellskred: NGU Report. (0800–3416) 67 p

Oppikofer T, Hermanns RL, Jaboydeoff M, Derron MH, Brideau MA, Jakob M, Sturzenegger M (2018) Comparison between three rock slope hazard assessment methodologies based on the Åknes case study from Norway. In: 7th Canadian Geohazards conference, Canmore, Alberta, Canada. 8 p

Oppikofer T, Hermanns RL, Roberts NJ, Böhme M (2019) SPLASH: semi-empirical prediction of landslide-generated displacement wave run-up heights. Geological Society, London, Special Publications 477(1):353–366

Oppikofer T , Hermanns RL, Jakobsen VU, Böhme M, Nicolet P, Penna I (2020) Forecasting dam height and stability of dams formed by rock slope failures in Norway: Natural Hazards and Earth System Sciences 1–24

Oppikofer T, Hermanns RL, Böhme M, Morken OA (2021) Rock avalanches in Norway–first complete hazard map in the county of Møre og Romsdal, Intraprevent, in press

Roberts NJ, McKillop R, Hermanns RL, Clague JJ, Oppikofer T (2014) Preliminary global catalogue of displacement waves from subaerial landslides. Science for a Safer Geoenvironment, Springer, pp 687–692

Vorren T, Mangerud J (2008) Glaciations come and go: the making of land: geology of Norway. Geological Society of Norway, Trondheim, pp 481–533

Landscape Formation and Large Rock Slope Instabilities in Manndalen, Northern Norway

Martina Böhme, Reginald L. Hermanns, and Tom R. Lauknes

Abstract

Manndalen valley in northern Norway has a high density of unstable rock slopes. Geomorphological characteristics, geological structures and slope activity have been mapped for all 10 rock slope instabilities. Foliation is sub-horizontal and joints almost vertical throughout the entire valley, leading to the conclusion that geological structures do not allow for an easy kinematic failure process. However Interferometric Synthetic Aperture Radar (InSAR) data from the TerraSAR-X satellite indicate that six of the sites are displacing with average rates in between 3 and 26 mm/yr. The most active unstable rock slope in Manndalen is Gamanjunni-3, showing displacements of locally up to 5 cm/yr and clearly developed limits of the displacing rock mass. Two rock avalanche deposits have been dated to 10.6 ± 0.6 ka and 4.1 ± 0.3 ka BP. This revealed a frequency of ~5000 years for rock avalanches in Manndalen in post glacial time, supporting earlier conclusions that the Gamanjunni-3 site is prone for a failure in close future.

Keywords

Terrestrial cosmogenic nuclides dating • InSAR • Landslide • Rock avalanche

M. Böhme (✉) · R. L. Hermanns
Geological Survey of Norway, P.O.Box 6315 7491 Torgarden, Trondheim, Norway
e-mail: martina.bohme@ngu.no

R. L. Hermanns
e-mail: reginald.hermanns@ngu.no

T. R. Lauknes
NORCE Norwegian Research Centre, Siva Innovasjonssenter, Sykehusvn 21, 9019 Tromsø, Norway
e-mail: tlau@norceresearch.no

Ž. Arbanas et al. (eds.), *Understanding and Reducing Landslide Disaster Risk*, ICL Contribution to Landslide Disaster Risk Reduction,
https://doi.org/10.1007/978-3-030-60713-5_30

Introduction

The municipality of Kåfjord in northern Norway is exposed to one of the highest densities of unstable rock slopes in Norway (Bunkholt et al. 2012). Out of 37 registered unstable rock slopes displaying postglacial deformation in this municipality, the catastrophic failure of 30 may have consequences for society. Several large unstable rock slopes may trigger displacement waves in adjacent fjords during a catastrophic collapse, endangering the population that is concentrating along the coastline. In addition, several are located on the valley slopes of the two major populated valleys Manndalen and Kåfjorddalen. Only in Manndalen, populated with 600 inhabitants, there are 10 unstable rock slopes located. They could have a direct impact on several dwellings or may also dam the river Manndalselva.

A strong effort was made in the past years to understand the hazard of the Gamanjunni-3 unstable rock slope located in Manndalen, as it shows one of the highest displacement rates in Norway (Böhme et al. 2016, 2019). The combination of morphological mapping of rock slope deformation and observations on slope activity lead to classifying this site as one of Norwegians most critical sites. In this contribution we put this site within the framework of the postglacial development of rock slope instabilities and failures. We thus compare the failure development of the remaining nine sites in the valley to Gamanjunni-3 and map the Manndalen valley for deposits of post glacial failures. Dating of the deposits gives a postglacial frequency of failures within the valley. The combined knowledge about the timing of deformation and previous failures is important in interpreting the hazard potential of certain unstable rock slopes.

Fig. 1 Examples of unstable rock slopes in Manndalen. **a** Overview of Ceallu and Liigavárri, both very large slope deformations. Vegetation and talus cover are masking deformation features. View towards S. **b** Backscarp of Storhaugen 2. A smaller still intact block is displaced at the back scarp. View towards S. **c** Gamanjunni 2. An open back crack is delimiting the unstable rock slope. View towards NE

Study Area

Manndalen is a north–south–striking, glacially U-shaped valley with steep valley walls on both sides. The lithology is dominated by mica schists belonging to Caledonian thrust sheets. Dated moraines indicate that the valleys in this part of Norway became ice free at ca. 12–11 ka (Hughes et al. 2016; Stroeven et al. 2016).

The eastern valley flank hosts nine unstable rock slopes within a distance of 13 km, while only one unstable rock slope is located on the western side of the valley (Figs. 1, 2). However, this instability is also facing westwards towards a tributary valley. The unstable rock slopes are showing quite different development and activity (Fig. 1). Several seem to be inactive or very slowly moving. However, the Gamanjunni-3 rock slope instability, is with a yearly average displacement of up to 5 cm among the five fastest-deforming rock slopes in Norway known so far.

Methods

Field mapping.

Most unstable rock slopes have been visited in the field. During fieldwork observations necessary for a hazard classification after Hermanns et al. (2012) are made. This includes morphological mapping in order to evaluate the development of the delimiting structures (Table 1) as well as signs of activity of each unstable rock slope. The orientation of metamorphic foliation, joints and fractures was measured in the field resulting in around 1000 measurements throughout the valley Manndalen (Table 2). Data collection was mainly undertaken at the upper part of the valley flanks or the plateau tops, but not at the slope toe. All structural data are presented using dip direction and dip angle.

Interferometric Synthetic Aperture Radar (InSAR).

Spaceborne InSAR can cover large areas and has also been applied to quantify deformation of slow-moving landslides (e.g., Colesanti and Wasowski 2006; Lauknes et al. 2010). Here, 11-day interval Stripmap mode TerraSAR-X satellite data covering the snow-free period from 2009 to 2014 are used. Average line-of-sight (LOS) displacement rates are estimated by using the interferogram stacking technique, utilizing short temporal baseline interferograms from each summer season. A spatial multilooking was applied producing a final pixel size of ca. 10 m. Only high-coherent pixels were selected. InSAR processing was done using the Norut GSAR software (Larsen et al. 2005).

Cosmogenic nuclide exposure dating.

Cosmogenic nuclide exposure dating was tested to be well suited to date ages of rock avalanche deposits or the development of sliding surfaces (e.g. Hermanns et al. 2001; Schwartz et al. 2017). This method was used to determine the age of two rock avalanche deposits (Fig. 2; Table 3) and the slide history of the unstable rock slope Gamanjunni-3. Three samples were collected on each deposit and processed following the method described by Hilger et al. (2018). At Gamanjunn-3, 10 samples were collected along the exposed back scarp. Details about the sampling and processing are described in Böhme et al. (2019). All samples were prepared and concentrated at the Dalhousie University CRISDal Laboratory in Halifax, Canada. Total $^{10}Be/^{9}Be$ ratios were determined at the Center for Accelerator Mass Spectrometry at Lawrence Livermore National Laboratory, Livermore, California. Ages were calculated with version 3 of the online exposure age calculator formerly known as the CRONUS-Earth online exposure age calculator (https://hess.ess.washington.edu).

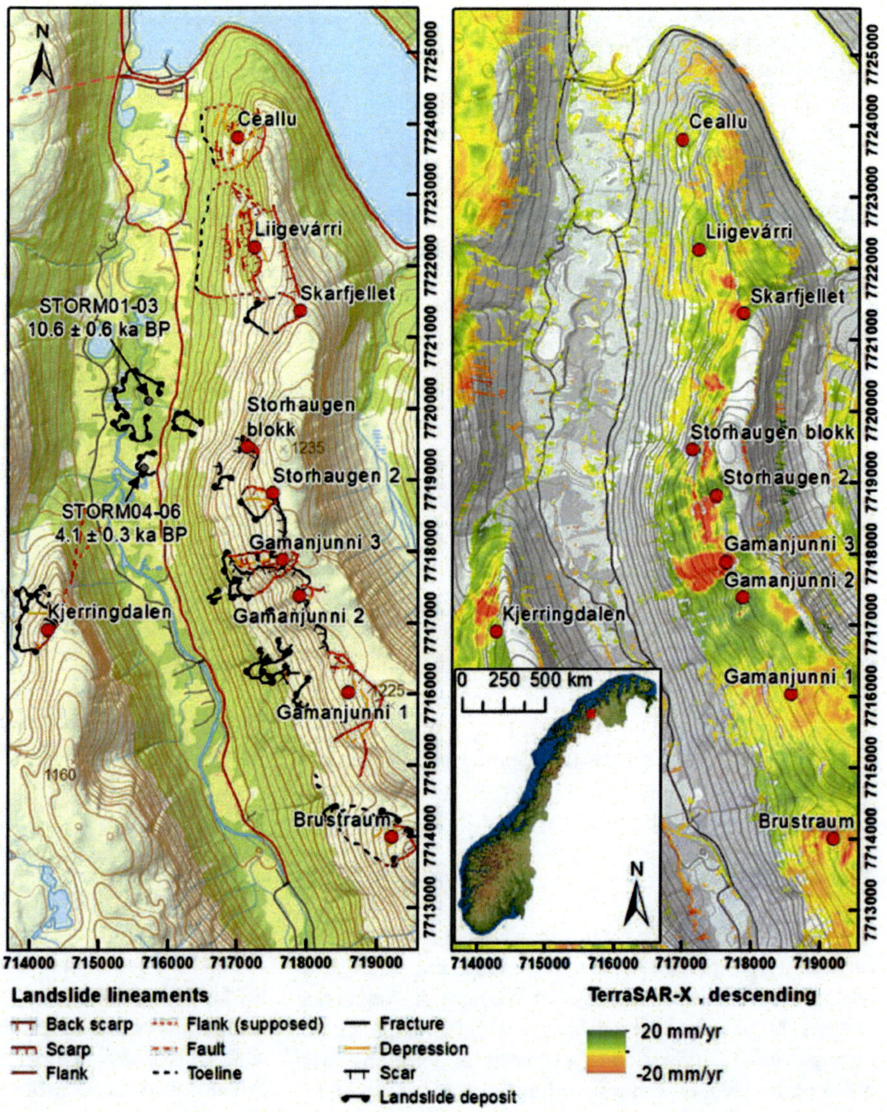

Fig. 2 Overview of the study area displaying the location of unstable rock slopes (red dots) in Manndalen valley. **A** Mapped gravitational lineaments and dated deposits on the valley bottom. Grey dots indicate the sample locations. **B** InSAR data from TerraSAR-X satellite, descending orbit. Negative values represent ground displacement in line-of-sight (LOS) away from the satellite. Downslope displacements are observed relating to several unstable rock slopes. Especially the limits of Gamanjunni-3 are clearly visible. In addition, there are some areas with surficial displacements of debris. The inset shows the location of the study area in Norway

Results

Rock slope instabilities.

There are 10 unstable rock slopes located on the valley flanks of Manndalen. However, only the 9 instabilities located on the eastern side of the valley are forming a hazard for society. The unstable rock slope on the western side of the valley is facing westwards towards a tributary uninhabited valley and will therefore not be discussed further.

All unstable rock slopes have their upper origin at the edge of the plateau, except the northernmost 2 instabilities. Observed geomorphological features are open cracks, large dislocated blocks obviously connected to detachment scarps, bulging slopes, sink holes and depressions. The main geomorphological properties for each unstable rock slope are summarized in Table 1. The surficial extend of the areas displaying signs of deformation varies in between 0.04 and 1.52 km^2 (Table 2).

Structural data.

The structural data reveal a very consistent subhorizontal foliation in entire Manndalen with an overall average of 288/06 ± 14°. However, the dip direction of the foliation varies locally due to Caledonian folding (Table 2). Fracture sets are dominantly subvertical. Two major fracture sets were observed with overall averages of 302/88 ± 16° and 194/90 ± 18°.

Table 1 Summary of main geomorphological characteristics of unstable rock slopes in Manndalen. These criteria are used to assess the state of failure development and the relative hazard

Unstable rock slope name	Development of back scarp		Development of lateral release surfaces					Morphologic expression of rupture surface
			General	morphology of slide mass				
Fully				Partly	Not developed			Developed/free on one side
Developed/free on both sides	No	Yes						
Ceallu (Fig. 1a)	x			x		x		Relatively fractured rock mass
Liigevárri (Fig. 1a)	x		x			x		Large grabens, no internal fracturing
Skarfjellet		x			x	x		Very fragmented rock mass
Storhaugen blokk*[1]	x			x		x		Open back scarp, almost no internal fracturing
Storhaugen 2 (Fig. 1b)		x	x			x		Very fragmented rock mass
Gamanjunni 3*[2]	x				x		x	Coherent block with some internal fracturing
Gamanjunni 2 (Fig. 1c)		x			x	x		Almost no internal fracturing
Gamanjunni 1*[3]	x			x		x		Almost no internal fracturing
Brustraum	x				x	x		Very fragmented rock mass
Kjerringdalen	x				x	x		Several coherent blocks

*[1]Böhme et al. (2015a, b), *[2]Böhme et al. (2016, 2019), *[3]Henderson et al. (2011)

Slope activity.

The slopes in Manndalen are in general very active regarding surficial slope processes, like rock falls and snow avalanches. Thus, visible activity for example in vegetation damage that is only due to deeper rock deformation are not distinguishable. InSAR data extracted at locations with outcropping bedrock can give information about the actual activity of the unstable rock slope. However, we used here averaged InSAR data over the entire deformation area of each site (Table 2). Four unstable rock slopes do not show displacement in the InSAR data. The other locations display a displacement in between 3 and 26 mm/yr averaged over the deformation area.

Cosmogenic nuclide exposure dating.

The two rock avalanche deposits have been dated with TCN at 10.6 ± 0.6 ka and 4.1 ± 0.3 ka BP respectively (Fig. 1; Table 3).

Ages from the back scarp of Gamanjunni-3 revealed that sliding started between 6.6 and 4.3 ka with a relatively fast or instantaneous displacement rate that decelerated a few hundred years after initiation (Böhme et al. 2019). The average paleo-slip rate is 28 mm/yr.

Discussion and Conclusions

Even if the foliation is sub horizontal its slight inclination towards SW-NW is likely the reason for the dominance of unstable rock slopes on the eastern valley side and their absence on the western side. Furthermore, this shallow foliation and the lack of other critically oriented structures does kinematically not allow for simple failure mechanisms. Mostly, detachment occurs along the steep fracture sets in combination with breaking through intact rock in order to develop a daylighting sliding plane. Breaking through intact rock is feasible within this rock type consisting of mica schists but formation of such structures depends on time dependent fracture growth.

This complex failure mechanism means in general a lower susceptibility for large rock slope failures to occur catastrophically. The only few rock avalanche deposits, compared to the number of unstable rock slopes are confirming the time dependent failure development. In addition, several objects, presently not showing displacement, are inactive or dormant and a fast failure development is unlikely. Others are having very low rates of displacement (few mm/yr), indicating that they are rather in an early stage of failure development. There are two unstable rock slopes with

Table 2 Summary of characteristics of unstable rock slopes in Manndalen. All structural data are presented using dip direction and dip angle. "Average LOS displacement" is the averaged InSAR displacement over the entire deformation area. Those are partially including fast displacing debris that is overlying the unstable rock mass

Unstable rock slope name	Foliation	Fracture set 1	Fracture set 2	Deformation area [km^2]	Average LOS displacement [mm/yr]
Ceallu	225/07 ± 3	283/81 ± 15	175/87 ± 14	0.53	0
Liigevárri	-	-	-	1.52	0
Skarfjellet	-	-	-	0.26	−6
Storhaugen blokk	250/08 ± 3	127/76 ± 12	-	0.04	0
Storhaugen 2	233/11 ± 13	240/80 ± 23	179/83 ± 11	0.11	−26
Gamanjunni 3	312/08 ± 13	124/89 ± 17	200/84 ± 18	0.41	−26
Gamanjunni 2	302/13 ± 7	-	016/79 ± 21	0.17	0
Gamanjunni 1	-	-	-	0.82	−3
Brustraum	199/04 ± 17	300/44 ± 13	027/82 ± 15	0.42	−6
Kjerringdalen	276/10 ± 7	300/85 ± 17	029/85 ± 20	0.17	−13

Table 3 Dating results of samples taken at boulders within rock-avalanche deposits on the valley bottom of Manndalen valley

Sample ID	Measured ^{10}Be concentration [atoms/g]	Age [ka]	Analytical uncertainty (Be)
STORM01	1.88E + 04	3.9	0.3
STORM02	2.17E + 04	4.5	0.4
STORM03	1.91E + 04	3.9	0.3
STORM04	4.70E + 04	9.9	0.7
STORM05	5.25E + 04	10.8	0.8
STORM06	5.26E + 04	11.0	0.8

larger displacements, Gamanjunni-3 and Storhaugen-2. Gamanjunni-3 has been defined as a high-risk object and is subject to permanent monitoring as a very large still relatively intact rock mass is displacing as a whole (Böhme et al. 2016, 2019). It is also the only unstable rock slope in Manndalen displaying a clear toe line and well-developed lateral limits (Table 1). Storhaugen-2 contrasts with Gamanjunni-3 being very fragmented, having not a completely developed back scarp and no clear toe line (Table 1).

The ages of the dated deposits as well as the start of deformation at Gamanjunni-3 are all different. The largest rock slope failure was dated to directly after deglaciation. Suggesting that glacial oversteepening and debuttressing by the retreating glaciers might have triggered the failure. A temporal peak of rock slope failures following deglaciation has been observed also in western Norway (Böhme et al. 2015a, b; Hermanns et al. 2017).

Initiation of sliding at Gamanjunni-3 and the age of the other deposit are several ka after deglaciation. A direct relation to it can thus be excluded. However, both the start of displacement and the second dated deposit occurred at the end of the Holocene thermal optimum, which lasted in northern Norway from 8 to 4 ka (Lilleøren et al. 2012; Mauri et al. 2015). The head of the valley slope lies at an altitude where permafrost conditions are expected today (Lilleøren et al. 2012). The initiation of sliding could thus be explained by effects of permafrost degradation on the rock mass.

To summarize, geological structures do not allow for an easy kinematic failure process and time dependent slope weakening is necessary before catastrophic failure of unstable rock slopes in Manndalen. Post glacially there has been a failure in Manndalen for about every 5000 years. Gamanjunni-3 is today one of the fastest moving rock slopes and expected to be one of the next failures in Norway. The temporal development of one event in 5000 years supports earlier conclusions that the Gamanjunni-3 site is prone for a failure in close future.

Acknowledgements This work was co-financed by the Norwegian Water Resources and Energy Directorate (NVE) through the national mapping program for unstable rock slopes in Norway. TerraSAR-X satellite data were provided by DLR (projects GEO0565, GEO0764, and GEO2497) and the development of the InSAR processing chain was supported by the Norwegian Space Centre via the Research Council of Norway projects PermaSAR, and by ESA via the PRODEX Permasat project.

References

Böhme M, Molina FY, Dehls J, Hermanns RL (2015a) Fare- og risikoklassifisering av det ustabile fjellpartiet Storhaugen blokk i Manndalen, Troms. NGU report 2015.050, Geological Survey of Norway, Trondheim

Böhme M, Oppikofer T, Longva O, Jaboyedoff M, Hermanns R, Derron M-H (2015b) Analyses of past and present rock slope instabilities in a fjord valley: implications for hazard estimations. Geomorphology 248:464–4

Böhme M, Bunkholt H, Dehls J, Oppikofer T, Hermanns R, Dalsegg E, Kristensen L, Lauknes TR, Eriksen HØ (2016) Geologisk modell og fare-og risikoklassifisering av det ustabile fjellpartiet Gamanjunni 3 i Manndalen, Troms. NGU report 2016.031. Geological Survey of Norway, Trondheim

Böhme M, Hermanns RL, Gosse J, Hilger P, Eiken T, Lauknes TR, Dehls JF (2019) Comparison of monitoring data with paleo–slip rates: cosmogenic nuclide dating detects acceleration of a rockslide. Geology 47(4):339–342

Bunkholt H, Redfield T, Osmundsen P, Oppikofer T, Hermanns R, Dehls J (2012) Landslide processes in hard rock in Troms, Norway. In: Proceedings of the 11th international & 2nd North American Symposium on Landslides, Banff, Canada, 3–8 June 2012

Colesanti C, Wasowski J (2006) Investigating landslides with space-borne synthetic aperture radar (SAR) interferometry. Eng Geol 88:173–199

Henderson IHC, Lauknes TR, Osmundsen PT, Dehls J, Larsen Y, Redfield TF (2011) A structural, geomorphological and InSAR study of an active rock slope failure development. Geological Society, London, Special Publications 351(1):185–199

Hermanns RL, Niedermann S, Garcia AV, Gomez JS, Strecker MR (2001) Neotectonics and catastrophic failure of mountain fronts in the southern intra-Andean Puna Plateau, Argentina. Geology 29:619–622

Hermanns R, Oppikofer T, Anda E, Blikra L, Böhme M, Bunkholt H, Crosta G, Dahle H, Devoli G, Fischer L (2012) Recommended hazard and risk classification system for large unstable rock slopes in Norway. NGU report 2012.029. Geological Survey of Norway, Trondheim

Hermanns RL, Schleier M, Böhme M, Blikra LH, Gosse J, Ivy-Ochs S, Hilger P (2017) Rock-avalanche activity in W and S Norway peaks after the retreat of the Scandinavian ice sheet. In: Proceedings workshop on world landslide forum 2017. Springer, pp 331–338

Hilger P, Hermanns RL, Gosse JC, Jacobs B, Etzelmüller B, Krautblatter M (2018) Multiple rock-slope failures from Mannen in Romsdal Valley, revealed from Quaternary geological mapping and 10Be exposure dating. The Holocene 28(12):1841–1854

Hughes AL, Gyllencreutz R, Lohne ØS, Mangerud J, Svendsen JI (2016) The last Eurasian ice sheets—A chronological database and time-slice reconstruction, DATED-1. Boreas 45:1–45. https://doi.org/10.1111/bor.12142

Larsen Y, Engen G, Lauknes TR, Malnes E, Høgda KA (2005) A generic differential interferometric SAR processing system, with applications to land subsidence and snow-water equivalent retrieval. In: Proceedings ESA Fringe 2005, ESA ESRIN, Frascati, Italy

Lauknes TR, Piyush Shanker A, Dehls JF, Zebker HA, Henderson IHC, Larsen Y (2010) Detailed rockslide mapping in northern Norway with small baseline and persistent scatterer interferometric SAR time series methods. Remote Sens Environ 114:2097–2109

Lilleøren KS, Etzelmüller B, Schuler TV, Gisnås K, Humlum O (2012) The relative age of mountain permafrost—estimation of Holocene permafrost limits. Glob Planet Change 92:209–223

Mauri A, Davis B, Collins P, Kaplan J (2015) The climate of Europe during the holocene: a gridded pollen-based reconstruction and its multi-proxy evaluation. Quatern Sci Rev 112:109–127

Schwartz S, Zerathe S, Jongmans D, Baillet L, Carcaillet J, Audin L, Dumont T, Bourlès D, Braucher R, Lebrouc V (2017) Cosmic ray exposure dating on the large landslide of Séchilienne: a synthesis to constrain slope evolution. Geomorphology 278:329–344

Stroeven AP, Hättestrand C, Kleman J, Heyman J, Fabel D, Fredin O, Goodfellow BW, Harbor JM, Jansen JD, Olsen L (2016) Deglaciation of Fennoscandia. Quat Sci Rev 147:91–121. https://doi.org/10.1016/j.quascirev.2015.09.016

Disaster Risk Assessment of the Silk Road

Peng Cui, Qiang Zou, Yu Lei, Zhengtao Zhang, and Shengnan Wu

Abstract

Disasters Risks, generated by the interaction of complex human and natural systems, are more significant, more complicated, and more difficult to foresee. Silk Road spans Asia, Europe, and Africa, involving more than 140 countries and nearly 66% of the world population. Relevant studies show that the countries along the Silk Road area suffered from the most frequent natural hazards and the most severe losses in the world. In order to promote social and economic development in Silk Road areas, it is urgent to carry out comprehensive research on disaster risk and disaster reduction. This compels new conceptual and analytical approaches to improve understanding of disaster risk at different scales. This paper presents the Silk Road disaster risk assessment at four scales, which could serve a different purpose. It is a robust response to one of the prioritized areas of Sendai Framework on Disaster Risk Reduction for people to understand disaster risk better.

Keywords

Silk road • Disaster • Risk assessment • Sendai framework • Disaster risk reduction

Introduction

The Silk Road, beginning in the Han Dynasty, has been playing an essential role in connecting Asia, Europe, and Africa through the exchanges of the trades, sciences, technologies, and cultures for centuries. Meanwhile, overwhelming disasters, such as earthquakes, floods, droughts, landslides, and debris flows, frequently occur, and bring significant challenges to sustainable development in the Silk Road area. The statistic showed that the average economic loss due to disasters against GDP for the Silk Road countries is twice the world average, while the annual mortality risks in this area are much higher than that of the rest of the world (Lei et al. 2018). Therefore, for a better understanding of disaster risks and more solid support in risk-informed sustainable development in the Silk Road area, disaster risk assessment of the entire Silk Road area is indispensable.

Disasters Overview of the Silk Road

As presented in the Global Assessment Report 2019 (GAR 2019), extreme changes in planetary and socio-ecological systems are happening now. The risks generated by the interaction of complex human and natural systems are escalating and greatly hindering our development. With the complex geological setting and climatic coverage, the Silk Road area has been plagued by natural hazards for a long time.

P. Cui (✉) · Z. Zhang · S. Wu
Institute of Geographic Sciences and Natural Resources Research, Chinese Academy of Sciences, Beijing, 100101, China
e-mail: pengcui@imde.ac.cn

Z. Zhang
e-mail: zhangzt@igsnrr.ac.cn

S. Wu
e-mail: wsn_zju@163.com

P. Cui · Q. Zou
The Centre for Excellence in Tibetan Plateau Earth Sciences (CETES) of the Chinese Academy of Sciences (CAS), University of Chinese Academy of Sciences, Beijing, China

Q. Zou · Y. Lei
Key Laboratory of Mountain Hazards and Earth Surface Process/Institute of Mountain Hazards and Environment (IMHE), Chinese Academy of Sciences (CAS), Chengdu, 610041, China
e-mail: zouqiang@imde.ac.cn

Y. Lei
e-mail: leiyu@imde.ac.cn

Ž. Arbanas et al. (eds.), *Understanding and Reducing Landslide Disaster Risk*, ICL Contribution to Landslide Disaster Risk Reduction,
https://doi.org/10.1007/978-3-030-60713-5_31

Earthquake

The Silk Road Economic Belt stretches along the Eurasia seismic belt. In the history, mega earthquakes taking place in the Silk Road area included the 1906 Manas earthquake (Ms8.0), the 1911 Alma-Ata earthquake (Ms 8.0), the 1920 Haiyuan (Ms 8.5) and the 1948 Ashgabat earthquake (Ms 7.8). The Maritime Silk Road crosses the subduction zone of the India Plate and the Eurasian Plate, the collision zone of the Arabian Plate, the Eurasian Plate and the African Plate and other plate boundaries, where a 9.3-magnitude earthquake occurred in 2004 and induced tsunami in the Indian Ocean (Cui et al. 2017).

Meteorological disaster

Global warming and the change of extreme weather events (such as extreme rainfall) have attracted wide attention. According to the fifth assessment report of the intergovernmental panel on climate change (IPCC 2013), from 1880 to 2012, the global average sea and land surface temperature showed a linear upward trend, raised by 0.85°C. As a result, a variety of extreme weather and climate events occur more frequently. The world meteorological organization (WMO) estimates that in the 25 years from 1967 to 1991, drought has affected 2.8 billion people worldwide, of whom 1.3 million died direct or indirect of drought. On the other hand, the flood also caused a huge problem in some regions. On October 31, 2012, heavy rains hit near Chennai, India, affecting 60,000 people, flooded many villages.

Geological disaster

Under global warming, extreme temperature and precipitation in high mountain areas increase in frequency and intensity. With the fragile environment in the Silk Road area, geological disasters in mountainous areas (such as landslide, debris flow, GLOF, melting glaciers) have also increased in frequency and scale, often led to substantial property losses and casualties. Take China-Pakistan Economic Corridor as an example, the Karakoram Highway (from China Xinjiang Kashgar to northern Pakistan city), of which there are 56 rockfall, 155 landslides, 21 debris flows, 2 landslide dammed lakes (Atabad landslide lake and Su Bing moraine lake), and 10 large glaciers.

Maritime Disaster

Most countries in the Silk Road area are affected by Maritime disasters, including storm surge, tsunami, coastal erosion, ocean acidification, and red tide. From the northern part of the South China Sea to the northern Indian Ocean coast, tropical cyclones are the most common disasters. In 2004, an 8.7 magnitude earthquake struck off the coast of Sumatra island in northern Indonesia, triggering a strong tsunami that directly or indirectly killed about 240,000 people in Sri Lanka, India, Indonesia, Thailand, Malaysia, and other countries, injuring more than 500,000 people and causing an economic loss of more than us $10 billion. Besides, sea-level rise caused by climate change will exacerbate Marine disasters such as increased coastal storm surges, coastal erosion.

Disaster Chain

Disaster chain refers to a series of disasters events induced by a primary disaster. It expands its influence in time and space, amplifies its damage and increases the disaster loss, which often leads to catastrophe or even transboundary disaster, increasing the complexity of disaster risk reduction. A typical example is the Yigong landslide—dammed lake—outburst flood in 2000 in China (Delaney and Evans 2015). It caused massive damage to both China and India.

Methodology

The principle of Disaster Risk

Landslide, debris flow, earthquake, tsunami, storm surge, and flood are natural phenomena, and they become disasters only when they interact with human society. In this regard, the disaster risk can be symbolized as the possibility of a specific disaster event causing casualties and property loss (UNISDR 2009). Disaster risk is a consequence of the interaction between the disaster and human and the properties exposed to hazards. Therefore, disaster and disaster risk are contributed by both elements of nature and society of great variety. Nowadays, the disaster risk model (Peduzzi et al. 2002; Granger 2003; ISDR 2004) extensively accepted to disaster assessment can be formulated as:

$$\text{Disaster Risk (R)} = \text{Hazard(H)} \times \text{Vulnerability (V)} \quad (1)$$

where, the Hazard (H) describing the probability or magnitude of occurrence of the disaster of given type and scale and reflecting the potential impact; the Vulnerability (V) describing the degree of damage of a specific element at risk under the influence of the disaster event, and mainly reflecting the capacity to resist disaster and value of this element at risk. The vulnerability can also be evaluated by exposure and fragility.

Multi-scale disaster risk assessment

The disaster risk assessment in different scale requires specific assessment methods and data support. It is an essential step to define the scale of assessment so that its end-user could adopt the risk assessment result. In this paper, we presented four scales of assessment based on our predefined objectives (Table 1). The data supporting this study are all open accessed from varies sources, such as USGS, NOAA, EMDAT, etc.

(1) Global Scale
The disaster risk assessment at this scale is to analysze the disaster fostering conditions and identify the general distribution of disaster in the Silk Road to provide support to the international cooperation in disaster risk reduction in the Silk Road area.
(2) Regional Scale
Disaster risk assessment at the regional scale evaluates the hazard zonation to facilitate countries in the Silk Road in establishing the inter-country collaborative risk reduction mechanism in accordance with the dominated disasters affecting their countries. Regional risk assessment address the potential disaster intensity and their possible consequences in the assessment area based on the analysis of the environment, disaster conditioning factors, and elemental risk (Cui et al. 2015).
(3) Local Scale
At the local scale, the risk assessment can be conducted incorporating the dynamic process of the disaster. A numerical simulation is a useful tool in this analysis. For example, we simulated the full process of a debris flow event and proposed a quantitative method of debris flow hazard assessment, based on kinetic energy. Using the value of elements at risk to represent the vulnerability, we propose a systematic and quantitative method for risk analysis of debris flow on a local scale (Cui et al. 2013; Zou et al, 2016). The assessment result is capable of supporting the design of early warning systems, emergency management plans, and disaster relief.
4. Infrastructure-oriented
Assessment at this scale focuses on the risk of essential infrastructures such as transportation line and hydropower project, to improve the construction and operation safety, reduce the maintenance cost of the infrastructure projects in their life cycle. This assessment takes a particular interest in the vulnerability of the element at risk in this case, the infrastructure (Zou et al. 2018). The structure response to disaster impact and failure process must be studied in this assessment.

Economic loss assessment

The total economic loss caused by the disaster is divided into Direct Economic Loss (DEL) and Indirect Economic Loss (IEL). The assessment method for DEL along the Silk Road mainly combines the disaster intensity, frequency, and spatial distribution, as well as the corresponding casualties and economic value exposed. The economic value is evaluated by the types of structures and their loss rate (or GDP distribution if it is assessment at a global scale).

Besides, with the development of social economy, acceleration of the process of global economic integration, the IEL (losses caused by production suspension and reduction due to direct economic loss, and the imbalance between supply and demand in the industrial linkage) hidden behind trade links cannot be ignored. The IEL caused by a catastrophic disaster could affect all the countries around the world and exceed the direct economic loss. Therefore, the assessment method of IEL is to build a comprehensive model that combined with the disaster mechanism and economic mechanism (e.g., Input–Output model) (Zhang et al. 2019):

$$YA_i(t) = Dem_i^{local}(t) + Dem_i^{recons}(t) \tag{2}$$

$$IEL = \sum_{i=1}^{n} \int_{1}^{p} \frac{YA_i(t)}{YA_i(0)} dt \tag{3}$$

where YA refers to the supply capacity of the ith sector at timestep t (e.g., day, month, etc.) after a disaster, Dem_i^{local} means the intermediate demand (e.g., government or residents demands for regular operation). Dem_i^{recons} Indicates the reconstruction demand (houses or infrastructures repair, rescue people). There will always be IEL until the end at time p. The IEL equals the sum of the cumulative losses (compare to pre-disaster level $YA(0)$) of each sector over time.

Table 1 Scales and objectives of disaster risk assessment

Assessment scale	Resolution	Objective
Global	1–10 km	Identify global disaster risk pattern, support international DRR
Regional	100 m	National disaster risk zonation, support the national strategy on collaborative disaster response
Local	1–5 m	Identify disaster influence and support urban plan
Infrastructure-oriented	Demand base, <1 m	Support Infrastructure design and safe operation

Assessment Results

Global scale

1. Risk Assessment of Mortality
   The risk of mortality is affected by the frequency of disasters and human vulnerability. Countries with a high-risk level include China and India, which is mainly caused by a high frequency of disaster and high population density. However, some other regions in Europe which have a low frequency but expose a medium or high risk since the population density is large or the capacity to resist disaster is weak (Fig. 1a).
2. Risk Assessment of Direct Economic Loss
   The risk of economic loss is mainly affected by the frequency and economic development level. Countries with high risk mainly locate in eastern Asia and Australia. These regions have more frequent disasters and high GDP, which leads to a high-risk level of economic loss. The regions with medium risk level distribute in central and western Europe and southwestern Asia(Fig. 1b).
3. Risk Assessment of Indirect Economic Loss
   Based on the statistics of the catastrophic flood in 2011, Thailand, the flood caused a DEL of 46,531 million USD (World Bank 2012), also an IEL of more than ten times larger than DEL in the next four years. Besides, many oversea invested factories built in Thailand (e.g., hard disk, automobile factories) were also damaged, resulting in large cross-boundary economic loss to other countries around the world through the trade network (Fig. 1c). The total cross-boundary economic loss has reached more than five times than its DEL. Japan suffers the most

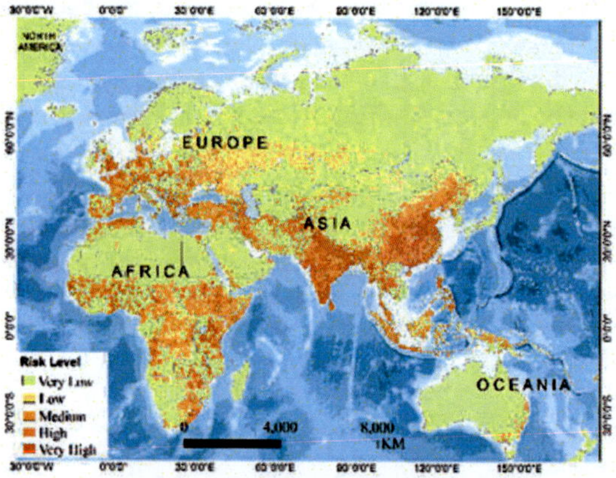

(a) Spatial distribution of mortality risk

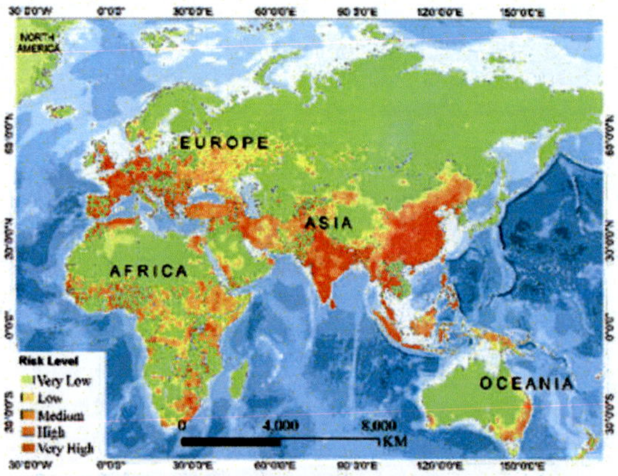

(b) Spatial distribution of economic loss risk

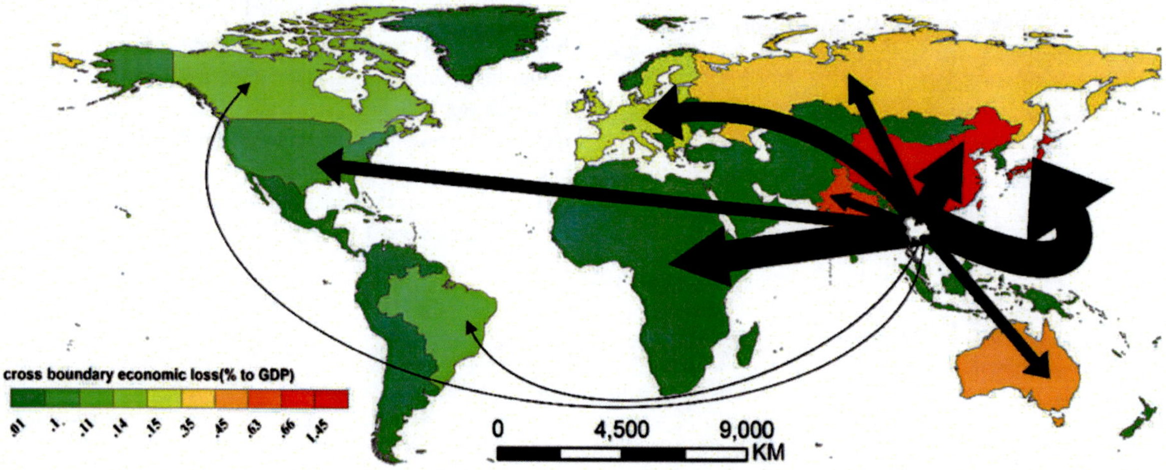

(c) Cross-boundary economic losses of the flood in 2011, Thailand (The width of arrows means the ranking of cross-boundary economic loss suffered by each country)

Fig. 1 Risk assessment results at globe scale

(178% of DEL), followed by China (86%), Europe (55%), USA(32%), Russia (12%), Australia (11%), India (10%), Brazil (5%), and Canada(5%). In addition, although the Europe suffers larger loss than Australia, it has the opposite effect on GDP, which shows that the actual cross-boundary economic loss suffered by different regions is closely related to the trade ties and their GDP. This analysis indicates that countries should not only pay attention to the DEL with the disaster area, but also the IEL and the cross-boundary loss of much wider region.

Regional scale: Tibet plateau disaster risk assessment

Base on the types of hazards (such as landslide, debris flow, flash flood, drought and snow calamity) at Qinghai-Tibet Plateau (Fig. 2)(Cui and Jia 2015), a hazardous and vulnerability analysis is carried out at this region and subsequently an integrated risk assessment is conducted. The assessment result shown in Fig. 3 indicates that the high-risk regions are mainly located at the fringe of western and southern Qinghai-Tibet Plateau while the moderate risk regions are in south-central, west-central and north-east area including southern Tibet, northern Qinghai and north-west of Sichuan. The low-risk regions are at the central of Qinghai-Tibet Plateau, including north-central of Tibet and south-west of Qinghai. The northern and north-west of Qinghai-Tibet plateau, including northern Tibet and southern Xinjiang, are under minimum risk (Cui et al. 2015).

Local scale: Dynamic Process-based Risk Assessment of Debris Flow

The proposed method on the local scale was applied to a case study of debris flow in Qipan gully, which caused severe damages for Duwen Highway and Qipan town of Sichuan Province in 2013 (Fig. 4a). Using debris flow simulation (Fig. 4b), remote sensing (RS), and GIS techniques, the proposed method studied the disaster movement and divided the debris flow areas into three zones as high-risk, medium-risk, and low-risk zone as shown in Fig. 4c. The result shows that the high-risk zone accounted for 54.2% of the total area (271,950 m^2), while the medium risk zone occupied 30.7% of the total area (154,070 m^2) and the low-risk zone contained 15.2% of total area (76,080 m^2). Comparing this result with the field survey data after the debris flow events, it showed that the calculated risk zones are very much consistent with the in-situ condition, which strongly suggests that this risk assessment method is able for debris-flow risk analysis and risk management of debris-flow alluvial fans.

Project-oriented risk assessment

In this case, we evaluated the potential hazard induced by debris flow in Ridi Gully (Zou et al. 2017). Considering the most damage from debris flow to the affected objects, we used the maximum simulated flow height to estimate the hazard of debris flow in Ridi Gully. By applying GIS spatial analysis techniques, we extracted the channel profile of Ridi Gully. Specifically, at the location where the Sichuan Tibet Railway crosses with a bridge. The depth of debris flow was 8.5 m with 100-year rainfall and 13.8 m with 200-year rainfall. Subsequently, we compared the simulated results with the 18 m headroom design of the Sichuan-Tibet Railway (Fig. 5a). It shows that half of the height of the designed headroom will be deposited by debris flow with a 100-year return period precipitation, and nearly two-thirds of the headroom will be taken by debris flow of a 200-year return

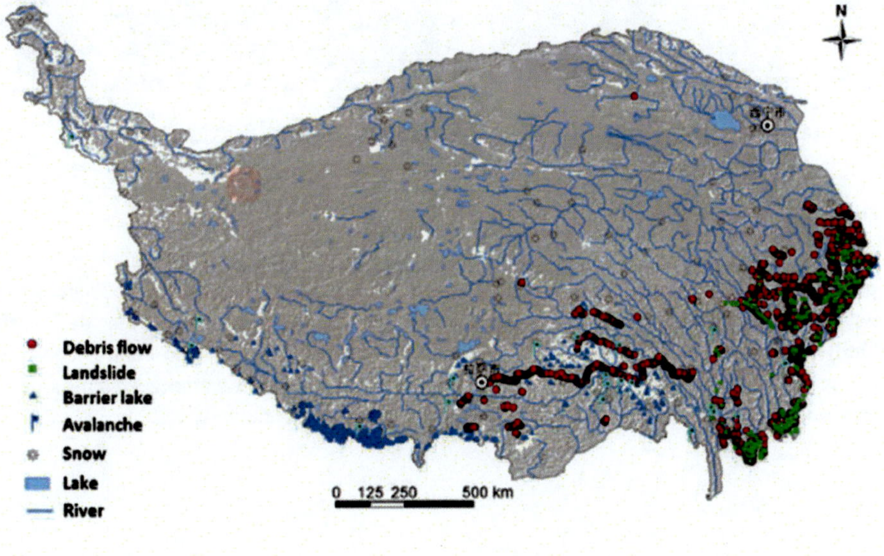

Fig. 2 Hazards distribution in the study area

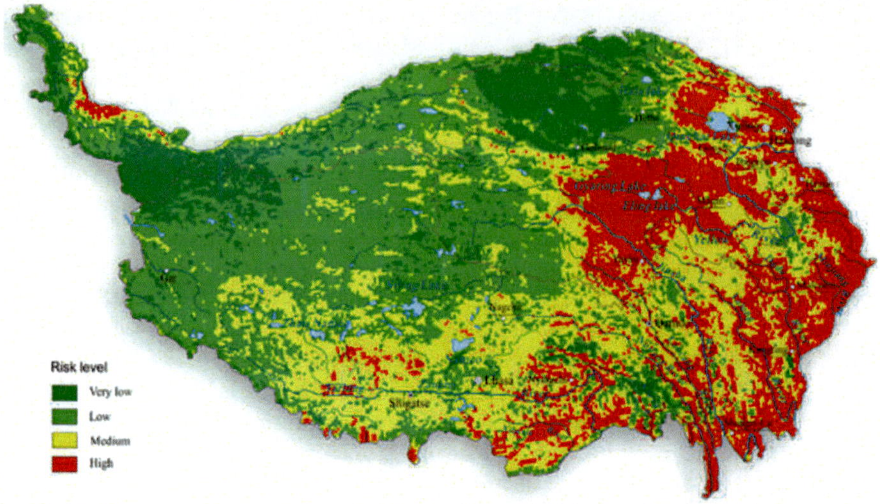

Fig. 3 Risk assessment of natural hazards

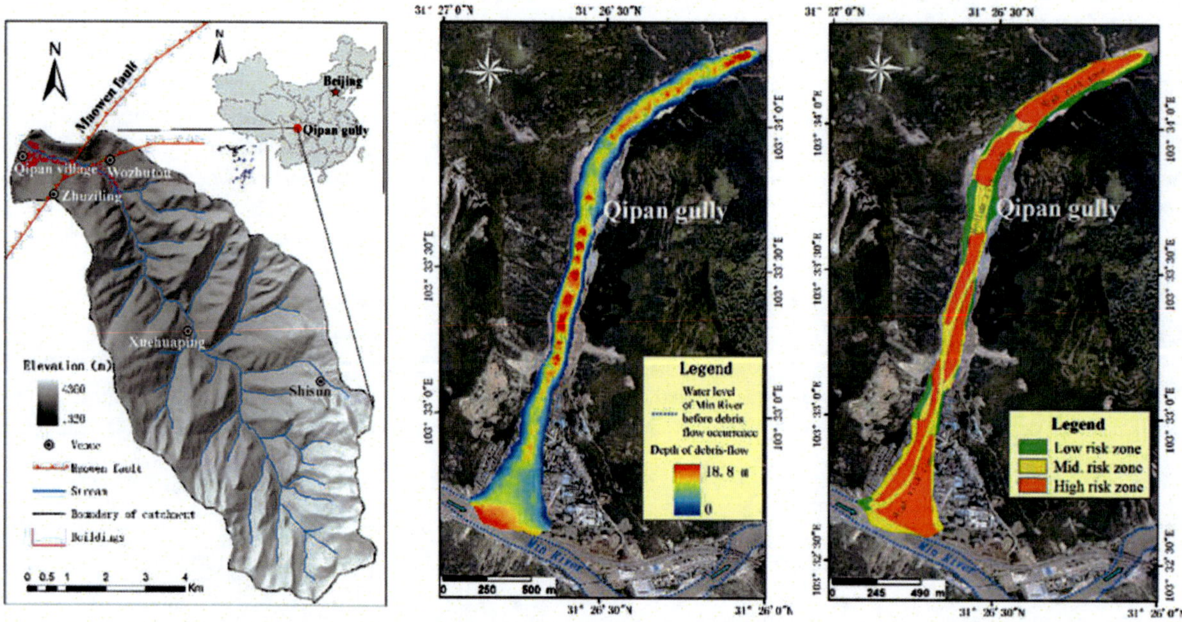

(a) Location of Qipan gully (b) Flow depth in the debris-flow affected area (c) Risk map of debris-flow in Qipan gully

Fig. 4 Risk assessment results in Qipan gully

period precipitation (Fig. 5b, c). As a result, the G318 national highway and villages in the alluvial area of the Ridi Gully are subjected to a high risk of debris flow. Considering the composite effects of debris flow impact, burying, and erosion, the risk of debris flow is high to very high around the Ridi Gully. Therefore, to reduce the potential damages induced by the debris flow, engineering measures are required to control debris flow, and a real-time monitoring and warning system is necessary to ensure the safety of the railway.

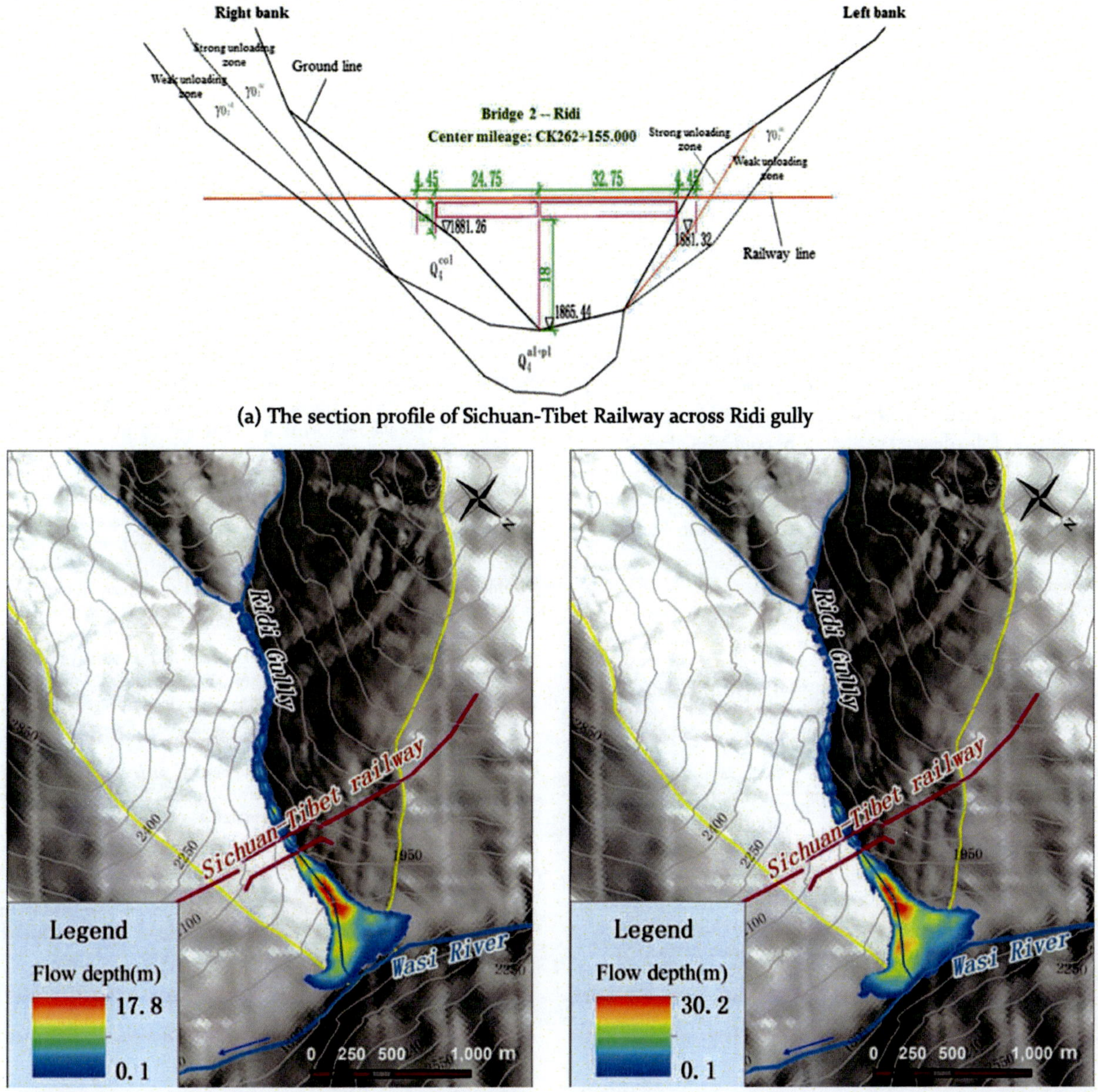

(a) The section profile of Sichuan-Tibet Railway across Ridi gully

(b) Simulation of a 100-year return period precipitation

(c) Simulation of a 200-year return period precipitation

Fig. 5 Simulated results of the Sichuan-Tibet Railway section in Ridi Gully

Discussion and Conclusions

The Silk Road area has a complex geological setting with a widely distributed earthquake, drought and waterlogging disaster, meteorological disaster, and geological disaster. Especially, South Asia is dominated by earthquakes, geological disasters, and flood disasters; Southeast Asia mainly suffers from earthquakes, geological disasters, and ocean disasters; and the Mediterranean is mainly influenced by earthquakes, geological disasters and flood disasters.

The paper carries out the disaster risk assessment of different scales, and the assessment results play an essential role in identifying global risk law, drawing inter-regional disaster risk maps, determining disaster impact and community planning, service engineering construction, safe operation and maintenance. However, due to the wide distribution and prominent uncertainty of natural disasters in the Silk Road area, it is necessary to:

1. understand the regional distribution of disaster and their risk;
2. combining the needs of real-time disaster monitoring and data transmission, developing affordable technologies and instruments, identify new indicators to develop a refined disaster monitor system and carrying out dynamic based disaster monitoring.
3. develop key technologies for disaster prevention and control of major projects, reduce the impact of natural hazards and improve the construction and operation safety toward a whole life circle project design.
4. enhance the disaster resilience of countries along the Silk Road, promote international cooperation on disaster risk and comprehensive disaster reduction and establish a inter-country disaster risk reduction mechanism for transboundary disasters.

Acknowledgements This work is supported by the International Partnership Program (Grant No. 131551KYSB20160002), the NSFC Research Fund for Young Scientists (41807509) and the Strategic Priority Research Program of Chinese Academy of Sciences(Grant No. XDA20030301).

References

Cui P, Jia Y (2015) Mountain hazards in the Tibetan Plateau: research status and prospects. Natl Sci Rev 2(4):397–402

Cui P, Zou Q, Xiang LZ, Zeng C (2013) Risk assessment of simultaneous debris flows in mountain townships. Prog Phys Geogr 37(4):516–542

Cui P, Fenghuan Su, Zou Q et al (2015) Risk assessment and disaster reduction strategies for mountainous and meteorological hazards in Tibetan Plateau. Chin Sci Bull 60:3067–3077 (in Chinese)

Cui P, Deep Regmi A, Zou Q et al (2017) Natural hazards and disaster risk in one belt one road corridors. In: 4th world landslide forum, Ljubljana, Slovenia, EU

Delaney KB, Evans SG (2015) The 2000 Yigong landslide (Tibetan plateau), rockslide-dammed lake and outburst flood: review, remote sensing analysis, and process modelling. Geomorphology 246:377–393

IPCC (2013). Climate change 2013: the physical science basis. Contribution of Working Group I to the Fifth Assessment Report of the Intergovernmental Panel on Climate Change. Cambridge University Press, Cambridge

ISDR (Intentional Strategy for Disaster Reduction). (2004). Living with risk. A global review of disaster reduction initiatives. United Nations publication, New York and Geneva.

Granger K (2003) Quantifying storm tide risk in cairns. In: Natural Hazards, vol 30. Kluwer Academic Publishers, pp 165–185

Peduzzi P, Dao Q-H, Herold C et al (2002) Global Risk and Vulnerability Index Trends per Year (GRAVITY), Phase II: Development, analysis and results, scientific report UNDP/BCPR. Switzerland, Geneva

UNDRR (2019) Global assessment report. [online] Available at: https://gar.unisdr.org/report-2019. Assessed 2 Dec 2019

UNISDR (2009) Terminology on disaster risk reduction. [online] Available at: https://www.unisdr.org/files/7817_UNISDRTerminologyEnglish.pdf. Assessed 13 Sep 2019

World Bank (2012) Thai Flood 2011. Rapid assessment for resilient recovery and reconstruction planning. Bangkok: The World Bank

Yu L, Cui P, Regmi AD et al (2018) An international program on Silk Road Disaster Risk Reduction-a Belt and Road initiative (2016–2020). Journal of Mountain Science 015(007):1383–1396

Zou Q, Cui P, Zeng C et al (2016) Dynamic process-based risk assessment of debris flow on a local scale. Phys Geogr 37(2):132–152

Zou Q, Zhou G, Li S et al (2017) Dynamic-process analysis and hazard prediction of debris flow in eastern Qinghai-Tibet Plateau area—a case study at RiDi gully. Arct Antarct Alp Res 49(3):373–390

Zou Q, Cui P, Zhou G et al (2018) A new approach to assessing vulnerability of mountain highways subject to debris flows in China. Prog Phys Geogr 42(3):305–329

Zhang Z, Li N, Cui P et al (2019) How to Integrate Labor Disruption into an Economic Impact Evaluation Model for Postdisaster Recovery Periods. Risk Anal 39(11):2443–2456

Analyzing the Characteristics of Glacial Debris Flow Activity in Parlung Tsangpo Basin, Tibet

2020 Kyoto Japan

Jiao Wang, Qiang Zou, Wen Jin, and Yanju Fu

Abstract

Parlung Tsangpo, which has a complex geological structure, active neotectonic movement, special hydrometeorological conditions and frequent glacier activity, has become the most dangerous segment of the Sichuan-Tibet highway due to the destruction caused by large-scale glacier debris flows. Evaluating the activity of glacier debris flow gullies is important for forecasting, assessing, and mitigating debris flows. The longitudinal profile of a debris flow valley is a fundamental morphological condition, and together with the rainfall and related hydraulic conditions, all three factors contribute to the initiation and transportation of glacial debris flows. To resolve the problems, such as the complexity of determination research and difficulty of data collection in the evaluation of debris flow gullies, the distribution characteristics of glacial debris flows are analyzed based on the statistical data in the study area. Meanwhile, an effective and convenient method for quantitatively analyzing the evolution stage and activity degree of glacier debris flow gullies is discussed. In this method, an index expressing the shape of the valley's longitudinal profile is used to link the excess entropy of the geomorphic system with the evolution stage for debris flow gullies. Then, the new evaluation standard for gully evolution can provide a quick activity assessment for debris flow-related early warning systems in the marine glacial area.

Keywords

Glacial debris flow • Shape of longitudinal profile • Excess entropy • Activity

J. Wang · Q. Zou (✉) · W. Jin · Y. Fu
Key Laboratory of Mountain Hazards and Earth Surface ProcessInstitute of Mountain Hazard and Environment, Chinese Academy of Sciences, Chengdu, 610041, China
e-mail: zouqiang@imde.ac.cn

J. Wang
e-mail: zouqiang@imde.ac.cn

W. Jin
e-mail: wenjin@imde.ac.cn

Y. Fu
e-mail: fuyanju@imde.ac.cn

W. Jin · Y. Fu
University of Chinese Academy of Sciences, Beijing, 100049, China

Introduction

The Parlung Tsangpo Basin in southeastern Tibet is one of the world's deepest gorges affected by the comprehensive action of plateau uplift and fluvial incision (King et al. 2016; Korup and Montgomery 2008). Large-scale glacial debris flows (DFs) may have severely impacted the Sichuan-Tibet highway (State Highway G318) along the right bank of the Parlung River over the past 60 years, causing the extended dysfunction of the highway (Wei et al. 2018), which seriously hinders the economic development of Tibet. Evaluating the activity of glacial DF gullies is a crucial and first step for forecasting, assessing, and mitigating debris flows (DFs). The evolution of a small watershed is an important part of geomorphologic evolution, and debris flow is its most significant manifestation. The occurrence of debris flow is closely linked with the geomorphologic evolution stage of the gully. Using geomorphologic evolution theory, the evolution stage of debris flow gullies is determined (Kumar et al. 2019; Procter et al. 2014; Schurch et al. 2016; Tang et al. 2019; Jiang 2001, 2003). The Parlung Tsangpo Basin has developed the largest and most typical marine glaciers on the Qinghai-Tibet Plateau, and glacier-related surface processes have imposed very strong control on the geomorphologic evolution of glacial debris flows; this information is essential for understanding the origin of DFs (Chiarle et al. 2007; Palacios and Marcos 1998). A large number of studies have described the characteristics, activity

Ž. Arbanas et al. (eds.), *Understanding and Reducing Landslide Disaster Risk*,
ICL Contribution to Landslide Disaster Risk Reduction,
https://doi.org/10.1007/978-3-030-60713-5_32

and risk assessment of DFs in the periglacial area (Baer 2017; Chiarle et al. 2007; Seinova et al. 2007), but only a few of them have specifically addressed DF activity and the related glacial landforms.

The basin where the debris flow occurs is a complex nonlinear and nonequilibrium system containing internal and external geological agent interactions and multiple positive or negative geomorphic evolutionary feedbacks (Jiang et al. 1999). The strong cutting effect of debris flow accelerates the change in landforms in the basin; meanwhile, the geomorphic conditions restrict the formation and movement of debris flow, which significantly affects the occurrence and scale of debris flow. The formation and activity characteristics of debris flows are closely related to the geomorphic stability of the basin (Qi 2000). Entropy represents the degree to which certain material system states may appear and is generally defined as a measure in science and technology. Excess entropy is a nonlinear and nonequilibrium form of entropy, and the stability of the region can be judged according to the positive and negative values (Yue et al. 1989). Therefore, the activity degree of the debris flow gully can be described by the hyperentropy value of the geomorphic system, and this information can be used to explore the relationship between the activity degree of debris flow and the stability of the geomorphic system.

General Regional setting

The Parlung Tsangpo Basin, which is located southeast of Tibet, is one primary tributary of Brahmaputra, with a total length of approximately 250 km. The geological features of this study area are shown in Fig. 1 and are characterized by strong active tectonic movement and special stratigraphic distribution due to the Eurasian plate colliding with the Indian plate (Ou et al. 2015). The lithology in this area includes granite and metamorphic rocks of granite, Devonian limestone, Carboniferous sandstone and Jurassic granite gneiss and slate, which were strongly affected by Quaternary geology and earthquake activities (Wei et al. 2018). The main channel of the Parlung Tsangpo developed along the Jiali fault, and its topography is affected by differential surface uplift and periodic climate change. Due to the effect of the strong compressional structure, the study area is also characterized by high elevation in the east and west and low middle terrain, and there is an interphase distribution of mountains, deep valleys, and wide valleys. The tributary branches are arranged in the shape of feathers, forming a typical lattice drainage on both sides of the main river or large tributary. In addition, the tributary valleys have large relief, with a maximum of 5744 m. The slopes on both sides of these valleys are greater than 30° while the maximal is up

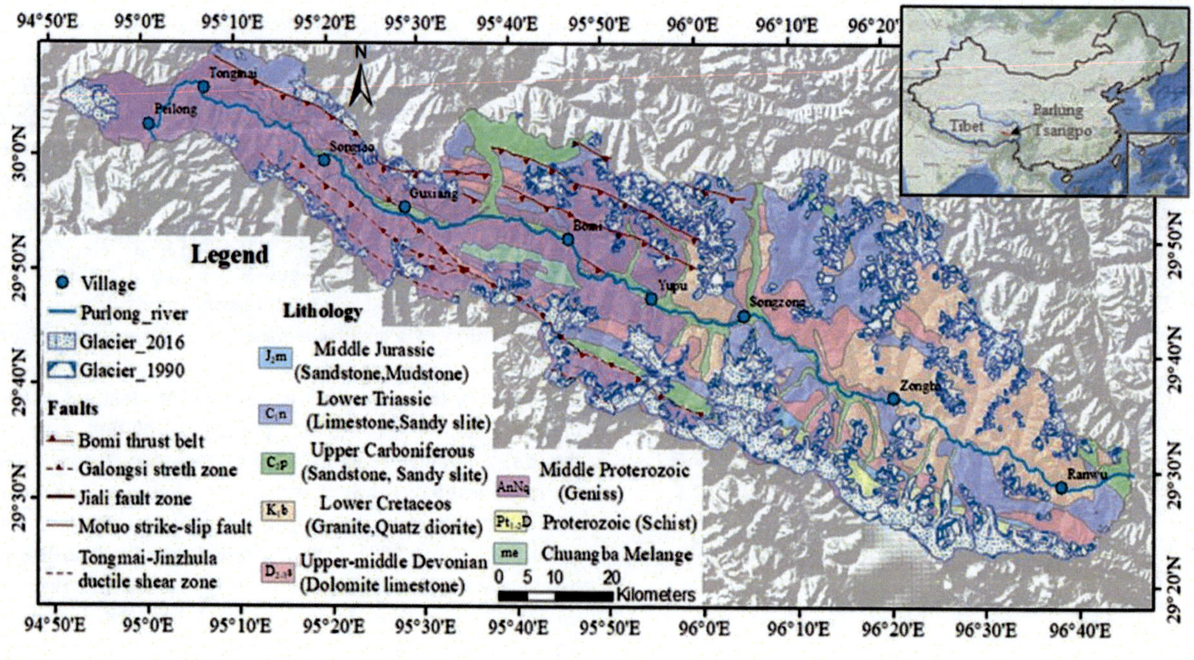

Fig. 1 Geological sketch map of the study area

to 80°. The high relief and steep slopes provide favorable terrain and potential energy conditions for the occurrence of mountain hazards such as landslides, avalanches, and debris flows.

Data and Methods

Data.

In this study, the ASTER Global Digital Elevation Models (two kinds of GDEMs; 5 and 25 m spatial resolution) were provided by the Tibet Bureau of Surveying and Mapping and used to evaluate the geomorphic condition of the study area. To enhance the landform characteristics, a shaded-relief image was generated from the DEM data. Gaofen-2 with 2 m resolution and Gaofen-1 with 1 m resolution images (from 2013 to 2017) were provided by the Geometrics Center of Sichuan Province, China, and confirmed using shaded-relief images with 5 m resolution. The information of most of the DFs was collected from field investigations; however, some DFs can be determined from high-resolution remote sensing images and high-resolution shaded relief when inaccessible in the field.

Methods.

Nanshan (1987) first introduced the concept of hyperentropy into the stability evaluation of river basin systems and proposed the mathematical expression formula of hyperentropy for a specific basin as follows:

$$\delta_x P = \frac{-\beta\alpha^3(\alpha - 1)(\alpha - 1)}{\alpha(-\beta - 1) - 3} \tag{1}$$

where α and β are the factors in the equation of Straler's hypsometric curve in a nonequilibrium state, α = 1–2/(1 + β). The sign $\delta_x P$ determines the stability of the watershed system, and its absolute value quantitatively represents the degree of stability. If $\delta_x P > 0$, the basin is stable, and the greater the absolute value is, the higher the basin stability is. If $\delta_x P < 0$, the basin is unstable, and the greater the absolute value is, the more unstable the basin is. Enough visible excess entropy is rich in geomorphic evolution information, which can quantitatively describe the degree of basin stability and can be used to determine the stability of the basin system.

Based on previous studies, another researcher (Jiang 2001, 2003) further proposed an excess entropy calculation formula expressed by the shape index of the gully longitudinal profile (Eq. 2), which was used to characterize the stability and DF activity of an open drainage basin:

$$\delta_x P = \frac{N^3(N^2 - 4)(N + 2)}{32(6 - N)} \tag{2}$$

where

$$\frac{h}{H} = \left(\frac{l}{L}\right)^N$$

where N is the shape index of the longitudinal profile in the DF basin; H and L are the height difference (m) and river length (m) of the basin, respectively; and h and l are the height difference (m) between a point on the longitudinal section of the gully and the gully mouth and the river length (m), respectively.
mm?>

Results

Distribution characteristics of glacier DFs.

The distribution of glacial DFs was sorted out in the study area according to the literature and combined with field investigations (Fig. 2). As mentioned above, wide valleys and valleys alternate in Parlung Tsangpo, and the region has distinct dry and wet seasons. The differences in geomorphology and meteorological conditions lead to obvious differences in the spatial and temporal distributions of DFs in the region.

The DFs are widespread and varies in the region. Among them, there were 61 DFs in the south bank with a distribution density of 0.28/km, which are less in the north bank than 78 DFs with a distribution density of 0.35/km. The DFs are the most densely distributed are in the upper and lower canyon section (I) of the river basin, and the density of DFs decreases in the sequence section of mesh valley (Fig. 2). Due to the steep terrain, small drainage and glacier area of gully in the canyon sections, where the scale of DFs was smaller but with high frequency. Whereas For the larger scale of the palaeo-glacier and drainage area of gullies in wide valley and segmented valley, the activity of DFs have the characteristics of a large scale but a small density.

The area of DF gully along the main river of Parlung Tsangpo was classified into the following categories: less than 1, 1–10, 10–20, 20–30, 30–50, 50–100 and greater than 100 km^2 (Fig. 3). The results show that DF mainly occurred in the small watershed with an area of 1–10 km^2, accounting

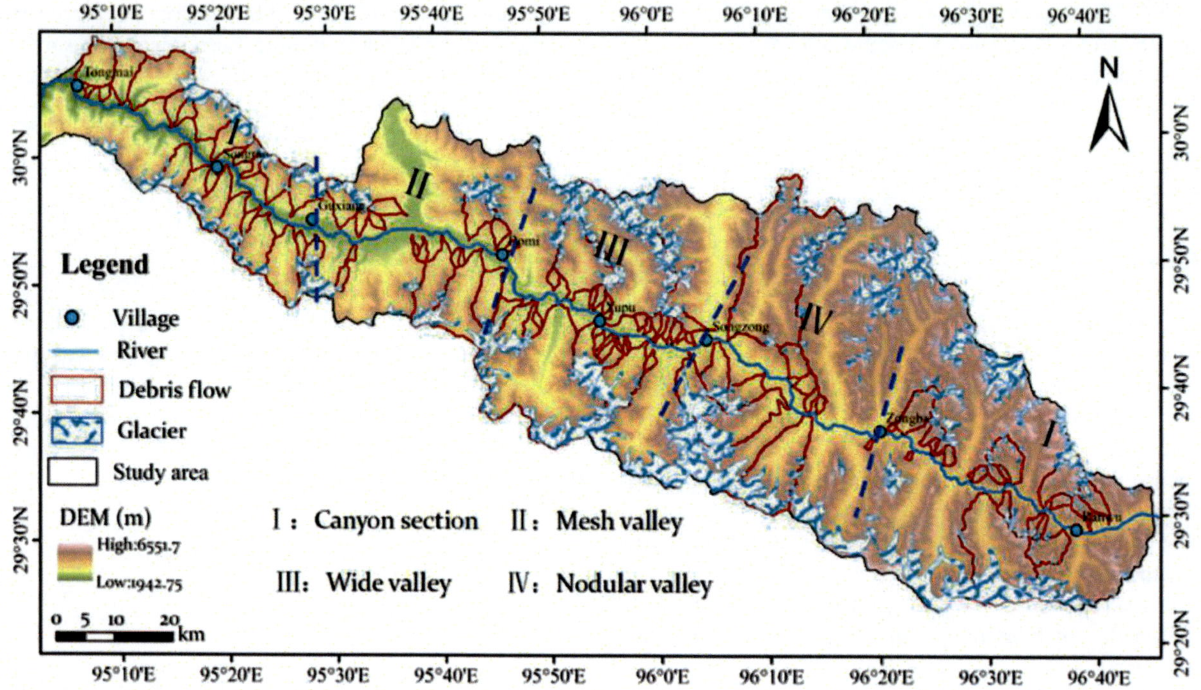

Fig. 2 Debris flow distribution map in the study area

Fig. 3 Statistical chart of glacier debris flow area

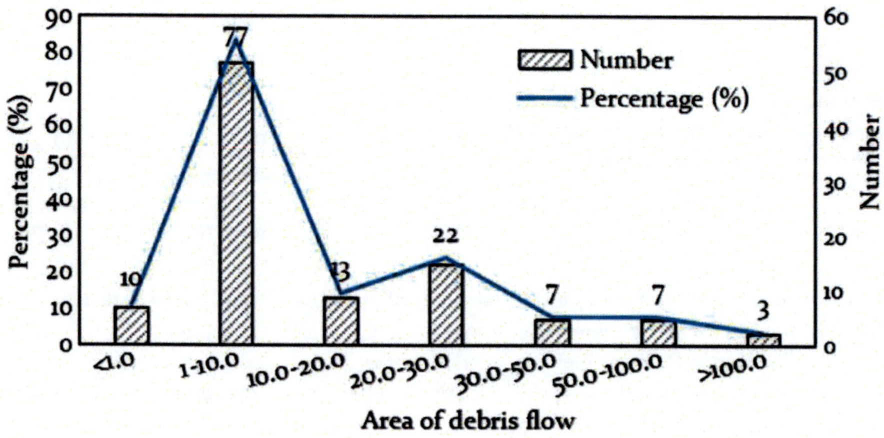

for 55%, followed by watersheds with an area of 20–30 km^2, which accounted for 15.71%. However, only 2.85% DFs occurred in the large watershed with an area greater than 100 km^2.

Negative correlation between activity frequency and catchment area.

In terms of the dynamic condition of glacial DF formation, the mean gully gradient was used to reflect the topographic characteristics of the basin and the dynamic condition of the basin. In other words, the larger the longitudinal slope was, the greater the movement energy of the loose materials in the channel (Xiang 2012). According this, 139 watersheds with DF events were counted in the study area, and their longitudinal slope was fitted to the subcatchment area (Fig. 4). There was an obvious negative power relationship between catchment area and channel slope ($S = 670.36A^{-0.334}$), which was similar to the results of the study on DF gullies in nonglacial areas. It is concluded that the average longitudinal slope of the larger watershed is smaller. The smaller the longitudinal slope of the basin is, the lower the potential energy loose debris source supplying DF activities is. Meaning that the initiation of DFs from these debris requires

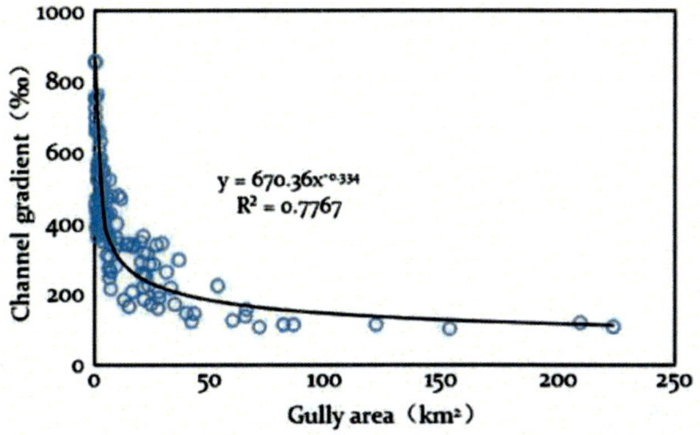

Fig. 4 The relationship between gully area and channel gradient of glacier debris flow

greater start-up and migration energy (From the hydrodynamic conditions of view) (Guo 2016). Therefore,under the same rainfall conditions, the frequency of DF outbreaks in the basins with a smaller area is higher than that in larger basins, the activity frequency of DFs decreases with increasing basin area.

Quantitative analysis the activity of DF gully.

The characteristics of the all gullies in the study, including the basin area (*S*), the length of the main gully (*L*), and the relative relief (*H*), were extracted by ArcGIS software. In addition, the shape of the longitude profile (*N*) and the excess entropy of the basin ($\delta_x P$) were selected to quantitatively analyze the gully activity.

The shape of the longitude profile (*N*) is an index that fits the power function ($y = x^N$) between the relative length ($x_i = l_i/L$) and relative height difference ($y_i = h_i/H$), which was obtained through the MATLAB platform. The excess entropy of the basin ($\delta_x P$) was calculated by Eq. 2.

The results showed that the shape of the gully longitudinal profile in the study area is ranged from 0.5705 to 2.9131, the fitting degree of each gully was greater than 0.90, and the value of hyperentropy ranged from -0.0151 to 5.5152. The range of these geomorphic indicator is of statistical significance.

To analyze the different stages of gully evolution in the study area, polynomial fitting was carried out between the shape of the longitudinal profile (*N*) of 159 gullies and the excess entropy of the basin ($\delta_x P$). The result of quintic polynomial fitting was the best, with a fitting degree of 0.99. The mathematical expression of the multinomial fitting is shown as Eq. 3.

$$f(\delta_x P) = 0.2278N^5 - 1.573N^4 + 3.7352N^3 - 4.6276N^2 + 2.2771N - 0.6414 \quad (3)$$

In order to determine the geomorphologic evolution stage boundary of watershed, the value of $\delta_x P$ was firstly used to separate the developing period and the decline period for the evolution of gullies. When $\delta_x P$ is less than 0, the gully is unstable, and vice versa. In addition, the delta valueof $\delta_x P$ and the corresponding N value on the curve (Fig. 5) were further analyzed to refine the evolution stage and activity of gullies according to the theory of geomorphic erosion cycle proposed by Davis. As shown in Fig. 5, there are three characteristic points on the curve in the development stage of DF: two minimum points are shown on the curve ($\delta_x P = -0.0174, N = 0.6595, \delta_x P = -0.1494$, $N = 1.6175$) where $0 < N < 2$, and one inflection point in the curve where the

Fig.5 The eigenvalue of, and the evaluation stages of the gully in the Parlung Tsangpo basin

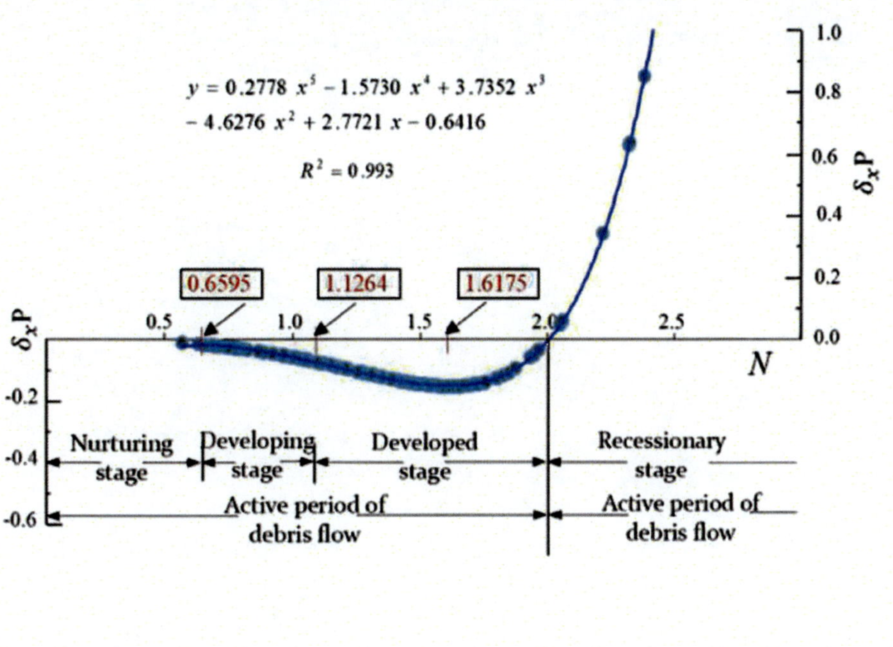

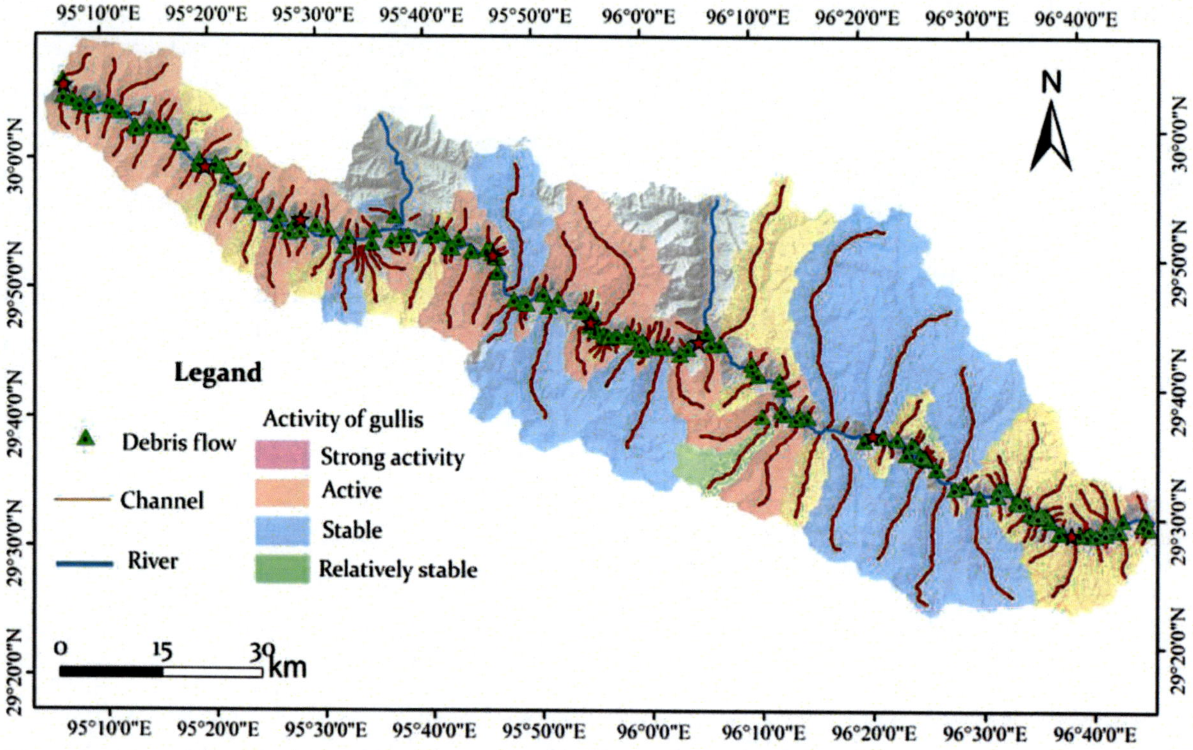

Fig. 5 Evaluation results of activity degree and distribution of debris flow

Table 1 The geomorphologic standard of the activity of debris flow gully in Parlung Tsangpo Basin

$\delta_x P$	N	Stage of gully evolution	Characteristic of gully	Activity of the debris flow
0 ~ −0.0174	0 ~ 0.6595	Nurturing phase	The negative hyperentropy increases slowly. The river system is in the expansion and branching stage. Meanwhile, the erosion is strengthened, and the river basin system begins to be unstable	These factors for triggering DF are emerging but not enough to generate the DF. This period is relatively stable
−0.0174 ~ −0.0806	0.6595 ~ 1.1264	Developing phase	The negative hyperentropy increases rapidly. The erosion is stronger, and the instability of the watershed system increases sharply	It is an active period in which the various conditions for DF are accelerated and can eventually promote the occurrence and development of DF
−0.0806 ~ 0	−0.0806 ~ 0	Developed phase	The negative hyperentropy increases to the extreme value at the decreasing speed and then accelerates to zero. The negative hyperentropy is the largest and the watershed system is the most unstable at this stage	The conditions are favorable for DF occurrence, The development of DF is vigorous, and the frequency of disasters is improved. The activity of DF is highest
> 0	> 2.0	Recessionary/stable phase	The entropy value continues to increase towards a positive value. The erosion is weakened, and the river system is at the end of the expansion. The system gradually tends to a relatively stable state	Debris flow can still occur sporadically. The activity of DF tends to be recessionary and stable

range of $\delta_x P$ from -0.0174 to -0.1494 ($\delta_x P = -0.0806$, $N = 1.1264$). According to the geomorphologic features of the study area, a new evaluatiOn standard for the activity of gully in the Marine glacier area was established according to these three feature points, as shown in Table 1. The results show that there were 80 basins in the strong active period (approximately 51.95%), 50 basins in the active period (approximately 32.47%), 26 basins in the stable period (16.88%), and 3 basins in the more stable period (approximately 2.15%). In conclusion, the evolution of gullies in the Parlung Tsangpo basin are in the developing-developed stage.

To assess the accuracy of this new evaluation standard, the historical data of DFs in the study area, which were collected by field investigation and from the literature, were superimposed on the activity evaluation results of all gullies in the study area (Fig. 5). The results showed that approximately 89.93% of the DFs occur mainly in a higher active gullies, including 80 DFs (accounting for approximately 58% of the total DFs number) in the developed period of gully belongd to the strong active type, and 45 DFs (accounting for 32.37%) in developing stage of gullies belonged to the active type,. In addition, there were 11 DFs accounting for 7.91% in the nurturing stage of gullies belonged to the relatively stable type and only 3 DFs (2.16%) in the recession period of gullies belong to recessionary type. According to the verification by field data, the evaluations of gullies quantified by the excess entropy and the shape of the longitudinal profile are consistent with the activity of DF in the region, and the result of the evaluation has objectivity and is reliable. This new criterion of the division of gullies along Parlung Tsangpo can be well used for DF activity assessment in the marine glacial area.

Discussion and Conclusions

Based on the field data about gullies along the Parlung Tsangpo basin, the distribution and characteristics of DF were analyzed. The differences in topographic, geomorphic and meteorological conditions have caused significant spatial and temporal differences in DF distribution and activity characteristics within the basin, and the distribution density increases from upstream to downstream. On the other hand, there was an obvious negative power relationship between catchment area and channel slope (y = 898.77A-0.528). The excess entropy and the shape of the longitudinal profile were used to qualitatively assess the characteristics of the gully and the activity of DF. Combining the value of excess entropy, the shape of the longitudinal profile and the field data, the activity of gullies in the study area could be divided into four levels: strong active ($0.0806 \leq \delta_x P < 0,1.1264 \leq N < 2.0$), active ($-0.0174 \leq \delta_x P < 0.\ 0806,0.6595 \leq N < 1.1264$), recessionary/stable ($\delta_x P \geq 0, N \geq 2.0$) and relatively stable ($0.0174 < < \delta_x P < 0, 0 < N < 0.6595$). It is verified by comparing the degree of DF gully activity by using the excess entropy of the catchment and the activity of DF in the region obtained by field assessment and literature. This evaluation standard for gully evolution can be applied to activity assessment of DF in the marine glacial area.

Acknowledgements This work is supported by the Major Program of National Natural Science Foundation of China (41941017) and Research Fund for Young Scientists (B) of Chinese Academy of Sciences (E0R2210210).

References

Ai Nanshan (1987) Information entropy of erosion watershed system. Journal of soil and waterconservation, 1987(2):1–8

Baer P (2017) Changing debris flow activity after sudden sediment input: a case study from the Swiss Alps. Geol Today 33(6):216–223

Chiarle M, Lannotti S, Mortara G et al (2007) Recent debris flow occurrences associated with glaciers in the Alps. Glob Planet Change 56(1–2):123–136

Guo XJ (2016) Debris flow formation model combining hydrological processes and stochastic material supplying simulation in watershed. Ph.d. theis, University of Chinese Academy of Sciences, Beijing, China

Jiang ZX (2001) A statistical analysis on longitudinal profile shape of debris flow valley along Palongzangbu River in Tibe. Chin J Geologic Hazard Control 12(4):41–47

Jiang ZX (2003) Models of shape and evolution on longitudinal profile of ice-snow melt-water valley. Chin J Geologic Hazard Control 04:22–28

Jiang ZX, Yao LK, Ai NS et al (1999) A new technique for nonlinear study and prevention of railway debris flow. Sichuan science and technology press, Chengdu

King GE, Herman F, Guralnik B (2016) Northward migration of the eastern Himalayan syntaxis revealed by OSL thermochronometry. Science 353(6301):800–804

Korup O, Montgomery DR (2008) Tibetan plateau river incision inhibited by glacial stabilization of the Tsangpo gorge. Nature 455 (7214):786

Kumar A, Bhambri R, Tiwari SK (2019) Evolution of debris flow and moraine failure in the Gangotri glacier region, Garhwal Himalaya: hydro-geomorphological aspects. Geomorphology 333:152–166

Palacios D, Marcos J (1998) Deglaciation on Mexico's Stratovolcanoes from 1994–95. J Glaciol 44:63–67

Procter JN, Cronin SJ, Zernack AV et al (2014) Debris flow evolution and the activation of an explosive hydrothermal system; Te Maari, Tongariro, New Zealand. J Volcanol Geoth Res 286:303–316

Ou XJ, Duller GAT, Roberts HM et al (2015) Single grain optically stimulated luminescence dating of glacial sediments from the Baiyu Valley, southeastern Tibet. Quat Geochronol 30(10):314–319

Qi L (2000) Evaluation method for activity degree of debris flow gully. J Mountain Sci 18(4):365–368

Schurch P, Densmore AL, Lvy-Ochs S et al (2016) Quantitative reconstruction of late Holocene surface evolution on an alpine debris-flow fan. Geomorphology 275:46–57

Seinova I B, Sidorova T L, Chernomorets S S. (2007) Processes of debris flow formation and the dynamics of glaciers in the Central Caucasus. In: International conference on debeis-flow hazards mitigation: mechanics. Chengdu, China, pp 77–85

Tang H, McGuire LA, Rengers FK et al (2019) Evolution of debris-flow initiation mechanisms and sediment sources during a sequence of postwildfire rainstorms. J Geophys Res Earth Surf 124 (6):1572–1595

Wei RQ, Zeng QL, Davies T et al (2018) Geohazard cascade and mechanism of large debris flows in Tianmo gully, SE Tibetan Plateau and implications to hazard monitoring. Eng Geol 233:172–182

Xiang LZ (2012) The risk assessment of debris flow along highway after the earthquake in Wenchuan county. Ph.d. theis, The Graduate University of Chinese Academy of Sciences, Beijing, China

Yue TX, Ai NS, Zhang YB (1989) Superentropy—a criterion for determining stability of drainage system. J Soil Water Conserv 3 (2):20–27

Rehabilitation of Gully-Dominant Hill Slopes by Using Low-Cost Measures—A Case Study in Nepal

Daisuke Higaki, Kishor Kumar Karki, Naoto Koiwa, Mio Takahashi, and Sohan Kumar Ghimire

Abstract

A study was carried out to find out the effectiveness of low-cost measures for the rehabilitation of gully-dominant hill slopes of the river terraces in the Central Nepal. The main process of gully development on these hill slopes involved crack formation in the laterite soil in dry season and subsequent soil block failure at gully heads followed by incising bed erosion. In order to stabilize these gullies and rehabilitate the hill slopes, low cost measures were adopted which mainly comprised gabion check dams and bamboo plantation. Some other measures such as catch drains and grass plantation were also implemented. As a collective result of these measures, gully erosion rates were minimized and also the gully-dominant slopes were rehabilitated as evident by increasing forest cover. Such low-cost measures were found effective for controlling gully erosion as well as for promoting land resource recovery.

Keywords

Gully erosion • laterite • Slope failure • Bamboo • Gabion check dam • Nepal

Introduction

Gully erosion is one of the major issues in land degradation in hilly and mountain terrain (Lal 1992; Sthapit 1996). In the Midlands of Nepal, gentle slopes or river terraces covered with thick soil are facing gully erosion problems (Higaki et al. 1998). Cyclic changes of crack formation in dry season and failure of soil blocks at the gully head followed by bed erosion due to monsoon rainfall induce such erosion processes every year (Higaki et al. 2005). Rehabilitation of gully-dominant slopes by low-cost measures are extremely important in the developing countries for the management of agricultural lands and natural resources such as water, soil and forests.

This paper aims to evaluate the effectiveness of the low-cost rehabilitation measures for controlling gully-dominant slopes through the monitoring of changes in land coverage conditions from a pilot project in Nepal.

Study Area

The study area, Pipaltar, is located on the slopes of fluvial terraces formed by Thadi River, a tributary of the Trishuli River in Nuwakot District, Central Nepal (Fig. 1). It consists of two monitoring sites called Upper Pipaltar and Lower Pipaltar (Fig. 1). Elevation of the sites is 500–600 m above sea level. Deforestation and over-grazing which mostly had taken place during 1960's has led to the formation of gullies and surface erosion (Fig. 2). The Water Induced Disaster Prevention Technical Centre (DPTC), Government of Nepal and Japan International Cooperation Agency (JICA) selected

D. Higaki (✉)
Land Infrastructure Development Operations, Nippon Koei Co. Ltd, Tokyo, 102-8539, Japan
e-mail: a9024@n-koei.co.jp

K. K. Karki
Former: Department of Water Induced Disaster Management, Ministry of Irrigation, Kathmandu, Nepal
e-mail: karki2000@hotmail.com

N. Koiwa
Hirosaki University, Hirosaki, 036-8560, Japan
e-mail: koiwa@hirosaki-u.ac.jp

M. Takahashi
Hirosaki Gakuin University, Hirosaki, 036-8577, Japan
e-mail: soumi.ho22@gmail.com

S. K. Ghimire
UK Flood Risk, London, DA1 2NL, UK
e-mail: sohan_kr@hotmail.com

Ž. Arbanas et al. (eds.), *Understanding and Reducing Landslide Disaster Risk*, ICL Contribution to Landslide Disaster Risk Reduction,
https://doi.org/10.1007/978-3-030-60713-5_33

this area for the pilot project for the evaluation of various measures to control gully and surface erosion in 1994 (Figs. 3, 4). Since the fluvial terraces are located near the confluence of the Trishili River and the Tadi River, thick fluvial sand and gravel deposits with more than 50 m in thickness formed by two rivers and overlaying laterite layer (5-8 m thick) are erodible by the gullies. The laterite soil is composed of 48% sand, 40% silt and 12% clay (DPTC 1999). Most plants hardly grow due to less contents of organic carbon and acidic condition of the laterite (0.4% in chemical composition) (DPTC 1999).

Adopted Measures and Monitoring of Erosion Control

Gully erosion generally occurred as a result of crack formation in laterite layer in dry season and subsequent soil block failure at gully heads followed by incising bed.

erosion in rainy season as indicated by the monitoring of gully head expansion (Higaki et al. 1998, 2005).

Adopted measures at each gully in Upper Pipaltar are as follows (Table 1; Fig. 3).

(1) Series of gabion check dams to prevent incising erosion on the gully bed (4 nos. at the G-4 main gully, 5 nos. at the gully G-4 LB 1) (Fig. 5).
(2) PNC panels with a check dam to protect the gully bed (Length = 20 m, the gully G-4 RB4)(Fig. 6).
(3) Series of check dams (3 nos.) with bamboo plantation (the gully G-4, RB-5).
(4) Bamboo plantation of 100 stems to protect fallen materials from the gully side slopes against bed erosion (the gully G-4, RB5A).
(5) A catch drain of 113 m in length at the gentle barren slopes on the ridge (the gully G-4, LB-1).

Fig. 1 Location of the study area

Fig. 2 Gully erosion on the slopes of a river terrace at Upper Pipaltar (Dec 1996)

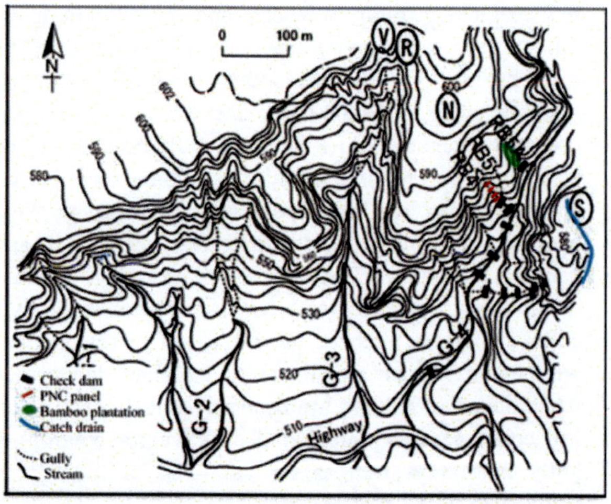

Fig. 3 Distribution of gullies and adopted measures at Upper Pipaltar

While steel wires of the gabion check dams had to be transported, stones were available from the riverbed of the Tadi River. Transportable PNC (Pipe Net Concrete) panels with the scale of length: 100 cm, height: 30 cm thickness: 3 cm were made of pre-cast concrete panel reinforced by steel net connected with a PVC pipe. PNC panels can be used to protect soil slopes from erosion (Hayashi and Fujita 1976). The catch drain was made of cement masonry (Fig. 7) in order to drain surface run-off from the barren slopes to the outside of the catchment of gullies.

Gully erosion monitoring before and after the adoption of the rehabilitation measures in 1995–2003 revealed that these measures reduced the gully erosion rates effectively and thus promoted vegetation recovery until 2003 (Higaki et al. 2005). Figure 8 shows the location of changes in the crown of gully heads during 1997–2003 which included major changes during maturation stage of the measures functioning (1994–97) and minor ones during full functioning stage

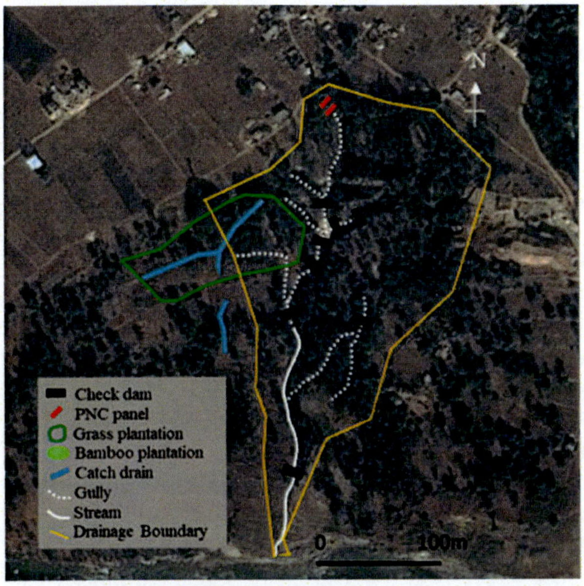

Fig. 4 Gullies and adopted rehabilitation measures in Lower Pipaltar (base image: Google earth (Nov 23, 2017))

(1997–2003) (Higaki et al. 2005). The gully head retreats during the former stage were 0.21, 3.32 and 1.26 m from RB4, RB5 and RB5A, The retreats remarkably decreased to 0.18, 0.91 and 0.49 m during the latter stage. The plane feature of gully heads in 1994, 1997 and 2003 indicated parallel retreat which was progressed by repeated small-scale failures of soil blocks in laterite layer. At the gully RB5A, well-growing bamboos originally introduced were 100 nos. which costed 5000 NRs (equivalent to eight thousands Japanese yen based on the exchange rate of April 1999) trapped the fallen materials from the gully side slopes.

Fig. 5 Gabion check dams at the G-4 main gully

Monitoring Results of Landscape Changes

In Lower Pipaltar, the pilot project adopted various rehabilitation measures with people's participation in the gully dominant Sukauri Khola watershed since 1995 after the Department of Soil Conservation and Watershed Management (DSCWM) had taken place in conservation of the devastated watershed (Table 2; Fig. 4). As this site is located near the confluence of the Tadi River and the sediment-loaded Trishuli River originating from High Himalaya and Tibet, thick layers of flood loam deposits consisting of the river terrace were expected to produce fine sediment toward the gully beds. Eight numbers of gabion check dams with the height of 1.5–3 m were constructed on the gully beds which trapped sand transported from the barren and grass-covered slopes in the upstream.

Table 1 Adopted measures and expenditure at the Upper Piplatar site

Item	Number	Length(m)	Width (m)	Height (m)	Location	Expenditure (NRs)	Year
PNC plate		ca. 20	ca. 2		G4 RB4, gully bed	145,414	1994–95
Gabion check dam	12	3–10	1–2	1.5–3	G4 RB4, 5, 5A, gully bed	158,982	1995–98
Gabion check dam	7	36 (total)	1–2	1.5–3	G4 LB1, gully bed	414,819	1998–99
Catch drain		113 (total)	ca.1		G4 LB1, flank slopes	659,000	1998
Bamboo plantation	ca. 1000				Along each gully bed and side slopes	63,995	1996–99
Others (survey map etc.)						90,822	1995–99
Total expenditure (NRs)						1,533,032	

Fig. 6 PNC plates installed on the gully bed of RB4

Fig. 7 Catch drain for diversion of run-off water from the Upper Slopes

Gullies also tend to be formed along foot trails due to run-off concentration and poor management (DSCWM 1991) (Fig. 9). Two stone-pitched catch drains were constructed (113 m long) to prevent the trail erosion in 1995 (Fig. 4). Grass harvesting in which not only locally available grass but improved grass such as Nepier grass (*Pennisetum Purpureum*) had also been introduced on the grass-covered and partially barren slopes. In addition, seed sowing was also executed. A Woman User's Group of the Sukauri Khola watershed participated in these initiatives in 1997–98 (Karki and Higaki 1998).

Since the monitoring of erosion rate could not be continued due to loss of monitoring pegs in Upper Pipaltar after 2003, inspection survey was undertaken during 2013–2016 by the authors. Comparison of the field monitoring photographs taken in different time indicated the effectiveness of rehabilitation measures in gully erosion control and the land resource recovery of degraded catchment. Though plants hardly recover on the head scarp of laterite soil layer, Ketuki (*Avaga Americana*), a low-height tree with dense roots were found growing in 2018 which prevents the crown from crack formation (Fig. 10). This resulted in the prevention of soil block failures which has enabled bamboo growing.

In Lower Pipaltar, a land coverage map prepared by Rizal et al. (1999) was compared with the Google Earth image taken in November 23, 2017 (Fig. 11). While the monitoring area of Sukauri Khola catchment with gully-dominant slopes is 8 ha in which tree covered area was 1.2 ha in 1999, forests have recovered on the slopes of 3.8 ha in area in November 2017. Figure 12 shows significant increase in tree-covered area on the gully-dominant slopes since July 1998 till March 2016. Sal trees (*Shorea Robsta*), a natural species of sub-tropical climate which are usable as woods for housing have grown very well (Fig. 12). These positive effects seem to be not only due to the adopted measures, but also due to the local people who followed the rules of user's group for natural resources management such as by stopping free-gazing in the gully-dominant catchment (Higaki et al. 1998). Millet cultivation as the traditional crop farming and mango plantation were combined on the terraced gentle slopes which had been partially barren by using Sloping Agricultural Land Technology (SALT) until 2013 (Partrap and Watson 1994) (Fig. 13).

Rehabilitation of Gully-Dominant Slopes by Low-Cost Measures

As gully erosion is a gradual geomorphic process continuing every year in Nepal, small-scale low-cost measures can be adopted using locally available materials including various plant species. Cyclic processes of crack formation with subsequent soil block failure in laterite layer and incising erosion on gully bed promoting instability of gully side slopes should be interrupted based on the monitored erosion

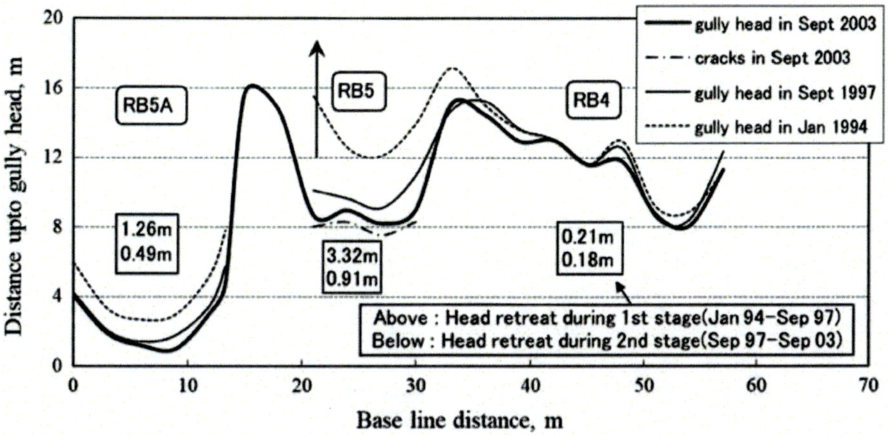

Fig. 8 Difference in the location of each crown of the gully head (1994–2003)

Table 2 Adopted measures and expenditure at the Lower Piplatar site

Item	Number	Length (m)	Width (m)	Height (m)	Location	Expenditure (NRs)	Year
PNC plate	180				Head of main gully	21,463	1994–95
Gabion check dam	8	3–8	1–2	1–3	Gully bed	1,021,952	1995–97
Catch drain		110 (total)	ca.1.5		Along eroded trail and flank slopes	301,446	1995–96
Bamboo plantation	ca. 1000				Along each gully bed and gully side slopes	58,996	1998–99
Grass harvesting and seed sowing		ca. 10.000 m^2			Natural grass-covered and partially barren slopes	18,400	1996–97
Others (water tank, fruit tree plantation)						134,180	1995–99
Total expenditure (NRs)						1,556,437	

rates. Reduction in run-off water concentration on the gullies will also be required.

Total expenditure of the pilot project for both Upper Pipaltar and Lower Pipaltar sites was 3,089, 470 NRs which was equivalent to five million Japanese yen during 1995–1999 (Rizal et al. 1999) in the total area of 35 ha monitored. Since gully erosion degrades the land resources such as soil, water and forests on which the local community depends for their livelihood, low-cost rehabilitation measures of gully-dominant slopes should also focus on the land resource recovery through various gully control measures especially in mountain agricultural areas of developing countries like Nepal. Low-cost rehabilitation measures of gully dominant slopes require (1) erosion process control by small structures including bio-engineering measures (Geo-environmental Unit, HMG. Nepal 1997) with easy maintenance, (2) direct benefit to villagers, and (3) sustainable natural resource management. The adopted measures such as series of low-height gabion check dams and bamboo plantation were very effective with people's participation in planning and implementation of the project.

Conclusions

Based on the erosion rates and landscape monitoring of the small catchments devastated by gully erosion, rehabilitation measures adopted by a pilot project were surveyed from the viewpoint of erosion control and land resource recovery in Nepal. Small gabion check dams and plantation of bamboos at gully beds effectively trapped fallen materials from the steep head and side slopes of the gullies thus promoting reforestation in gully-dominant slopes. As slopes consisting of laterite soil hardly recovered only by plantation, drainage of surface run-off from such slopes are also recommended. Not only physical measures mentioned above, but

Fig. 9 A gully formed by trail erosion (Mar 1997)

Fig. 10 Stabilization of the gully head of G-4 RB5 by Ketuki, a low-height tree (right side) and bamboos on the gully bed (photo, Apr 2018)

participation of local people in the rehabilitation projects of gully-dominant slopes should be introduced thus creating awareness on the sustainable management of degraded lands.

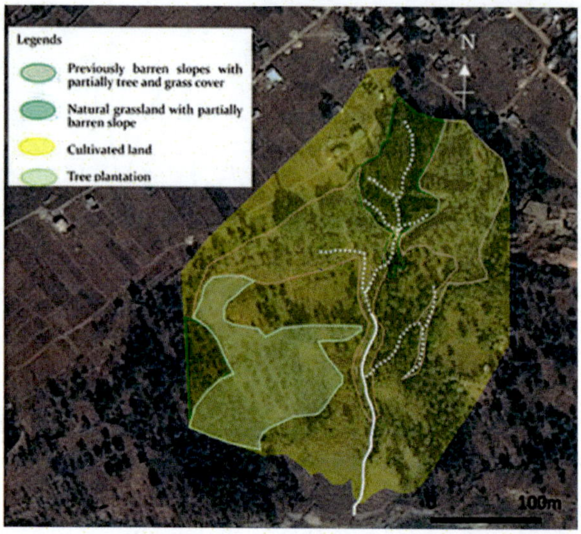

Fig. 11 Comparison of the land coverage at lower Pipaltar in 1999 and 2017 (Land coverage in 1999 was drawn based on Rizal et al. (1999), Base image: Google earth

Fig. 12 Reforestation of the gully-dominant hill slopes **a** July 1999, **b** Mar 2016

Fig. 13 Change in landuse on the previously barren and grass-covered gentle slope on the leftbank of the gully **a** July 1998, **b** Oct 2013

Acknowledgements This study was based on the monitoring of the pilot projects for which the author had worked by Japan International Cooperation Agency and DPTC, DMSP, Government of Nepal and author's research. Part of the study was funded by the Grant-in—aid from the Ministry of Education, Science, Sports and Culture (Kiban-C, No.25450200) to Hirosaki University. Mr. Shambu Prasad and Mr. Shanmukesh Chandra Amatya contributed in monitoring works. We are thankful to these organization and persons.

References

Department of Soil Conservation and Watershed Management, Ministry of Forests and Soil Conservation, HMG/N (1991) Soil conservation and watershed management activities. Kathmandu Sajha Press, 54 p

DPTC (1999) Final report on Udayapur model site. Water Induced Disaster Prevention Technical Centre, 41 p

Geo-environmental Unit, Ministry of Works and Transport, Department of Roads, HMG/N (1997) Guidelines, use of bi-engineering in the road sector, 20 p

Hayashi M, Fujita T (1976) Examples of PNC plate for soil slope protection of farmlands. J Jpn Soc Irrigat Drain Reclam Eng 44 (2):89–95 (in Japanese)

Higaki D, Karki KK, Gautam CS (1998) Soil erosion problems in the midland of Nepal with reference to the Trishuli model site, Nuwakot district, central Nepal. News Bull Nepal Geol Soc 15:33–38

Higaki D, Karki KK, Gautam CS (2005) Soil erosion control measures on degraded sloping lands: a case study in Midlands of Nepal. Aquat Ecosyst Health Manage 8(3):243–250

Karki KK, Higaki D (1998) Significance of participatory soil conservation in Nepal. Sabo-to-Chisui (Japan), 122:59–62 (in Japanese)

Lal R (1992) Restoring land degraded by gully erosion in the Tropics. In Lal R, Stewart BA (eds) Soil restoration—advances in soil science, vol 17. Springer, New York. (ISBN 978-1-4612-2820-2), pp 123–152

Partrap T, Watson HR (1994) Sloping agricultural land technology (SALT), A regenerative option for sustainable mountain farming. ICIMOD Occasional Paper 23, Kathmandu, Nepal. (ISBN 92-9115-323-0), 140 p

Rizal KP, Bhattarai D, Pantha RH (1999) Final report on Pipaltar Sabo model works. Report submitted to DPTC/JICA, Nepal, 40 p (unpublished)

Sthapit KM (1996) Soil erosion problems in Nepal. In: Paper for Second space informatics seminar for sustainable development—mountain resource management, Kathmandu, Nepal, 20 p

Site Suitability Analysis for Nature-Based Landslide Risk Mitigation

2020 Kyoto Japan

G. A. Chinthaka Ganepola, Udeni Priyantha Nawagamuwa, Anurudda Kumara Karunarathna, Senaka Basnayake, Lilanka Kankanamge, and Dhanushka Jayathilake

Abstract

Nature-based landslide risk mitigation involves use of vegetation in mitigating slope instabilities and has become popular in Asian countries such as Philippines, Malaysia, Hong Kong, Nepal, India and Thailand. They are considered as cost effective, environmental friendly and sustainable solutions for landslide risk mitigation. However, not all unstable sites, with landslide risks can be mitigated with nature-based techniques. The failure mechanism and the level of risk must be studied in detail before implementing such solutions. Hence, there is a need for a site selection process in order to select the most appropriate sites for applying nature-based techniques. In this study, five factors were selected in determining the level of appropriateness of vegetation in slope stabilization. Relative importance (weightage) of each individual factor towards applicability of nature-based techniques were determined using the Analytic Hierarchy Process (AHP). Site characteristics can be evaluated against the five factors and can be given marks on a scale of 4 which is multiplied by corresponding weightage of factors. The results of the multiplication are added to arrive at the final score for a particular site. This score can then be used to conclude the level of appropriateness of nature-based techniques at site-specific level. The proposed site selection process can be applied as a tool to rank candidate sites and decide on suitable landslide risk mitigation solutions.

Keywords

Nature-based • Vegetation • Landslides • Site suitability

G. A. C. Ganepola (✉)
Climate Resilience Department, Asian Disaster Preparedness Center (Sri Lanka Country Office), Bangkok, 10400, Thailand
e-mail: chinthaka.ganepola@adpc.net

S. Basnayake
Climate Resilience Department, Asian Disaster Preparedness Center, Bangkok, 10400, Thailand
e-mail: senaka_basnayake@adpc.net

U. P. Nawagamuwa
Department of Civil Engineering, University of Moratuwa, Moratuwa, 10400, Sri Lanka
e-mail: udeni@uom.lk

A. K. Karunarathna
Department of Agricultural Engineering, University of Peradeniya, Peradeniya, 20400, Sri Lanka
e-mail: akk@pdn.ac.lk

L. Kankanamge
Geotechnical Engineering Division, National Building Research Organization, Colombo - 05, Colombo, Sri Lanka
e-mail: lilanka.kankanamge@nbro.gov.lk

D. Jayathilake
Human Settlements Planning and Training Division, National Building Research Organization, Colombo - 05, Colombo, Sri Lanka
e-mail: danushkatcp@gmail.com

Introduction

The records of International Disaster Database (EM-DAT), indicated that landslides account for 4.9% of all-natural disaster events and 1.3% of all natural hazard fatalities between 1990 and 2015. 54% of these landslide events had occurred in Asia (Guha-Sapir as cited in Froude and Petley 2018). Further, the potential for loss of life and destruction of assets through landslides is increasing in many mountainous and hilly areas of Asia (Forbes and Broadhead 2013).

Different civil engineering measures can be implemented to mitigate the landslide risk. They include retaining structures, slope reinforcing measures, and surface & sub-surface drainage structures. However, it has been demonstrated in many countries in Asia that nature-based solutions can be effective in reducing the occurrence and impacts of landslides.

Nature-based Solutions (NbS) are interventions which make use of natural processes and ecosystem services to address hazards such as floods, erosion and landslides. NbS

Ž. Arbanas et al. (eds.), *Understanding and Reducing Landslide Disaster Risk*, ICL Contribution to Landslide Disaster Risk Reduction,
https://doi.org/10.1007/978-3-030-60713-5_34

can be completely "Green" (consisting of only ecosystem elements) or "Hybrid" (combination of ecosystem elements and conventional engineering measures) (Natural Hazards—Nature-based Solutions 2017).

Nature-based landslide risk management involves use of vegetation in mitigating slope instabilities and have become popular in Asian countries such as Philippines, Malaysia, Hong Kong and Nepal. They are seen as cost effective, environmental friendly and sustainable solution providers for landslide risk management practices.

Numerous research studies affirm the positive role of vegetation in preventing shallow landslides (Forbes and Broadhead 2013; Eab et al. 2015; Ni et al. 2017; Coppin and Richards 2007; Schwarz et al. 2010). Howel (2002) had mentioned six main functions of vegetation that contribute to increase the slope stability. They are:

1. Catch eroding material moving down the slope;
2. Armour the slope against surface erosion from both runoff and rain splash;
3. Reinforce the soil by providing a network of roots that increases the soils' resistance to shear;
4. Anchor the surface material by extending roots through potential failure planes into firmer strata below;
5. Support the soil mass by buttressing and arching;
6. Drain excess water from the slope by having different planting configurations that enhance drainage avoiding saturation and slumping of earth materials. Vegetation also help reduce pore-water pressure within the slope by extracting water from the roots and transpiring through the leaves.

Forbes and Broadhead (2013) indicated that the role of vegetation in relation to deep-seated landslides is considerably smaller although reducing moisture content of soils by tree roots can still help avoid excessive soil water pressures. Fatahi et al. (2010) reported that South-east Asian countries such as Phillipines, Malaysia and Hong Kong employ vegetation in ground treatment more than other countries in particular for stability of slopes in those regions.

Landslide risks can be mitigated using only civil engineering solutions or vegetation alone (Green Nature-based Solutions) or a combination of civil engineering solutions with vegetation (Hybrid solutions). More sustainable solutions can be introduced to sites with landslide hazard employing nature-based techniques. In this view, selection of optimum sites for nature-based landslide risk management remains a challenge.

This paper discusses a site selection process which can be employed to rank candidate sites on the suitability of application of vegetation in order to mitigate the landslide risk on a site-specific level.

Need for a Site Selection Process

Unstable sites with landslide symptoms must be studied in detail based on site-specific aspects. Factors such as failure mechanism, subsurface material characteristics and conditions for growth of vegetation needs to be assessed before implementing nature-based solutions. Campbell et al. (2008) stated that such solutions would not be very effective for sites where soil thickness is minimum, with deep seated failures or on steep slopes (over about 35–50°).

Hence, there is a need for developing a process for shortlisting sites in order to select most appropriate locations for implementing nature-based techniques for landslide risk mitigation.

Method of Study

A systematic literature review was conducted to identify the most influential factors determining the appropriateness of nature-based techniques for landslide risk mitigation. The limitations of such techniques too were annotated.

Having identified the important factors, the relative importance of each of them in determining site suitability for application of nature-based techniques were determined. All identified factors would not have an equal importance, when considering the application of nature-based techniques to mitigate landslide risk at sites.

The Analytic Hierarchy Process (AHP) was applied to determine the relative importance of each factor. According to Banai-Kashani (1989), the AHP had been effective in evaluating problems involving multiple and diverse criteria with limited data. The AHP exhibits flexibility in dealing with both the qualitative (intangible) and quantitative (tangible) factors in a multi-criteria evaluation problem. Furthermore, the AHP provides a methodological framework within which the inconsistency in judging the relative importance of factors in a site suitability analysis can be both detected and corrected.

The final site selection process proposed in this study enables calculation of a composite score for each site based on the suitability of application of nature-based techniques.

Factors Determining the Appropriateness of Nature-Based Techniques for Landslide Risk Management

Effective application of nature-based techniques needs careful studying of site-specific characteristics. Information on factors such as depth to failure plane, slope, environmental conditions and elements at risk should be assessed.

In order to determine the most suitable sites for application of nature-based techniques, five factors were selected in this study and same are summarized as follows.

i. Depth to failure plane

Nature-based techniques are more suitable to reduce shallow slope instabilities (Schwarz et al. 2010; Greenwood et al. 2004; Leung et al. 2015). Marks based on a scale of four were allocated to each category as shown in the Table 1. Campbell et al. (2008) stated that effectiveness of vegetation-based techniques is estimated as not to exceed 2 m which is the maximum rooting depth of most available plant species.

However, in situations where the depth to failure plane is greater than 2 m, vegetation can be used to control surface soil erosion rate and to prevent gullying effects. Further, vegetation has the ability to maintain shallow soil layers in an unsaturated condition and to increase the soil shear strength by establishing sufficient matric suction.

ii. Slope range (in degrees)

Choi and Cheung (2013) mentioned that in Hong Kong, vegetation was used as a slope surface cover in the upgrading of existing man-made slopes which are not steeper than 55 degrees. Further, it must be noted that with the increase of slope angle, the soil thickness tends to decrease which is an unfavourable factor for the growth of vegetation.

Further, Campbell et al. (2008) suggested that vegetation works can be carried out cost-effectively and relatively safely at slopes less than 35°.

Hence, considering the factors described above, marks are suggested for each slope range as shown in Table 2. Slope angles indicated in Table 2 refer to overall slope gradients within the sites.

iii. Suitability for creating a vegetation cover

The ability of plant species to grow in a certain site is governed by factors such as soil thickness, presence of boulders and the climatic conditions of the present ecosystem. Sufficient thickness of soil is essential for vegetation to be applied at site.

Campbell et al. (2008) stated that sites containing more than 50% rock outcrop are considered to be unsuitable for installation of vegetation measures, due to their increased installation cost and their reduced effectiveness. Existence of favourable rainfall conditions at site favour the application of nature-based techniques. Table 3 summarizes the different categories considered in allocating marks under this factor.

iv. Maintainability

In the implementation of landslide risk mitigation strategies using vegetation, it is vital to consider maintenance challenges. Vegetation must be properly maintained in order to make a positive contribution towards slope stability. Therefore, marks can be assigned under this factor (Table 4) considering the possibility of implementing a maintenance schedule after application of vegetation.

v. Probable loss considering the exposure elements at risk together with its socio-economic and environmental aspects

It is important to prioritize areas where the landslide risk is higher in carrying out risk mitigation activities. An idea on probable loss can be gauged by analysing exposure elements together with its socio-economic and environmental aspects. Marks can be assigned under this factor as given in Table 5.

Table 1 Marks allocated for depth to failure plane

Depth to failure plane	Appropriateness of nature-based techniques	Marks
(<0.5 m)	High	4.00
(0.5–1 m)	Medium	3.00
(1–2 m)	Low	2.00
(>2 m)	Very Low	1.00

Table 2 Marks allocated for slope range

Slope range in degrees	Appropriateness of nature-based techniques	Marks
(0–15)	High	4.00
(16–30)	Medium	3.00
(31–40)	Low	2.00
(>40)	Very low	1.00

Table 3 Marks allocated depending on site suitability for creating a vegetation cover

Suitability for creation of vegetation cover	Appropriateness of nature-based techniques	Marks
Soil thickness greater than 0.5 m, less than 25% of the site covered with rock outcrops and with annual rainfall >2,500 mm	High	4.00
Soil thickness greater than 0.5 m, less than 50% of the site covered with rock outcrops and with annual rainfall between 1,250 mm & 2,500 mm	Medium	3.00
Soil thickness less than 0.5 m, more than 50% of the site covered with rock outcrops and with annual rainfall between 1,250 mm & 2,500 mm	Low	2.00
Soil thickness less than 0.5 m, more than 50% of the site covered with rock outcrops and with annual rainfall <1,250 mm	Very low	1.00

Table 4 Marks allocated for categories under probable loss

Possibilities of Maintainability	Appropriateness of nature-based techniques	Marks
Maximum	High	4.00
Average	Medium	3.00
Medium	Low	2.00
Minimum	Very low	1.00

Table 5 Marks allocated for probable loss

Probable loss	Appropriateness of nature-based techniques	Marks
Maximum loss	High	4.00
Medium loss	Medium	3.00
Average loss	Low	2.00
Minimum loss	Very low	1.00

Determination of Relative Importance of Factors Using Analytic Hierarchy Process (AHP)

All the above five factors determine the level of appropriateness in applying nature-based techniques for landslide risk management. However, their relative importance in accessing the suitability for application of nature-based techniques would not be equal. The priority weight of each factor based on their importance was assessed quantitatively by using Analytic Hierarchy Process (AHP) developed by Saaty (1987).

Banai-Kashani (1989) mentioned that the AHP exhibited flexibility in dealing with both qualitative (intangible) and quantitative (tangible) factors in a multi-criteria evaluation problem. Further, it was reported that the AHP provided a methodological framework within which the inconsistency in judging the relative importance of factors in a site suitability analysis could be both detected and corrected.

Saaty (1987) mentioned that pairwise comparisons were fundamental in the use of the AHP. In assigning priorities to each factor by judging them in pairs considering their relative importance, numbers presented in Table 6 were used.

The consistency of the judgment was evaluated by calculating the consistency ratio as elaborated by Banai-Kashani (1989). It was mentioned that a consistency ratio value of less than 10% demonstrated a good consistency judgment.

The pair wise comparison matrix shown in Table 7 was developed after analysing the importance of the five factors in determining the level of appropriateness of nature-based techniques for landslide risk mitigation.

Priority weights of each of the five factors, derived by applying the AHP are given in Table 8. The judgment presented a consistency ratio of 7% which was within the acceptable limits.

Table 6 The AHP scale for paired comparisons (Adapted from Saaty as cited in Banai-Kashani 1989)

Intensity of importance	Definition and explanation
1.00	Equal importance–two activities contribute equally to the objective
3.00	Moderate importance–experience and judgment slightly favour one activity over another
5.00	Essential or strong importance–experience and judgment strongly favour one activity over another
7.00	Demonstrated importance–an activity is strongly favoured and its dominance is demonstrated in practice
9.00	Extreme importance–the evidence favouring one activity over one another is of the highest possible order of affirmation
2.00, 4.00, 6.00, 8.00	Intermediate values between two adjacent judgments–when compromise is needed
Reciprocal of numbers	If an activity has one of the above numbers assigned to it when compared with the second activity has the reciprocal value when compared to the first

Table 7 Matrix comparing the relative importance of each factor

	Depth to failure plane	Slope range	Suitability for creation of vegetation cover	Maintainability	Probable loss
Depth to failure plane	1.00	3.00	3.00	2.00	1.50
Slope range	0.33	1.00	2.00	2.00	2.00
Suitability for creation of vegetation cover	0.33	0.50	1.00	2.00	1.50
Maintainability	0.50	0.50	0.50	1.00	1.50
Probable loss	0.67	0.50	0.67	0.67	1.00

Table 8 Relative weightage assigned to factors

Name of factor	Weight	Weight (%)
Depth to failure plane	0.36	36.00
Slope range	0.22	22.00
Suitability for creation of vegetation cover	0.16	16.00
Maintainability	0.13	13.00
Probable loss	0.13	13.00

Calculation of Final Score for Each Site

A final score needs to be calculated for each site in order to rank them according to the level of appropriateness of nature-based techniques.

Marks allocated under each factor was multiplied by the corresponding priority weightages shown in Table 8. The results of the multiplication are added to arrive at the final score which is expressed based on a scale of 4 (Eq. 1).

$$\text{Final Score} = \sum_{i=1}^{n} [W_i * S_i] \qquad (1)$$

where

n—no. of factors.

W—Weightage assigned to each factor.

S—Marks assigned under each factor.

The final score is then categorized into three categories of landslide risk mitigation measures based on the level of

Table 9 Different categories of landslide risk mitigation measures

Final score	Landslide risk mitigation measures
4.00 ≤ S_{final} <3.00	Landslide risk can be mitigated using nature-based techniques alone or together with small scale engineering structures only if required
3.00 ≤ S_{final} <2.00	Landslide risk can be mitigated using hybrid solutions (combination of nature-based techniques and conventional engineering measures)
S_{final} ≤2.00	Landslide risk can be mitigated using large scale engineering structures. Nature-based techniques can be used to control surface erosion and to improve aesthetic appearance

appropriateness of nature-based techniques. The different categories are shown in Table 9.

Discussion

Nature-based techniques in landslide risk mitigation are cost effective, environmentally friendly and contributing to formulate sustainable solutions. Identifying optimal sites for application of nature-based techniques is very vital since, all unstable sites with landslides threats cannot be mitigated using only vegetation. In this study, a site selection process was developed to determine the level of appropriateness of nature-based techniques established on five criterion factors. The Analytic Hierarchy Process (AHP) was used to quantify the priority weightages of each factor.

The assigned weightages are in a common relative scale which ranges from 0 to 100%. According to Table 8, "Depth to failure plane" is very important in deciding the appropriateness of nature-based techniques. It is followed by "Slope range" and "Suitability for creation of vegetation cover". "Maintainability" and "Probable loss" have been assigned equal weightages. These two factors have a relatively less importance when compared with first three factors listed in Table 8 in determining the level of appropriateness of nature-based techniques.

Conclusions

This study identified the most influential factors in determining site suitability for nature-based landslide risk mitigation which is cost effective, environmentally friendly and provide sustainable solutions. The Analytic Hierarchy Process (AHP) was used to determine the relative importance of each individual factor. The proposed site selection process enables calculation of a composite score for the interested site and can be used as a tool to rank candidate sites in order to decide on suitable landslide risk mitigation solutions.

Acknowledgements The authors wish to acknowledge thankfully the invaluable inputs and guidance by: Eng. N. M. S. I. Arambepola (Team Leader of NBLRM Project, Asian Disaster Preparedness Center), and Eng. (Dr.) Asiri Karunawardena (Director General, National Building Research Organization). Further, they wish to convey sincere thanks to the World Bank for providing financial support for the study.

References

Banai-Kashani R (1989) A new method for site suitability analysis: the analytic hierarchy process. Environ Manage 13(6):685–693

Campbell SDG, Shaw R, Sewell RJ, Wong JCF (2008) Guidelines for soil bioengineering applications on natural terrain landslide scars. Geo Report No. 227. Geotechnical Engineering Office. Hong Kong

Choi KY, Cheung RWM (2013) Landslide disaster prevention and mitigation through works in Hong Kong. J Rock Mech Geotech Eng 5:354–365. https://doi.org/10.1016/j.jrmge.2013.07.007

Coppin NJ, Richards IG (2007) Use of vegetation in civil engineering. CIRIA London. (ISBN_0–86017–711–4). 52 p

Eab KH, Likitlersuang S, Takahashi A (2015) Laboratory and modelling investigation of root-reinforced system for slope stabilization. Soils Found 55(5):1270–1281. https://doi.org/10.1016/j.sandf.2015.09.025

Fatahi B, Khabbaz H, Indraratna B (2010) Bioengineering ground improvement considering root water uptake model. Ecol Eng 36:222–229. https://doi.org/10.1016/j.ecoleng.2008.12.027

Forbes K, Broadhead J (2013) Forests and landslides. The role of trees and forests in the prevention of landslides and rehabilitation of landslide-affected areas in Asia. FAO Bangkok. (ISBN_978–92–5–106994–3). 49 p

Froude MJ, Petley DN (2018) Global fatal landslide occurrence from 2004 to 2016. Nat Hazards Earth Syst Sci 18:2161–2181. https://doi.org/10.5194/nhess-18-2161-2018

Greenwood JR, Wint NJEJ (2004) Assessing the contribution of vegetation to slope stability. Geotech Eng 157(4):199–207. https://doi.org/10.1680/geng.2004.157.4.199

Howell J (2002) Roadside bio-engineering, reference manual, his majesty's government of Nepal, Department of Roads, Kathmandu. (ISBN_1 86192 170 5). 218 p

Leung FTY, Yan WM, Hau BCH, Tham LG (2015) Root systems of native shrubs and trees in Hong Kong and their effects on enhancing slope stability. CATENA 125:102–110. https://doi.org/10.1016/j.catena.2014.10.018

Natural Hazards—Nature-based Solutions (2017) https://naturebasedsolutions.org/. Last accessed 01 Mar 2020

Ni JJ, Leung AK, Ng CWW, So PS (2017) Investigation of plant growth and transpiration-induced matric suction under mixed grass-tree conditions. Can Geotech J 54:561–573. https://doi.org/10.1139/cgj-2016-0226

Saaty R (1987) The analytic hierarchy process—hat it is and how it is used. Math Model 9(3–5):161–176

Schwarz M, Preti F, Giadrossich F, Lehmann P Or D (2010) Quantifying the role of vegetation in slope stability. A case study in Tuscany (Italy). Ecol Eng 36(3):285–291. https://doi.org/10.1016/j.ecoleng.2009.06.014

Study on the Application of Nature Based Landslide Mitigation in Sri Lanka

2020 Kyoto Japan

Udeni P. Nawagamuwa and Markandu Dishan

Abstract

Slope failures could be considered as one of the major disasters in Sri Lanka, causing damages to properties and lives of people. Many precautions and studies have been done regarding the prevention and risk assessment of landslides. However, role of vegetation on stabilizing the slope is yet to be properly modelled. This study identifies the issues related to role of vegetation in slope stability analysis. The main objective of this study is to assess the influence of roots in mechanically stabilizing the soil. Root cohesion and root morphology of ten selected native plants are derived and analysed for different slopes with different soil properties complying with Sri Lankan conditions. SLOPE/W software is used to analyse the Factor of Safety for each case in accordance with Spencer method. Resulted FOS are then tabulated to identify the angle of stable slope. Derived root cohesion values varied with different types of plants with different root characteristics. The value of root cohesion ranged from 4.99 for *Lantana Camara* to 36.7 for *Eucaliptus* sp. resulting different safe slope angle ranging from 26 to 29° for soil conditions with low cohesion and friction angle and 42–57° for soils with high cohesion and friction angle. It was also identified that type of root morphological approximation and the location of a tree on a slope plays important role in slope stability. Moreover, usage of plants in hybrid solution of combining traditional retaining structures with vegetation can be considered as a starting point for stabilizing slope, since vegetation may take longer time to get matured.

Keywords

Root cohesion • Safe and critical slope angle • Root morphology • Slope stability

Introduction

Landslide is one of the major catastrophic events that causes loss of human life and damages to the properties. Concerning the risk and impacts of landslides, a deep understanding on main factors influencing slope failure is essential to identify the prevention measures. Geotechnical Engineering studies on slope stability emphasises the importance of study and investigation on landslide through observation and analysis. Moreover, traditional solutions such as soil nailing and retaining walls are being used in Sri Lanka to stabilize any slope instability. Even though they provide support against slope failure and control over erosion, their negative effects on existed nature and higher cost are considered as some drawbacks (Nawagamuwa et al. 2014). When considering environmental and geotechnical problems, interaction between the presence of vegetation and slope, especially the influence of roots in soil are considered. Different roles of vegetation on slope stability adopted from Menashe and Box (1993) are presented in Fig. 1.

This study presents a measure on nature-based landslide mitigation using Bio-engineering aspects for Sri Lankan conditions regarding the risk of happening a landslide. Vegetation can influence in slope stability in different ways. Considering the tensile strength of the roots, a resisting force

U. P. Nawagamuwa (✉) · M. Dishan
Department of Civil Engineering, University of Moratuwa, Katubedda, Moratuwa, 10400, Sri Lanka
e-mail: udeni@uom.lk

M. Dishan
e-mail: dishanandrew@yahoo.com

Ž. Arbanas et al. (eds.), *Understanding and Reducing Landslide Disaster Risk*, ICL Contribution to Landslide Disaster Risk Reduction,
https://doi.org/10.1007/978-3-030-60713-5_35

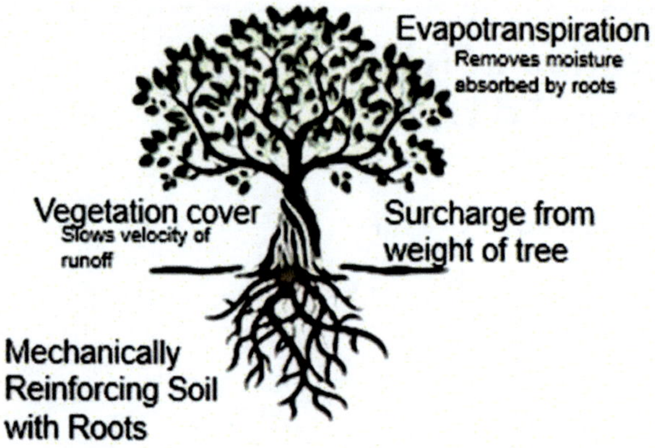

Fig 1 Different roles of Vegetation on Slope Stability using influence of roots and moisture removal from soil (Modified from Menashe and Box 1993)

Class of root system	Real morphology	Geometrical approximation
Heart root system		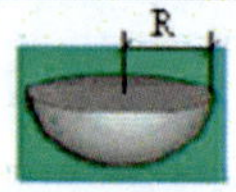
Tap root system		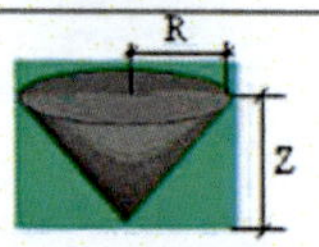
Plate root system		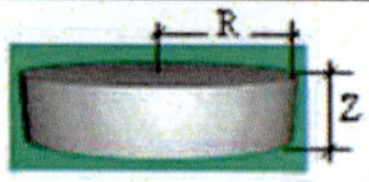

Fig. 2 Indication of different Root morphology and their relevant geometrical approximations (Shang 2016). 'R' indicated Radius and 'Z' indicates depth of considered are

against shear failure can develop due to the anchorage properties of roots into soil which is commonly known as mechanically reinforcing property of roots (Greenwood et al. 2004). Likewise, the moisture absorption by plant roots can have some effect on lowering the water content in soil. This helps create a matric suction which enforces the stability of soil from failing (Hoogland et al. 1981). In this study the main area of concern will be on tensile strength of local plants roots, relationship between root cohesion and slope stability and modelling the real application of local plants on stabilizing the soil types in Sri Lanka for local conditions.

Roots of Vegetation

Roots are an important part of trees which support the whole body by anchoring deep into soil and provide proper water and nutrient supply to process photosynthesis. These properties of roots are significant on slope stability and erosion control (Genet et al. 2010). Therefore, it is necessary to study about roots to understand the support provided for slope stability by vegetation. The classification of roots in accordance with diameter is important while analysing the root strength and depth verifications. The root diameters are characterised into main six groups, very fine (<0.5 mm), fine (0.5–2 mm), small (2–5 mm), medium (5–10 mm), large (10–20 mm) and very large (>20 mm) (Glinski and Lipiec 2018). Root length density is another important factor need to be considered for identification of tree root influence zone on a slope. Usually it is maximum at 20-cm depth and decreases with depth, however, also it should be considered that few trees in nature have high root density below 20 cm depth as well (Akinnifesi et al. 2004).

Root morphology is also a significant factor needed for analysis of the effects of tree roots on slope stability by mechanical reinforcement, considering the mass and different cohesion it exerts on slope (Shang 2016). Main types of root morphology and respective geometrical approximation are represented in Fig. 2.

According to Gray and Harukazu (1983), greatest reinforcement by fibres in soil occurs when the fibres are positioned at 60° with respect to deformed zone. Thus the Eq. 1 is modified by Schmidt et al. (2001) to relate root cohesion with tensile strength of root tread and the ratio between cross sectional area of examined ground with cumulative cross sectional area of roots in that zone. This relationship helps to determine or estimate the total root cohesion by root reinforcement on a given area.

$$Cr = 1.2\sum_{i=1}^{n} Tri\{\frac{Ari}{As}\} \quad (1)$$

C_r = Root cohesion by root reinforcement by a given vegetation species.

T_{ri} = Tensile strength of a root fibre.

A_{ri} = Cross sectional area of the selected root fibre.

A_s = Cross sectional area of examined ground area.

n = number of roots found in the examined ground area.

According to this relationship, to obtain higher root cohesion, the value of either root thread tensile strength or diameter of root threads or root density should be relatively high enough.

Moreover, a study was done to create a Plant Manuel with the title of Nature Based Landslide Risk Management Project in Sri Lanka by National Building Research Organization with the Asian Disaster Preparedness Center (2019). This manual presents the relevant details accounting for plant characteristics, Vegetation-Climate propagation method and Engineering significance. In this study roots of eleven plant species were selected and tested for tensile strength. These root samples were collected close to Badulusirigama in Badulla which is a landslide prone area. Many countries have performed this type of experiments for their local conditions, implying the necessity to carry out this study in Sri Lankan context to create a cost effective and efficient way to avoid slope failures.

Root Cohesion of Native Vegetation

Root cohesion of selected ten native plants were derived using Eq. 1. Necessary details for computation of root cohesion were gathered and derived or estimated from details presented on various studies. These details are summarised in Table 1.

These relevant data of each vegetation and the tensile strength of root thread of those vegetation presented in Plant Manuel by National Building Research Organization with the Asian Disaster Preparedness Centre (2019) are used in deriving the root cohesion.

The calculated values for root cohesion are used in the analysis for different soil types relevant to Sri Lankan soil characteristics.

The root morphologies of each plants were selected in accordance with their relevant root types and considering their possible overlapping. The derived values for root cohesion of those trees with respect to their selected root morphological approximation are presented in Table 2.

Numerical Simulation of Mechanical Reinforcement of Roots

Method of slices, Spencer method is used in the analysis as the considered hypothetical slopes consist of different soil properties (Cho 2007). This analysis is based on Mohr–Coulomb failure criteria. However, it must be noted that the water table is not considered in this study as this study mainly focused on mechanical strengthening by tree roots.

Other major assumption in this study is that existing slope soil is homogeneous along the profile of slope and the top organic layer is ignored. Also, it was assumed that the soil density and friction angle do not change for the soil around tree roots as mentioned by Falkiner et al. (2006) and the root distribution does not change with different soil types. This was taken into consideration therefore that the mentioned tree root details do not vary with site condition.

Table 1 Root distribution details

Name	Depth of rooted (m)	No. of Root (m^{-2})	Fine root % (%)	Radius of rooted area (m)	References
Artocarpus hetarophyllus	1	70	25.80	2.25	Jamaludheen et al. (1996), Prakash et al. (2009)
Azadirachta indica	2.33	51	68.80	1.76	Das et al. (2008), Toky and Bisht (1992)
Bauhinia racemosa	4.78	51	45	1.1	Das et al. (2008), Verma et al. (2014)
Calliandra calothyrsus	0.55	63.8	50	0.9	Shang (2016)
Clidermia hirta	0.4	60	70	0.5	Miyanishi and Kellman (1988), Shang (2016)
Coffea arabica	1.64	57	65	1.5	Carducci et al. (2015), Shang (2016)
Eucaliptus sp.	2	52	26	2	Laclau et al. (2001), Falkiner et al (2006)
Grevillea banksii	1	82	50	6.95	Smith et al. (1999), Shang (2016)
Lantana camara	0.41	60	45	0.96	Fensham et al. (1994), Shang (2016)
Osbeckia sp.	0.3	60	65	0.5	Ramakrishnan and Ram (1988), Shang (2016)

Table 2 Root cohesion and Root morphology of selected vegetations

Name of the plant	Root cohesion	Root morphology
Artocarpus hetarophyllus	14.6	Heart
Azadirachta indica	28.49	Heart
Bauhinia racemosa	21.33	Tap
Calliandra calothyrsus	7.86	Cylindrical
Clidermia hirta	4.9	Cylindrical
Coffea arabica	19.59	Heart
Eucaliptus sp.	36.7	Heart
Grevillea banksii	6.72	Cylindrical
Lantana camara	4.99	Cylindrical
Osbeckia sp.	7.22	Cylindrical

Case Study 1, Stability of a Slope for Different Root Morphology

The analysis of stability of slope for different root morphology is carried out using slope/w software with Spencer method. The first part of the analysis was done considering typical material properties of Sri Lankan Laterite (γ = 18 kNm^{-3}) and c = 5 kPa and ϕ = 20°. The main three types of root morphologies are analysed with this hypothetical slope of 45° as slope angle and with three consecutive trees to check the effect of different vegetation on a slope. Each morphological shape represents the root influence soil area on the slope. The soil properties of Sri Lankan soil and material properties root influenced soil sockets with their morphological shapes are entered in software analysis as shown in Fig. 3.

For material properties of root influenced area, values calculated for Eucalyptus were applied. The factor of safety resulted from the numerical analysis is then tabulated and compared in Table 3.

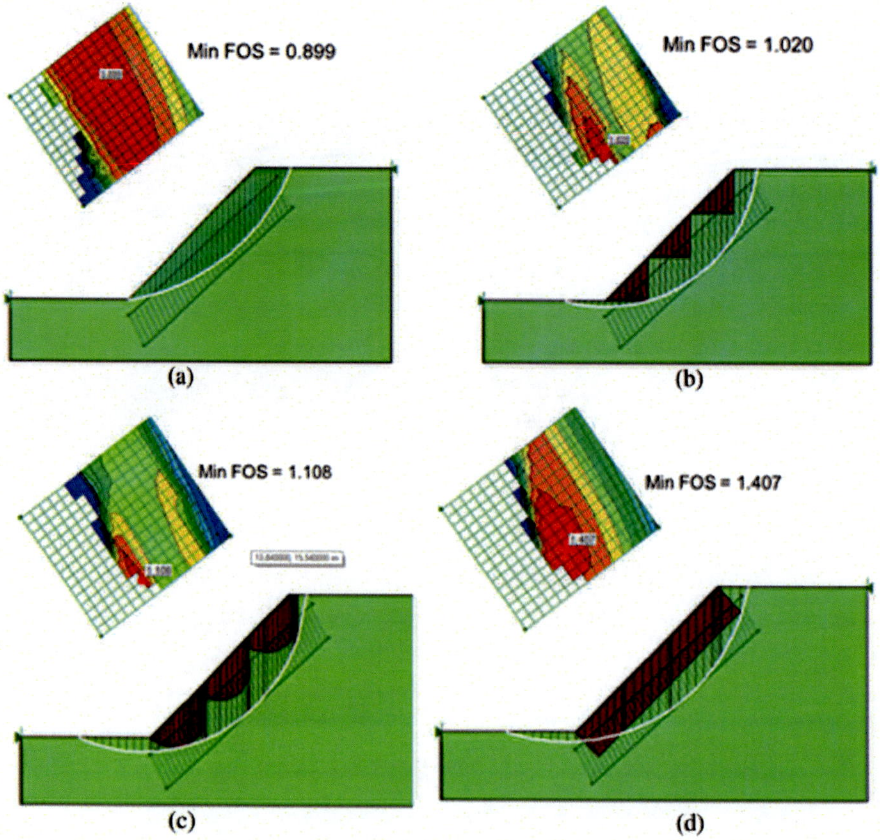

Fig. 3 Slope stability for a fully vegetated slope for with 1:1 gradient Considering different root morphology. **a** non-vegetated slope; slope vegetated by 3 consecutive; **b** cone root; **c** semi-spherical; **d** cylindrical, root morphological roots

Table 3 Representation of FOS for each geometric approximation shapes

Geometrical approximation	Min. FOS	Percentage increment in FOS
None	0.899	0
Tap	1.020	13.4%
Heart	1.108	23.5%
Cylinder	1.407	56.51%

The results show that the cylinder-shaped root morphology will stabilize the slopes comparatively higher than the rest.

Case Study 2, Stability for Locating Trees at Different Location in the Slope

Another hypothetical analysis is carried out to check the stability of a slope after placing the trees at different locations in the slopes. The same assumption and the parameters mentioned above were used in the analysis. The factor of safety for each scenario were analysed through the numerical analysis and compared in Fig. 4.

Results from different plants have shown that a significant increase in factor of safety when plant is near the toe compared to the rest. The reason is due to difference in compression level from top to bottom surcharged by the weight of the tree (Rees and Ali 2006).

Case Study 3, Stability for Different Types of Native Vegetation

The analysis of stability provided by ten selected native vegetation were carried out for the cases where vegetation cover provided by each plant on slope of different slope angles such as 30, 45, 60 and 70° and constant vertical height of 10 m. The same assumptions mentioned in the above case study were used here. The root morphology of each plants presented in Table 2 and root influenced area from Table 1 are taken into consideration for the parameters entered in the numerical analysis. The relative soil properties of Sri Lankan soil and material properties of root influenced soil sockets with their morphological shapes are applied in the analysis. The safest angle and critical angle are taken as where the slope angle in which the factor of safety is equal to 1.2 and 1.0, respectively. The safe slope angle and critical slope angles for cases where friction angle is constant at 20° and cohesion of soil varies as 5, 10 and 15 kPa as well as for

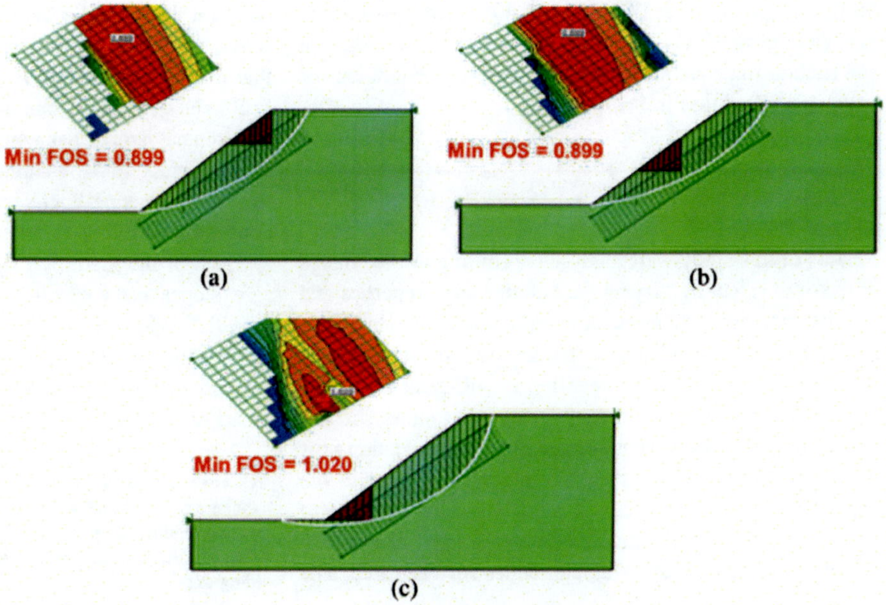

Fig. 4 Slope stability for a fully vegetated slope for 1:1 gradient considering different location of vegetation. Different locations of the trees are **a** near the top of slope; **b** middle of the slope; **c** near the toe of the slope

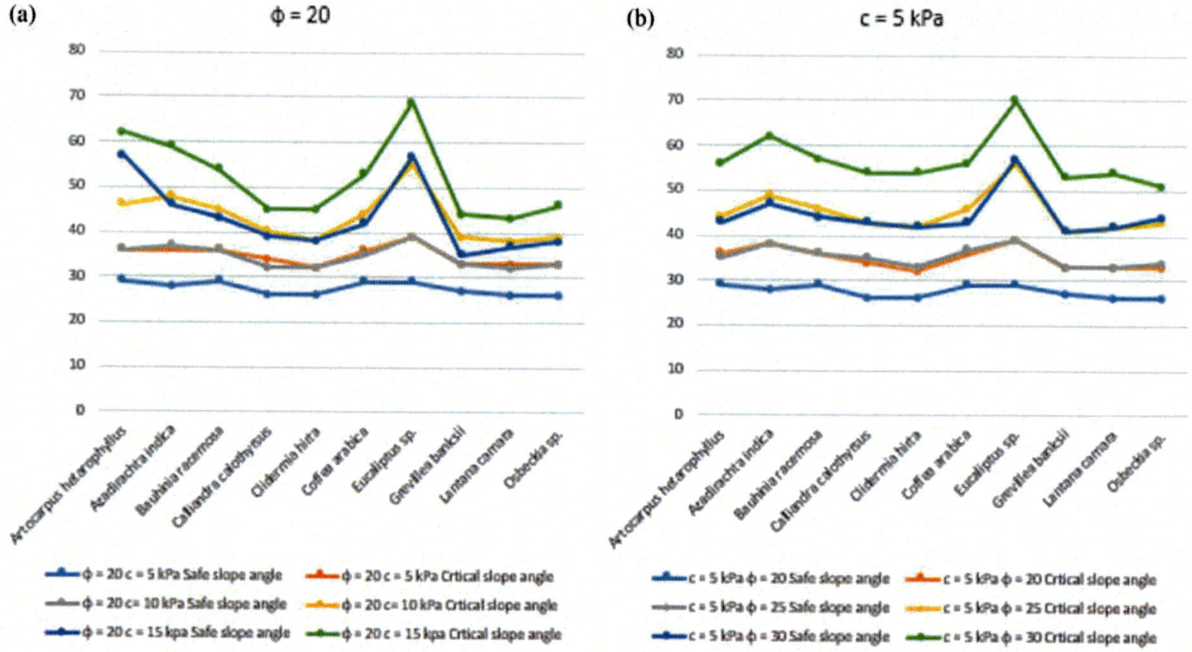

Fig. 5 **a** and **b** Safe and critical Slope angle for different soil cohesion and friction Angle of slopes vegetated with (Artocarpus hetarophyllus, Azadirachta indica, Bauhinia racemose, Calliandra calothyrsus, Clidermia hirta, Coffea arabica, Eucaliptus sp., Grevillea banksia, Lantana camara, Osbeckia sp.) for different soil condition. Soil conditions are, **a** Safe slope angle:- (light blue) c = 5 kPa and ϕ = 20; (grey) c = 10 kPa and ϕ = 20; (dark blue) c = 15 kPa and ϕ = 20; Critical slope angle:- (red) c = 5 kPa and ϕ = 20; (yellow) c = 10 kPa and ϕ = 20; (green) c = 15 kPa and ϕ = 20, **b** Safe slope angle:- (light blue) c = 5 kPa and ϕ = 20; (grey) c = 5 kPa and ϕ = 25; (dark blue) c = 5 kPa and ϕ = 30;, Critical slope angle:- (red) c = 5 kPa and ϕ = 20; (yellow) c = 5 kPa and ϕ = 25; (green) c = 5 kPa and ϕ = 30

the case where the cohesion is constant at 5 kPa and the friction angle varies as 20°, 25° and 30° are represented in Fig. 5a, b respectively. The derived values for soil cohesion and friction angle are representing different soil conditions of vegetated hill slopes in Sri Lanka.

Conclusion

Results of this study were analysed according to the change in factor of safety by varying the different soil properties and for different root cohesion. Ten different native trees are selected to perform the analysis as they can withstand safe in local conditions. Analysis regarding the different root morphological shapes and different location on slope showed that cylindrical shaped root morphology and vegetation near toe of the slope are significant in slope stability with the vegetation effect. The analysis regarding reinforcement of soil by different trees which are native to Sri Lanka on different soils showed different safe angles for each tree. The range of safe angle lies between 26–57° for c = 5 kPa and different ϕ values lie between 26–69° for ϕ = 20° and different c values. While comparing the results, it was observed that tree type vegetation such as Eucalyptus sp. have higher FoS values for safe slope angles and shrub type plants such as Lantana camara has comparatively lower values for safe slope angles. Table 4 summarises the suggested vegetation for different slope angles considering critical cases on soil conditions available in most of the Sri Lankan slopes.

Each of the identified slope angles can help identify the safe slopes on vegetated hill slopes and can be used to provide slope stability for manmade slopes. Moreover, the above finding can later be used in mixed construction method in which traditional retaining structures are combined with vegetation related aspects.

Acknowledgements Authors would like to acknowledge the staff members of Soil Mechanics Laboratory of University of Moratuwa for their continuous help. National Building Research Organization and Asian Disaster Preparedness Centre are kindly acknowledged for providing data.

Table 4 Safe and Critical slope angles for different vegetation for different soils types

Name of the plant	ϕ = 20						c = 5 kPa					
	c = 5 kPa		c = 10 kPa		c = 15 kpa		ϕ = 20		ϕ = 25		ϕ = 30	
	Safe slope angle	Critical slope angle	Safe slope angle	Critical slope angle	Safe slope angle	Critical slope angle	Safe slope angle	Critical slope angle	Safe slope angle	Critical slope angle	Safe slope angle	Critical slope angle
Artocarpus hetarophyllus	29	36	36	46	57	62	29	36	35	44	43	56
Azadirachta indica	28	36	37	48	46	59	28	38	38	49	47	62
Bauhinia racemcsa	29	36	36	45	43	54	29	36	36	46	44	57
Calliandra calothyrsus	26	34	32	40	39	45	26	34	35	43	43	54
Clidermia hirta	26	32	32	38	38	45	26	32	33	42	42	54
Coffea arabica	29	36	35	44	42	53	29	36	37	46	43	56
Eucaliptus sp.	29	39	39	55	57	69	29	39	39	56	57	70
Grevillea banksii	27	33	33	39	35	44	27	33	33	41	41	53
Lantana camara	26	33	32	38	37	43	26	33	33	42	42	54
Osbeckia sp.	26	33	33	39	38	46	26	33	34	43	44	51

References

Akinnifesi FK, Rowe EC, Livesley SJ, Kwesiga FR, Vanlauwe B, Alegre JC (2004) Tree root architecture. In: van Noordwijk M, Cadisch G, Ong CK (eds) Below-ground interactions in tropical agroecosystems: concepts and models with multiple plant components. CABI, pp 61–81. https://doi.org/10.1079/9780851996738.0061

Carducci C, Oliveira G, Curi N, Heck R, Rossoni D, de Carvalho T, Costa (2015) Gypsum effects on the spatial distribution of coffee roots and the pores system in oxidic Brazilian Latosol. Soil Tillage Res 145:171–180. https://doi.org/10.1016/j.still.2014.09.015

Cho S (2007) Effects of spatial variability of soil properties on slope stability

Das D, Chaturvedi OP (2008) Root biomass and distribution of five agroforestry tree species. Agrofor Syst 74:223–230. https://doi.org/10.1007/s10457-008-9159-9

Falkiner RA, Nambiar EKS, Polglase PJ, Theiveyanathan S, Stewart LG (2006) Root distribution of Eucalyptus grandis and Corymbia maculata in degraded saline soils of south-eastern Australia. Agrofor Syst 67(3):279–291. https://doi.org/10.1007/s10457-005-5258-z

Fensham RJ, Fairfax RJ, Cannell RJ (1994) The invasion of *Lantana camara L.* in forty mile scrub National Park, north Queensland. Aust J Ecol 19:297–305

Genet M, Stokes A, Fourcaud T, Norris JE (2010) The influence of plant diversity on slope stability in a moist evergreen deciduous forest. Ecol Eng 36(3):265–275. https://doi.org/10.1016/j.ecoleng.2009.05.018

Glinski J, Lipiec J (2018) Soil physical conditions and plant roots. https://doi.org/10.1201/9781351076708

Gray DH, Harukazu O (1983) Mechanics of fiber reinforcement in sand. J Geotech Eng 109(3):335–353. https://doi.org/10.1061/(ASCE)0733-9410(1983)109:3(335)

Greenwood J, Norris J, Wint J (2004) Assessing the contribution of vegetation to slope stability. Proceedings of the Institution of civil engineers-geotechnical engineering - Proc Inst Civil Eng-Geotech E. 157:199–207. https://doi.org/10.1680/geng.2004.157.4.199

Hoogland JC, Feddes RA, Belmans C (1981) Root water uptake model depending on soil water pressure head and maximum extraction rate. Acta Hortic 119:123–131

Jamaludheen V, Kumar BM, Wahid PA (1996) Root distribution pattern of the wild jack tree (*Artocarpus hirsutus Lamk.*) as studied by 32P soil injection method. Agrofor Syst 35:329–336

Laclau JP, Bouillet JP, Ranger J (2001) Spatial localization of roots in a clonal plantation of Eucalyptus in Congo. Influence on the ability of the stand to take up water and nutrients. Tree Physiol 21:129–136. https://doi.org/10.1093/treephys/21.2-3.129

Menashe E, Box PO (1993) Vegetation management: a guide for puget sound bluff property owners, 54

Miyanishi K, Kellman M (1988) Ecological and simulation studies of the responses of Miconia albicans and Clidemia sericea populations to prescribed burning. For Ecol Manage 23:121–137

Nature Based Landslide Risk Management Project in Sri Lanka, No 139699, pp 1–5. The World Bank (2019). https://documents.worldbank.org/curated/en/733271564052371375/Nature-Based-Landslide-Risk-Management-Project-in-Sri-Lanka

Nawagamuwa UP, Sarangan S, Janagan B, Neerajapriya S (2014) Study on the effect of plant roots for stability of slopes. In: Proceedings of the 3rd world landslide forum, Bejing, China, 2–6 June 2014

Prakash O, Kumar R, Mishra A, Gupta R (2009) Artocarpus heterophyllus (Jackfruit): an overview. Pharmacognosy Rev 3:353–358

Ramakrishnan PS, Ram SC (1988) Vegetation, biomass and productivity of seral grasslands of Cherrapunji in north-east India. Vegetation 74:47–53

Rees SW, Ali N (2006) Seasonal water uptake near trees: A numerical and experimental study. Geomech Geoeng Int J 1(2):129–138

Schmidt KM, Roering JJ, Stock JD, Dietrich WE, Montgomery DR, Schaub T (2001) The variability of root cohesion as an influence on shallow landslide susceptibility in the Oregon Coast Range. Can Geotech J 38(5):995–1024. https://doi.org/10.1139/t01-031

Shang KY (2016) The simulation of water uptake by vegetation and its impact on slope stability using an image–based model of plant root architecture

Smith DM, Jackson NA, Roberts JM, Ong C (1999) Root distributions in a Grevillea robusta-maize agroforestry system in semi-arid Kenya. Plant Soil 211:191–205. https://doi.org/10.1023/A:1004635414462

Toky OP, Bisht RP (1992) Observations on the rooting patterns of some agroforestry trees in an arid region of north-western India. Agrofor Syst 18:245–263

Verma P, Yadav AN, Kazy S, Saxena A, Suman A (2014) Evaluating the diversity and phylogeny of plant growth promoting bacteria associated with wheat (Triticum aestivum) growing in central zone of India. Int J Curr Microbiol Appl Sci 3:432–447

Slope Stability Around the Northern Part of the Tegucigalpa Basin, Honduras: A Case of Landslide Process at Residential Development Areas

Kiyoharu Hirota, Cincy Rosa, and Koichi Hasegawa

Abstract

This paper describes the landslide, terrain, and geology which focused to residential area around the basin occupying Tegucigalpa, the capital of Honduras. The basin has a large residential area in the northern part, and landslides sometimes affect the collapse of houses. Landslides in the studied area (*see* quadrangle in this paper's figures) were investigated to identify the external and internal factors of the three landslides, El Bambú, El Reparto, and Guillén, and examined the landslide process. In the studied area, the landslides are essentially derived from Slide-type landslide, followed by Flow-type landslide as debris flows, and finally undergo a termination process. Landslides in El Bambú and El Reparto have been stabled due to countermeasures. The Guillén landslide is always active due to the supply of water from residential areas. Internal factors such as topography and geology are generally considered to be landslide characteristics, whereas in the northern part of Tegucigalpa, wastewater from residential areas was found to be causing landslides on the slope located houses as an external factor.

Keywords

Landslide • Catchment • Geology • Residential area • Honduras • JICA

Introduction

This paper pesents slope stability around the northern part of Tegucigalpa, Honduras, specially treats landslides of residential area. Selecting geologic and geomorphic characters from the studied area to find the trigger factors of landslide, we approached three landslides of El Bambú, El Reparto, and Guillén. Three landslides ultimately move as Flow-type in Vernes' landslide classification. Even though the hillside moved as Slide-type landslide initially. This phenomenon is often seen in residential areas on the northern slope of Tegucigalpa and is an external factor that tends to destabilize the slope. This project considered the relation to geology and geomorphology and their impact upon landslides in Tegucigalpa, Honduras a part of a project for Japan International Cooperation Agency (hereafter referred to as JICA) named "Proyecto para el Control y Mitigación de Desastres en Laderas del Distrito Central (in *English*: Project for Control and Mitigation of Slope Disasters in the Central District in Republic of Honduras)".

Geological Background

Geology of Honduras

The geology of Honduras (Fig. 1A) is composed of a Quaternary deposit (Alluvium), metamorphic rocks, sedimentary rocks, and volcanic rocks respectively, from the north to the south.

K. Hirota (✉)
International Consulting Department, Kokusai Kogyo Co., Ltd., 2 Rokubancho, Chiyoda-ku, Tokyo, 102-0085, Japan
e-mail: sbhirota@gmail.com

C. Rosa
Alcaldía Municipal del Distrito Central, UMGIR, Tegucigalpa, 11101, Honduras
e-mail: cincy.rosa@amdc.hn

K. Hasegawa
OYO International Corporation, OYO Corporation Group, Tokyo, 112-0002, Japan
e-mail: hasegawa@oyointer.com

Ž. Arbanas et al. (eds.), *Understanding and Reducing Landslide Disaster Risk*,
ICL Contribution to Landslide Disaster Risk Reduction,
https://doi.org/10.1007/978-3-030-60713-5_36

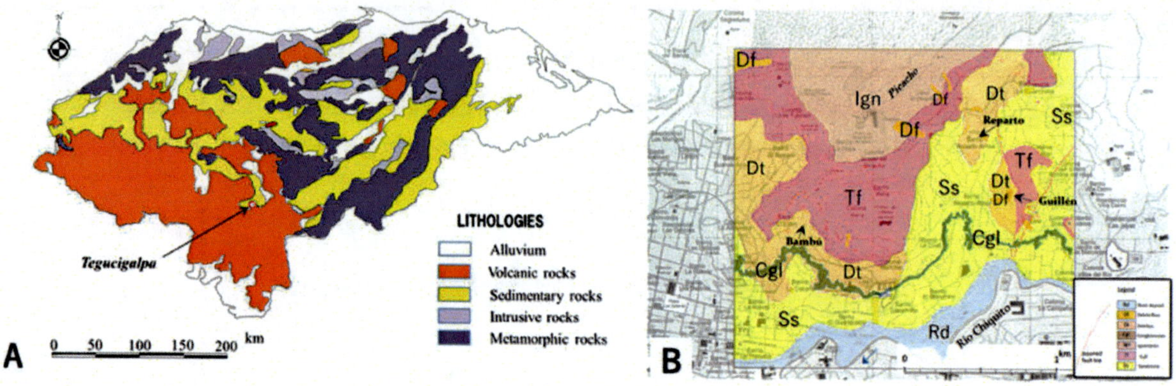

Fig. 1 Geologic maps of Honduras (**A** Hirota 2016, simplified from Rogers et al. 2007; **B** Studied area in the Northern part of Tegucigalpa, Honduras)

Geology of Investigated Area in Tegucigalpa

Focusing the geology in Tegucigalpa, rhyolitic tuff and pumice tuff lie over the sandstone, and ignimbrite that forms the ridge called El Picacho, along the northern margins of the basin (Fig. 1B). The sandstone in the Valle de Ángeles Group (Rodgers and O'Conner 1990), is characterized by reddish fine-grained sandstone in which strata facies are partially be changed siltstone. Tuff is characterized by various strata facies, pumice tuff, rhyolitic tuff, which includes andesitic rock and dacitic rock. Rocks are partially weathered at the boundary between sandstone and ignimbrite. Ignimbrite usually has vertical joints. Ignimbrite is widely broken on the surface of the rock and tilted by joints and faults on the steep slope. It is easy for water from rainfall to penetrate through the vertical joints.

Landslide in the Studied Area

Landslide Study

The landslide study in Tegucigalpa has started in earnest after Hurricane Mitch 1998, which affected tremendous disaster of landslide and inundation (JICA 2002; Harp et al. 2002a, b). Landslide studies related the JICA project include, making landslide distribution map, hazard map, landslide education, and planning for protection against future landslides (Grupo de Mapeo de Deslizamientos de Tierra de Tegucigalpa 2014; Hirota and Kamiya 2014; Hirota 2015, 2016; Mancada and Yamagishi 2016; Bernhard et al. 2018).

To find out the landslide distribution in Tegucigalpa, JICA conducted a survey based on which landslide countermeasures would be optimal after the hurricane Mitch. JICA (2002) selected seventeen landslides for focusing the places where countermeasures were needed against further landslide. Then three landslides, El Berrínche, El Bambú, and El Reparto, were selected. In those day, there was no landslide at Guillén, Nueva Santa Rosa.

Re-evaluation of seventeen previous landslides was conducted in order to create a Landslide Inventory Map of the Tegucigalpa region (Hirota and Kamiya, 2014). Landslide types are divided mainly into two types, Flow-type and Slide-type. The exception is Fall-type in Tegucigalpa according to Varnes' landslide classification (Varnes 1978). Next, Grupo de Mapeo de Deslizamientos de Tierra de Tegucigalpa (2014) compiled a detailed landslide distribution map (Fig. 2, Left) as a result of the JICA project 2011–2013 to find out the landslide distribution throughout the Tegucigalpa Basin. According to this distribution map, we comprehended there are many landslide areas to the north part of Tegucigalpa, which are concentrated in narrow places geographically along the channel of the river Río Cholteca, and in areas where steep slopes are formed at the southern side of the hill named Picacho.

As part of JICA project 2013–2015, a hazard map was created based on a more detailed landslide distribution map, and analysis of it. Based on the hazard map created during this project, we investigated landslides in the area we are currently studying (Fig. 2, white quadrangle). El Bambú, El Reparto, and Guillén are examples of active landslides, including both unstable and stable statea. The team considered why landslides occured in the northern slopes of Tegucigalpa.

Basement Geology in Micro-Catchment

The Río Choluteca is the main river through Tegucigalpa and adjacent areas. The main tributaries of the Río Choluteca are the Río Chiquito and the Río San Jose on the right bank

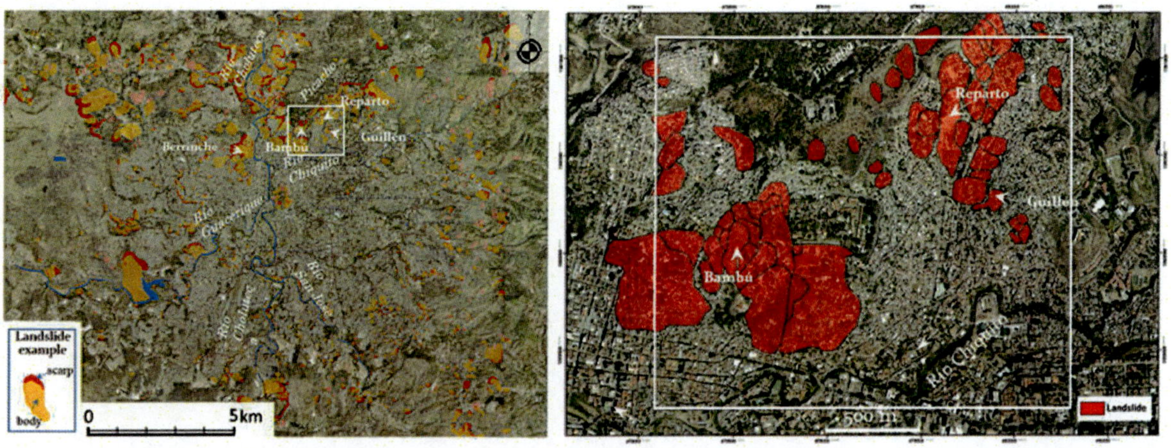

Fig. 2 Left: Landslides distribution around the Tegucigalpa basin (modified from Grupo de Mapeo de Deslizamientos de Tierra de Tegucigalpa 2014), Right: Landslide distribution of studied area. Quadrangle: studied area

and the Río Guacerique on the left bank. The tributaries that flow through the main part of Tegucigalpa are the Río Chiquito and the Río San Jose (Fig. 2, Left). The basement geology of the catchment basin formed by the Río Chiquito (Fig. 2, Right: a portion of the catchment) is mainly sandstone of the Valle de Ángeles Group (Mesozoic), and the basement geology of the catchment formed by the Río San Jose is tuff of the Padre Miguel Group (Cenozoic) mainly (*from* Rodgers and O′Conner 1990). Tuff of the Padro Miguel Group, which lays a micro-catchment area in the Río San Jose, is hard, and the sandstone distributed in the micro-catchment area scoured by the Río Chiquito is more weathered and softer and more susceptible to erosion. Clay occurs near the geologic boundaries between sandstone and tuff.

Landslides in Micro-Catchments of the Río Chiquito

Landslides distributed in the studied area are concentrated on the right bank of the Río Chiquito, and landslide distribution is observed upstream of each small catchment area (Fig. 3).

The main river, the Río Choluteca flows north from Tegucigalpa, and when it makes a large detour to the east, then it flows to the south and falls into the Pacific Ocean. The concentration of landslides in the upper tributaries of the river system are on the right bank of the Río Chiquito. Upper tributaries correspond to the zone of soil erosion in the catchment area.

How to Find Vulnerable Landslide

Landslide distribution in the studied area was confirmed, and topographical analysis was performed by combining the classification results of "Relief Energy" and "Slope Inclination" using GIS.

[Relief Energy: RE]

The relief energy map shows the difference in elevation of the terrain. Here, the difference between the highest and lowest altitudes is measured on a 10 m square grid. There are four categories: $0\ \text{m} \leq \text{RE} < 5\ \text{m}$, $5\ \text{m} \leq \text{RE} < 10\ \text{m}$ (green), $10\ \text{m} \leq \text{RE} < 20\ \text{m}$ (yellow), and $20\ \text{m} \leq \text{RE} < 40\ \text{m}$ (red) (Fig. 4). As one of the classification numbers, the classification of slope failure of 5 m or more, used in Japan, has been adopted.

[Slope Inclination: SI]

Classification of the slope inclination map is divided into 4 categories: $0° \leq \text{SI} < 30°$, $30° \leq \text{SI} < 45°$ (yellow), $45° \leq \text{SI} < 60°$ (red), $60° \leq \text{SI} < 90°$ (dark pink) express the distribution (Fig. 5). As one of the classification degrees, the classification of slope failure of 30° or more is the Japanese standard been used. As one of classification degrees for the slope direction, the classification of slope failure of 30° or more was also used.

Fig. 3 Landslides distribution and small catchments of the Río Chiquito at the northern part of Tegucigalpa by GIS (quadrangle: studied area)

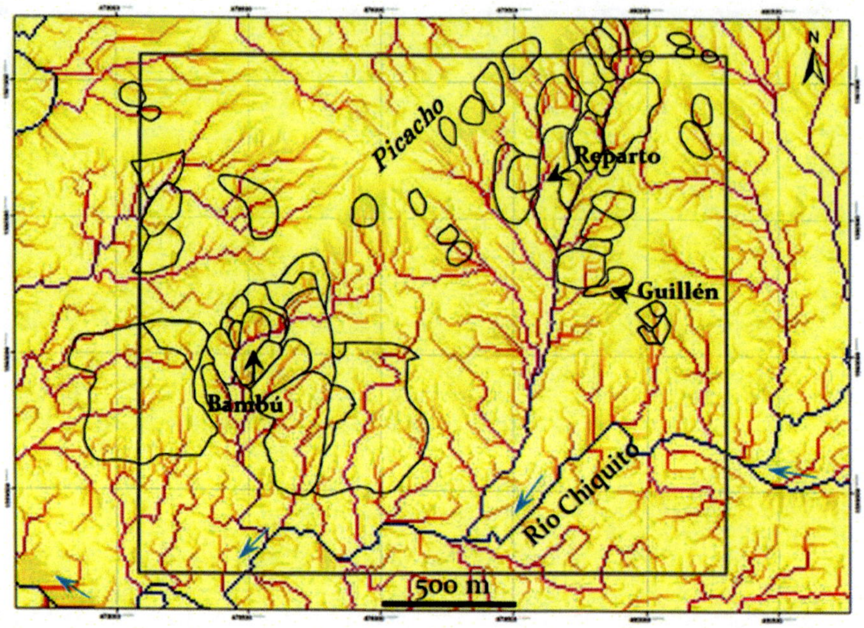

Fig. 4 Relief energy map at the northern part of Tegucigalpa (quadrangle: studied area)

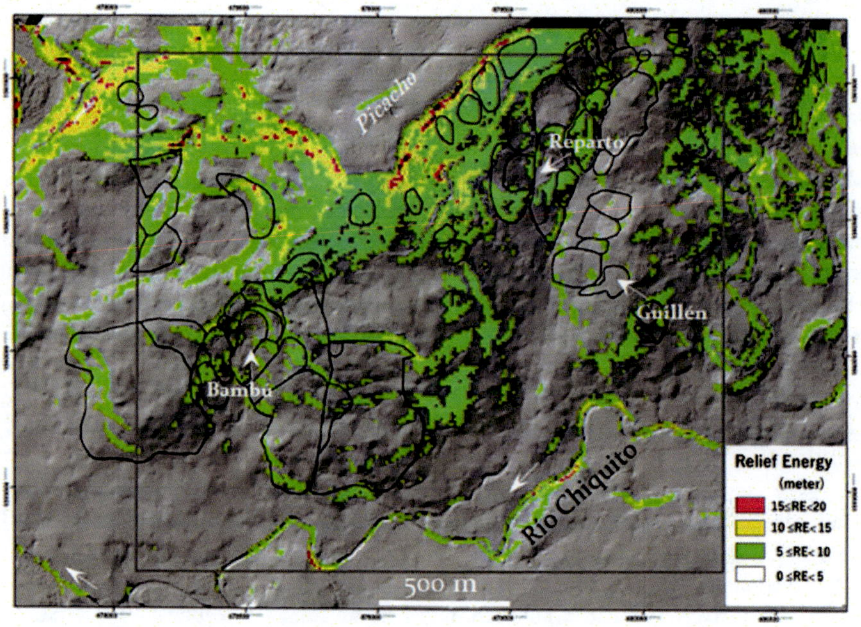

Results and Discussion

Landslides and Geologic/Geomorphic Attributes

Depending on the classification of categories in the relief energy map, the distribution area can be largely divided into those where landslide bodies are in the range of $0\text{ m} \leq RE < 5\text{ m}$ and those in the range of $5\text{ m} \leq RE < 10\text{ m}$. In addition, some landslide bodies are in the area of, but some near the scarp of head and the end are $5\text{ m} \leq RE < 10\text{ m}$. There are no landslides on steep slopes of the division $15\text{ m} \leq RE$ area, but there is a few landslides including $15\text{ m} \leq RE$ area within the landslide body. In such an area, a scarp is formed near the slope divisional boundary.

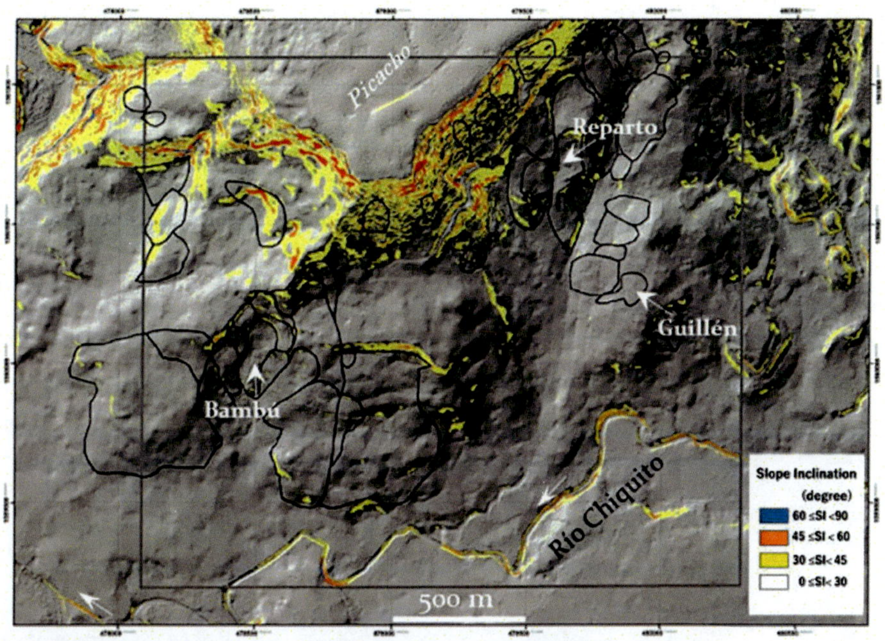

Fig. 5 Slope inclination map at the northern part of Tegucigalpa (quadrangle: studied area)

In the slope division, the area of 0° ≤ SI < 30° is found on the cliff extending to the north-northeast-southwest on the southside of El Picacho. The area of 30° ≤ SI < 45° is partially above the southern slope of the hill El Picacho and on both sides of the bank slope where the Río Chiquito is dissected. There are head of a landslide and/or foot of a landslide in the area of 30° ≤ SI < 45°. The 45° ≤ SI < 60° area is concentrated on the southern slope of the hill El Picacho. The area of 60° ≤ SI < 90° is partially observed in the dissected valley of the Río Chiquito. Based on the analysis of geologic and geomorphic features, Fig. 6 shows the results of applying the geologic and geomorphic elements to the three landslide areas, El Bambú, El Reparto, and Guillén, which correspond to each category.

The geologic attributes common to the three landslides of El Bambú, EL Reparto, and Guillén are "sandstone as basement rock", "the landslide body is in the Detritus distribution area", and "geological discontinuity exists". However, the geologic boundary of the three landslides, Bambú, Reparto, and Guillén, has the boundary between sandstone and tuff near the slip surface of the landslide body. In the landslide of Guillén, the top tuff has moved due to a fault, forming a scarp landslide. The existence of the fault is added as an attribute. Stage 1 of Guillén is a Slide-type landslide with tuff remaining on the surface. In the next stage 2, the lower sandstone boundary is used for the sliding surface. In stage 3, it slides as a debris flow due to being next to stage 2, the lower sandstone boundary is used for the sliding surface, and in stage 3, it slides as a debris flow due to surface soil.

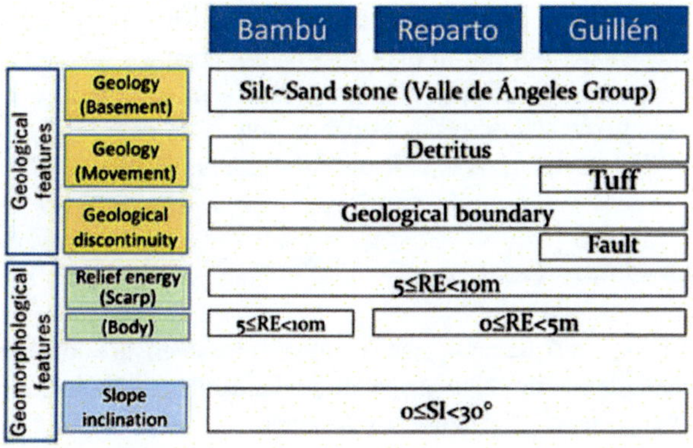

Fig. 6 Attribute among three landslides: Bambú, Reparto and Guillén

As a geomorphic attribute, the three landslides have in common that the slope inclination division is 0° ≤ SI < 30°, and the relief energy division is 5 m ≤ RE < 10 m at the landslide slope. In Bambú, an attribute of the entire slope including scarps and bodies of landslide, the relief energy division is in the range of 5 m ≤ RE < 10 m. By being in the range of 5 m ≤ RE < 10 m, a gentle slope is formed.

Landslide Process

We discuss how the three landslides slide (Figs. 7, 8 and 9).

(1) **Bambú** (Fig. 7)

The Bambú landslide was caused as debris flow and affected to the El Edén region enormously within detritus (Dt) deposited area in the aftermath of Hurricane Mitch (Hurricane

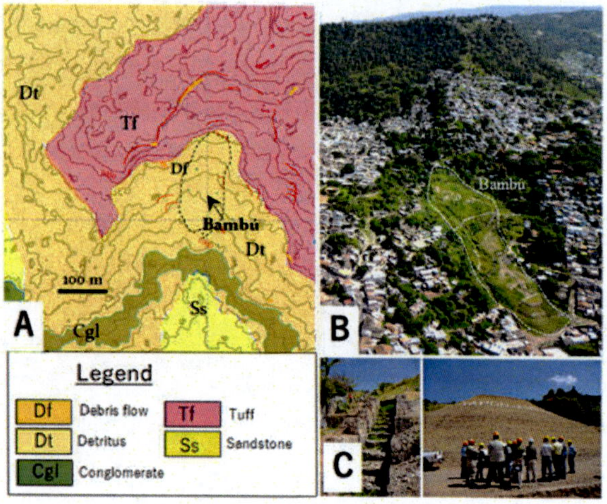

Fig. 7 El Bambú. A: Geologic map, B: Landslide view, C: Drainage, Embankment

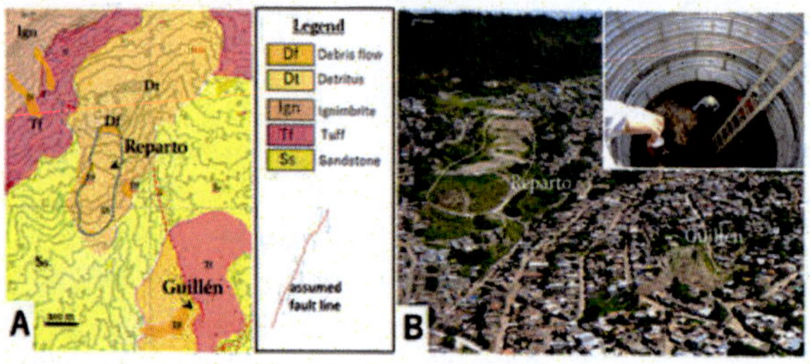

Fig. 8 El Reparto. **A** Geologic map, **B** Landslide view of Reparto and Guillén, photo at right upper: drainage well

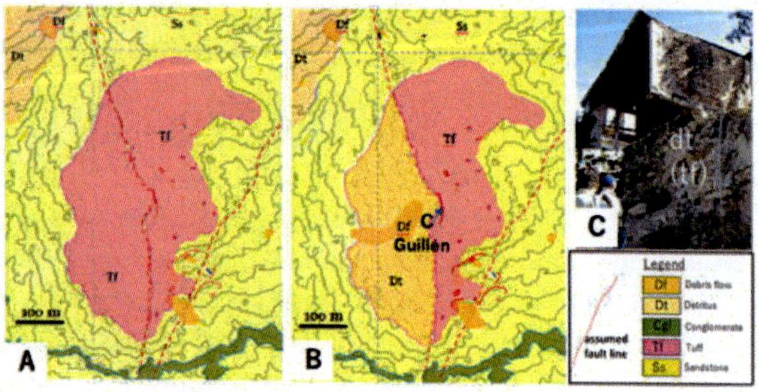

Fig. 9 Guillén (A: before movement of tuff, B: after movement of tuff), C: Scarp of landslide

mitch affected El Edén not directly, according to interviews at site).

There is a geologic boundary between Tuff (Tf) and Sandstone (Ss) near the scarp of landslide. We consider the following process for El Bambú landslide.

Stage 1: Water flows inside of terrain along potential cracks (faults and/or joints in tuff) at high-angle geological discontinuity vertically, destabilizing and depositing sediment on the surface.

Stage 2: Debris flow is generated due to precipitation in addition to daily residential wastewater.

(2) **Reparto** (Fig. 8)

The Reparto landslide occupies in the detritus (Dt) distribution area and is a re-movement of some previously deposited detritus or debris flow deposits.

The process of the Reparto landslide is as follows:

Stage 1: Tuff (Tf) and ignimbrite (Ign) forming the hill El Picacho accumulate along the slope due to slope failure. Here, rockfalls, debris and rock avalanches, and debris flows are possible.
Stage 2: The geological interface between tuff (Tf) and sandstone (Ss) promotes debris flow (Df), and detritus (Dt) is relatively widespread.
Stage 3: The thick detritus (Dt) partially causes debris flow (Df). Here, mountain stream scouring is also a trigger for landslides.

(3) **Guillén** (Fig. 9)

The landslide at Guillén, Nueva Santa Rosa was reported in a newspaper a Tribuna in Honduras on October 12, 2011. Thirteen years have passed since Hurricane Mitch struck Honduras.

The process of the Guillén landslide is as follows:

Stage 1: Residential wastewater constantly flows in the rock where there is a high-angle crack vertically due to the fault, and the landslide body rotates and slips.
Stage 2: After that, the unstable block slides with geologic boundaries between tuff (Tf) and sandstone (Ss) due to differences in physical properties. The landslide body slides until stratum of tuff in the upper part is gone.
Stage 3: After sliding of stratum, sediment accumulated in the surface soil descends as debris flow (Df).

We summarized three landslides process (Table 1).

Landslide Factors

Taking into account the regional characteristics of the studied area in Tegucigalpa Basin, landslide factors include "steep slope", "geological discontinuity", "surface soil" as

Table 1 Landslide process with combination of geology

Combination of geology Strata: Upper/ Lower	Expected Landslide classification	Applicable Landslide
Detritus/ Tuff	Slide Flow	
Detritus/ Detritus	Slide Flow	Flow → Bambú Stage 2, Reparto Guillén Stage 3
Detritus/ Sandstone	Slide Flow	Slide → Guillén Stage 2
Ignimbrite/ Tuff	Rock fall Rock avalanche	
Tuff/ Tuff	Rock fall Slide	
Tuff/ Sandstone	Slide	Slide → Bambú Stage 1, Guillén Stage 1

Table 2 Possible factors for landslide in the studied area

Landslide factors	
Internal factor	External factor
Steep slope	Precipitation
Geological discontinuity	Residential wastewater
Surface soil	Land use change

internal factors and "precipitation", "residential wastewater", and "land use change" as external factors. Among these factors, the intrinsic steep slope element is included, the geological discontinuity, which is a geologic element, is regarded as the formation of a cliff, and surface soil is regarded as the material for final movement (Table 2).

The essential factor is that the siltstone to sandstone of the Valle de Ángeles Group is the catchment of the Río Chiquito, and the loaded balance on the slope is lost until tuff overlying sandstone has collapsed or slid down.

On the other hand, the Guillén landslide was not collapsed by the rainfall (max a day: 129.3 mm/day, 27–31 October, Tegucigalpa) when Hurricane Mitch struck. Although the precipitation is considered an external factor as a trigger, but the landslide did not occur. Earthquakes and earthquake damages were not recorded during nor after Hurricane Mitch.

Residential wastewater is constantly used during the day for meals, showers, toilets, etc., so that water is always flowing at locations where slopes are formed. For this reason, in the northern residential areas, the daily wastewater is supplied to geological discontinuities by residents until the strata of tuff disappears and debris flow occurs. Then, the landslide continues to move to a stable state.

Conclusion

(1) Landslides in residential area of Tegucigalpa is affected more by geologic factors than by geomorphic factors. In summary, it is caused by the following characteristics.

1) There is sandstone as the basement rock of the catchment area of Rio Chiquito.
2) Geology in the studied area is composed of tuff and sandstone in the catchment area of Río Chiquito.
3) Residential wastewater always flows on the slope's surface, and penetrates into geological discontinuities (faults, joints, etc.).

(2) The slope stability in the northern part of Tegucigalpa supplies the equilibrium in the process of landslides combined of tuff and sandstone, and debris flow occurrence. Phenomena of penetration to underground from residential wastewater, and also from precipitation affects landslide. Due to the findings of this project it is recommended that investigation of residential wastewater and its influence on landslides be conducted in order for the residential area to maintain slope stability.

Acknowledgements This investigation was conducted through the JICA project: "Proyecto para el Control y Mitigación de Desastres en Laderas del Distrito Central". We are grateful to members of JICA project, the JICA Headquarters, the JICA Honduras Office, our counterparts: UNAH (Universidad National Autónoma de Honduras), AMDC (Alcaldía Municipal del Distrito Central), and COPECO (Comisión Permanente de Contingencias).

References

Bernhard LT, Sevilla N, Hirota K, Yamagishi H, Tsukamoto S (2018) How to make a database of landslides in Tegucigalpa, Honduras. TXT-tool 4.504–1.1, Landslide Dynamica: ISDR-ICL landslide interactive teaching tools, pp 743–759

Grupo de Mapeo de Deslizamientos de Tierra de Tegucigalpa (2014) Manual para elaboración de mapa de inventarios de Deslizamientos de tierra. Caso de aplicación: ciudad de Tegucigalpa (ISBN 978-99926-701-1), p 80

Harp EL, Castañeda M, Held MD (2002) Landslides triggered by Haricane Mitch in Tegucigalpa, Honduras USGS Open-File Report 02-33, pp 1–11

Harp EL, Held MD, Castañeda M, McKenna JRW (2002) Landslide Hazard Map of Tegucigalpa Honduras. USGS Open-File Report 02-219, p 9

Hirota K, Kamiya S (2014) Re-evaluation of landslide occurred by Hurricane Mitch 1998. Tegucigalpa Honduras. In: Sassa K (ed) Landslide science for a safer geo-environment, vol 3, pp 393–400

Hirota K (2015) Landslides and geology education in Honduras, as activities of Japan International Cooperation Agency senior volunteer. 52(4):36–42

Hirota K (2016) Preliminary Report: Relationship between 9geology and landform in landslides of the Tegucigalpa Basin, Honduras. Second Central American and Caribbean Congress, pp 18–20

JICA (2002) The study on flood control and landslide prevention in Tegucigalpa metropolitan area of the republic of Honduras (official document). Final Report in English version

Moncada R, Yamagishi H (2016) Educational methodologies implemented in Latin America for landslide inventory and analysis. Second Central American and Caribbean Congress, pp 18–20

Rodgers RD, O'Conner EA (1990) Geologic map of the Tegucigalpa quadrangle, Honduras (scale 1:50,000), Instituto Geográfico Nacional

Rogers RD, Mann P, Emmet PA (2007) Tectonic terranes of the Chortis black based on integration of regional aeromagnetic and geologic data. Geological Society of America Special Papers, vol 428, pp 65–88

Varnes DJ (1978) Slope movement types and processes. In: Shuster RL, Krizek RJ (eds) Special Report 176: landslides: analysis and control. Transportation and Road Research Board, National Academy of Science, Washington DC, pp 11–33

Classification of Cryogenic Landslides and Related Phenomena (by Example of the Territory of Russia)

Oleg V. Zerkal and Alexander L. Strom

Abstract

Large part of Russian territory (more than 60%) is located within permafrost zone. Cryogenic landslides and related phenomena in the permafrost zone are quite specific and include some types of slope processes that are absent in "warm" regions (e.g. block fields, very thin earth flows with diurnal cycle of activity, etc.). However, common types of landslides in the permafrost zone also have some specific peculiarities. The classification of cryogenic landslides and related phenomena is proposed. Compiling this classification, we considered both seven types of the frozen (permafrost) soil and six well-known landslide types. Role of the permafrost surface in the formation of the sliding surface was considered too, as well as the thermodynamic instability of the permafrost when temperature changes from positive to negative and vise-versa, which is a specific factor of slopes instability typical of the permafrost zone. Peculiarities of various types of the cryogenic landslides evolution and related phenomena at the territory of Russian Federation are described.

Keywords

Permafrost • Cryogenic landslides • Classification

Introduction

Large part of Russian territory is covered by the permafrost almost completely and is characterized by arctic and sub-Arctic landscapes (arctic deserts, polar tundra, forest-tundra and mountain tundra). The natural climatic sectoral distribution and altitude zonality play an important role in localization of the southern boundary of the continuous permafrost. Within the sector with the Siberian atmospheric circulation the southern boundary of this engineering-geological zone in the highlands might pass south from 52°–54° N, while within the sector influenced by the Atlantic type of circulation (areas located west from the Ural mountains and in northern part of Western Siberia) this boundary is shifted northward significantly—up to 65°–66° N (Zerkal and Strom 2017).

Classification of Cryogenic Landslides and Related Phenomena

Types of Permafrost

The specific feature of the engineering-geological permafrost zone is wide distribution of frozen rocks and soils of different types that have negative temperature for three years or more. The following types can be selected (Euro-Asian Council 2011):

- Ice soil that contains more than 90% of ice;
- frozen soil and rock–deposits that have negative or zero temperature and contain visible icy inclusions and/or icy cement and characterized by the cryogenic structural bonds;
- frosty soil—soil with negative temperature that, however, does not contain ice and melt water;
- plastic frozen soil consolidated with ice and characterized by viscous-ductile properties;
- cooled soil—salt-containing soil, which negative temperature is still higher than its freezing temperature;
- frosty debris—coarse-grained and sandy soil that has negative temperature but is not cemented by ice;
- quasi-solid frozen soil strongly cemented by ice and subjected to the relatively brittle failure.

O. V. Zerkal (✉)
Geological Department, Lomonosov Moscow State University, Moscow, 119991, Russia
e-mail: igzov@mail.ru

A. L. Strom
Geodynamics Research Center LLC, 3rd Novomikhailovsky passage 9, Moscow, 125008, Russia
e-mail: strom.alexandr@yandex.ru

Ž. Arbanas et al. (eds.), *Understanding and Reducing Landslide Disaster Risk*,
ICL Contribution to Landslide Disaster Risk Reduction,
https://doi.org/10.1007/978-3-030-60713-5_37

Approaches to Classification of the Cryogenic Landslides and Related Phenomena

Slopes composed of various types of the permafrost soils listed above can be subjected to both well-known landslide types and to the slope processes that are absent in "warm" regions composed of thawed soils. Significant influence on the activity of slope processes in the permafrost areas is also provided by the thermodynamic instability of the permafrost when temperature changes from positive to negative and vise-versa.

No universal classification of the cryogenic slope processes that would take all aspects of their variability and the complexity of motion mechanisms has been elaborated up to now. One of the general classifications that considers influence of ice on slopes stability was proposed in Sharpe (1938). He selected such types of slope processes as "rock glacier cap", "solifluction" and "debris avalanche". Most well-known local permafrost landslides classifications elaborated for the specific types of the phenomena in frozen soil were proposed by T.N. Kaplina, L.S. Garagulia, V.F. Bolighovsky and V.V. Kiuntsel, and recently by M.O. Leibman and Kiziakov and by E.A. Kozyreva with her co-authors (Kaplina 1965; Garagulia 1983; Bolikhovsky and Kiuntcel 1990; Leibman and Kiziakov 2007; Kozyreva et al. 2011; Dokuchaev and Dokuchaeva 2015). Landslides and related phenomena in frozen rock were not considered by these classifications. Classification of slope processes in rock permafrost were elaborated for rock streams and rock seas (Tiurin 1979) and for rock glaciers (Romanovsky et al. 1989; Galanin 2009). D. Schneider with co-authors developed the classification of how ice can be involved in large rapid mass movements (Schneider et al. 2011). D. Barsch and W. Haeberli had previously addressed the problem of the investigation of rock glaciers (Barsch 1988; Haeberli 1985).

The proposed classification of the specific cryogenic landslides and related phenomena in the permafrost zone is shown in Table 1. It is based on the well-known types of landslides described in (Cruden and Varnes 1996; Hungr et al. 2014).

Types of the Cryogenic Landslides and of the Related Phenomena and Their Systematization

Generally, two major zones can be selected within the overall permafrost area—zone of the continuous permafrost and zone where both frozen and thawed soils are developed. Here we consider the first one, where two sub-zoned can be identified—the presently glaciated sub-zone and the sub-zone of seasonal thawing (Trofimov 2002; Zerkal and Strom 2017).

The presently glaciated subzone embraces areas of the modern polar ice shields, as well as high-altitude areas of modern glaciation whose lower boundary depends on the latitude.

In the polar islands modern glaciation occupies up to 50% of the Novaya Zemlia Northern Island, up to 2% of its Southern Island and up to 85% of the Frantz Josef Land. It is well developed in the high mountain systems of the Cherskyi and Stanovoi Ranges in Eastern Siberia with sub-arctic climatic conditions as well as in high mountain systems located in the middle latitude zone (Central Caucasus, Altai Mountains, etc.). The nival zone of the presently glaciated areas is characterized by the slow movement of glaciers and rock glaciers in the valleys and by snow and ice avalanches displacing clastic material on the steep slopes.

The formation of block fields (rock streams and rock seas) is typical of the study mountainous areas. In particular, very thin (0.3–0.4 m thick only) "film coating" block fields composed of the frosty debris gradually moving downslope are widespread on the Vrangel, Novaya Zemlia, Severnaya Zemlya Islands at altitudes of 50–160 m a.s.l. (Ershov 2001).

Farther to the south at elevations where the permafrost is developed well but locally (the alpine tundra belt) formation of the block fields is also typical at altitudes of 400–450 m a. s.l in the North and up to 700 m a.s.l. in the South of the region. Here they are up to 1 m thick and became thicker southward (Ershov 2001).

Another permafrost-related mass-wasting phenomenon is the formation of rock glaciers. About 6500 rock glaciers of different morpho-dynamic types (slope lobate, cored tongue-like, complex) have been identified in the north-east of Asia (Galanin 2009). Areas where rock glaciers are well-developed are characterized, mainly, by mean annual precipitation exceeding 500 mm/year (for Eastern Siberia with the Pacific type of atmospheric circulation). No rock glaciers have been found in the area of medium-altitude mountainous areas with sharply continental climate (mean annual precipitation less than 300 mm/year).

It should be noticed that role of the freezing–thawing of the fissure water as an important factor of rocky slopes destabilization increases in the areas with the sporadic permafrost in the sharply continental climate. Above all else freezing of the ground water leads to the increase of the hydrodynamic pressure heads that also have negative effect on rocky slopes stability.

Intensive physical rock weathering in the high-altitude nival zone predetermines intensive formation of rock falls

Table 1 Classification of the Cryogenic Landslides and Related Phenomena (classical types of landslides are not included), modified after Zerkal and Strom (2017)

		Material									
		Frozen									
		Ice and soil-ice mix	Frozen and frosty rock	Frozen debris and soil					Seasonal thawing		
				Quasi-solid frozen soil	Frozen soil	Frosty debris	Plastic frozen soil	Cooled soil	Rock	Debris	Soil
Movement type (after Hungr et al. 2014)	Creep					Kurum			Do not differ from the "classical" types slope failures		
	Fall	Ice fall and rock-ice fall[a]	Rockfall and rock-ice fall[b]								
	Topple										
	Slide	Ice-rock slide	Rock slide[c]							Fast kurum[d]	Cryo-Earth slide[e]
	Spread					Rock sea		Cooled earth spread			Axtive-layer detachments spread[f]
	Flow	Slow movement of glaciers and rock glaciers	Slow movement of ice-rock glacier		Slow movement of ice-earth glacier	Rock stream	Ice-earth flow				Slow and fast solifluction (cryogenic earth flow) and filmy earth flow[g]
	Complex	Entrainment of frozen debris by snow- or ice avalanche	Ice-rock avalanche, Ice-frozen debris/soil avalanche							Ice-Debris avalanche	

[a]Including failure of steep glaciers, after (Schneider et al. 2011)
[b]Including failure within bedrock or rock debris overlying glaciers, after (Schneider et al. 2011)
[c]Including translational and rotational slides. During warm period of the year deposits of rockslides that originated from the quasi-solid frozen soil and frozen soil, thaw and spread
[d]Motion of debris (including pieces of ice) over the surface of frozen rock/soil and (or) ice (boundary of the frozen and thaw material)
[e]Predominantly block (translational) motion of the seasonal thawed soil over the surface of ice and/or ice soils (slip surface—boundary of the frozen and thaw material) including retrogressive thaw slump. All cryo-earth slides are shallow since their thickness is limited by the seasonal thaw thickness. During warm period of the year deposits of the cryo-earth slides thaw and spread
[f]By M.O. Leibman (Leibman and Kiziakov 2007)
[g]Flow of the water-saturated thawed soil (including daily thawed) over the boundary between the thawed and frozen soils, including solifluction lobes, formation of thermal circuses and thermo-erosional cauldrons with filmy earth flows

and ice-rock falls. In these areas with the extremely rugged topography large ice-rock avalanches occur rather often being induced by falls and slides in the frozen and frosty rock. Such phenomenon can be exemplified by the 2002 Kolka-Karmadon ice-rock avalanche that was triggered by rockfall of the frosty shale on the Kolka glacier (Petrakov et al. 2008; Evans et al. 2009b). Similar phenomena—the 1962 and 1970 Huaskaran ice-rock avalanches (Plafker and Eriksen 1978; Evans et al. 2009a).

Within the sub zone with seasonal thawing, which southern boundary depends on the atmospheric circulation significantly, evolution of the modern slope processes takes place where the low-temperature permafrost is developed. Here the characteristic feature of slope processes has a significant countrywide decrease of their intensity and smaller variability of the landslide types comparative with the non-permafrost regions. Some inactive rockslides in mountainous regions are, likely, relict features. Anticipated climate change, and change of the temperature at first, should not change frozen state of rock massifs in general.

At the same time, thermal effects that increase effects of ice thawing in the permafrost zone subjected to seasonal thaw is a leading factor of the increase of landslide activity. Such thermodynamic instability of the permafrost provokes development of thermal erosion and thermal abrasion associated with slope deformations, predominantly in the form of cryo-earthflows and retrogressive thaw slumps (Leibman and Kiziakov 2007). Filmy earthflows with daily cycle of the activity when thin soil layer thaw every day and moves downslope over the frozen base, are the specific type of the cryo-earthflows. They originate during “polar day” when some parts of slopes are in the shadow when sun is low. Development of filmy earthflows leads to formation of the thermal circuses (“thermal kettles”) with typical earthflows below them typical of the tundra and forest-tundra areas.

The syngenetic and epigenetic ice soils with high content of ice are widespread in the polar areas (Trofimov 2002). Significant development of such soil in the seashore zone predetermine intensive thawing of its icy component that decreases the overall stability of the slopes. Such phenomenon intensifies at present due to global climate change.

The long-term annual average rate of the bank edge retreat caused by slope deformations produced by thermal erosion reaches 2–3 m/yr at the Barents and Kara Sea shores, ranges from 1.9–2.0 to 4–6 m/yr at the Laptev Sea shore (up to more than 20 m/yr on the islands’ shores), and from 1.6 to 5–7 m/yr at the Eastern-Siberian Sea shores (Zerkal 2013; Zhelezniak et al. 2014; Zerkal and Strom 2017). At the same time at the sea shore sections without such landslides mean rate of the thermal erosion does not exceed 0.3 m/yr (Zhelezniak et al. 2014). The permafrost-related landslides and related phenomena affect up to 38% of the arctic seas shorelines (except the eastern part of the Chukchee Sea with different geological conditions) (Zerkal and Strom 2017).

Further evolution of the seashore cliffs destruction by thermal-gravitational processes will depend on the climate change tendencies, and will intensify with global warming and decrease with relative cooling.

The long-term annual average rate of the seasonal bank edge retreat induced by thermal erosion and associated slope processes in the valleys of large Eastern Siberian Rivers (Lena. Indigirka, Yana, Kolyma) and of their tributaries can reach several tens of meters. In the continental areas, besides the cryogenic earthflows, the active-layer detachment spreads are widespread as well. They take place not only near river-bank cliffs, but also in the middle parts of the slopes with inclination less than 5°. At steeper slopes dipping more than 5–7°, where the seasonal thaw depth reaches 0,4–1,0 m, the slow-moving and fast-moving solifluction earthflows form. The former are plastic and viscous-plastic flows moving from 8–10 to 30 cm/yr; the latter are more or less viscous flows moving up to several meters per day, sometimes even faster (Garagulia 1983; Zerkal 2013; Zhelezniak et al. 2014; Zerkal and Strom 2017). Slopes in the continuous permafrost zone can be divided into three classes according to the hazard of the cryogenic earthflows: low solifluction-hazard slopes (5–7°), (7–15°), and maximal solifluction-hazardous slopes (>15°) (Garagulia 1983).

Slow and fast solifluction earthflows create special landforms on the affected slopes—flows and tongue-shape solifluction lobes (Leibman and Kiziakov 2007) (local pseudo-terraces). Such landforms are several meters or tens of meters wide and their height rarely exceeds 2 m. In the small valleys the solifluction flows can be up to 100 m long and 25–30 m wide with a frontal lobe up to 1–1.5 m high.

Solifluction earthflows are developed at the hilly and low mountainous areas covered by tundra where they can cover up to 16% of the territory (Leibman and Kiziakov 2007). Further to the south, within the middle altitude mountainous areas, similar landforms originate at the uppermost parts of the slopes, above the upper boundary of the taiga forest. For the southern tundra with more intensive vegetation such cryogenic phenomena are not typical and corresponding landforms, are, likely, relict.

The solifluction earthflows evolve seasonally, during the warm season of the year, and their activity depends on maximal monthly average and on the long-time annual average temperature at the surface, and on its annual variation. Activity of the solifluction earthflows can lead to their formation at new sites and to the increase of the size of the existing features. According to (Garagulia 1983), possible

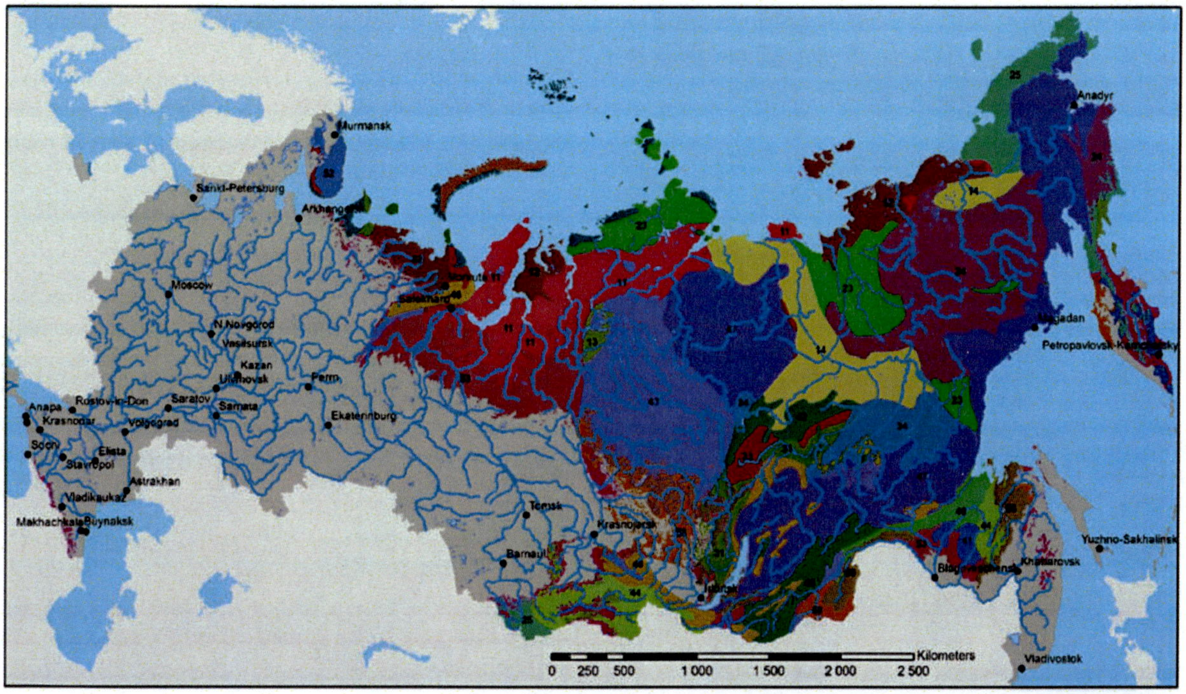

Fig. 1 Map showing the distribution of the cryogenic landslides and related phenomena in Russia. Legend—see Fig. 2 on the next page

rise of the annual average temperature for 1°C together with similar rise of the monthly average temperature will result in 0.2–0.25 m increase of the seasonally thawed layer thickness for the areas with the 7,5°C annual variation of the soil surface temperature (marine type if climate) and in its 0,15–0,2 m increase in the areas with such annual temperature variations more than 20°C (harsh continental climate) (Garagulia 1983). One can expect that maximal variation of the activity of the cryogenic earthflows (fast and slow solifluction) will be in the areas with continental and harsh continental climate (with the Siberian type of atmospheric circulation (Zerkal and Strom 2017).

Formation of the block fields within the poorly developed sporadic permafrost sub-zone has been reported at the altitudes up to 1100–1200 m a.s.l. at the west of the sub-zone and up to 1200–1300 m a.s.l. at the east in the alpine tundra belt. Further to the south such belt is located at altitudes from ~1600–2300 m a.s.l. (Kuznetsk Alatau, South-east of Trans-Baikal region) to 3500 m a.s.l. (Tyva). Block fields cover up to 60–70% of these areas (e.g. in the Khamar-Daban Range) (Ershov 2001, Zerkal and Strom 2017). The suffusion extraction of fines in the areas with rugged terrain facilitates formation of the coarse carapace, which thickness in Southern Yakutia and in Northern part of the Trans-Baikal region increases up to 2.0–2.5 m.

In the sub-zone of poorly developed sporadic permafrost the intensity of slope processes increases gradually. It is caused by the significant decrease of the role of the cryogenic component in the strength of the soil massifs along with disintegration of their subsurface parts due to active seasonal freezing–thawing processes and gradual development of not only physical, but also chemical weathering. Role of the solifluction in the slope processes gradually decreases southward, being transformed into typical earthflows. The proportion of shallow block slides increases, contrary, such as, e.g. in the Belogorsk area on the right bank of the Ob River near the Khanty-MansyiskCity in the Western Siberia lowland (Zerkal and Strom 2017).

Spatial distribution of the cryogenic landslides of different types within the territory of Russian Federation is demonstrated in Fig. 1.

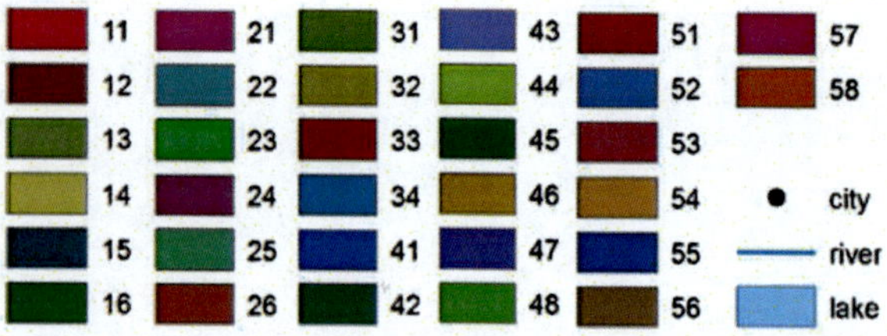

Fig. 2 Legend of the cryogenic landslides distribution map shown in Fig. 1. The two-digit code was used in ArcGIS to describe regions with different combinations of the landslide types: 11–16—areas with prevailing landslides in frozen soils; 21–26—areas where, besides soil and rock, ise is widely involved in slope processes; 31–34—areas with prevailing kurums of various types; 41–48—areas with widely developed permafrost-related rock falls and rock slides; 51–58—areas with sparsely distributed cryogenic landslides and related phenomena. Their description is in the following list: 11 Cryo-earth slides, slow and fast solifluction. 12 Cryo-earth slides, slow and fast solifluction, cooled earth spreads. 13 Cryogenic earth slides, cryogenic earth flows. 14 Cryogenic earth slides, cryogenic earth flows, filmy earth flows. 15 Cryogenic earth slides, cryogenic earth flows, rarely active-layer detachments spreads. 16 Slow and fast solifluction, cooled earth spreads. 21 Icefalls, rock-ice falls, ice-rock slides, ice-rock avalanches. 22 Ice-frozen debris/soil avalanches. 23 Ice-rock slides, ice-rock and ice-frozen debris/soil avalanches, cryo-earth slides. 24 Ice-rock slides, ice-rock and ice-frozen debris/soil avalanches, cryo-earth slides, slow moving rock glaciers. 25 Ice-rock slides, ice-rock and ice-frozen debris/soil avalanches, cryogenic earth slides and cryogenic earth flows. 26 Rock-ice falls, ice-rock slides, ice-rock avalanches, kurums and filmy kurums, cryogenic earth slides, slow moving rock glaciers. 31 Kurums, rock stream, cryogenic earth slides and cryogenic earth flows. 32 Kurums, rock streams. 33 Kurums, rock streams, rock seas. 34 Kurums, rock streams, rock seas, slow moving rock glaciers. 41 Permafrost-related rock slides, kurums, rarely, ice-rock avalanches. 42 Permafrost-related rock slides, kurums, rock seas, rock streams, cryogenic earth slides. 43 Permafrost-related rock slides, kurums, rock streams, cryogenic earth slides. 44 Permafrost-related rock slides, kurums, rock streams, cryogenic earth slides, ice-debris flows. 45 Permafrost-related rockfalls and rockslides, kurums and fast kurums, rock streams, cryogenic earth slides. 46 Permafrost-related rockfalls and rockslides, kurums, rock streams, cryogenic earth slides. 47 Permafrost-related rockslides, kurums, rock seas, rock streams, cryogenic earth slides, slow moving rock glaciers. 48 Permafrost-related rockslides, kurums, rock seas, sow moving rock glaciers, cryogenic earth slides, ice-debris flows. 51 Rarely, cryogenic earth slides, cryogenic earth flows, and ice-frozen debris/soil avalanches. 52 Rarely, ice-frozen debris/soil avalanches. 53 Rarely, permafrost-related earth slides and earth flows. 54 Rarely, permafrost-related rock falls and rock slides, and cryogenic earth slides. 55 Rarely, permafrost-related rock falls and rock slides, cryogenic earth slides and cryogenic earth flows, and ice-rock avalanches. 56 Rarely, permafrost-related rock falls, rocks slides, cryogenic earth slides and cryogenic earth flows. 57 Rarely, permafrost-related rockslides, and cryogenic earth slides. 58 Rarely, permafrost-related rockslides, rarely, cryogenic earth slides

Conclusions

The permafrost is widely developed in arctic and sub-arctic climatic zones of Russia. Due to the specific conditions, slope deformations in the permafrost areas are quite special and can be classifies as a group of the cryogenic landslides and related phenomena, whose local classifications were proposed by S. Sharpe, T.N Kaplina, L.S. Garagulia, M.O. Leibman (Sharpe 1938; Kaplina 1965; Garagulia 1983; Leibman and Kiziakov 2007). However, no uniform classification of the cryogenic slope processes taking into account their general variability and peculiarities of their formation has been elaborated till now.

Such classification could be proposed based on the combined analysis of types of the permafrost soils affected by slope processes and of various mechanisms of their evolution.

At the same time, the main goal of this study was not to elaborate the universal classification of the cryogenic slope processes, but to demonstrate wide spectrum of their special types that are absent in "warm" regions. The proposed classification, however, can be useful for generalization of the available data on cryogenic landslides and related phenomena and their comparative analysis with the general concepts of landslides evolution, and we hope that it can be useful for researchers and practitioners.

References

Barsch D (1988) Rock glaciers.In: Clark MJ (ed) Advances in periglacial geomorphology. Wiley, Chichester, pp 69–90

Bolikhovsky VF, Kiuntcel VV (1990) Evolution of landslides in the permafrost of the Western Siberian tundra. Enginernaya Geologiya [Eng Geology]. 1:65–70 (in Russian)

Cruden DM, Varnes DJ (1996) Landslide types and processes. In: Turner AK, Schuster RLE (eds) Landslides: investigation and Mitigation. National Academy Press, Washington, pp 36–75

Dokuchaev NE, Dokuchaeva VB (2015) Cryogenic translational landslides in the Upper Kolyma River basin (the Magadan Region). J North-Eastern Sci Center Far East Branch Russ Acad Sci 3:25–29 (in Russian)

Ershov ED (ed) (2001) Basic of the geocryology. Part 4. Dynamic geocryology. Moscow State University Press. Moscow, p 688. (in Russian)

Evans SG, Bishop NF, Smoll LF, Murill PV, Delaney KB, Oliver-Smith A (2009b) A reexamination of the mechanism and human impact of catastrophic mass flows originating on Nevado Huascaran, Cordillera Blanca, Peru in 1962 and 1970. Eng Geol 108:96–118

Evans SG, Tutubalina OV, Drobyshev VN, Chernomorets SS, McDougall S, Petrakov D, Hungr O (2009b) Catastrophic detachment and high-velocity long-runout flow of Kolka Glacier, Caucasus Mountains, Russia in 2002. Geomorphology 105:314–321

Galanin AA (2009) Rock Glaciers of North-Eastern Asia: mapping and geographical analysis. Earth's Cryosphere 13(4):49–61

Garagulia LS (1983) Classification of solifluction types for the estimation mapping of the territory. Enginernaya Geologiya [Eng Geology] 4:10–17 (in Russian)

Haeberli W (1985) Creep of mountain permafrost: internal structure and flow of alpine rock glaciers. Mitteilungen der Versuchsantalt für Wasserbau, Hydrologie und Glaziologie, vol 77, p 142

Hungr O, Leroueil S, Picarelli L (2014) The Varnes classification of landslide types, an update. Landslides 11(2):167–194

Interstate standard GOST 25100–2011 Soils. Classification (2011) Euro-Asian Council for Standardization, Metrology and Certification (EASC). (in Russian)

Kaplina TN (1965) Kriogenic slope processes. Nauka (Science) Publishing House, Moscow, p 296. (in Russian)
Kozyreva EA, Rybchenko AA, Shipek T, Pellinen VA (2011) Solifluction landslides of the Olhon Island coast. J Irkutsk State Techn Univ 4:41–49 (in Russian)
Leibman MO, Kiziakov AI (2007) Cryogenic landslides of Yamal and Yugorsky Peninsula. Moscow-Tiumen, p 205 (in Russian)
Petrakov DA, Chernomorets SS, Evans SG, Tutubalina OV (2008) Catastrophic glacial multi-phase mass movements: a special type of glacial hazard. Adv Geosci 14:211–218
Plafker G, Eriksen GE (1978) Nevados Huascaran avalanches, Peru. In: Voight B (ed) Rockslides and avalanches, vol 1. Elsevier, Amsterdam, pp 277–314
Romanovsky NN, Turin AI, Sergeev DO (1989) Kurums of the buld mountain belt. Novosibirsk, Nauka Publishers, p 152 (in Russian)
Schneider D, Huggel C, Haeberli W, Kaitna R (2011) Unraveling driving factors for large rock-ice avalanche mobility. Earth Surf Proc Land 36:1948–1966
Sharpe CFS (1938) Landslides and related phenomena: a study of mass movements of soil and rock. Columbia University Press, New York, p 137
Trofimov VT (2002) Zonality of the engineering-geological conditions of the continents of Earth. Moscow, Moscow State University Publishers., p 348 (in Russian)
Tiurin AI (1979) The genetic classification of kurums. Bulletin of Moscow State Universiy, Geological series No 3:74–82
Zerkal OV (2013) Regularities of modern natural geological processes distribution. In: Trofimov VT, Kalinin EV (eds) Engineering geology of Russia, vol 2. Engineering Geodynamics of Russian Territory, Moscow, pp 674–696 (in Russian)
Zerkal OV, Strom AL (2017) Overview of landslides distribution in Russian Federation and variation of their activity due to climatic change. In: Slope safety preparedness for impact of climatic change. Balkema, London, pp 253–288
Zheleznіak MN, Grigoriev MN, Shepeliev YV, Skachlcov YB (2014) Modern state and evolution of Arctic Criolitozone; Ecological aspects of development. In: Proceedings of the conference «Arctic-2014». (CD-ROM). Yakutsk (in Russian)

Cutting-Edge Technologies Aiming for Better Outcomes of Landslide Disaster Mitigation

2020 Kyoto Japan

Kazuo Konagai

The International Consortium on Landslides (ICL) and The Global Promotion Committee of the International Programme on Landslides (GPC/IPL) have been responsible for organizing the World Landslide Forums (WLFs) every three years since 2008. Ever since the 1st WLF, the forums have long been the arena for landslide researchers and practitioners to exchange up-to-date information of recent devastations caused by landslides, cutting-edge technologies for landslide disaster mitigations and early warnings etc. to establish synergies among all participants worldwide.

Though the upcoming WLF5 has officially been postponed by one year to 2–6 November 2021 due to the global disruption caused by the coronavirus pandemic, the WLF5 will be all the more important with the Kyoto Landslide Commitment 2020 (KLC2020) to be launched as planned in the final online signatory meeting on 5 November 2020; the KLC 2020 is intended to be our action goals as the further advanced successor of the 'Sendai Landslide Partnerships 2015–2025 for Global Promotion of Understanding and Reducing Landslide Disaster Risk' in line with some of 17 Sustainable Development Goals (SDGs), particularly SDG 11, "Make cities and human settlements inclusive, safe, resilient and sustainable," of the United Nations.

For these important goals, the ICL has been inviting sponsorship from industries, businesses, and government agencies; all leading players in landslide science and technologies. They have been supporting a variety of the ICL/IPL activities such as publishing the International full-color journal "Landslides (Journal of the International Consortium on Landslides)", full-color books for WLFs, exhibiting their cutting-edge technologies in WLFs, etc. Here follow short introductions of their activities with their names, addresses and contact information.

Marui & Co. Ltd.

1-9-17 Goryo, Daito City, Osaka 574-0064, Japan
URL: https://marui-group.co.jp/en/index.html
Contact: hp-mail@marui-group.co.jp

Marui & Co. Ltd. celebrates its 100th anniversary in 2020. Marui, as one of the leading manufacturers of testing apparatuses in Japan, has been constantly striving to further improve its service since its foundation in 1920, thus contributing to the sustainable development of our nation and society. Our main products cover a wide variety of destructive and non-destructive testing apparatuses in the fields of geotechnical engineering, concrete engineering (mortar, aggregates, etc.), and ceramic engineering. Of special note is that Marui has been helping manufacture ring-shear apparatuses half-century long based on the leading-edge idea of Dr. Kyoji Sassa, Professor Emeritus at the Kyoto University. Marui has delivered total 7 ring-shear apparatuses to the Disaster Prevention Research Institute, Kyoto University, and 2 to the International Consortium on Landslides. Also the apparatuses were exported to the United States of America, China, Croatia and Vietnam.

Marui & Co. Ltd. takes great pleasure in developing, manufacturing, and providing new products of high value sharing the delight of achievement with our customers, and thus contributing to the social development. The whole staff of Marui & Co. Ltd. are determined to devote ceaseless effort to keep its organization optimized for its speedy and high-quality services, by the motto "Creativity and Revolution", and strive hard to take a step further, as a leading

K. Konagai (✉)
Organizing Committee of the Fifth World Landslide Forum,
International Consortium On Landslides, Kyoto, 606-8226, Japan
e-mail: konagai@iclhq.org

Ž. Arbanas et al. (eds.), *Understanding and Reducing Landslide Disaster Risk*,
ICL Contribution to Landslide Disaster Risk Reduction,
https://doi.org/10.1007/978-3-030-60713-5_38

manufacturer of testing apparatuses, to answer our customer's expectations for the twenty-second century to come.

Nippon Koei Co., Ltd.

5-4 Kojimachi, Chiyoda-ku, Tokyo 102-8539, Japan
URL: https://www.n-koei.co.jp/english/
Contact: https://www.n-koei.co.jp/english/contact/input

Nippon Koei Co., Ltd. and its group companies conduct many projects to support the growth of developing countries in Asia, Africa, the Middle and Near East, Latin America and other regions. Examples of their efforts include environmental measures to combat global warming, development of regional transportation infrastructure to support the rapid growth of emerging economies, and reconstruction assistance for regions affected by conflict and/or natural disasters.

OSASI Technos, Inc.

65-3 Hongu-cho, Kochi City, Kochi 780-0945, Japan
URL: https://www.osasi.co.jp/en/
Contact: cs@osasi.co.jp

OSASI Technos, Inc. has been making its best efforts to develop its cutting-edge technologies for landslide early warning. Its unique compact and lightweight sensors making up the Landslide Early Warning System enable long-term monitoring of unstable landslide mass movements, precipitations, porewater pressure buildups, etc., in a remote mountainous area where commercial power is often unavailable. OSASI Technos, Inc. is also proud of its advanced technology to transfer observed data even in areas with poor telecom environments as proven in the successful implementations in South Asia.

All stuff members of OSASI Technos work together for mitigation of landslide disasters worldwide.

Godai Corporation

1-35 Kuroda, Kanazawa, Ihikawa 921-8051, Japan
URL: https://soft.godai.co.jp/En/Soft/Product/Products/LS-RAPID/
Contact: pp-sales@godai.co.jp

Ever since its foundation in 1965, Godai Kaihatsu Co., Ltd., a civil engineering consulting firm, has long been providing a variety of software and measures particularly for natural disaster mitigation. With its rich expertise in both civil engineering and information technology (IT), the company has its primary goal to address real world needs of disaster mitigation. All the staff of Godai Kaihatsu Co., Ltd. feel it more than happy that their cutting-edge technologies help mitigate natural disasters.

Japan Conservation Engineers & Co., Ltd.

3-18-5 Toranomon, Minato-ku, Tokyo 1,050,001, Japan
URL: https://www.jce.co.jp/en/
Contact: go_info@jce.jp

Japan Conservation Engineers & Co., Ltd. (JCE) is a general consulting firm working on landslide prevention research and consulting. JCE provides various disaster prevention technologies for debris flows, landslides, slope failures, rockfalls, etc. In addition, JCE is proud of its expertise having been conducting surveys and consulting works on coastal erosions and tsunami countermeasures for about 20 years. JCE contributes to the world through its activities in the realm of both structural and non-structural measures to build a resilient society.

OYO Corporation

7 Kanda-Mitoshiro-cho, Chiyoda-ku, Tokyo 101-8486, Japan
URL: https://www.oyo.co.jp/english/
Contact: https://www.oyo.co.jp/english/contacts/

OYO Corporation, the top geological survey company in Japan established in Tokyo in 1957, is well known as one of leading companies providing cutting-edge technologies and measures for natural disasters such as landslides, earthquakes, tsunamis, and floods. Not just developing and selling measuring instruments related to disaster prevention, OYO also delivers a market-leading services in 3D ground/geological modeling and 3D exploration technologies.

Kokusai Kogyo Co., Ltd.

2 Rokubancho, Chiyoda-ku, Tokyo 102-0085, Japan
URL: https://www.kkc.co.jp/english/index.html
Contact: overseas@kk-grp.jp

Kokusai Kogyo Co., Ltd. as a leading company of geospatial information technologies, has long been providing public services with its comprehensive expertise to address real world needs and cutting-edge measurement technologies. Kokusai Kogyo Co., Ltd. helps rebuild "Green Communities," which has been of our great concern in terms of "environment and energy," "disaster risk reduction" and "asset management". Kokusai Kogyo Co., Ltd. offers the advanced and comprehensive analyses of geospatial information for developing new government policies, maintaining and operating social infrastructures safe and secure, and implementing low-carbon measures in cities.

Influenced by the recent global climate change, extreme rainfall events have become more frequent worldwide and resultant hydro-meteorological hazards are creating more deaths and devastations particularly in many developing countries where effective advanced countermeasures are not readily available. Kokusai Kogyo Co., Ltd. is proud of its achievements in establishing resilient infrastructure systems and implementing effective monitoring/early warning systems in developing countries, which have long been helping reduce the risks from natural hazards.

Geobrugg AG

Aachstrasse 11, 8590 Romanshorn, Switzerland

URL: www.geobrugg.com
Contact: info@geobrugg.com

Swiss company Geobrugg is the global leader in the supply of high-tensile steel wire safety nets and meshes—with production facilities on four continents, as well as branches and partners in over 50 countries. True to the philosophy "Safety is our nature" the company develops and manufactures protection systems made of high-tensile steel wire. These systems protect against natural hazards such as rockfall, landslides, debris flow and avalanches. They ensure safety in mining and tunneling, as well as on motorsport tracks and stop other impacts from falling or flying objects. More than 65 years of experience and close collaboration with research institutes and universities make Geobrugg a pioneer in these fields.

Ellegi srl

Via Petrarca, 55 I-22070 Rovello Porro (CO) Italy

URL: https://www.lisalab.com/engl/?seze=1
Contact: info@lisalab.com

Ellegi srl provides worldwide monitoring services and produces Ground Based synthetic aperture radar (GBInsAR) for remote measurement of displacements and deformations on natural hazards and manmade buildings using its own designed and patented LiSALab system.

Its activities started in 2003 as a spin off project to exploit commercially the Ground Based Linear Synthetic Aperture Radars technology developed by European Commission's Ispra Joint Research Centre and based on the results of more than 10 years of research. Since then Ellegi has industrialized and developed the core technology of the LiSALab system and latest LiSAmobile system represents the 5th generation of development.

In 2003 it was the first commercial company in the world to provide GBInSAR measurements of natural hazards and structure.

Ellegi srl offers:

- Displacement fields measurement, control and monitoring of the deformation caused by natural hazards, like landslides, rockslides, sinkhole, volcanic deformation in every operative condition, including emergencies.
- Structural strain fields measurement, control, monitoring and diagnosis of the deformation affecting buildings, bridges, viaducts, dams.
- GBInSAR monitoring systems, installation, management and maintenance in order to provide information about natural hazards or anthropic activity, that can generate or cause slopes failures or buildings instabilities.

In all the above-mentioned activities Ellegi srl uses the GBInSAR LiSALab technology that represents a real "break-through".

Chuo Kaihatsu Corporation

3-13-5 Nishi-waseda, Shinjuku-ku, Tokyo 169-8612, Japan

URL: https://www.ckcnet.co.jp/global/
Contact: https://www.ckcnet.co.jp/contactus/

Chuo Kaihatsu Corporation (CKC) was founded in 1946, and has been aiming to become the "Only One" consultant for our customers. We engage in the hands-on work that will "Remain with the earth, Remain in people's hearts, and Lead to a prosperous future". We focus on road, river and dam engineering to flesh out industrial infrastructures specifically by means of geophysical/geotechnical/geological investigations, civil engineering surveys and project implementations. In recent years, we make significant efforts on earthquake disaster mitigation, sediment disaster prevention/mitigation and ICT information services. Many achievements of ours have already contributed to mitigation of natural disasters such as landslides, earthquakes and slope failures in Japan, Asia and the Pacific Region.

IDS GeoRadar s.r.l.

Via Augusto Righi, 6, 6A, 8, Loc. Ospedaletto, Pisa, Italy, 56,121

URL: https://idsgeoradar.com/
Contact: info@idsgeoradar.com

IDS GeoRadar, part of Hexagon, provides products and solutions, based on radar technology, for monitoring applications including landslides, rockfalls, complex structures, mining and civil engineering. The company is a leading provider of Ground Penetrating Radar (GPR) and Interferometric Radar solutions worldwide. IDS GeoRadar is committed to delivering best-in-class performance solutions and to the pursuit of product excellence, through the creation of application-specific, innovative and cost-efficient systems for a wide range of applications.

METER Group, Inc.

2365 NE Hopkins Court, Pullman, WA 99,163, USA

URL: metergroup.com/wlf5
Contact: bryan.wacker@metergroup.com

METER Group provides accurate, rugged, and dependable instrumentation to monitor moisture in all its phases within an unstable slope. METER specializes in instrumentation for near real-time monitoring of incoming moisture in the form of rain and weather. In addition, we provide real-time below-surface monitoring of existing moisture conditions like moisture content and soil suction which show how the soil profile is filling with water to saturation, including the transition to positive pore water pressure.

The ZL6 advanced cloud data logger works together with ZENTRA Cloud data software to simplify and speed up data collection, management, visualization, and alerting. Our well-published instrumentation is used worldwide in universities, research and testing labs, government agencies, and industrial applications.

For almost four decades, scientists and engineers have relied on our instrumentation to understand critical moisture parameters. We've even partnered with NASA to measure soil (regolith) moisture on Mars. Wherever you measure, and whatever you're measuring, rely on METER for accuracy, affordability, and simplicity that will make your job easier.

Asia Air Survey Co., Ltd.

Shinyuri 21 BLDG 3F, 1-2-2 Manpukuji, Asao-Ku, Kawasaki, Kanagawa 215-0004, Japan

URL: https://www.ajiko.co.jp/en/
Contact: service@ajiko.co.jp

Asia Air Survey (AAS), as one of the leading engineering and consulting companies, has long been providing disaster prevention and mitigation services for over 65 years, particularly in the fields of landslide, debris flow, erosion control, etc. AAS is proud of being the inventor of Red Relief Image Map (RRIM), which is a cutting-edge 3D terrain visualization method allowing great geomorphological details to be visualized in one glance, thus has been used in various facets of disaster prevention and mitigation.

Kiso-Jiban Consultants Co., Ltd.

Kinshicho Prime Tower 12 Floor, 1-5-7 Kameido, Koto-ku, Tokyo 36-8577, Japan

URL: https://www.kisojiban.com/
Contact: kisojiban-contactus@kiso.co.jp

Kiso-Jiban Consultants, established in 1953, is an engineering consulting firm especially well known in the field of geotechnical engineering. The areas of its comprehensive services are listed below:

- Geological and Geotechnical Survey
- Geotechnical Analysis and Design
- Disaster Prevention and Management
- Geographic Information Systems (GIS)
- Soil and Rock Laboratory Tests
- Instrumentation and Monitoring
- Geophysical Exploration and Logging
- Distribution of Geosynthetics Products.

Much-talked-about new service is Kiso-SAR System allowing accurate estimation of both extent and rate of landslide movements based upon a comprehensive interpretation of InSAR results from geotechnical and landslide engineering viewpoint (see the one-page introduction of Kiso-Jiban Consultants Co., Ltd.). With Kiso-SAR system, the following pieces of important geotechnical information can be provided:

(1) Extent of a deforming landslide mass (and the rate of its movement)
(2) Consolidation buildup in soft clay underlying a fill
(3) Deformation buildups induced by slope cutting.

Okuyama Boring Co., Ltd.

10-39 Shimei-cho, Yokote City, Akita 013-0046, Japan

URL: https://okuyama.co.jp/en/
Contact: info@okuyama.co.jp

Okuyama Boring Co., Ltd. is proud of its achievements in various projects to help solve many landslide problems. The company has been offering services in geological surveys and analyses, developing rational countermeasures against various geotechnical problems as well as safe workflow diagrams, and providing necessary pieces of advice for ensuring safety during landslide countermeasure works. For this purpose, Okuyama Boring Co., Ltd. works on monitoring, observations, field surveys, numerical analyses, countermeasure works, etc. of landslides.

Kawasaki Geological Engineering Co. Ltd.

Mita-Kawasaki Bldg, 2-11-15 Mita, Minato-ku, Tokyo108-8337, Japan

URL: https://www.kge.co.jp/
Contact: post-master@kge.co.jp

Kawasaki Geological Engineering Co., Ltd. as one of the leading members of SAAM Research Group, has proactively been involved in developing "Sustainable Asset Anchor Maintenance (SAAM, hereafter) System," enabling easy maintenance of ground anchors. Its unique jack, weighing about half the weight of a conventional jack, together with a newly developed jig, can be applied to any type of anchor even with a short extra length, thus allowing for in-situ lift-off tests on these anchors. The SAAM system also has an optional weight meter that can be installed after performing a lift-off test.

Nissaku Co., Ltd.

4-199-3 Sakuragi-cho, Omiya-ku, Saitama 330-0854, Japan

URL: https://www.nissaku.co.jp/
Contact: survey@nissaku.co.jp

Nissaku Co., Ltd., founded in 1912 as a well drilling company, provides services for far-flung fields of not only groundwater exploitation but also measures for landslides. Having its rich expertise in these fields, Nissaku Co., Ltd. offers general reliable one-stop technical services including designs, investigations, analyses, constructions, and maintenances.

Full-color presentations from the above seventeen exhibitors focusing on their landslide technologies are shown on the following pages. Their cutting-edge technologies have of course been instrumental in the progress that we have made in landslide risk-reduction worldwide, and we want to exert even greater effort to aim high given the KLC 2020 as our new action goals. The International Consortium on Landslides seeks volunteers willing to support our activities introducing their brand-new technologies for landslide disaster mitigation in our international journal "Landslides," full color books for WLFs, exhibitions at WLFs, etc. If you are interested in being engaged in supporting ICL activities, please contact the ICL secretariat. secretariat@iclhq.org.

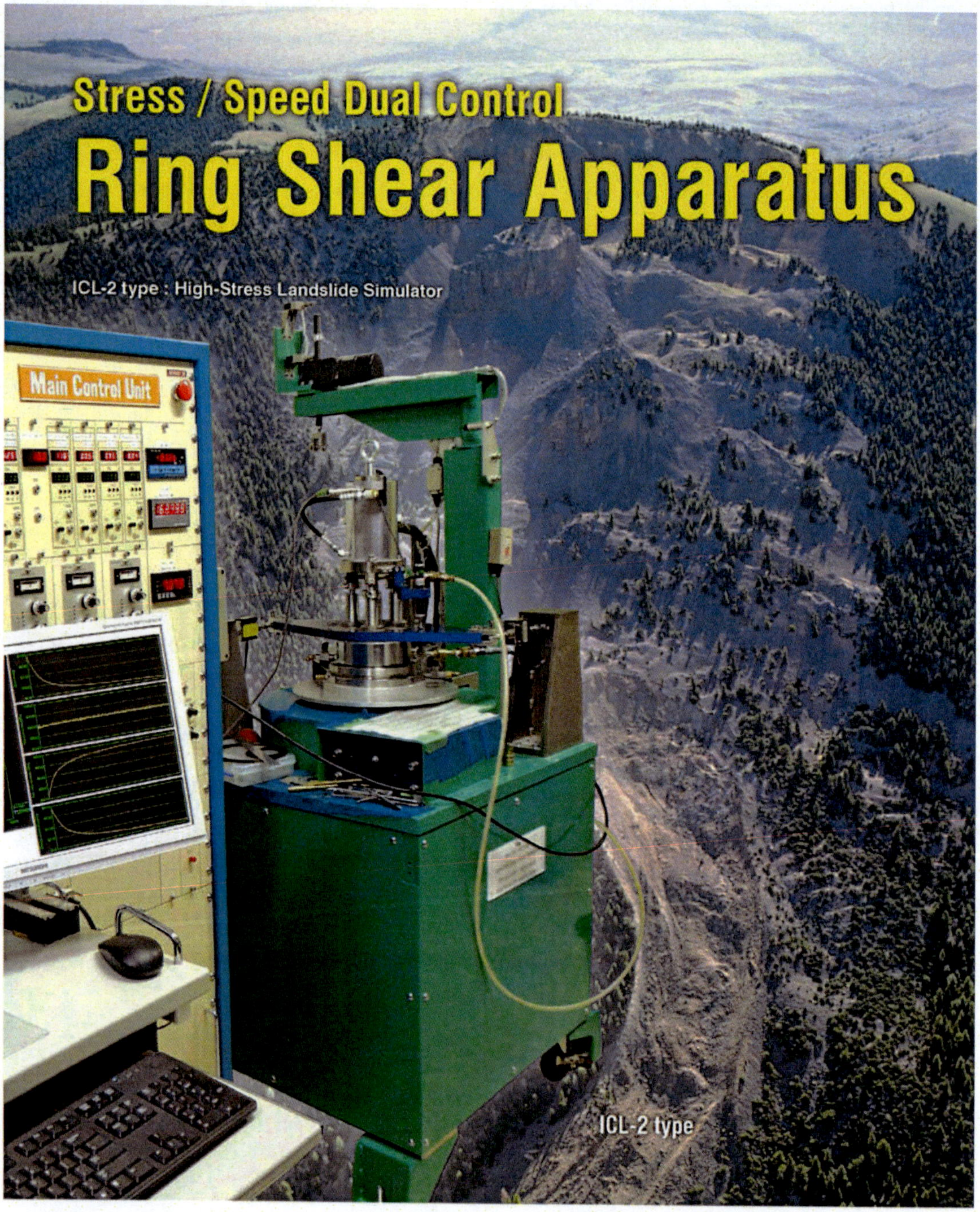
Stress / Speed Dual Control
Ring Shear Apparatus
ICL-2 type : High-Stress Landslide Simulator
Main Control Unit
ICL-2 type

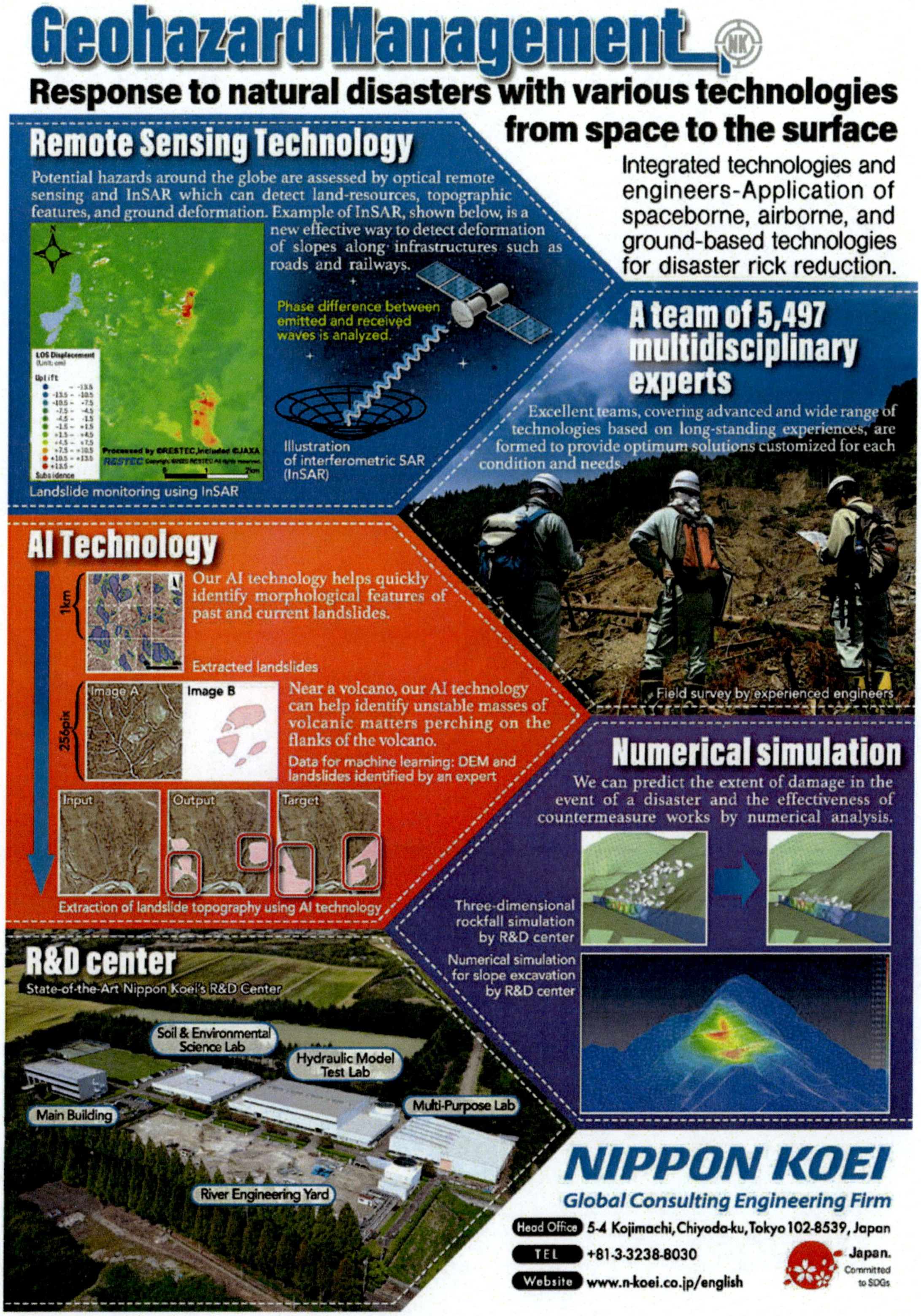
Geohazard Management
Response to natural disasters with various technologies from space to the surface
Integrated technologies and engineers-Application of spaceborne, airborne, and ground-based technologies for disaster rick reduction.
Remote Sensing Technology
Potential hazards around the globe are assessed by optical remote sensing and InSAR which can detect land-resources, topographic features, and ground deformation. Example of InSAR, shown below, is a new effective way to detect deformation of slopes along infrastructures such as roads and railways.
Phase difference between emitted and received waves is analyzed.
Illustration of interferometric SAR (InSAR)
Landslide monitoring using InSAR
A team of 5,497 multidisciplinary experts
Excellent teams, covering advanced and wide range of technologies based on long-standing experiences, are formed to provide optimum solutions customized for each condition and needs.
Field survey by experienced engineers
AI Technology
Our AI technology helps quickly identify morphological features of past and current landslides.
1km
Extracted landslides
Image A
Image B
256pix
Near a volcano, our AI technology can help identify unstable masses of volcanic matters perching on the flanks of the volcano.
Data for machine learning: DEM and landslides identified by an expert
Input
Output
Target
Extraction of landslide topography using AI technology
Numerical simulation
We can predict the extent of damage in the event of a disaster and the effectiveness of countermeasure works by numerical analysis.
Three-dimensional rockfall simulation by R&D center
Numerical simulation for slope excavation by R&D center
R&D center
State-of-the-Art Nippon Koei's R&D Center
Soil & Environmental Science Lab
Hydraulic Model Test Lab
Main Building
Multi-Purpose Lab
River Engineering Yard
NIPPON KOEI
Global Consulting Engineering Firm
Head Office 5-4 Kojimachi, Chiyoda-ku, Tokyo 102-8539, Japan
TEL +81-3-3238-8030
Website www.n-koei.co.jp/english
Japan. Committed to SDGs

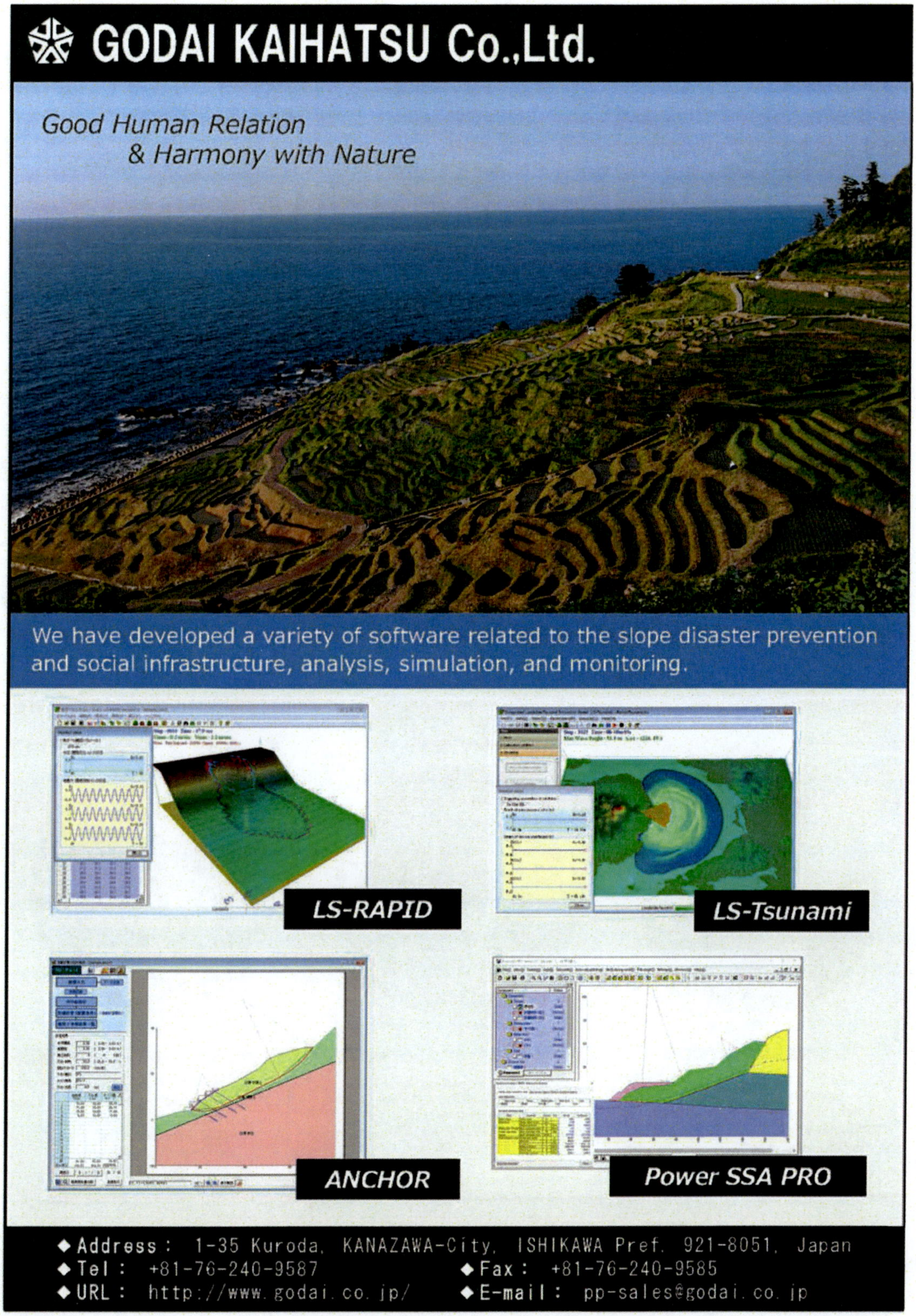
GODAI KAIHATSU Co.,Ltd.
Good Human Relation
& Harmony with Nature
We have developed a variety of software related to the slope disaster prevention and social infrastructure, analysis, simulation, and monitoring.
LS-RAPID
LS-Tsunami
ANCHOR
Power SSA PRO
◆Address : 1-35 Kuroda, KANAZAWA-City, ISHIKAWA Pref. 921-8051, Japan
◆Tel : +81-76-240-9587
◆Fax : +81-76-240-9585
◆URL : http://www.godai.co.jp/
◆E-mail : pp-sales@godai.co.jp

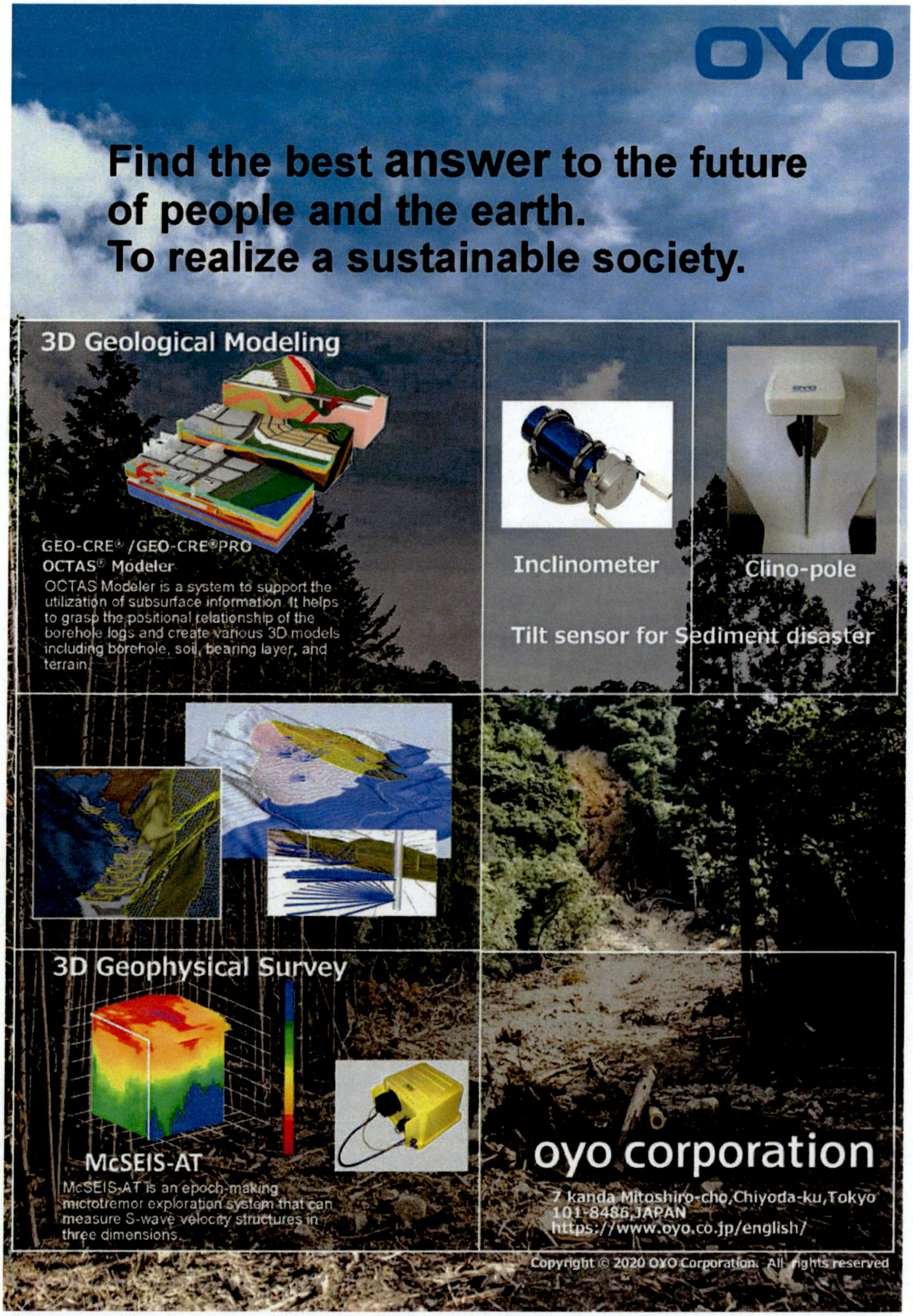
OYO
Find the best answer to the future
of people and the earth.
To realize a sustainable society.
3D Geological Modeling
GEO-CRE® /GEO-CRE®PRO
OCTAS® Modeler
OCTAS Modeler is a system to support the utilization of subsurface information. It helps to grasp the positional relationship of the borehole logs and create various 3D models including borehole, soil, bearing layer, and terrain.
Inclinometer
Clino-pole
Tilt sensor for Sediment disaster
3D Geophysical Survey
McSEIS-AT
McSEIS-AT is an epoch-making microtremor exploration system that can measure S-wave velocity structures in three dimensions.
oyo corporation
7 kanda Mitoshiro-cho,Chiyoda-ku,Tokyo
101-8486,JAPAN
https://www.oyo.co.jp/english/
Copyright © 2020 OYO Corporation. All rights reserved

Green Communities
~Towards a better future,
for people and the environment~
Participations
SUSTAINABLE DEVELOPMENT GOALS
SENDAI FRAMEWORK
PARIS2015 COP21·CMP11
Japan Asia Group
KOKUSAI KOGYO
Kokusai Kogyo as a leading company of geospatial information technologies, has been contributing to the improvement of public services with advanced measurement technologies and a wide range of consulting technologies. Kokusai Kogyo supports the development of "Green Communities" representing the new era public concerns on "environment and energy," "disaster risk reduction" and "asset management". Kokusai Kogyo offers the advanced analysis of geospatial information consultancy for developing new government policies, maintaining and operating social infrastructures with safe and secure city planning, and building low-carbon cities.
3D-GIV (3D-Geomorphic Image Velocimetry)
3D-GIV can help grasp the ground surface displacements caused by natural phenomena such as landslides by analyzing differences between digital geomorphic images obtained through ad hoc Airborne Laser Surveys.
ELSAMAP (Elevation and Slope Angle Map)
ELSAMAP is our cutting-edge 3D terrain visualization method allowing great geomorphological details to be visualized in one glance with gray-scaled slope inclinations and colored altitudes. ELSAMAP has been used to interpret micro-topographies, landslides and some other things.
Realtime Hazard Map System
Our Realtime Hazard Map reflects up-to-date information of soil natures and precipitations at landslide hazard sites, etc. that can constantly be changing, and evaluates area-wide hazard risk in real time.
shamen-net
(a network system using GNSS for slope disaster detection)
"shamen-net" is a total monitoring system integrating GNSS and other monitoring devices (Measurement precision: +/-1mm, on a real time basis)
2 Rokubancho,Chiyoda-ku,Tokyo,Japan 102-0085 https://www.kkc.co.jp/english/ info_overseas@kk-grp.jp

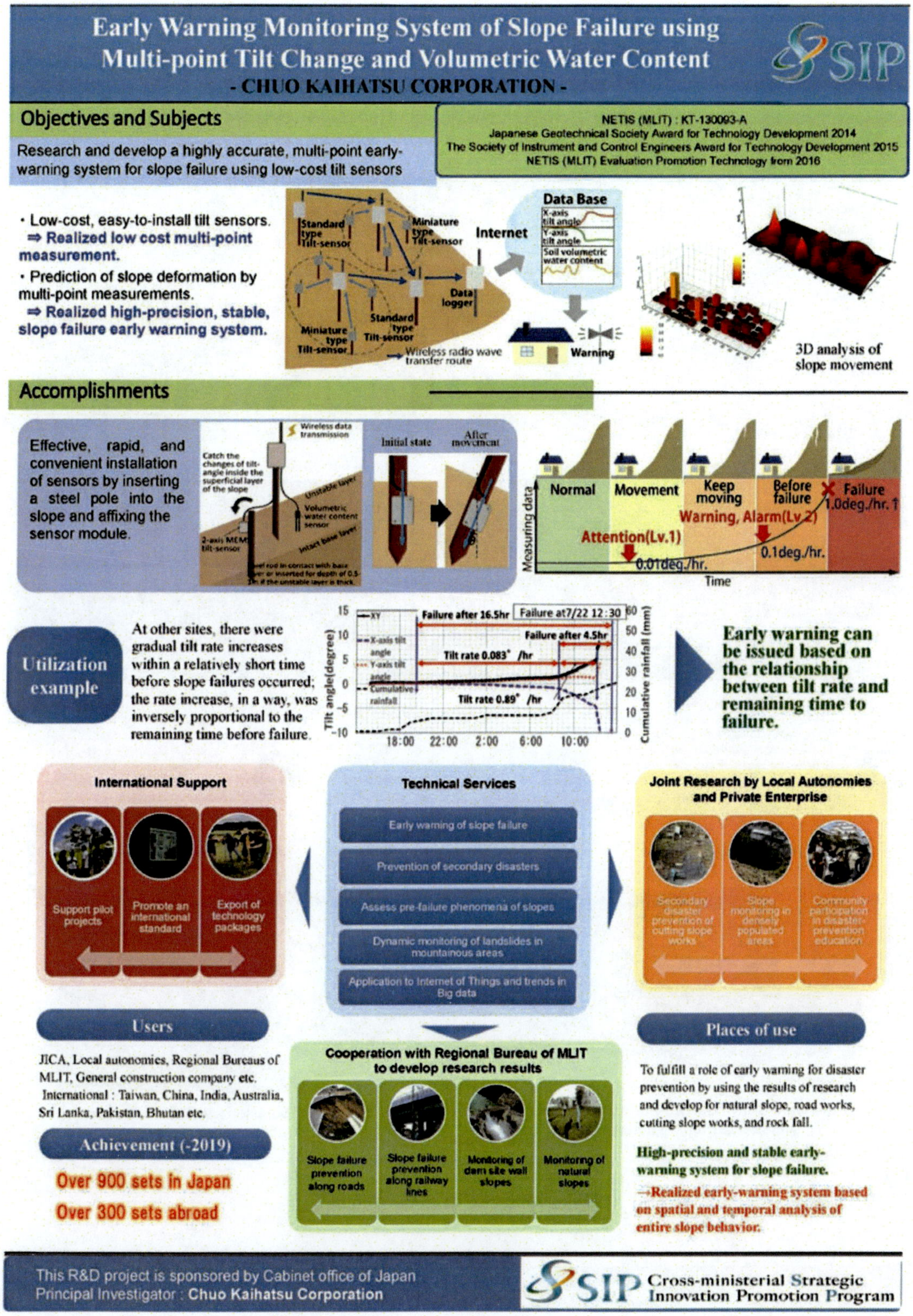

Early Warning Monitoring System of Slope Failure using Multi-point Tilt Change and Volumetric Water Content
- CHUO KAIHATSU CORPORATION -
SIP
Objectives and Subjects
Research and develop a highly accurate, multi-point early-warning system for slope failure using low-cost tilt sensors
NETIS (MLIT) : KT-130093-A
Japanese Geotechnical Society Award for Technology Development 2014
The Society of Instrument and Control Engineers Award for Technology Development 2015
NETIS (MLIT) Evaluation Promotion Technology from 2016
• Low-cost, easy-to-install tilt sensors.
⇒ Realized low cost multi-point measurement.
• Prediction of slope deformation by multi-point measurements.
⇒ Realized high-precision, stable, slope failure early warning system.
Standard type Tilt-sensor
Miniature type Tilt-sensor
Data logger
Wireless radio wave transfer route
Internet
Data Base
X-axis tilt angle
Y-axis tilt angle
Soil volumetric water content
Warning
3D analysis of slope movement
Accomplishments
Effective, rapid, and convenient installation of sensors by inserting a steel pole into the slope and affixing the sensor module.
Wireless data transmission
Catch the changes of tilt-angle inside the superficial layer of the slope
Unstable layer
Volumetric water content sensor
2-axis MEMS tilt-sensor
Intact base layer
Initial state
After movement
Normal
Movement
Keep moving
Before failure
Failure
Attention(Lv.1)
Warning, Alarm(Lv.2)
0.01deg./hr.
0.1deg./hr.
1.0deg./hr.
Measuring data
Time
Utilization example
At other sites, there were gradual tilt rate increases within a relatively short time before slope failures occurred; the rate increase, in a way, was inversely proportional to the remaining time before failure.
Failure after 16.5hr
Failure at7/22 12:30
Failure after 4.5hr
Tilt rate 0.083° /hr
Tilt rate 0.89° /hr
X-axis tilt angle
Y-axis tilt angle
Cumulative rainfall
Tilt angle(degree)
Cumulative rainfall (mm)
18:00 22:00 2:00 6:00 10:00
Early warning can be issued based on the relationship between tilt rate and remaining time to failure.
International Support
Support pilot projects
Promote an international standard
Export of technology packages
Technical Services
Early warning of slope failure
Prevention of secondary disasters
Assess pre-failure phenomena of slopes
Dynamic monitoring of landslides in mountainous areas
Application to Internet of Things and trends in Big data
Joint Research by Local Autonomies and Private Enterprise
Secondary disaster prevention of cutting slope works
Slope monitoring in densely populated areas
Community participation in disaster-prevention education
Users
JICA, Local autonomies, Regional Bureaus of MLIT, General construction company etc.
International : Taiwan, China, India, Australia, Sri Lanka, Pakistan, Bhutan etc.
Achievement (-2019)
Over 900 sets in Japan
Over 300 sets abroad
Cooperation with Regional Bureau of MLIT to develop research results
Slope failure prevention along roads
Slope failure prevention along railway lines
Monitoring of dam site wall slopes
Monitoring of natural slopes
Places of use
To fulfill a role of early warning for disaster prevention by using the results of research and develop for natural slope, road works, cutting slope works, and rock fall.
High-precision and stable early-warning system for slope failure.
→Realized early-warning system based on spatial and temporal analysis of entire slope behavior.
This R&D project is sponsored by Cabinet office of Japan
Principal Investigator : Chuo Kaihatsu Corporation
SIP Cross-ministerial Strategic Innovation Promotion Program

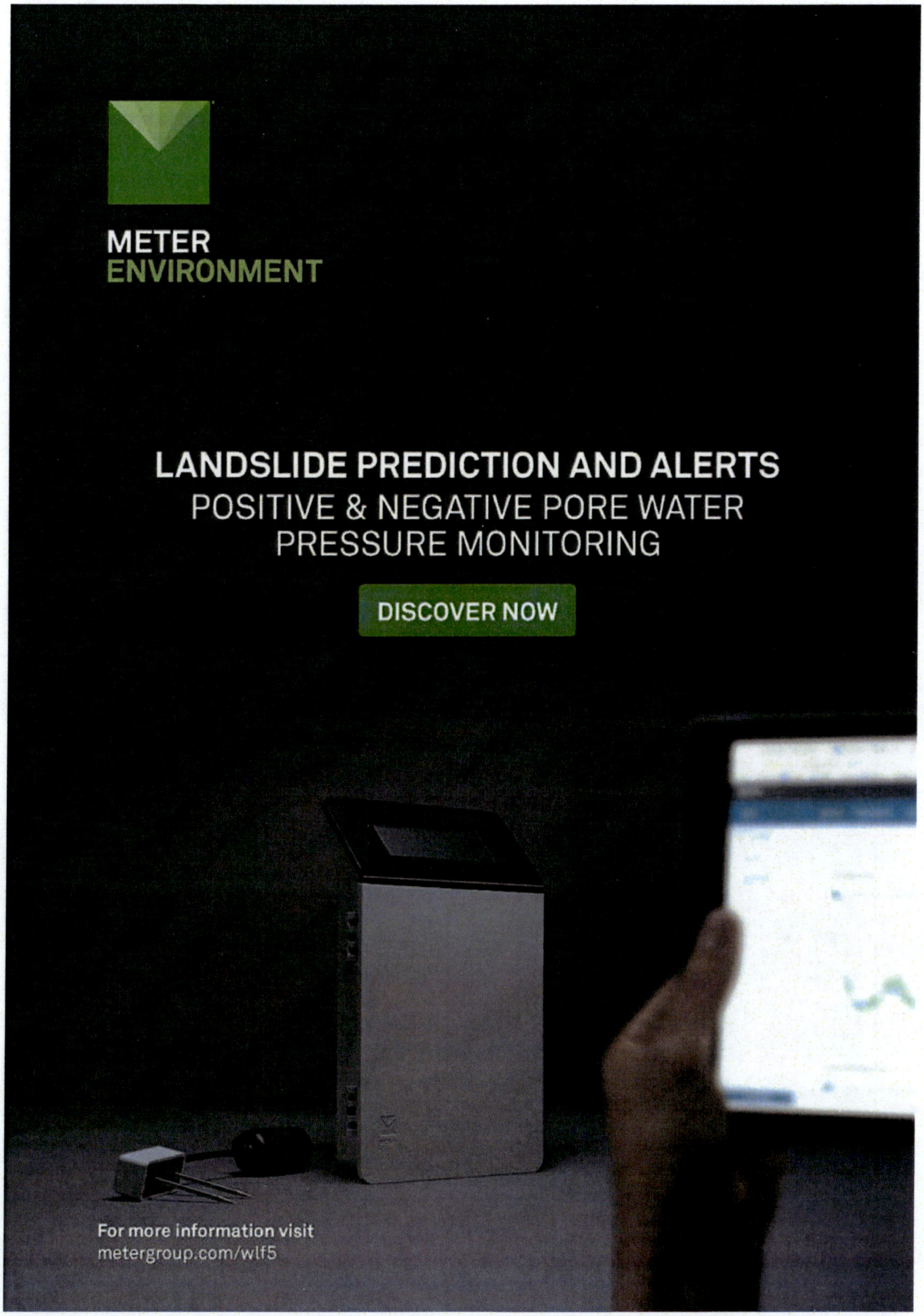

Kisojiban

Kiso-Jiban Consultants Co., Ltd.

Monitoring Service for Landslide by Kiso-SAR System

What's SAR ?

SAR (Synthetic Aperture Radar) is a technique that utilizes interference of radio waves for precise determination of distance.
InSAR (interferometric SAR), the phase of the received backscattered signal from two images of the same scene are used to measure path length differences with an accuracy of a few millimeters.

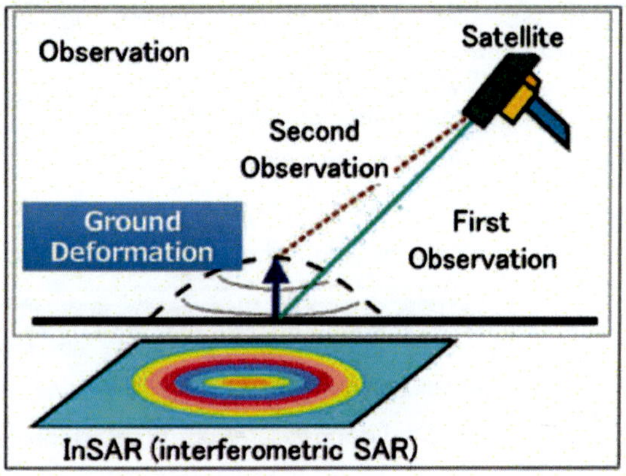

Ground Deformation Estimation in Landslides

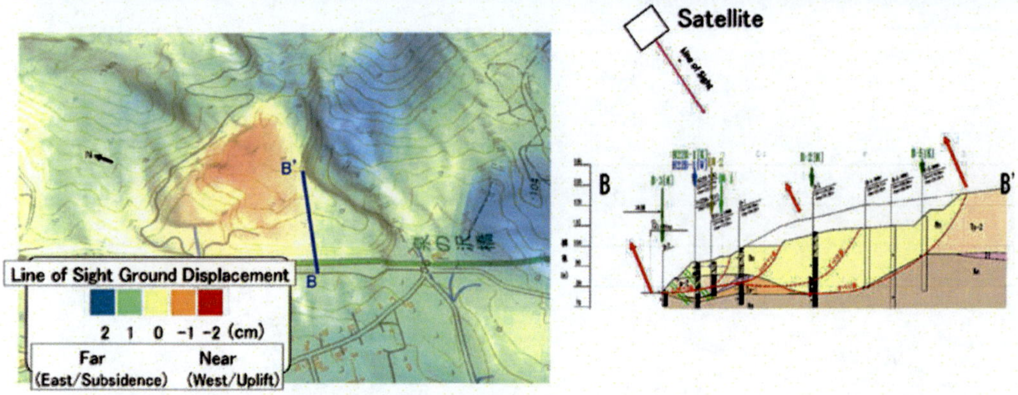

Ground Deformation of Landslide Observed by Kiso-SAR System

Kisojiban
Kiso-Jiban Consultants Co., Ltd.
Kinshicho Prime Tower 12 Floor,1-5-7
Kameido,Koto-ku,Tokyo 136-8577,Japan
Tel.:+81-3-6861-8800

International Consortium on Landslides

International Consortium on Landslides

An international non-government and non-profit scientific organization promoting landslide research and capacity building for the benefit of society and the environment

President: Peter T. Bobrowsky (Geological Survey of Canada)
Vice Presidents: Matjaz Mikos (University of Ljubljana, Slovenia), Dwikorita Karnawati (Agency for Meteorology, Climatorology, and Geophysics, Indonesia), Nicola Casagli (University of Florence, Italy), Binod Tiwari (California State University, USA), Zeljko Arbanas (University of Rijeka, Croatia)
Executive Director: Kaoru Takara (Kyoto University, Japan), Treasurer: Kyoji Sassa (Prof. Emeritus, Kyoto University, Japan)

ICL Full Members:

Geotechnical Engineering Office, Hong Kong Special Administrative Region, China
UNESCO Chair for the Prevention and the Sustainable Management of Geo-hydrological Hazards - University of Florence, Italy
Korea Institute of Geoscience and Mineral Resources (KIGAM)
University of Ljubljana, Faculty of Civil and Geodetic Engineering (ULFGG), Slovenia

Albania Geological Survey / The Geotechnical Society of Bosnia and Herzegovina / Center for Scientific Support in Disasters – Federal University of Parana, Brazil/ Geological Survey of Canada / University of Alberta, Canada / Northeast Forestry University, Institute of Cold Regions Science and Engineering, China / China University of Geosciences / Chinese Academy of Sciences, Institute of Mountain Hazards and Environment / Tongji University, College of Surveying and Geo-Informatics, China / The Hong Kong University of Science and Technology, China / Shanghai Jiao Tong University, China / The University of Hong Kong, China / Universidad Nacional de Colombia / Croatian Landslide Group (Faculty of Civil Engineering, University of Rijeka and Faculty of Mining, Geology and Petroleum Engineering, University of Zagreb) / City of Zagreb, Emergency Management Office, Croatia / Charles University, Faculty of Science, Czech Republic / Institute of Rock Structure and Mechanics, Department of Engineering Geology, Czech Republic / Brown Coal Research Institute, Czech Republic / Cairo University, Egypt / Technische Universitat Darmstadt, Institute and Laboratory of Geotechnics, Germany / National Environmental Agency, Department of Geology, Georgia / Universidad Nacional Autonoma de Honduras (UNAH), Honduras / Amrita Vishwa Vidyapeetham, Amrita University / Vellore Institute of Technology, India / National Institute of Disaster Management, India / Agency for Meteorology, Climatology, and Geophysics of the Republic of Indonesia (BMKG Indonesia) / University of Gadjah Mada, Center for Disaster Mitigation and Technological Innovation (GAMA-InaTEK), Indonesia / Parahyangan Catholic University, Indonesia / Building & Housing Research Center, Iran / Italian Institute for Environmental Protection and Research (ISPRA) - Dept. Geological Survey, Italy / University of Calabria, DIMES, CAMILAB, Italy / Istituto de Ricerca per la Protezione Idrogeologica (IRPI), CNR, Italy / DIA–Universita degli Studi di Parma, Italy / University of Torino, Dept of Earth Science , Italy / Centro di Ricerca CERI - Sapienza Università di Roma, Italy / Kyoto University, Disaster Prevention Research Institute, Japan / Japan Landslide Society / Korean Society of Forest Engineering / National Institute of Forest Science, Korea / Korea Infrastructure Safety & Technology Corporation / Korea Institute of Civil Engineering and Building Technology / Slope Engineering Branch, Public Works Department of Malaysia / Institute of Geography, National Autonomous University of Mexico (UNAM) / International Centre for Integrated Mountain Development (ICIMOD), Nepal / University of Nigeria, Department of Geology, Nigeria / Moscow State University, Department of Engineering and Ecological Geology, Russia / JSC "Hydroproject Institute", Russia / University of Belgrade, Faculty of Mining and Geology, Serbia / Comenius University, Faculty of Natural Sciences, Department of Engineering Geology, Slovakia / Geological Survey of Slovenia / University of Ljubljana, Faculty of Natural Sciences and Engineering (ULNTF), Slovenia / Central Engineering Consultancy Bureau (CECB), Sri Lanka / National Building Research Organization, Sri Lanka / Landslide group in National Central University from Graduate Institute of Applied Geology, Department of Civil Engineering, Center for Environmental Studies, Chinese Taipei / National Taiwan University, Department of Civil Engineering, Chinese Taipei / Asian Disaster Preparedness Center, Thailand / Ministry of Agriculture and Cooperative, Land Development Department, Thailand / Institute of Telecommunication and Global Information Space, Ukraine / California State University, Fullerton & Tribhuvan University, Institute of Engineering, USA & Nepal / Institute of Transport Science and Technology, Vietnam / Vietnam Institute of Geosciences and Mineral Resources (VIGMR).

ICL Associates

State Key Laboratory of Geohazard Prevention and Geoenvironment Protection (Chengdu University of Technology), China / Czech Geological Survey, Czech Republic / Department of Earth and Environmental Sciences, University Aldo Moro, Bari, Italy / Department of Sciences and Technologies, University of Sannio, Italy / Department of Earth and Environmental Sciences – University of Pavia, Italy / Geotechnical Engineering Group (GEG), University of Salerno, Italy / Niigata University, Research Institute for Natural Hazards and Disaster Recovery, Japan / Ehime University Center for Disaster Management Informatics Research, Japan / Tian-Shan Geological Society, Kyrgyzstan / Institute of Environmental Geoscience RAS (IEG RAS), Russia / Russian State Geological Prospecting University n.a. Sergo Ordzhonikidze (MGRI-RSGPU) / TEMPOS, environmental civil engineering Ltd., Slovenia / Institute of Earth Sciences – Faculty of Geoscience and Environment, University of Lausanne, Switzerland / Middle East Technical University (METU), Turkey / North Dakota State University, USA

ICL Secretariat:

Secretary General: Kyoji Sassa
International Consortium on Landslides, 138-1 Tanaka Asukai-cho, Sakyo-ku, Kyoto 606-8226, Japan
Web: http://icl.iplhq.org/, E-mail: secretariat@iclhq.org
Tel: +81-75-723-0640, Fax: +81-75-950-0910

Ž. Arbanas et al. (eds.), *Understanding and Reducing Landslide Disaster Risk*,
ICL Contribution to Landslide Disaster Risk Reduction,
https://doi.org/10.1007/978-3-030-60713-5